Springer Series in

MATERIALS SCIENCE 140

Springer Series in

MATERIALS SCIENCE

Editors: R. Hull R. M. Osgood, Jr. J. Parisi H. Warlimont

The Springer Series in Materials Science covers the complete spectrum of materials physics, including fundamental principles, physical properties, materials theory and design. Recognizing the increasing importance of materials science in future device technologies, the book titles in this series reflect the state-of-the-art in understanding and controlling the structure and properties of all important classes of materials.

For other titles published in this series, go to
www.springer.com/series/856

Lorena Pardo • Jesús Ricote

Multifunctional Polycrystalline Ferroelectric Materials

Processing and Properties

Springer

Piezo INSTITUTE
The European Institute of Piezoelectric Materials and Devices

Professor Dr. Lorena Pardo
Instituto de Ciencia de Materiales
de Madrid
Consejo Superior de Investigaciones
Científicas
(ICMM-CSIC)
Cantoblanco
28049 Madrid
España

Dr. Jesús Ricote
Instituto de Ciencia de Materiales
de Madrid
Consejo Superior de Investigaciones
Científicas
(ICMM-CSIC)
Cantoblanco
28049 Madrid
España

Series Editors:

Professor Robert Hull
University of Virginia
Dept. of Materials Science and Engineering
Thornton Hall
Charlottesville, VA 22903-2442, USA

Professor Jürgen Parisi
Universität Oldenburg, Fachbereich Physik
Abt. Energie- und Halbleiterforschung
Carl-von-Ossietzky-Straße 9-11
26129 Oldenburg, Germany

Professor R. M. Osgood, Jr.
Microelectronics Science Laboratory
Department of Electrical Engineering
Columbia University
Seeley W. Mudd Building
New York, NY 10027, USA

Professor Hans Warlimont
DSL Dresden Material-Innovation GmbH
Pirnaer Landstr. 176
01257 Dresden, Germany

Published by Springer,
P.O. Box 17, 3300 AA Dordrecht, The Netherlands
In association with
Canopus Academic Publishing Limited,
15 Nelson Parade, Bedminster, Bristol, BS3 4HY, UK

www.springer.com and www.canopusbooks.com

Springer Series in Materials Science ISSN 0933-033X
ISBN 978-90-481-2874-7 e-ISBN 978-90-481-2875-4
Springer Dordrecht Heidelberg London New York

Library of Congress Control Number: 2009940448

Printed on acid-free paper

Springer is part of Springer Science+Business Media (www.springer.com)

Preface

Most of the recent efforts to produce books on ferroelectric materials have focused on issues such as the integration of ferroelectrics into different types of devices (Ferrolectric and Dynamic Random Access Memories; Piezoelectric Devices), mostly in thin film form, with intrusions into the realm of nanoscale phenomena. Although some attempts have been made to cover more fundamental topics, such as mechanical fatigue or phase transitions, which are essential to understand the performance of polycrystalline ferroelectrics in applications, an overview of the recent advances in processing and properties of both ferroelectric bulk ceramics and thin films is still lacking, despite its direct impact on the improvement or development of new applications. We think that this book can fill such gap. Here the reader will find in one book updated information on the preparation and properties of this technologically relevant range of materials – information that is currently scattered throughout a number of publications.

Basic concepts of polycrystalline ferroelectrics processing and properties are found, together with references to their multiple applications, in the introductory sections of the chapters. On the other hand, research topics that arose in the recent past and are nowadays the focus of intense activity are also addressed in this book. Such is the case for the environmentally friendly polycrystalline ferro-piezoelectric materials, seen from the point of view of elimination of hazardous components, such as the commonly used lead oxide, or the development of clean processing routes for lead-based ferroelectrics. The challenges in the processing and characterization of crystallographically oriented bulk ferroelectric ceramics and nanosized ferroelectrics are also analysed here. All chapters were written by leading authorities on the topics with reference to the basics and to recent advances.

C. Galassi (ISTEC, Faenza, Italy) has written *Advances in Processing of Bulk Ferroelectric Materials,* using both classical and non-conventional techniques. M. Kosec, D. Kuscer and J. Holc (Institute Jožef Stefan, Ljubljana, Slovenia) have written *Processing of Ferroelectric Ceramic Thick Films,* a topic at the first stage of the integration of ferroelectrics with other hybrid and microelectronic technologies. Following the integration steps that require even higher reduction of the dimensions of the ferroelectric material, some chapters are devoted to thin-film issues and nano-sized ferroelectrics. K. Kato (National Institute of Advanced Industrial Science and Technology (AIST), Nagoya, Japan) has written *Tailored Liquid Alkoxides for the Chemical Solution Processing of Pb-free Ferroelectric Thin Films*. M. L. Calzada (ICMM-CSIC, Madrid, Spain) has written *Ferroelectrics onto Substrates Prepared by Chemical Solution Deposition: From the Thin Film to the Self-Assembled Nano-sized Structures* and I. Bretos and M. L. Calzada (ICMM-CSIC. Madrid, Spain) have written *Approaches Towards the Minimisation of Toxicty in Chemical Solution Deposition Processes of Lead-Based Ferroelectric Thin Films.*

Ferroelectricity and crystal structure are closely related, and the detailed analysis of this requires the use of singular and advanced techniques. L. E. Fuentes-Cobas (Centro de Investigación de Materiales Avanzados, Chihuahua, México) has written about *Synchrotron Radiation Diffraction and Scattering in Ferroelectrics*; M. E. Montero Cabrera (Centro de Investigación de Materiales Avanzados. Chihuahua, Mexico) – *X-Ray Absorption Fine Structure Applied to Ferroelectrics*; D. Chateigner (CRISMAT-ENSICAEN, Caen, France) and J. Ricote (ICMM-CSIC, Madrid, Spain) – *Quantitative Texture Analysis of Polycrystalline Ferroelectrics; and* V. V. Svartsman (Duisburg-Essen University. Duisburg, Germany); and A. L. Kholkin (Aveiro University. Aveiro, Portugal) –*Nanoscale Investigation of Polycrystalline Ferroelectric Materials Via Piezoresponse Force Microscopy.*

Frequently ferro-piezoelectric ceramic materials in devices are subjected to high mechanical loads and must present a high resistance to fatigue under electromechanical vibrations. D. Lupascu, J. Schröder (University of. Duisburg-Essen, Essen, Germany), C. Lynch (UCLA, Los Angeles, USA), W. Kreher (University of Dresden, Dresden, Germany) and I. Westram (Darmstadt University of Technology, Darmstadt, Germany) have written about *Mechanical Properties of Ferro-Piezoceramics*. C. Chima-Okereke, W. L. Roberts, A. J. Bushby and M. J. Reece (Queen Mary College, University of London, UK) have written about *The Elastic Properties of Ferroelectric Thin Films Using Nanoindentation.*

A glimpse of the multifunctionality of ferro-piezoelectric ceramics, also mentioned in other chapters, is provided by R. Jiménez and B. Jiménez (ICMM-CSIC, Madrid, Spain), writing on *Pyroelectricity in Polycrystalline Ferroelectrics.* Special attention was given to issues related to the piezoelectric properties of polycrystalline ferroelectrics which are far from being fully explored, and nowadays face important challenges. L. Pardo (ICMM-CSIC, Madrid, Spain) and K. Brebøl (Limiel ApS, Langebæk, Denmark) cover *Properties of Ferro-Piezoelectric Ceramic Materials in the Linear Range: Determination from Impedance Measurements at Resonance* and J. Erhart (Technical University of Liberec, Liberec, Czech Republic) describes *Domain Engineered Piezoelectric Resonators.* A. Albareda and R. Pérez (Politechnic University of Catalonia, Barcelona, Spain) have written about *Non-linear Behaviour of Piezoelectric Ceramics.* Finally, also as a glimpse into the many possible applications of polycrystalline ferroelectrics, in particular in the field of ultrasonic transducers,

Y. Gómez-Ullate Ricón and F. Montero de Espinosa Freijo (Acoustics Institute, CSIC, Madrid, Spain) have written *Piezoelectric Transducers for Structural Health Monitoring: Modelling and Imaging*.

This book offers interesting content for the beginner from academia or industry who is curious about the possibilities of polycrystalline ferroelectric materials; they will find here a wide range of information. But, also, researchers involved in the study of ferroelectric materials or end-users of ferro-piezoelectric ceramics will find some recent developments in the field and some topics that are not commonly discussed in books devoted to ferroelectrics.

L. Pardo
J. Ricote

Contents

List of Contributors

Alfons Albareda
Applied Physics Department,
Universitat Politècnica de Catalunya,
08034 Barcelona,
Spain

K. Brebøl
Limiel ApS,
DK - 4772 Langebæk,
Denmark

Iñigo Bretos,
Instituto de Ciencia de Materiales de Madrid (CSIC),
Sor Juana Inés de la Cruz, 3,
Cantoblanco,
28049 Madrid,
Spain

A. J. Bushby,
Centre for Materials Research and School of Engineering and Materials Science,
Mile End Road,
London E1 4NS,
UK

Maria Elena Montero Cabrera
Centro de Investigación en Materiales Avanzados, S.C.
Miguel de Cervantes 120,
Complejo Industrial Chihuahua,
31109 Chihuahua,
Mexico

M. Lourdes Calzada
Instituto de Ciencia de Materiales de Madrid (CSIC),
Sor Juana Inés de la Cruz, 3,
Cantoblanco,
28049 Madrid,
Spain

D. Chateigner
Laboratoire de CRIstallographie et Science de MATériaux CRISMAT-ENSICAEN,
Institut Universitaire de Technologie (IUT),
Université de Caen Basse Normandie,
6 Boulevard du Maréchal Juin,
F-14050 Caen,
France

C. Chima-Okereke,
Centre for Materials Research and School of Engineering and Materials Science,
Mile End Road,
London E1 4NS
UK

Jiří Erhart
Department of Physics and International Centre for Piezoelectric Research,
Technical University of Liberec,
Studentská 2,
CZ-461 17 Liberec 1,
Czech Republic

Luis E. Fuentes-Cobas
Centro de Investigación en Materiales Avanzados, S. C.
Complejo Industrial Chihuahua,
Miguel de Cervantes 120,
31109 Chihuahua,
México

Francisco Montero de Espinosa Freijo
Instituto de Acústica,
Serrano 144,
28006 Madrid,
Spain

Carmen Galassi
Institute of Science and Technology for Ceramics ISTEC-CNR,
Via Granarolo, 64,
I-48018 Faenza (RA),
Italy

Janez Holc
Jožef Stefan Institute,
Jamova cesta 39,
SI-1000 Ljubljana,
Slovenia

B. Jiménez
Instituto de Ciencia de Materiales de Madrid (CSIC),
Consejo Superior de Investigaciones Científicas,
28049, Madrid,
Spain

R. Jiménez
Instituto de Ciencia de Materiales de Madrid (CSIC),
Consejo Superior de Investigaciones Científicas,
28049, Madrid,
Spain

Kazumi Kato
National Institute of Advanced Industrial Science and Technology (AIST),
2266-98 Anagahora,
Shimoshidami,
Moriyama-ku,
Nagoya 463-8560,
Japan

A. L. Kholkin
Departent of Ceramic and Glass Engineering, CICECO,
University of Aveiro,
Aveiro,
Portugal

Marija Kosec
Jožef Stefan Institute,
Jamova cesta 39,
SI-1000 Ljubljana,
Slovenia

Wolfgang Kreher
Institute for Materials Science,
Technische Universität Dresden,
Germany

Danjela Kuscer
Jožef Stefan Institute,
Jamova cesta 39,
SI-1000 Ljubljana,
Slovenia

Doru C. Lupascu
Institut für Materialwissenschaft,
Universität Duisburg-Essen,
Essen,
Germany

Christopher S. Lynch
Department of Mechanical and Aerospace Engineering,
University of California,
Los Angeles,
USA

L. Pardo
Instituto de Ciencia de Materiales de Madrid (ICMM-CSIC),
Cantoblanco,
28049 – Madrid,
Spain

Rafel Pérez
Applied Physics Department,
Universitat Politècnica de Catalunya,
08034 Barcelona,
Spain

M. J. Reece
Centre for Materials Research and School of Engineering and Materials Science,
Mile End Road,
London E1 4NS,
UK

Yago Gómez-Ullate Ricón
Instituto de Acústica,
Serrano 144,
28006 Madrid,
Spain

J. Ricote
Instituto de Ciencia de Materiales de Madrid (CSIC),
Sor Juana Inés de la Cruz 3,
Cantoblanco,
E-28049 Madrid,
Spain

W. L. Roberts
Centre for Materials Research and School of Engineering and Materials Science,
Mile End Road,
London E1 4NS,
UK

Jörg Schröder
Institute of Mechanics,
Universität Duisburg-Essen,
Essen,
Germany

V. V. Shvartsman
Angewandte Physik,
University of Duisburg-Essen,
Duisburg,
Germany

Ilona Westram
Institute for Materials Science,
Darmstadt University of Technology,
Darmstadt,
Germany

Chapter 1
Advances in Processing of Bulk Ferroelectric Materials

Carmen Galassi

1.1 Introduction

The development of ferroelectric bulk materials is still under extensive investigation, as new and challenging issues are growing in relation to their widespread applications. Progress in understanding the fundamental aspects requires adequate technological tools. This would enable controlling and tuning the material properties as well as fully exploiting them into the scale production. Apart from the growing number of new compositions, interest in the first ferroelectrics like $BaTiO_3$ or PZT materials is far from dropping. The need to find new lead-free materials, with as high performance as PZT ceramics, is pushing towards a full exploitation of barium-based compositions. However, lead-based materials remain the best performing at reasonably low production costs. Therefore, the main trends are towards nano-size effects and miniaturisation, multifunctional materials, integration, and enhancement of the processing ability in powder synthesis. Also, in control of dispersion and packing, to let densification occur in milder conditions. In this chapter, after a general review of the composition and main properties of the principal ferroelectric materials, methods of synthesis are analysed with emphasis on recent results from chemical routes and cold consolidation methods based on the colloidal processing.

1.2 Ferroelectric Materials

Ferroelectric materials are a subgroup of spontaneously polarised pyroelectric crystals, and are characterised by the presence of a spontaneous polarisation. This polarisation is reversible under the application of an electric field of mag-

---

Institute of Science and Technology for Ceramics ISTEC-CNR, Via Granarolo, 64, I-48018 Faenza (RA) Italy, Phone + 39 0546 699750, carmen.galassi@istec.cnr.ita nitude less than the dielectric breakdown of the material itself [1, 2, 3]. Ferroelectric materials are divided into four categories: the perovskite group (ABO_3) that is the most important one, the bismuth layer structure group, the tungsten bronze group (Fig. 1.1) and the pyroclore group. Most ferroelectric materials undergo a structural phase transition from a high temperature paraelectric phase to a low temperature ferroelectric phase. The temperature of the phase transition is called the Curie temperature (T_C). In the ferroelectric state, the displacement of the central B ion, when an electric field is applied to the unit cell, causes the reversal of polarisation. The areas with the same polarisation orientation are referred to as domains, with domain walls existing between areas of unlike polarisation orientation. The switching of many adjacent unit cells is referred to as domain reorientation or switching. When this ionic movement occurs, it leads to a macroscopic change in the dimensions of the unit cell and the ceramic as a whole.

In ferroelectric ceramics, domains are randomly oriented and thus the net polarisation is zero because of their cancellation effect. Therefore, the as-prepared ferroelectric ceramics are neither piezoelectric nor pyroelectric. To show piezoelectric and pyroelectric properties, polycrystalline ferroelectric ceramics must be poled at a strong external DC electric field (1–10 kV/mm). This must also be done at elevated temperatures to make the domains more easily switchable.

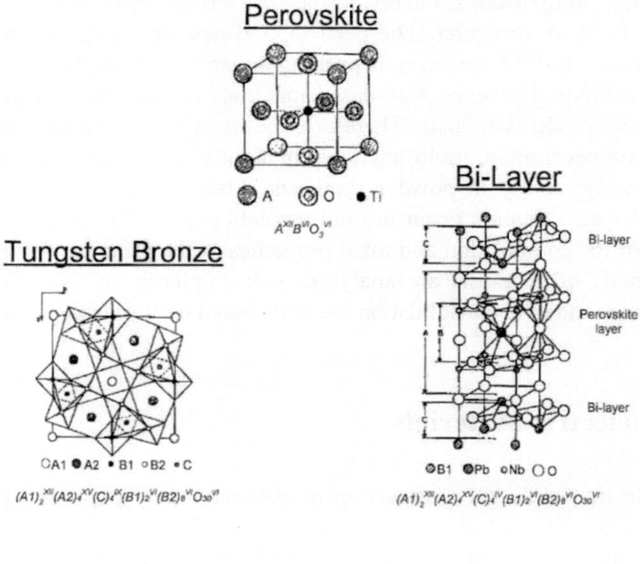

Fig. 1.1 Variants of the perovskite structure (from [4])

For a given composition, the T_C and the electrical, mechanical and optical properties strongly depend on the microstructure. This means density, grain size, shape, size distribution, porosity size and distribution, and anisotropy of grains or porosity. For most of the applications, fully dense materials are required to maximise the dielectric constant, the electrical breakdown strength, and the mechanical strength. These are also needed to minimise the dielectric loss tangent. Porosity is introduced in the materials, for example, when the reduction of the acoustic impedance is required. The value of the piezoelectric coefficients in ferroelectric materials at room temperature ranges from several pC N^{-1} in the $Sr_2Nb_2O_7$ family of layer structure perovskites, to more than 2000 pC N^{-1} in single crystals of relaxor-based ferroelectrics. This covers three orders of magnitude. Ferroelectricity was first discovered in the Rochelle salt (sodium potassium tartrate tetrahydrate, $KNa(C_4H_4O_6)_4H_2O$). But only after the discovery of ferroelectric ceramics, during the Second World War, did the number of applications grow rapidly.

1.2.1 Perovskite Type Materials

Perovskite crystals are represented by the general formula ABO_3 where the valence of A cations is from +1 to +3 and of B cations from +3 to +6. The perovskite unit-cell consists of a corner-linked network of oxygen octahedra, creating an octahedral cage (B-site), and the interstices (A-sites). Various A-site substitutions result in a large family of simple perovskite ferroelectrics (more than 100). In many ferroelectric ceramics, represented by the families of $BaTiO_3$ and lead-based solid solutions, Ti^{4+} (Zr^{4+}) ions occupy the B-site while Pb^{2+} (Ba^{2+}) ions occupy the A-site [4]. Variations of the corner linked octahedral-like tilt or rotations result in new families of ferroelectrics. Among those, the tungsten bronze and the bilayer structures are the most important (Fig. 1.1).

1.2.1.1 Barium Titanate

Barium titanate ($BaTiO_3$ or BT) [5] is a ferroelectric and piezoelectric material with a variety of commercial applications. These applications include multilayer ceramic capacitors (MLCCs), embedded capacitance in printed circuit boards, underwater transducers (sonars), thermistors with positive temperature coefficient of resistivity (PTCR), and electroluminescent panels. It shows relatively low T_C (120°C) and low electromechanical coupling factor (0.35). The grain size of $BaTiO_3$ plays a major role in ferroelectric properties. Much attention has therefore been paid to the synthesis of single-phase $BaTiO_3$ ceramics with a controlled microstructure (grain size > critical size). In addition, both stoichiometry and composition control are important parameters for the control of the ferroelectric properties [6]. $BaTiO_3$ is often combined with additives to modify and improve properties. Sr^{2+} reduces the T_C while Pb^{2+} increases it. Ca^{2+} enhances the temperature range of stability of the tetragonal phase (T_C shifters are $SrTiO_3$, $CaZrO_3$, $PbTiO_3$, and $BaSnO_3$). Co^{2+} reduces the high electric field losses. For example, in the system $BaTiO_3$–$SrTiO_3$, the high Tc value of pure $BaTiO_3$ ceramics can be decreased by increasing the Sr content. The dependence of the permittivity on the electric field can also be accurately tailored (tunability). The barium strontium titanate solid solutions ($Ba_xSr_{1-x}TiO_3$ or BST) are then very attractive for tunable resonators, filters, phase-shifters and variable frequency oscillators.

Depressors, such as $Bi_2(SnO_2)_3$, $MgZrO_3$, $CaTiO_3$, $NiSnO_3$, and the shifters, are added in small (1–8 wt%) quantities to the base $BaTiO_3$ composition. This is to lower or depress the sharpness of the dielectric constant peak at the T_C, thus giving a flatter dielectric constant–temperature profile.

Solid solutions of $BaTiO_3$ and non-ferroelectric $BaSnO_3$, $Ba(Ti_{1-x}Sn_x)O_3$ (BTS) exhibit ferroelectric properties. These are used for capacitors and ceramic boundary layer capacitors, bolometers, actuators and microwave phase shifters [7]. In BTSx ceramics, the isovalent Sn-substitution on the Titanium (Ti) site makes it possible to reduce the temperature dependence. It is also possible to control the room-temperature values of macroscopic properties, such as dielectric characteristics, relaxor behaviour, and sensor performance. With increasing Tin (Sn) content between 10% and 20%, the Tc of the paraelectric–ferroelectric phase transition decreases considerably. For $x > 0.05$, deviations from the Curie–Weiss law for the temperature dependence of the permittivity increase significantly. The phase transition of BTSx becomes increasingly diffuse.

1.2.1.2 Lead-Based Materials

Lead Titanate ($PbTiO_3$ or PT) is a ferroelectric material with a phase transition temperature of 490°C. It has unique properties like high transition temperature, low dielectric constant, low ratio for the planar-to-thickness coupling factor, and a low aging rate of the dielectric constant. PT ceramics are good candidates as stable pyroelectric and piezoelectric devices for high temperature or high frequency applications. Anisotropic thermal expansion during cooling from a high sintering temperature creates large internal stresses in the material, which is destroyed by microcracking. The expansion is caused by the phase transition from cubic paraelectric to tetragonal ferroelectric (with a relatively large c/a ratio of ~1.065). Therefore, PT materials can be prepared via the conventional solid state reaction only after modification with proper dopants.

Lead Zirconate Titanate ($Pb(Zr_xTi_{1-x})O_3$ or PZT) ceramics [8], solid solutions of PT and $PbZrO_3$ (PZ) possess high electromechanical coupling coefficients (K_p=0.70). They have higher T_C values than $BaTiO_3$, which permit higher operation and processing temperatures. They can be easily poled. They possess a wide range of dielectric constants. They are relatively easy to sinter. And, most important, they form solid-solution compositions with many different constituents. This allows a wide range of achievable properties. Although PZT ceramics of different compositions have various functions, a salient feature of the phase diagram for this solid solution system is the existence of the almost temperature-independent phase boundary around x = 0.52–0.53. This separates a rhombohedral Zr-rich phase from a tetragonal Ti-rich phase. The dielectric constant, piezoelectric constant and electromechanical coupling coefficient exhibit a pronounced maximum value for the composition corresponding to this phase boundary. This is generally referred to as the morphotropic phase boundary (MPB). The position of the MPB is not exactly defined as it is associated to a phase coexistence region for which different models have been proposed: coexistence of the two phases, metastability of one of two phases, and unwanted compositional fluctuations [9]. An inversely proportional dependence of the width of the coexistence phase was proposed on the grain size [10] as the result of the thermal fluctuations during cooling. Recently, for compositions close to the MPB, the existence of a low symmetry monoclinic (M) phase bridging the tetragonal (T) and rhombohedral (R) ones was revealed in the temperature range 20-300K [11]. The T–M transition gradually changes from first to second order increasing the Ti content from the boundary with the R phase to x = 0.48. This confirms that the coexistence of T and M phases is intrinsic and not due to compositional fluctuations [12]. The PZT materials are almost always used with a dopant, a modifier, or other chemical constituents to tailor their basic properties to specific applications. Donor doping (Nb^{5+} replacing Zr^{4+} or La^{3+} replacing Pb^{2+}) increases the electrical resistivity of the materials by at least three orders of magnitude. The donors are usually compensated by A-site vacancies. These additives (and vacancies) enhance domain reorientation. Ceramics produced with these additives are characterised by high dielectric constants, maximum coupling factors, square hysteresis loops, low coercive fields, high remnant polarisation, higher dielectric loss, high mechanical compliance, and reduced aging. Acceptor doping (Fe^{3+} replacing Zr^{4+} or Ti^{4+}) is compensated by oxygen vacancies and usually has limited solubility in the lattice. Domain reorientation is limited. Hence, ceramics with acceptor additives are characterised by poorly developed hysteresis loops, lower dielectric constants, low dielectric losses, low compliances, and higher aging rates. Further, the substituting ions can be of the same valency and approximately the same size as the replaced ion. Isovalent substitution such as Ba^{2+} or Sr^{2+} replacing Pb^{2+} or Sn^{4+} replacing Zr^{4+} or Ti^{4+}, usually produce inhibited domain reorientation and poorly developed hysteresis loops, lower dielectric loss, low compliance, and higher aging rates.

Lead Lanthanum Zirconate Titanate (PLZT) ceramics, with variable lanthanum concentrations and different Zr/Ti ratios, exhibit a variety of ferroic phases and embrace all compositional aspects of the dielectric, piezoelectric, pyroelectric, ferroelectric, and electro-optic ceramics. The addition of lanthanum lets the system maintain extensive solid solution. It also decreases the stability of the ferroelectric phases in favour of the paraelectric and anti-ferroelectric phases. T_C reduces with increasing lanthanum. At a 65/35 ratio of Zr/Ti, a concentration of 9% lanthanum (designated as 9/65/35) is sufficient to reduce the temperature of the stable ferroelectric polarisation to slightly below room temperature. This results in a material that is non-ferroelectric and cubic in its virgin state. The cross-hatched area existing along the FE–PE phase boundary denotes a region of diffuse, metastable relaxor phases that can be electrically induced to a ferroelectric phase. Materials within this region exhibit electro-optic behaviour.

1.2.1.3 Relaxor Ferroelectrics

Some ferroelectric ceramics exhibit significantly large electrostrictive effect mostly just above their Tc. This effect is where an electric field can enforce the energetically unstable ferroelectric phase related to the presence of nanoscale ordered regions in a disordered matrix [13]. This effect is utilised in relaxor ferroelectrics (RFE) that show extraordinarily high dielectric constants and a diffuse Tc in a moderate temperature range. Among those materials, lead magnesium niobate $Pb(Mg_{1/3}Nb_{2/3})O_3$ (PMN) based relaxor ceramics have been thoroughly investigated. They have been successfully applied as high-strain (0.1%) electrostrictive actuators and high dielectric constant (>25 000) capacitors. The most popular specific composition in this system is PMN–0.1PT. This increases the T_m (the temperature of maximum dielectric constant for relaxors, equivalent to T_C for normal ferroelectrics) of PMN to ~40°C. For this composition, the temperature of polarisation loss (T_d) is ~10°C. Hence, the material is a relaxor at room temperature (25°C). An addition of ~28% PT causes the material to revert to a normal ferroelectric tetragonal phase with T_C ~130°C.

1.2.1.4 Alkaline Niobates

Alkaline niobates are one of the families of materials under investigation as a possible alternative to lead-based piezoelectric materials [14]. In Fig. 1.2, the Tc against the piezoelectric constant is reported for comparison with PZT and other lead-free materials.

Potassium niobate ($KNbO_3$) (KN) exhibits the same sequence of phase transitions as $BaTiO_3$. These transitions are from the cubic paraelectric to the tetragonal phase at 435°C, from the tetragonal to the orthorhombic phase at 225°C, and from the orthorhombic to the rhombohedral phase at −10°C. The tetragonal, orthorhombic and rhombohedral phases are all ferroelectric. Potassium niobate ceramics exhibit weak piezoelectric properties. Solid solutions of $KNbO_3$ with $NaNbO_3$ (NN) lead to a system with many MPBs, showing ferroelectricity up to about 90% $NaNbO_3$. $K_{0.5}Na_{0.5}NbO_3$ (KNN) [15] ceramics, fabricated by conventional sintering, show relatively good piezoelectric properties (Table 1.1). Further improvement of the properties is being made by continuous improvement of processing routes for powder preparation, doping, shaping and sintering methods.

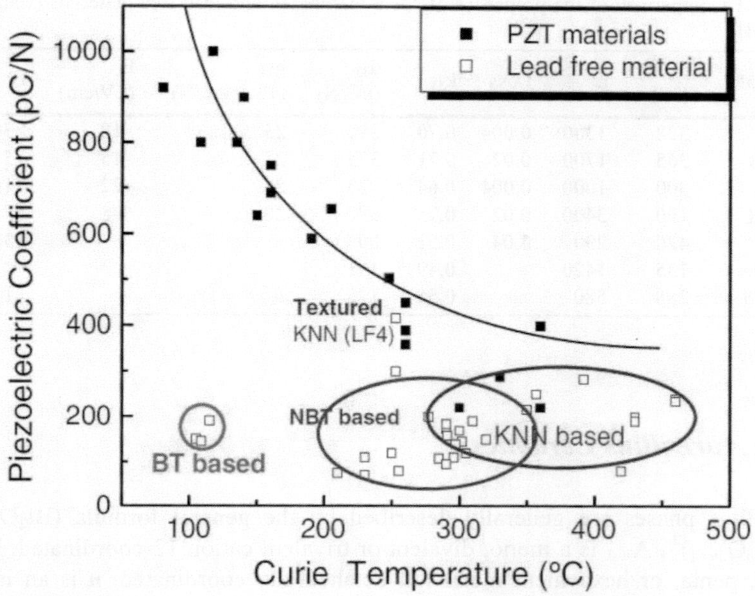

Fig. 1.2 Room temperature values of d_{33} as a function of T_C for various Piezoceramics (from [14]).

1.2.1.5 Bismuth-Based Materials

Bi-based compounds have similar or larger levels of ion off-centering than Pb-based compounds, leading to large ferroelectric polarisations [16]. In addition, they have considerably higher transition temperatures to the paraelectric phase. This results in reduced temperature dependence of the properties under room-temperature operating conditions [17]. $(Bi_{0.5}Na_{0.5})TiO_3$ (BNT) is the most important among the bismuth containing ferroelectric perovskites. The BNT ceramic exhibits a large remnant polarisation, a Curie temperature Tc=320°C and a phase transition point from ferroelectric to anti-ferroelectric Tp=200°C. However, data on piezoelectric properties of the BNT ceramic are scarce because it is difficult to pole this ceramic with its large coercive field, (Ec=73 kV/cm). Therefore, BNT-based solid solutions that can be poled easily have recently been studied. Particularly, a large piezoelectricity is expected for the BNT-based solid solutions with a morphotropic phase boundary (MPB). A morphotropic phase boundary separating ferroelectric tetragonal and rhombohedral phases exists in the $Na_{0.5}Bi_{0.5}TiO_3$-$BaTiO_3$ and related ternary system with $K_{0.5}Bi_{0.5}TiO_3$ [18, 19]. In Table 1.1, the main piezoelectric properties of some representative PZT materials are reported for comparison with lead-free materials.

Table 1.1 Comparison of the properties of some lead-based and lead-free materials (elaborated from [14]).

Material	T_C (°C)	ε_r	Loss	k_{33}	d_{33} (pC/N)	g_{33} (10^{-3} Vm/N)	E_C (kV/cm)	Q
PZT4	328	1300	0.004	0.70	290	25	~18	>500
PZT5A	365	1700	0.02	0.71	375	25	~15	75
PZT8	300	1000	0.004	0.64	225	25	~22	>1000
PZT5H	190	3400	0.02	0.75	590	20	6–8	65
KNN	420	290	0.04	0.51	80			130
$BaTiO_3$	135	1420		0.49	191			
BNBT6	288	580		0.55	125			81

1.2.2 Aurivillius Ceramics

Aurivillius phases are generally described by the general formula $(Bi_2O_2)^{2+}$ – $(A_{n\text{-}1}B_nO_{3n+1})^{2-}$. A_{n-1} is a mono, divalent or trivalent cation 12-coordinated. B is a quadri, penta, or hexavalent metal ion octahedrally coordinated. *n* is an integer representing the number of perovskite layers and can range from 1 to 8 [20]. These bismuth layer-structured ferroelectric (BLSF) compounds, first studied by Aurivillius, belong to the family of bismuth titanate ($Bi_4Ti_3O_{12}$ or BiT) [21]. Except for BiT that is monoclinic, they possess pseudoperovskite layers (A_{n-1} B_nO_{3n+1}) $^{2-}$ stacked between $(Bi_2O_2)^{2+}$ layers (Fig. 1.1). They are of great interest due to their high Curie temperatures (up to 980°C) and good piezoelectric properties. Several critical issues concern the processing, like reproducibility of the properties, narrow range of sintering temperature. Due to the layer structure, the compositions exhibit a very high anisotropy of properties [15]. The microstructure of such ceramic materials consists of plate-like shaped grains. From the point of view of piezoelectric properties, $SrBi_4Ti_4O_{15}$ (SBiT) is of special interest because of its high Curie temperature (~530°C) and its remarkably stable properties on the driving field amplitude and frequency [22]. Wang *et al.* [23] recently succeeded in producing Potassium Bismuth Titanate ($K_{0.5}Bi_{4.5}Ti_4O_{15}$) ~95% dense with Tc= 555°C, Qm 1602 and d_{33}= 21.2 pC/N. Mixed Aurivillius phases are of interest for their potentially enhanced properties [24].

1.2.3 Tungsten Bronze Ceramics

Tungsten bronze (TB) type ferroelectric materials exhibit interesting electro-optic, non-linear optic, piezoelectric and pyroelectric properties. TB structure has a general formula of $(A1)_4(A2)_2(C)_4(B1)_2(B2)_8O_{30}$. B-type cations occupy A1, A2 and C sites. B-type cations occupy the B1 and B2 octahedral sites [25]. In the formula,

A1, A2, C, and B are 15-, 12-, 9- and 6-fold coordinated sites in the crystal lattice structure (Fig. 1.1). Generally, A1 and A2 sites can be filled by Na, Li, K, Ca, Sr, Ba, Pb, Bi and some rare earth (Sm, Nd, Dy, Ce) cations. B1 and B2 sites can be filled by W, Nb, Ta. The smallest interstice C is often empty, and hence a formula is $A_6B_{10}O_{30}$ for the filled TB structure. The metal cations distribution in the different sites of the TB structure plays a crucial role in tailoring physical and functional properties. Moreover, the properties of the TB structure could be modified in a wide scale, by coupling the most important members of the TB family. These members are barium sodium niobates (BNN), potassium lanthanum niobates (KLN), strontium barium niobates (SBN), strontium sodium niobates (SNN), etc. $PbNb_2O_6$ (PN) with a Curie temperature close to 570°C is ferroelectric orthorhombic for sintering temperature above 1250°C. It can be obtained in a single phase only upon appropriate thermal treatments [26].

1.2.4 Pyrochlore

The pyrochlore structure is shown by materials of the stoichiometry $A_2B_2O_7$. B is a tetravalent or pentavalent species and A trivalent or divalent, respectively [27]. Cadmium pyroniobate, $Cd_2Nb_2O_7$ (CNO) is ferroelectric at low temperatures. It exhibits three dielectric anomalies in the narrow temperature range from 195 to 205K, above which it is cubic [28]. The ferroelectric behaviour disappears above 185K, and at the same temperature there are anomalies in the dielectric constant and specific heat [29]. The pyrochlore structure is commonly described as composed of two interpenetrating networks without common constituents. The frequency dependence of the dielectric constant in this temperature regime is similar to that seen in typical relaxor materials. This indicates the presence of polar clusters in CNO.

1.2.5 Multiferroics

Recently, materials that combine ferroelectric and magnetic properties are triggering scientific and technological interest for application in novel multifunctional devices [30]. Multiferroics and magnetoelectric materials can be single phase or two phase materials where magnetisation can be induced by an electric field and electrical polarisation can be induced by a magnetic field. Single phase materials include the anti-ferromagnetic relaxor ferroelectrics like $Pb(Fe_{1/2}Nb_{1/2})O_3$ (PFN) and $Pb(Fe_{1/2}W_{1/3})O_3$ (PFW). They are also orthorhombic manganites, $REMnO_3$ or $REMn_2O_5$, where RE is a rare earth element, $LiCoPO_4$ and $BiFeO_3$, and its solid solutions with $BaTiO_3$ [31]. Indirect coupling, via strain, between two materials such as a ferromagnet and a ferroelectric can be introduced because of the low value of the magnetoelectric coefficient or the low temperature range of the magnetoelectric effect. Intimate contact between a piezomagnetic (or magnetostrictive) material and a piezoelectric (or electrostrictive) material can be achieved in the form of composites [32]. Particulate ceramic composites have been studied by combining BT and ferrites like (Ni(Co,Mn,Zn)Fe_2O_4–$BaTiO_3$ [33], $CoFe_2O_4$–$BaTiO_3$ [34] $NiFe_2O_4$–$BaTiO_3$, $LiFe_5O_8$–$BaTiO_3$, or PZT and ferrites like $NiFe_2O_4$–PZT, (Ni,Zn)Fe_2O_4–PZT, $CuFeCrO_4$–$BaPbTiO_3$, or $CoFe_2O_4$–$Bi_4Ti_3O_{12}$, or laminates (as PZT or PMN-PT) with high magnetostrictive materials, such as Ni–Co–Mn ferrite (NCMF) [35].

1.3 Powder Synthesis

1.3.1 Solid State Reaction (SSR)

The most commonly used process for the powder synthesis is based on the thorough mixing of the starting oxides or carbonates. This is followed by solid-state reaction at high temperatures. The successful production of powders for advanced electronic ceramics depends on the control of the synthesis parameters and purity and morphology of the raw materials.

1.3.1.1 Barium Titanate

Barium titanate (BT) is produced from the reaction between TiO_2 and $BaCO_3$ [36]. The reactants are mixed in order to reduce agglomerates, to increase the homogeneity and to reduce the particle size. After mixing, the raw materials are treated at high temperatures and then, the $BaTiO_3$ is produced. According to Beauger *et al.* [37], the reaction between $BaCO_3$ and TiO_2 proceeds through the following stages:

1. Formation of $BaTiO_3$ at the expense of TiO_2:

$$BaCO_3 \rightarrow BaO + CO_2 \quad (1)$$

$$BaO + TiO_2 \rightarrow BaTiO_3 \quad (2)$$

The reaction proceeds rapidly at the surface of contact between the reactants.

2. When $BaTiO_3$ is formed at the surface, the reactants are separated by a product layer; then the course of the reaction becomes diffusion-controlled. Barium ions must diffuse through $BaTiO_3$ and penetrate into TiO_2 grains. However, when reaching the $BaTiO_3$ interface, barium can react according to:

$$BaTiO_3 + BaO \rightarrow Ba_2TiO_4 \tag{3}$$

Hence, the formation of Ba_2TiO_4 proceeds by the reaction between BaO and the prior-formed $BaTiO_3$.

3. Finally, TiO_2 and Ba_2TiO_4 react to produce $BaTiO_3$, which is the final phase:

$$Ba_2TiO_4 \rightarrow BaTiO_3 + BaO \tag{4}$$

$$TiO_2 + BaO \rightarrow BaTiO_3 \tag{5}$$

To obtain single-phase $BaTiO_3$ temperatures above 900°C are required and the powders are often aggregated. This affects the grain growth during sintering. In order to avoid or minimise this problem, it is important to select an appropriate milling method and control the chemical impurities introduced into the powders from the grinding medium (Al_2O_3, ZrO_2, etc.). Moreover, it is difficult to mix and to maintain chemical homogeneity in the final product, especially when one of the reactants is present in a minor proportion. Small-grained and well-crystallised pure ferroelectric materials are often required as a consequence of the evolution towards miniaturisation, while keeping the highest dielectric constant and low production costs. Therefore, a renewed interest into the formation of BT nanoparticles by a solid state reaction has grown recently [38]. By a solid-state reaction at 700°–800°C of mixtures of nanocrystalline raw materials, Buscaglia *et al.* [39] obtained single-phase $BaTiO_3$ powders with a specific surface area up to 15 m^2/g (particle size: 70 nm). This resulted in highly dense materials after sintering. When nanocrystalline $BaCO_3$ and TiO_2 are chosen as starting powders, they react directly at a temperature lower than the air decomposition of $BaCO_3$. This prevents the formation of the side product Ba_2TiO_4. The same group developed a two-step method for the fabrication of hollow $BaTiO_3$ ferroelectric particles [40]. It involves the suspension of barium carbonate powder in the aqueous solution of peroxy titanium. Amorphous titania precipitates on the barium carbonate crystals by slowly heating the suspension up to 95°C and keeping the temperature constant for five hours. The resulting $BaCO_3$@TiO_2 core–shell particles are then converted into $BaTiO_3$ hollow particles by calcination at 700°C. The out-diffusion of the core phase is faster than in-diffusion of the shell. This leads to the formation of the cavity in the material. Initially, the strong imbalance of the diffusion fluxes determines the formation of Kirkendall porosity close to the original $BaCO_3/TiO_2$ interface without intermediate decomposition. Here, the $BaCO_3$ crystals behave as sacrificial templates. The starting $BaCO_3$ elongated crystals produced empty shells with an average thickness of about 70 nm composed of equiaxed nanocrystals. The size of the core crystals and the reaction temperature are critical in the solid-state fabrication of hollow structures. Calcination at 900–1000°C results in the collapse of the empty shell, with the formation of aggregates of small $BaTiO_3$ particles.

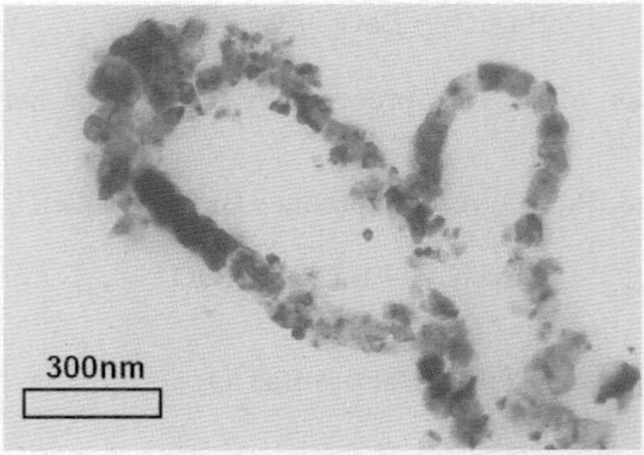

Fig. 1.3 TEM image of the cross-section of $BaTiO_3$ hollow particles obtained after 24 h calcination at 700°C of the $BaCO_3$@TiO_2 powder (from [40]).

$BaTiO_3$ is a stable perovskite end member and is a good candidate to produce lead-free materials. This is because of its high degree of solubility into other perovskite phases forming solid solution [41], and its stable tetragonal crystal structure at room temperature. A few examples are $Bi(Zn_{1/2}Ti_{1/2})O_3$–$BaTiO_3$ [42], $CaTiO_3$–$BaTiO_3$ [43], $BaTiO_3$– $Ba_5Nb_4O_{15}$ [44].

1.3.1.2 Lead-Based Perovskites

PZT powders are still mostly produced by the conventional mixed-oxide route [8]. This has been extensively investigated for compositions close to the MPB. The investigations were to enable an understanding of how the processing parameters and the dopants added affect the compositional fluctuations and phase coexistence. As in the binary mixtures, PT is formed firstly (450°-600°C) with an exothermic reaction and a large volume expansion. PZ is formed at 700°-800°C with endothermic reaction and large volume increase. The formation of the PZT solid solution [45] proceeds with expansion of more that 12% depending on the particle size [46,47]. Therefore, direct reaction sintering is not a means to obtain dense materials. Owing to the enlarged phase coexistence region [48], the equilibrium state is not reached in the calcined powders, even for long heat treatments. This results in poor reproducibility of the process. Moreover, even if the as-reacted powders consist of a mixture of phases, where the degree of homogenisation of the starting oxides depends on the milling conditions [49], reactivity and dispersion of the raw materials [50, 51], they homogenise at the sintering temperature and form monophase materials. A recent study [52], in terms of mutual interaction of the phases in the composition PZT53/47 calcined at 1000°C, confirms the presence of the rhombohedral phase in the calcined powders. The materials are generally prepared by homogenisation and milling of the starting oxides in a liquid media, drying and calcination at temperatures between 600°C and 900°C. A two-stage solid state reaction (reactive calcination) has been developed for PZT ceramics to achieve finer-grained starting powder. The method is based on a pre-reaction of ZrO_2 and TiO_2 at high temperature (1400°C) to form a rutile structure. This exploits the molar volume expansion of the B-site precursor phase. PbO is then added to form the perovskite [53,54]. Being mechanically weak calcined agglomerates, as a consequence of the volume expansion, they are fractured in nano-sized particles (70 nm) by high energy milling. The powder can be densified at 950-1000°C to obtain grain size in the range 0.1 to 10 μm [55]. The milling step becomes more critical when dopants are added or multicomponent compositions are produced. Galassi *et al.* investigated the influence of milling introduced at different steps of the powder treatments (Table 1.2), on the microstructure and dielectric and piezoelectric properties of a multicomponent complex PZT system ($[Pb(Li_{0.25}Nb_{0.75})]_{0.06}O_3$-$PMN_{0.06}$-$PZT_{0.88}$) [56]. Simple milling in agata mortar or long ball milling of the raw materials altogether, or pre-grinding of the coarser oxide, resulted in sintered materials with comparable density but different microstructure and quality factor (Q_m) ranging from 380 to more than 2000.

Table 1.2 Density and piezoelectric properties of the PZT samples (elaborated from [56]).

Sample	Grinding procedure	Green Density (%)	Sintered Density (%)	k_p	d_{31} (10^{-12}m/V)	Q_m
A	Agate mortar, after calcination	64.2	99.7	0.32	-50.0	380
B	Wet milling for 100 h, after calcination	60.4	100.0	0.54	-64.7	2132
C	Sample A, with pre-ground MnO_2	57.4	99.5	0.54	-69.0	994

In the solid state synthesis of PMN powders, the formation of lead niobate based pyrochlore is a critical issue. Among the several methods of powder preparation developed to reduce the undesirable pyrochlore phase, the process that has been more successful is the so-called columbite precursor method [57]. In this technique, MgO and Nb_2O_5 are first reacted to form the columbite structure ($MgNb_2O_6$), with high volume expansion. This is then reacted with PbO and eventually TiO_2 to form the PMN or PMN–PT compositions. Recently, Kwon *et al.* [58] produced 0.65PMN-0.35PT by reactive calcination of precursor mixture of fine and coarse raw materials after prolonged milling.

Lead Metaniobate ($PbNb_2O_6$) has been produced by SSR [26] by adding excess PbO to the nominal formula as the required calcining temperature is 1050°C and the sintering one is higher than 1250°C, to avoid the formation of the low temperature rhombohedral structure. For polycrystalline PN materials, obtaining a single orthorhombic phase is a difficult task. Generally, these materials are formed from a mixture of rhombohedral and orthorhombic phases. Nevertheless, an appropriate thermal treatment followed by rapid cooling can yield a PN material with a single orthorhombic phase.

Microwave irradiation of reactants is a means of accelerating a variety of chemical reactions and to lower the reaction temperature [59, 60]. Microwave heating is affected by mainly two factors; permittivity (dielectric constant) (ε'), and dielectric loss (ε'') [61]. The dependence of the heating rate is given by ($\varepsilon'/\varepsilon''$), defined as loss tangent (tan δ). The dielectric loss tangent (tan δ) depends on the temperature, composition, and physical state of the reactants and the frequency of the electromagnetic waves. Single phase PZT at temperature as low as 600°C can be obtained by microwave-assisted SSR procedure with enhancement of the reaction rates if one of the constituent precursors is a non-stoichiometric oxide [62].

1.3.2 Mechanochemical Synthesis

It consists of the activation of the reactions of the oxide precursors by mechanical energy rather than heat energy, like in the conventional SSR. The main issue is that it skips the calcination step leading to nano-sized powders with better sinterability. Kong *et al.* recently published a comprehensive review on this technique [63] where they show many successful examples of the production of ferroelectric powders via the high energy mechanical milling. They did this by direct synthesis of compounds, mainly in the PT, PZT, PLZT, PMN, and PZN, PFW and BiT systems, or by improved reaction, to form BT, or by amorphisation of precursors for Aurivillius family compounds. Different equipments are used including vibration shake mills, planetary mills or attritor mills. Important parameters are the type of mill, the materials used for the milling vial and media, the milling speed and time, ball to powder weight ratio (BPR), milling environment, process additives, temperature control and the application of an electrical or magnetic field during milling. The comparison of planetary mill and shaker mill for the activation of PZT 53/47 powder in different conditions [64] showed that the BPR has a marked influence on the phase formation. Under a certain value, even prolonged (120 h) milling does not cause the formation of the single perovskitic phase. BPR of 20 at 500 rpm were the necessary conditions for the planetary mill to produce the single phase after 65 hours. This suggests that the milling intensity is related to the shock power injected on the mass of powder trapped in the collision. This could be related to localised temperature increase in the collision point that influences the rate of grain boundary and lattice diffusion processes. In the early stage of milling, the starting oxide powders are refined in both particle and crystallite size. A certain degree of amorphisation takes place that is lower at the higher BPR and milling speed. The mechanical activation makes dynamic processes of the diffusion and atomic rearrangements, enhanced by repeated fracturing and rebinding. Rojac *et al.* [65] proposed the use of a milling map as a tool for determining the critical or minimum cumulative kinetic energy for the formation of the amorphous phase or the intermetallic compound. This was to compare different milling equipments or process conditions. They designed and tested low energy and high energy milling experiments and showed that a critical cumulative kinetic energy between 7 and 12 kJ/g is necessary to induce the formation of $NaNbO_3$. This milling energy strongly changes for different compositions up to values as high as 150 kJ/g. By mechanochemical synthesis, Xue *et al.* synthesised the PZT 52/48 single phase at room temperature [66] after 20 hours shaking at 900 rpm (SPEX shaker-mill) of the pre-ball milled mixture of the starting oxides. The fine powders partially mechanically activated for 10 hours show high sinterability [67]. Single phase PMN-PT ceramics at different PT content were produced in batches of 200 g each by mechanically activating the whole mixture of the starting oxides or by pre-activating two of the starting powders. The different particle size distribution influenced the final microstructure [68].

An amorphous mixture of Aurivillius compounds of the composition $(Bi_3TiNbO_9)_x(SrBi_2Nb_2O_9)_{1-x}$ with x=1.00, 0.65 and 0.35 was obtained upon mechanical activation for 336 hours in a vibrating mill Fritsch Pulverisette [69], of the starting materials that are transformed in the Aurivillius structure at 600 °C. This is a considerably lower temperature than the one needed in the conventional process. The highly reactive powder can be sintered at 1000°C or even at 700°C by hot pressing. Even $Bi_4Sr_{n-3}Ti_nO_{3n+3}$ compounds at increasing n become amorphous after 168 hours mechanical activation in vibration mill [70].

1.3.3 Chemical Methods

Several wet chemical routes are investigated to produce ultrafine starting powders, with improved chemical homogeneity, reduced agglomerate hardness and higher reactivity. They are transformed in the single phase at lower temperature, and result in better control of the stoichiometry and of the final material microstructure. Methods based on precipitation-filtration, such as co-precipitation and hydrothermal or sol-gel synthesis, were extensively applied.

1.3.3.1 Co-Precipitation

PZT and PZT-based powders and their composition modifications with one and more dopants have been produced by co-precipitation routes. The lower calcination temperature of powders produced by hydroxide and oxalate co-precipitation was shown to have strong effects on the sintered material [71]. Non-stoichiometric $PbTiO_3$ perovskites were obtained when an initial equimolar mixture of both oxides precursors was used [72]. Based on solubility calculations, Choy [73] analysed the optimum co-precipitation conditions to obtain PZT 52/48 from metal hydroxides precipitation. He found consistent experimental results (homogeneous precipitation at pH 9 and single perovskitic phase on heating at 900°C for two hours). Optimising the calcination temperature at 500°C, PZT 52/48 powder pellets, isostatically pressed, were densified in one step at 1050°C [74]. A two-step method was investigated to produce $(PbLa)(ZrSnTi)O_3$

composition, by adding the solution of the (Pb,La) ions to the as-washed (ZrSnTi) precipitates [75].

Residual alides when metal chlorides were used proved detrimental for the final properties of the materials. The peroxide-based route was proposed as an alternative method. Camargo [76] proposed the dissolution of Ti metal in the hydrogen peroxide and aqueous ammonia solution to form the peroxotitanato solution. Lead nitrate and zirconil nitrate solutions are added to this. PZT 53/47 coprecipitated starting from nitrites by using urea directly crystallises at 550°C [77]. Fine-grained $BaTi_{0.87}Sn_{0.13}O_3$ (BTS13) powder was synthesised from an oxalate precursor and used to prepare sintered ferroelectric BTS13 ceramics [78]. The precipitate powders show the coexistence of the BTS13 and $BaCO_3$ phases at heating till 1100°C, and the pure BTS13 phase at 1300°C. A small dielectric loss and low frequency dispersion of dielectric characteristics were found owing to high permittivity of the very fine particles. The BTS13 ceramic showed a diffuse paraelectric–ferroelectric phase transition due to the Sn substitution on Ti-sites and the fine grain size of the material. The advantages of this technique used to prepare BTS13 solid solution powder are the mild sintering conditions needed to densify.

1.3.3.2 Sol-Gel Synthesis

Sol-gel processing is widely used to synthesise multicomponent oxides with an intermediate stage including a sol or a gel state. It is a colloidal route based on the hydrolysis/condensation reaction of metal alcoxide salts or complexes (metal carboxilate complexes). Despite the potential for cation mixing at the molecular level, gels usually do not directly crystallise into the equilibrium oxide phase. However, intermediate phases are formed that require high temperature solid state reaction to form the pure phase systems [79]. From the analysis of the local structure of partially heat treated gels [80], it was found that heterogeneity exists at the molecular level. This is related to differences in the hydrolysis and condensation rates of alkoxides of different metals. Carboxylate gels involve the reaction of metal cations with carboxylate ligands to form carboxylate complexes, depending on the nature of the ligand, pH and temperature that form a crosslinked network. The carboxylate complexes and their subsequent crosslinking are formed as a result of deprotonation, complexation and polymerisation sequence. The rise of viscosity during concentration of the solution prevents the precipitation of the metal carboxylate complexes.

In order to keep as low as possible the amount of free metals in the solution, an excess of chelating agent is necessary. Large constant complex formation is preferable. Citric acid has been used as a chelating agent for barium and lead oxide and their multicomponent systems. The citrato-metal complexes are commonly used because they are stable to hydrolysis and ionically crosslink in concentrated solution. This prevents precipitation during gelation. Excess ligand and pH control are necessary to form homogenous gel. During the gel thermolysis, several reactions take place. These can be discerned in scission of the different bonds in the amorphous gel at T<300°C and decomposition of the carboxylato-metal bonds at higher temperature, which gives rise to the formation of the oxide. Phase separation [79] can arise owing to the different thermal stability of the bonds or to the formation of carbonates with the decomposition products. During crystallisation, an intermediate pyrochlore phase is often formed before the formation of the stable perovskitic phase. This is attributed to the pyrochlore's higher nucleation site density and is related to its much disordered structure in comparison with the perovskite phase. The sol-gel chemistry, combined with the combustion process, leads to homogeneous, fine and highly reactive powders in a shorter time and at lower temperatures than traditional processes. Many nitrate citrate gels, when heated on a hot plate, burn in a self-propagating reaction, thus converting the precursor mixture directly into the products.

1.3.3.3 Self-Combustion Synthesis

The self-combustion synthesis technique consists in the heating of a saturated aqueous solution or gel of the desired metal salts or complexes, and a suitable organic fuel to boil. This is done until the mixture ignites and a self-sustaining and rather fast combustion reaction takes off. This results in a dry, usually crystalline, fine oxide powder. It exploits an exothermic, usually very rapid and self sustaining redox-type chemical reaction. The heat required to drive the chemical reaction is provided by the reaction itself. The combustion of metal nitrates-urea mixtures, for example, usually occurs as a self-propagating and non-explosive exothermic reaction. The large amounts of gases formed can result in the appearance of a flame that can reach temperatures up to 1000°C [81]. It shows many advantages (inexpensive precursors, short preparation time, modest heating, and simplicity). The main parameters that affect the course of the reaction are composition of starting mixture (fuel to oxidiser ratio), pH of the sol, modality of gel formation, and starting combustion temperature. Aqueous-based citrate-nitrate sol-gel combustion process can be adopted to reduce the reaction and sintering temperatures, and to improve chemical homogeneity in PZT bulk materials. The presence of nitrate ions is fundamental to the desired oxidation-reduction reaction. In this case, the exothermic reaction between the citrate as organic fuel and the nitrates as oxidiser can be initiated in a muffle furnace or on a hot plate at temperature lower than 500°C. Banerjee [82] synthesised free standing nanocrystalline PZT 52/48 particles using this method. He found that the increase of the citrate to nitrate ratio (C/N) enhances the coagulation of particles and tends to form nanoporous particles. Niobium-doped lead zirconate titanate of the composition titanate (PZTN) precursor solution was prepared starting from Zr-oxynitrate, Pb-citrate, Nb- and Ti-peroxo-citrate precursors [83]. Citric acid monohydrate was the source of citrate anion that was used as both a chelating agent for metal cations and a fuel for combustion. The combustion reactions are highly exothermic (Fig. 1.4). They are autocatalytic in nature and lead to the complete decomposition of the organic com-

compounds. There is also a slow crystallization of PZTN from the amorphous gel to the perovskite structure (Fig. 1.5), with a limited quantity of pyrochlore phase that disappears upon heating at 500°C. The crystallite size was in the range of 20–30 nm but the particles aggregated into hard agglomerates, which preserve the lamellar structure deriving from the gel polymeric network. The influence of the molar ratio of citrate to nitrate onto chemical composition, phase evolution and morphology of $(Pb_{0.93}La_{0.07})[(Zr_{0.60}Ti_{0.40})]_{0.9825}Nb_{0.0175}O_3$ (PZTLN) powders produced by the sol-gel combustion was investigated [84]. Stable sol was obtained by adding nitric acid as oxidiser at C/N molar ratio varying from the stoichiometric (1.30) to an increasing nitrate excess (0.36 and 0.09). PH of the sol was maintained at 7. The polymeric gels resulting from these sols were pyrolysed by an exothermic reaction which becomes an "explosive" process for C/N = 0.09. According to the velocity of the decomposition reaction, the average size of PZTLN powder decreased with lowering C/N ratio. Thus, the powder with the lower C/N ratio presented a uniform microstructure with grain size of about 50 nm diameter bonded in chains and randomly oriented.

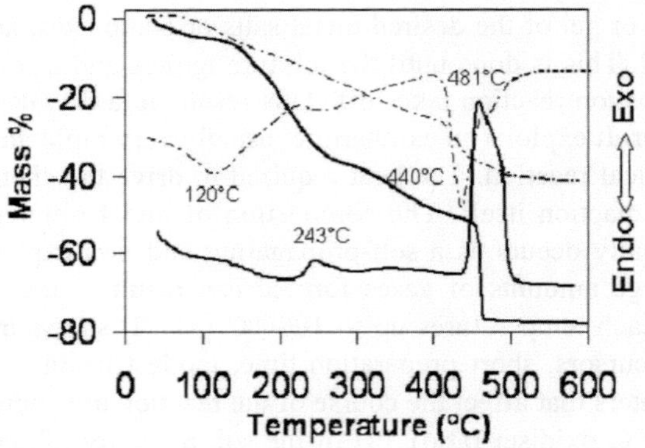

Fig. 1.4 Thermal analysis plots of gel and mixed metal nitrates solution (dot line) (composition PZTN). DTA and TGA analysis of the citrate gel is compared with that of a Pb, Zr, Ti, Nb nitrate mixture prepared by spray-drying the aqueous solution of salts mixed in the same stoichiometric ratio (from [83]).

A significant lead loss was identified that required an excess amount of lead in the starting lead solution. A residual amount of lead carbonate is found at the sintering temperature. Fine Ba-modified bismuth sodium titanate with composition $0.94[(Bi_{0.5}Na_{0.5})TiO_3]$-$0.06BaTiO_3$ (BNBT) powders were prepared by optimising C/N and the gel firing temperature. Decomposition of the precursors and the direct formation of the pure BNBT perovskitic phase in a single step [85]. The best conditions to obtain the desired phase are (C/N) = 0.2, and a combustion temperature of 500°C. Ball milled powders were densified at temperature 100°C lower than that one used in the conventional mixed oxide route. The addition of urea (urea:metal cations = 1.4:1) prevents the precipitation of metal-citrate compounds as a consequence of pH variations induced by the evaporation of NH_3. The stoichiometry is maintained from gel formation to the sintered body.

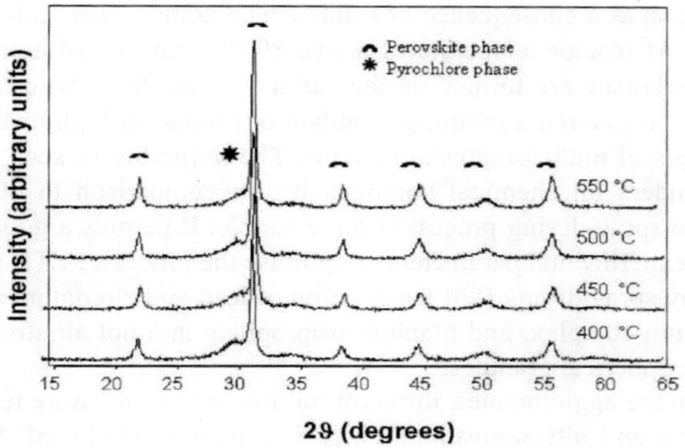

Fig. 1.5 XRD spectra of gel combusted powders calcined at different temperatures (composition PZTN) (from [83]).

1.3.3.4 Direct Synthesis from Solution

In the direct synthesis from solution (DSS) methods, the precipitation of components, consequent to the solvent evaporation, takes place inside the droplets. This ensures a control of stoichiometry at this level. Together with a better control of stoichiometry, DSS methods produce finer particles ensuring a better powder sinterability. In spray-pyrolysis, an aerosol of the precursor solution is directly pyrolised inside the reaction chamber [86]. The aerosol is generated by nozzle atomisation or by ultrasonic atomisation. In the nozzle atomisation, the droplet sizes are relatively large, giving particle sizes of tens of microns. Ultrasonic atomisation is capable of producing aerosols with much smaller droplet sizes, but is generally a low-throughput process.

Recently, $BaTiO_3$ nanoparticles with tunable size (from about 23 to 71 nm) were successfully prepared by flame-assisted spray pyrolysis (FASP) [87]. This was from an aqueous solution of barium acetate and titanium-tetra-isopropoxide, by varying the concentration of precursors and methane flow rate. The flame temperature is a key factor in producing particles with a narrow size distribution. When high methane flow rates are applied, the structure of $BaTiO_3$ nanoparticles is cubic. Hexagonal phase is formed. Nimmo [88] investigated the processing conditions to produce PZT 52/48 powders by twin-fluid nozzle atomisation of aqueous sol–gel precursor solutions of lead acetate, zirconium acetate, and titanium isopropoxide modified by acetylacetone.

The initial drying and low-temperature decomposition are critical stages for particle morphology. This results in a mixture of spherical and larger collapsed particles with size up to 10 μm. In fact, the rapid solvent evaporation causes surface precipitation as a consequence of solute concentration distribution during the condensation. At reactor temperatures above 800°C, nano-sized elongated particles of lead carbonate are formed on the surface of the PZT particles. The spray drying (SD) is considered a promising method of producing high quality powders for the synthesis of multicomponent systems. The properties of such powders are strongly dependent on chemical homogeneity. In comparison to other solution techniques, the spray-drying process is quite simple. It permits a good theoretical simulation of experimental parameters to optimise the process. PZTN powders can be produced by spray drying [89] the solution of lead and zirconium nitrates, niobium ammonium complex, and titanium isopropilate in a hot air stream. This results in minute spherical granules.

To break up the agglomerates, different milling procedures were tested, including ball milling and ultrasonication. There is a high reactivity of the very fine powder that leads to the formation of the perovskitic phase at 550°C. But the hollow sphere morphology of the agglomerates results in inhomogeneous microstructure even after sintering (Fig. 1.6). The efficiency of the synthesis by spray drying a solution is strongly limited by the maximum concentration of the salts in the precursor solutions. In this case, as the four salts show the maximum solubility in different conditions, the concentration of the multi-component solution is governed by the solubility of lead and zirconium nitrates in nitric acid. The maximum concentration reachable resulted 0.2M on lead basis at pH=1.

By carefully controlling the stoichiometry and the milling procedure, Bezzi [90] obtained a homogeneous material with as good properties as the one produced by the solid state reaction. However, this was at a 100°C lower sintering temperature. PMN powders were prepared [91] starting from an aqueous nitrate solution containing Pb^{2+}, Mg^{2+} and Nb^{5+} cations. On heating at 450°C, the pyrochlore phase is formed. The perovskite phase coexists with pyrochlore and unreacted PbO in the temperature range 700–800°C. They are almost completely transformed into the perovskite phase at 800°C. Above 800°C, the PbO partially evaporates and the pyrochlore phase is formed again. The direct synthesis of the perovskite phase is possible during densification when the amorphous starting powder calcined at 350°C is used. The densification takes place simultaneously to the formation of the perovskite phase, when the samples are sintered at 1050°C through a reaction sintering mechanism.

Fig. 1.6 SEM morphology of SD powders: powder As sprayed **a** after ball milling, **b** and after sintering **c**; bar =5 μm (from [89]).

1.3.3.5 Hydrothermal Synthesis

A solvothermal process can be defined "a chemical reaction in a closed system in the presence of a solvent (aqueous and non aqueous solution) at a temperature higher than that of the boiling point of such a solvent" [92]. Consequently, a solvothermal process involves high pressures. Hydrothermal synthesis is a process that utilises single or heterogeneous phase reactions in aqueous media at elevated temperature (T>25°C). Pressure (P>100 kPa) is used to crystallise anhydrous ceramic materials directly from solution [93]. The precursors can be solutions, gels, and suspensions. Inorganic or organic additives are often used at high concentrations to control pH or to promote solubility. Other additives are used to promote particle dispersion or control crystal morphology. For large scale production of materials, typical temperature and pressure fall around 350°C and 100 MPa (saturated vapour pressure of water at this temperature is 16 MPa), while milder or more severe reaction conditions are also considered. A thermodynamic model was developed to determine the formation conditions of the lead titanate zirconate, starting from the thermodynamic data of the solids and aqueous species [94]. This led to the synthesis of phase pure PZT at minimum temperatures (150-200°C) and optimum reagent conditions.

Also adjusted were KOH concentration and the amount of PbO [95], and the size and the morphology of the PZT powder. PZT 52/48 nanocrystals were prepared [96] with morphology changing from particle to rod and wire. For this, the ratio of polyvinyl alcohol to polyacrylic acid used as surfactants and reaction time was adjusted. Nano-sized BT powders have been produced by several low-temperature methods and hundreds of paper have been published. Recently, Wei *et al.* [97] successfully synthesised highly dispersed $BaTiO_3$ nanocrystals of 5–20 nm in size. This was done via a solvothermal method using the mixture of ethylenediamine and ethanolamine as the solvent. Pure perovskite KTN particles have been solvothermally synthesised under a milder condition, such as a lower reaction time and [KOH] in comparison to the hydrothermal route [98]. The sol-gel hydrothermal process was applied to produce NKBT nanowires at T below 200°C [99]. First, a dry gel is prepared by heating at 80°C the sol obtained. A mixture of bismuth nitrate dissolved in acetic acid and potassium and sodium nitrates dissolved in water is added into the solution of tetrabutyl titanate in ethanol. The gel is suspended in the 10 M solution of NaOH and hydrothermally treated at T 100°–200°C under autogenerated pressure for 48 hours. At 160°C, crystalline nanowire are formed and the NKBT is formed that show upon densification superior properties to that prepared by the conventional sol-gel or solid state reaction.

The conventional hydrothermal method is a time-consuming process that could take several days. It can be enhanced by introducing microwaves into the reaction vessels (microwave hydrothermal process) to reduce synthesis times and temperatures as shown for PZT 52/48 [100].

The influence of microwave frequency, bandwidth sweep time, and processing time on the particle size, phase, microstructure, and porosity of barium titanate prepared by microwave hydrothermal processing has been systematically investigated [101]. These have been compared with results obtained with conventional hydrothermal synthesis (170°C for 2 to 40 hours). The $BaTiO_3$ nanoparticles show global cubic structure with local tetragonal clusters. Increasing the sweep time results in hexagonal phase impurity. This indicates that the transient heating patterns were not uniform, and that the particle growth was slower than for conventional hydrothermal synthesis.

Well crystallised BST nanopowders have been synthesised under supercritical conditions [102] through a single step continuous synthesis over the entire range of composition. This is done in the temperature range of 150°C–380°C at 26 MPa using a continuous process from a mixing of barium, strontium and titanium isopropoxides in ethanol (feed solution) [103]. The synthesis was carried out in an 8 m tubular coiled reactor fitted with an external heater. The first part of the reactor was heated at 150°C and the last one at 380°C.

1.4 Colloidal Processing

1.4.1 Slurry Formulation

Submicrometer-sized (with a tendency towards nano-sized) powders are generally employed in the processing of ferroelectric materials to reduce the densification temperature. This is also to obtain fine grained microstructures with significant improvements in properties. A major problem of fine particles is that they spontaneously agglomerate due to van der Waals attractive forces. This results in inhomogeneous particle packing and pore-size distribution in the green body. The inhomogeneities, being introduced in a larger scale than the primary particles themselves, will control the microstructure evolution during sintering. Therefore, they must be minimised during the cold consolidation treatments. Colloidal processing [104] is a means to address the problem. It prevents the agglomerate formation by controlling and manipulating the interparticle forces. It eliminates by sedimentation the hard agglomerates formed by partial sintering of primary particles in previous calcining steps [105].

The green body properties are significantly influenced by controlling the state of the powder dispersion (colloidally stable or flocculated). In general terms, colloidally stable suspensions result in more densely packed green bodies with more narrow pore distribution that strongly flocculated suspensions. In contrast, weakly flocculated suspensions have been shown to result in optimal green bodies with some advantages [106]. Many studies have attempted to find the optimal dispersion conditions to maximise the properties of the sintered body. The powders are preferably dispersed in aqueous media. However, one of the problems is the solubility of the powder, which makes the interparticle interaction more complex. The resulting stoichiometry alterations can be detrimental for the final properties of the material. The dissolution of $BaTiO_3$ is favoured in acidified water and accounts for the dependence of the electrokinetic behaviour on the solids concentration [107]. Pb dissolution in aqueous PZT suspensions occurs both in acidic and in strong basic conditions and is influenced for example by the dispersants [108].

Although less environmentally compatible, non-aqueous media are often preferred because the wider choice of organic additives makes easier the optimisation of the formulations for several ceramic fabrication methods. The interparticle interaction can be controlled by introducing electrostatic or steric repulsion or a combination of the two. By adding electrolytes, short chain polymers or polyelectrolytes, long or short range repulsive interaction arises, which results in stable or weakly flocculated suspensions. Nano-sized $BaTiO_3$ powder suspensions in decane were cold consolidated by pressure casting and sintered. The final microstructure can be correlated with the dispersion state. The stable or weakly flocculated ones result in good sinterability and better microstructure than those strongly flocculated [109].

The colloidal processing of PZT powder added with Nb_2O_5 in different organic dispersants like toluene, heptane or methyl ethyl ketone (MEK) resulted in high green and sintered densities. This was when suspensions stabilised by electrosteric hindrance [110] were used. The study of the electrokinetic and rheological behaviour of aqueous PLZT suspensions showed the dependence on pH and amount of ammonium polymethacrylate dispersant. However, lot-to-lot variations, the order of dispersant addition, and pH adjustment [111] also influenced the outcome. Polyacrilic acid (PAA) is a good dispersant for many ceramic powders in water, including $BaTiO_3$ [112].

Green bodies of 62% theoretical density can be achieved by slip casting at high pH and very high PAA coverage. Nevertheless, PAA exhibits both passivation and sequestration effects at the $BaTiO_3$ solution interface [113]. Sequestering occurs at pH>8 and Ba^{2+} dissolution increases linearly with PAA concentration, while a relatively strong passivation was found for PAA in acidic solution. Recently, Yoshikawa *et al.* [114] showed that comb polymer architectures like poly methacrylic acid) (PMAA), are less dependent on ionic strength in comparison to the pure polyelectrolyte. This was when they were used as a backbone, combined with poly ethilene oxide (PEO) as a charge neutral teeth. Upon increasing the molecular weight of the teeth, they are less susceptible to conformational changes induced by counter-ions in the dispersion of $BaTiO_3$ fine particles in water. Further, comb copolymers can associate the binding effect to the dispersing one [115].

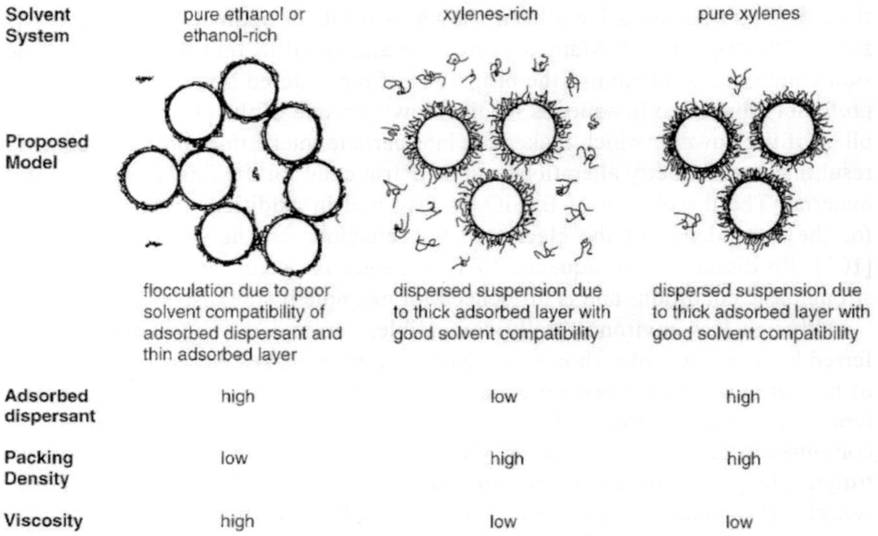

Fig. 1.7 Models for dispersion state of particles in different solvent mixtures containing the same amount of dispersant (from [118]).

1.4.2 Suspension-Based Shaping Techniques

Colloidal processing is the basic approach for the formulation of suspensions for most of the wet cold consolidation routes. This is for routes such as tape casting, screen printing, electrophoretic deposition, ink jet printing, and other direct writing methods.

1.4.2.1 Tape Casting

Tape casting is largely used [116] for the preparation of thin bulk sheets with thickness in the range 50 – 800 μm. These are the basic building blocks in many electro-ceramic components including multilayer ceramic packages. The slurry is deposited on a carrier surface via the doctor blade technique. The coating dries and forms a flexible layer consisting of a polymeric matrix filled with the ceramic powder [117]. After a debonding treatment to eliminate the organic additives, the material is directly sintered. Traditionally, the suspensions were based on organic solvents due to their low latent heat of evaporation and low surface tension. Binary solvent systems are commonly used because of their increased ability to dissolve the organic additive and faster drying. Often, they are based on ethanol (EtOH) (with methyl-ethyl-ketone, or toluene or xylene), and the mixing ratio can be a critical parameter for the optimisation of the slurry performance (Fig. 1.7) [118]. The dispersant can combine electrostatic or steric effects while binders and plasticisers confer to the tape the necessary flexibility for the further handling. Effective deflocculants in organic medium are glycerol tryoleate and phosphate ester [119]. Polyvinylbutyral, eventually at different molecular weights, is typically used as the binder, combined with butylbenzylphtalate as the plasticiser. Nano-sized PZT powders were dispersed in organic medium [120]. Various ferroelectric materials (BST, PZT, etc.) have been successfully tape cast by adjusting the combination of the same organic components [121]. Then, they were laminated to produce co-fired multimaterial structures with well defined interfaces without delamination.

More critical parameters for successful processing are the ratio between the binder and the plasticiser, the order of the addition of the ingredients, and the milling procedure (that is typically prolonged for 24 hours). Recently, interest has been focused on the production of aqueous suspensions [122] that result in far more sensitive process perturbations. These are drying conditions, casting composition or film thickness, pH, and dissolution of the ceramic powder. Therefore, a careful control of the compositional and process variables is required. Two different types of binders can be selected: water soluble binders (like cellulose ethers, polyvinyl alcohol), or water-emulsion binders (dispersions of non-soluble binder particles in water) like acrylics, vinyls, or polyurethane. The deflocculants are usually polyelectrolytes, as mentioned previously. A wetting agent is often added as the wetting behaviour on the carrier film is a critical issue. The complexity of the variables to be controlled is well represented in the work of Smay *et al.* [123].

Laminated PZT multilayers were produced by tape casting water-based slurry. Aqueous dispersion of nano-sized $BaTiO_3$ dispersed with NH_4-PAA with PVA as the binder, and glycerol as plasticiser, has produced homogeneous green body and 95% rel density upon sintering at 1200°C [124]. Feng *et al.* studied the effect of molecular size and chemical structures of plasticisers and binders (PVA with varying molecular weights and hydrolysis percentages), on the mechanical properties of the green tapes of PLZT powder [125]. During drying, a stress evolution occurs because of constrained volume shrinkage. The latter stage of drying is linked mainly to the polymer phase that undergoes coalescence and affects the residual stresses depending on the polymer chain mobility or on the hydrolysis level of PVA. The higher hydrolysis generates larger stresses in the green tapes. Water-soluble PMAA-b-PEO polyelectrolyte was used to disperse PLZST powder, with PVA as binder and PEG as plasticiser. The green tape resulted in very flexible and almost fully dense sintered body [126]. Deliormanli *et al.* [127] showed that the use of comb polymer PAA-PEO with a non-ionic acrylic latex emulsion, and hydroxypropyl methylcellulose and the preparation by a multi-step process, led to high solid loading PMN suspensions. The final microstructure is homogeneous as a consequence of the careful control of the rheological properties and wetting and drying behaviour of the slurry.

1.4.2.2 Electrophoretic Deposition

Electrophoretic deposition (EPD) is usually carried out in a two-electrode cell in two steps [128]. In the first step, an electric field is applied between two electrodes and charged particles suspended in a suitable liquid move toward the oppositely charged electrode (electrophoresis). In the second step, the particles accumulate at the deposition electrode and create a relatively compact and homogeneous film (deposition). Therefore, it is necessary to produce a stable suspension containing charged particles free to move when an electric field is applied. After the deposition, a heat-treatment step is normally needed to further densify the deposit. The technique is mainly applied to produce coatings and films. Thick films of ferroelectric materials have consequently been developed [129, 130]. Moreover, it is of interest to produce monolithic ceramics, ceramic laminates, fibre reinforced composites and functionally graded materials [131]. The powders can be dispersed in organic medium (often ethanol) or water. PH, zeta potential, and conductivity are critical parameters in the control of the deposit homogeneity [132]. Recently, magnetic field-assisted EPD has been applied to obtain grain-oriented $Bi_4Ti_3O_{12}$–$BaBi_4Ti_4O_{15}$ (BiT-BBTi) [133]. The powder dispersed in ethanol with phosphoric ester and polyethylenimine is aligned along a specific crystallographic orientation by applying a magnetic field among the facing electrodes and an electric voltage. The dried compacts were sintered to 97% relative density. It showed that the alignment of the particles induced by the magnetic field in the suspension, remains in the green compacts and is further enhanced by densification followed by grain growth. The piezoelectric properties are significantly improved in comparison to the randomly oriented samples.

1.4.2.3 Inkjet Printing

Inkjet printing is a mean of fabricating 3D ceramics solid structures. It is a near net shape, tool less manufacturing technique, consisting in the deposition of ceramic ink micro-droplets ejected via nozzles to build the successive layers [134]. Drop-on demand printers are frequently used and very fluid inks are required that rapidly solidify by evaporation of the carrier vehicle (EtOH or MEK/EtOH). Sometimes, the jetting is done at high temperature. PZT parts were fabricated by Noguera *et al.* [135] with a printing head by adjusting the fluid properties of the ceramic suspension (size distribution in relation to the aperture of the nozzle, organic fraction composition – in order to avoid sedimentation) and control drying, viscosity and surface tension. They also controlled the velocity, initial size and path of the droplets before spreading. Ceramic pillar arrays were fabricated with pillar diameter 110 μm starting from a 10 volume percentage suspension. Minimising the dimensional change between the printed part and the sintered part requires high solid loading suspensions. Only a limited range of viscosity (10-50 mPas) is compatible with this technology. One variant of inkjet printing is phase change or hot melt printing that makes use of a particulate suspension in a low melting point vehicle. The printing occurs at a temperature above its melting point. PZT powder was dispersed by ball milling at 110°C at 40 volume percentage in paraffin wax (melting point 57°C) added with paraffin oil, to lower the viscosity, with a combination of stearylamine and polyester [136]. The parts were printed at the same temperature with a nozzle of 70 μm diameter. The debonding is a critical and time consuming step. The wax is removed by capillary action, by keeping the samples in a powder bed for two days, instead of burning the residual wax by a controlled heating followed by sintering.

1.4.2.4 Electro-hydrodynamic Deposition

Electro-hydrodynamic deposition is an alternative method to downsize the structure, for example for preparing micrometer and sub-micrometer scale composites. It is a method of liquid atomisation. By means of electrical forces, the liquid flowing out of a capillary nozzle, which is maintained at high electric potential, is forced by the electric field to be dispersed into fine droplets [137]. The droplet size can range from hundreds micrometer to tens of nanometer and can be nearly monodisperse. They are electrically charged and are driven by means of an electric field. Electrospray is mostly used for micro- and nano-thin film deposition, micro- or nano-particle production, and micro- or nano-capsule formation. PZT columnar structures have been grown as thick as 35 μm from the 0.6 M sol in 1-propanol and glacial acetic acid, deposited at flow rate $2.5x10^{-11}$ m^3s^{-1} and applied voltage 4.2 kV [138].

1.5 Templated Grain Growth

Microstructure control is a key issue in optimising the performance of the ferroelectric materials. Microstructure evolution has been extensively investigated. The sintering process involves both densification and grain growth. The densification process is the replacement of solid and vapour interfaces either by solid and solid interfaces (solid-state sintering), or by solid and liquid interfaces (liquid phase sintering). Grain growth is related to the minimisation of total interfacial area by interface migration. The driving force and kinetics of sintering are related to interfacial energy evolution. Capillarity effect drives grain coarsening or Ostwald ripening of solid grains dispersed in a liquid matrix. Grains smaller than average dissolve and larger ones grow. In the solution-reprecipitation process, surface structure strongly affects the growth of the new grains. Pore and grain size are closely related because of geometrical constraints that impose a critical pore to size ratio above which the pore is stable. In most ferroelectric materials, a liquid phase is present during densification, at least as a thin intergranular film. Grain growth occurs by the migration of the solid-liquid interface. Sintering in oxygen is a further key step in obtaining fully dense ferroelectrics. Pores filled with oxygen can close more easily due to the faster diffusion of oxygen in comparison to the slower diffusion of nitrogen that is the major component of air.

The crystallographic texturing of polycrystalline ferroelectric ceramics has been intensively investigated since the finding by Park and Shrout [139] that relaxor-based ferroelectric single crystals show remarkably higher piezoelectric strains than ceramics, primarily in the (001) crystal orientation. Similar increased properties have been found even in $BaTiO_3$ and Zr-doped $BaTiO_3$. The efforts to produce single crystals have resulted in a significant progress in the production of larger single crystals. But the difficulties in controlling the uniformity of the concentration, particularly for MPB crystals, and the intrinsically high costs still limit the diffusion of the single crystal technology. Therefore, the scientific and commercial interest in the processing and properties of textured ceramics has increased significantly. Templated Grain Growth (TGG) consists of the nucleation and growth of the desired polycrystalline material on aligned single crystal template particles. This results in an increased fraction of oriented material upon heating. The template particles must be anisometric, to promote the alignment during cold consolidation. Single crystals must act as preferred growth sites and chemically stable up to the densification temperature. In a reactive matrix they can be the seed sites for the phase formation (Reactive Templated Grain Growth RTGG) [140].

In a comprehensive review, Messing *et al.* [141] showed that more than one to more than three times higher d_{33} values are obtained for all the compositions investigated. Texturing degree as high as 90% can be obtained. TGG can be homoepitaxial when the growing single crystal has the same composition and crystal structure of the template material. It can be heteroepitaxial when the template has a different composition but the same crystal structure, or when there is matching between the lattice and the matrix. A small amount of larger template particles is dispersed in a finer and equiaxed particles matrix. It is subsequently oriented, usually by shear forming processes like tape casting or extrusion. Anisometric particles can be induced to align under a gated doctor blade. During densification, once the ceramic exceeds 95% density, texture evolves by growth of the template particle. They can act as nucleation sites and seed the phase transformation of the matrix [142].

The growth process is sustained by a size ratio between the template particles and the matrix grains larger than approximately 1.5. It is often enhanced by a liquid phase. The template particles must have similar crystal structure and less than 15% lattice parameter mismatch with the phase to be templated. High aspect ratio with axis matching the crystallographic expectation is preferable, like stability in the presence of the liquid phase at the growth temperature. Therefore, the most preferred are whisker or platelet shaped particles with a small grain size that after growth should not exceed the mean size of 40 μm. They are produced by molten flux or hydrothermal synthesis [143,144]. Perovskites like $PbTiO_3$, $BaTiO_3$, and $SrTiO_3$ are used as templates for complex peroskitic systems like PMN-based relaxors, because of the small lattice mismatch. Brosnan *et al.* recently used tabular $SrTiO_3$ [145] to optimise the processing conditions for highly textured PMN-28PT composition. 20 μm thick green tapes were stacked to 60-70 layers, laminated, debonded and isostatically pressed to about 54% TD. After sintering at 1150°C, (15 hours soaking) in lead atmosphere and flowing oxygen, an almost fully dense material was obtained with templated grains with average size of 40 μm. No residual $SrTiO_3$ grains were found as they were rather fine. A soaking at 750°C was necessary to stabilise the templates against dissolution in the PbO rich intergranular phase before the beginning of the TGG process. An increase of the piezo coefficient by a factor of up to 1.8 was found [146] by templating PMN-32PT and PMN-37PT-21PZ materials with $BaTiO_3$ platelets, following similar processing conditions. Plate-like $NaNbO_3$ was used as reactive template to texture compositions in the system $(KNa)NbO_3$-$LiTaO_3$ [147] by RTGG method. This led to obtaining a Lotgering factor of orientation higher than 90% and excellent piezoelectric properties. The $NaNbO_3$ particles were synthesised by a topochemical reaction in which the particle morphology was preserved and a <001> plane of the perovskite was developed by ion exchanging the Na for Bi ions on bismuth layer-structured plate-like composition. Recently, materials in the BNT and BNBT system were textured with templates of the composition $Bi_4Ti_3O_{12}$ (BiT) by TGG [148], and through RTGG by varying the Na content from 2 mol% excess to 2 mol% deficient [149].

The evolution of the microstructure of the sample with Na -excess shows texturing already at temperature lower than 800°C. This was attributed to the formation of a liquid phase (that promotes the dissolution of the polycrystalline matrix grains and deposition on the lowest energy surface that is the single crystal template). A large degree of orientation remains at the densification temperature, but some porosity remains in the final microstructure (final density about 93%). Gao *et al.* [150] measured the piezoelectric properties of NKBT textured with BiT template particles, in the perpendicular and parallel to the tape casting direction. He found a significant improvement in comparison to the conventional process.

1.6 Conclusions

In recent years, the research on the processing of ferroelectric materials has been extended to new compositions and routes to produce powders with controlled morphology, mostly reduced size, and controlled size distribution towards the nano scale. Some of the most investigated processes to produce fine powders were reviewed in the present work, together with a few of the processing routes to produce cold consolidated bodies. Control of the powder agglomeration is a critical issue both during the powder synthesis and in the shaping step. Wet forming methods are widely used to control the green body homogeneity. Further, templated grain growth has been focused as a means to enhance the properties of new lead-free materials.

References

1. Jaffe B, Cook WR and and Jaffe H (1971) Piezoelectric Ceramics Academic Press, London and New York
2. Rabe KA, Dawber M, Lichtensteiger C, Ahn CH and Triscone JM (2007) Modern physics of ferroelectrics: Essential background Physics of Ferroelectrics: a Modern Perspective (Topics in Applied Physics), pp 1-30
3. Haertling GH (1999) Ferroelectric ceramics: History and technology, Journal of the American Ceramic Society 82: 797-818
4. Bhalla AS, Guo RY and Roy R (2000) The perovskite structure – a review of its role in ceramic science and technology. Materials Research Innovations 4: 3-26
5. Lee T and Aksay IA (2001) Hierarchical structure-ferroelectricity relationships of barium titanate particles. Crystal Growth & Design 1: 401-419
6. Lee S, Randall CA and Liu ZK (2007) Modified phase diagram for the barium oxide-titanium dioxide system for the ferroelectric barium titanate. Journal of the American Ceramic Society 90: 2589-2594
7. Markovic S, Mitric M, Cvjeticanin N and Uskokovic D (2006) Structural and dielectric properties of $BaTi_{1-x}Sn_xO_3$ ceramics In: Uskokovic DPMSKRDI (ed), pp 241-246
8. Jaffe B, Roth RS and Marzullo S (1954) Piezoelectric properties of lead zirconate-lead titanate solid-solution ceramics. Journal of Applied Physics 25: 809
9. Kakegawa K, Mohri J, Takahashi T, Yamamura H and Shirasaki S (1977) Compositional fluctuation and properties of $Pb(Zr,Ti)O_3$. Solid State Communications 24: 769-772
10. Cao WW and Cross LE (1993) Theoretical-model for the morphotropic phase-boundary in lead zirconate lead titanate solid-solution. Physical Review B 47: 4825-4830
11. Noheda B, Cox DE, Shirane G, Gonzalo JA, Cross LE and Park SE (1999) A monoclinic ferroelectric phase in the $Pb(Zr_{1-x}Ti_x)O_3$ solid solution. Applied Physics Letters 74: 2059-2061
12. Cordero F, Craciun F and Galassi C (2007) Low-temperature phase transformations of $PbZr_{1-x}Ti_xO_3$ in the morphotropic phase-boundary region. Physical Review Letters 98255701
13. Bokov AA and Ye ZG (2006) Recent progress in relaxor ferroelectrics with perovskite structure. Journal of Materials Science 41: 31-52
14. Shrout TR and Zhang SJ (2007) Lead-free piezoelectric ceramics: Alternatives for PZT? Journal of Electroceramics 19: 111-124
15. Maeder MD, Damjanovic D and Setter N (2004) Lead free piezoelectric materials. Journal of Electroceramics 13: 385
16. Mehring M (2007) From molecules to bismuth oxide-based materials: Potential homo- and heterometallic precursors and model compounds. Coordination Chemistry Reviews 251: 974-1006
17. Baettig P, Schelle CF, LeSar R, Waghmare UV and Spaldin NA (2005) Theoretical prediction of new high-performance lead-free piezoelectrics. Chemistry of Materials 17: 1376-1380

18. Takenaka T, Maruyama K and Sakata K (1991) $(Bi_{1/2}Na_{1/2})TiO_3$-$BaTiO_3$ system for lead-free piezoelectric ceramics. Japanese Journal of Applied Physics Part 1-Regular Papers Short Notes & Review Papers 30: 2236-2239
19. Nagata H, Yoshida M, Makiuchi Y and Takenaka T (2003) Large piezoelectric constant and high Curie temperature of lead-free piezoelectric ceramic ternary system based on bismuth sodium titanate-bismuth potassium titanate-barium titanate near the morphotropic phase boundary. Japanese Journal of Applied Physics Part 1-Regular Papers Short Notes & Review Papers 42: 7401-7403
20. Isupov VA (2006) Systematization of Aurivillius-type layered oxides. Inorganic Materials 42: 1094-1098
21. Shulman HS, Testorf M, Damjanovic D and Setter N (1996) Microstructure, electrical conductivity, and piezoelectric properties of bismuth titanate. Journal of the American Ceramic Society 79: 3124-3128
22. Maeder MD, Damjanovic D, Voisard C and Setter N (2002) Piezoelectric properties of $SrBi_4Ti_4O_{15}$ ferroelectric ceramics. Journal of Materials Research 17: 1376-1384
23. Wang CM and Wang JF (2008) Aurivillius phase potassium bismuth titanate: $K_{0.5}Bi_{4.5}Ti_4O_{15}$. Journal of the American Ceramic Society 91: 918-923
24. Sanson A and Whatmore RW (2005) Phase diagram of the $Bi_4Ti_3O_{12}$-$BaTiO_3$-$(Na_{1/2}Bi_{1/2})TiO_3$ system. Journal of the American Ceramic Society 88: 3147-3153
25. Jamieson P.B., Abrahams S.C. and Bernstein J.L. (1968) Ferroelectric tungsten bronze – type crystal structures. I. Barium Strontium Niobate $Ba_{0.27}$ $Sr_{0.75}$ $Nb_2O_{5.78}$. The Journal of Chemical Physics 48: 5048-5067
26. Venet M, Vendramini A, Garcia D, Eiras JA and Guerrero F (2006) Tailoring of the lead metaniobate ceramic processing. Journal of the American Ceramic Society 89: 2399-2404
27. Weller MT, Hughes RW, Rooke J, Knee CS and Reading J (2004) The pyrochlore family – a potential panacea for the frustrated perovskite chemist. Dalton Transactions: 3032-3041
28. Ang C and Yu Z (2004) Phase-transition temperature and character of $Cd_2Nb_2O_7$. Physical Review B 70134103
29. Fischer M, Malcherek T, Bismayer U, Blaha P and Schwarz K (2008) Structure and stability of $Cd_2Nb_2O_7$ and $Cd_2Ta_2O_7$ explored by ab initio calculations. Physical Review B 78014108
30. Eerenstein W, Mathur ND and Scott JF (2006) Multiferroic and magnetoelectric materials. Nature 442: 759-765
31. Liu XH, Xu Z, Wei XY and Yao X (2008) Ferroelectric and Ferromagnetic Properties of $0.7BiFe_{1-x}Cr_xO_3$–$0.3BaTiO_3$ Solid Solutions. Journal of the American Ceramic Society 91: 3731-3734
32. Nan CW, Bichurin MI, Dong SX, Viehland D and Srinivasan G (2008) Multiferroic magnetoelectric composites: Historical perspective, status, and future directions. Journal of Applied Physics 103031101
33. Testino A, Mitoseriu L, Buscaglia V, Buscaglia MT, Pallecchi I, Albuquerque AS, Calzona V, Marre D, Siri AS and Nanni P (2006) Preparation of multiferroic composites of $BaTiO_3$-$Ni_{0.5}Zn_{0.5}Fe_2O_4$ ceramics. Journal of the European Ceramic Society 26: 3031-3036
34. Duong GV and Groessinger R (2007) Effect of preparation conditions on magnetoelectric properties of $CoFe_2O_4$-$BaTiO_3$ magnetoelectric composites, pp E624-E627
35. Babu SN, Srinivas K, Suryanarayana SV and Bhimasankaram T (2008) Magnetoelectric properties in NCMF/PZT particulate and bulk laminate composites. Journal of Physics D-Applied Physics 41165407
36. Amin A, Spears MA and Kulwicki BM (1983) Reaction of anatase and rutile with barium carbonate. Journal of the American Ceramic Society 66: 733-738

37. Beauger A, Mutin JC and Niepce JC (1983) Synthesis reaction of metatitanate $BaTiO_{3\,2}$. Study of solid-solid reaction interfaces. Journal of Materials Science 18: 3543-3550
38. Yanagawa R, Senna M, Ando C, Chazono H and Kishi H (2007) Preparation of 200 nm $BaTiO_3$ particles with their tetragonality 1.010 via a solid-state reaction preceded by agglomeration-free mechanical activation. Journal of the American Ceramic Society 90: 809-814
39. Buscaglia MT, Bassoli M and Buscaglia V (2008) Solid-state synthesis of nanocrystalline $BaTiO_3$: Reaction kinetics and powder properties. Journal of the American Ceramic Society 91: 2862-2869
40. Buscaglia MT, Buscaglia V, Viviani M, Dondero G, Rohrig S, Rudiger A and Nanni P (2008) Ferroelectric hollow particles obtained by solid-state reaction. Nanotechnology 19225602
41. Lemanov VV (2007) Barium titanate-based solid solutions. Ferroelectrics 354: 69-76
42. Huang CC and Cann DP (2008) Phase transitions and dielectric properties in $Bi(Zn_{1/2}Ti_{1/2})O_3$-$BaTiO_3$ perovskite solid solutions. Journal of Applied Physics 104024117
43. Lee S and Randall CA (2008) A modified Vegard's law for multisite occupancy of Ca in Ca in $BaTiO_3$-$CaTiO_3$ solid solutions. Applied Physics Letters 92, 111904, 1-3
44. Maso N, Beltran H, Cordoncillo E, Sinclair DC and West AR (2008) Polymorphism and dielectric properties of b-doped $BaTiO_3$. Journal of the American Ceramic Society 91: 144-150
45. Chandratreya SS, Fulrath RM and Pask JA (1981) Reaction Mechanisms in the Formation of PZT Solid Solutions. Journal of the American Ceramic Society 64: 422-425
46. Nakamura Y, Chandratreya SS and Fulrath RM (1980) Expansion During the Reaction Sintering of PZT. Ceramurgia international 6: 57-60
47. Shrout TR, Papet P, Kim S and Lee GS (1990) Conventionally prepared submicrometer lead-based perovskite powders by reactive calcination. Journal of the American Ceramic Society 73: 1862-1867
48. Isupov VA (1980) Reasons for Discrepancies Relating to the Range of Coexistence of Phases in Lead Zirconate-Titanate Solid Solutions. Soviet Physics, Solid State (English translation of Fizika Tverdogo Tela) 22: 98
49. Kingon AI, Terblanche PJ and Clark JB (1982) Effect of Reactant Dispersion on Formation of PZT Solid Solutions. Ceramics International 8: 108-144
50. Hiremath BV, Kingon AI and Biggers JV (1983)Reaction sequence in the formation of lead zirconate-lead titanate solid-solution – role of raw-materials. Journal of the American Ceramic Society 66: 790-793
51. Galassi C, Roncari E, Capiani C and Costa A (2000) Influence of processing parameters on the properties of PZT materials In: Galassi C, Dinescu M, Ukino K and Sayer M (eds) Piezoelectric materials: advances in science, technology and applications. Kluwer Academic Publishers, pp 75-86
52. Medvecky L, Kmecova M and Saksl K (2007) Study of $PbZr_{0.53}Ti_{0.47}O_3$ solid solution formation by interaction of perovskite phases. Journal of the European Ceramic Society 27: 2031-2037
53. Shrout TR, Papet P, Kim S and Lee G-S (1990) Conventionally prepared submicrometer lead-based perovskite powders by reactive calcination. Journal of the American Ceramic Society 73: 1862-1867
54. Amer AM, Ibrahim SA, Ramadan RM and Ahmed MS (2005) Reactive calcination derived PZT ceramics. Journal of Electroceramics 14: 273-281
55. Randall CA, Kim N, Kucera J-P, Cao W and Shrout TR (1998) Intrinsic and extrinsic size effects in fine-grained morphotropic-phase- boundary lead zirconate titanate ceramics. Journal of the American Ceramic Society 81: 677-688

56. Galassi C, Roncari E, Capiani C and Craciun F (1999) Processing and characterization of high Qm ferroelectric ceramics. Journal of the European Ceramic Society 19: 1237
57. Swartz SL and Shrout TR (1982) Fabrication of perovskite lead magnesium niobate. Materials Research Bulletin 17: 1245-1250
58. Kwon S, Sabolsky EM and Messing GL (2001) Low-temperature reactive sintering of 0.65PMN – 0.35PT. Journal of the American Ceramic Society 84: 648-650
59. Sutton WH (1989) Microwave processing of ceramic materials. American Ceramic Society Bulletin 68: 376-386
60. Rao KJ, Vaidhyanathan B, Ganguli M and Ramakrishnan PA (1999) Synthesis of inorganic solids using microwaves. Chemistry of Materials 11: 882-895
61. Gabriel C, Gabriel S, Grant EH, Halstead BSJ and Mingos DMP (1998) Dielectric parameters relevant to microwave dielectric heating. Chemical Society Reviews 27: 213-223
62. Vaidhyanathan B, Singh AP, Agrawal DK, Shrout TR, Roy R and Ganguly S (2001) Microwave Effects in Lead Zirconium Titanate Synthesis: Enhanced Kinetics and Changed Mechanisms. Journal of the American Ceramic Society 84: 1197-1202
63. Kong LB, Zhang TS, Ma J and Boey F (2008) Progress in synthesis of ferroelectric ceramic materials via high-energy mechanochemical technique. Progress in Materials Science 53: 207-322
64. Beitollahi A and Moravej M (2004) Phase formation study of PZT nanopowder by mechanical activation method at various conditions. Journal of Materials Science 39: 5201
65. Rojac T, Kosec A, Malic B and Holc J (2006) The application of a milling map in the mechanochemical synthesis of ceramic oxides. Journal of the European Ceramic Society 26: 3711-3716
66. Xue J, Wan D, Lee SE and Wang J (1999) Mechanochemical synthesis of lead zirconate titanate from mixed oxides. Journal of the American Ceramic Society 82: 1687-1692
67. Lee SE, Xue JM, Wan DM and Wang J (1999) Effects of mechanical activation on the sintering and dielectric properties of oxide-derived PZT. Acta Materialia 47: 2633-2639
68. Alguero M, Moure A, Pardo L, Holc J and Kosec M (2006) Processing by mechano synthesis and properties of piezoelectric $Pb(Mg_{1/3}Nb_{2/3})O_3$-$PbTiO_3$ with different compositions. Acta Materialia 54: 501-511
69. Moure A, Pardo L, Alemany C, Millan P and Castro A (2001) Piezoelectric ceramics based on Bi_3TiNbO_9 from mechano chemically activated precursors. Journal of the European Ceramic Society 21: 1399-1402
70. Ferrer P, Alguero M, Iglesias JE and Castro A (2007) Processing and dielectric properties of $Bi_4Sr_{n-3}Ti_nO_{3n+3}$ (n=3, 4 and 5) ceramics obtained from mechanochemically activated precursors. Journal of the European Ceramic Society 27: 3641-3645
71. Villegas M, Moure C, Jurado JR and Duran P (1993) Influence of the calcining temperature on the sintering and properties of PZT ceramics. Journal of Materials Science 28: 3482-3488
72. Calzada ML and Defrutos J (1994) Nonstoichiometric lead titanate ceramics prepared by a wet-chemical processing. Journal of Materials Science-Materials in Electronics 5: 13-16
73. Choy JH, Han YS and Kim JT (1995) Hydroxide coprecipitation route to the piezoelectric oxide $Pb(Er,Ti)O_3$ (PZT). Journal of Materials Chemistry 5: 65-69
74. Junmin X and Wang J (1999) Lead zirconate titanate via reaction sintering of hydroxide precursors. Journal of Materials Research 14: 1503-1509
75. Chen M, Yao X and Zhang L (2001) Preparation of (Pb, La)(Zr, Sn, Ti)O_3 antiferroelectric ceramics using colloidal processing and the field induced strain properties. Journal of the European Ceramic Society 21: 1159

76. Camargo ER, Frantti J and Kakihana M (2001) Low-temperature chemical synthesis of lead zirconate titanate (PZT) powders free from halides and organics. Journal of Materials Chemistry 11: 1875-1879
77. Roy S, Bysakh S and Subrahmanyam J (2006) Crystallization kinetics of homogeneously precipitated lead zirconate titanate using urea: Comparison with the conventional ammonia precipitated sample. Journal of Materials Research 21: 856-863
78. Cernea M, Manea A, Piazza D, Galassi C and Vasile E (2007) Ba $(Ti_{1-x}Sn_x)O_3$ (x=0.13) dielectric ceramics prepared by coprecipitation. Journal of the American Ceramic Society 90: 1728-1732
79. Narendar Y and Messing GL (1997) Kinetic analysis of combustion synthesis of lead magnesium niobate from metal carboxylate gels, Journal of the American Ceramic Society 80: 915-924
80. Sengupta SS, Ma L, Adler DL and Payne DA (1995) Extended x-ray-absorption fine-structure determination of local-structure in sol-gel-derived lead titanate, lead zirconate, and lead-zirconate-titanate. Journal of Materials Research 10: 1345-1348
81. Fumo DA, Jurado JR, Segadaes AM and Frade JR (1997) Combustion synthesis of iron-substituted strontium titanate perovskites. Materials Research Bulletin 32: 1459-1470
82. Banerjee A and Bose S (2004) Free-standing lead zirconate titanate nanoparticles: Low-temperature synthesis and densification. Chemistry of Materials 16: 5610-5615
83. Montanari G, Costa AL, Albonetti S and Galassi C (2005) Nb-doped PZT material by sol-gel combustion. Journal of Sol-Gel Science and Technology 36: 203-211
84. Cernea M, Montanari G, Galassi C and Costa AL (2006) Synthesis of la and Nb doped PZT powder by the gel-combustion method. Nanotechnology 17: 1731-1735
85. Mercadelli E, Galassi C, Costa AL, Albonetti S and Sanson A (2008) Sol-gel combustion synthesis of BNBT powders. Journal of Sol-Gel Science and Technology 46: 39-45
86. Messing GL, Zhang SC, Jayanthi GV and Narendar Y (1993) Ceramic powder synthesis by spray pyrolysis. Journal of the American Ceramic Society 76: 2707-2726
87. Purwanto A, Wang WN, Lenggoro IW and Okuyama K (2007) Formation of $BaTiO_3$ nanoparticles from an aqueous precursor by flame-assisted spray pyrolysis. Journal of the European Ceramic Society 27: 4489-4497
88. Nimmo W, Ali NJ, Brydson R, Calvert C and Milne SJ (2005) Particle formation during spray pyrolisis of lead zirconate titanate. Journal of the American Ceramic Society 88: 839-844
89. Galassi C, Roncari E, Capiani C, Fabbri G, Piancastelli A, Silvano F (2002) Spray-Dried PZT from Precursors: Cold Consolidation and Sintering. Journal of Materials Synthesis and Processing 9: 213-221
90. Bezzi F, Costa AL, Piazza D, Ruffini A, Albonetti S and Galassi C (2005) PZT prepared by spray drying: From powder synthesis to electromechanical properties. Journal of the European Ceramic Society 25: 3323-3334
91. Costa AL, Galassi C and Roncari E (2002) Direct synthesis of PMN samples by spray-drying. Journal of the European Ceramic Society 22: 2093-2100
92. Demazeau G (2008) Solvothermal reactions: an original route for the synthesis of novel materials. Journal of Materials Science 43: 2104-2114
93. Riman RE, Suchanek WL and Lencka MM (2002) Hydrothermal crystallization of ceramics. Annales De Chimie-Science Des Materiaux 27: 15-36
94. Lencka M, Anderko A and Riman RE (1995) Hydrothermal precipitation of lead zirconate titanate solid solutions: thermodynamic modelling and experimental synthesis. 78: 2609
95. Ohba Y, Rikitoku T, Tsurumi T and Daimon M (1996) Precipitation of lead zirconate titanate powders under hydrothermal conditions. Journal of the Ceramic Society of Japan 104: 6-10

96. Ren ZH, Xu G, Wei X, Liu Y, Shen G and Han GR (2007) Shape evolution of Pb $(Zr,Ti)O_3$ nanocrystals under hydrothermal conditions. Journal of the American Ceramic Society 90: 2645-2648
97. Wei X, Xu G, Ren ZH, Wang YG, Shen G and Han GR (2008) Synthesis of highly dispersed barium titanate nanoparticles by a novel solvothermal method. Journal of the American Ceramic Society 91: 315-318
98. Wei N, Zhang DM, Han XY, Yang FX, Zhong ZC and Zheng KY (2007) Synthesis and mechanism of ferroelectric potassium tantalate niobate nanoparticles by the solvothermal and hydrothermal processes. Journal of the American Ceramic Society 90: 1434-1437
99. Hou YD, Hou L, Zhang TT, Zhu MK, Wang H and Yan H (2007) $(Na_{0.8}K_{0.2})_{0.5}Bi_{0.5}TiO_3$ nanowires: Low-temperature sol-gel-hydrothermal synthesis and densification. Journal of the American Ceramic Society 90: 1738-1743
100. Abothu IR, Lin SF, Komarneni S and Li QH (1999) Processing of $Pb(Zr_{0.52}Ti_{0.48})O_3$ (PZT) ceramics from microwave and conventional hydrothermal powders. Materials Research Bulletin 34: 1411-1419
101. Nyutu EK, Chen CH, Dutta PK and Suib SL (2008) Effect of microwave frequency on hydrothermal synthesis of nanocrystalline tetragonal barium titanate. Journal of Physical Chemistry C 112: 9659-9667
102. Aymonier C, Loppinet-Serani A, Reveron H, Garrabos Y and Cansell F (2006) Review of supercritical fluids in inorganic materials science. Journal of Supercritical Fluids 38: 242-251
103. Reveron H, Elissalde C, Aymonier C, Bousquet C, Maglione M and Cansell F (2006) Continuous supercritical synthesis and dielectric behaviour of the whole BST solid solution. Nanotechnology 17: 3527-3532
104. Lewis JA (2000) Colloidal processing of ceramics. Journal of the American Ceramic Society 83: 2341-2359
105. Lange FF (1989) Powder processing science and technology for increased reliability. Journal of the American Ceramic Society 72: 3-15
106. Velamakanni BV, Chang JC, Lange FF and Pearson DS (1990) New method for efficient colloidal particle packing via modulation of repulsive lubricating hydration forces. Langmuir 6: 1323-1325
107. Paik U and Hackley VA (2000) Influence of solids concentration on the isoelectric point of aqueous barium titanate. Journal of the American Ceramic Society 83: 2381-2384
108. Hsu WT, Yu BY and Wei WCJ (2006) Colloidal processing of $Pb(Zr,Ti)O_3$, part I-Pb dissolution. Journal of Ceramic Processing Research 7: 206-210
109. Bergstrom L, Shinozaki K, Tomiyama H and Mizutani N (1997) Colloidal processing of a very fine $BaTiO_3$ powder – Effect of particle interactions on the suspension properties, consolidation, and sintering behavior. Journal of the American Ceramic Society 80: 291-300
110. Meng CH, Wei WCJ, Shieh J and Chen CS (2006) Colloidal processing of $Pb(Zr,Ti)O_3$ targets part II – Effect of $NbO_{2.5}$ additive. Journal of Ceramic Processing Research 7: 281-287
111. Cho JM and Dogan F (2001) Colloidal processing of lead lanthanum zirconate titanate ceramics. Journal of Materials Science 36: 2397-2403
112. Chen ZC, Ring TA and Lemaitre J (1992) Stabilization and processing of aqueous $BaTiO_3$ suspension with polyacrylic-acid. Journal of the American Ceramic Society 75: 3201-3208
113. Paik U, Hackley VA, Lee J and Lee S (2003) Effect of poly(acrylic acid) and poly(vinyl alcohol) on the solubility of colloidal $BaTiO_3$ in an aqueous medium. Journal of Materials Research 18: 1266-1274

114. Yoshikawa J, Lewis JA and Chun B-W (2009) Comb Polymer Architecture, Ionic Strength, and Particle Size Effects on the $BaTiO_3$ Suspension Stability. Journal of the American Ceramic Society 92:S42-S49
115. Chen LP and Hsu KC (2008) Synthesis of an amide/carboxylate copolymer for barium titanate suspensions. I. As a dispersant. Journal of Applied Polymer Science 108: 2077-2084
116. Mistler RE (1990) Tape casting – the basic process for meeting the needs of the electronics industry. American Ceramic Society Bulletin 69: 1022-1026
117. Mistler RE and Twiname ER (2000) Tape Casting: Theory and practice The American ceramic Society, Westerville
118. Feng JH and Dogan F (2000) Effects of solvent mixtures on dispersion of lanthanum-modified lead zirconate titanate tape casting slurries. Journal of the American Ceramic Society 83: 1681-1686
119. Galassi C, Roncari E, Capiani C and Pinasco P (1997) PZT-based Suspensions for Tape Casting. Journal of the European Ceramic Society 17: 367-371
120. Reddy SB, Singh PP, Raghu N and Kumar V (2002) Effect of type of solvent and dispersant on NANO PZT powder dispersion for tape casting slurry. Journal of Materials Science 37: 929-934
121. Jantunen H, Hu T, Uusimaki A and Leppavuori S (2004) Tape casting of ferroelectric, dielectric, piezoelectric and ferromagnetic materials, Journal of the European Ceramic Society 24: 1077-1081
122. Hotza D and Greil P (1995) Aqueous tape casting of ceramic powders. Materials Science and Engineering a-Structural Materials Properties Microstructure and Processing 202: 206-217
123. Smay JE and Lewis JA (2001) Structural and Property Evolution of Aqueous-Based Lead Zirconate Titanate Tape-Cast Layers. Journal of the American Ceramic Society 84: 2495
124. Song YL, Liu XL, Zhang JQ, Zou XY and Chen JF (2005) Rheological properties of nanosized barium titanate prepared by HGRP for aqueous tape casting. Powder Technology 155: 26-32
125. Feng JH and Dogan F (2000) Aqueous processing and mechanical properties of PLZT green tapes. Materials Science and Engineering a-Structural Materials Properties Microstructure and Processing 283: 56-64
126. Zeng YP, Zimmermann A, Zhou LJ and Aldinger F (2004) Tape casting of PLZST tapes via aqueous slurries, Journal of the European Ceramic Society 24: 253-258
127. Sakar-Deliormanli A, Çelik E and Polat M (2009) Preparation of the $Pb(Mg_{1/3}Nb_{2/3})O_3$ films by aqueous tape casting. Journal of the European Ceramic Society 29: 115-123
128. Sarkar P and Nicholson PS (1996) Electrophoretic deposition (EPD): Mechanisms, kinetics, and application to ceramics. Journal of the American Ceramic Society 79: 1987-2002
129. Nagai M, Yamashita K, Umegaki T and Takuma Y (1993) Electrophoretic deposition of ferroelectric barium-titanate thick-films and their dielectric-properties. Journal of the American Ceramic Society 76: 253-255
130. Doungdaw S, Uchikoshi T, Noguchi Y, Eamchotchawalit C and Sakka Y (2005) Electrophoretic deposition of lead zirconate titanate (PZT) powder from ethanol suspension prepared with phosphate ester. Science and Technology of Advanced Materials 6: 927-932
131. Corni I, Ryan MP and Boccaccini AR (2008) Electrophoretic deposition: From traditional ceramics to nanotechnology. Journal of the European Ceramic Society 28: 1353-1367
132. Ma J and Cheng W (2002) Electrophoretic deposition of lead zirconate titanate ceramics. Journal of the American Ceramic Society 85: 1735-1737

133. Suzuki M, Miyayama M, Noguchi Y and Uchikoshi T (2008) Enhanced piezoelectric properties of grain-oriented $Bi_4Ti_3O_{12}$-$BaBi_4Ti_4O_{15}$ ceramics obtained by magnetic-field-assisted electrophoretic deposition method. Journal of Applied Physics 104: 014102-1-6
134. Calvert P (2001) Inkjet printing for materials and devices. Chemistry of Materials 13: 3299-3305
135. Noguera R, Lejeune M and Chartier T (2005) 3D fine scale ceramic components formed by ink-jet prototyping process. Journal of the European Ceramic Society 25: 2055-2059
136. Wang T and Derby B (2005) Ink-jet printing and sintering of PZT. Journal of the American Ceramic Society 88: 2053-2058
137. Jaworek A and Sobczyk AT (2008) Electrospraying route to nanotechnology: An overview. Journal of Electrostatics 66: 197-219
138. Sun D, Rocks SA, Wang D, Edirisinghe MJ and Dorey RA (2008) Novel forming of columnar lead zirconate titanate structures. Journal of the European Ceramic Society 28: 3131-3139
139. Park SE and Shrout TR (1997) Ultrahigh strain and piezoelectric behavior in relaxor based ferroelectric single crystals. Journal of Applied Physics 82: 1804-1811
140. Tani T (1998) Crystalline-oriented piezoelectric bulk ceramics with a perovskite-type structure. Journal of the Korean Physical Society 32: S1217
141. Messing GL, Trolier-McKinstry S, Sabolsky EM, Duran C, Kwon S, Brahmaroutu B, Park P, Yilmaz H, Rehrig PW, Eitel KB, Suvaci E, Seabaugh M and Oh KS (2004) Templated grain growth of textured piezoelectric ceramics. Critical Reviews in Solid State and Materials Sciences 29: 45-96
142. Suvaci E and Messing GL (2000) Critical factors in the templated grain growth of textured reaction-bonded alumina. Journal of the American Ceramic Society 83: 2041-2048
143. Yoon KH, Cho YS and Kang DH (1998) Molten salt synthesis of lead-based relaxors. Journal of Materials Science 33: 2977-2984
144. Mao Y, Park TJ, Zhang F, Zhou H and Wong SS (2007) Environmentally friendly methodologies of nanostructure synthesis. Small 3: 1122-1139
145. Brosnan KH, Poterala SF, Meyer RJ, Misture S and Messing GL (2009) Templated Grain Growth of < 001 > Textured PMN-28PT Using $SrTiO_3$ Templates. Journal of the American Ceramic Society 92 : S133-S139
146. Richter T, Denneler S, Schuh C, Suvaci E and Moos R (2008) Textured PMN-PT and PMN-PZT. Journal of the American Ceramic Society 91: 929-933
147. Saito Y, Takao H, Tani T, Nonoyama T, Takatori K, Homma T, Nagaya T and Nakamura M (2004) Lead-free piezoceramics. Nature 432: 84-87
148. Jones JL, Iverson BJ and Bowman KJ (2007) Texture and anisotropy of polycrystalline piezoelectrics. Journal of the American Ceramic Society 90: 2297-2314
149. Motohashi T and Kimura T (2007) Development of texture in $Bi_{0.5}Na_{0.5}TiO_3$ prepared by reactive-templated grain growth process. Journal of the European Ceramic Society 27: 3633-3636
150. Gao F, Zhang CS, Liu XC, Cheng LH and Tian CS (2007) Microstructure and piezoelectric properties of textured $(Na_{0.84}K_{0.16})_{0.5}Bi_{0.5}TiO_3$ lead-free ceramics. Journal of the European Ceramic Society 27: 3453-3458

Chapter 2
Processing of Ferroelectric Ceramic Thick Films

Marija Kosec, Danjela Kuscer, Janez Holc

2.1 Introduction

The rapid development of the electronics industry has created the need for high-performance, high-reliability, miniaturised electronic components integrated into various electronic devices. Additional requirements, such as the desired size and weight, low cost, low power consumption, and portability, should be considered to make the devices user friendly and widely accessible. Attempts to miniaturise discrete elements have generally failed due to the difficulty in handling and assembly. A lot of waste material and high costs are also involved. In this approach, the ceramic parts are manufactured as a bulk ceramic, followed by a reduction in size by cutting, polishing, etc., to specified dimensions. The final step is the assembling of a thin layer of ceramic with the other components. This top-down approach imposes limits on the minimum dimensions of the manufactured parts. It constrains the geometry of the parts to simple shapes, like discs, plates, rings, cylinders, etc.

The bottom-up approach, where a layer is built on the substrate, has been shown to be an effective way to produce a thick-film component. The paste, consisting of a fine powder mixed with an organic phase, is deposited on a substrate and fired at a high temperature. The obtained layers, with thicknesses from a few to a few tens of micrometres, are the basis for the thick-film component. This technology enables the direct integration of layers onto the substrates. It therefore eliminates the difficulty of handling the thin, bulk ceramic. Thick-film technology was first successfully applied during the Second World War when conductive silver and resistive carbon inks were deposited on a ceramic substrate. This was to miniaturise an electronic part of a mortar's proximity fuse. The invention of the transistor in the late 1940s initiated an intensive development of this technique. This resulted in more reliable and lower-priced electronic

---

Jožef Stefan Institute, Jamova cesta 39, SI-1000 Ljubljana, Slovenia components. These subsequently became widely used in everyday products such as radios and televisions.

Nowadays, thick-film technology is well established. The implementation of an extra element in the form of films is a natural development. Ceramic ferroelectric thick-film structures are incorporated in micro-electro-mechanical systems (MEMS). These integrate the mechanical elements and the electronics on a particular substrate. This integration results in the production of miniaturised, high-power and highly sensitive sensors, actuators and transducers.

Ferroelectric thick films are planar structures that generally consist of a substrate, a bottom electrode, a ferroelectric film and a top electrode (Fig. 2.1). The thickness of the ferroelectric layer is typically between 1 and 100 μm.

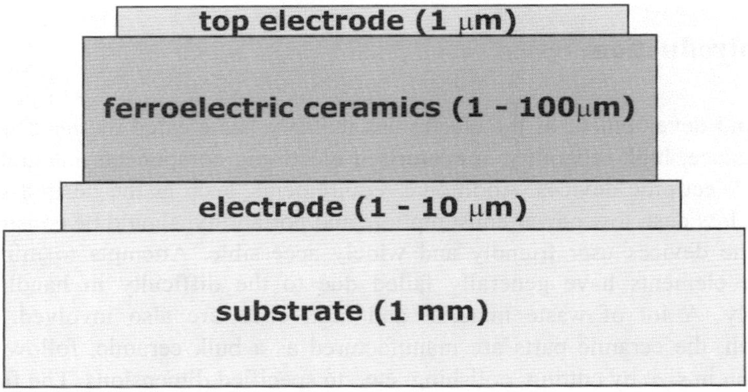

Fig. 2.1. Structure of a ferroelectric thick film.

Ferroelectric thick films are based either on lead-containing or lead-free perovskite materials. The intensive development of ferroelectric materials began in the 1950s with the widespread use of ceramics based on barium titanate ($BaTiO_3$) for capacitors and piezoelectric transducer applications. Later, many lead-based ferroelectric ceramics, including lead titanate ($PbTiO_3$), lead zirconate titanate (PZT), lead lanthanum zirconate titanate (PLZT), and relaxor ferroelectrics such as compositions based on lead magnesium niobate (PMN), were developed and used in a variety of applications. A number of lead-free ferroelectric compositions received attention in the 1990s due to increased environmental awareness. However, only selected compositions have been realised in thick-film form.

Ferroelectric films have been prepared on various substrates like metals (silicon, stainless steel, nickel) and ceramics (alumina, zirconia and, more recently, low-temperature co-fired ceramic (LTCC)). Commonly used electrodes include sputtered or screen-printed metals (platinum, gold, silver, their alloys), oxide-based electrodes (manganites, cobaltites, ruthenates), or composite materials (ruthenate-PZT electrode).

The processing of a thick film involves the synthesis of the powder, the formation of the ink, the deposition of the ink onto the substrate using suitable deposition methods, and the sintering of the deposit to obtain a layer with a good functional response. The processing steps of thick films and bulk ceramics are compared and schematically shown in Fig. 2.2.

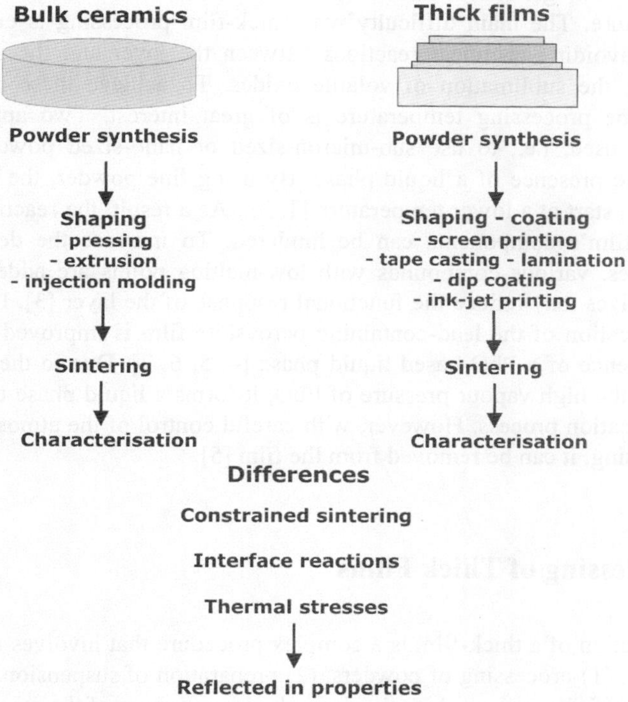

Fig. 2.2. Schematic representation of the processing of thick films as compared to bulk ceramics.

The formation of the thick film involves numerous steps that are characteristic not only for thick films, but also for the manufacturing of bulk ceramics. In both cases, the process begins with the powder synthesis. To obtain a bulk ceramic, the powder-forming processes are chosen on the basis of technical requirements. These include the shape and the size of the product, its microstructure, its properties, its cost, etc. The common forming processes for bulk ceramics include pressing, extrusion, injection moulding and casting. The most common techniques for patterning thick films on various substrates start from powder and include screen printing, pad printing, ink jet, dip coating, electrophoretic deposition, tape casting, and lamination. All these shaping methods require the preparation of suitable suspensions or slurries. After the deposition, the drying and the removal of the organic phases, the green film is fired at a temperature that is sufficient to develop useful properties in the ceramic. In this processing step, referred to as sintering, the shrinkage of the material occurs. The sintering mechanism in a thick film is different from that in a bulk ceramic. Since ceramic compacts shrink approximately uniformly in the x, y and z directions, the thick films are clamped to the substrate. As a consequence, the film shrinks exclusively in the direction perpendicular to the substrate and is subjected to constrained conditions.

To obtain good functional properties, the ferroelectric layer has to be chemically homogeneous without secondary phases and with a uniform, dense microstructure. The main difficulty with thick-film processing is ensuring good adhesion, avoiding chemical reactions between the layer and the substrate and minimising the sublimation of volatile oxides. To achieve these requirements, lowering the processing temperature is of great interest. Two approaches are commonly used, i.e., to use sub-micron-sized or nano-sized powder and/or to sinter in the presence of a liquid phase. By using fine powder, the densification process can start at a lower temperature [1, 2]. As a result, the reactivity between the thick film's components can be hindered. To increase the density at low temperatures, various compounds with low-melting points are added. However, these additives may reduce the functional response of the layer [3]. It is clear that the densification of the lead-containing perovskite film is improved significantly in the presence of a PbO-based liquid phase [4, 5, 6, 7]. Due to the low melting point, and the high vapour pressure of PbO, it forms a liquid phase that improves the densification process. However, with careful control of the atmosphere during the processing, it can be removed from the film [5].

2.2 Processing of Thick Films

The fabrication of a thick-film is a complex procedure that involves the following basic steps: (1) processing of powders, (2) preparation of suspension, (3) shaping, and (4) densification to produce the desired microstructure of the ceramic.

2.2.1 Processing of the Powder

The starting powder is a crucial factor for processing high-quality ferroelectric thick films. The performance characteristics of a sintered thick film are significantly influenced by the precursor powder's characteristics. Among the most important characteristics are the particle size, the particle size distribution, the chemical composition, and the chemical homogeneity of the powder. Agglomerate-free powders with a narrow size distribution can be compacted with a high green density. When they are in the nano-sized region, these powders can be sintered at reduced temperatures.

A powder that is used for processing a thick film is generally prepared by solid-state synthesis. After the homogenisation, the powder mixture is annealed at elevated temperatures to ensure the formation of the desired compound. The obtained powder consists of micron-sized particles. Ferroelectric materials have a regularly complex chemical composition, and the processing of a single-phase composition is challenging.

The solid-state processing of perovskite powders with a complex chemical composition may result in a chemically non-homogeneous distribution of the elements in a single-phase calcined powder. One such example is screen-printed $(Pb,La)(Zr,Ti)O_3$ (denoted PLZT) thick films on an alumina substrate [8]. The paste for screen printing was prepared from PLZT powder calcined at 900°C. The chemically inhomogeneous distribution of all the elements was observed after sintering at 1050°C. By increasing the sintering temperature to 1150°C, the chemical homogeneity was generally improved. However, the loss of PbO was observed as a result of the intensified sublimation of the PbO.

One possible approach to processing chemically homogeneous thick films, with complex chemical compositions at temperatures as low as possible, is to use a chemically homogeneous starting powder with nano-sized particles.

Mechano-Chemical Synthesis, also known as high-energy milling, may serve as an alternative to solid-state synthesis. It enables the synthesis of nano-sized powders with a complex chemical composition at close to room temperature by introducing mechanical energy into the powder mixture.

A chemically homogeneous 0.65 $Pb(Mg_{1/3}Nb_{2/3})O_3$ – 0.35 $PbTiO_3$ (denoted PMN-PT) thick film screen printed on a platinised alumina substrate was prepared at 950°C [9, 10]. The suspension was prepared from high-energy-milled powder mixed with an organic phase. The obtained PMN-based powder was characterised by a nano-sized particle and a high chemical homogeneity [11, 12, 13]. PMN-PT thick films were sintered in the presence of a PbO-based liquid phase and this resulted in a single-phase, dense thick film with a good functional response [13]. In addition, by using a low sintering temperature, the chemical reactivity between the thick film and the substrate was minimised.

The Hydrothermal Method is of considerable interest for the synthesis of nano-structured powders and thin- and thick-film layers on metal substrates. This is so because it is a low-cost and environmentally friendly technique. The hydrothermal method utilises the chemical reactions among different ions dissolved in solution and exposed to high temperatures and elevated pressures. Nano-sized PZT powders with spherical particles between 5 and 10 nm have been synthesised from inexpensive metal salts at 160°C [14]. In addition, this method enables not only the synthesis of the compounds from a precursor solution but also a simultaneous deposition on the substrate. A deposit up to a few tens of micrometres thick on a complex shaped substrate can be obtained. PZT- and PMN-based thick films have been grown *in-situ* on a titanium substrate from oxide-based precursors [15, 16, 17, 18].

The Coprecipitation [19, 20, 21] and sol-gel [22] methods have been used for processing ferroelectric thick films from powders.

2.2.2 Shaping Methods

There are a number of suitable methods for making a film in the range of few to a hundreds of μm. These methods can be divided into two groups: those methods primarily designed for thick films (screen printing, tape casting), and those developed for thin films and 3D structures.

2.2.2.1 Screen Printing

The most common method for processing thick films is screen printing. It is a relatively simple process that makes it possible to deposit various materials on a wide variety of substrates, such as ceramics, metals, glass, textile, organic flexible substrates, etc. It is equally well suited to small-scale batches and to high-volume production. This process makes it possible to produce films with thicknesses from a few to several tens of μm with a reliable lateral resolution of above 100 μm.

Screen printing is well suited to the production of electronic components. Its great advantage is its ability to realise the whole structure, ranging from the bottom electrode to the ferroelectric film and the upper electrode, with the same technology and in some cases even with co-firing.

The method requires suspensions with a relatively high viscosity (ink, paste) containing the powder and the "organic vehicle" that is prepared under shear-mode mixing in a three-roll mill. The composition of the suspension should be carefully designed to obtain pseudo-plastic properties in order to moderate the thixotropic properties. In addition, a solvent with a relatively high boiling point is required to avoid drying during the printing. For laboratory experiments the well-known α-terpineol and butylcarbitole are used as the solvents and ethyl-cellulose is used as a binder [23]. The suspension is squeezed through the screen onto the substrate, either manually or automatically. After the deposition, the film is dried to remove the solvents. The desired thickness of the structure is ensured by multiple screen-printing and drying processes. Finally, the organic components, such as polymers and modifiers, are removed from the layer using a thermal treatment between 300°C and 600°C. This is subsequently densified by heating to an appropriate temperature. The technical details of the equipment and the technology can be found elsewhere [3, 24, 25]. The resulting film suffers from a relatively low green density. However, it can be improved by additional isostatic pressing of the green deposit [5, 26].

An adapted screen-printing method was used to prepare PZT films with a thickness up to 3 μm on stainless steel or silicon substrates. To increase the density of the thick film, a sol-infiltration procedure was used [27]. After the screen printing and the burning out of the organic phase, the films were coated with a sol and then sintered [28, 29]. Another process involves the screen-printed and subsequently sintered films being coated with a solution. They are then thermally treated again, but at a lower temperature [30]. After the annealing of each deposit, the sol was infiltrated into a porous structure and thermally treated [31]. In order to decrease the sintering temperature of the PZT thick films, low-melting point additives were added to the basic dispersed phase [32]. The so-called ComFi technology uses a slurry containing a sol, the powder and low-melting-point additives deposited on a substrate, and the subsequent infiltration of the sol into each deposited layer [1, 33, 34].

2.2.2.2 Tape Casting

Tape casting is a forming technique for producing thin, flat ceramics. The method was originally developed for producing electronic ceramics, including substrates, packages and multilayer capacitors. The tape thickness that can be achieved is generally in the micrometre-to-millimetre range.

The process starts with the preparation of a concentrated suspension containing the deflocculated powder in an organic solvent or water, mixed with several additives, such as dispersing agents, a binder and a softener. The suspension is subsequently cast by means of the tape-casting facility. During the process, the suspension flows from a storage container onto a plastic foil, which is continuously moved with a controlled velocity under the container. A deposit is formed on the plastic foil. The height of this deposit is controlled by a doctor blade, which determines the final thickness of the green tape. After casting the suspension, the green ceramic foil passes into a drying chamber, in which the foil is dried [35].

The green tapes can be used for various applications. They can be laminated and consequently sintered to form ceramic substrates, which have been applicable in thick-film technologies [36, 37, 38]. Lamination is widely used for producing low-temperature, cofired-ceramic (LTCC) tapes. LTCC technology is a three-dimensional ceramic technology utilising the third dimension (z) for the interconnecting layers, the electronic components, and the different 3D structures, such as cantilevers, bridges, diaphragms, channels and cavities. Thick-film technology contributes the lateral and vertical electrical interconnections, and the embedded and surface passive electronic components (resistors, thermistors, inductors, capacitors).

2.2.2.3 Electrophoretic Deposition

Electrophoretic deposition (denoted EPD) is a processing method that enables the shaping of various materials in a variety of shapes and dimensions. For example, it can be used to produce coatings, films, and free-standing objects. EPD is a process in which, in the first step, charged particles, suspended in a liquid medium, migrate towards an electrode when applying an electric field. In the second step, they deposit on an electrode. The suspension for EPD consists of the particles suspended in a solvent with some additives [39].

Aqueous and non-aqueous suspensions can be used for EPD [40]. The solvents should be inert with respect to the powder. The main advantage of aqueous suspensions is that high deposition rates can be obtained by the application of relatively low electric field strengths. However, when applying DC voltages, the electrolysis was found to induce the decomposition of the water and cause gas formation. The deposit may contain pinholes in addition to a lack of adhesion with the substrate. Organic solvents are commonly used for EPD, but they need severe safety precautions, leading to increased processing costs. When compared to other thick-film processing routes, the advantage of EPD is the possibility to process a deposit in a wide range of thicknesses, from a few tens of nanometres to hundreds of millimetres. There is also the possibility to deposit not only on flat but also on curved substrates. In addition, the method is fast, inexpensive, requires simple equipment, and is suitable for mass production.

A comprehensive overview of the electrophoretic deposition of ceramic materials has been published [41]. The PZT deposits were prepared by EPD from water-based suspensions using hydrothermally synthesised PZT powder [42, 43]. It is reported that after sintering at 1100°C, micro-cracks are formed in PZT layers thicker than 5 μm. At this temperature, the PZT chemically interacts with the substrate. A range of non-aqueous colloids have been studied in order to deposit the PZT onto various substrates. PZT films about 20 micrometres thick were prepared from micron-sized PZT powder dispersed in acetyl-acetone with the addition of iodine [44]. PZT particles dispersed in glacial acetic acid have been deposited on an electroded alumina substrate [45], and on metal foils [46, 47]. It is reported that when the sintering occurred in the presence of the liquid phase, by the addition of both Li compounds and PbO, a 10-μm-thick film on Al_2O_3 exhibited a polarisation switching behaviour similar to bulk ceramics. When PZT is deposited on a Cu foil, the formation of a Cu_xPb alloy and, consequently, the deterioration of the functional properties of the PZT thick-film was reported [47]. PZT has been deposited on SiC fibres using a coprecipitated PZT powder suspended in a mixture of water, ethanol, acetone and acetylacetone solvent [19]. High-quality crack-free layers with a thickness up to 40 μm have been reported.

2.2.2.4 Inkjet Printing

Direct-write assembly techniques offer the possibility of fabricating ceramic materials with complex 2D and 3D structures. Inkjet printing involves the direct deposition of colloidal inks in a desired pattern via a layer-by-layer build sequence. The printing information is created directly from a computer and stored digitally. Inkjet printing has several key advantages. It is a simple, non-contact technique that deposits the material in the desired pattern via a nozzle. It avoids the use of screens, printing plates or photolithography. It has good resolution, being capable of depositing tracks with a 50-μm width, with the potential for future improvement. The process is potentially compatible with many rigid and flexible substrates. It enables rapid development and manufacture and is well suited to high-speed, multilayer processing. It offers the possibility of producing a range of electronic components and three-dimensional structures. It is a low-waste, highly flexible process, equally suited to mass production and small development batches.

The inkjet printing method relies on the formulation of suitable colloidal inks with the desired degree of colloidal stability and rheological behaviour. Colloid inks for direct ink-jet printing typically contain 5 to 40 vol. % of solids. They must be agglomerate-free to avoid clogging of the print-head nozzle and they must form a consistent droplet. This successful droplet formation requires careful control of the surface tension and the rheological parameters of the fluid, such as viscosity, yield stress under shear and compression, and viscoelastic properties. The droplet's spreading influences the lateral resolution and the thickness of the deposit. To minimise it, a high drying rate for the fluid is desired, in addition to the proper surface tension and viscosity.

Numerous successful inkjet printings of PZT have been demonstrated. For example, an aqueous suspension of PZT particles has been successfully deposited on paper [48], and a paraffin-oil/wax-based suspension of PMN–PT has been studied for deposition [49]. A PZT self-standing 3D structure has also been demonstrated [50], and a lot of attention has been given to the inkjet printing of PZT pillar structures, used mainly for ultrasonic transducer applications. The successful fabrication of PZT 1-3 composites has also been demonstrated [51, 52, 53].

2.2.2.5 High Density Deposition Methods

Screen printing, tape casting, inkjet printing, and electrophoretic deposition produce green layers of low density as a result of the low stresses applied to the powder particles during shaping. Deposits with a high density can be obtained by using high energy deposition methods. These techniques rely on the direct deposition of the powder on a certain substrate and are referred to as a jet printing [22, 54], aerosol deposition [55, 56, 57], and airflow deposition [58]. A submicron-sized powder is used, and it is mixed with a high pressure carrier gas to form an aerosol flow. It is then injected into the deposition chamber. The accelerated particles collide with the substrate to form a dense ceramic film at room temperature. These methods enable the low-temperature fabrication of high-quality complex structures, such as mono-morphs, bimorphs, multilayer stacks, and compositionally graded elements [59].

PZT-PMN-based thick films with thicknesses from 5 to 200 μm have been deposited on nickel substrates. A density of ~80 % of the theoretical value (TD) for various thick films has been obtained after the deposition. The green density of aerosol-deposited films is higher than the one obtained for isostatically pressed films, i.e., 67 % TD. After the sintering between 800 and 1000°C, films with more than 98 % of TD and good ferroelectric properties were reported [58]. Lebedev *et al.* [57] reported that a PZT thick film, aerosol deposited on a platinised silicon substrate at 550°C and post-annealed at 600°C, possessed a density higher than 95 % and good functional properties.

2.2.2.6 Thin-Film Adapted Methods

Films with thicknesses of 1 to 10 μm have been prepared using the chemical solution deposition (CSD) method. Two approaches have been used. The first one is multiple spin or dip coating and intermediate firing of the deposit. With this method, PZT films with a thickness of 1 to 4 μm have been prepared on a silicon substrate [60]. It was also reported that 10-μm-thick PZT films prepared by numerous deposition methods exhibited d_{33} values of 220 pm/V. This is comparable to the values of PZT bulk ceramic [61]. Due to the necessity for numerous deposition layers, and the time-consuming processing, the deposition was run automatically. A 15-μm-thick PZT layer has been demonstrated using an automatic dip-coating procedure [62].

The second approach is an adapted thin-film method. The CSD precursor for the coatings is modified with the addition of nano-sized particles with the same composition to increase the solids load in the slurry. Consequently, the viscosity and the density of the slurry are increased. A higher viscosity leads to a thicker deposited layer in a single step. A higher density reduces the subsequent shrinkage and prevents the film from the cracking. When cracks are formed in the deposit, the powder inhibits their propagation. After the deposition process, the film is fired at a typical CSD-processing temperature. This approach is suitable for large-area deposition processes and is compatible with silicon technology.

Deposits with thicknesses of a few tens of micrometers can be processed by the interfacial polymerisation method and a composite precursor. This one-step method used an alkoxide precursor solution that is put into the reaction vessel containing water and the substrate. At the interface of two immiscible liquids, a gel layer is formed. After draining the water, the gel layer is placed on the substrate. After the sintering, a PZT layer with a thickness of 23 μm and a good functional response was obtained [63].

2.2.2.7 Other Methods

Thick films can also be made using micropen writing [64, 65], robocasting [66, 67], the micro-stereo-lithographic process, gelcasting [68, 69], electrohydrodynamic deposition [70] and others.

2.2.3 Densification of Thick Films

To obtain a suitable functional response, the film has to be chemically homogeneous without an undesirable phase and with a uniform microstructure.

The film deposited on a substrate has to be sintered. The film is clamped to the substrate and therefore it densifies in constrained conditions. In contrast to a bulk ceramic that shrinks isotropically, the thick-film structure shrinks exclusively in the direction perpendicular to the substrate. This behaviour results in different microstructural characteristics of the film when compared to a bulk ceramic processed under identical sintering conditions (temperature, time, and atmosphere).

During the processing, the thick film is in direct contact with chemically different materials. Therefore, the main difficulty with thick-film processing is ensuring good adhesion as well as avoiding chemical reactions between the film and the substrate. To achieve these requirements, lowering the processing temperature is of great interest. The sintering temperature can be lowered by the addition of low-melting-point compounds or glass frit. This allows sintering in the presence of a liquid phase. The particle size of the powder also influences the sinterability of the thick films. The higher surface/volume ratio of the fine particles consequently leads to a higher density at lower sintering temperatures. By using fine powder, the sintering temperature is lowered. Consequently, the chemical reactivity between the components of the thick-film structure is hindered.

The formation of undesirable reaction products is observed when the layer and the substrate are not chemically compatible phases under particular sintering conditions. The chemical reactivity of the substrate and the thick films may be hindered by incorporating an additional layer between them. This acts as a diffusion barrier that hinders the formation of undesired reaction products. This may therefore improve the adhesion between the film and the substrate.

2.2.3.1 Constrained Sintering

The substrate and the as-deposited layer expand during the thermal treatment of a ceramic thick-film structure deposited on a substrate. At a particular temperature, the film tends to shrink in all directions due to the driving force for sintering. The film is clamped to the rigid substrate and therefore cannot shrink in the plane, but only in the direction perpendicular to the substrate. This results in a tensile stress in the plane of the substrate because the sintering occurs in constrained conditions. The constrained sintering of a thick-film structure on a rigid substrate is schematically shown in Fig. 2.3.

The sintering behaviour of the thick-film structure under constrained sintering conditions has been widely discussed [71, 72, 73]. Under constrained conditions, a slower densification rate and the generation of processing defects have been observed in thick-film structures. This may originate from the stresses that are present in the thick-film structure during the sintering. The in-plane tensile stresses reduce the driving force for sintering and promote the formation of cracks in the structure. The magnitude of the tensile stress depends on the shear rates and the densification rate of the film. When the thick-film structure responds in an appropriate way to the stresses, the number of defects is reduced and the density is improved.

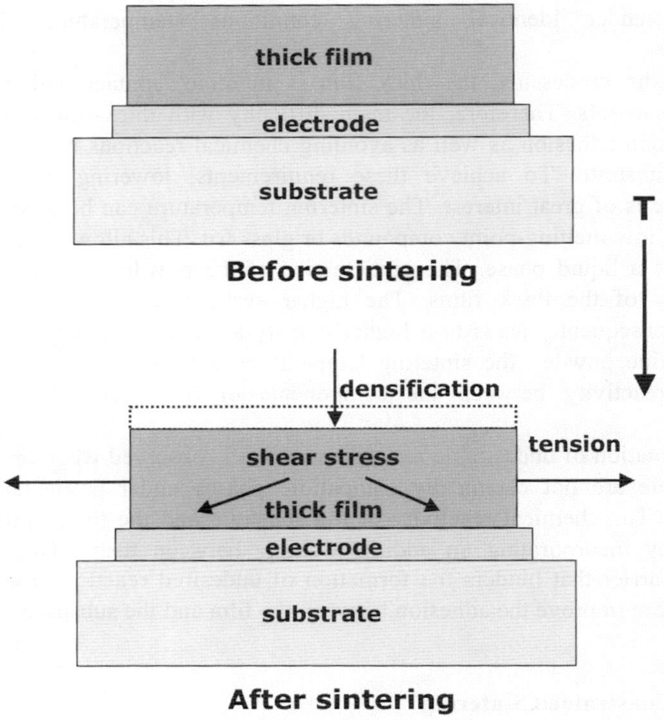

Fig. 2.3. Schematic representation of the constrained sintering of a thick film on a rigid substrate.

The presence of a liquid seems to be beneficial, since it helps release the shear stress through a process of particle rearrangement. This may also be due to enhanced transport via a stress-supported dissolution-precipitation mechanism. The need for a liquid phase to obtain high-density films was confirmed by sintering PLZT thick films [5]. The presence of a liquid phase (PbO based) was ensured by a PbO-saturated atmosphere that prevented the sublimation of the lead oxide, which was initially added to the PLZT in excess [74]. For a comparison, the other samples were kept in an atmosphere that allowed switching from liquid- to solid-phase sintering. The densification was retarded with the disappearance of the liquid phase. The amount of liquid phase can, however, be kept low [5].

The shear stress can be released via particle rearrangement (sliding) and/or enhanced transport processes involving a stress-supported dissolution-precipitation mechanism. This leads to coarsening and reshaping of the grains. The densification of the film is enhanced in comparison with the bulk [5]. Even a minor amount of liquid phase, provided by the capillary condensation of PbO in the pellets with an initially stoichiometric composition, supports densification [75, 76].

2.2.3.2 Sintering in the Presence of a Liquid Phase

The densification of thick-film structures is enhanced in the presence of a liquid phase. Several requirements should be taken into account when designing the sintering in the presence of a liquid phase. First, the melting point of the additive should be lower than the onset sintering temperature of the material. This compound has to be thermodynamically stable in equilibrium with other thick-film components under the processing conditions (temperature, atmosphere). The additive is usually not a ferroelectric material and may reduce the ferroelectric response of the layer. When the additive forms a solid solution with the ferroelectric material, it can change the functional response due to doping. The amount of liquid should be optimised to minimise its influence on the functional properties. It can also be transient and disappear during the processing as a result of evaporation and/or incorporation into the film.

PbO is often used for the processing of lead-based ferroelectric thick films. The addition of a few weight percent of excess PbO to the PZT starting powder gives a corresponding amount of PbO-based liquid phase at temperatures above the melting point of the ternary eutectic in $PbO–ZrO_2–TiO_2$ [77].

PbO liquid-phase-assisted sintering was exploited in several lead-based thick-film structures [5, 6, 7, 13, 78, 79, 80, 81, 82, 83]. Typically, 1 to 5 mol % of PbO is added to lead-based ferroelectric material to ensure sintering in the presence of the liquid phase. However, due to the high vapour pressure of PbO under the sintering conditions, it tends to sublimate from the film. It is necessary, therefore, to ensure a PbO-rich atmosphere around the sample during the sintering course to prevent the loss of PbO from the layer. It is highly desirable that the film does not contain a secondary phase after the sintering. An attempt has been made to remove the PbO-based liquid phase from the film during the final stage of the process.

It was shown that a range of microstructural properties and a very different functional response can be obtained for a 0.65 $Pb(Mg_{1/3}Nb_{2/3})O_3$–0.35 $PbTiO_3$ thick film sintered at an identical temperature and time. This needed a different PbO-atmosphere, determined by the amount of packing powder [13]. When a PMN-PT thick film was surrounded by a small amount of packing powder, all the excess PbO sublimated from the PMN–PT in the initial stage of the sintering. The period when the liquid PbO was present in the PMN–PT was not sufficiently long to obtain a dense PMN–PT layer. The film was then characterised by poor functional properties. When using a larger amount of packing powder around the sample, the excess PbO remained in the PMN–PT for a longer period of the thermal treatment. This resulted in liquid-phase sintering. The film sintered with an optimal amount of packing powder was thinner and denser, with a significantly better functional response. When using a large amount of packing powder, the PbO remained in the sample. Even though the PMN–PT film is dense, its functional response was lower due to the presence of a thin PbO dielectric layer at the grain boundaries of the PMN-PT.

PbO and various compounds or glass frits with a low melting point have been added to the starting powder, to lower the processing temperature of ferroelectric thick films. For lead-based ferroelectrics, the sintering temperature can be lowered with binary and ternary mixtures of low-melting oxides such as $Pb_5Ge_3O_{11}$ [63, 84, 85], Cu_2O-PbO [1, 32, 33], Li_2CO_3-Bi_2O_3 [86, 87], Bi_2O_3-ZnO [30], Li_2CO_3-Bi_2O_3-CuO [88], $Pb_5Ge_2SiO_{11}$ [88], PbO-PbF_2 [89], $Ba(Cu_{0.5}W_{0.5})O_3$, $BiFeO_3$ [90] and Li_2CO_3 [91]. The additives for lead-free ferroelectric materials are CuO [92], $K_{5.4}Cu_{1.3}Ta_{10}O_{29}$ [93], (Na,K)-germanate [94] and $BiFeO_3$ [95].

Glass frits are also used as additives for lowering the sintering temperature. The borosilicate glass phases [96], B_2O_3-Bi_2O_3-CdO [89, 97] or a mixture of boron oxide and silica [34, 98, 99, 100] have been added to lead-based ferroelectric thick films.

2.3 Processing of Ferroelectric Thick Films on Various Substrates

2.3.1.1 Ferroelectric Thick Films on Silicon

The piezoelectric thick-film structures deposited on silicon substrates have been intensively studied for various applications such as sensors, accelerometers and transducers. The advantage of using a silicon substrate is that it is a well-developed technology that offers the possibility of Si micromachining. It also enables the integration of the thick-film material and the electronics within a single chip.

To obtain a good functional response of the ferroelectric layer, lead-based materials require thermal treatment at a temperature between 800°C and 900°C. At these temperatures the volatility of the PbO and the interdiffusion of the lead oxide and Si through the bottom electrode are significant. As a result, SiO_2, which is commonly used as a passivation layer for a Si wafer, chemically reacts with the PbO and forms lead-silicate compounds at the substrate/electrode interface [101, 102]. Consequently, they lead to the delamination of the electrode from the silicon [103].

Various approaches have been used to improve the adhesion of the lead-based ferroelectric layer on the Si substrate. It was shown, for example, that the bottom electrode may act as a buffer layer. The interdiffusion of Si and PbO has not been observed through a continuous and dense gold electrode. Consequently, the delamination of the PZT from the Si substrate was avoided [104]. A PZT thick film with the addition of low-melting-point compounds was deposited on silicon, and a bilayer Au/Pt structure was used as an electrode. Good dielectric properties and a high d_{33} value of the thick-film structure were reported [97].

To improve the adhesion of the layer on the silicon substrate, and to minimise the chemical reactivity between the thick-film components, various structures have been reported. Si/SiO_2/Cr/Pt [54], Si/SiO_2/Ir/Pt [57, 63], Si/SiO_2ZrO_2/Pt [105], Si/SiO_2/YSZ/TiO_2/Pt [28, 29], Si/SiO_2/Ti/Pt [88], Si/SiO_2/Si_3N_4/TiO_2/Pt [103] and Si/SiO_2/Al_2O_3/Au [106] have been tested successfully.

2.3.1.2 Ferroelectric Thick Films on Ceramic Substrates

There are several potential applications of ferroelectric thick films on ceramic substrates, including gravimetric sensors [107] pyroelectric sensors [78, 83], ultrasound medical transducers [6, 85, 108, 109], microbalances [80], pressure sensors [36, 110] and electrophoretic printing [5, 30].

Alumina is a widely used material as a ceramic substrate because it is thermally stable and chemically inert. However, the presence of a small amount of oxides such as MgO, SiO_2, CaO in the alumina substrate causes the formation of secondary phases at the thick-film/substrate interface.

An example is a PLZT thick film on a platinised alumina substrate [111]. It was observed that during the processing the PLZT together with the bottom Pt electrode always peeled off the alumina substrate. The main reason was the formation of beta-alumina crystals at the interface between the alumina and the platinum.

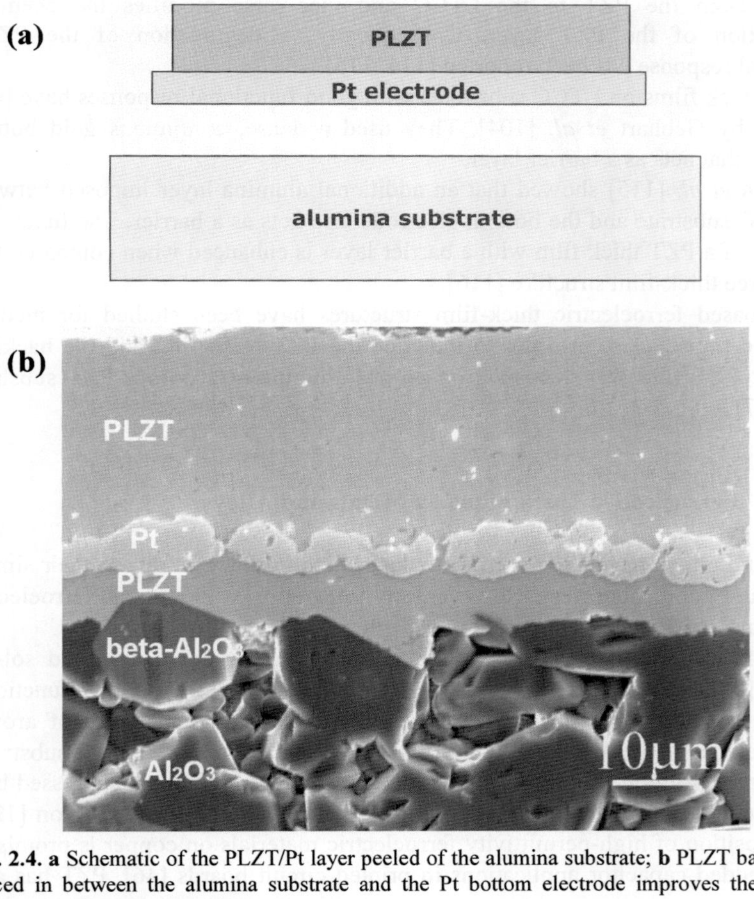

Fig. 2.4. a Schematic of the PLZT/Pt layer peeled of the alumina substrate; **b** PLZT barrier layer placed in between the alumina substrate and the Pt bottom electrode improves the adhesion between the PLZT layer and the substrate.

To avoid the delamination of active PLZT/Pt from the alumina substrate, the formation of undesirable large crystals at the Pt/alumina interface should be eliminated. The incorporation of an additional PLZT barrier layer between the alumina and the Pt electrode hinder the diffusion of Al_2O_3 into the active PLZT. It minimised the reaction products and, therefore, the undesirable reaction products are not formed at the alumina/Pt interface (Fig. 2.4) [5].

The weak adhesion between the bottom Pt electrode and the alumina substrate was applied for processing substrate-free PMN-PT bending-type actuators. Composites were peeled off from the substrates [112] after screen printing and firing the PMN-PT/Pt.

For some applications, such as micro-sensors or actuators, the ferroelectric thick-film structures tend to be integrated into low-temperature co-fired ceramic (LTCC). This is a key substrate material in micro-system technologies [104, 113]. LTCC exhibits a high chemically reactivity with ferroelectric materials at a typical temperature of 850°C required for the processing. The considerable inter-diffusion of ions from the PZT to the LTCC and vice-versa modifies the chemical composition of the PZT layer. Consequently, a degradation of the PZT's functional response has been reported [114, 115].

PZT thick films on LTCC substrates with good functional responses have been reported by Gebhart *et al.* [104]. They used a dense, continuous gold bottom electrode that acts as a barrier layer.

Hrovat *et al.* [115] showed that an additional alumina layer imposed between the LTCC substrate and the bottom electrode also acts as a barrier. The functional response of a PZT thick film with a barrier layer is enhanced when compared to a barrier-free thick-film structure [116].

PZT-based ferroelectric thick-film structures have been studied for medical ultrasonic transducers. In order to integrate the PZT thick film and the backing, the thick PZT film was deposited on porous alumina or a porous PZT substrate [108, 117].

2.3.1.3 Ferroelectric Thick Films on Metals and Alloys

Metals and alloys have been used to replace silicon substrates due to their simple tooling and good robustness. However, any integration with ceramic ferroelectric thick-films is difficult.

PZT films have been processed on stainless steel using a modified sol-gel method [27], and the aerosol deposition technique [118, 119]. Good functional response of PZT films on stainless steel prepared at a temperature of around 600°C has been reported. PZT thick-films have also been made on Ni substrates using screen printing. The excessive oxidation of Ni was partially suppressed by a double screen-printed Au electrode, with the first one being fired in argon [120]. The deposition of high-permittivity ferroelectric materials on copper is promising for embedded capacitor applications in printed circuit boards [46]. PZT has also been deposited by electrophoretic deposition on copper foil. After sintering, a Cu_xPb alloy was formed, which deteriorated the properties of the PZT thick film.

A PZT ferroelectric thick film with an oxide bottom electrode has been demonstrated. A composite lead-ruthenate-PZT electrode was screen printed and fired on an alumina substrate, followed by screen-printed and sintered PZT [121].

2.4 Summary

A brief review of the processing features of ferroelectric thick films has been presented here. It is clear that the processing of ferroelectric thick films has advanced significantly in the past 15 years. This has been a typical application-driven research that has resulted in a number of practical solutions. More effort is needed in powder synthesis and in novel deposition methods. A better understanding of phenomena like constrained sintering, the physics and chemistry of interfaces, the general properties of thick films and, in particular, the processing-properties relationship is required.

2.5 Acknowledgment

The financial support of the Slovenian Research Agency and EU 6FP Network of Excellence MIND (NoE 515757-2) is gratefully acknowledged.

References

1. Dorey RA, Whatmore RW (2004) Electrical properties of PZT and PMN-PT/PZT thick films produced using ComFi technology. J. Eur. Ceram. Soc. 24:1091- 1094.
2. Kwon S, Sabolsky E, Messing GL (2001) Low-temperature re sintering of 0.65PMN-0.35PT. J. Am. Ceram. Soc. 84:648-650.
3. Dorey R A, Whatmore R W (2004) Electroceramic thick film fabrication for MEMS. J.Electroceram., 12:19-32
4. Guha JP, Hong DJ, Anderson, HU (1998) Effect of excess PbO on the sintering characteristic and dielectric properties of $Pb(Mg_{1/3}Nb_{2/3})O_3 - PbTiO_3$-based ceramics. J. Am. Ceram. Soc. 71:C152-154.
5. Kosec M, Holc J, Malič B, Bobnar V (1999) Processing of high performance lead lanthanum zirconate titanate thick films. J. Eur. Ceram. Soc. 19: 949-954.
6. Kosec M, Murko D , Holc J, Malič B, Čeh M, Hauke T, Beige H (2001a) Low-temperature processing of $(Pb,La)(Zr,Ti)O_3$ thick films on alumina substrates. Z. Met. Kd. 92: 97-104

7. Saha D , Sen A, Maiti H S (1999) Low temperature liquid phase sintering of lead magnesium niobate. Ceram. Int. 25:145-151
8. Bernik S, Marinenko RB, Holc J, Samardžija Z, Čeh M, Kosec M (2003) Compositional homogeneity of ferroelectrics (Pb,La)(Ti,Zr)O_3 thick films. J. Mat. Res. 18:515-523.
9. Kosec M, Holc J, Kuscer D, Drnovšek S (2007) $Pb(Mg_{1/3}Nb_{2/3})O_3$ – $PbTiO_3$ thick films from mechanochemically synthesized powder. J. Eur. Ceram. Soc. 27:3775-3778
10. Kuscer D, Kovač J, Tchernychova Šturm E, Kosec M (2009a) Characterisation of the amorphous phase and the nano-sized crystallites in high-energy-milled lead-magnesium-niobate powder, J. Am. Ceram. Soc. 92:1224-1229.
11. Kuscer D, Meden A, Holc J, Kosec M (2006) Mechano-Synthesis of Lead-Magnesium-Niobate Ceramics. J. Am. Ceram. Soc. 89:3081-3088
12. Kuscer D, Holc J, Kosec M (2007) Formation of $Pb(Mg_{1/3}Nb_{2/3})O_3$ -0.35 $PbTiO_3$ using a high-energy milling process. J. Am. Ceram. Soc. 90:29-35
13. Kuscer D, Skalar M, Holc J, Kosec M (2009) Processing and properties of 0.65 $Pb(Mg_{1/3}Nb_{2/3})O_3$ -0.35 $PbTiO_3$ thick films. J. Eur. Ceram. Soc. 29:105-113
14. Deng Y, Liu L, Cheng Y, Nan CW, Zhao S, (2003) Hydrothermal synthesis and characterisation of nanocrystalline PZT powders. Mat. Lett. 57:1675-78.
15. Chen X, Fan H, Ke S (2006) Low-temperature synthesis of (Pb,La)(Zr,Ti)O_3 thick-film on Ti substrates by the hydrothermal method using oxide precursors. Appl. Phys. Lett., 88:012901
16. Chen X, Fan H, Liu L, Ke S (2008) Low-temperature growth of lead-magnesium niobate thick films by hydrothermal process. Ceram. Int. 34:1063-66
17. Ndiaye PA, Loiseau B, Minaut S, Pernot P, Tricot JC (1999) $PbZr_xTi_{1-x}O_3$ hydrothermal synthesis on titanium substrate for actuators. Myrosystem Techn. 6:15-18.
18. Ndiaye PA, Loiseau B, Minaut S, Pernot P, Tricot JC (1999) $PbZr_xTi_{1-x}O_3$ hydrothermal synthesis on titanium substrate for actuators. Myrosystem Techn. 6:15-18.
19. Heinrich JG, Kim JW (2005) Influence of processing parameters on microstructure and ferroelectric properties of PZT-coated SiC fibers J. Eur. Ceram. Soc. 25:1637-45
20. Rodrigues JB and Eiras JA (2002) Preparation and characterization of PLT thick-films produced by a chemical route J. Eur. Ceram. Soc. 22:2927-2932
21. Simon L, Le Dren S, Gonnard P (2001) PZT and PT screen-printed thick films. J. Eur.Ceram.Soc 21:1441-1444
22. Lee DY, Yu JH, Shin YS, Park D, Yu TU, Hwang J (2008) Formation of ceramic nano-particle patterns using electrohydrodinamic jet printing with pin –to-pin electrodes. J. Appl. Phys. 47:1723-25
23. Prudenziati M (1994) Thick Film Sensors, Handbook of Sensors and Actuators 1, Elsevier, Amsterdam
24. Harper C A (1974) Handbook of Thick Film Hybrid Microelectronics, McGraw-Hill, New York
25. Torah RN, Beeby SP, Tudor MJ, White NM (2007) Thick film piezo ceramics and devices. J. Electroceram. 19:95-110.
26. Lucat C, Menil F, Von Der Mühll R. (1997) Thick-film densification for pyroelectric sensors. Meas. Sci. Technol. 8:38-41.
27. Zhou Q F, Chan H L W, Choy C L, (2000) PZT ceramic/ceramic 0–3 nanocomposite films for ultrasonic transducer applications. Thin Solid Films, 375:95-99
28. Kim Y B, Kim T S, Choi K S, Choi D J (2001) Densification method of screen printed PZT(52/48) thick films. Integrated Ferroelectrics 35:199-208
29. Kim YB, Kang JY, Kim TS (2002), Fabrication and Resonant Behavior of PZT Thick Film Cantilever for BioChip. Integrated Ferroelectrics 50:11-20
30. Gebhardt S, Seffner L, Schönecker A, Rödel J, Beckert W, Kreher W, Sotnikov A, Häßler W, Reuter S, Hübler A (2004) Bi-layered PZT films by combining thick and thin film technology. J. Eur. Ceram. Soc. 24:1101-1105

31. Wang Z, Zhu W, Zhao C, Tan O K (2003) dense PZT thick films derived from sol-gel based nanocomposite process. Materials Science and Engineering B 99:56-62
32. Corker D L, Zhang Q, Whatmore R W, Perrin C (2002) PZT composite ferroelectric thick films, J. Eur. Ceram. Soc. 22:383-390
33. Dorey RA, Stringfellow S B, Whatmore R W, (2002) Effect of sintering aid and repeated sol infiltration on the dielecrric and piezoelectric properties of PZT composite of a PZT composite thick film. J. Eur. Ceram. Soc., 22:2921-2926
34. Koch M, Harris N, Evans AGR, White NM, Brunnschweiler A, (1998) A novel micromachined pump based on thick-film piezoelectric actuation. Sensors and Actuators A 70:98-103
35. Reed J S (1995) Principles of Ceramics Processing, Second Addition, John Wiley & Sons, Inc. New York
36. Belavič D, Hrovat M, Holc J, Santo-Zarnik M, Kosec M, Pavlin M (2008) The application of thick-film technology in C-MEMS. J. Electroceram. 19:363-368.
37. Golonka LJ (2003) Low temperature cofired ceramics (LTCC) technology in microelectronics. In: Kosec M, Malič B, Kuščer D (eds) Conference Notes, Processing of Electroceramics, Bled, August 31-September 3, 2003, pp 313-329.
38. Wolny WW (2000) Proc. 12th IEEE Int. Symp. on Applications of Ferroelectrics, ISAF 2000, Honolulu, July 21-August 2, pp 257-262.
39. Zhitomirsky I (2002) Cathode electro deposition of ceramic and organoceramic materials. Fundamental aspects. Advances in Colloid and Interface Science 97:279-317
40. Tabellion J, Clasen R (2004) Electrophoretic deposition from aqueous suspension for near shape manufacturing of advanced ceramics and glasses-applications. Journal of Material Science 39:803-811.
41. Sarkar P, Nicholson P S (1996) Electrophoretic deposition (EPD): mechanisms, kinetics, and application to ceramics. J. Am. Ceram. Soc. 79:987-2002
42. Kaya C, Kaya F, Su B, Thomas B, Boccaccini AR (2005) Structural and functional thick ceramic coatings by electrophoretic deposition. Surface and Coatings Techn. 191:303-310
43. Su B., Ponton CB, Button TW (2001) Hydrothermal and electrophoretic deposition of lead zirconate titanate (PZT) films. J. Eur. Ceram. Soc., 21:1539-42
44. Ng SY, Boccaccini AR (2005) Lead zirconate titanate films on metallic substrates by electrophoretic deposition. Mater. Sci. Eng. B 116:208-214
45. Van Tassel J, Randall CA (1999) Electrophoretic deposition and sintering of thin/thick PZT films. J. Eur. Ceram. Soc. 19:955-958
46. Wu A, Vilarinho PM, Kingon AI (2006) Electrophoretic deposition of lead zirconate titanate films on metal foils for embedded components. J. Am. Ceram. Soc. 89:575-581
47. Wu A, Vilarinho PM, Srinivasan S, Kingon AI, Reanay IM, Woodward D, Ramos AR, Alvis E (2006a) Microstructural studies of PZT thick films on Cu foils. Acta Materialia 54:3211-3220
48. Windle J, Derby B (1999) Ink-jet printing of PZT aqueous ceramic suspensions, J. Mater. Sci. Lett. 18:87-89.
49. Senslis G, Dubarry M, Lejeune M, Chartier T (2002) 3D piezoelectric structure made by ink-jet printing. Ferroelectrics 273:279-284
50. Wang T, Derby B (2005) Ink-jet printing and sintering of PZT. J. Am. Ceram. Soc. 88:2053-58
51. Bhati AB, Mott M, Evans JRG, Edirishinge MJ (2001) PZT pillars for 1-3 composited prepered by ink-jet printing, J. Mater. Sci. Lett. 20:245-48
52. Lejeune M, Cartier T, Dossou-Yovo C, Noguera R (2009) Ink-jet printing of ceramic micro-pillar arrays. J. Eur. Ceram. Soc. 29:905-911
53. Noguera R, Lejeune M, Cartier T (2005) 3D fine scale ceramic components formed by ink-jet prototyping process. J. Eur. Ceram. Soc. 25:2055-59

54. Adachi H, Kuroda Y, Imahashi T, Yanagisawa K (1997) Preparation of Piezoelectric Thick Films using a Jet Printing System. Jpn. J. Appl. Phys. 36:1159- 1163
55. Akedo J, Lebedev M, Sato H, Park J (2005) High-speed optical microscanner driven with resonation of lam waves using $Pb(Zr,Ti)O_3$ thick films formed by aerosol deposition. Jap. J. Appl. Phys. 44:7072-7
56. Choi JJ, Hahn BB, Ryu J, Yoon WH, Park DS (2007) Effect of PZN addition and post annealing temperature on electrical properties of PZT thick films prepared by aerosol deposition method. J. Appl. Phys. 102:044101
57. Lebedev M, Akedo J, Akiyama Y (2000) Actuation Properties of Lead Zirconate Titanate Thick Films Structured on Si Membrane by the Aerosol Deposition Method. Jpn. J. Appl. Phys. 39:5600-5603
58. Stytsenko E, Ryan MJ, Daglish M (2004) Airflow deposition of oxide electroceramic films. J. Eur. Ceram. Soc. 24:999-1003
59. Schubring N W, Mantese J V, Micheli A L, Catalan A B, Mohammed M S, Naik R, Auner G (1999) Graded ferroelectrics: a new class of steady state thermal/electrical/mechanical elergy interchange devices. Integrated Ferroelectrics 24:155-168
60. Ledermann N, Muralt P, Gentil S, Mukati K, Cantoni M, Baborowski J, Settter N, (2003) {1 0 0}-Textured, piezoelectric $Pb(Zr_xTi_{1-x})O_3$ thin films for MEMS: integration, deposition and properties. Sensors and Actuators A 105:162-170
61. Takashi I, Imai S, Matsuda H, (2004) fabriaction of lead zirconate titanate thick film discs for micro transducers devices. Mat. Res. Soc. Proc. 785: D4.5.1-D4.5.6.
62. He XY, Ding AL, Zheng XS, Qiu PS, Luo WG (2003) Preparation of PZT(53/47) thick films deposited by a dip-coating process. Microelectronic Engineering, 66:865-871
63. Tsurumi T, Ozawa S, Abe G, Ohashi N, Wada S, Yamane M, (2000) Preparation of PZT(53/47) thick films deposited by a dip-coating process. Jpn. J. Appl. Phys. 39:5604-5608
64. King B H, Dimos D, Yang P, Morissette S L (1999) Direct-write fabrication of integrated, multilayer ceramic components. Journal of Electroceramics 3:173-78
65. Morissette S L, Lewis J A, Clem P G, Cesarano J, Dimos D B (2001) Direct-Write Fabrication of $Pb(Nb,Zr,Ti)O_3$ Devices: Influence of paste rheology on print morphology and component properties. J. Am. Ceram. Soc. 84: 2462-2468
66. Sarkar-Deliormanh A, Celik E, Polat M (2008) Rheological behavior of PMN gels for solid freeform fabrication. Colloids and Surfaces A: Physicochemical and Engineering Aspects 324:159-66
67. Smay J E, Cesarano III J, Tuttle B A, Lewis J A, (2004) Directed Colloidal Assembly of Linear and Annular Lead Zirconate Titanate Arrays. J. Am. Ceram. Soc. 87:293-295
68. Guo D, Cai K, Li L, Zhilun Gui Z, (2003) Application of gelcasting to the fabrication of piezoelectric ceramic parts. J. Eur. Ceram. Soc. 23:1131-37
69. Guo D, Li L, Cai K, Gui Z, Nan C (2004) Rapid Prototyping of Piezoelectric Ceramics via Selective Laser Sintering and Gelcasting. J. Am. Ceram. Soc. 87:17-22
70. Sun D, Rocks SA, Wang D, Edirisinghe MJ, Dorey RA (2008) Novel forming of columnar lead zirconate titanate structures. J. Eur. Ceram. Soc. 28:3131-39
71. Bordia R K, Raj R, (1985) Sintering behaviour of ceramic films constrained by a rigid substrate. J. Am. Ceram. Soc. 68:287-292.
72. Mohanram A, Lee S, Messing GL, Green DJ (2006) Constrained sintering of Low-Temperature Co-Fired Ceramics. J. Am. Ceram. Soc. 89: 1923-1929
73. Tzeng SY, Jean JH (2002) Stress development during constrained sintering of alumina/glass/alumina sandwich structure J. Am. Ceram. Soc. 85:335-340
74. Härdtl KH, Rau H (1969) PbO vapour pressure in the $Pb(Ti_{1-x})O_3$ system. Solid State Commun. 7:41-45

75. Kosec M, Holc J, Levassort F, Tran-Huu-Hue P, Lethiecq M (2001) Screen-printed Pb(Zr,Ti)O_3 thick films for ultrasonic medical imaging applications. International Symposium on Microelectronics, Baltimore, Maryland. Proceedings. Washington: IMAPS, pp 195-200
76. Kuscer D, Korzekwa J, Kosec M, Skulski R (2007a) A- and B-compensated PLZT x/90/10: Sintering and microstructural analysis. J. Eur. Ceram. Soc. 27 :4499-4507
77. Fushimi S, Ikeda T (1967) Phase Equilibrium in the System PbO-TiO_2–ZrO_2. J. Am. Ceram. Soc., 50:129-132
78. De Cicco G, Morten B, Dalmonego D, Prudenziati M (1999) Pyroelectricity of PZT-based thick-films. Sensors and Actuators A, 76:409-415.
79. Ferrari V, Marioli D, Taroni A (1997) Thick-film resonant piezo-layers as new gravimetric sensors. Meas. Sci. Technol. 8:42-48.
80. Ferrari V, Marioli D, Taroni A, Ranucci E (2000) Multisensor array of mass microbalances for chemical detection based on resonant piezo-layers of screen-printed PZT. Sensors and Actuators, B 68:81-87
81. Futakuchi T, Nakano K, Adachi M (2000) Low-Temperature Preparation of Lead-Based Ferroelectric Thick Films by Screen-Printing. Jpn. J. Appl. Phys. 39:5548-5551
82. Lee BY, Cheon CI, Kim JS, Bang KS, Kim JC, Lee HG (2002) Low temperature firing of PZT thick films prepared by screen printing method. Materials Letters 56:518-521
83. Lozinski A, Wang F, Uusimäki A, Leppävuori, (1997) PLZT thick films for pyroelectric sensors. Meas. Sci. Technol. 8:33-37
84. Belavič D, Santo Zarnik M, Holc J, Hrovat M, Kosec M, Drnovšek S, Cilenšek J, Maček S (2006) Properties of Lead Zirconate Titanate Thick-Film Piezoelectric Actuators on Ceramic Substrates. Int. J. Appl. Ceram. Techn. 3: 448-454
85. Tran-Huu-Hue P, Levassort F, Meulen FV, Holc J, Kosec M, Lethiecq M (2001) Preparation and electromechanical properties of PZT/PGO thick films on alumina substrate. J. Eur. Ceram. Soc. 21:1445-1449
86. Akiyama Y, Yamanaka K, Fujisawa E, Kowata Y (1999) Development of lead zirconate titanate family thick films on various substrates. Jpn. J. Appl. Phys. 38:5524-5527
87. Chen HD, Udayakumar KR, Cross LE, Bernstein JJ, Niles LC (1995) Dielectric, ferroelectric, and piezoelectric properties of lead zirconate titanate thick films on silicon substrates. J. Appl. Phys. 77:3349-3353
88. Simon-Seveyrat L, Gonnard P (2003) Processing and characterization of piezoelectric thick films screen-printed on silicon and glass-ceramic substrates. Integrated Ferroelectrics 51:1-18
89. Le Dren S, Simon L. Gonnard P, Troccaz M, Nicolas A (2000) Investigation of factoers affecting the preparation of PZT thick films. Mat. Res.Bull. 35:2037-45.
90. Kaneko S, Dong D, Murakami K (1998) Effect of Simultaneous Addition of $BiFeO_3$ and $Ba(Cu_{0.5}W_{0.5})O_3$ on Lowering of Sintering Temperature of Pb(Zr,Ti)O_3 Ceramics. J. Am. Ceram. Soc. 81: 1013-1018
91. Gentil S, Damjanovic D, Setter N (2005) Develpoment of relaxor ferroelectric materials for screen-printing on alumina and silicon substrates. J. Eur. Ceram. Soc. 25:2125-2128
92. Matsubara M, Toshiaki Y, Sakamoto W, Kikuta K, Yogo T, Hirano S (2005) processing and piezoelectric propeorties of lead-free (K,Na)(Nb,Ta)O_3 ceramics. J. Am. Ceram. Soc. 88:1190-96
93. Matsubara M, Yamaguchi Y, Kikuta K, Hirano S (2005a) Sintering and piezoelectric properties of potasium niobate ceramics with newly developed syntering aid. Jpn. J. Appl. Phys. 44:258-63.
94. Bernard J, Benčan A, Rojac T, Holc J, Malič B, Kosec M (2008) Low-temperature sintering of $K_{0.5}Na_{0.5}NbO_3$ ceramics. J. Am. Ceram. Soc. 91:2409-2411.
95. Zuo R, Ye C, Fang X (2008) $Na_{0.5}K_{0.5}NbO_3$-$BiFeO_3$ lead-free piezoelectric ceramics. J. Phys. Chem. Solids, 69:230-235

96. Thiele ES, Setter N (2000) Lead zirconate titanate particle dispersion in thick film ink formulation. J. Am. Ceram. Soc. 83:1407-12.
97. Thiele ES, Damjanović D, Setter N J. (2001) Processing and Properties of Screen-Printed Lead Zirconate Titanate Piezoelectric Thick Films on Electroded Silicon. J. Am. Ceram. Soc. 84:2863-2868
98. Beeby SP, Blacburn A, White NM (1999) Processing of PZT piezoelectric thick films on silicon for microelectromechanical systems. J. Michromech. Microeng. 9:218-229
99. Jones GJ, Beeby SP, Dargie P, Papakostas T, White N (2000) An investigation into the effect of modified firing profiles on the piezoelectric properties of thick-film PZT layers on silicon Measurement Science and Technology 11:526-531
100. Koch M, Harris N, Maas R, Evans AGR, White NM, Brunnschweiler A, (1997) A novel micropump design with thick-film piezoelectric actuation. Meas. Sci. Technol. 8:49-57.
101. Glynne -Jones P, Beeby SP, Dargie P, Papakostas T, White NM (2000) An investigation into the effect of modified firing profiles on the piezoelectric properties of thick-film PZT layers on silicon. Meas. Sci. Technol. 11:526-531.
102. Smart RM, Glasser FP (1974) Compound Formation and Phase Equilibria in the System $PbO-SiO_2$. J. Am. Ceram. Soc. 57:378-382.
103. Duval FCC, Dorey RA, Haigh RH, Whatmore RW (2003) Stable TiO_2/Pt electrode structure for lead containing ferroelectric thick films on silicon MEMS structures. Thin Solid Films, 444:235-240.
104. Gebhardt S, Seffner L, Schlenkrich F, Schonecker A (2007) PZT thick films for sensor and actuator applications. J. Eur. Ceram. Soc. 27:4177-80.
105. Jeon Y, Kim DG, No K, Kim S J, Chung J (2000) Residual stress analysis of Pt bottom electrodes on $ZrO_2/SiO_2/Si$ and SiO_2/Si substrates for $Pb(ZrTi)O_3$ thick films. Jap. J. Appl. Phys. 39:2705-2709
106. Kosec M, Holc J, Hauke T, Beige H (2001b) PZT based thick films on silicon. Abstracts of the 10th International Meeting on Ferroelectricity, IMF 10, Madrid, (Spain)
107. Huang Z, Zhang Q, Corkovic S, Dorey R A, Duval F, Leighton G, Wright R, Kirby P, Whatmore R W (2006) Piezoelectric PZT films for MEMS and their characterisation by interferometry. J. Electroceram. 17:549-556
108. Maréchal P, Haumesser L, Tran-Huu-Hue LP, Holc J, Kuscer D, Lethiecq M, Feuillard G (2008) Modeling of a high frequency ultrasonic transducer using periodic structures. Ultrasonics 2:141-149.
109. Maréchal P, Levassort F , Holc J, Tran-Huu-Hue LP, Kosec M, Lethiecq M (2006) High-frequency transducers based on integrated piezoelectric thick films for medical imaging. IEEE Trans. Ultrason. Ferroelectr. Freq. Control 53:1524-1533.
110. Santo Zarnik M, Belavič D, Maček S, Holc J (2009) Feasibility study of a thick-film PZT resonant pressure sensor made on a prefired 3D LTCC structure. Int. J. Appl. Ceram. Techn. 6:9-17.
111. Holc J, Hrovat M, Kosec M (1999) Interactions between alumina and PLZT thick films. Mat.Res.Bul., 34:2271-78.
112. Uršič H, Hrovat M, Holc J, Santo-Zarnik M, Drnovšek S, Maček S, Kosec M (2008b) A large-displacement $65Pb(Mg_{1/3}Nb_{2/3}O_3$-$35PbTiO_3$/Pt bimorph actuator prepared by screen printing. Sensors and Actuators B 133:699-704.
113. Golonka LJ, Buczek M, Hrovat M, Belavic D, Dziedzic A, Roguszczak H, Zawada T (2005) Properties of PZT thick films made on LTCC. Microelectronics International 22:13-16
114. Hrovat M. Holc J, Drnovšek S, Belavic D, Bernard J, Kosec M, Golonka L, Dziedzic A, Kita J (2003) Characterization of PZT thick films fired on LTCC substrates. J. Mat. Sci. Lett. 22:1193-1195.
115. Hrovat M, Holc J, Drnovšek S, Belavič D, Cilenšek J, Kosec M (2006) PZT thick films on LTCC substrate with imposed alumina barrier layer. J. Eur. Ceram. Soc. 26:897-900.

116. Uršič H, Hrovat M, Belavič D, Cilenšek J, Drnovšek S, Holc J, Santo-Zarnik M, Kosec M (2008a) Microstructural and electrical characterisation of PZT thick films on LTCC substrates. J. Eur. Ceram. Soc. 28:1839-44.
117. Levassort F, Holc J, Ringaard E, Bove T, Kosec M, Lethiecq M (2007) Fabrication, modelling and use of porous ceramics for ultrasonic transducer applications. J. Electroceram. 19:125-137.
118. Akedo J, Lebedev M (2000) Piezoelectric properties and poling effect of $Pb(Zr,Ti)O_3$ thick films prepared for microactuators by aerosol deposition. Appl. Phys. Lett. 77:1710-1712.
119. Park JH, Akedo J, Sato H. (2007) High-speed metal-based optical microscanners using stainless-steel substrate and piezoelectric thick films prepared by aerosol deposition method Sensors and Actuators A 135:86-91
120. Benčan A, Holc J, Hrovat M, Dražić G, Kosec M (2002) Interactions between PZT thick films and Ni substrates. Key Eng. Mater. 206-213:1301-1304.
121. Holc J, Hrovat M, Kuščer D, Kosec M (2002) The preparation and properties of a PZT thick film on an alumina substrate with a $Pb_2Ru_2O_{6.5}$ electrode. Ferroelectrics, 270:87-92.

Chapter 3
Tailored Liquid Alkoxides for the Chemical Solution Processing of Pb-Free Ferroelectric Thin Films

Kazumi Kato

3.1 Tailored Alkoxides

The chemical solution deposition method has been applied widely for multicomponent thin films. Generally, its high potential is attributable to the homogeneity of solutions. However, low affinity among conventional raw materials such as metalorganic compounds and organic solvents sometimes remains problematic. Tailoring the molecular structure of a liquid source and optimising its solubility and reactivity for hydrolysis, condensation, and combustion would yield the following: a precisely controlled composition, low-temperature crystallisation, high phase purity, and uniform microstructure in a thin film deposited using a liquid source [1, 2, 3]. In particular, low-temperature crystallisation is essential for integration of ferroelectric thin films into semiconductors.

3.2 $Sr[BiTa(OR)_9]_2$ and $Sr[BiNb(OR)_9]_2$ for $SrBi_2Ta_2O_9$ and $SrBi_2Nb_2O_9$

3.2.1 Chemistry in Solutions of Sr-Bi-Ta and Sr-Bi-Nb Complex Alkoxides [3, 4]

Precursors for thin films of layer-structured perovskite $SrBi_2Ta_2O_9$ and $SrBi_2Nb_2O_9$ were prepared by the reactions of a strontium-bismuth double methoxyethoxide and tantalum or niobium methoxyethoxide in methoxyethanol,

National Institute of Advanced Industrial Science and Technology (AIST), 2266-98 Anagahora, Shimoshidami, Moriyama-ku, Nagoya 463-8560, Japan followed by partial hydrolysis [4]. Several spectroscopic techniques, such as ^{1}H-, ^{13}C-, and ^{93}Nb-NMR, and Fourier-transform infrared spectroscopy were used to analyse the arrangement of the metals and oxygen in the precursor molecules.

The bismuth alkoxide was insoluble in methoxyethanol because of its own strong association. However, it easily dissolved in methoxyethanol containing strontium alkoxide in $Sr(OR)_2$:$Bi(OR)_3$ molar ratios of > 0.5 [5]. As previously reported, the strong association of the bismuth alkoxide can be destroyed by the formation of a strontium-bismuth double alkoxide. The formation of the strontium-bismuth double alkoxide is discussed next. Also, the bismuth alkoxide was insoluble in strontium-tantalum or strontium-niobium double alkoxide solutions. Therefore, the strontium-bismuth double alkoxide has formed and then reacted with the tantalum or niobium alkoxide to prepare uniform alkoxide complex solutions [4]. $Sr(OCH_2CH_2OCH_3)_2$ showed ^{1}H-NMR signals at 3.92, 3.50, and 3.34 ppm assigned to the hydrogen of the methylene group that is bonded to the hydroxyl group ($CH_3OCH_2C\underline{H}_2O$-Sr), the hydrogen of the methylene group ($CH_3OC\underline{H}_2CH_2O$-Sr), and the hydrogen of the terminal methyl group ($C\underline{H}_3OCH_2CH_2O$-Sr), respectively. Where identified, hydrogen and carbon atoms are shown in bold face. $Sr(OCH_2CH_2OCH_3)_2$ exhibited three ^{13}C-NMR signals at 77.1, 61.8, and 58.8 ppm ascribed to the carbon of the methylene group ($CH_3O\underline{C}H_2CH_2O$-Sr), the carbon of the methylene group that is bonded to strontium ($CH_3OCH_2\underline{C}H_2O$-Sr) via the adjacent oxygen, and the carbon of the terminal methylene group ($\underline{C}H_3OCH_2CH_2O$-Sr), respectively. The differences in chemical shifts between the methoxyethoxy groups of $Sr(OCH_2CH_2OCH_3)_2$ and pure methoxyethanol are caused by the substitution of a strontium atom for the hydrogen atom of the hydroxyl group. $Sr[Bi(OCH_2CH_2OCH_3)_4]_2$ had ^{1}H-NMR signals at 3.92, 3.50, and 3.34 ppm, and three ^{13}C-NMR signals at 77.6, 62.4, and 58.5 ppm. The small differences in chemical shifts in comparison to $Sr(OCH_2CH_2OCH_3)_2$ are indicative of the formation of strontium-bismuth double alkoxide, wherein the hydrogens and carbons are in a chemically equivalent environment. This explanation agrees with a previous report [5] on formation of $Sr[Bi(OEt)_4]_2$ double alkoxide.

$Ta(OCH_2CH_2OCH_3)_5$ exhibited ^{1}H-NMR signals at 4.70, 3.56, and 3.28 ppm, and three ^{13}C-NMR signals at 75.0, 71.6, and 58.6 ppm. The ^{1}H-NMR for $Sr[Ta(OCH_2CH_2OCH_3)_6]_2$ as a reference also consisted of signals at 4.55, 3.48, 3.37 ppm. The ^{13}C-NMR consisted of three signals at 76.2, 68.7, and 59.0 ppm [4]. These results are indicative of the formation of strontium-tantalum double alkoxide. Previously, a $Sr[Ta(OPr^{i})_6]_2$ double alkoxide was reported to comprise two TaO_6 octahedra connected by a strontium atom [6]. $Sr[Ta(OCH_2CH_2OCH_3)_6]_2$ is considered to have the same structure as $Sr[Ta(OPr^{i})_6]_2$.

The unhydrolysed $SrBi_2Ta_2O_9$ precursor showed ^{1}H-NMR signals at 4.54, 3.48, and 3.37 ppm, strong ^{13}C-NMR signals at 76.1, 68.6, and 59.0 ppm, and satellite signals at 62.2 and 58.5 ppm [4]. The main ^{1}H and ^{13}C signals closely correspond to the signals of $Sr[Ta(OCH_2CH_2OCH_3)_6]_2$. The satellite signals closely correspond to the signals of $Sr[Bi(OCH_2CH_2OCH_3)_4]_2$. The close correspondences or small differences indicate that the unhydrolysed $SrBi_2Ta_2O_9$ precursor is not a mixture of the two double alkoxides. It has a homogeneous ternary solution structure. The chemical shifts of the hydrolysed $SrBi_2Ta_2O_9$ precursors did not change when the water:alkoxy group molar ratios, R, were varied between 1:18 and 1:6. A sol with a large amount of water for hydrolysis (i.e., R=1:3) neither showed significant deviation of the chemical shifts with respect to the unhydrolysed $SrBi_2Ta_2O_9$ precursor.

Analogous results were observed on the chemical environments of the hydrogens and carbons of methoxyethoxy groups in the $SrBi_2Nb_2O_9$ system. ^{93}Nb is an appropriate nucleus for NMR investigations. Therefore, the chemical shifts for unhydrolysed and hydrolysed $SrBi_2Nb_2O_9$ precursors were measured to further probe the chemical environment of niobium in the octahedral unit [4]. Two signals appeared at -1240 and -1160 ppm in the ^{93}Nb-NMR for $Nb(OEt)_5$. In the ^{93}Nb-NMR of $Nb(OCH_2CH_2OCH_3)_5$, a broad signal, indicative of an asymmetric pentafold environment of the niobium atom in a monomeric unit, appeared at -1170 ppm. The broadening behaviour of the ^{93}Nb signal for niobium alkoxides, with ligand substitution, also had been identified in a previous report. In contrast, the ^{93}Nb-NMR for the unhydrolysed $SrBi_2Nb_2O_9$ precursor exhibited a sharp signal at -1140 ppm with a half-width value of 14000 Hz. This could be explained by niobium atoms in a chemically equivalent environment but in a symmetric octahedron site in the molecule, as reported for $K[Nb(OEt)_6]_2$ double alkoxide [7, 8]. The ^{93}Nb-NMR signal of the $SrBi_2Nb_2O_9$ precursor after partial hydrolysis was identical to the unhydrolysed precursor. In the ^{93}Nb-NMR for $Sr[Nb(OCH_2CH_2OCH_3)_6]_2$, which was prepared for a reference. A signal at -1150 ppm with a half-width value of 11400 Hz also was observed.

In the FT-IR spectra of $Sr[Bi(OCH_2CH_2OCH_3)_4]_2$, $Ta(OCH_2CH_2OCH_3)_5$, the unhydrolysed $SrBi_2Ta_2O_9$ precursor, and $Sr[Ta(OCH_2CH_2OCH_3)_6]_2$, the peaks in the range of 1000-800 cm^{-1} and of 600-400 cm^{-1} were due to C-O vibrations within the methoxyethoxy groups and M-O bonds, respectively [4]. In the FT-IR spectrum of $Sr[Bi(OCH_2CH_2OCH_3)_4]_2$, peaks due to C-O vibrations appeared at 982, 962, 895, and 837 cm^{-1}, and peaks due to Bi-O vibrations and Sr-O vibrations appeared at 591, 555, 525, and 461 cm^{-1}. The peak shifts, and an additional absorption peak at 555 cm^{-1} in the latter (M-O vibrations), with respect to individual strontium alkoxide and bismuth alkoxide, are due to the formation of $Sr[Bi(OCH_2CH_2OCH_3)_4]_2$ double alkoxide. A similar shift and an additional absorption peak at 559 cm^{-1} have been reported for the formation of $Sr[Bi(OEt)_4]_2$ double alkoxide containing Sr-O-Bi bonds [5]. In the FT-IR spectrum of $Ta(OCH_2CH_2OCH_3)_5$, peaks due to C-O vibrations appeared at 981, 964, 926, 893, 837, and 801 cm^{-1}, and peaks due to Ta-O vibrations appeared at 580 and 490 cm^{-1}. The peaks due to Ta-O vibrations in $Ta(OCH_2CH_2OCH_3)_5$ are in a similar frequency range to those reported for $Ta(OEt)_5$ [9]. In the FT-IR spectrum of the $SrBi_2Ta_2O_9$ precursor, absorption peaks appeared at 982, 966, 918, 897, 838, 587, and 474 cm^{-1}. Comparison of the spectrum of the $SrBi_2Ta_2O_9$ precursor to the spectra of $Sr[Bi(OCH_2CH_2OCH_3)_4]_2$ and $Ta(OCH_2CH_2OCH_3)_5$ indicated that the peaks around at 587 andn474 cm^{-1} are due to vibration modes resulting from the M-O bonds in the complex molecule. These peaks are identified easily when a comparison is made with the reference $Sr[Ta(OCH_2CH_2OCH_3)_6]_2$, which exhibited absorption peaks, due to Ta-O vibrations, at 583 and 475 cm^{-1}. Analogous results concerning the chemical bonds in the $SrBi_2Nb_2O_9$ system were obtained.

The ^{1}H-, ^{13}C-, and ^{93}Nb-NMR and FT-IR spectra indicate that $SrBi_2Ta_2O_9$ and $SrBi_2Nb_2O_9$ precursors contain Sr-O-M (where M is Ta or Nb) bonds comprising two MO_6 octahedra connected by a strontium atom, and Sr-O-Bi links. These precursors are structurally identical. ^{93}Nb-NMR also confirms a symmetric niobium site within the NbO_6 octahedra, which are connected by a strontium atom and bonded to two bismuth atoms. A possible molecular structure of $SrBi_2Ta_2O_9$ and $SrBi_2Nb_2O_9$ precursors is proposed from these results. The features of $SrBi_2Ta_2O_9$ and $SrBi_2Nb_2O_9$ precursors are identical to the sublattices of $SrBi_2Ta_2O_9$ and $SrBi_2Nb_2O_9$ crystals. The sublattice units also are considered to be preserved in the cross-linked oligomeric species as condensation proceeds in the case of the hydrolysis with small amounts of water [4].

3.2.2 $SrBi_2Ta_2O_9$ and $SrBi_2Nb_2O_9$ Thin Films [*4*]

The sol-gel-derived $SrBi_2Ta_2O_9$ thin films from the former precursors were crystallised by rapid thermal annealing in an oxygen atmosphere below 550°C. They exhibited preferred (115) orientation [4]. The crystallinity improved and the crystallite size increased with increasing temperature up to 700°C. In the case of $SrBi_2Nb_2O_9$ thin films, a low heating rate (2°C/min) was necessary for the control of the crystallographic (115) orientation. A rate of 200°C/s (rapid thermal annealing) produced films that exhibited c-axis orientation. The (115) $SrBi_2Ta_2O_9$ thin film, heated to 700°C, exhibited improved ferroelectric properties. The $2P_r$ and E_c values at an applied voltage of 5 V were 8.9 $\mu C/cm^2$ and 36 kV/cm, respectively. Moreover, upon 10^{10} cycles of switching at an applied voltage of 3 V, the thin film exhibited no change in polarisation. The low crystallisation temperature can be explained as follows. Because the molecular structure of the $SrBi_2Ta_2O_9$ and $SrBi_2Nb_2O_9$ precursors is preserved upon hydrolysis, the homogeneous gels that form after spin coating may have low-energy amorphous structures. Moreover, because of the structural similarity of the precursors to that of the crystalline sublattice, the activation energies for the amorphous-gel to crystalline phase transformation are expected to be low. A similar process, with low activation energies, has been reported for the crystallisation of homogeneous gels of $LiNbO_3$ and $PbTiO_3$ derived by the alkoxide route [10, 11]. Also, low crystallisation temperatures in $Ba(Mg_{1/3}Ta_{2/3})O_3$ ceramics using alkoxy-derived precursor have been reported [12]. The difference in the heating-rate dependence of the crystallographic orientation between the $SrBi_2Ta_2O_9$ and $SrBi_2Nb_2O_9$ thin films may have resulted from the higher refractoriness of $SrBi_2Nb_2O_9$ over $SrBi_2Ta_2O_9$.

The crystallinity and crystallographic orientation of the $SrBi_2Ta_2O_9$ thin films crystallised at 650°C, were improved by the UV irradiation under appropriate conditions at the stage of the as-deposited and noncrystalline thin films [13]. By the spectroscopic analyses, it was found that the chemical structure of the noncrystalline Sr-Bi-Ta-O thin films on Pt layer on Si changed by UV irradiation under the appropriate conditions. Additionally, the UV irradiation was considered to accelerate the polymerisation and to change the interaction between the noncrystalline Sr-Bi-Ta thin films and Pt layer on Si.

Thin films of $CaBi_2Ta_2O_9$, which has a similar crystal structure of $SrBi_2Ta_2O_9$ and $SrBi_2Nb_2O_9$, were prepared on Pt-passivated silica glass substrates using the Ca-Bi-Ta triple alkoxide solution [14]. The 750°C-crystallised thin film was a single phase of layer-structured perovskite $CaBi_2Ta_2O_9$ and showed random orientation. The 750°C-crystallised $CaBi_2Ta_2O_9$ thin film exhibited P-E hysteresis loop at relatively high voltage of 13 V. It showed no fatigue after 2×10^{10} switching cycles. The dielectric constant of the randomly-crystallised $CaBi_2Ta_2O_9$ thin film was not so low as the c-axis oriented $CaBi_2Nb_2O_9$ thin film. However, the loss factor was enough lower with respect to the randomly-crystallised $CaBi_2Ta_2O_9$ thin film [14].

3.3 $CaBi_4Ti_4(OCH_2CH_2OCH_3)_{30}$ for $CaBi_4Ti_4O_{15}$

3.3.1 Chemistry in Solution of Ca-Bi-Ti Complex Alkoxide [15]

Preparation of Ca-Bi-Ti complex alkoxide is schematically drawn in Fig. 3.1. Calcium metal was dissolved in ethanol (C_2H_5OH) by reaction at the boiling point of 78°C. Separately, bismuth triethoxide ($Bi(OC_2H_5)_3$) and titanium tetraisopropoxide ($Ti(Oi\text{-}C_3H_7)_4$) were dissolved together in methoxyethanol ($CH_3OC_2H_4OH$) at 124°C in a molar ratio of 1:1. The two solutions were mixed to adjust the atomic ratios of calcium, bismuth, and titanium to be stoichiometric 1:4:4. Then, they were heated at 78°C for two hours. Next, deionised water diluted in $CH_3OC_2H_4OH$ was added to the mixture in a molar ratio of 1:30 (H_2O:alkoxy group). The solution was stirred at room temperature for an hour after hydrolysis. The concentration of the hydrolysed alkoxide solutions was 0.02 M. The volume ratio of C_2H_5OH to $CH_3OC_2H_4OH$ was about 1:4.5.

Chemical shifts of ^{1}H- and ^{13}C-NMR for the Ca-Bi-Ti complexes and related alkoxides are obtained [15]. In the ^{1}H-NMR for $BiTi(OCH_2CH_2OCH_3)_7$, signals at 4.77, 3.59, and 3.30 ppm were caused by hydrogen of the methylene group adjacent to O-M bond ($CH_3OCH_2C\underline{H}_2O$-M, M: Bi or Ti), hydrogen of the methylene group ($CH_3OC\underline{H}_2CH_2O$-M, M: Bi or Ti), and hydrogen of the terminal methyl group ($C\underline{H}_3OCH_2CH_2O$-M, M: Bi or Ti), respectively. In the ^{13}C-NMR for $BiTi(OCH_2CH_2OCH_3)_7$, signals at 76.2, 69.1, and 58.3 ppm were caused by carbon of the methylene group ($CH_3O\underline{C}H_2CH_2O$-M, M: Bi or Ti), carbon of the methylene group adjacent to O-M bond ($CH_3OCH_2\underline{C}H_2O$-M, M: Bi or Ti), and carbon of the terminal methyl group ($\underline{C}H_3OCH_2CH_2O$-M, M: Bi or Ti), respectively. Differences in chemical shifts of the methoxyethoxy group of $BiTi(OCH_2CH_2OCH_3)_7$ from the individual $Bi(OCH_2CH_2OCH_3)_3$ and $Ti(OCH_2CH_2OCH_3)_4$ indicated the formation of Bi-Ti double alkoxide, wherein the hydrogen and carbon were in a chemically equivalent environment. Formation of Bi-Ti double alkoxide was reported [16]. However, no structural analysis details have been presented so far.

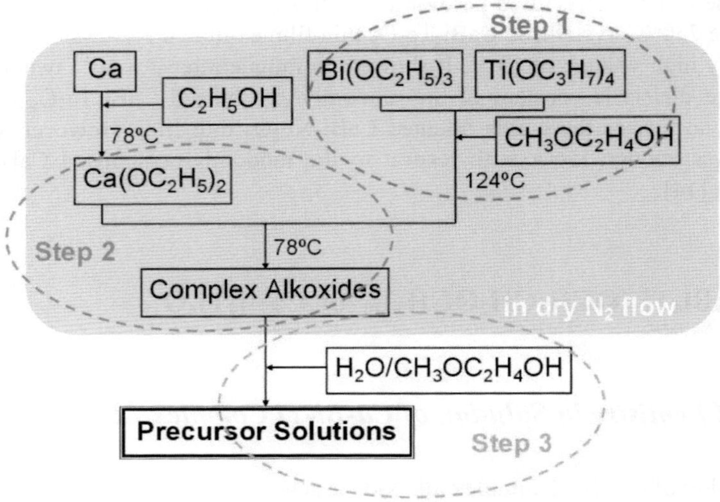

Fig. 3.1 Scheme for preparation of Ca-Bi-Ti complex alkoxide.

Signals at 4.72, 4.44, 3.52, and 3.32 ppm in the ^{1}H-NMR for $CaBi_4Ti_4(OCH_2CH_2OCH_3)_{30}$ were attributable to hydrogen of the methylene group adjacent to the O-M bond ($CH_3OCH_2C\underline{H_2}O$-M, M: Ca, Bi or Ti), hydrogen of the methylene group ($CH_3OC\underline{H_2}CH_2O$-M, M: Ca, Bi or Ti), and hydrogen of the terminal methyl group ($C\underline{H_3}OCH_2CH_2O$-M, M: Ca, Bi or Ti), respectively. Signals at 76.6, 75.0, 72.1, 69.1 and 58.5 ppm in the ^{13}C-NMR for $CaBi_4Ti_4(OCH_2CH_2OCH_3)_{30}$ were caused by carbon of the methylene group ($CH_3O\underline{C}H_2CH_2O$-M, M: Ca, Bi or Ti), carbon of the methylene group adjacent to the O-M bond ($CH_3OCH_2\underline{C}H_2O$-M, M: Ca, Bi or Ti), and carbon of the terminal methyl group ($\underline{C}H_3OCH_2CH_2O$-M, M: Ca, Bi or Ti), respectively. The main ^{1}H and ^{13}C signals corresponded closely to the Bi-Ti double alkoxide signals. The small signals corresponded closely to signals of $Ti(OCH_2CH_2OCH_3)_4$. Those close correspondences or small differences confirmed that the Ca-Bi-Ti complex alkoxide was not a mixture. It instead consisted of some interactions between the two alkoxides. Furthermore, Ca atom is considered to be linked to

O atoms of the Bi-O bonds of Bi-Ti double alkoxide, because isolated $Ca(OCH_2CH_2OCH_3)_2$ was not detected in the Ca-Bi-Ti alkoxide systems and for the $Ca(OCH_2CH_2OCH_3)_2$. A Ca-O-Bi link was easy to form as $Ca[Bi(OCH_2CH_2OCH_3)_4]_2$. A previous report on the formation of $Ca[Bi(OC_2H_5)_4]_2$ double alkoxide [17] concurs with this explanation.

In the chemical shifts of partially hydrolysed $CaBi_4Ti_4(OCH_2CH_2OCH_3)_{30-x}(OH)_x$, the ^1H and ^{13}C signals corresponding to the Bi-Ti double alkoxide signals disappeared preferentially. However, we observed no marked deviation of chemical shifts with respect to the unhydrolysed $CaBi_4Ti_4(OCH_2CH_2OCH_3)_{30}$. Moreover, the chemical shifts of a partially hydrolysed and then six-month-aged complex were not distinguishable from those of a partially hydrolysed fresh complex. This indicates that the complex retains the structure persistently and therefore offers good stability.

In the FT-IR spectrum of $BiTi(OCH_2CH_2OCH_3)_7$, peaks caused by Bi-O and Ti-O vibrations appeared at 593, 517, and 462 cm^{-1}. The shift and strengthening of the absorption peak at 593 cm^{-1}, with respect to the individual $Bi(OCH_2CH_2OCH_3)_3$ and $Ti(OCH_2CH_2OCH_3)_4$, were attributable to formation of Bi-Ti double alkoxide. Three typical peaks appeared in the FT-IR spectra of $CaBi_4Ti_4(OCH_2CH_2OCH_3)_{30}$. Those peaks shifted depending on the chemical composition. In the FT-IR spectrum of partially hydrolysed $CaBi_4Ti_4(OCH_2CH_2OCH_3)_{30-x}(OH)_x$, the appearance of the three peaks was confirmed to be similar. However, the individual peaks shifted slightly with respect to the unhydrolysed $CaBi_4Ti_4(OCH_2CH_2OCH_3)_{30}$.

Spectroscopic analysis results indicate that the Ca-Bi-Ti complex alkoxide such as $CaBi_4Ti_4(OCH_2CH_2OCH_3)_{30}$, had similar local structures that contained the Bi-O-Ti bonds as in the $BiTi(OCH_2CH_2OCH_3)_7$ and Ca atoms linking to O atoms of the Bi-O bonds. Results showed that the partial hydrolysis of the complex proceeded preferentially at sites by the Bi-O-Ti bonds, but that no compositional deviation occurred. Although no significant change in viscosity was observed, the complexes are considered to proceed to a condensation reaction forming oligomers in the partially hydrolysed sources. Further investigations of the development of amorphous gel structure from the Ca-Bi-Ti complex and the relationship between the gel structure and layer-structured crystal must be conducted.

3.3.2 $CaBi_4Ti_4O_{15}$ Thin Films Integrated on Pt-Coated Si for FeRAM Application [15, 18, 19]

Fig. 3.2 shows XRD profiles of 650°C-crystallised $CaBi_4Ti_4O_{15}$ thin films on the three kinds of Pt-coated substrates. The crystallographic appearance of the thin films was distinctive in comparison with the XRD profile of the alkoxy-derived $CaBi_4Ti_4O_{15}$ powders crystallised at 850°C [18]. The $CaBi_4Ti_4O_{15}$ thin film on

Pt:200 nm/Ti:20 nm/SiO_2:1000 nm/Si shows a Pt layer with a high (111) orientation. That film was a mixture of perovskite and pyrochlore phases (Fig. 3.2(a)). The pyrochlore phase showed (111) orientation and the perovskite phase showed random orientation. Nevertheless, it is noteworthy that it has rather high intensity of the (200)/(020) diffraction line compared to the single-phase $CaBi_4Ti_4O_{15}$ thin film on Pt:300 nm/TiO_X:10 nm/SiO_2:10 nm/Si, with Pt showing (111) orientation and relatively lower crystallinity (Fig. 2(c)). In contrast, the $CaBi_4Ti_4O_{15}$ thin film on Pt:150 nm/TiO_X:20 nm/SiO_2:200 nm/Si with Pt, showing (200) orientation and relatively lower crystallinity, was almost a single phase of perovskite and showed c-axis orientation (Fig. 3.2(b)). Differences between the Pt bottom electrodes were shown not only in terms of the orientation, but also in the strain and crystallinity. We analysed the d_{111} and full width at half-maximum (FWHM) values of the (111) diffraction lines [19]. The d_{111} and FWHM values were: 0.2258 nm and 0.118°, respectively, for Pt:200 nm/Ti:20 nm/SiO_2:1000 nm/Si; 0.2255 nm and 0.208°, respectively, for Pt:150 nm/TiO_X:20 nm/SiO_2:200 nm/Si; and 0.2252 nm and 0.271°, respectively, for Pt:300 nm/TiO_X:10 nm/SiO_2:10 nm/Si. All of the Pt bottom electrodes were found to be strained because the d_{111} of Pt bulk metal is 0.2265 nm. The strain increased, the FWHM value increased and, consequently, the crystallinity decreased. These results suggest that the lattice matching governs the $CaBi_4Ti_4O_{15}$ thin films' orientation to a greater extent than the strain and crystallinity. The phase transition may be suppressed, especially in the interface region, because the atomic arrangement in the (111) plane of the cubic pyrochlore structure matches that in the (111) plane of the highly oriented Pt bottom. The residual pyrochlore phase in the interface region may affect the orientation of the upward perovskite phase. On the other hand, the atomic arrangement in the (00l) plane of the orthorhombic perovskite structure matches that in the local (h00) plane of the Pt bottom, which consequently promotes the phase transition. Misfits of pyrochlore (111)/Pt (111) and perovskite (00l)/Pt (h00) were calculated as about 5.4% and 3.4%, respectively.

Fig. 3.3 shows cross-section TEM and electron diffraction profiles of 650°C-crystallised $CaBi_4Ti_4O_{15}$ thin films on the three kinds of Pt-coated Si. Both $CaBi_4Ti_4O_{15}$ thin films had columnar structures. The electron diffraction profiles of the grains 1–6 in the $CaBi_4Ti_4O_{15}$ thin films on various Pt-coated Si show that the crystalline spots were indexed as the perovskite structure and the grains had individually differing orientations. In the cross-section profile of the thin film on Pt:200 nm/Ti:20 nm/SiO_2:1000 nm/Si (Fig. 3.3(a)), small grains with diameters of several tens of nanometres typically coexist at the Pt bottom surface. The small grains seem to have a pyrochlore structure, which is suppressed to transit under the effect of the Pt bottom. In contrast, well-developed and uniform columnar grains of perovskite single phase were observed for the thin films both on Pt:150 nm/TiO_X:20 nm/SiO_2:200 nm/Si (Fig. 3.3(b)) and Pt:300 nm/TiO_X:10 nm/SiO_2:10 nm/Si (Fig. 3.3(c)).

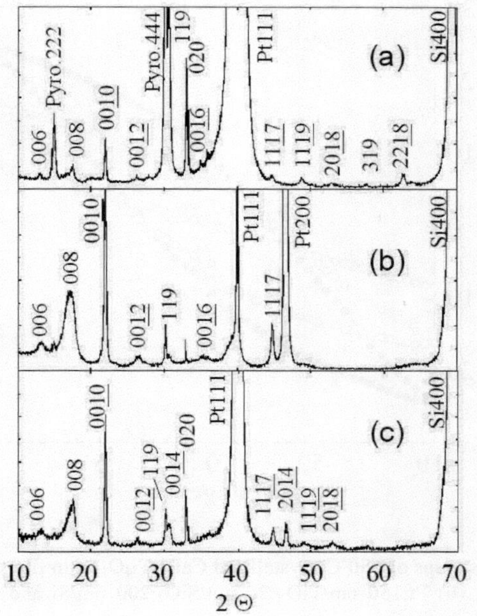

Fig. 3.2 XRD profiles of 650°C-crystallised $CaBi_4Ti_4O_{15}$ thin films on **a** Pt:200 nm/Ti:20 nm/SiO_2:1000 nm/Si, **b** Pt:150 nm/TiO_X:20 nm/SiO_2:200 nm/Si and **c** Pt:300 nm/TiO_X:10 nm/SiO_2:10 nm/Si [2].

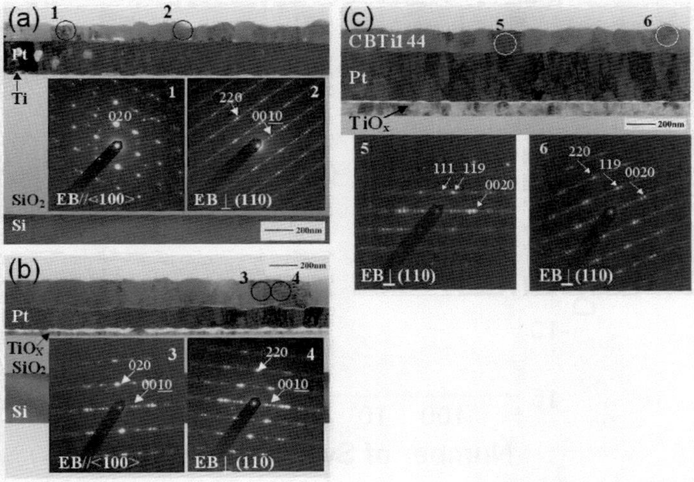

Fig. 3.3 Cross-section TEM and electron diffraction profiles of 650°C-crystallised $CaBi_4Ti_4O_{15}$ thin films on **a** Pt:200 nm/Ti:20 nm/SiO_2:1000 nm/Si, **b** Pt:150 nm/TiO_X:20 nm/SiO_2:200 nm/Si and **c** Pt:300 nm/TiO_X:10 nm/SiO_2:10 nm/Si [2].

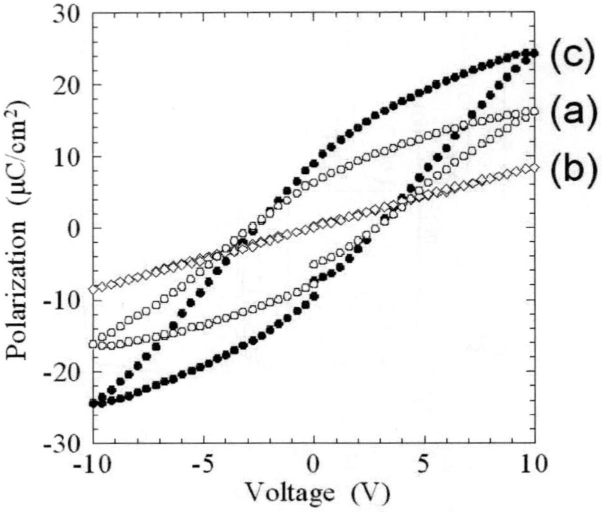

Fig 3.4 P-V hysteresis loops of 650°C-crystallised $CaBi_4Ti_4O_{15}$ thin films on (a) Pt:200 nm/Ti:20 nm/SiO_2:1000 nm/Si, (b) Pt:150 nm/TiO_X:20 nm/SiO_2:200 nm/Si and (c) Pt:300 nm/TiO_X:10 nm/SiO_2:10 nm/Si [2].

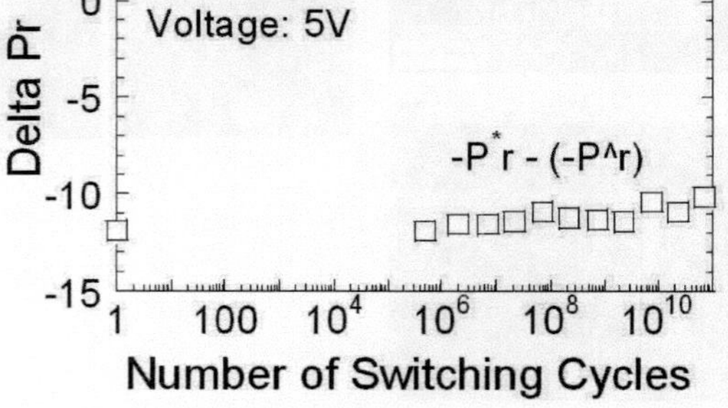

Fig. 3.5 Changes of the polarisation with switching cycles for 650°C-crystallised $CaBi_4Ti_4O_{15}$ thin film on Pt:300 nm/TiO_X:10 nm/SiO_2:10 nm/Si [18].

Fig 3.4 shows P-V hysteresis properties of the 650°C-annealed thin films on the three kinds of Pt-coated Si. The thin film on Pt:200 nm/Ti:20 nm/SiO_2:1000 nm/Si, which consisted of a mixture of randomly-crystallised perovskite columnar grains showing rather high (200)/(020) diffraction line and pyrochlore small grains, exhibited hysteresis loops (Fig 3.4a). The Pr and Ec were 7.1 $\mu C/cm^2$ and 140 kV/cm, respectively, at an applied voltage of 10 V. Another c-axis oriented thin film on Pt:150 nm/TiO_X:20 nm/SiO_2:200 nm/Si showed no ferroelectric loop (Fig 3.4b). The Pr and Ec of the single phase thin film with random orientation on Pt:300 nm/TiO_X:10 nm/SiO_2:10 nm/Si were the highest, at 9.3 $\mu C/cm^2$ and 140 kV/cm, respectively (Fig 3.4c). The values of the Pr were associated with both crystallographic orientation and phase purity. Fig. 3.5 shows the endurance behaviours of the randomly-crystallised single phase $CaBi_4Ti_4O_{15}$ thin film against a number of switching cycles with voltages of 5 V and a pulse width of 10^{-6} s. The polarisation did not change after 10^{11} switching cycles. The P-E hysteresis and fatigue properties of the stoichiometric $CaBi_4Ti_4O_{15}$ thin film were comparable to those of $SrBi_2Ta_2O_9$ thin film [20, 21], of which the stoichiometry had to be deviated precisely to enhance the properties. However, relatively high voltages were required with respect to the $CaBi_4Ti_4O_{15}$ thin film. Fig. 3.6 shows the dielectric properties of the 650°C-annealed $CaBi_4Ti_4O_{15}$ thin film. The dielectric constant (ε) and loss factor (tanδ) were almost constant in the frequency range of 10 kHz to 1 MHz. They were 340 and 0.033, respectively, at 100 kHz. The ε and tanδ of $CaBi_4Ti_4O_{15}$ ceramics have been reported as about 150 and 0.1, respectively [22, 23]. The ε of the thin film was much higher than that of the ceramics. The enhanced ε may be stem from the crystallinity relating to residual stress in the thin film.

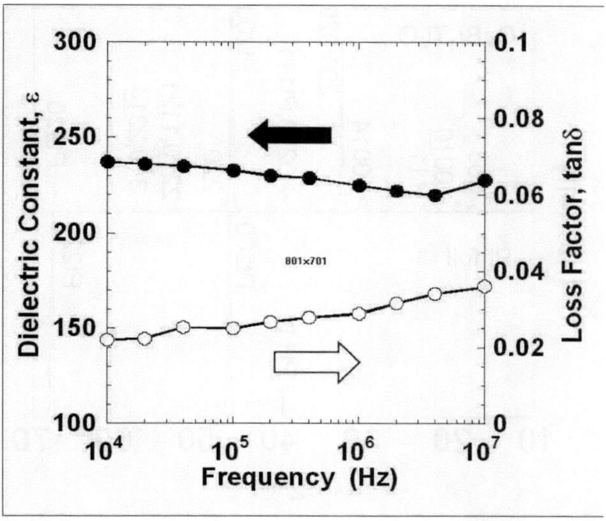

Fig. 3.6 Dielectric properties of 650°C-crystallised $CaBi_4Ti_4O_{15}$ thin film on Pt:300 nm/TiO_X:10 nm/SiO_2:10 nm/Si [2].

Fig. 3.7 SEM cross-section profiles of $CaBi_4Ti_4O_{15}$ film crystallised on Pt foil [24].

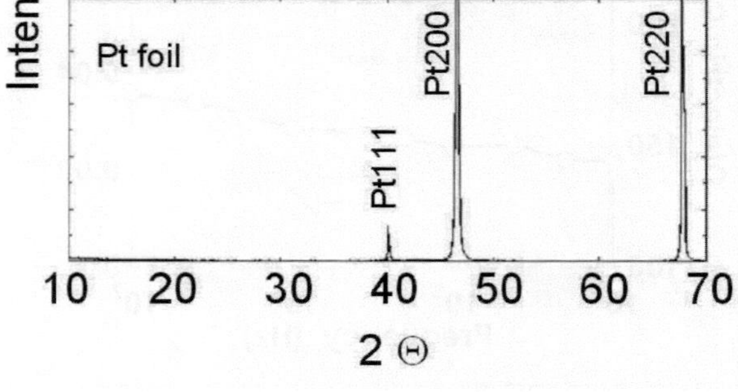

Fig. 3.8 XRD profiles of 700°C-crystallised $CaBi_4Ti_4O_{15}$ film on Pt foil [24].

3.3.3 $CaBi_4Ti_4O_{15}$ films integrated on both sides of Pt foils for piezoelectric application [24, 25]

Fig. 3.7 shows a SEM cross-section profile of $CaBi_4Ti_4O_{15}$ film on Pt foil [24]. The film had a columnar structure and consisted of well-developed grains with diameters of about 200 nm; the column heights were about 500 nm. The film surface was affected by that of Pt foil. It was rather rough compared to the thin films on Pt-coated Si. Fig. 3.8 shows XRD profiles of $CaBi_4Ti_4O_{15}$ film and Pt foil [24]. The Pt foil had high intensities of (200) and (220) diffraction lines and low intensity of (111) diffraction line. The orientation of the Pt foil differed markedly from that of the Pt layers coated on Si. $CaBi_4Ti_4O_{15}$ films on the Pt foil showed a high intensity of (200)/(020) diffraction line when compared to the other lines. However, the (200) and (020) diffraction lines were indistinguishable from each other. The lattice constants of *a* (or *b*) and *c* were estimated as 0.5417 nm and 4.086 nm, respectively, by precise measurement. The degree of the *a* or *b*-axis orientation *f* was calculated as 0.58 using the following definition proposed by Lotgerling [26]:

$$f = (P - P_0)/(1 - P_0) \quad (1)$$

$$P = I_{(h00)/(0k0)}/[I_{\Sigma(hkl)}] \text{ (}P_0\text{: same value for powders [18])} \quad (2)$$

Characteristics differed from those of $CaBi_4Ti_4O_{15}$ thin films on Pt-coated Si and are attributable to good matching of atomic arrangements in $CaBi_4Ti_4O_{15}$ (100)/(010) and Pt(110) planes. The lattice mismatch, which is calculated between c of $CaBi_4Ti_4O_{15}$ lattice and 15 times of $2^{1/2}$a of Pt lattice, is remarkable: 1.1%. The good matching between $SrBi_2Ta_2O_9$(001) and Pt(110) planes has already been mentioned [27]. However, it has not been realised in synthesis of a/b-axis oriented $SrBi_2Ta_2O_9$ or any other layer-structured perovskite thin film to date.

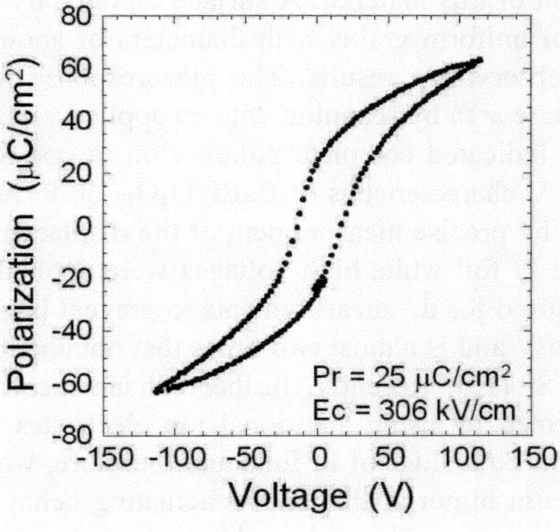

Fig. 3.9 P-V hysteresis characteristic of Pt/ $CaBi_4Ti_4O_{15}$/Pt-foil capacitor [24].

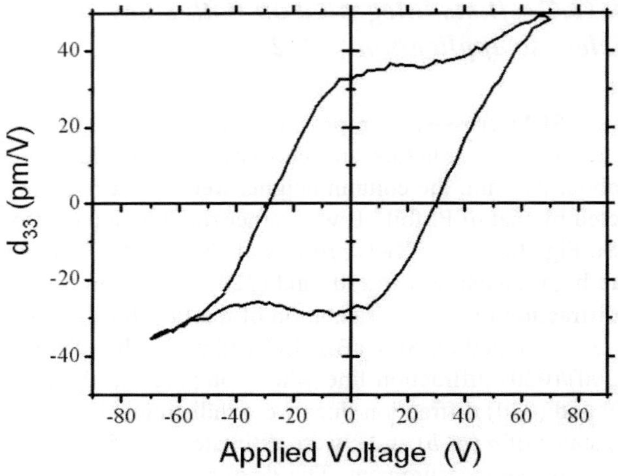

Fig. 3.10 d_{33}-V characteristics of 700°C-crystallised $CaBi_4Ti_4O_{15}$ film on Pt foil [24].

Fig. 3.9 shows P-V characteristics of Pt/ $CaBi_4Ti_4O_{15}$/Pt capacitor [24]. The P-V hysteresis loop showed an improved shape. The remnant polarisation (Pr) and coercive electric field (Ec) were 25 μC/cm^2 and 306 kV/cm, respectively, at an applied voltage of 115 V and a frequency of 500 Hz. The Pr and Ec values were enhanced to more than two times that of the values obtained for $CaBi_4Ti_4O_{15}$ thin films with random orientation. The Pr value was higher than that of stoichiometric $SrBi_2Ta_2O_9$ single crystal [28]. This is 18 μC/cm^2, and is almost identical to that of non-stoichiometric $SrBi_2Ta_2O_9$ single crystal [28], which is 28 μC/cm^2. The relatively high Ec value is attributable to the high Tc. It may also be related to oxygen vacancies. Further investigation for improvement of these properties will be necessary for application of this material. A surface topography image indicated that the film consisted of uniform grains with diameters of about 200 nm [24]. This agreed with SEM observation results. The piezoresponse image obtained after polling an 8-μm-square area by scanning with an applied voltage of 60 V had high uniform contrast. It indicated complete polarisation in one direction: downward. Fig. 3.10 shows d_{33}-V characteristics of $CaBi_4Ti_4O_{15}$ on Pt foil [24]. The d_{33} values were calculated by precise measurement of the displacement between the top of cantilever tip and Pt foil while high voltages were applied. The additional top electrodes were not used for d_{33} measurements to prevent bending effects. The d_{33} value is about 30 pm/V and is almost two times that obtained for $CaBi_4Ti_4O_{15}$ thin films on Pt-coated Si [15]. Recently, further enhancement of the piezoelectric property was confirmed by using additional top electrodes [29]. The films had uniform properties on both sides of Pt foils and therefore, would be applicable to piezoelectric devices in bimorph shape. The actuating behaviours and piezoelectric constant of d_{31} were primarily evaluated by using a laser vibrometer for application into microelectromechanical systems [30].

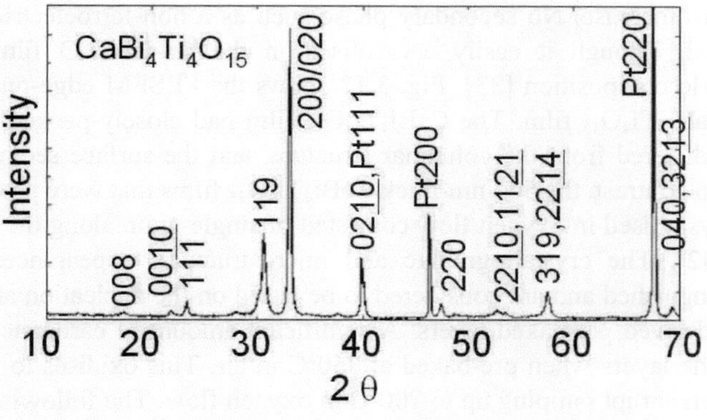

Fig. 3.11 XRD profile of 1 μm-thick CBTi144 film crystallised on Pt foil [25].

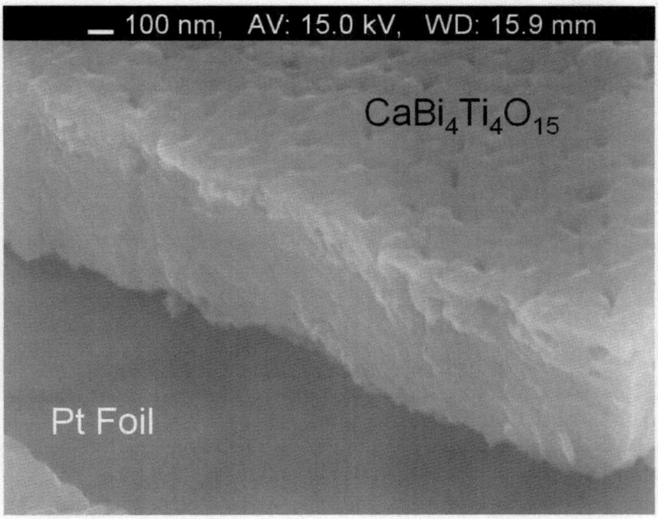

Fig. 3.12 FESEM edge-on profile of 1 μm-thick CBTi144 film on Pt foil [25].

Fig. 3.11 shows XRD profile of 1 μm-thick $CaBi_4Ti_4O_{15}$ film crystallised on Pt foil. The $CaBi_4Ti_4O_{15}$ film showed high intensities of (200)/(020) diffraction lines compared to the other lines, although the (200) and (020) diffraction lines could not be distinguished. The characteristic is considered to be due to good matching between the c-axis of $CaBi_4Ti_4O_{15}$ film and the 110 direction in the (100) plane of Pt foil [24, 31]. The relative intensities of the (200)/(020) diffraction lines were higher with respect to the present 1 μm-thick film when compared with the $CaBi_4Ti_4O_{15}$ thin films with a thickness of 500 nm [32]. The thin films were pre-baked in air and then crystallised in oxygen flow.. The number of grains with the polar axis orientation was found to increase. No secondary phase such as a non-ferroelectric pyrochlore was detected. Though it easily crystallised in the Ca-Bi-Ti-O film with non-stoichiometric composition [33]. Fig. 3.12 shows the FESEM edge-on profile of 1 μm-thick $CaBi_4Ti_4O_{15}$ film. The $CaBi_4Ti_4O_{15}$ film had closely-packed dense structure which differed from the columnar structure, and the surface seems to be relatively flat. In contrast, the 500 nm-thick $CaBi_4Ti_4O_{15}$ films that were pre-baked in air and then crystallised in oxygen flow consisted of single grain along the out-of-plane direction [32]. The crystallographic and microstructural appearances should be clearly distinguished and are considered to be based on the nucleation and growth in the alkoxy-derived pre-baked layers. A significant amount of carbonaceous residue may be in the layers when pre-baked at 350°C in air. This oxidises to gas phase to remove when abrupt ramping up to 700°C in oxygen flow. The following nucleation and growth proceed rapidly and almost simultaneously. The out of control in crystallisation resulted in the individual grain growth with large voids and undesirable rough surface.

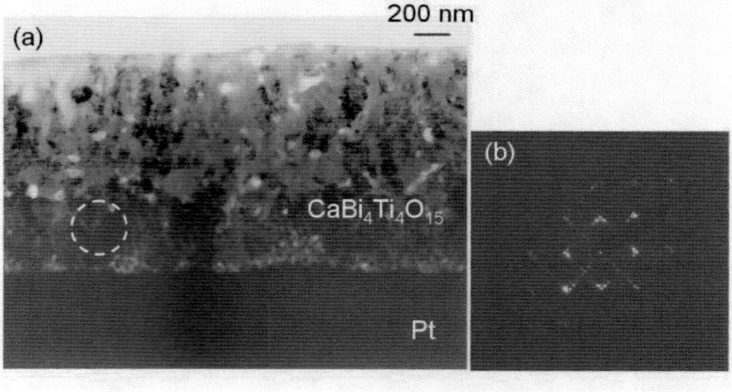

Fig. 3.13 a TEM cross section profile of 1 μm-thick CBTi144 film on Pt foil and **b** electron diffraction patterns of the selected area as indicated in a [25].

Fig. 3.13 shows cross-section TEM profile of 1 μm-thick $CaBi_4Ti_4O_{15}$ film and electron diffraction patterns of the selected area. It was found that the film pre-baked in oxygen flow and then crystallised in oxygen flow was completely densified along both of the in-plane and out-of plane directions. There have been no crack and void in the film. The structure is considered to be based on controlled nucleation at the bottom sites close to the interface and the gradual growth to the upper in each deposition layer. The electron diffraction patterns indicated that the film exhibited (100)/(010) orientation. The result is in good agreement with the XRD data as shown Fig. 3.11.

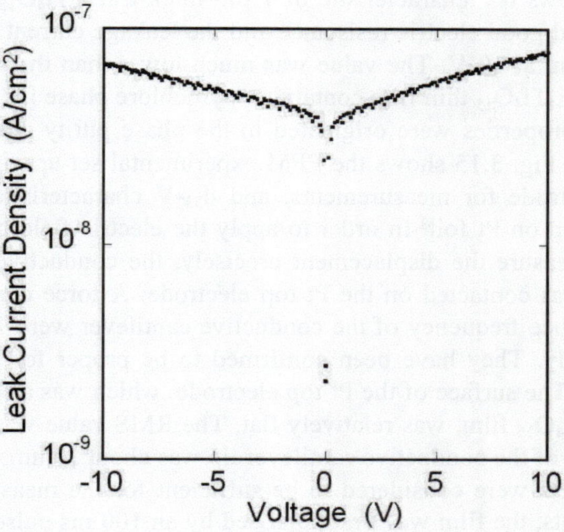

Fig. 3.14 I-V characteristic of 1 μm-thick CBTi144 film on Pt foil [25].

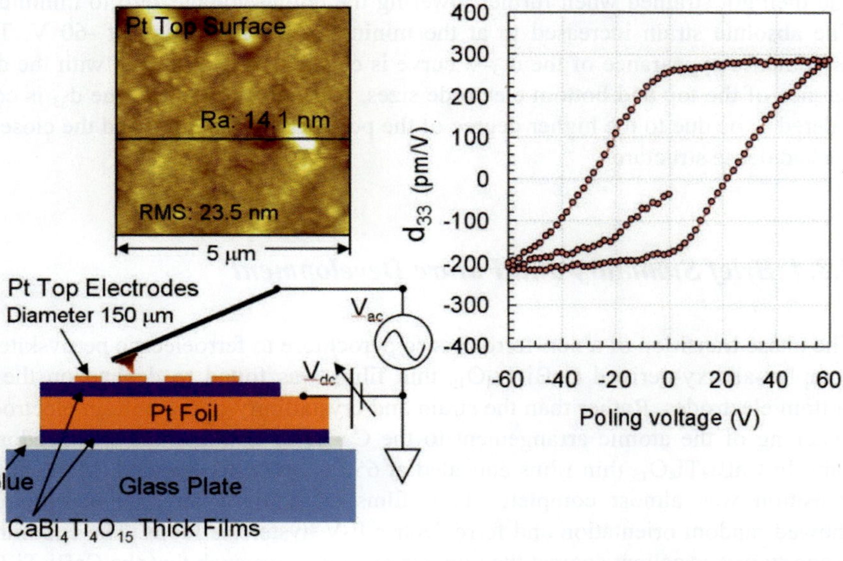

Fig. 3.15 PFM experimental set-up, the surface topography of Pt top electrode, and the d_{33}-V characteristic of 1 μm-thick CBTi144 film on Pt foil [25].

Fig. 3.14 shows I-V characteristic of 1 μm-thick $CaBi_4Ti_4O_{15}$ film on Pt foil. The film showed good electric resistance and the leakage current density was low as 7 x 10^{-8} A/cm^2 at 10 V. The value was much lower than the recently reported one for the $CaBi_4Ti_4O_{15}$ thin film containing pyrochlore phase [33]. The high electric resistance properties were originated in the phase purity and closely-packed dense structure. Fig. 3.15 shows the PFM experimental set-up, the topography of the Pt top electrode for measurements, and d_{33}-V characteristic of 1 μm-thick $CaBi_4Ti_4O_{15}$ film on Pt foil. In order to apply the electric field homogenously to the film and measure the displacement precisely, the conductive W_2C-coated Si cantilever tip was contacted on the Pt top electrode. A force constant and a mechanical resonance frequency of the conductive cantilever were 7.0 N/m and 180 kHz, respectively. They have been confirmed to be proper for the piezoelectric measurements. The surface of the Pt top electrode, which was affected by the surface of $CaBi_4Ti_4O_{15}$ film, was relatively flat. The RMS value was about 23.5 nm. As the diameter of the conductive cantilever tip was about 30 nm, the physical and electrical contacts were considered to be sufficient for the measurements. In the d_{33} measurements, the film was first polarised by an 100 ms pulse at various voltages. Then, measurement was performed within the following 500 ms. As on-top electrode measurements, the depolarisation field has been completely compensated. The observed value represents the quasistatic state of the film. Therefore, no relaxation was observed in the d_{33} loop measurement when lowering the poling voltage from maximum to zero. The d_{33} at the maximum voltage of 60 V was about 260 pm/V. This is much higher than the recently reported value (180 pm/V) of 500 nm-thick polar axis oriented $CaBi_4Ti_4O_{15}$ films [29]. The domains reversed and then got strained when further lowering the poling voltage zero to minimum. The absolute strain increased to at the minimum poling voltage of -60 V. The asymmetric appearance of the d_{33}-V curve is considered to associate with the difference of the top and bottom electrode sizes. The enhancement of the d_{33} is considered to be due to the higher degree of the polar axis orientation and the closely-packed dense structure.

3.3.4 Brief Summary and Future Development

The phase transition of a non-ferroelectric pyrochlore to ferroelectric perovskite in complex-alkoxy-derived $CaBi_4Ti_4O_{15}$ thin films was found to depend on the Pt bottom electrodes. Rather than the strain and crystallinity of the bottom electrode, matching of the atomic arrangement to the Ca-Bi-Ti-O thin films was predominant. In $CaBi_4Ti_4O_{15}$ thin films annealed at 650°C on (200)-oriented Pt, the phase transition was almost complete. Thin films crystallised on (111)-oriented Pt showed random orientation and ferroelectric P-V hysteresis loops. The endurance property was excellent against the number of switchings such that the $CaBi_4Ti_4O_{15}$ thin films are anticipated for application to FeRAM.

Polar-axis oriented $CaBi_4Ti_4O_{15}$ films were fabricated on Pt foils using the complex metal alkoxide solution. The 500-nm-thick film had a columnar structure and consisted of well-developed grains. The *a/b*-axis orientation of the ferroelectric films is considered to associate with the preferred orientation of Pt foil. The film showed improved ferro- and piezoelectric properties. The Pr, Ec and d_{33} values were enhanced two times compared to those of the $CaBi_4Ti_4O_{15}$ thin film with random orientation. Polar-axis oriented $CaBi_4Ti_4O_{15}$ films are expected to enhance development of novel devices with Pb-free piezoelectric materials.
1 μm-thick polar-axis oriented $CaBi_4Ti_4O_{15}$ films were fabricated by control of nucleation and growth in the alkoxy-derived non-crystalline layers on Pt foils. The oxygen ambient during pre-baking impacted the cross-section microstructure and crystallographic orientation. The oxygen ambient during crystallisation impacted the crystallite size and oxygen stoichiometry. The resultant 1 μm-thick film showed relatively higher degrees of polar axis orientation and simultaneously had non-columnar closely-packed dense structure. The leakage current density was about 7×10^{-8} A/cm^2 at 10 V. Piezoelectric constant d_{33} was determined to be 260 pm/V at the maximum poling voltage of 60 V by PFM measurements.

3.4 $BaTi(OR)_6$ for $BaTiO_3$

3.4.1 Chemistry in Solutions of Ba-Ti Double Alkoxides

For preparation of $BaTiO_3$ precursor solution, barium metal and Ti-isopropoxide were used as starting materials, and 2-methoxyethanol as solvent. The binary system of Ba-Ti alkoxides has been first reported for use of simple mother alcohol as solvent such as isopropanol and aminoalcohol [34]. Instead of the simple alcohol, 2-methoxyethanol was found to substitute functional groups of the alkoxide and form the metaloxane bond of Ba-O-Ti. In the FT-IR spectrum of the Ba-Ti alkoxide, the absorption peaks in the range of 800-400 cm^{-1} indicated the formation of the bond [35].

3.4.2 $BaTiO_3$ Films Deposited on $LaNiO_3$ Seeding Layers on Si [36, 37]

$BaTiO_3$ films deposited using the solution of 0.5 M had smaller effects of $LaNiO_3$ seeding layers on orientation degree and crystallinity. However, by lowering the concentration of the solution to 0.2 M, a highly (100)-oriented with high crystallinity $BaTiO_3$ films can be deposited on the obtained $LaNiO_3$ seeding layer. In order to promote the (100) orientation of $BaTiO_3$ film and to reduce the number of time-consuming steps, the highly (100)-oriented $BaTiO_3$ thinner layer was adopted as a buffer layer [36]. Fig. 3.16 shows the XRD pattern of thicker $BaTiO_3$ film deposited onto the thinner $BaTiO_3$ layer with the solution of 0.5 M. It was found that the obtained $BaTiO_3$ films with a thickness of 1 μm are still (100)-predominant. The crystalline quality of the $BaTiO_3$ films is significantly improved as compared to the $BaTiO_3$ films directly deposited onto $LaNiO_3$ layer with the solution of 0.5 M. It indicated that highly (100)-oriented thinner $BaTiO_3$ layer with high crystallinity is effective as a buffer layer to grow (100)-oriented $BaTiO_3$ films with high concentration solution. Therefore, it is obvious that, except for the low lattice match between the seeding layer and grown films, the high crystallinity of the identical composition buffer layer should also be important for oriented growth.

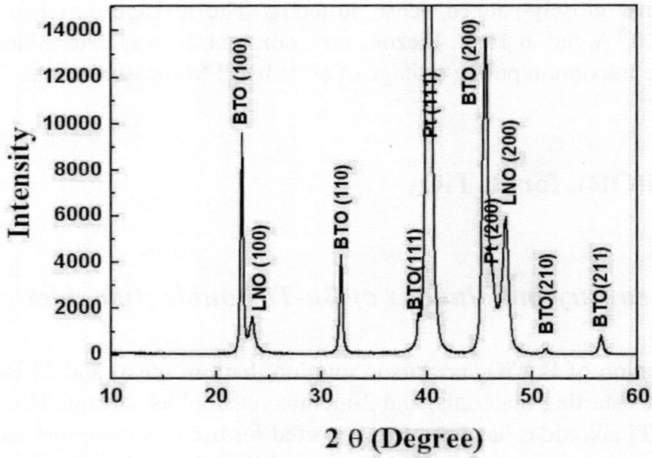

Fig. 3.16 XRD profile of 1 μm-thick $BaTiO_3$ film [36].

The cross-sectional FE-SEM micrograph of the $BaTiO_3$ film on $LaNiO_3/Pt/TiO_x/SiO_2/Si$ substrate is shown in Fig. 3.17. It is obvious that the thinner $BaTiO_3$ layer (□160 nm) has a polycrystalline columnar growth characteristic of the grains. However, predominant columnar grains, together with granular grains were observed in the thicker $BaTiO_3$ layer. In $BaTiO_3$ films, columnar grain growth only can be realised when the heterogeneous nucleation events at the interface overwhelm the homogenous nucleation of oxo-carbonate [38]. Thus, with increasing the thickness of a single layer, the granular grains cannot be overgrown by the columnar grains due to the limitation of the growth rate. The appearing of voids should also be related to the two kinds of grain growth. In order to further improve the orientation degree, deposition conditions, i.e., the solution concentration and heat-treatment, will be optimised in the near future. The interface between the two $BaTiO_3$ layers can be observed clearly and is denoted by the arrows. The total thickness of the $BaTiO_3$ film is found to be approximately 1 μm. The surface morphology of the $BaTiO_3$ film was observed by AFM. It is noticed that the $BaTiO_3$ film deposited on the smooth surface of the $BaTiO_3$ buffer layer (RMS=2.673 nm) [39] shows an increasing surface roughness (RMS=7.070 nm). The average grain size of these films is ~50 nm, which is also smaller than that of $BaTiO_3$ buffer layer. These further display both interface heterogeneous and homogenous nucleation within the bulk of the films.

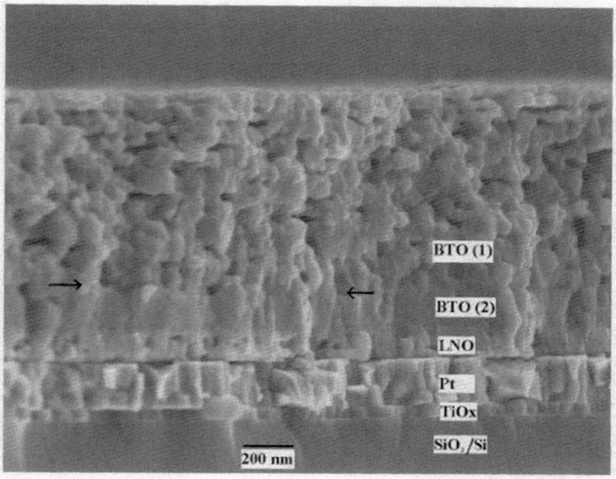

Fig. 3.17 FE-SEM cross section profile of $BaTiO_3$/ $LaNiO_3$ films on Pt-coated Si [36].

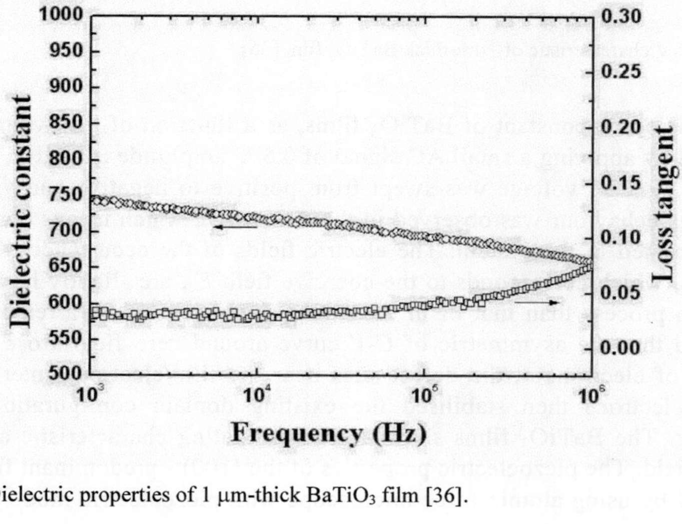

Fig. 3.18 Dielectric properties of 1 μm-thick $BaTiO_3$ film [36].

The dielectric constant ε_r and dielectric loss tan δ were measured by applying a small AC signal of 0.5 V amplitude, as a function of frequency in the range of 1 kHz-1 MHz. as shown in Fig. 3.18. The ε_r decreases from 745 to 665 with increasing the frequency from 1 kHz to 1 MHz. A loss of 0.028 is obtained at 1 kHz, and then it gradually increases with increasing frequency. Above 100 kHz, it increases with a rapidity that might be attributed to the extrinsic resonance behaviour. This is also observed in the random oriented or (100)-oriented $BaTiO_3$ thin films prepared on $LaNiO_3$/Si substrate [39, 40, 41].

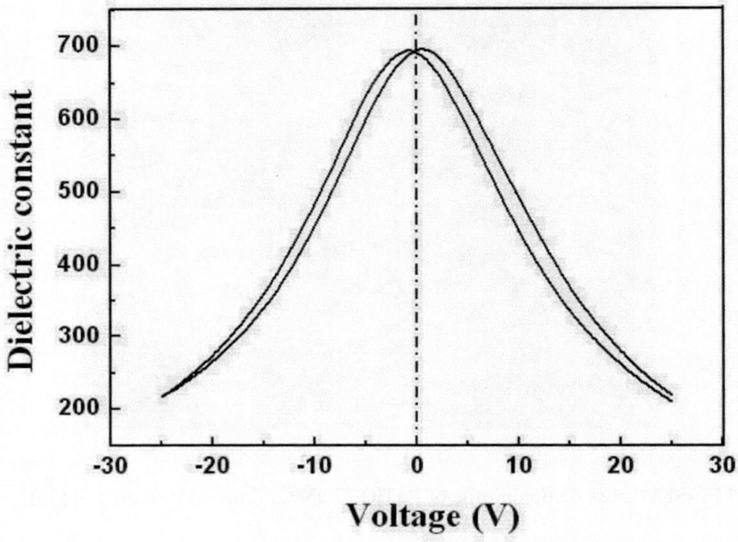

Fig. 3.19 C-V characteristic of 1μm-thick $BaTiO_3$ film [36].

The dielectric constant of $BaTiO_3$ films, as a function of bias voltage, were measured by applying a small AC signal of 0.5 V amplitude a 1 MHz, as shown in Fig. 3.19. The voltage was swept from positive to negative and vice-versa. Hysteresis behaviour was observed in the *C-V* curve, which means domain wall motion existed in the system. The electric fields of the occurrence of ε_r maximum, E_m, which corresponds to the coercive field E_c, are slightly higher in the sweep up process than that of in the sweep down process. Warren *et al.* [41] suggested that the asymmetric of *C-V* curve around zero field arose from the trapping of electrons at the defect sites near the film/electrode interface. The trapped electrons then stabilised the existing domain configuration against switching. The $BaTiO_3$ films show a good insulating characteristic against an applied field. The piezoelectric properties of the (100) - predominant films were measured by using atomic force microscope with piezoelectric mode. An average local piezoelectric coefficient higher than 50 pm/V (Fig. 3.20a) has been obtained on the bare films. These d_{33} values of $BaTiO_3$ films are comparable to those of $Pb(Zr,Ti)O_3$ films [42, 43]. They are higher than those of lead-free piezoelectric films when they are deposited on Si substrates [44, 45]. Therefore, the (100)-predominant $BaTiO_3$ thicker films should be a promising candidate as the lead-free piezoelectric films for MEMS applications.

$BaTiO_3$ thin films deposited on $LaNiO_3$ seeding layers using the dilute 0.2 M solution have been clarified to show the high crystallintiy and high degree of (100) orientation [39, 40]. The high quality $BaTiO_3$ thin film worked well as a buffer layer on the $LaNiO_3$ seeding layer for fabrication of thicker $BaTiO_3$ films, with (100) preferred orientation. Fig. 3.21 shows XRD profile of $BaTiO_3$ film with a thickness of 1 μm deposited on the identical buffer and $LaNiO_3$ seeding layers, using the concentrated 0.4 M Ba-Ti double alkoxide solution. It was found that the $BaTiO_3$ thick film was still (100)-predominant with the high orientation degree ($I_{100}/(I_{100}+I_{110}+I_{111})$) of 0.67, although the degree was lower than that of the thin buffer layer as 0.97 [37]. The crystalline quality of the thick film was significantly improved as compared to the $BaTiO_3$ films directly deposited onto the $LaNiO_3$ sole seeding layer with the concentrated solutions. It indicated that highly (100)-oriented thinner $BaTiO_3$ layer with high crystallinity is effective as a buffer to grow (100)-oriented $BaTiO_3$ films with the concentrated solutions. Therefore, instead of the low lattice match between the seeding layer and grown films, the high crystallinity of the identical composition buffer layer is obviously important for highly oriented growth.

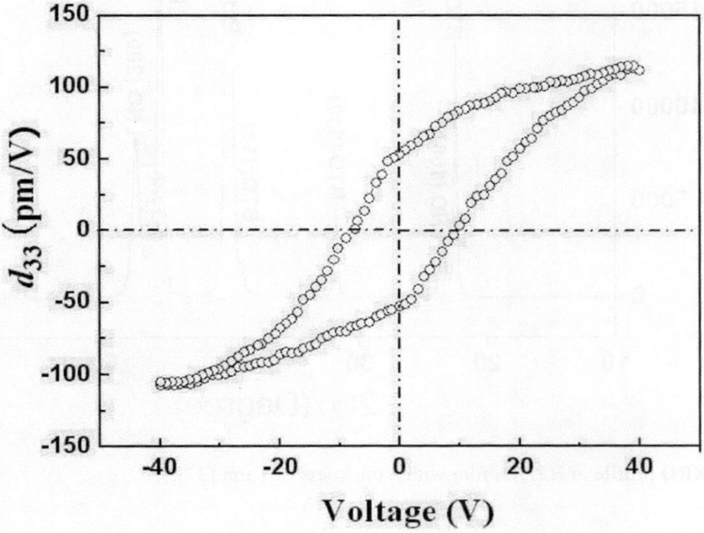

Fig. 3.20 d_{33}-V characteristics of 1μm-thick $BaTiO_3$ film [36].

Fig. 3.22 a, b, c and d show the cross-section TEM and electron diffraction profiles of the 280 nm-thick $BaTiO_3$ film deposited on the $LaNiO_3$ seeding layer. The 0.2 M solution was used, which can work as the identical buffer for textured $BaTiO_3$ thick films. The thin film had a columnar structure and was densified without inner voids. This often appeared in the films deposited by using organic compounds as starting materials. In the highly magnified profile as Fig. 3.22b, the layer-like structure was observed in the one columnar grain. The HR-TEM image of the selected area indexed as X (Fig. 3.22c) indicated there is no discontinuity in the atomic arrangement. The electron diffraction profile revealed that the film was $BaTiO_3$ and showed highly (100) preferred orientation. Fig. 3.23a, b, c, d and e show the cross section TEM and electron diffraction profiles of the 1 μm-thick $BaTiO_3$ film fabricated on the identical $BaTiO_3$ buffer and $LaNiO_3$ seeding layers using the 0.4 M solution. The thick film had a combined structure, which was a gradient of columnar and granular structures in the region of interface to surface. The constituent grains were single phase of $BaTiO_3$ with high crystallinity and preferred orientation in diameters of 40 nm to 100 nm.

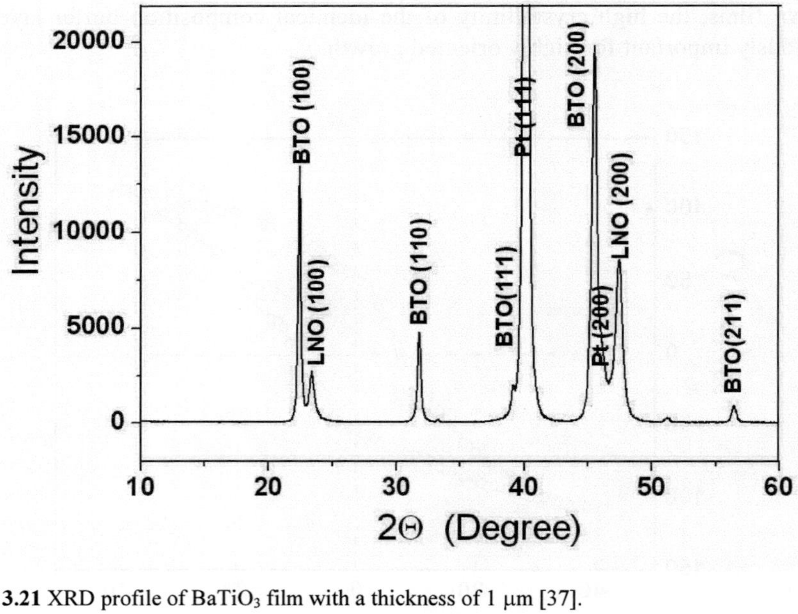

Fig. 3.21 XRD profile of $BaTiO_3$ film with a thickness of 1 μm [37].

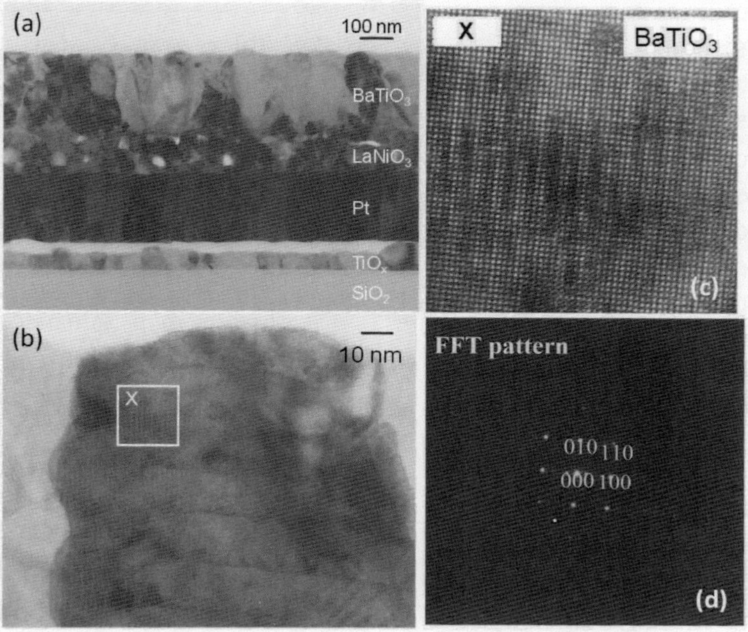

Fig. 3.22 Cross-section TEM micrographs **a** and **b**, high-resolution image **c**, and electron diffraction pattern **d** of the 280 nm-thick $BaTiO_3$ film deposited on $LaNiO_3/Pt/TiO_x/SiO_2/Si$ substrate [37].

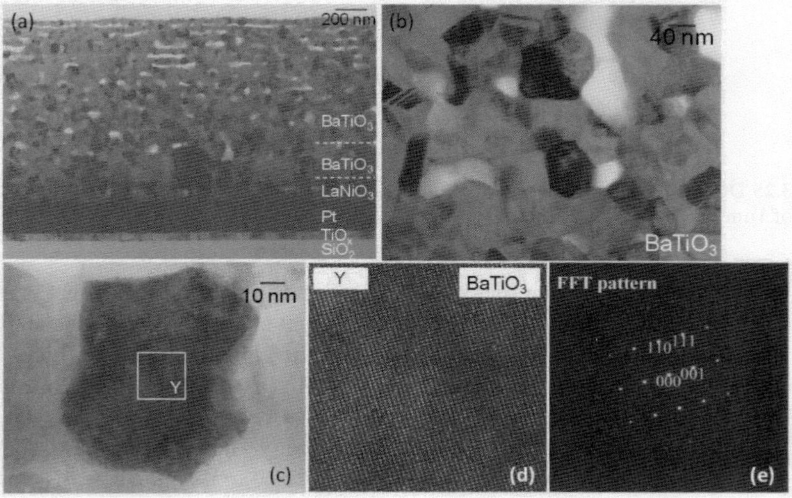

Fig. 3.23 Cross-section TEM micrographs **a**, **b** and **c**, high-resolution image **d**, and electron diffraction pattern **e** of the 1 μm-thick $BaTiO_3$ film deposited on $LaNiO_3/Pt/TiO_x/SiO_2/Si$ substrate [37].

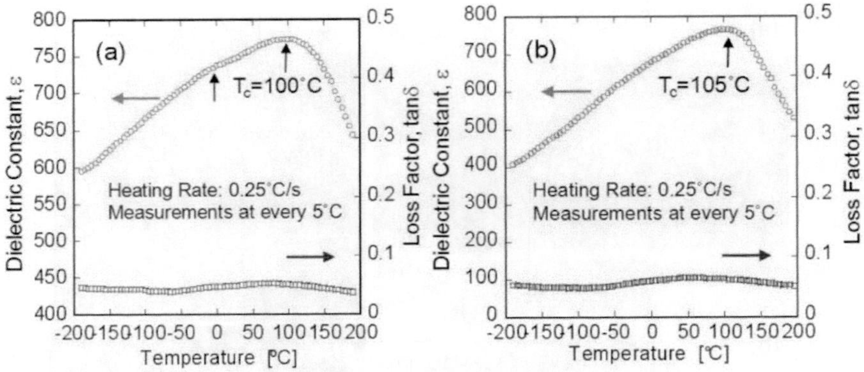

Fig. 3.24 Temperature dependences of the dielectric constant and loss factor of **a** the 280 nm-thick and **b** 1 μm-thick $BaTiO_3$ films [37].

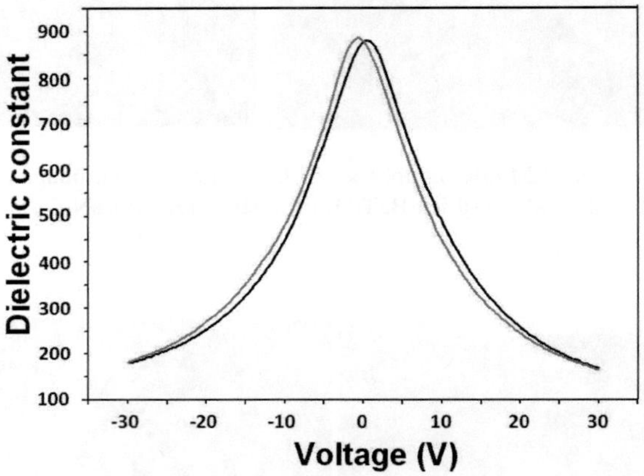

Fig. 3.25 DC bias dependence of the dielectric constant at 1 MHz of $BaTiO_3$ film with a thickness of 1μm [37].

Fig. 3.24a and b show the dielectric properties as a function of temperature for $BaTiO_3$ thin and thick films. The dielectric constant showed broad peaks at 100°C and 105°C as the Curie temperature, and T_c for the 280 nm and 1 μm-thick films, respectively. The film showing higher crystallinity and (100) orientation degree exhibited the lower T_c. The lowered T_c was considered to associate with the high crystallinity, crystallographic orientation and, thereby, induced stress [39, 42, 46]. The broad transition may originate in nanometre sized grains, the size distribution and the gradient structure. In the 280 nm-thick film, an additional shoulder appeared at around 0°C. This is considered to relate to the phase transition of tetragonal to orthorhombic. The detail analysis of the later transition is in progress and the characteristic will be clarified.

The dielectric constant of the 1 μm-thick $BaTiO_3$ film, as a function of bias voltage, was measured by applying a small AC signal of 0.5 V amplitude a 1 MHz, as shown in Fig. 3.25. The voltage was swept from positive to negative and vice-versa. Hysteresis behaviour was observed in the *C-V* curves, which mean domain wall motion existed in the system. The electric fields of the occurrence of ε_r maximum, E_m, which corresponds to the coercive field E_c, are slightly higher in the sweep up process than that of in the sweep down process. Warren *et al.* [41] suggested that the asymmetric of *C-V* curve around zero field arose from the trapping of electrons at the defect sites near the film/electrode interface. The trapped electrons then stabilised the existing domain configuration against switching. If one were to consider these films for varactor application, the dielectric tunability (delta $\varepsilon = (\varepsilon m_{ax}$-$\varepsilon m_{in})/\varepsilon m_{ax}$) is approximately calculated as 0.80 at 300 kV/cm, measured at 1 MHz.

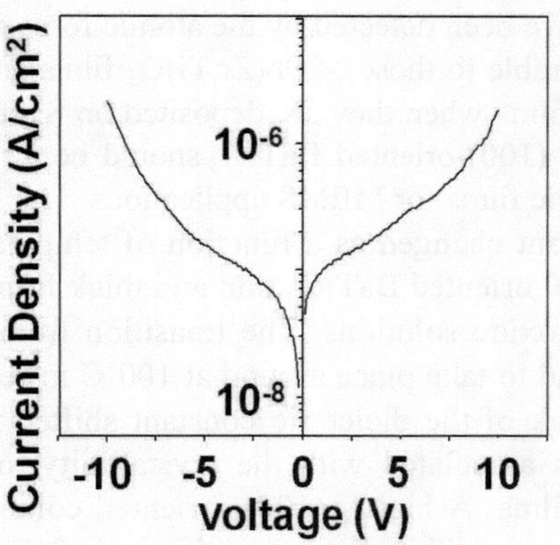

Fig. 3.26 J-V curve of $BaTiO_3$ film with a thickness of 1 μm [37].

Fig. 3.26 shows the dependence of leakage current on applied voltage for the 1 μm-thick $BaTiO_3$ film. The $BaTiO_3$ film shows a good insulative characteristic against applied field. The magnitude of resistivities (ρ = VA/It, where V is an applied voltage, A is the area of the capacitor, I is the leakage current, and t is the thickness of $BaTiO_3$ film) of the 1 μm-thick $BaTiO_3$ film is around the order of 10^{11}Ùcm. The insulating performance is good enough for actuator or sensor applications. The leakage current in the capacitor with the bilayer $BaTiO_3$ film in the positive bias voltage region (with $LaNiO_3$ conductive seeding layer under the positive bias voltage) is higher than those of in negative one. Our previous measurements in columnar $BaTiO_3$ films showed that the leakage current in the positive bias voltage region was always higher than those in the negative bias voltage region [39]. It is known that the positive leakage current is limited by the interface between the bottom electrode and the films (bottom interface). The negative leakage current is limited by the interface between the top electrode and the films (top interface) [47, 48, 49, 50]. The enhancement of the positive leakage current was relatively suppressed in the present bilayer $BaTiO_3$ film. This indicates the improvement of the interface between the $BaTiO_3$ film and $LaNiO_3$ seeding layer. The results are in good agreement of the microstructure as shown in Fig. 3.22 and Fig. 3.23.

3.4.3 Brief Summary and Future Development

Bilayer $BaTiO_3$ films with thickness of 1 μm were prepared on $LaNiO_3/Pt/TiO_x/SiO_2/Si$ substrates using two kind of concentration solution. The $BaTiO_3$thinner layer is found to be effective as a buffer layer for the crystallisation and oriented grain growth of the upper $BaTiO_3$ layer. The dielectric constant of ~745 and loss tangent of ~0.028 (1 kHz) were obtained. The $BaTiO_3$ films still show a good insulating characteristic against applied field. Local piezoelectric coefficients higher than 50 pm/V have been detected by the atomic force microscope on the bare films. These are comparable to those of $Pb(Zr,Ti)O_3$ films and higher than those of lead-free piezoelectric films when they are deposited on Si substrates. These results indicate that the highly (100)-oriented $BaTiO_3$ should be a promising candidate as the lead-free piezoelectric films for MEMS applications.

The dielectric constant changed as a function of temperature in the range of -200°C to 200°C in (100) oriented $BaTiO_3$ thin and thick films deposited on Si substrate using double alkoxide solutions. The transition from paraelectric to ferroelectric phase was found to take place around at 100°C instead of 130°C for single crystals. The broad peak of the dielectric constant shifted to lower temperatures and the behaviour was associated with the crystallinity, orientation degree and microstructure of the films. A highly (100) -oriented columnar $BaTiO_3$ thin film with thickness of 280 nm exhibited two transitions at 0°C and 100°C. The additional lower temperature transition was considered to be tetragonal to orthorhombic. The 1μm-thick $BaTiO_3$ film with a combined structure consisted of columnar and granular grains showed a transition at 105°C.

References

1. R. C. Mehrotora, J. Non-Cryst. Solids, 100, 1 (1988).
2. K. Kato, D. Fu, K. Tanaka, K. Suzuki, T. Kimura, K. Nishizawa, T. Miki, Int. J Appl. Ceram. Technol., 2, 64 (2005).
3. S. Sakka, Handook of Sol-Gel Science and Technology: Processing Characterization and Applications, Kluwer Academic Publishers, Vol.1, p.41 (2004).
4. K. Kato, C. Zhang, J. M. Finder, S. K. Dey, J. Am. Ceram. Soc., 81, 1869 (1998).

5. S. Katayama, M. Sekine, J. Mater. Res., 6, 36 (1991).
6. S. Govil, P. N. Kapoor, R. C. Mehrotra, J. Inorg. Nucl. Chem., 38, 172 (1976).
7. M. M. Amini, M. D. Sacks, Mater. Res. Soc. Symp. Proc., 180, 675 (1990).
8. T. Yogo, K. Kikuta, Y. Ito, S. Hirano, J. Am. Ceram. Soc., 78, 2175 (1995).
9. D. C. Bradley, R. C. Holloway, J. Chem. Soc., A, 219 (1968).
10. S. Hirano, K. Kato, 2, 142 (1987).
11. P. R. Coffman, C. K. Barlingay, A. Guota, S. K. Dey, J. Sol-Gel Sci. Technol., 6, 83 (1996).
12. S. Katayama, I. Yoshinaga, N. Yamada, T. Nagai, J. Am. Ceram. Soc. 79, 2059 (1996).
13. K. Nishizwa, T. Miki, K. Suzuki, K. Kato, J. Mater. Res. 18, 899 (2003).
14. K. Kato, K. Suzuki, K. Nishizawa, T. Miki, J. Appl. Phys., 88, 3779 (2000).
15. D. Fu, K. Suzuki and K. Kato, Jpn. J. Appl. Phys., 42, 5994 (2003).
16. M. Toyoda, Y. Hamaji, K. Tomono and D. A. Payne, Jpn. J. Appl. Phys., 32, 4158 (1993).
17. S. Katayama and M. Sekine: J. Mater. Res., 6, 36 (1991).
18. K. Kato, K. Suzuki, K. Nishizawa and T. Miki, Appl. Phys. Lett. 78, 1119 (2001).
19. K. Kato, K. Suzuki, D. Fu, K. Nishizawa and T. Miki, Appl. Phys. Lett. 81, 3227 (2002).
20. S. H. Kim, D. J. Kim, K. M. Lee, M. Park, A. I. Kingon, R. J. Nemanich, J. Im and S. K. Streiffer, J. Mater. Sci., 14, 4395 (2000).
21. K. Aizawa, E. Tokumitsu, K. Okamoto and H. Ishiwara, Appl. Phys. Lett., 76, 2609 (2000).
22. T. Takeuchi, T. Tani and Y. Saito, Jpn. J. Appl. Phys., 38, 5553 (1999).
23. H. Yan, C. Li, J. Zhou, W. Zhu, L. He and Y. Song, Jpn. J. Appl. Phys., 39, 6339 (2000).
24. K. Kato, D. Fu, K. Suzuki, K. Tanaka, K. Nishizawa, T. Miki, Appl. Phys. Lett., 84, 3771 (2004).
25. K. Kato, K. Tanaka, S. Kayukawa, K. Suzuki, Y. Masuda, T. Kimura, K. Nishizawa, T. Miki, Appl. Phys. A-Mater. Proc., 87, 637 (2007).
26. F. K. Lotgerling, J. Inorg. Nucl. Chem., 9, 113 (1959).
27. J. S. Lee, H. H. Kim, H. J. Kwon and Y. W. Jeong, Appl. Phys. Lett., 73, 166 (1998).
28. Y. Noguchi, K. Murata, M. Takahashi and M. Miyayama, private communication.
29. D. Fu, K. Suzuki and K. Kato, Appl. Phys. Lett., 85, 3519 (2004).
30. F. Arai, K. Motoo, T. Fukuda and K. Kato, Appl. Phys. Lett., 85, 4217 (2004).
31. J. S. Lee, H. H. Kim, H. J. Kwon and Y. W. Jeong, Appl. Phys. Lett., 73, 166 (1998).
32. K. Kato, K. Tanaka, K. Suzuki, T. Kimura, K. Nishizawa, T. Miki, Appl. Phys. Lett., 86, 112901 (2005).
33. A. Z. Simoes, M. A. Ramirez, A. Ries, J. A. Varela, E. Longo and R. Ramesh, Appl. Phys. Lett., 88, 072916 (2006).
34. K. S. Mazdiyasni, R. T. Dolloff, J. S. Smith II, J. Am. Ceram. Soc., 52, 523 (1969).
35. K. Tanaka, K. Suzuki, D. Fu, K. Nishizawa, T. Miki, K. Kato, Jpn. J. Appl. Phys., 43, 6525 (2004).
36. Y. Guo, K. Suzuki, K. Nishizawa, T. Miki, K. Kato, Acta Mater., 54, 3893 (2006).
37. K. Kato, K. Tanaka, K. Suzuki, S. Kayukawa, Appl. Phys. Lett., 91, 172907 (2007).
38. S. Hoffmann and R. Waser, J. Euro. Ceram. Soc. 19, 1339 (1999).
39. Y. Guo, K. Suzuki, K. Nishizawa, T. Miki, and K. Kato, J. Cryst. Growth, 284, 190 (2005).
40. Y. Guo, K. Suzuki, K. Nishizawa, T. Miki, and K. Kato, Jpn. J. Appl. Phys. 45, p (2006).
41. W. L. Warren, D. Dimos, and R. Waser, MRS Bull. 21 40 (1996)
42. P. Gerber, C. Kügeler, U. Böttger, and R. Waser, J. Appl. Phys. 98, 124101 (2006).
43. C. Chao, Z. Wang, and W. Zhu, Thin Solid Films, 493, 313 (2005).

44. G. Hu, J. Xu, and I. Wilson, Appl. Phys. Lett., 75, 1610 (1999).
45. H. Maiwa, N. Iizawa, D. Togawa, T. Hayashi, Appl. Phys. Lett., 82, 1760 (2003).
46. B. Dkhil, E. Defay and J. Guillan, Appl. Phys. Lett., 90, 022908 (2007).
47. S. Zafar, R. E. Jones, B. Jiang B. White, V. Kaushik, and S. Gillespie, Appl. Phys. Lett., 73, 3533 (1998).
48. J. C. Shin, J. Park, C. S. Hwang, and H. J. Kim, J. Appl. Phys., 86, 506 (1999).
49. G. W. Dietz, M. Schumacher, R. waser, S. K. Streiffer, C. Basceri, and A. I. Kingon, J. Appl. Phys., 82, 1455 (1999).
50. S. Maruno, T. Kuroiwa, N. Mikami, K. Sato, S. Ohmura, M. Kaida, T. Yasue, and T. Koshikawa, Appl. Phys. Lett., 73, 954 (1998).

Chapter 4
Ferroelectrics onto Substrates Prepared by Chemical Solution Deposition: From the Thin Film to the Self-Assembled Nano-sized Structures

M. L. Calzada

4.1 Introduction

Ferroelectrics are high dielectric permittivity materials with spontaneous polarisation, P_s, in a range of temperature. This is due to the lack of symmetry of the crystal structure with the higher symmetry of the paraelectric phase. [1, 2] The wide range of properties of these materials has been used in electronic devices since the 1940s, when $BaTiO_3$ capacitors were commercialised [3]. Nowadays, devices for information storage can be fabricated with ferroelectric materials making use of their high permittivity values (Dynamic Random Access Memories, DRAM). They can even be fabricated by the switching of polarisation with the electric field (Non-Volatile Ferroelectric Random Access Memories, NVFRAM). The piezoelectric or the pyroelectric activity of ferroelectrics (polarisation changes with pressure or with temperature, respectively) is exploited in sensors and actuators. [4, 5, 6] At the beginning, these devices were fabricated with single crystals and bulk ceramics. But, the trend towards miniaturisation of electronic devices has been the driving force for the reduction of the size of the ferroelectric material. [7] Fig. 4.1 shows this trend in ferroelectrics during the past decades. [8] The progressive miniaturisation of the ferroelectric material and their integration with the IC (Integrated Circuit) technology has additional benefits in the electronic industry. Some of these benefits are an increase of the integration density, and the low voltage operation leading to low power consumption. Therefore, new areas of application can be found for ferroelectric materials in the next generation of electronic devices, if integrated with semiconductor substrates as thin, ultra-thin film, or isolated nano-sized structures. Fig. 4.2 shows some of the possible applications of ferroelectrics onto substrates. [4, 9]

Inst. Ciencia de Materiales de Madrid (CSIC), Cantoblanco, 28049 – Madrid, Spain.
lcalzada@icmm.csic.es

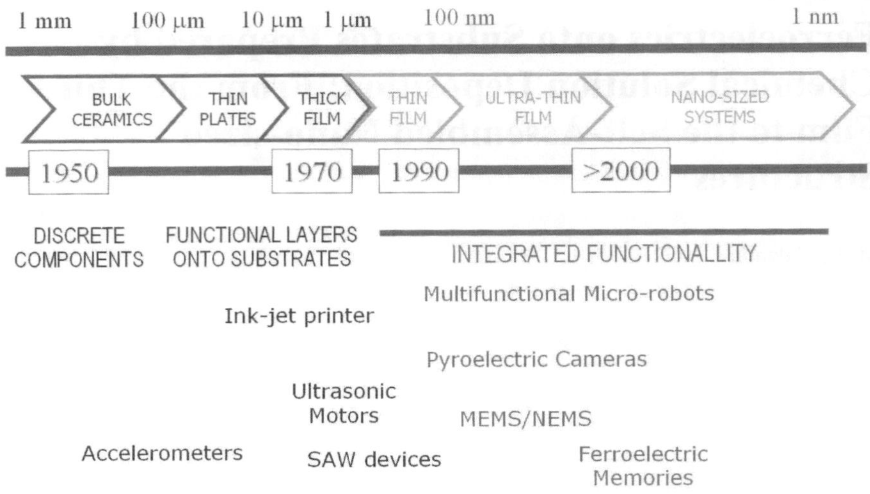

Fig. 4.1 Trends in ferroelectric polycrystalline materials: from bulk ceramics to nano-sized systems.

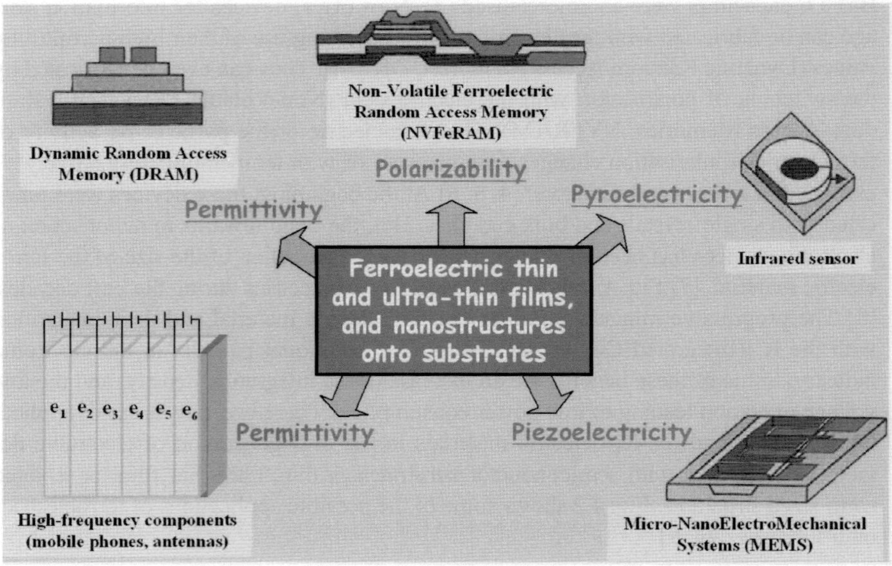

Fig. 4.2 Applications of ferroelectrics supported onto substrates. [4, 9] .

Table 4.1 Ferroelectric perovskite compositions prepared onto substrates and possible applications in micro-nanoelectronic devices

Ferroelectric compositions	Property used in the device	Device applications	Material form	References
$BaTiO_3$ (BT), $SrTiO_3$ (ST) $(Ba,Sr)TiO_3$ (BST), $Pb(Zr,Ti)O_3$ (PZT)	High permittivity, high dielectric non-linearity	Dynamic Random Access Memories (DRAM), Microwave devices (MW)	Thin and ultrathin films Isolated nanostructures onto substrates	[10,11]
$SrBi_2Ta_2O_9$ (SBT), $Bi_{3.25}La_{0.75}Ti_3O_{12}$ (BLT) $Pb(Zr,Ti)O_3$ (PZT)	Switchable polarisation, low fatigue	Non-Volatile Ferroelectric Random Access Memories (NVFERAM)]	Thin and ultrathin films Isolated nanostructures onto substrates	[4,12]
$PbTiO_3$ (PT), $(Pb,La)TiO_3$ (PTL) $(Pb,La)(Zr,Ti)O_3$ (PLZT) $Pb(Zr,Ti)O_3$ (PZT), $Pb(Sc,Ta)O_3$ (PST)...	Pyroelectric response	Infrared sensors	Thin films	[13,14]
$Pb(Zr,Ti)O_3$(PZT), $PbTiO_3$ (PT), $(Pb,Ca)TiO_3$ (PTC), $(Pb,La)TiO_3$ (PTL)...	Piezoelectric response	MicroElectroMechanical Systems (MEMS)	Thin and ultrathin films	[15,16]
$(Pb,La)(Zr,Ti)O_3$, (PLZT), $(Pb,La)TiO_3$ (PTL), $LiNbO_3$, $KNbO_3$,...	Electrooptic effect	Optical waveguides	Thin films	[17,18]

Ferroelectric perovskite-related compounds have been most studied as thin-ultrathin films and nanostructures onto substrates. They have been prepared by different deposition techniques and have shown to be potentially useful for applications in micro-and nano-electronic devices (Table 4.1). [4, 10, 11, 12, 13, 14, 15, 16, 17, 18] As shown in Table 4.1, ferroelectric composition determines the application in devices of these materials. $PbTiO_3$ is the first member of the lead-containing perovskites. It has a large tetragonality and spontaneous polarisation at room temperature. Pure $PbTiO_3$ thin and ultrathin films have been prepared and characterised, [19, 20]. Lately, basic studies have been carried out on $PbTiO_3$ nanostructures supported onto substrates. [21, 22] The partial substitution of Pb^{2+} by isovalent (Ca^{2+}, Sr^{2+}, ...) or off-valent (La^{3+}, Sm^{3+}, ...) cations have provided materials in thin film form with competitive piezo and pyroelectric responses. [23, 24, 25, 26, 27, 28, 29] The substitution of Ti(IV) by Zr(IV) in $PbTiO_3$ leads to the $Pb(Zr,Ti)O_3$ (PZT) solid solution. This has been considered for a long time the best candidate for NVFERAM, piezoelectric devices or infrared sensors. [4, 30, 31, 32] In the mid-1990s, the competitiveness of layered perovskites, such as $SrBi_2Ta_2O_9$ (SBT) [33] or $Bi_{3.25}La_{0.75}Ti_3O_{12}$ (BLT) [34], for non-volatile computer memories was proved. This showed these compositions have low fatigue and large retention when deposited onto semiconductor substrates with conventional Pt bottom electrodes. Other lead-free compounds based on the well-known $BaTiO_3$ ferroelectric perovskite are also of interest in micro-nanoelectronic devices. Thus, the solid solution resulting from the substitution of Ba^{2+} by Sr^{2+}, $Ba_xSr_{1-x}TiO_3$ (BST), has been intensively studied for its use in applications where high charge storage densities (DRAM) in thin film capacitors and high dielectric non-linearity for microwave (MW) devices are required. [10, 35] Also, lead-free perovskites of pure and doped $LiNbO_3$ are the most used ferroelectric compositions for applications in optical thin film waveguides. [36, 37] The concern about toxicity of lead-related compounds with new government directives on the control of the content of this element in commercial products [38] has stressed the use of ferroelectric lead-free compositions in devices. Thus, solid solutions of niobate orthorhombic perovskite-type compounds ($(K,Na)NbO_3$) with $LiTaO_3$ or $LiSbO_3$, show stable piezoelectric characteristics over a wide temperature range. They seem to be promising materials for piezoelectric devices, although, until now, mainly the fabrication and properties of bulk ceramics have been reported. [39]

Fabrication of ferroelectrics onto substrates can be carried out using different deposition techniques, which are grouped into physical and chemical methods. Table 4.2 summarises some of these methods, indicating if thin or ultrathin films or isolated nanostructures have been obtained by these techniques. The requirements of each type of device or integration issues, such as coating conformity, processing temperature or epitaxy, dictate the suitability of the deposition technique for the preparation of ferroelectrics onto substrates. [40, 41, 42, 43, 44, 45, 46, 47, 48, 49, 50, 51, 52]

Table 4.2 Techniques used for the deposition of ferroelectrics onto substrates.

Techniques used for the deposition of ferroelectrics onto substrates		Material form	Advantages	Handicaps	Devices
Physical deposition techniques*	RF Sputtering	Thin films [41] ultrathin films [42], isolated nanostructures [43]	Epitaxial coatings	High cost, negative ions, low stoichiometric control	Capacitors, memories, MEMS, pyroelectric devices, substrate acoustic waves (SAW), ...
	Magnetron sputtering	Thin films [41],	Epitaxial coatings	High cost, target surface, low stoichiometric control	Capacitors, memories, MEMS, pyroelectric devices, SAW, ...
	Ion beam sputtering	Thin films[41]	Good stoichiometric control	High cost, uniformity	Capacitors, memories, MEMS, SAW, pyroelectric devices, ...
Chemical deposition techniques**	Laser ablation	Thin films [44], ultrathin films [45] isolated nanostructures [46]	Expitaxial coatings	High cost, uniformity, low stoichiometric control	Capacitors, infrared sensors, waveguides, ...
	MOCVD	Thin films [47], ultrathin films [48], isolated nanostructures [49]	Conformal coatings, good stoichiometric control	High cost, high substrate temperature	Memories, capacitors, MEMS, pyroelectric devices, ..
	CSD	Thin films [50], ultrathin films [51], isolated nanostructures [52]	Low cost, good stoichiometric control, uniformity, tailoring of the solution chemistry	Multiple coatings	MEMS, infrared sensors, capacitors, ...

* Fabrication of isolated ferroelectric nanostructures onto substrates by physical deposition techniques usually involves the carving of a thin film using a lithography based technique (top-down technology).

** Also denoted self-assembly methods (bottom-up technology) in the case of the fabrication of isolated nanostructures.

Chemical Solution Deposition (CSD) is an attractive technique for the preparation of these materials. It presents problems, for instance the fabrication of conformal coatings or ordered arrays of nanostructures, which are needed in memory devices. However, its advantages make it a commercially viable technique for producing ferroelectric materials of interest in specific applications. These advantages are: low cost, the possibility of tailoring the chemistry of the solutions for designing special materials, and the well-adjusted stoichiometry and uniformity of the coating.

Here, a review of the recent work on CSD-processed thin and ultra-thin films, and how they can evolve towards nanostructures supported onto substrates is shown. The emphasis is on the ferroelectric activity of these materials that make them useful in potential micro or nanodevices. The work carried out at the Materials Science Institute of Madrid (ICMM-CSIC) over the years on thin films prepared by CSD is shown. Special attention is paid to the effect of the processing on the properties of the films. Following the trends of reduction of the microelectronic device sizes, our work has focused on the fabrication by CSD of nano-sized ferroelectrics. First, in the approach as continuous ultra-thin films (film thickness <100 nm) and second, in the approach as nano-sized structures in the sub-100nm size regime (ferroelectric nano-capacitors).

4.2 Chemical Solution Deposition (CSD) of Ferroelectric Materials

The initial development of the CSD technique for the fabrication of ferroelectric materials (films) onto substrates came towards the end of the 1970s. Fukushima *et.al.* [53] reported on the preparation of $BaTiO_3$ thin films by the deposition of partially hydrolised solutions formed by alkoxides and organometallic compounds. In the early 1980s, alternative solution routes were proposed for the preparation of $BaTiO_3$ thin films, and for $PbTiO_3$ and $PbZrO_3$ thin films. [54, 55, 56] Since then, CSD has been widely used for the preparation of films with different ferroelectric compositions.

The CSD technique involves three steps (Fig. 4.3) [40, 50, 57]:

1. Preparation of the precursor solution. It has to contain reagents of the cations that are going to form the crystalline ferroelectric composition, dissolved in appropriate solvents to get a homogeneous and stable solution.

2. Deposition of the solution onto a substrate. Usually spin or dip-coating are used for obtaining a wet layer that is dried and partially pyrolysed to get an amorphous coating.

3. Crystallisation of the ferroelectric phase. Thermal treatment of the amorphous coating converts the amorphous film into a crystalline ferroelectric film.

The different strategies used for the preparation of the precursor solution give rise to the three main subgroups in which CSD methods are classified: Metal-Organic Decomposition (MOD), Sol-Gel, and hybrid routes.

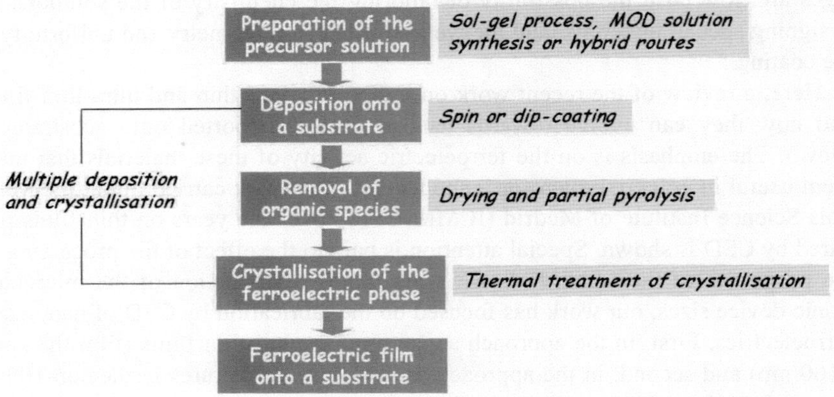

Fig. 4.3 Chemical Solution Deposition (CSD) of ferroelectrics onto substrates.

MOD is based on the use of large carboxylate or β-diketonates compounds dissolved in a common solvent. There is no interaction among the starting compounds or with the solvent. As a result, the MOD solution is a simple mixture of components.

Sol-gel uses metal alkoxides dissolved in alcohols as synthesis reagents. Contrary to MOD, hydrolysis and condensation are the key reactions occurring in sol-gel chemistry. This leads to the formation of a precursor sol containing metal-oxygen-metal (M-O-M) bonds:

Hydrolysis

$$M(OR)_z + H_2O \rightarrow HO\text{-}M(OR)_{z-1} + ROH \quad (1)$$

$$M(OR)_z + zH_2O \rightarrow M(OR)_z + zROH \quad (2)$$

Condensation

$$(OR)_{z-1}M\text{-}OH + HO\text{-}M(OR)_{z-1} \rightarrow$$

$$(OR)_{z-1}M\text{-}O\text{-}M(OR)_{z-1} + H_2O \textit{ oxolation} \quad (3)$$

$$(OR)_{z-1}M\text{-}OH + RO\text{-}M(OR)_{z-1} \rightarrow$$

$$(OR)_{z-1}M\text{-}O\text{-}M(OR)_{z-1} + ROH \textit{ alcoxolation} \quad (4)$$

These reactions occur simultaneously during synthesis, leading to the formation of a continuous solid structure with liquid phase trapped, which is the gel (visco-elastic or elastic solid). Therefore, these reactions have to be minimised in the synthesis of the sols used in CSD. This is also because a spinnable precursor (viscous liquid or sol) is needed for the deposition of the wet layer. As sol-gel involves reaction among the starting compounds, it offers good opportunities for tailoring the chemistry of the precursor sol and thus, the properties of the derived materials.

Ferroelectric compositions contain two or more cations. Therefore, the precursors (reagents) have to be all together in the stable and homogeneous multicomponent solution needed for the CSD deposition. However, alkoxides of some metals (e.g., alkaline or earth alkaline elements) are difficult to synthesise. Transition metal alkoxides (e.g., titanium, zirconium or tantalum) react vigorously with water to produce metal-oxo precipitates. As a consequence, almost all the solution processes involved in the CSD deposition of ferroelectric materials should be strictly classified as *hybrid processes*. Here, metal short-chain carboxylates (e.g., acetates) or inorganic salts (e.g., nitrates) are used together with metal alkoxides dissolved in a solvent. Besides, chemical additives (e.g., acetic acid, acetylacetone or amines) are generally used to decrease the reactivity of the metal alkoxides and to obtain easier handled solutions.

In all the cases, the solutions (sols and/or mixtures of components dissolved in a solvent) are highly homogeneous. They provide good stoichiometric control and uniformity to the deposited coating. Besides, the chemistry of the solutions can be easily tailored to obtain precursors and materials with designed properties. Examples of how to tailor the CSD processing for getting ferroelectric thin and ultrathin films, and nanostructures onto substrates with improved properties, are shown in the next sections.

4.3 Tailoring the Chemistry of the Precursor Solutions

Some examples are shown here on how to tailor the chemistry of the solutions used in the CSD fabrication of ferroelectric materials onto substrates. This is to obtain precursors with specific characteristics such as: reduction of the moisture sensitivity or the toxicity of the solution, increase of the solution homogeneity, photoactivity of the solution, the formation of micelles working as micro-nanoreactors, etc. Although many other examples can be found in the literature, [58, 59] those presented here are enough to illustrate the importance of the control of the solution chemistry for the preparation of these materials.

4.3.1 Control of the Hydrolysis of the Solutions

Alcohols are the most used solvents in CSD processing of electronic multi-metal oxide films. Among them, 2-methoxyethanol, $CH_3OCH_2CH_2OH$ (MOE), has been widely used in the preparation of alkoxide based solutions since the 1980s. AT that point, the synthetic method known as the "*methoxyethanol route*" [55, 56, 60] was developed for the fabrication of $PbTiO_3$, $PbZrO_3$ and related thin film ferroelectric compositions. The use of this solvent has advantages such as the easy dissolving of metal carboxylates or the alcohol-exchange reaction with the starting metal alkoxides (5). This reduces the hydrolysis sensitivity of the precursor solutions.

$$M(OR)_n + H_3C - O - CH_2 - CH_2 - OH \rightarrow$$

$$M - [O - CH_2 - CH_2 - O - CH_3]_n (MOE) + n\,ROH \qquad (5)$$

The stabilisation towards hydrolysis is due to the formation of bonds among the metal centre of the alkoxide and the polar centres of the MOE. As a consequence, solutions are less reactive with water, although their synthesis and handling has to be carried out under a dry atmosphere.

One of the first approaches to get solutions with a relatively low reactivity with moisture was the use of metal alkoxides modified with β-diketonate ligands. These are molecules with two keto groups separated by one methylene group. An internal exchange reaction is produced in them, converting the keto groups into an alcohol group (6). This alcohol group can react with metal alkoxides, stabilising them by the chelating effect and the delocalisation of the electrons in the 6-membered ring (7). [60]

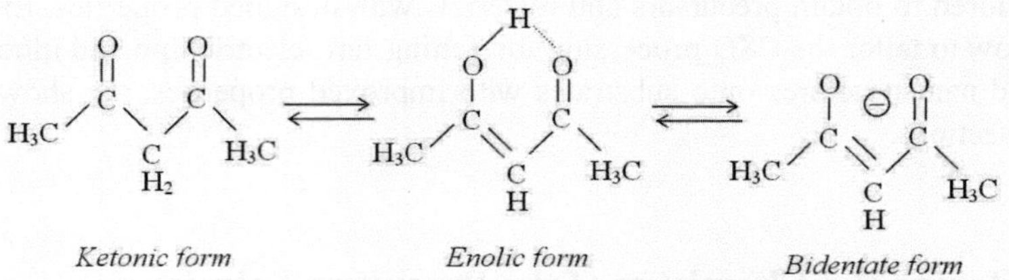

(6)

Modified titanium tetraisopropoxide with acetylacetone (titanium di-isopropoxide bis-acetylacetonate) was used in the synthesis of $PbTiO_3$ precursor solutions synthesised by the MOE route. The blocking effect of the acetylacetonate ligands increased the stability of the resulting solutions and led to single-layer crack-free perovskite films with thickness over 0.5 μm. [61]

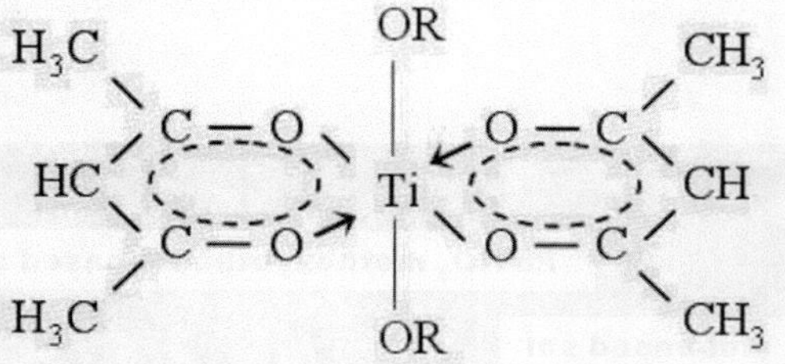

(7)

A further improvement in the solution synthesis of ferroelectric thin films was the development of the "*diol route*". Here, besides using the former titanium reagent, the toxic MOE solvent was changed by a diol. [19, 62] The most used diol solvent was 1,3-propanediol ($OH(CH_2)_3OH$), although the use of the 2, 4 and 5-C chain diols (1,2-ethanediol, $OH(CH_2)_2OH$, 1,4-butanediol, $OH(CH_2)_4OH$ and 1,5-pentanediol, $OH(CH_2)_5OH$) have also been reported. Air-stable $PbTiO_3$ based solutions were obtained (Fig. 4.4). Nuclear Magnetic Resonane (NMR) studies demonstrated [63] that in the diol synthesised solutions, the two isopropoxide groups initially bonded to the titanium di-isopropoxide bis-acetylacetonate (see (7)) were changed during reflux by two diol groups, working these as unidentate ligands. As a consequence, polymerised structures are formed in solution by bridging through the remaining hydroxyl groups (8).

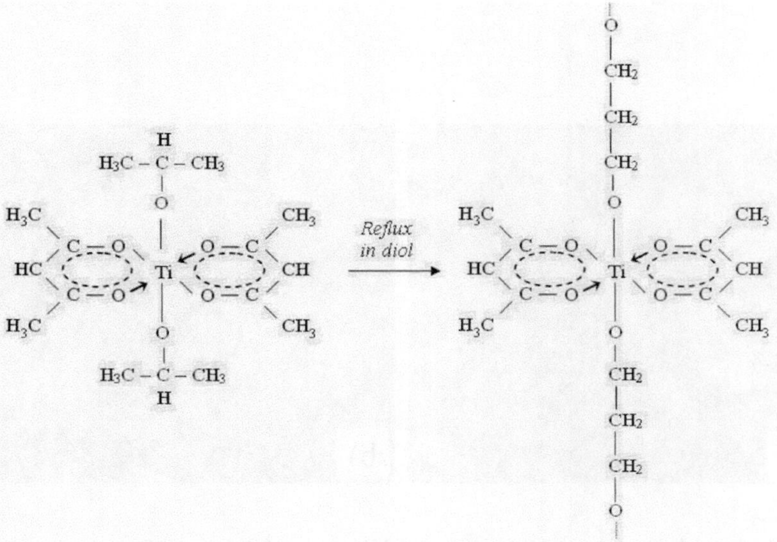

(8)

Because of the steric hindrance, the inorganic polymer formed in the sol has a low sensitivity towards moisture. The resulting sols can be stored and handled in air (Fig. 4.4).

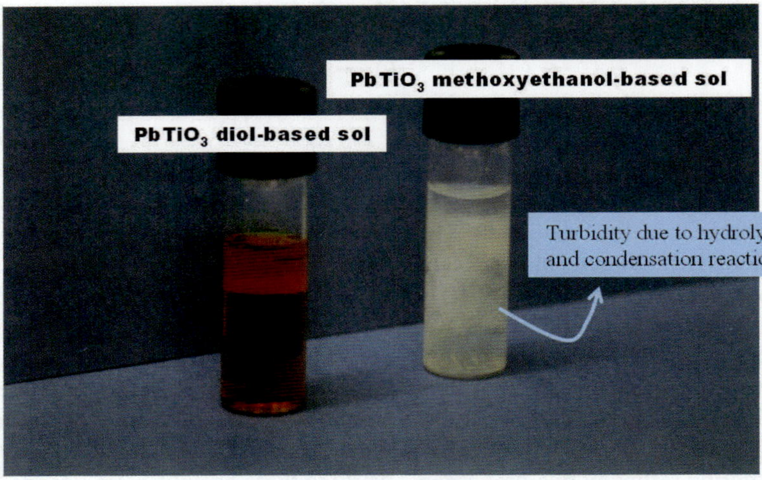

Fig. 4.4 Sols synthesised by the diol and methoxyethanol routes.

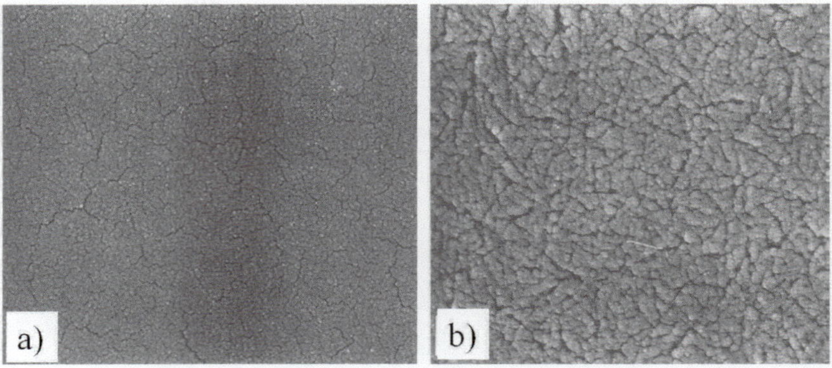

Fig. 4.5 Surface microstructure of Ca-modified $PbTiO_3$ perovskite prepared from the diol route, but with different length of the C-chain of the diol: a) 1,2-ethanediol ($OH(CH_2)_2OH$ and b) 1,4-butanediol ($OH(CH_2)_4OH$) [64 65].

In addition, crack-free single layer perovskite films with a thickness over a micron were obtained [19]. A clear effect of the length of the C-chain of the diol on the film microstructure (grain size) was observed (Fig. 4.5). [64, 65] This route also permits the use of mixtures of a diol and water as solvent. This makes easy the preparation of multicomponent solutions containing other cations, different from Ti(IV), Zr(IV) or Pb(II), such as Ca(II), Sm(III) or La(III). [23]

The versatility of this route has also been proved with ferroelectric compositions containing elements of the group Vb, such as Ta(V) and Nb(V). For compositions containing Ta(V), strontium bismuth tantalate ($Sr_2Bi_2Ta_2O_9$, SBT) films with a layered-perovskite structure were prepared (Fig. 4.6). Here, NMR also demonstrated that the synthesised sol contained the Ta(V) cation bonded to five diol groups. This was due to the result of the interchange reaction among the ethoxide groups of the alkoxide reagent (tantalum pentaethoxide, $Ta(OC_2H_5)_5$), and the diol that works at the same time as reagent and solvent (Fig. 4.6). Crystalline films derived from these sols have the characteristic microstructure of SBT films. They are formed by grains with anisotropic shape, and also show well-defined ferroelectric responses. [66, 67] Using this knowledge, air-stable precursor solutions of the relaxor-ferroelectric $Pb(Mg_{1/3}Nb_{2/3})O_3$-$PbTiO_3$, with nominal compositions close to the Morphotropic Phase Boundary (MPB), were also synthesised. The resulting crystalline films were single perovskite and had appropriate dielectric and ferroelectric responses. [68]

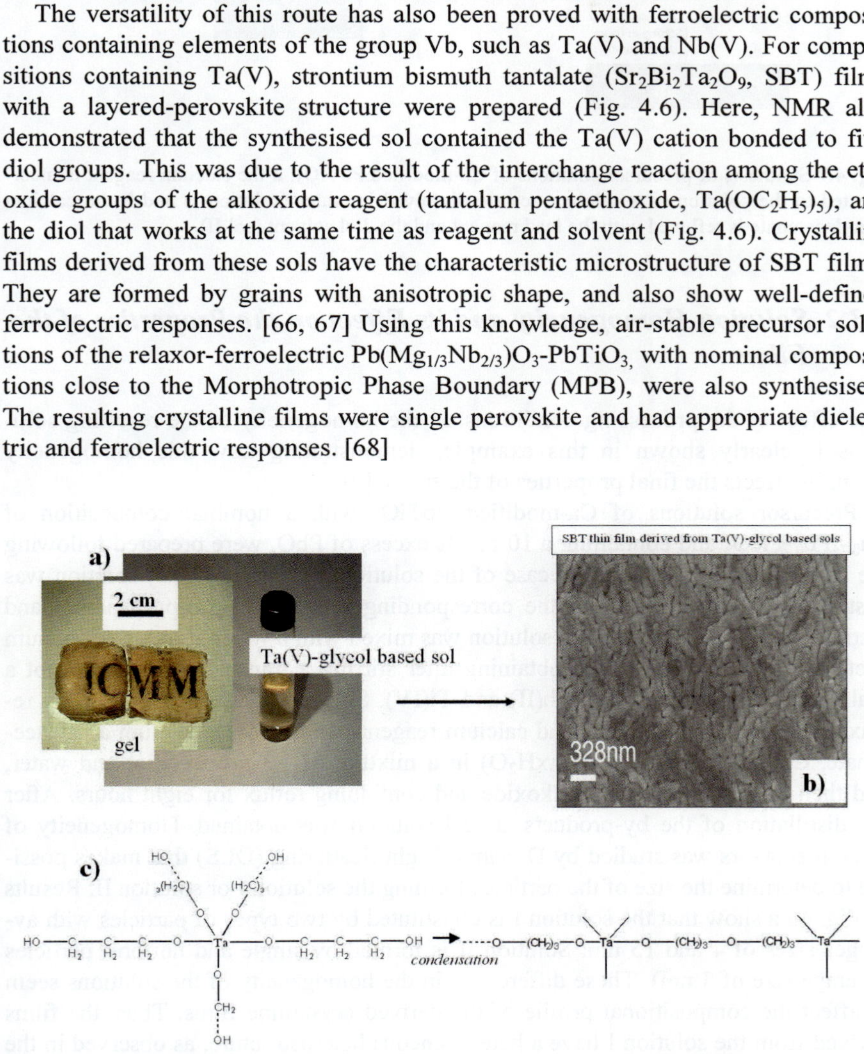

Fig. 4.6 a $SrBi_2Ta_2O_9$ (SBT) precursor sols and gels synthesised by the diol route and **b** the corresponding crystalline SBT thin films. **c** The polymer structure developed in the sol and gel and determined by Nuclear Magnetic Resonance (NMR) is shown (see [66])

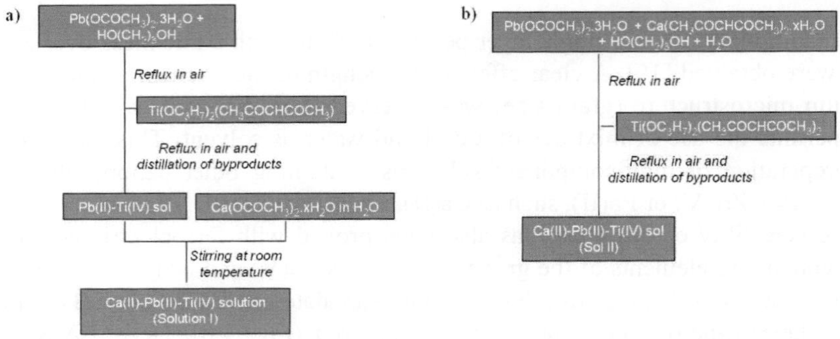

Fig. 4.7 Synthesis of precursor solutions of Ca-modified $PbTiO_3$: **a** the calcium reagent is incorporated to the solution as a calcium acetate dissolved in water (solution I) and **b** the calcium acetylacetonate is refluxed with the lead reagent and the diol solvent (sol II).

4.3.2 Solution Homogeneity and its Effect on the Properties of the Films

The effect of the processing conditions on the homogeneity of the resulting solutions is clearly shown in this example, demonstrating how this homogeneity strongly affects the final properties of the films. [9]

Precursor solutions of Ca-modified $PbTiO_3$ with a nominal composition of $Ca_{0.24}Pb_{0.76}TiO_3$ and containing a 10 mol% excess of PbO, were prepared following the schemes of Fig. 4.7. In the case of the solution I, a Pb(II)-Ti(IV) solution was first synthesised by refluxing the corresponding reagents in 1,3-propanediol and distillation of by-products. This solution was mixed with a water solution of calcium acetate, $Ca(OCOCH_3)_2.xH_2O$, obtaining after stirring a transparent solution (not a real sol) that contains Ca(II), Pb(II) and Ti(IV). Solution II was synthesised by refluxing for one hour the lead and calcium reagents. In this case, calcium acetylacetonate, $Ca(CH_3COCHCOCH_3)_2.xH_2O$) in a mixture of 1,3-propanediol and water, and then adding the titanium alkoxide and continuing reflux for eight hours. After the distillation of the by-products, a real solution was obtained. Homogeneity of these precursors was studied by Dynamic Light Scattering (DLS) that makes possible to determine the size of the particles forming the solution I or solution II. Results of Fig. 4.8a show that the solution I is constituted by two types of particles with averages sizes of 4 and 15 nm. Solution II is formed by single and uniform particles (average size of 3 nm). These differences in the homogeneity of the solutions seem to affect the compositional profile of the derived crystalline films. Thus, the films derived from the solution I have a heterogeneous heterostructure, as observed in the corresponding RBS spectrum (Fig. 4.8b). Lead-deficient surfaces (6% or the film thickness) and bottom interfaces (13% of the film thickness) are developed. The thickness of the film has the expected perovskite composition of an 81% of the total film thickness. However, the films derived from solution II have a homogeneous compositional profile that fits well with that of the expected perovskite.

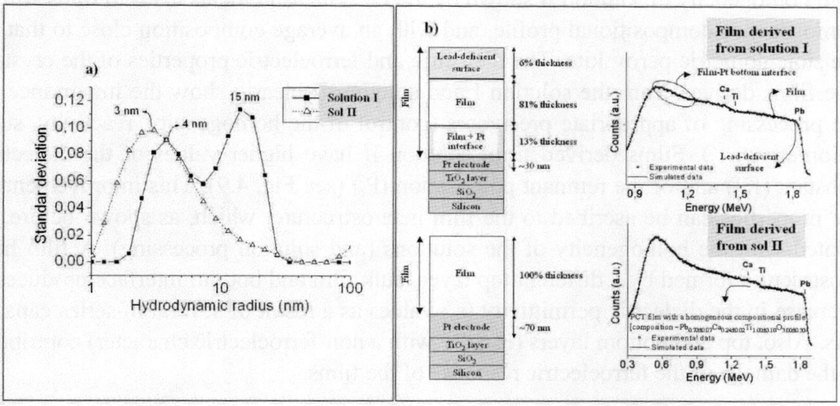

Fig. 4.8 a Bimodal and lognormal distributions of the particle size of the solution I and solution II, measured by Dynamic Light Scattering (DLS). **b** Experimental and simulated Rutherford Backscattering Spectroscopy (RBS) spectra of the films derived from the solution I and solution II along with the heterostructure deduced from these experiments [9].

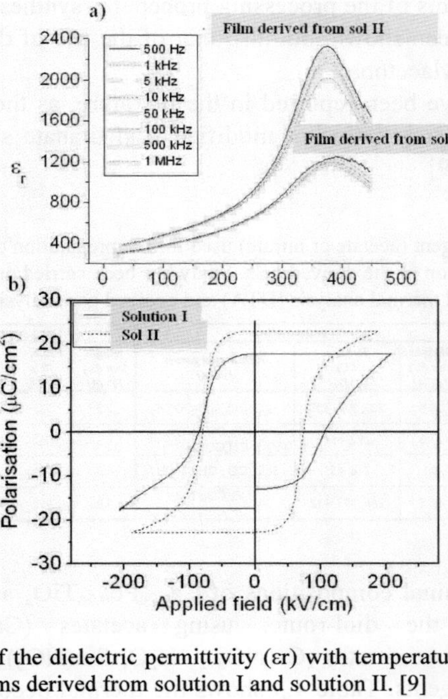

Fig. 4.9. a Variation of the dielectric permittivity (εr) with temperature and **b** ferroelectric hysteresis loops for the films derived from solution I and solution II. [9]

Therefore, these RBS results indicate that the characteristics of the heterostructure of the films are related with the solution processing and solution homogeneity. The high homogeneity of solution II shown by the DLS measurements leads to films with a homogeneous compositional profile, and with an average composition close to that of the stoichiometric perovskite. The dielectric and ferroelectric properties of the crystalline films derived from the solution I and solution II clearly show the importance of the processing of appropriate precursors (control of the homogeneity, reactivity, stoichiometry, ….). Films derived from solution II have higher values of the dielectric constant (K') and of the remnant polarisation (P_r) (see Fig. 4.9). This improvement in the properties can be ascribed to the film heterostructure, which, as shown before, is related with the homogeneity of the solutions (and solution processing). A film heterostructure formed by a different top layer, bulk film and bottom interface produces a decrease in the dielectric permittivity (ε_r) values as a result of several in-series capacitors. Also, top and bottom layers (usually with a non-ferroelectric character) contribute to the damage of the ferroelectric response of the films.

4.3.3 Effect of the Chemical Reagents Used for the Preparation of the Precursor Solutions

A clear example of how the chemicals involved in solution preparation can affect the properties of the crystalline films has just been shown in the previous section. Not only have the effects of the processing procedure (synthesis of a sol or preparation of a solution) been shown, but also that of the use of different calcium reagents (acetate or acetylacetonate).

Other examples have been reported in the literature, as those related with the use of nitrates in the preparation of modified lead titanate solutions, instead of acetate compounds. [69]

Table 4.3 Effect of the reagent (acetate or nitrate) used for the preparation of the precursor solution on the thermal decomposition of the derived gels. Study has been carried out by thermogravimetric analysis (TGA), differential thermal analysis (DTA) and evolved gas analysis (EGA) [70]

Film composition (reagent)	25-200°C			200-400°C			450-600°C			>600°C		
	Weight loss (%) [TGA]	Peaks* (°C) [DTA]	Eliminated gasses [EGA]	Weight loss (%) [TGA]	Peaks** (°C) [DTA]	Eliminated gasses [EGA]	Weight loss (%) [TGA]	Peaks** (°C) [DTA]	Eliminated gasses [EGA]	Weight loss (%) [TGA]	Peaks** (°C) [DTA]	Eliminated gasses [EGA]
$Ca_{0.24}Pb_{0.76}TiO_3$ (calcium acetate)	5.5	100	H_2O	20.6	280, 374, 433	H_2O, CO_2, CH_3COCH_3	2.5	583	H_2O, CO_2	1.8	748	CO_2
$Ca_{0.24}Pb_{0.76}TiO_3$ (calcium nitrate)	6.5	100	H_2O	15.3	297, 454	H_2O, CO_2, NO, CH_3COCH_3	4.3	547	H_2O, CO_2, NO	1.7	725	CO_2
$Sm_{0.08}Pb_{0.88}TiO_3$ (samarium acetate)	1.3	100	H_2O	8.1	310, 435	H_2O, CO_2, CH_3COCH_3	4.9	525	H_2O, CO_2	---	---	---
$Sm_{0.08}Pb_{0.88}TiO_3$ (samarium nitrate)	5.0	100	H_2O	16.8	266, 293, 412	H_2O, CO_2, NO, CH_3COCH_3	3.0	512	H_2O, CO_2, NO	---	---	---

Solutions with nominal compositions of $Ca_{0.24}Pb_{0.76}TiO_3$ and $Sm_{0.08}Pb_{0.88}TiO_3$ were prepared by the diol-route, using acetates ($Ca(CH_3COO)_2.xH_2O$, $Sm(CH_3COO)_3.xH_2O$) or nitrates ($Ca(NO_3)_2.4H_2O$, $Sm(NO_3)_3.xH_2O$), as the calcium or samarium reagents. Table 4.3 shows the thermal decomposition (thermogravimetric and differential thermal analysis, TGA/DTA, and evolved gas analysis, EGA) of these solutions. The combustion of the organic compounds is detected in the acetate precursors by the elimination of water (H_2O) and carbon dioxide (CO_2). There are also small amounts of acetone (CH_3COCH_3) as a side-product resulting from the two-step decomposition of the acetate groups [70]:

$$2CH_3COO^- \rightarrow CO_3^{2-} + CH_3COCH_3$$

$$CO_3^{2-} \rightarrow CO_2 + O^{2-}$$

However, in the case of the nitrate precursors, the TGA/DTA/EGA study also shows the elimination of NO gasses, which are due to the pyroysis of the nitrates.

Films deposited from these solutions, and crystallised at 650°C with a heating rate of 10°C/min, contain the perovskite phase along with a second pyrochlore phase. But, the content of the latter is larger in the films derived from the nitrate precursors. This is observed in the scanning electron microscopy (SEM) images of Fig. 4.10 (a, c and d). It is related to the more complicated thermal decomposition of the nitrate precursors than that of the acetate precursors. Note, in Fig. 4.10e and Fig. 4.10f, how the elimination of organic and nitrate groups enlarges up to higher temperatures in the case of the nitrate precursor than in the acetate precursor. Besides, for the $Ca_{0.24}Pb_{0.76}TiO_3$ solutions, the EGA analysis (Table 4.3) shows the evolution of CO_2 gas above 700°C. This indicates the decomposition of residual CO_3^{2-} groups coming from the formation of the intermediate $CaCO_3$. This is not detected for the $Sm_{0.08}Pb_{0.88}TiO_3$ solutions, probably due to the highest negative enthalpy of formation of the $CaCO_3$ than of the $Sm_2(CO_3)_3$. [71]

The use of calcium nitrate in the preparation of the $Ca_{0.24}Pb_{0.76}TiO_3$ solutions produces additional problems in the fabrication of the crystalline films derived from these precursors. Large craters appear in the surface microstructure of the films (Fig. 4.10b). These craters may be the result of the exothermic reactions occurring during the water desorption of the crystal water of nitrates deposited onto a solid substrate. In the special case of the calcium nitrate, this desorption is accompanied by a surface explosion process. [72]

These results show the important effect that can have the chemical reagents used in solutions on the characteristics of the resulting films. This indicates, in this particular case, that acetates are better candidates than nitrates for the incorporation of modifier cations into the lead titanate films. However, a deeper study of the processing conditions needs to be made for improving the crystallinity of the films, avoiding the formation of the second pyrochlore phase.

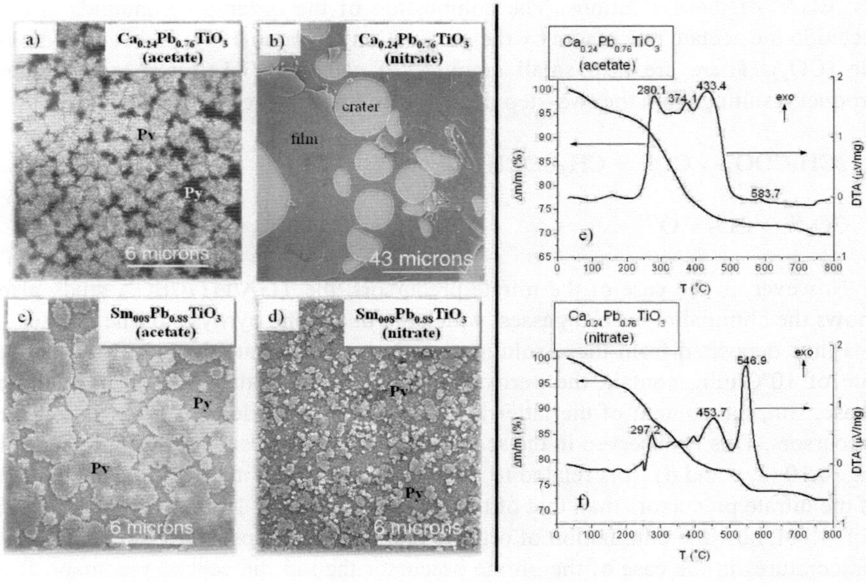

Fig. 4.10 Scanning electron microscopy images of crystalline films derived from the solutions with nominal compositions of a) $Ca_{0.24}Pb_{0.76}TiO_3$ and calcium acetate reagent, b) $Ca_{0.24}Pb_{0.76}TiO_3$ and calcium nitrate reagent, c) $Sm_{0.08}Pb_{0.88}TiO_3$ and samarium acetate reagent and d) $Sm_{0.08}Pb_{0.88}TiO_3$ and samarium nitrate reagent. e) TGA and DTA curves corresponding to the thermal decomposition of the dried powders derived from the $Ca_{0.24}Pb_{0.76}TiO_3$ solutions prepared with the calcium acetate and nitrate reagents. Pv: perovskite, Py: pyrochlore.

4.3.4 Stoichiometry of the Precursor Solution

Thin films have a high surface to volume ratio. This can promote losses by volatilisation of some elements forming the deposited amorphous layer during its thermal treatment of crystallisation. As a consequence, an unbalance in the composition of the crystalline film can result. This is detrimental for its properties.

In lead titanate based films, lead is easily lost by volatilisation at the temperatures usually used for the crystallisation of the films. This favours the formation of second non-ferroelectric phases with pyrochlore or fluorite structure. If the content of this second phase is noticeable, it can be easily observed as a morphologically differentiated phase on the surface microstructure of the films (see Fig. 4.10).

The incorporation of excess lead in the $PbTiO_3$ based solutions is an easy way to compensate for lead losses during the film crystallisation. This is widely used by the groups working on the fabrication of thin films by CSD. [73, 74] However, this is not always enough for getting ferroelectric films with good properties. Other processing parameters should be properly addressed.

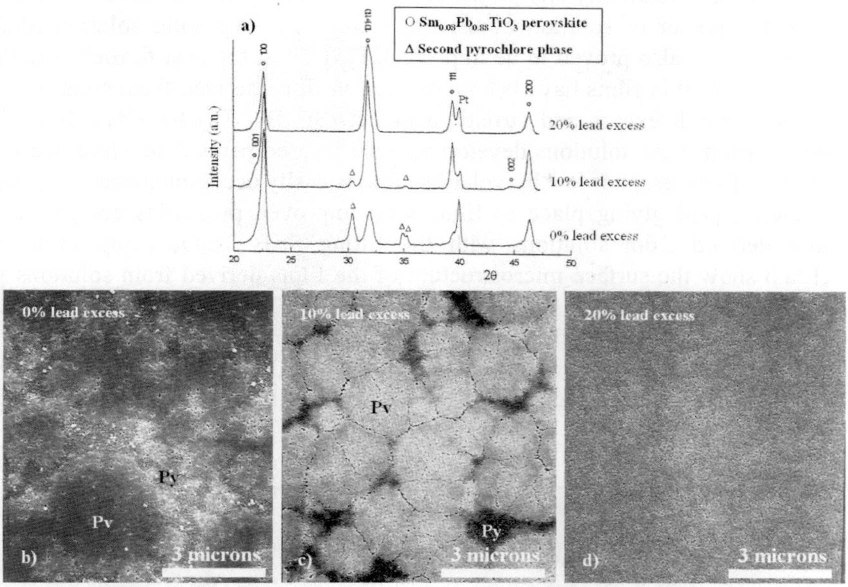

Fig. 4.11 a) Grazing incidence x-ray diffraction (GIXRD) patterns of $Sm_{0.08}Pb_{0.88}TiO_3$ thin films derived from solutions with different lead excesses and crystallised at 650ºC for 60 min with a heating rate of 10ºC/min. SEM images of the surfaces of the films prepared with: a) 0 mol % lead excess, b) 10 mol % lead excess and c) 20 mol % lead excess. Pv: $Sm_{0.08}Pb_{0.88}TiO_3$ perovskite, Py: second pyrochlore phase.

Fig. 4.11 shows the effect of the lead content of the precursor solution on the formation of the crystal phases of $Sm_{0.08}Pb_{0.88}TiO_3$ thin films. [65] A second pyrochlore phase is formed in these films along with the perovskite phase, when the films are deposited from solutions without excess lead. As the content of lead is increased in the precursor solution, the second phase decreases. Single perovskite films are only obtained from solutions containing a 20 mol% of lead excess (see Fig. 4.11a). The coexistence of pyrochlore and perovskite is observed in the film surfaces as phases with different morphology (grain size, porosity, …) (Fig. 4.11b,c). Only the film prepared with a 20 mol% of lead excess has a homogeneous microstructure (Fig. 4.11d). This confirms its full crystallisation into the perovskite phase, as detected by x-ray diffraction.

But, other ferroelectric compositions of solid solutions containing elements with a low volatilisation temperature also develop second non-ferroelectric phases when prepared as thin films. For instance, bismuth in $SrBi_2Ta_2O_9$ (SBT) compounds.. In SBT thin films, a second crystalline phase with a fluorite structure is formed during annealing. This can also be easily detected by x-ray diffraction and distinguished from the perovskite phase on the surface microstructure of this type of films. Bismuth excess in the precursor solution minimises the formation of this second phase. However in these films, not only does the bismuth content have effect on the crystallinity and properties of the resulting films, but the stoichiometry in the precursor solution of the other cations of the solid solution (mainly strontium) has also proved to be important. [75] Thus, the best ferroelectric properties of SBT thin films have been measured in films derived from solutions containing bismuth excess and strontium defect ($Sr_{0.8}Bi_{2.2}Ta_2O_9$). Crystalline films derived from these solutions develop a single layered perovskite. Here, the residual bismuth excess, not lost by volatilisation, is easily accommodated in strontium vacancies, [76] giving place to films with improved properties compared with those derived from solutions with the fitting $Sr_2Bi_2Ta_2O_9$ composition. Fig. 4.12a,b show the surface microstructure of the films derived from solutions with nominal compositions of $SrBi_2Ta_2O_9$ and $Sr_{0.8}Bi_{2.2}Ta_2O_9$. The ferroelectric hysteresis loops obtained in these films are also shown in Fig. 4.12c,d. Larger remnant polarisations (P_r) are measured in the $Sr_{0.8}Bi_{2.2}Ta_2O_9$ film than in the $SrBi_2Ta_2O_9$ one. This is a clear indication of a better ferroelectric response.

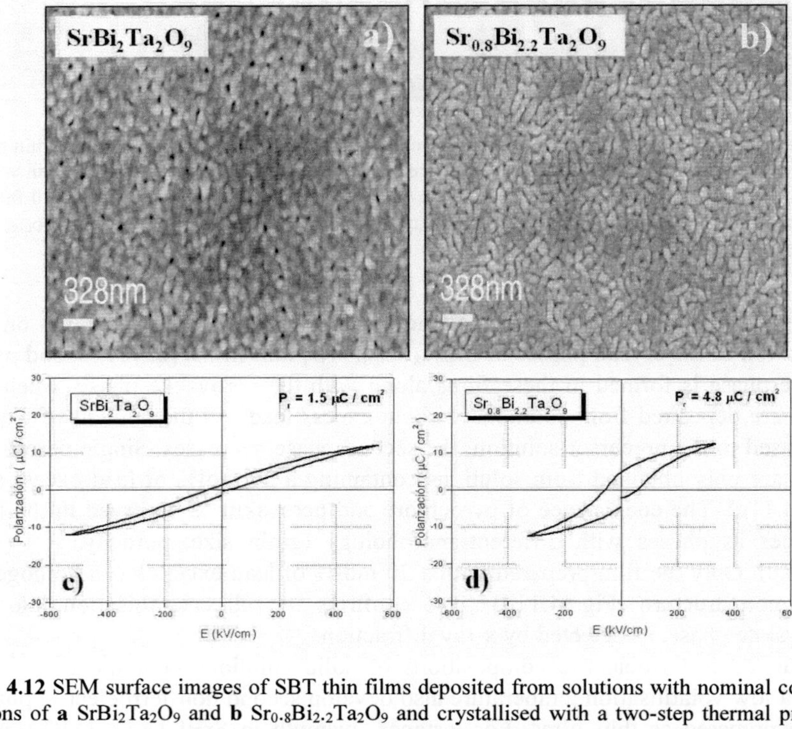

Fig. 4.12 SEM surface images of SBT thin films deposited from solutions with nominal compositions of **a** $SrBi_2Ta_2O_9$ and **b** $Sr_{0.8}Bi_{2.2}Ta_2O_9$ and crystallised with a two-step thermal process (550°C for 2 h in O2 + Rapid Thermal Process at 650°C for 1 h in O_2 with a heating rate of 200°C/s). **c** and **d** show the ferroelectric hysteresis loops of the former films.

4.3.5 Photo-Activation of the Precursor Solutions

The preparation of light-sensitive materials by using sol-gel chemistry with UV-irradiation has been used for the fabrication of oxide films. [77] Energetic photons of UV lights can activate chemical bounds of organic compounds. This produces $\pi \rightarrow \pi^*$ transitions that give place the dissociation of the alkyl group – O bonds and the further formation of metal – O – metal bonds. The UV activation of sol-gel precursors has been used for increasing densification of films [78], the reduction of materials [79], the hydroxylation of surfaces [80] and the crystallisation of thin films [81]. In the case of ferroelectric thin films, it has been used for the photo-patterning of films. [82] However, it has been scarcely used for the fabrication of films at low temperatures. [83] But this is a key topic for ferroelectric materials, because low processing temperatures are demanded for the preparation of ferroelectric thin films onto semiconductor substrates (compatibility with the Si-technology). Or onto metallic or polymer substrates.

The photo-sensitivity of sol-gel precursors depends on the chemical nature of their components. The chemistry of sols and gels can be, in principle, easily tailored for getting precursors containing UV absorbing species. External photo-sensitive generators can be added to the solution, or inherent photo-sensitive solutions can be synthesised. Both, external and internal photo-activation of solutions are here shown for the fabrication of ferroelectric thin films at low temperatures. [84, 85, 86]

Synthesis of calcium modified lead titanate precursor solutions ($Ca_{0.24}Pb_{0.76}TiO_3$) has been carried out via a previously reported diol-based sol-gel process. [23] Titanium di-isopropoxide bis-acetylacetonate (see Fig. 4.13a) was used as the titanium reagent. These solutions are denoted Ti-Pr-Ac. External compounds that can absorb light in the UV-range were added to this Ti-Pr-Ac solution. This resulted in precursor solutions that contain acetylacetone ($CH_3COCH_2COCH_3$) (Fig. 4.13b), o-nitrobenzyl alcohol ($O_2NC_6H_4CH_2OH$) (Fig. 4.13c) and cyclohexyl-phenyl-ketone ($HOC_6H_{10}COC_6H_5$) (Fig. 4.13d). These solutions are denoted AcH, NBA and HPK, respectively. All the solutions have maximum absorption at wavelengths between 200 and 300 nm, but the solutions containing HPK and NBA have highest absorptions (Fig. 4.13e). These are well-known UV-generators, mainly due to the benzyl group of both compounds. Films were deposited from these solutions, irradiated at 250ºC in air, and crystallised by rapid thermal processing (heating rate of 30ºC/s) in air at 550ºC. All the films presented ferroelectric response as observed in the hysteresis loops of Fig. 4.13f. However, the highest response was measured in the films derived from the Ti-Pr-Ac solution, where external photo-activators were not incorporated. The inherent photo-sensitivity of the Ti-Pr-Ac solution is due to the β-diketonate complex used as the titanium reagent in the synthesis of the solution. Although its absorption is lower than those of the other solutions with external activators (AcH, NBA and HPK), it is enough for getting the prompt elimination of the organics and formation of the ferroelectric perovskite phase at a low temperature. The role of the AcH, NBA and HPK activators is not clear. They increase the UV sensitivity of the solutions. But, they do not improve the ferroelectric properties of the resulting films, probably due to a more complicated thermal decomposition as reported in [85].

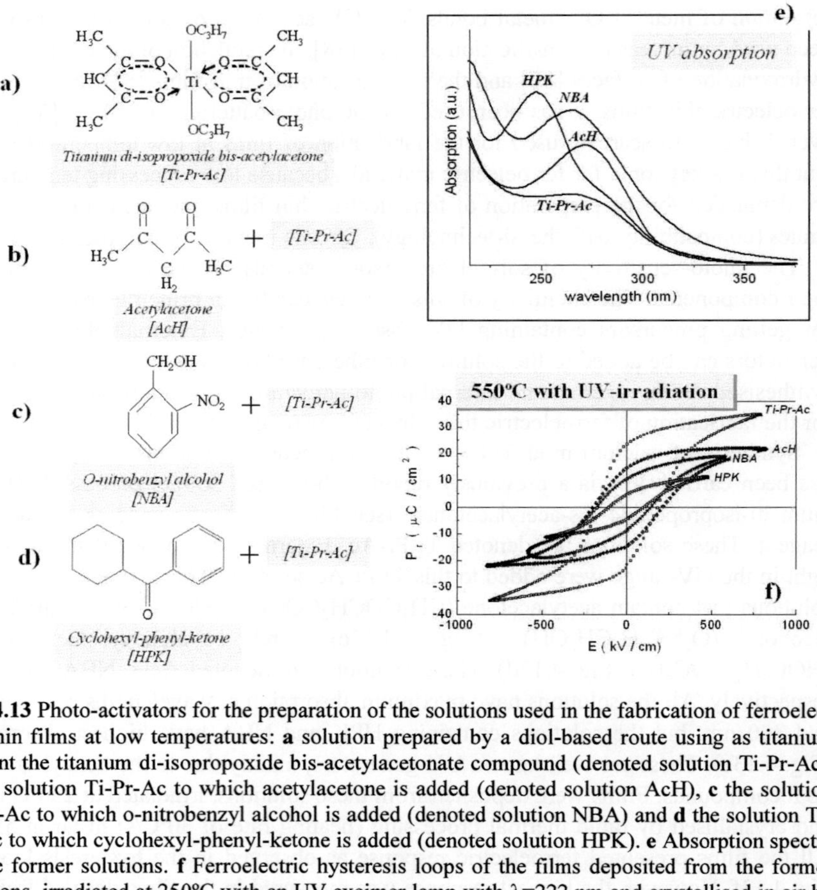

Fig. 4.13 Photo-activators for the preparation of the solutions used in the fabrication of ferroelectric thin films at low temperatures: **a** solution prepared by a diol-based route using as titanium reagent the titanium di-isopropoxide bis-acetylacetonate compound (denoted solution Ti-Pr-Ac), **b** the solution Ti-Pr-Ac to which acetylacetone is added (denoted solution AcH), **c** the solution Ti-Pr-Ac to which o-nitrobenzyl alcohol is added (denoted solution NBA) and **d** the solution Ti-Pr-Ac to which cyclohexyl-phenyl-ketone is added (denoted solution HPK). **e** Absorption spectra of the former solutions. **f** Ferroelectric hysteresis loops of the films deposited from the former solutions, irradiated at 250°C with an UV excimer lamp with λ=222 nm and crystallised in air by Rapid Thermal Processing at 550°C.

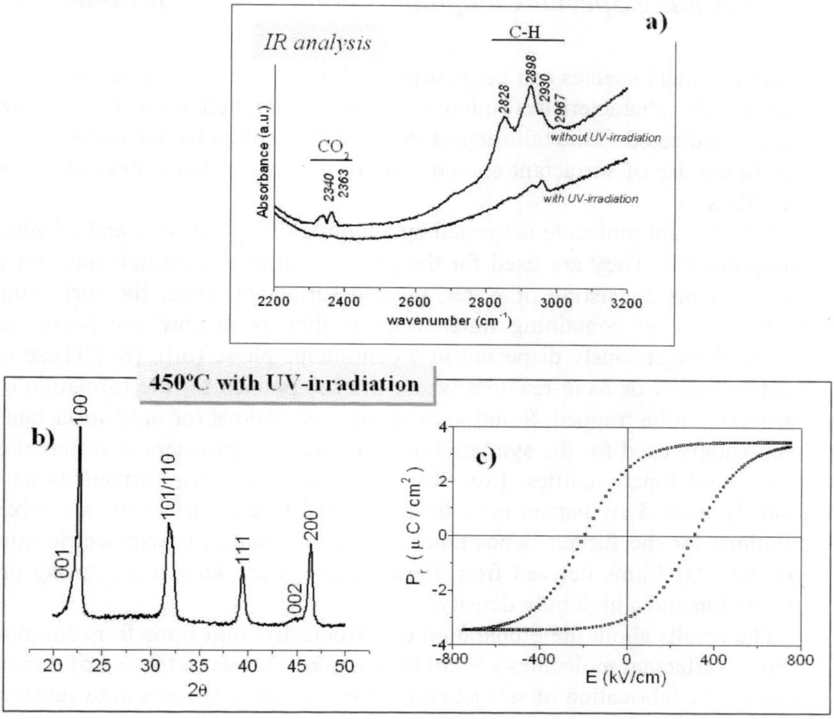

Fig. 4.14 a IR spectra of the gel layers deposited from the solution Ti-Pr-Ac and treated at 250°C with and without UV-irradiation. **b** X-ray diffraction pattern of the film deposited from the solution Ti-Pr-Ac, UV-irradiated at 250°C and crystallised in air by Rapid Thermal Processing at 450°C. **c** Ferroelectric hysteresis loop of the former film [83]

The IR spectra of Fig. 4.14a of a non-irradiated film derived from the Ti-Pr-Ac solution and treated at 250°C shows the vibrations of C – H chains. These bands have almost disappeared in the UV-irradiated film. This indicates that elimination of organics occurs at a lower temperature under UV-light. As a consequence, prompt formation of the ferroelectric perovskite is produced. Even films derived from these solutions could be crystallised after irradiation at a low temperature of 450°C. These films were single perovskite phase, as observed by x-ray diffraction (Fig. 4.14b). They have an appropriate ferroelectric response (Fig. 4.14c) for using in devices. [87]

These results prove the competitiveness of using photo-active solutions for the fabrication of ferroelectric thin films at temperatures compatible with the Si-technology; or at temperatures required for the integration of these films with non-traditional substrates used in the development of new materials.

4.3.6 Adding Special Compounds to the Precursor Solutions

Other chemical species can be incorporated to the precursor solutions to get films with specific characteristics (microstructure, texture, bulk density, …). To end this section dedicated to the tailoring of the precursor solutions, we make a short mention to the use of surfactant compounds for the sol-gel fabrication of ferroelectric thin films.

A surfactant molecule is formed by a hydrophilic polar head and a hydrophobic non-polar tail. They are used for the preparation of microemulsions that are ternary systems consisting of water, oil and surfactant. Here, the surfactant forms reverse micelles containing water drops in their polar core and being these micelles homogeneously dispersed in a continuous phase (oil). [88] These micelles work as micro or nano-reactors where the compounds for the formation of oxide particles can be trapped. Solutions and microemulsions (or only surfactants) have been widely used for the synthesis of inorganic nanopowders of different compositions and functionalities. [89, 90] In the case of films, surfactants have been scarcely used. Few papers have been reported where surfactants are mixed with solutions for the further deposition and crystallisation of ferroelectric thin films. [91, 92, 93] Films derived from these solutions are known for strong preferred orientation and a high bulk density.

The results about the preparation of ferroelectric thin films from solutions containing surfactant molecules are not shown here. However, the use of these precursors for the fabrication of self-assembled ferroelectric systems onto substrates will be presented below.

4.4 Tailoring the Conversion of the Solution Deposited Layer into a Ferroelectric Crystalline Thin Film

Once the precursor solution has been prepared, films onto a substrate are usually deposited by spin or dip-coating, resulting in a wet layer. This layer is first treated at a low temperature (drying), at which evaporation of solvents and pyrolysis of organics occurs. After this, an amorphous film is obtained containing M – O – C and M – O – H bonds. The amorphous-to-crystalline perovskite conversion breaks the former bonds and forms M – O – M bonds. This is obtained by heat treatment. Fig. 4.15 illustrates some of the phenomena occurring during the thermal treatment of crystallisation of the as-deposited amorphous film. During this treatment, the ferroelectric phase is formed and other non-ferroelectric phases can crystallise. Non-stoichiometry can occur due to volatility of some elements (e.g. Pb, Bi, …), shrinkage of the film takes place generating stresses, interfaces are formed as a result of the diffusion between the film and the substrate, etc. These drawbacks have to be minimised during heat treatment, to obtain the full transformation of the amorphous layer into a ferroelectric film with appropriate properties. For this, different approaches may be used, in which temperature, soaking time and heating rate are the major parameters controlling the thermal process. However, other factors, not directly involved in the thermal schedule, such as the type of substrate or the firing atmosphere also affect the characteristics and the properties of the crystalline films. Some of these factors are discussed with the control of the heterostructure, microstructure, and ferroelectric response of the resulting film.

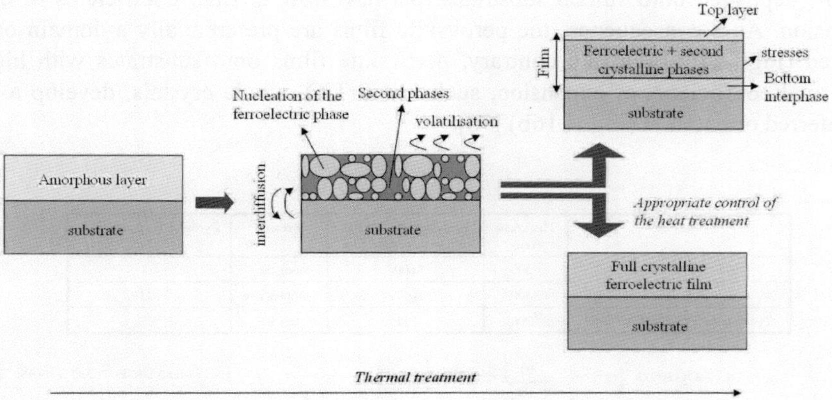

Fig. 4.15 Conversion of the amorphous layer deposited from the solution into a ferroelectric perovskite film by means of a heat treatment.

4.4.1 Effect of the Substrate during the Heat Treatment

The substrate and the deposited amorphous layer form the material that is subjected to the thermal treatment. The aim of this treatment is the conversion of the amorphous layer into an active (ferroelectric) crystalline film. But, besides crystallisation of the ferroelectric perovskite, other phenomena related with the interaction between the substrate and the film also occur during heating. These affect the properties of the final material.

Two examples about the effect of the substrate on the film characteristics are shown here. The first is how the texture of the ferroelectric perovskite is determined by the underlying substrate. The second is the formation of bottom interfaces related with the chemical nature of the surface substrate.

Orientation of a film on a substrate is affected by the nucleation mechanisms of the crystals on the substrate surface, and by the stresses.

Substrate surface is a site for heterogeneous nucleation. However, during the crystallisation of the films prepared by CSD, a competition between the homogeneous nucleation of crystals in the bulk layer and the heterogeneous nucleation of crystals at the film/substrate interface occurs. The film orientation enhances by depositing thin layers, due to the reduction of the volume where homogeneous nucleation can occur. Therefore, the films shown here have been prepared by multiple deposition and crystallisation. [94]

The stresses in films prepared by CSD mainly result from the wetting of the substrate by the solution, the pyrolysis of organics during the amorphous to crystalline conversion, and the thermal expansion mismatch between the film and the substrate during cooling down. [95] Tensile stresses occur in $PbTiO_3$ based thin films deposited onto silicon substrates that have low thermal coefficients of expansion. As a consequence, the perovskite films are preferentially a-domain oriented (Fig. 4.16a). On the contrary, perovskite films onto substrates with high thermal coefficients of expansion, such and $SrTiO_3$ single crystals, develop a c-preferred orientation (Fig. 4.16b) [96].

Substrate	Thickness (nm)	Preferred orientation of the perovskite film	stresses	Pyroelectric coefficient ($\gamma \times 10^9$ C cm^{-2} K^{-1})
Pt/TiO_2/SiO_2/(100)Si	~300	<100>	tensile	~5
Pt/$SrTiO_3$	~300	<001>	compressive	~11
Ti-template/Pt/Ti/SiO_2/(100)Si	~200	<111>	tensile	~7

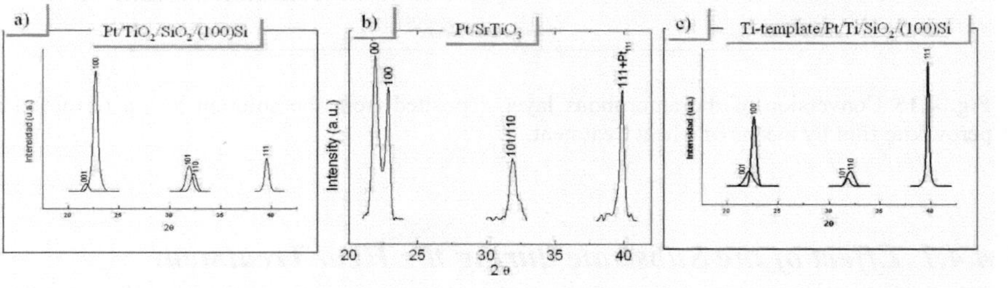

Fig. 4.16 Perovskite ($Ca_{0.24}Pb_{0.76}TiO_3$) thin films onto different substrates: **a** platinised silicon substrate, **b** platinised strontium titanate substrate and **c** Ti template layer onto a platinised silicon substrate. Table shows the preferred orientation of the films related with the type of substrate and the stresses developed during heating. The value of the pyroelectric coefficient is an indication of the spontaneous polarisation of the film, which is related with its orientation along the polar direction of the perovskite. [2, 97]

The orientation of these perovskite films affects their functional response. Films with an appreciable c-preferred orientation are desired for pyro and piezoelctric applications, due to their high spontaneous polarisation (P_s). See the higher values of the pyroelectric coefficient of the c-axis oriented film of Fig. 4.16b, than that of the a-axis oriented film of Fig. 4.16a.

The preferred orientation of perovskite films onto silicon substrates may also be modified by the use of template layers that change the chemical nature and structure of the substrate surface on which the perovskite nucleates. [97] In this case, it is necessary to get a good mismatch between the film and the substrate surface that prevails over the tensile stresses developed during the heating. It must promote the preferred orientation of the crystalline film along its polar direction. Fig. 4.16c shows a perovskite film onto a platinised silicon substrate with a Ti template layer. Authors report the formation of intermediate phases at the substrate surface during heating (Pt_3Ti, TiO_2 enriched perovskites, ...) [98, 99], that have a good matching with the perovskite crystal structure, facilitating the growing of the film with a <111> preferred orientation.

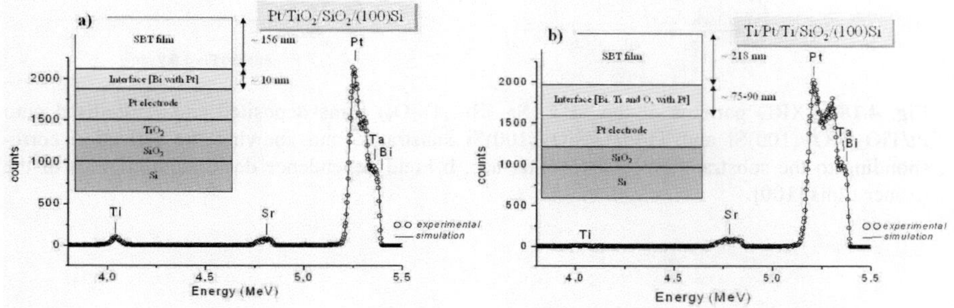

Fig. 4.17 Experimental and simulated Rutherford Backscattering Spectroscopy (RBS) spectra of the SBT ($Sr_{0.8}Bi_{2.2}Ta_2O_9$) films deposited and crystallised onto **a** $Pt/TiO_2/SiO_2/(100)Si$ and **b** $Ti/Pt/Ti/SiO_2/(100)Si$ substrates. The heterostructures of both type of films deduced from the RBS study are shown [100].

But during the heating and the interaction (chemical reaction, interdiffusion, ...), the substrate and the film is produced. This affects the nucleation of the perovskite onto the substrate surface. This also promotes chemical reaction between the substrate and the film and the formation of interfaces. These interfaces determine the ferroelectric response of the material. Fig. 4.17 shows the RBS analysis of two SBT thin films prepared onto silicon substrates with different heterostructure electrodes ($Pt/TiO_2/SiO_2/(100)Si$ and $Ti/Pt/Ti/SiO_2/(100)Si$). It has just been shown that the reaction between Ti and Pt, and/or the film during heating (see Fig. 4.16c), seems to produce chemical species at the interface that favour the preferred growth of $PbTiO_3$ perovskite thin films. In the case of SBT thin films, the high mobility of Ti compared with the high stability of TiO_2 results in a thicker interface of reaction for the films onto Ti/Pt/Ti electrodes, than onto Pt/TiO_2, and in interfaces with different chemical nature. SBT films onto Pt/TiO_2 develop interfaces formed by a Pt-Bi alloy. On Ti/Pt/Ti electrodes, the interface contains Bi, Ti, O and Pt, with a composition and structure close to that of the $Bi_4Ti_3O_{12}$ layered perovskite. As a consequence, by controlling the substrate surface, it is possible to modify the characteristics of the metal-ferroelectric junctions of the films obtained after the thermal treatment (Fig. 4.18). These junctions work on the leakage current behaviour of both films, and thus can influence their performance in devices (NVFERAM) [100].

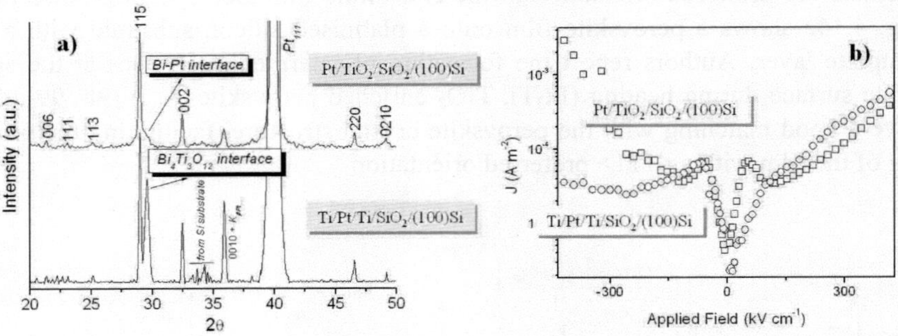

Fig. 4.18 a XRD patterns of the SBT ($Sr_{0.8}Bi_{2.2}Ta_2O_9$) films deposited and crystallised onto $Pt/TiO_2/SiO_2/(100)Si$ and $Ti/Pt/Ti/SiO_2/(100)Si$ substrates, and showing the reflection corresponding to the substrate/ferroelectric interface. **b** Field dependence dc-leakage currents of the former films [100].

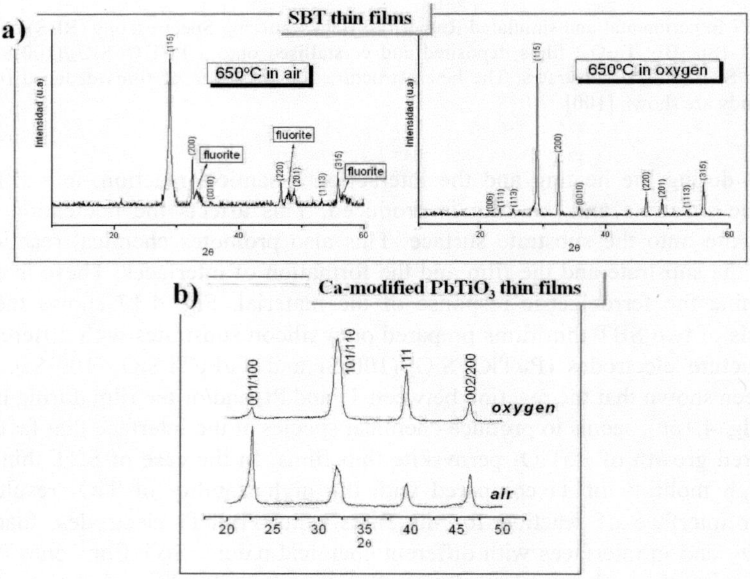

Fig. 4.19 Effect of the firing atmosphere on the films obtained after the heat treatment of crystallisation. **a** SBT ($Sr_{0.8}Bi_{2.2}Ta_2O_9$) films onto silicon substrates and treated at 650ºC in air or in oxygen. The second fluorite phase appears in the films treated in air. **b** Ca-modified $PbTiO_3$ thin films onto silicon substrates and treated in air or in oxygen. The firing atmosphere affects the texture of the crystalline film.

4.4.2 Firing Atmosphere

The atmosphere during the heat treatment of the films also affects their final characteristics. Usually, films prepared by CSD are crystallised in air or oxygen atmosphere, since these facilitate the combustion and elimination of organics during the thermal treatment. Besides, second phases formed in ferroelectric films during crystallisation are usually oxygen-deficient phases with pyrochlore or fluorite structure. So, oxygen-rich atmospheres during firing minimise the formation of these second phases and facilitate the full-crystallisation of the film into the ferroelectric perovskite phase. This is shown in Fig. 4.19a, where the XRD patterns of SBT thin films crystallised at 650ºC in air and in oxygen are shown. Reflections corresponding to the second fluorite phase are observed in the diffractogram of the films treated in air, where it is not detected in the films prepared in oxygen. [67]

Firing atmosphere also can affect the nucleation mechanisms of the ferroelectric phase during its crystallisation. Fig. 4.19b shows two Ca-modified $PbTiO_3$ thin films that have been heat-treated in air and in oxygen. The films prepared in the oxygen atmosphere have a random orientation. However, those treated in air have an a-axis preferred orientation. Both films are onto silicon substrates. This must favour the preferred orientation of the perovskite along the a-axis, due to the tensile stresses developed during annealing. However, the use of an oxygen atmosphere during firing enhances the combustion of the organic compounds and their elimination. This makes possible the homogeneous nucleation of the perovskite crystals in the bulk layer and thus, the obtention of random oriented thin films. [9]

4.4.3 Conventional Heating versus Rapid Heating

The Rapid Thermal Processing (RTP) technique uses heat infrared lamps that make possible the heating up to the desired temperature in short rise times. This reduces the processing time of the material to few seconds.

Thermal treatments using rapid heating rates have been traditionally used in the semiconductor technology. Here, Si or GaAs wafers are treated at relatively low temperatures (<500ºC) with rapid heating rates (10-300 ºC/s). This avoids the oxidation of the wafer and favours the obtention of the low resistance ohmic contacts needed in the fabrication of ultra-thin gate dielectrics. [101]

For CSD derived ferroelectric thin films, the thermal treatment used for the crystallisation of the film has a strong influence on the properties. Initially, these films were fabricated at temperatures in excess of 650ºC for several hours in conventional ovens (conventional heating). This produces the crystallisation of the ferroelectric phase, but also other undesired side effects. Thus, volatilisation of some elements from the film (mainly lead in lead-perovskite thin films) can occur. This leads to an unbalance in the composition of the crystalline film and to the formation of non-ferroelectric second phases. Also, long annealings favour the interdifussion between the substrate and the film. All of these phenomena spoil the ferroelectric response of the final material. Therefore, alternative methods of crystallisation of solution that solve the problems have been explored. In the 1990s, the RTP technique began to be used for the fabrication of ferroelectric thin films. [102]

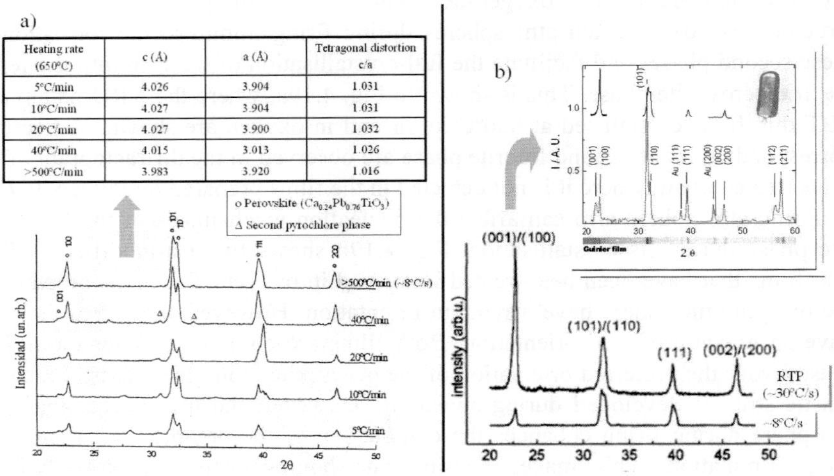

Heating rate (650°C)	c (Å)	a (Å)	Tetragonal distortion c/a
5°C/min	4.026	3.904	1.031
10°C/min	4.027	3.904	1.031
20°C/min	4.027	3.900	1.032
40°C/min	4.015	3.013	1.026
>500°C/min	3.983	3.920	1.016

Fig. 4.20 a Effect of the heating rate on the development of the crystal phases in the film. Increase of the heating rate minimises the formation of the second pyrochlore phase. Table shows the cell parameters and the tetragonal distortion of the perovskite formed in the films, for different heating rates. The perovskite tetragonality decreases with the increase of the heating rate. **b** Effect of the heating rate on the preferred orientation of the perovskite film onto a $Pt/TiO_2/SiO_2/(100)Si$ substrate. Preferred orientation along the a-axis of the perovskite is promoted by rapid heating rates. Inset shows the X-ray diffraction pattern of a perovskite film onto a silicon substrate and that of the same film after separation from the substrate. Stresses are realised and the tetragonality of the perovskite is recovered in the separated film [103]

Fig. 4.20 shows the effect of the heating rate on the development of crystalline phases, on crystal structure and texture of Ca-modified $PbTiO_3$ thin films deposited from solutions onto $Pt/TiO_2/SiO_2/(100)Si$ substrates and treated at 650°C in air. [65] The films contain an appreciable amount of the second pyrochlore phase when they are treated with slow heating rates (5°C/min). The content of this second phase decreases as the heating rate increases. This results in single perovskite films for films treated with a rapid heating rate of ~8°C/s (Fig. 4.20a). Rapid heating minimises the formation of second phases by circumventing the temperature at which they are stable. The tetragonality of the perovskite film also decreases with the heating rate, indicating an increase of the film stresses produced during the thermal treatment.

Also, the <100> preferred orientation of the film increases with the heating rate. Rapid heating promotes the heterogeneous nucleation of the first perovskite crystals onto the surface substrate, over the homogeneous nucleation produced in films crystallised with conventional heating (Fig. 4.20b). The stresses supported by the perovskite film onto the substrate when prepared by a rapid heating can be inferred from the inset of Fig. 4.20b. It shows a film that has been electrolytic separated from the substrate. [103] The separated layer recovers the tetragonality of the perovskite and losses the perovskite texture. This indicates that part of the stresses developed during the processing of the film has been released.

Heating rate (650°C)	Film thickness	Thickness of the film – Pt electrode interface
~8 °C/s	~4400 Å	~700 Å
~30 °C/s	~4000 Å	~600 Å

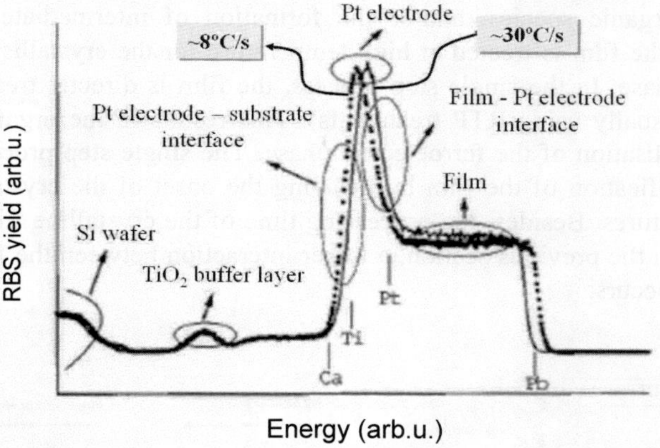

Fig. 4.21 RBS spectra of two $Ca_{0.24}Pb_{0.76}TiO_3$ thin films deposited onto $Pt/TiO_2/SiO_2/(100)$ substrates and treated with heating rates of ~8°C/s and ~30°C/s. The table shows the thickness of the film/substrate interface and of the bulk ferroelectric layer of both films. Note that interface thickness and layer thickness decrease with the increase of the heating rate, indicating a lower interaction between the film and the substrate and a higher bulk density of the film.

The duration of the thermal treatment also increases the interdiffusion between the substrate and the film, producing their chemical reaction and the formation of detrimental interfaces. Therefore, RTP treatments are preferable for the crystallisation of CSD derived films. The effect of the heating rate on the interaction between the film and the substrate can be observed in the RBS study of Fig. 4.21. [2, 104] This figure shows the multilayer structure of the material, which is formed by: the ferroelectric film, the film/electrode (Pt) interface, the electrode (Pt)/substrate interface, and the substrate (TiO_2 buffer layer onto a SiO_2/(100)Si wafer). The results indicate that the increase of the heating rate produces a decrease in the thickness of the film/electrode interface. This also decreases the thickness of the bulk film (ferroelectric layer) due to its lower porosity (higher bulk density). As a consequence of the lower porosity of the bulk film, and the thinner thickness of the film/substrate interface, the dielectric and ferroelectric properties of the films crystallised by RTP are improved. This shows the RTP technique as an adaptable method for the integration of ferroelectric layers with the semiconductor technology.

4.4.4 Two Step Heating versus Single Step Heating

Two approaches are normally reported for the transformation of the solution deposited layer into a ceramic film: the two step and the single step processes. [40, 57, 105] In the former, the as-deposited film is subjected to thermal treatment at an intermediate temperature (100-500°C). This results in the pyrolysis and thermolysis of the organic species, and/or the formation of intermediate crystalline phases. Then, the film is treated at high temperature for the crystallisation of the ferroelectric phase. In the single step process, the film is directly treated at high temperature (usually using RTP treatments). This results in the organic removal and the crystallisation of the ferroelectric phase. The single step process by RTP enhances densification of the film by delaying the onset of the crystallisation to higher temperatures. Besides, the processing time of the crystalline film is lower. As discussed in the previous section, a lower interaction between the film and the substrate also occurs.

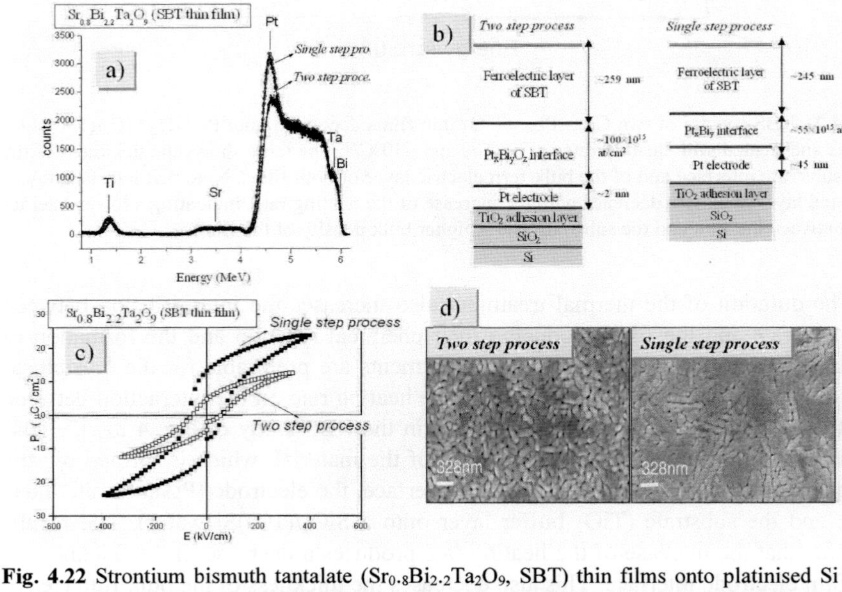

Fig. 4.22 Strontium bismuth tantalate ($Sr_{0.8}Bi_{2.2}Ta_2O_9$, SBT) thin films onto platinised Si substrates and crystallised at 650°C using a two step or a single step process. **a** Experimental and simulated RBS spectra, **b** heterostructure of the films inferred from the simulation of the RBS experimental spectra, **c** ferroelectric hysteresis loops and **d** microstructure of the film surfaces.

Fig. 4.22 shows the characteristics of two SBT thin films prepared by a two step and a single step process. [67] In the two step process, the deposited and dried film is subject to intermediate treatment at 550ºC for two hours., to stabilise a second fluorite phase. A subsequent treatment at 650ºC produces the transformation of the fluorite phase into the layered SBT perovskite. In the single step process, the deposited and dried film was directly heated by RTP (heating rate of 200ºC/s) at 650ºC.

The type of heating process affects the grain size, grain shape, porosity and interface thickness of the films. [67, 75, 106, 107] Larger and more elongated grains are obtained in the SBT films here shown, crystallised by a single step process. These films also have a lower porosity and thinner interfaces than the films prepared with a two step treatment. As a consequence, the single step SBT films exhibit less slanted ferroelectric hysteresis loops with higher values of the remnant polarisation (P_r), than the two step SBT films.

In the example here shown, the characteristics and properties of the SBT thin films crystallised with a single step process are clearly better than those of the films prepared with a two step process. However, the use of a single step treatment does not always enhance the ferroelectric response of the films. This is related with the precursor chemistry, its pyrolysis and thermolysis. It is indicated in some cases that intermediate heating previous to the crystallisation assures the elimination of organics or intermediate crystalline phases. It also controls the bulk density of the final film.

4.4.5 UV-Assisted Rapid Thermal Processing

The integration of the ferroelectric thin films with the Si-based readout integrated circuits (ROIC) requires the processing of the film at a low temperature. This is compatible with those used in the semiconductor industry (<550ºC). Higher temperatures would damage the substrate and other elements of the microelectronic device. However, the crystallisation of the ferroelectric perovskites usually occurs at higher temperatures. Therefore, the reduction of the processing temperature is one of the main challenges for the integration of ferroelectric materials. Different methods have been reported for this.. [108] One of the successful approaches in the low-temperature CSD processing of films is the so-called UV sol-gel photoannealing or PhotoChemical Solution Deposition (PCSD). This technique was first used for the preparation of single oxide films (Ta_2O_5, ZrO_2 or SiO_2) [109, 110]. However, it has scarcely been explored for the fabrication of ferroelectric perovskite thin films.

PCSD requires the use of photo-sensitive solutions (see section 3.5) and high power UV-irradiation. Over the years, the development of new excimer lamps with wavelengths in the range of 108-354 nm have increased the research effort on the low-temperature photoannealing of thin films. This is mainly deposited from liquids (CSD) or vapours (Chemical Vapour Deposition, CVD) that contain species that absorb UV-light. [111] These excimer lamps have been incorporated in RTP equipments and CVD chambers for getting the photoexcitation of the deposited film. This results in quick elimination of the organic compounds and in prompt formation of the crystalline phase.

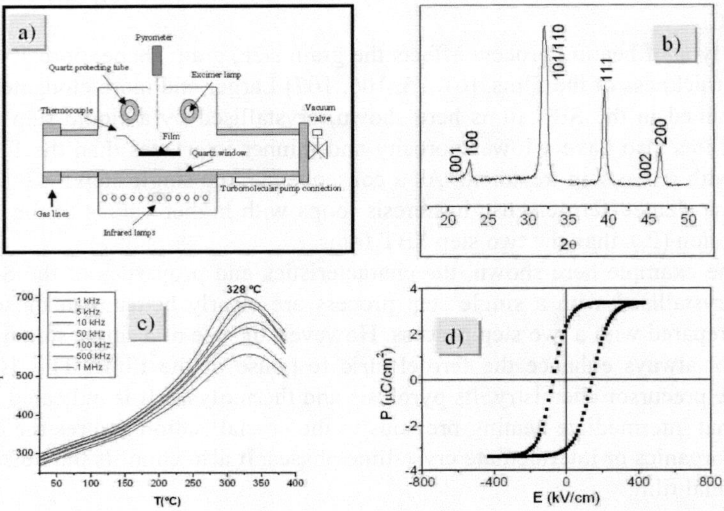

Fig. 4.23 a Scheme of an UV-assisted RTP processor commercialised by QUALIFLOW-JIPELEC [112]. Crystalline structure, dielectric and ferroelectric behaviour of $PbTiO_3$ based film UV-irradiated and heated at 450ºC in the former equipment: **b** X-ray diffraction pattern showing the perovskite structure, b) variation of the dielectric permitivitty with the temperature and **c** ferroelectric hysteresis loop [87]

Fig. 4.23 shows a scheme of an UV-assisted RTP processor that, besides the infrared lamps for the rapid heating (RTP), has two excimer lamps for the irradiation of the sample before the heating up to the crystallisation temperature. This equipment is a commercial processor [112].However, laboratory-scale equipment has also been used for the preparation of ferroelectric thin films. [113]

Using this technique, we have successfully prepared thin films based on ferroelectric $PbTiO_3$ at temperatures as low as 450ºC. Previously, some works had shown how the formation of the ferroelectric perovskite in these films started at a relatively low temperature (~500ºC). However, the films only presented a clear ferroelectric response when processed at higher temperature. Here, the use of the PCSD technique makes possible the formation of the ferroelectric phase at a very low temperature of 450ºC (Fig. 4.23b). It also allows the obtention of films with an appropriate dielectric and ferroelectric response. [83, 87] The characteristic ferroelectric-paraelectric transition of a ferroelectric material is observed in Fig. 4.23c. The ferroelectric hysteresis loop measured in these materials (Fig. 4.23d) is another indication of the ferroelectricity of the films processed by PCSD.

These results prove the competitiveness of the UV-assisted RTP for fabrication at low temperatures of solution-derived ferroelectric thin films. This overcomes traditional problems associated with the integration of these films into silicon devices.

4.5 Scaling down the Ferroelectric Thin Film

Ferroelectrics in thin film form started to be integrated with semiconductor substrates in the 1990s. Since then, the trend in microelectronics has been the continuous reduction of the device size below the micron (nanometre sizes). This would make possible the increase of density of integration in the device. This would also allow low voltage consumption. Therefore, the integration of nano-sized ferroelectrics with semiconductor substrates is of high interest in the new and future micro and nanoelectronic devices. This has been the driving force for the miniaturisation of the ferroelectric material. Thus, thin films (thickness over 100 nm) have moved to the sub-100 nm regime size, where ferroelectric ultra-thin films (thickness below 100 nm) and isolated ferroelectric nanostructures (lateral dimensions below 100 nm) are included (Fig. 4.1).

In this last section of this chapter, different strategies for the fabrication of ferroelectric ultra-thin films and nanoislads will be shown. All of them are based on the use of CSD methods, trying to show the competitiveness of this technique for the preparation of nano-sized ferroelectrics onto substrates with appropriate performances in devices.

4.5.1 Ultra-Thin Films

The reduction of thickness of the ferroelectric layer in devices below 100 nm would decrease the power consumption of the devices. But, the active (ferroelectric) layer should preserve properties (functionality) even for small thickness. For instance, the film should maintain remnant polarisation values enough to yield large piezoelectric activity (MEMS/NEMS). Or, it should have sufficient capacity for storing charge (information) in DRAM memories. Or, it may have polarisation that would be not lost with the reading and writing of the memory, for storing information in non-volatile FERAM memories.

For ferroelectric thin films, the critical thickness, below which spontaneous polarisation disappears, depends on the nature of the material (observing ferroelectricity in polymer films with thickness of only 10 Å.) [114] Stable ferroelectric states have been also demonstrated by direct or indirect measurements in perovskite films less than 10 Å thick. [115, 116]

For ferroelectric perovskite-based films, sub-100 nm thick layers have been prepared by different deposition techniques. They are: pulsed laser deposition [117], rf magnetron sputtering [118], or metalorganic chemical vapour deposition [119].

CSD methods have also been used for the fabrication of continuous ultra-thin ferroelectric films. As shown in the previous sections, CSD provides uniform coatings with a high degree of composition control and good properties. However, CSD was believed to be limited for the deposition of very thin layers, since very thin films usually suffered from incomplete coating of the substrate. Therefore, electrical characterisation of these films is commonly carried out by Piezoresponse Force Microscopy (PFM). This provides fundamental knowledge about this type of films. However, from the application point of view, macroscopically addressable device-ready ultrathin capacitors with appropriate properties are desired.

Table 4.4 CSD methods used for the preparation of ferroelectric ultra-thin films

CSD method	Film composition	Film thickness	Size of the top electrode	Reference
Multiple deposition and low-molarity solutions	$Pb(Zr,Ti)O_3$ (PZT)	~75 nm	---	[120,121]
	$Ba_{0.7}Sr_{0.3}TiO_3$ (BST)	~25 to 100 nm	---	[122]
	$Bi_4Ti_3O_{12}$ (BTO)	~30 nm	---	[123]
	$Sr_{0.8}Bi_{2.2}Ta_2O_9$ (SBT)	~40 nm	~ 560×10^2 μm²	[51,124,125]
	$Sr_{0.8}Bi_{2.5}Ta_2O_9$ (SBT)	~35 nm	~ 308×10^2 μm²	[126]
	$Ca_{0.50}Pb_{0.50}TiO_3$ (PTC)	~80 nm	~ 560×10^2 μm²	[9]
	$PbTiO_3$ (PT)	~35 nm	~ 220×10^2 μm²	[20,127]
Multiple deposition and low-molarity solutions, combining excess Pb in solution and PbO cover coats	$Pb_{0.88}La_{0.12}Zr_{0.485}Ti_{0.485}O_3$ (PLZT)	~50 nm	~ 10^6 μm²	[128]
Solid solutions of ferroelectrics with Bi_2SiO_5	$Bi_4Ti_3O_{12}$ (BTO) $SrBi_2Ta_2O_9$ (SBT) $Pb(Zr,Ti)O_3$ (PZT)	~13 to 25 nm	~ 78×10^2 μm²	[129]
Heat treatment by Rapid Thermal Annealing of the capacitor prepared by the deposition of the top electrode on the amorphous film	$SrBi_2Ta_2O_9$ (SBT)	~25 to 60 nm	---	[130]

Different CSD methods have been reported for fabrication of ferroelectric perovskite ultra-thin films of sufficient quality for the reliable preparation of macroscopic capacitors. Table 4.4 summarises some of these methods [20, 51, 120, 121, 122, 123, 124, 125, 126, 127, 128, 129, 130]. This indicates the film thickness and the size of the top electrode on which ferroelectric hysteresis loops were measured. This is a good indication of the continuity and quality of the crystalline ultra-thin film.

In general, all the CSD methods used for the deposition of ultra-thin films are based on the multiple deposits of highly-diluted (low-molarity) solutions. However, it has been reported that for thickness below a critical value, the polycrystalline film break up into islands. This gives place to the loss of the continuity of the film. [131] This critical value is related with the grain size to the thickness ratio.

Therefore, attempts to prepare very thin continuous films have been directed to the reduction of the grain size of the film. Conventional annealing of films leads to grain growth. Rapid thermal processing (RTP) of the films has shown to be an efficient method for getting ultra-thin films. This is so because it provides sufficient energy for the crystallisation of the CSD derived film, but leaves insufficient time for completing grain growth. [51, 127, 130] Decreased film thickness also enhances the volatility of some compounds from the film (Pb, Bi), due to the large surface to volume ratio. This damages the properties of the film. Thus, authors use excess Pb in solution and PbO cover coats (for lead-based perovskites), or excess Bi (for SBT compositions), to enable the fabrication of phase-pure films with thickness below 50 nm. [126, 128] Finally, a successful CSD method based on the deposition of Si-substituted ultrathin ferroelectric films has been reported. [129] Here, precursor solutions of the solid solutions of the ferroelectric phase and Bi_2SiO_5 are deposited. By this way, temperature of crystallisation of the films was lowered by ~200°C in comparison with films without Bi_2SiO_5. As a consequence, small grain size and flat surfaces were obtained in the films. This makes possible the fabrication of very thin continuous films (13-25 nm) with good ferroelectric responses. [129]

Therefore, strategies for decreasing the layer thickness of CSD films have to be directed to the reduction of the average grain size. RTP treatments or reduction of the crystallisation temperature avoid grain growth. Thus, the UV-assisted RTP process described above may be a promising technology for the fabrication of ultra-thin films. Additionally, as described in previous sections, CSD offers the opportunity of tailoring the chemistry of the solutions for getting films with controlled microstructures. Thus, grain size can be modified in the films prepared by the diol-route by changing the carbon length of the diol used for the solution synthesis. Short C-chain diols lead to films with smaller grain size than long C-chain diols (see Fig. 4.5). [64, 65] This is being used in our laboratory for the fabrication of continuous films with thickness below 50 nm. [127]

As examples, two types of ultra-thin films (Ca-modified $PbTiO_3$ and $Sr_{0.8}Bi_{2.2}Ta_2O_9$) prepared by the diol-route are shown. Both with properties of interest for memories; the first one ($Ca_{0.50}Pb_{0.50}TiO_3$) for DRAMs, and the second one ($Sr_{0.8}Bi_{2.2}Ta_2O_9$) for NVFERAMs.

Ferroelectric oxides are being investigated as possible candidates to replace the SiO_2 gate oxide traditionally used in DRAMs. The projected SiO_2 thickness in these devices is less than 2 nm. This makes fabrication and reliability of the device extremely difficult. [132] Ferroelectrics have high dielectric permittivities. They are thus candidates to replace SiO_2 in DRAM devices, by allowing the use of a thicker insulating layer than those of SiO_2, but with a similar effective capacitance. The storage of charge in DRAM devices increases with the decrease of film thickness. This is because capacitance increases as thickness of the capacitor decreases. The major problem, as thickness of the capacitor decreases, is the increase of the leakage current density of the device. Leakages below 10^{-7} are required for charge storage devices.

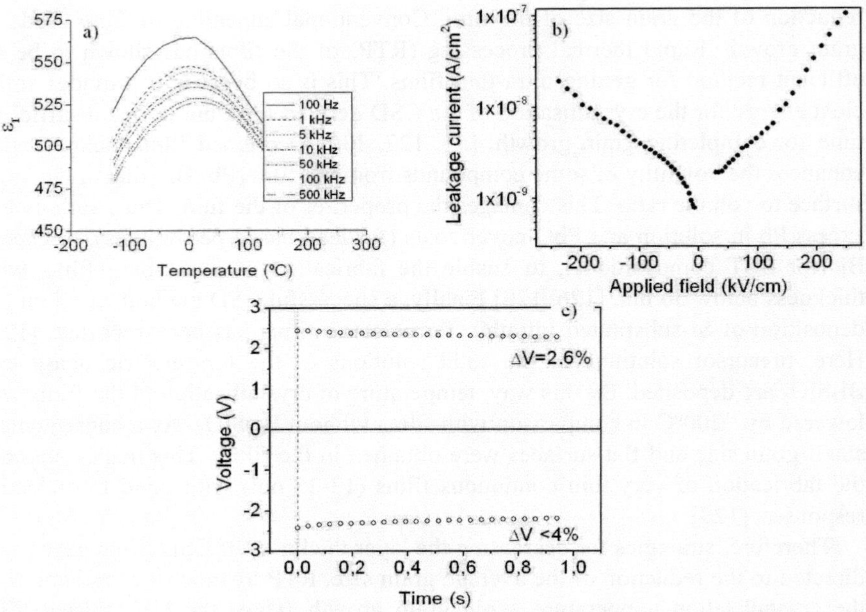

Fig. 4.24 Dielectric and electrical properties of a ~80 nm thick $Ca_{0.50}Pb_{0.50}TiO_3$ film prepared by CSD. **a** Variation of the dielectric permittivity (εr) with temperature. **b** Leakage current density. **c** Time voltage drop after the writing pulse, for the two signs of voltage charge. [9]

Here, ~80 nm thick $Ca_{0.50}Pb_{0.50}TiO_3$ films were prepared by multiple deposition and crystallisation (650ºC) of low-molarity precursor solutions synthesised by the diol-route. [9] Continuous ultra-thin crystalline films were obtained with relatively high dielectric permittivities at room temperature, ~525 (Fig. 4.24a). Additionally, low leakages currents, below 10^{-8} A/cm^2, are obtained in this film for the range of applied electric field used in the measurements (Fig. 4.24b). The testing of the voltage drop of the film after a writing pulse was also performed (Fig. 4.24c). This measurement shows the reliability of the film in a DRAM that is successively charged-released between zero and the refreshment voltage of the memory (real operation of the film in the memory device). The voltage drop of this film is less than 5%, which fulfils the recommendations for using a capacitor in a DRAM device. [133]

Therefore, the indications of the reliability of these CSD derived films for DRAM applications are: high dielectric permittivity at room temperature of these ultra-thin $Ca_{0.50}Pb_{0.50}TiO_3$ films, their low leakage current density, and the ability to retain above 95% of the written voltage after 200 ms (Fig. 4.24). [133].

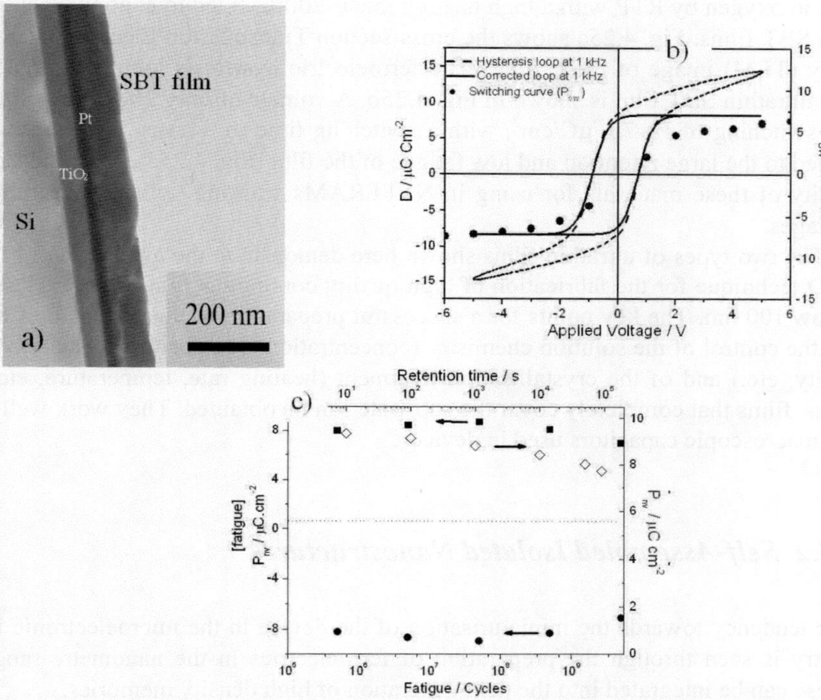

Fig. 4.25 Microstructure and ferroelectric properties of a ~40 nm thick $Sr_{0.8}Bi_{2.2}Ta_2O_9$ film prepared by CSD. **a** Cross-section image of the film observed by Transmission Electron Microscopy (TEM). **b** Ferroelectric hysteresis loops and switched polarisation curve of the film. **c** Fatigue for positive (■) and negative (●) voltage reading pulses. ◇ Retention for positive reading. [51, 67]

SBT ultra-thin films are good candidates for using in NVFERAMs because they can be fatigue-free materials (up to 10^{12} cycles) with the conventional platinum electrodes. [32] We have fabricated continuous ~40 nm thick films by CSD, by using diol-based solutions resulting from the mixing of a tantalum-glycolate solution, solutions of strontium and bismult carboxylates dissolved in 2-ethylhexanoic acid, and 1,3-propanediol solvents. [134] Multiple spin-coating of low molarity solutions (~0.04 equivalent moles per litre) and multiple crystallisation in oxygen by RTP with a high heating rate (~200°C/s) led to continuous ultrathin SBT films. Fig. 4.25a shows the cross-section Transmission Electron Microscopy (TEM) image of these films. The ferroelectric hysteresis loop measured in this ultrathin SBT film is shown in Fig. 4.25b. A voltage of only 2V is enough for the switching of P_r~7.5 μC/cm^2, with a switching time of ~1.3μs. These results, joined to the large retention and low fatigue of the film (Fig. 4.25c), show the suitability of these materials for using in NVFERAMs working with low operation voltages.

The two types of ultrathin films shown here demonstrate the availability of the CSD technique for the fabrication of high quality continuous films with thickness below 100 nm. The key points for a successful preparation of these films by CSD are the control of the solution chemistry (concentration, composition, density, viscosity, etc.) and of the crystallisation treatment (heating rate, temperature, etc.). Thus, films that completely cover the substrate can be obtained. They work well in the macroscopic capacitors used in devices.

4.5.2 Self-Assembled Isolated Nanostructures

The tendency towards the miniaturisation of the device in the microelectronic industry is seen through the preparation of ferroelectrics in the nanometre range. These can be integrated into the next generation of high density memories.

Successful techniques have been used to prepare nanostructures of simple systems onto substrates (metals, semiconductors, etc). [135, 136, 137] Isolated nanostructures of complex oxides (e.g., ferroelectric perovskites) integrated with semiconductor substrates are more difficult to prepare. Additionally, the physical properties of these nano-sized structures are dramatically different from those of bulk materials, which is a handicap for applications.

Top-down and bottom-up techniques have been used for the fabrication of ferroelectric nanostructures. [138, 139] Top-down techniques are lithography-based patterning methods that have two major advantages: good spatial resolution and positioning precision. However, they are expensive techniques limited to structures with lateral sizes not much smaller than ~100 nm. They are invasive methods that can produce the damage of the ferroelectric structure and thus, of its physical properties. Bottom-up (also called self-assembly methods) approaches are low cost techniques that allow relatively easy fabrication of structures with sizes below 100 nm. However, the major handicap of these techniques is the difficulty for getting nanoscale islands with a periodic order onto the substrate. Only the use of a bottom-up method, assisted by a lithography technique (hybrid approaches), has permitted the definition of regular arrays of ferroelectric perovskite nanostructures.

Table 4.5 Methods used for ferroelectric nanostructures onto substrates

	PROCESSING METHOD	NANOSTRUCTURE COMPOSITION	LATERAL SIZE (L) AND HEIGHT (H) OF THE NANOSTRUCTURES	DEMONSTRATION OF FERROELECTRICITY	REFERENCES
TOP-DOWN APPROACHES	Focused ion milling	$Pb(Nb,Zr,Ti)O_3$ (PNZT) $SrBi_2Ta_2O_9$ (SBT)	L~70nm L~90nm	Yes (by PFM) Yes (by PFM)	[140,141,142]
	Electron-beam direct writing	$SrBi_2Ta_2O_9$ (SBT) $Pb(Zr,Ti)O_3$ (PZT)	L~100nm	Yes (by PFM)	[143]
	Electron-beam lithography + lift-off process + dry etching	$Pb(Zr,Ti)O_3$ (PZT)	L~100nm, H~ 200nm	Yes (by PFM)	[144]
	Nanoimprint lithography	$Pb(Zr,Ti)O_3$ (PZT)	L~350nm, H~ 50nm	Yes (by PFM)	[145]
BOTTOM-UP APPROACHES	Using the microstructural instability of ultra-thin films [131,146]	Nano-electroded cells of $Bi_4Ti_3O_{12}$ (BIT)	L~125-200nm, H~180nm	Yes (conventional ferroelectric hysteresis loops using Hg-ball contacts and by PFM)	[147,148]
		$PbTiO_3$ (PT) $BaTiO_3$ (BT)	L~70nm	Yes (by PFM)	[149] [150]
		$Pb(Zr,Ti)O_3$ (PZT)	L~20-90nm, H~ 10-20nm	Yes (by PFM)	[151,152]
		$PbTiO_3$ (PT)	L~100nm	Yes (by PFM)	[8]
	Hydrothermal growth	$PbTiO_3$ (PT)	L~200nm	No	[153]
	Block-copolymer templated assisted growth	$SrTiO_3$ (ST)	L~10-20nm	---	[154]
	Nanosphere lithography	$BaTiO_3$ (BT)	L~230nm, H~ 9nm	Yes (by PFM)	[155,156,157]
		$SrBi_2Ta_2O_9$ (SBT)	L~250nm, H~ 30nm	Yes (by PFM)	[158]
	Microemulsion aided sol-gel	$BaTiO_3$ (BT)	L~50-60nm, H~20-30nm	No	[159]
		$PbTiO_3$ (PT)	L~40nm	Yes (by PFM)	[22]
HYBRID APPROACHES	Lithography-modulated self-assembly	$Pb(Zr,Ti)O_3$ (PZT)	L~50nm, H~20nm	Yes (by PFM)	[160]
	Tempate-controlled growth of solution deposited nanostructures	$PbTiO_3$ (PT)	L~50nm	Yes (by PFM)	[161]
			L~30nm	No	[52]
	Soft-electron beam lithography	$Pb(Zr,Ti)O_3$ (PZT)	L~500nm L~80nm ring structures	No	[162]
			L~70-80nm, H~10nm	Yes (by PFM)	[163]
		$BaTiO_3$ (BT)	L~70-100nm, H~100nm	Yes (by PFM)	[164]

Table 4.5 shows some of the methods reported for the fabrication of ferroelectric isolated nanostructures. [8, 22, 52, 131, 140, 141, 142, 143, 144, 145, 146, 147, 148, 149, 150, 151, 152, 153, 154, 155, 156, 157, 158, 159, 160, 161, 162, 163, 164, 165] All of these methods may involve the use of the CSD technique.

This could be for the deposit of a continuous thin film that is later carved for drawing the isolated structures (top-down approaches). Or, it could be as a self-assembly method for building structures from the bottom to the top (bottom-up approaches). Here, some CSD-based bottom-up methods will be detailed as techniques that allow the inexpensive fabrication of ferroelectric nanostructures with sizes down to 100 nm.

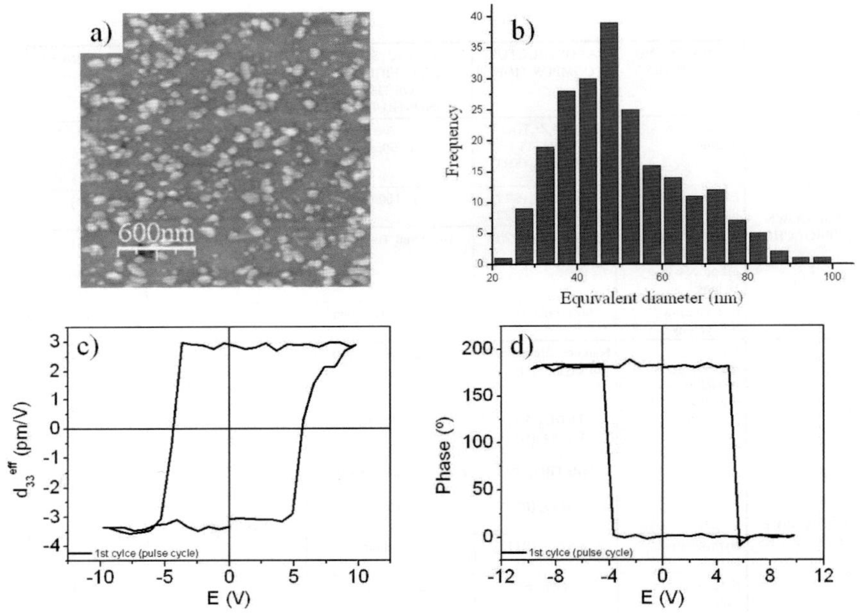

Fig. 4.26 $PbTiO_3$ isolated nanostructures onto $Pt/TiO_2/SiO_2/(100)Si$ substrates, prepared by the deposition and crystallisation of low molarity precursor solutions. **a** Atomic Force Microscopy (AFM) topography image, **b** distribution of the nanostructure sizes, **c** Local d_{33}^{eff} piezoelectric hysteresis loop obtained in one of the islands and **d** phase hysteresis loop obtained in one of the islands. [8]

The most common CSD-bottom-up approach to get these materials is based on the use of the phenomenon of the microstructural instability of ultra-thin films. In this case, low molarity solutions are deposited onto the substrate. Crystallisation of the ferroelectric phase is carried out at relatively high temperatures. Under these conditions, the continuity of the film is lost [131, 146] and the coating is formed by islands of different sizes. In this method, the size distribution and shape of the ferroelectric nanostructures depend on: the equivalent film thickness (or solution concentration equivalent to the quantity of the deposited material), and obtaining lognormal or bimodal distributions. [166, 167, 168] Fig. 4.26 shows the AFM image of $PbTiO_3$ isolated nanostructures obtained by the deposition of a solution with a concentration of 0.04 equivalent moles of $PbTiO_3$ per litre of solution onto a $Pt/TiO_2/SiO_2/(100)Si$ substrate, and the crystallisation at 650°C. The size distribution of these nanostructures is shown in Fig. 4.26b. The type of distribution (lognormal) indicates that complex mechanisms are involved in the formation of the nanostructures. The coalescence of particles or/and the surface energies between substrate and deposited material plays an important role in the processing and characteristics of these nanostructures. The ferroelectric response of these materials was tested by Piezoresponse Force Microscopy (PFM). Figs.26c and 26d shows the d_{33}^{eff} piezoelectric and the phase hysteresis loops measured at local scale in one of the islands. This demonstrates the ferroelectric nature of these isolated nanostructures.

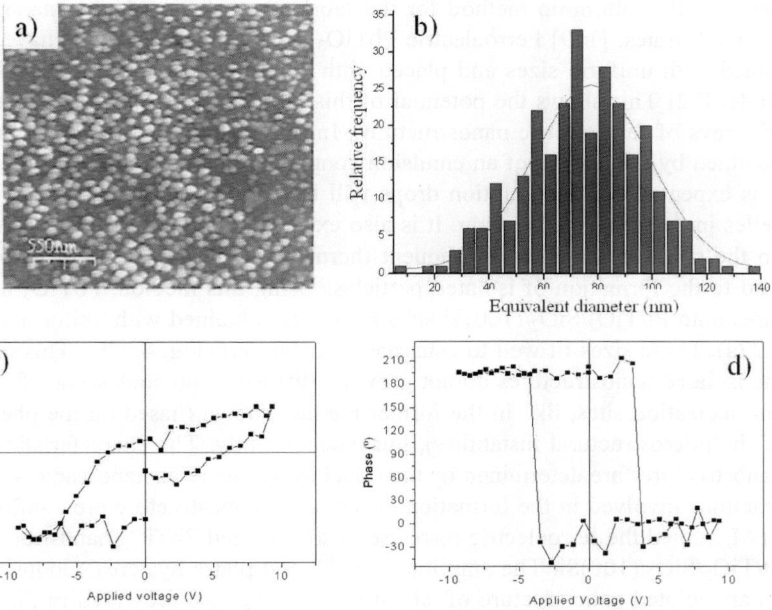

Fig. 4.27 $PbTiO_3$ isolated nanostructures onto $Pt/TiO_2/SiO_2/(100)Si$ substrates, prepared by the microemulsion aided sol-gel deposition. **a** Atomic Force Microscopy (AFM) topography image, **b** distribution of the nanostructure sizes, **c** Local d_{33}^{eff} piezoelectric hysteresis loop obtained in one of the islands and **d** phase hysteresis loop obtained in one of the islands. [169]

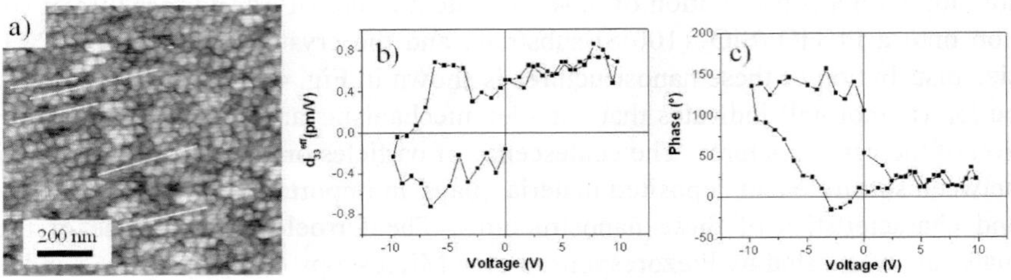

Fig. 4.28 $PbTiO_3$ isolated nanostructures onto (100)$SrTiO_3$ substrates, prepared by the microemulsion aided sol-gel deposition. **a** Atomic Force Microscopy (AFM) topography image (some lines showing the short-range periodical order of the nanostructures have been drawn) , **b** local d_{33}^{eff} piezoelectric hysteresis loop and **c** phase hysteresis loop [22].

The microemulsion mediated synthesis has been recently reported as an alternative CSD-bottom-up method for the fabrication of ferroelectric nanostructures onto substrates. [159] Ferroelectric $PbTiO_3$ isolated nanostructures have been obtained with uniform sizes and placed with periodicity in some zones of the substrate. [22] This shows the potential of this bottom-up method for the fabrication of arrays of ferroelectric nanostructures. In this method, the precursor solution is obtained by the mixing of an emulsion containing reverse micelles and a solution. It is expected that the solution drops will be placed in the polar core of the micelles in the resulting solution. It is also expected that the deposit of this solution on the substrate, and the subsequent thermal treatment of crystallisation, would lead to the formation of isolated particles. Using this method, $PbTiO_3$ nanostructures onto $Pt/TiO_2/SiO_2$/(100)Si substrates were obtained with uniform sizes (Fig. 4.27a). These sizes fit well to Gaussian distributions (Fig. 4.27b). This means that the isolated nanostructures do not grow by diffusion and coalescence from different nucleation sites, like in the former method shown (based on the phenomenon of the microstructural instability), but independently. The characteristics of these nanostructures are determined by the micelles that work as nanoreactors, where all reactions involved in the formation of the oxide nanostructure are confined. [168] PFM studied the ferroelectric response of the isolated $PbTiO_3$ nanostructures onto $Pt/TiO_2/SiO_2$/(100)Si. The amplitude (d_{33}^{eff}) and phase hysteresis loops measured on an isolated nanostructure of lateral size <80 nm are presented in Figs.27c and 27d. These results indicate that the polarisation of these nanostructures can be switched, thus demonstrating their ferroelectric nature. However, one of the major drawbacks of the bottom-up approaches, the ordering of the islands onto the substrate, is not solved. This can be due to incorrect selection of the substrate. $PbTiO_3$ nanostructures of Fig. 4.27 are prepared onto a policrystalline surface ($Pt/TiO_2/SiO_2$/(100)Si substrate). [169] This promotes the nucleation of the crystalline structures onto the defects of the Pt surface (grain boundaries). [170] It does not preserve the expected arrangement of the micelles in the solution deposited layer. [22] Therefore, substrates with smoother surfaces were used. Thus, the solution formed by the microemulsion and the solution was deposited onto (100)$SrTiO_3$ substrates polished on one side. It is also subjected to thermal treatment of crystallisation. Crystalline coatings as those shown in Fig. 4.28a were obtained. Particles were obtained with an average size of ~40 nm and with a short-range periodical order (nanostructures place along lines) in some regions of the substrate. The ordered nanostructures are not extended all over the coating. Areas of ~3.0×3.0 μm^2 have been observed in the AFM topography images, where a lateral order is detected. The ferroelectric character of these ordered nanostructures onto Nb-doped $SrTiO_3$ substrates is proved by PFM (see local hysteresis loops of Figs.28b and 28c). Structural analysis with synchrotron radiation, reported for these nanostructures, [22] demonstrates the $PbTiO_3$ perovskite crystal structure, which is the origin of their ferroelectricity.

Therefore, the results make these oxide nanostructures attractive for applications in nanoelectronic devices. They point to microemulsion aided sol-gel preparation as a promising method for self-assembly and self-ordering of ferroelectric isolated nanostructures.

4.6 Final Remark

Chemical Solution Deposition (CSD) is shown as a powerful and versatile technique for the preparation of ferroelectric thin and ultrathin films, and self-assembled nano-sized isolated structures onto substrates. Properties of these materials can be controlled by the appropriate tailoring of the solution chemistry and of the thermal conversion into a ceramic, of the deposited precursor. Additionally, CSD fulfils two of the major requirements for the fabrication of active (ferroelectric) elements in micro/nanoelectronic devices. They are the downscaling and the integration of the ferroelectric material. CSD offers the possibility of producing continuous ultrathin films and sub-100 nm ferroelectric structures, below the sizes accessible by other techniques. This, along with the integration of these materials with semiconductor substrates at temperatures compatible with those used in the microelectronic industry, place CSD among the most competitive fabrication techniques of ferroelectrics supported onto substrates.

Acknowledgments

The author thanks Dr. R. Jiménez, Dr. R. Sirera, Dr. A. González, Dr. I. Bretos and Dr. M. Torres for their contributions to some of the results presented here. The work collected in this chapter has been financed by different Spanish and European projects, and special mention has to be made to the projects now in progress: EU VI FP NoE MIND "Multifunctional and Integrated Piezoelectric Devices", NMP3–CT2005–515757 and Spanish MAT2007–61409.

References

1. Pardo L, Ricote J, Algueró M and Calzada ML (1999) Ferroelectric materials based on lead titanate, in Handbook of low and high dielectric constant materials and their applications. Vol.1. Materials and processing. Ed.by H.S.Nalwa. Hitachy Research Laboratory. Academic Press. San Diego, CA-USA. Chapter 10: 457
2. Mendiola J and Calzada ML (2002) Ferroelectric thin films of modified lead titanate, in Handbook of thin films materials. III. Ferroelectric and dielectric thin films. Ed. by H.S.Nalwa. Standord Scientific Corporation. Academic Press. Los Angeles, CA-USA. Chapter 6: 369.
3. Gray RB (1949) Transducer and method of making same. US Patent No.2486560.
4. Auciello O, Scott JF and Ramesh R (1998) The physics of ferroelectric memories. Physics Today, 7: 22.
5. Kingon AI and Streiffer SK (1999) Ferroelectric films and devices. Curr. Opin. Solid State Mater. Sci., 4: 39.
6. Muralt P (2000) Ferroelectric thin films for micro-sensors and actuators: a review. J. Micromech. Microeng., 10: 136.
7. Semiconductor Industry Association. International Technology Roadmap for Semiconductors (ITRS). Available at http://public.itrs.net on 2008.
8. Calzada ML, Ricote J, Jiménez R, Bretos I, Ramos P, Mendiola J, Algueró M and Pardo L (2006). Ferroelectrics onto silicon prepared by chemical solution deposition methods: from the thin film to the self-assembled systems. Bol.Soc.Esp.Cerám.Vidrio, 45(3): 126.
9. Bretos I (2006) Low-toxic chemical solution deposition methods for the preparation of multifunctional $(Pb_{1-x}Ca_x)TiO_3$ thin films. Ph.D Thesis. Univ. Autónoma de Madrid. Spain.
10. Kingon AI, Maria JP and Streiffer SK (2000) Alternative dielectrics to silicon dioxide for memory and logic devices. Nature, 406: 1032.
11. Kingon AI, Streiffer SK, Basceri C and Summerfelt SR (1996) High-permittivity perovskite thin films for dynamic random-access memories. Mater.Res.Bull., 21(7): 46.
12. Takasu H (2000) The ferroelectric memory and its applications. J.Electroceram., 4:2/3: 327.
13. Whatmore RW (1986) Pyroelectric devices and materials. *Rep.Prog.Phys., 49(12):* 1335.
14. Sixmith A, Johnson N and Whatmore RW (2005) Pyroelectric IR sensor arrays for fall detection in the older population. J.Phys.IV, 128: 153.
15. Trolier-Mckinstry S and Muralt P (2004) Thin film piezoelectric for MEMS. J.Electroceram., 12(1-2): 7.
16. Craighead HG (2000) Nanoelectromechanical systems. Science, 290: 1532.
17. Egger P, Hulliger J (1999) Optical materials for short wavelength generation. Coord.Chem.Rev., 183: 101.
18. Ouaddari M, Delprat S, Vidal F, Chaker M, Wu K (2005) Microwave characterization of ferroelectric thin-film materials. IEEE Trans.Micr.Theor.Techn., 53(4) Part 1: 1390.
19. Phillips NJ, Calzada ML and Milne SJ (1992) Sol-gel derived lead titanate films. J.Non-Cryst.Solids, 147&148: 285.
20. Ricote J, Holgado S, Ramos P and Calzada ML (2006) Piezoelectric ultrathin lead titanate films prepared by deposition of aquo-diol solutions. IEEE Trans.Ultr.Ferr.Freq.Contr., 53(12): 2299.
21. Clemens S, Schneller T, Waser R, Rudiger A, Peter F, Kronholz S, Schmitz T and Tiedke S (2005) Integration of ferroelectric lead titanate nanoislands for direct hysteresis measurements. Appl.Phys.Let., 87 (14): 142904.

22. Calzada ML, Torres M, Fuentes-Cobas LE, Mehta A, Ricote J and Pardo L (2007) Ferroelectric self-assembled $PbTiO_3$ perovskite nanostructures onto (100)$SrTiO_3$ substrates from a novel microemulsion added sol-gel preparation method. Nanotechnology, 18: 375603.
23. Sirera R and Calzada ML (1995) Multicomponent solutions for the deposition of modified lead titanate films. Mater.Res.Bull., 30(1): 11.
24. Kholkin AL, Calzada ML, Ramos P, Mendiola J and Setter N (1996) Piezoelectric properties of Ca-modified $PbTiO_3$ thin films. Appl.Phys.Lett., 69(23): 3602.
25. Algueró M, Kholkin A, Calzada ML and Pardo L (2007) Dependence of functional properties on processing in sol-gel (Pb,La)TiO_3 thin films. Appl.Phys.A, 71: 195.
26. Kholkin AL and Calzada ML (1998) Calcium modified lead titanate thin films for electromechanical applications. J.Phys.IV France, 8: 195.
27. Ramos P, Mendiola J, Jiménez R, Calzada ML, González A and Tejedor P (2002) Detector de IR de lámina ferroeléctrica de (Pb,Ca)TiO_3. Bol.Soc.Esp.Cerám.Vidr., 41(1): 5.
28. Poyato R, Calzada ML and Pardo L (2005) Tailoring of the functional properties of sol-gel films on Pt/TiO_2/SiO_2/(100)Si substrates: (Pb,La)TiO_3/(Pb,Ca)TiO_3 multilayer heterostructures. Appl.Phys.A, 80: 369,
29. Poyato R, Calzada ML and Pardo L (2005) (Pb,Ca)TiO_3/(Pb,La)TiO_3/(Pb,Ca)TiO_3 heterostructure characterized as ferroelectric multifunctional materials. J.Appl.Phys., 97: 11.
30. Shimizu M, Fujisawa H and Niu H (2005) Pb-based ferroelectric thin films prepared by MOCVD. Ferroelectric thin films topics in applied physics, 98: 59.
31. Yu X, Zhang ZS and Hu H (2005) PZT ferroelectrics thin films used in MEMS. J.Inorg.Mater., 20(1): 27.
32. Kang DH, Kim KW, Lee SY, Kim YH and Gil SK (2005) Influencing factors on the pyroelectric properties of Pb(Zr,Ti)O_3 thin film for uncooled infrared detector. Mater.Chem.Phys., 2-3: 411.
33. DeAraujo CAP, Cuchiaro JD, Mcmillan LD, Scott MC and Scott JF (1995) Ferroelectric capacitors with platinum-electrodes. Nature, 374: 627.
34. Park BH, Kang BS, Bu SD, Noh TW, Lee J and Jo W (1999) Lanthanum-substituted bismuth titanate for use in non-volatile memories. Nature, 401: 682.
35. Kozyrev A, Ivanov A, Samoilova T, Soldatenkov O, Astafiev K and Sengupta LC (2000) Nonlinear response and power handling capability of ferroelectric $Ba_xSr_{1-x}TiO_3$ film capacitors and tunable microwave devices. J.Appl.Phys., 88(9): 5334.
36. Schrofel J and Hradilova J (1994) Basic principles and materials in integrated optics. Chemicke Listy, 88(11): 686.
37. Jelínek M (2003) Progress in optical waveguiding thin films. Czechoslovak J.Phys., 53(5): 365.
38. European Union Directive 2002/95/EC on the Restriction Of the use of certain Hazardous Substances in electrical and electronic equipment (RoHS directive).
39. Saito Y, Takao H, Tani T, Nonoyama T, Kakatori K, Homma T, Nagaya T and Nakamura M (2004) Lead-free piezoceramics. Nature, 432: 84.
40. Schwartz RW (1997) Chemical solution deposition of perovskite thin films. Chem.Mater., 9: 2325.
41. Auciello O, Kingon AI and Krupanidhi SB (1996) Sputter synthesis of ferroelectric films and heterostructures. MRS Bull., 21(6): 25.
42. Tybell T, Ahn CH and Triscone JM (1999) Ferroelectricity in thin perovskite films. Appl.Phys.Lett., 75(6): 856.
43. Muralt P (2006) Texture control and seeded nucleation of nanosize structures of ferroelectric thin films. J.Appl.Phys., 100 (5): 051605.

44. Auciello O and Ramesh R (1996) Laser-ablation deposition and characterisation of ferroelectric capacitors for nonvolatile memories. MRS Bull., 21 (6): 31.
45. Nagarajan V, Prasertchoung S, Zhao T, Zheng H, Ouyang J, Ramesh R, Tian W, Pan XQ, Kim DM, Eom CB, Kohlstedt H and Waser R (2004) Size effects in ultrathin epitaxial ferroelectric heterostructures. Appl.Phys.Lett., 84 (25): 5225.
46. Zheng H, Straub F, Zhan Q, Yang PL, Hsieh WK, Zavaliche F, Chu YH, Dahmen U and Ramesh R (2006) Self-assembled growth of $BiFeO_3$-$CoFe_2O_4$ nanostructures. Adv.Mater., 18: 2747.
47. DeKeijser M and Dormans GJM (1996) Chemical vapour deposition of electroceramic thin films. MRS Bull., 21(6): 37.
48. Fong DD, Stephenson GB, Streiffer SK, Eastman JA, Auciello O, Fuoss PH and Thompson C (2004) Ferroelectricity in ultrathin perovskite films. Science, 304(5677): 1650.
49. Nonomura H, Nagata M, Fujisawa H, Shimizu M, Niu H and Honda K (2005) Structural control of self-assembled $PbTiO_3$ nanoislands fabricated by metalorganic chemical vapour deposition. Appl.Phys.Lett., 86: 163106.
50. Tuttle BA, Schwartz RW (1996) Solution deposition of ferroelectric thin films. MRS Bull., 21(6): 49.
51. González A, Jiménez R, Mendiola J, Alemany C and Calzada ML (2002) Ultrathin ferroelectric strontium bismuth tantalate films. Appl.Phys.Lett., 81(14): 2599.
52. Kronholz S, Rathgeber S, Karthäuser S, Kohlsedt H, Clemens S and Schneller T (2006) Self-assembly of diblock-copolymer micelles for template-based preparation of $PbTiO_3$ nanograins. Adv.Funct.Mater., 16: 2346.
53. Fukushima J (1976) Preparation of $BaTiO_3$ films by hydrolysis of organometallic compounds. Am.Ceram.Soc.Bull., 55(12): 1064.
54. Fukushima J, Kodaira K and Marsushita T (1984) Preparation of ferroelectric PZT films by thermal decomposition of organometallic compounds. J.Mater.Sci., 19 : 595.
55. Budd KD, Dey SK and Payne DA (1985) Sol-gel processing of $PbTiO_3$, $PbZrO_3$, PZT and PZLT thin films. Brit.Ceram.Proc., 36: 107.
56. Dey SK, Budd KD and Payne DA (1988) Thin-film ferroelectrics of PZT by sol-gel processing. IEEE Trans.Ultras.Ferr., 35(1): 80.
57. Schwartz RW, Schneller T and Waser R (2004) Chemical solution deposition of electronic oxide films. C.R.Chimie, 7 : 433.
58. Hubert-Pfalzgraf LG (2004) To what extent can design of molecular precursors control the preparation of high tech oxides?. J.Mater.Chem., 14(21): 3113.
59. Kato K, Tanaka K, Suzuky K, Fu D, Nishizawa K and Miki T (2003) Chemical solution processing for electroceramic films. Proceeding on the symposium processing of electroceramics, as an initiative of POLECER (European thematic network on polar electroceramics). Ed.by M.Kosec, D.Kuscer and B.Malic. Jozef Stefan Institute, Ljublana, Slovenia. p.155.
60. Blum JD and Gurkovich SR (1985) Sol-gel-derived $PbTiO_3$. J.Mater.Sci., 20(12) : 4479.
61. Milne SJ and Pyke SH (1991) Modified sol-gel process for the production of lead titanate films. J.Am.Ceram.Soc., 74(6): 1407.
62. Milne SJ and Phillips NJ (1991) Diol-based sol-gel system for the production of thin-films of $PbTiO_3$. J.Mater.Chem., 1(5): 893.
63. Calzada ML, Sirera R, Carmona F and Jiménez B (1995) Investigations of a diol-based sol-gel process for the preparation of lead titanate materials. J.Am.Ceram.Soc., 78(7): 1802.
64. Calzada ML and Sirera R (1996) Chemically derived ferroelectric calcium modified lead titanate thin films deposited from aquo-diol solvent solutions. J.Mater.Sci.-Mater.Electr., 7: 39.

65. Sirera R. Ph.D Thesis (1997) Síntesis por sol-gel de soluciones de titanato de plomo modificado para la preparación de láminas delgadas ferroeléctricas. Univ.Autónoma de Madrid, Spain.
66. Calzada ML and González A (2005) Tantalum penta-glycolate sol as a precursor of strontium bismuth tantalate ferroelectric thin films. J.Am.Ceram.Soc., 88(10), 2702.
67. González A (2002) Láminas delgadas y ultradelgadas de tantalato de estroncio y bismuto obtenidas a partir de derivados de glicolato de tántalo para su uso en memorias ferroeléctricas no volátiles. Ph.D Thesis. Univ.Autónoma de Madrid (Spain).
68. Calzada ML, Algueró M, Ricote J, Santos A and Pardo L (2007) Preliminary results on sol-gel processing of <100> oriented $Pb(Mg_{1/3}Nb_{2/3})O_3$-$PbTiO_3$ thin films using diol-based solutions. J.Sol-Gel Sci.Techn., 42: 331.
69. Calzada ML, Malic B, Sirera R and Kosec M (2002) Thermal decomposition chemistry of modified lead-titanate aquo-diol gels used for the preparation of thin films. J.Sol-Gel Sci.Techn., 23: 221.
70. Malic B, Kosec M, Smolej K and Stavber S (1999) Effect of precursor type on the microstructure of $PbTiO_3$ thin films. J.Eur.Ceram.Soc., 19: 1344.
71. Dictionary of Inorganic Compounds (1992). Ed.by J.E. Macintyre. Chapman & Hall, London.
72. Asada H, Udaka M and Kawano H (1994) Explosive vaporization of crystal water on temperature-programmed heating of a thin-film of hydrated calcium nitrate. Thin Solid Films, 252: 49.
73. Calzada ML, Martin MJ, Ramos P, Mendiola J, Sirera R, Da Silva MF and Soares C (1997) Effect of composition and annealing conditions on the properties of sol-gel prepared calcium-modified lead titanate thin films. J.Phys.Chem.Solids, 58(7): 1033.
74. Aggarwal S, Madhukar S, Nagaraj B, Jenkins IG, Ramesh R, Boyer L and Evans JT (1999) Can lead nonstoichiometry influence ferroelectric properties of $Pb(Zr,Ti)O_3$ thin films?. Appl.Phys.Lett., 75(5): 716.
75. Jimenez R, Alemany C, Calzada ML, Gonzalez A, Ricote J and Mendiola J (2002) Processing effects on the microstructure and ferroelectric properties of strontium bismuth tantalate thin films. Appl.Phys.A., 75: 607.
76. Takemura T, Noguchi T, Hase T, Kimura H and Miyasaka Y (1999) Composition dependence of dielectric anomaly in strontium bismuth tantalate thin films. Symposium on Ferroelectric Thin Films VII. Ferroelectric thin films VII. 541: 235.
77. Imai H (2005) Ultraviolet (UV) irradiation. Handbook of sol-gel science and technology. Processing characterization and applications. Ed.by S.Sakka. Kluwer academic publishers, London, Vol.1, Chp.27: p.639.
78. Ohishi T, Mackawa S and Katoh A (1992) Synthesis and properties of tantalum oxide films prepared by the sol-gel method using photo-irradiation. J.Non-Cryst.Solids, 147&148: 493.
79. Asakuma N, Hirashima H, Fukui T, Toki M, Awazu K and Imai H (2002) Photoreduction of amorphous and crystalline ZnO films. Jpn.J.Appl.Phys., 41, 3909.
80. Asakuma N, Fukui T, Toki M, Awazu M and Imai (2003) Photoinduced hydroxylation at ZnO surface. Thin Solid Films, 445: 284.
81. Imai H, Tominaga A, Hirashima H, Aizawa M and Toki M (1998) Ultraviolet-laser-induced crystallization of sol-gel derived indium oxide films. J.Sol-Gel Sci.Tech., 85(13): 994.
82. Soyama N, Sasaki G, Atsuki T, Yonezawa T and Ogi K (1994) The formation of a fine-patterned ferroelectric thin film from a sol-gel solution containing a photo-sensitive water-generator. Proceeding of the ninth international symposium on applications of ferroelectrics. Eds. R.K.Pandey, M.Liu and A.Safari. IEEE, Piscataway, NJ: p.408.

83. Calzada ML, Bretos I, Jiménez R, Guillon H and Pardo L (2004) Low-temperature processing of ferroelectric thin films compatible with silicon integrated circuit technology. Adv.Mater., 16(18): 1620.
84. Bretos I, Jiménez R, Rodríguez-Castellón E, García-López J and Calzada ML (2008) Heterostructure and compositional depth profile of low-temperature processed lead titanate-based ferroelectric thin films prepared by photochemical solution deposition. Chem.Mater., 20: 1443.
85. Calzada ML, Gonzalez A, Poyato R and Pardo L (2003) Photo-sensitive sol-gel solutions for the low-temperature UV-assisted processing of $PbTiO_3$ based ferroelectric thin films. J.Mater.Chem., 13: 1451.
86. Pardo L, Poyato R, Gonzalez A and Calzada ML (2003) Low temperature preparation of piezoelectric thin films by ultraviolet-assisted rapid thermal processing. Mater.Sci.Semic.Proc., 5: 77.
87. Calzada ML, Bretos I, Jimenez R, Guillon H, Ricote J and Pardo L (2007) Low-temperature ultraviolet sol-gel photoannealing processing of multifunctional lead titanate based thin films. J.Mater.Res., 22(7): 1824.
88. Pileni MP (1993) Reverse micelles as microreactors. J.Phys.Chem., 97: 6961.
89. Herrig H and Hempelmann R (1997) Microemulsion mediated synthesis of ternary and quaternary nanoscale mixed oxide ceramic powders. Nanostr.Mater., 9: 241.
90. Malic B (2003) Solution processing of electroceramics in Proceeding on the symposium processing of electroceramics, as an initiative of POLECER (European thematic network on polar electroceramics). Ed.by M.Kosec, D.Kuscer and B.Malic. Jozef Stefan Institute, Ljublana, Slovenia. p.41.
91. Kim SS, Choi EK, Kim JK, Kim JS, Song TK and Kim J (2003) Effects of surfactant on surface morphology and orientation of Nb-doped $Bi_4Ti_3O_{12}$ thin films. J.Korean Phys.Soc., 42: S1126.
92. Kim J, Kim JK, Heo S and Lee HS (2006) Ferroelectric properties of sol-gel prepared La- and Nd-substituted, and Nb-co-substituted bismuth titanate using polymeric additives. Thin Solid Films, 503: 60.
93. Pithan C, Schneller T, Shiratory Y, Majurader SB, Haege FH, Dornseiffer J and Waser R (2006) Microemulsion mediated synthesis of nanocrystalline $BaTiO_3$: possibilities, potential and perspectives. Int.J.Mater.Res., 9(7), 499.
94. Ricote J, Poyato R, Algueró M, Pardo L and Calzada ML (2003) Texture development in modified lead titanate thin films obtained by chemical solution deposition on silicon-based substrates. J.Am.Ceram.Soc., 86(9): 1571.
95. Mendiola J, Calzada ML, Ramos P, Martín MJ and Agulló-Rueda F (1998) On the effects of stresses in ferroelectric $(Pb,Ca)TiO_3$ thin films. Thin Solid Films, 315: 195.
96. Tuttle BA, Garino TJ, Voigt JA, Headley TJ, Dimos D and Eatough MO (1995) Relationships between ferroelectric 90° domain formation and electrical properties of chemically prepared $Pb(Zr,Ti)O_3$ thin films. in Science and technology of electroceramic thin films, ed.by O.Auciello and R.Waser. Kluwer Academic, Netherlands. p.117.
97. González A, Poyato R, Jiménez R, Mendiola J, Pardo L and Calzada ML (2000) Oriented growth of sol-gel-modified $PbTiO_3$ thin films on Si-based substrates. Surf.Interface Anal., 29 : 325.
98. Song YJ, Zhu YF and Desu SB (1998) Low temperature fabrication and properties of sol-gel derived (111) oriented $Pb(Zr_{1-x}Ti_x)O_3$ thin films. Appl.Phys.Lett., 72(21): 2686.
99. Muralt P, Maeder T, Sagalowicz L, Hiboux S, Scalese S, Naumovic D, Agostino RG, Xauthopoulos N, Mathieu HJ, Patthey L and Bullock EL (1998) Texture control of $PbTiO_3$ and $Pb(Zr,Ti)O_3$ thin films with TiO_2 seeding J.Appl.Phys., 83(7): 3835.
100. Calzada ML, Jiménez R, González A, García-López J, Leinen D and Rodríguez-Castellón E (2005) Interfacial phases and electrical characteristics of ferroelectric stron-

tium bismuth tantalate films on Pt/TiO_2 and Ti/Pt/Ti heterostructure electrodes. Chem.Mater., 17 : 1441.

101. Hori T and Iwasaki H (1989). IEEE Electron Device Letters, 10(2): 64.
102. Hu H, Shi L, Kumar V and Krupanidhi SB (1992) Rapid thermal annealing processed ferroelectric Pb(Zr,Ti)O_3 films. Ceramic Transactions, 25 : 113.
103. Jiménez R, González A, Calzada ML and Mendiola J (2000) Study of electrolytic laminated ferroelectric thin films from electrode substrates. J.Mater.Res., 15(5): 1041.
104. Sirera R, Leinen D, Rodríguez-Castellón E and Calzada ML (1999) Processing effects on the compositional depth profile of ferroelectric sol-gel Ca-$PbTiO_3$ thin films. Chem.Mater., 11: 3437.
105. Bhuiyan MS, Paranthaman M and Salama K (2006) Solution-derived textured oxide thin films - a review Supercond.Sci.Techn., 19 : R1.
106. Calzada ML, González A, Jiménez R, Alemany C and Mendiola J (2001) Rapid thermal processing of strontium bismuto tantalate ferroelectric thin films prepared by a novel chemical solution deposition method. J.Eur.Ceram.Soc., 21(10-11): 1517.
107. Calzada ML, González A, García-López J and Jiménez R (2003) Crystallization, heterostructure, microstructure and properties of ferroelectric strontium bismuth tantalate films derived from tantalum glycolate solutions. Chem.Mater., 15: 4775.
108. Calzada ML, Pardo L, Guillon H, Bretos I, González A, Poyato R and Jiménez R (2003) Low temperature CSD processing of ferroelectric thin films in Proceeding on the symposium processing of electroceramics, as an initiative of POLECER (European thematic network on polar electroceramics). Ed.by M.Kosec, D.Kuscer and B.Malic. Jozef Stefan Institute, Ljublana, Slovenia. p.167.
109. Maekawa S and Ohishi T (1994) Evaluation of SiO_2 thin-films prepared by sol-gel method using photoirradiation. J.Non-Cryst.Solids, 169, 207.
110. Hirai S, Shimakage K, Sekiguchi M, Wada K and Nukui A (1999) Zirconium oxide coating on anodised aluminium by the sol-gel process combined with ultraviolet irradiation at ambient temperature. J.Am.Ceram.Soc., 82: 2011.
111. Van de Leest RE (1995) UV photo-annealing of thin sol-gel films. Appl.Surf.Sci., 86: 278.
112. Qualiflow/Jipelec. http://www.jipelec.com/news/index.htm on 2009.
113. Bretos I, Jiménez R, García-López J, Pardo L and Calzada ML (2008) Photochemical solution deposition of lead-based ferroelectric films: avoiding the PbO-excess at least. Chem.Mater., 20: 5731.
114. Bune AV, Fridkin VM, Ducharme S, Blinov LM, Palto SP, Sorokin AV, Yudin SG and Zlatkin A (1998) Two-dimensional ferroelectric films. Nature, 391: 874.
115. Sai N, Kolpak AM and Rappe AM (2005) Ferroelectricity in ultrathin perovskite films. Phys.Rev.B. B72: 020101(R).
116. Lichtensteiger C, Triscone JM, Junquera J and Ghosez P (2005) Ferroelectricity and tetragonality in ultrathin $PbTiO_3$ films. Phys.Rev.Lett., 94: 047503.
117. Kim YS, Jo JY, Kim DJ, Chang YJ, Lee JH, Noh TW, Song TK, Yoon JG, Chung JS, Balk SI, Kim YW and Jung CU (2006) Ferroelectric properties of $SrRuO_3/BaTiO_3/SrRuO_3$ ultrathin film capacitors free from passive layers. Appl.Phys.Lett., 88: 072909.
118. Kelman MB, Schloss LF and McIntyre PC (2002) Thickness-dependent phase evolution of polycrystalline Pb(Zr0.35Ti0.65)O-3 thin films. Appl.Phys.Lett., 80, 1258.
119. Nonomura H, Fujisawa H, Shimizu M and Niu H (2002) Epitaxial growth and ferroelectric properties of the 20-nm thick Pb(Zr,Ti)O_3 film on $SrTiO_3$ (100) with an atomically flat surface by metalorganic chemical vapour deposition. Jpn.J.Appl.Phys., Part 1. 41(11B): 6682.

120. Nouwen R, Mullens J, Franco D, Yperman J, Van Poucke LC, Norga G, Wouters D and Maes H (1999) Sol-gel processing of piezoelectric and ferroelectric films US patent, application number 09-255,916.
121. Wouters DJ, Norga GJ and Maes HE (1999) Ultra-thin PZT films for low-voltage ferroelectric non-volatile memories. Ferroelectric thin films VII. MRS Symp.Proc., 541: 581.
122. Kügeler C, Liedtke R and Waser W (2002) Leakage current properties of $Ba_{0.7}Sr_{0.3}TiO_3$ thin films depending on the film thickness. ISAF 2002: Proceedings of the 13th IEEE international symposium on applications of ferroelectrics: 215.
123. Schneller T and Waser R (2002) Chemical solution deposition of ferroelectric thin films - State of the art and recent trends. Ferroelectrics, 267: 293.
124. Jiménez R, Calzada ML, González A, Mendiola J, Shvartsman VV, Kholkin AL and Vilarinho PM (2004) Nanoscale properties of ferroelectric ultrathin SBT films. J.Eur.Ceram.Soc., 24: 319.
125. Ramos P, Jiménez R, Alemany C, Calzada ML, González A, Alonso CE and Mendiola J (2004) Testing SBT ferroelectric thin films for non-volatile RAM. Integrated Ferroelectrics, 60: 33.
126. Aizawa K and Ishiwara H (2005) Reduction of pyrochlore phase and pronounced improvement of ferroelectric properties in ultrathin $SrBi_2Ta_2O_9$ films derived from Bi-rich sol-gel solution. Jpn.J.Appl.Phys., 44(8): 6221.
127. Ricote J (2007). MIND Internal workshop on films prepared by chemical solution deposition. Madrid (Spain).
128. Brennecka GL, Parish CM, Tuttle BA and Brewer LN (2008) Multilayer thin and ultra-thin film capacitors fabricated by chemical solution deposition. J.Mater.Res., 23(1): 176.
129. Kijima T and Ishiwara H (2002) Si-substituted ultrathin ferroelectric films. Jpn.J.Appl.Phys., 41: L716.
130. Celinska J, Joshi V, Narayan S, McMillan L and Paz de Araujo C (2003) Effects of scaling the film thickness on the ferroelectric properties of $SrBi_2Ta_2O_9$ ultrathin films. Appl.Phys.Lett., 82(22): 3937.
131. Miller KT, Lange FF and Marshall DB (1990) The instability of polycrystalline thin films: experiment and theory. J.Mater.Res., 5: 151.
132. Schulz M (1999) The end of the road for silicon?. Nature, 399: 729.
133. Hoffman S and Waser RM (1997) Dielectric properties, leakage behaviour and resistance degradation of thin films of the soli solution series $Ba(Ti_{1-y}Zr_y)O_3$. Integrated Ferroelectrics, 17: 141.
134. Calzada ML, Jiménez R, González A and Mendiola J (2001) Air-stable solutions for the low-temperature crystallization of strontium bismuth tantalate ferroelectric films. Chem.Mater., 13: 3.
135. Aggarwal S, Ogale SB, Ganpule CS, Shinde SR, Novikov VA, Monga AP, Burr MR and Ramesh R (2001) Oxide nanostructures through self-assembly. Appl.Phys.Lett., 78: 1442.
136. Zhang Z, Wei BQ and Ajayan PM (2001) Self-assembled patterns of iron oxide nanoparticles by hydrothermal chemical-vapour deposition. Appl.Phys.Lett., 79(25): 4207.
137. Guarini KW, Black CT, Zhang Y, Kim H, Sikorski EM and Babich IV (2002) Process integration of self-assembled polymer templates into silicon nanofabrication. J.Vac.Sci.Tehnol., 20: 2788.
138. Alexe M and Hesse D (2006) Self-assembled nanoscale ferroelectrics. J.Mater.Sci., 41: 1.
139. Gruverman A and Kholkin A (2006) Nanoscale ferroelectrics: processing, characterization and future trends. Rep.Prog.Phys., 69: 2443.
140. Stanishevsky A, Aggarwal S, Prakash AS, Melngailis J and Ramesh R (1998) Focused ion-beam patterning of nanoscale ferroelectric capacitors. J.Vac.Sci.Techn., 16(6): 3899.

141. Ganpule CS, Stanishevsky A, Su Q, Aggarwal S, Melngailis J, Williams E and Ramesh R (1999) Scaling of ferroelectric properties in thin films. Appl.Phys.Lett., 75(3): 409.
142. Ganpule CS, Stanishevsky A, Su Q, Aggarwal S, Melngailis J, Williams E and Ramesh R (1999) Scaling of ferroelectric and piezoelectric properties in Pt/$SrBi_2Ta_2O_9$/Pt thin films. Appl.Phys.Lett., 75(24): 3874.
143. Alexe M, Harnagea C, Hesse D and Gösele U (1999) Patterning and switching of nanosize ferroelectric memory cells. Appl.Phys.Lett., 75(12): 1793.
144. Bühlmann S, Ewir B, Baborowski J and Muralt P (2002) Size effect in mesoscopic epitaxial ferroelectric structures: increase of piezoelectric response with decreasing feature size. Appl.Phys.Lett., 80(17): 3195.
145. Harnagea C, Alexe M, Schilling J, Choi J, Wehrspohn RB, Hesse D and Gösele U (2003) Mesoscopic ferroelectric cell arrays prepared by imprint lithography. Appl.Phys.Lett., 83(9): 1827.
146. Seifert A, Vojta A, Speck JS and Lange FF (1996) Microstructural instability in single-crystal thin films. J.Mater.Res., 11: 1470.
147. Alexe M, Scott JF, Curran C, Zakharov ND, Hesse D and Pignolet A (1998) Self-patterning nano-electrodes on ferroelectric thin films for gigabit memory applications. Appl.Phys.Lett., 73(11): 1592.
148. Alexe M, Gruverman A, Harnagea C, Zakharov CD, Pignolet A, Hesse D and Scot JF (1999) Switching properties of self-assembled ferroelectric memory cells. Appl.Phys.Lett., 75(8): 1158.
149. Waser R, Schneller T, Hoffmann-Eifert S and Ehrhart P (2001) Advanced chemical deposition techniques – from research to production. Integrated Ferroelectrics, 36(1-4): 3.
150. Schneller T and Waser R (2002) Chemical solution deposition of ferroelectric thin films – state of the art and recent trends. Ferroelectrics, 267: 293.
151. Szafraniak I, Bhattacharyya S, Harnagea C, Scholz R and Alexe M (2004) Self-assembled ferroelectric nanostructures. Integrated Ferroelectrics., 68: 279.
152. Chu MW, Szafraniak I, Scholz R, Harnagea C, Hesse D, Alexe M and Gösele U (2004) Impact of misfit dislocations on the polarization instability of epitaxial nanostructured ferroelectric perovskites. Nature Mater., 3: 87.
153. Szafraniak I and Alexe M (2003) Hydrothermal growth of nano-sized ferroelectrics. Ferroelectrics, 291: 19.
154. Grosso D, Boissiere C, Smarsly B, Brezesinski T, Pinna N, Albouy PA, Amenitsch H, Antonietti M and Sanchez C (2004) Periodically ordered nanoscale islands and mesoporous films composed of nanocrystalline multimetallic oxides. Nature Mater., 3: 787.
155. Ma W, Harnagea C, Hesse D and Gösele U (2003) Well-ordered arrays of pyramid-shaped ferroelectric $BaTiO_3$ nanostructures. Appl.Phys.Lett., 83(18): 3770.
156. Ma W and Hesse D (2004) Polarization imprint in ordered arrays of epitaxial ferroelectric nanostructures. Appl.Phys.Lett., 84(15): 2871.
157. Ma W, Hesse D and Gösele U (2005) Formation of ferroelectric perovskite nanostructure patterns using latex sphere monolayers as masks: an ambient gas pressure effect during pulsed laser deposition. Small. 1(8-9): 837.
158. Ma W and Hesse D (2004) Microstructure and piezoelectric properties of sub-80 nm high polycrystalline $SrBi_2Ta_2O_9$ nanostructures within well-ordered arrays. Appl.Phys.Lett., 85(15): 3214.
159. Bhattacharyya S, Chattopadhyay S and Alexe M (2003) Fabrication of isolated ferroelectric nanostructures. Mater.Res.Soc.Symp.Proc., 740: I10.3.1.
160. Calzada ML, Torres M, Ricote J and Pardo L (2009) Ferroelectric $PbTiO_3$ nanostructures onto Si-based substrates with size and shape control. J.Nanopart.Res., 11; 1227.

161. Bühlmann S, Muralt P and Von Allmen S (2004) Lithography-modulated self-assembly of small ferroelectric Pb(Zr,Ti)O3 single crystals. Appl.Phys.Lett., 84(14): 1614.
162. Clemens S, Schneller T, Vand der Hart A, Peter F and Waser R (2005) Registered deposition of nanoscale ferroelectric grains by template-controlled growth. Adv.Mater., 17: 1357.
163. Donthu S, Pan Z, Myers B, Shekhawat G, Wu N and Dravid V (2005) Facile scheme for fabricating solid-state nanostructures using e-beam lithography and solution precursors. Nano Letters, 5(9): 1710.
164. Nagarajan V, Stanishevsky A and Ramesh R (2006) Ferroelectric nanostructures via a modified focused ion beam technique. Nanotechnology, 17: 338.
165. Pan Z, Alem N, Sun T and Dravid VP (2006) Site-specific fabrication and epitaxial conversion of functional oxide nanodisk arrays. Nano Letters, 6(10), 2344.
166. Dawer M, Szafraniak I, Alexe M and Scott JF (2003) Self-patterning of arrays of ferroelectric capacitors: description by theory of substrate mediated strain interactions. J.Phys.:Condens.Matter., 15: L667.
167. Clemens S, Röhrig S, Rüdiger A, Schneller T and Waser R (2006) Variable size and shape distribution of ferroelectric nanoislands by chemical mechanical polishing. Small, 4: 500.
168. Landfester K (2001) The generation of nanoparticles in miniemulsions. Adv.Mater., 13(10): 765.
169. Torres M, Ricote J, Pardo L and Calzada ML (2008) Nanosize ferroelectric $PbTiO_3$ structures onto substrates. Preparation by a novel bottom-up method and nanoscopic characterisation. Integrated Ferroelectrics, 99: 95.
170. Rüdiger A, Schneller T, Roelofs A, Tiedke S, Schmitz T and Waser R (2005) Nanosize ferroelectric oxides – tracking down the superparaelectric limit. Appl.Phys.A., 80: 1247.

Chapter 5
Approaches Towards the Minimisation of Toxicity in Chemical Solution Deposition Processes of Lead-Based Ferroelectric Thin Films

Iñigo Bretos, M. Lourdes Calzada

Abstract

The ever-growing environmental awareness in our lives has also been extended to the electroceramics field during the past decades. Despite the strong regulations that have come up (RoHS directive), a number of scientists work on ferroelectric thin film ceramics containing lead. Although the use of these materials in piezoelectric devices is exempt from the RoHS directive, successful ways of decreasing toxic load must be considered a crucial challenge. Within this framework, a few significant advances are presented here, based on different Chemical Solution Deposition strategies. Firstly, the UV sol-gel photoannealing technique (Photochemical Solution Deposition) avoids the volatilisation of hazardous lead from lead-based ferroelectric films, usually observed at conventional annealing temperatures. The key point of this approach lies in the photo-excitation of a few organic components in the gel film. There is also a subsequent annealing of the photo-activated film at temperatures low enough to prevent lead volatilisation, but allowing crystallisation of the pure perovskite phase. Ozonolysis of the films is also promoted when UV-irradiation is carried out in an oxygen atmosphere. This is known to improve electrical response. By this method, nominally stoichiometric solution (i.e., a solution without PbO-excess) derived films with reliable properties, and free of compositional gradients, may be prepared at temperatures as low

Instituto de Ciencia de Materiales de Madrid, C.S.I.C., Sor Juana Inés de la Cruz, 3, Cantoblanco, 28049 Madrid, Spain
Phone: +34 91 334 9000
Fax: +34 91 372 0623
Email: ibretos@icmm.csic.es, lcalzada@icmm.csic.es

as 450°C. A Pt_xPb interlayer between the ferroelectric film and the Pt silicon substrate is observed in the heterostructure of the low-temperature processed films. This is when lead excesses are present in their microstructure. The influence of this interface on the compositional depth profile of the films will be discussed. We will evaluate the feasibility of the UV sol-gel photoannealing technique in fabricating functional films while fulfilling environmental and technological aspects (like integration with silicon IC technology). The second part of this chapter is focused on the solution chemistry involved in the synthesis of lead-based precursor solutions. The toxicity associated with the use of hazardous organic solvents (e.g., 2-methoxyethanol) in the CSD methodology is shown to be significantly lowered (even suppressed) by i) an entirely aqueous solution-gel route, and/or by ii) a diol-based sol-gel method. They use water and 1,3-propanediol respectively as solvents for the reactions. The physicochemical characteristics of the solutions (rheology, sol-network, aging phenomenon) obtained through these methods are observed to be highly influenced by the different synthetic routes followed, and the chemical reagents used for their preparation. Consequently, films with tailored microstructures and heterostructures are obtained that lead to a notable optimisation in the functionality of the material. This shows its potential for a wide range of microelectronic devices.

5.1 Introduction

The global trend towards the protection and sustainability of environment is considered to have begun with the United Nations Conference on Environment and Development (UNCED) in Rio de Janeiro (Brazil) in 1992. The so-called *Earth Summit* [1] brought together the main heads of State or Government, and transmitted an ambitious message to the rest of the planet. It said worldwide states should cooperate towards international agreements that respect the interests of all citizens and protect the integrity of the global environmental and developmental system. It was not long before the consequences of this agreement were also felt in the electroceramic materials field. D. Q. Xiao [2, 3] was the first author to propose the concept of *Ecoferroelectrics* (Environmentally Conscious Ferroelectrics) as a clear extension of the word *Ecomaterial* [4], previously adopted during the early 1990s. These materials should comprise a series of strategies on composition, processing technology and manufacturing process. This is to enable a small environmental load to be achieved during the life-cycle of the product. As a result, several environmental laws and regulations have come up, particularly in Europe, to support the aforementioned goals.

One of these directives is considered as probably the most disruptive event in the history of electronic manufacturing. It has also played an important role in research of the ferroelectric community in the past decade. Directive 2002/95/EC, *on the restriction of the use of certain hazardous substances in electrical and electronic equipment* [5], popularly known as the "RoHS directive", entered into force across all EU member-states (together with Iceland, Lichtenstein and Norway) on July 1, 2006. From that date, electrical and electronic equipment was not to contain lead, mercury, cadmium, hexavalent chromium, polybrominated biphenyls (PBB) or polybrominated diphenyl ethers (PBDE), because of the associated health and safety risks. Successful piezoelectric ceramics, which are widely used in sensors, actuators and other electronic components, are based on the lead zirconate [$Pb(Zr_xTi_{1-x})O_3$, PZT] and lead titanate ($PbTiO_3$, PT) perovskites. Therefore, the major environmental problem currently concerning electroceramic materials may be considered as the design and fabrication of microelectronic devices based on ferroelectric compositions containing lead. There has been a substantial effort to develop lead-free piezoelectric ceramics, such as the research on compounds based on alkali niobates ($LiNbO_3$, $NaNbO_3$, $KNbO_3$), barium titanate ($BaTiO_3$, $Ba(Ti,Zr)O_3$), bismuth sodium titanates $(Bi,Na)TiO_3$ and/or Aurivillius bismuth oxides (e.g., $SrBi_2Ta_2O_9$). However, these compositions display much smaller ferroelectric, pyroelectric and piezoelectric responses compared to lead-based ceramics [6]. No effective alternative to PZT was found until 2004 with the work of Saito *et al.* [7]. Despite this, *lead in electronic ceramic parts (e.g., piezoelectric devices)* is exempt from the RoHS directive. The original exemptions were published within the annex to this directive on 27 January 2003, before the work of Saito and others was published. They were promoted under situations where science or technology cannot yet offer alternative solutions, or when the practical benefits of the non-substitution outweigh the negative environmental and health impacts. In the case of lead in piezoelectric devices, it is not clear which was the main reason that motivated its exemption from the directive. Evidence of reliable lead-free piezoceramics has been proved at last [7]. But, the many amendments to the original RoHS directive until July 2006 neither examined nor modified point 7 of the directive, which referred to lead in electronic ceramic parts. On the other hand, several manufacturing companies [8] deal with this exemption on the basis that the lead concentration in a ferroelectric capacitor device (e.g., 200 nm-thick PZT layer) is similar to that found in living environment soil, and even of the order of that present in natural foods (~30 ppm).

This chapter has been structured without going into the technical aspects of the RoHS directive. We also assume that there is plenty of research on multifunctional polycrystalline ferroelectric materials containing lead. Here, several approaches based on Chemical Solution Deposition (CSD) methods are presented with the purpose of minimising the toxicity generally involved in the fabrication of lead-based ferroelectric thin films. One of them is particularly demonstrated as avoiding the volatilisation of hazardous lead into the atmosphere during the thermal processing of the films [9]. The key process of this approach consists in the photo-excitation of a few organic components in the gel film. Subsequently, the photo-activated film is annealed at low temperatures by means of the ultraviolet (UV) sol-gel photoannealing technique [10, 11]. This is also referred to as Photochemical Solution Deposition, PCSD. The use of such annealing temperatures (450°C) prevents the loss of lead from the nominal composition of the ceramic. It also obtains single-phase crystalline films with optimum dielectric and ferroelectric properties, which are compatible with the thermal budgets required in the CMOS process employed within the silicon semiconductor industry (≤500°C). We will also discuss the main effects produced by the low-temperature processing on the microstructure and heterostructure of the films (e.g., grain size, interfaces, and composition gradients). We will end with important remarks about the influence of excess lead in the precursor solutions on the crystallisation kinetics of lead-based multioxide compositions.

Another cause of concern associated with the technology of most CSD methods is the toxicity of the solvents and reagents normally used. These methods are considered low-energy processes compared to solid- and/or vapour-state reactions. This is one of the key advantages within the soft solution processing of advanced materials [12]. However, the majority of works reporting on chemical solution preparation (mainly through sol-gel and Metallorganic Decomposition, MOD) of multimetal precursors use organic solvents that involve both safety and health risks. Some of these compounds have been recognised as carcinogens (benzene), reproductive hazards (2-methoxyethanol, 2-ethoxyethanol), neurotoxins (toluene, n-hexane), or depressants of the central nervous system (xylene, buthoxyethanol). Their use in specific manufacturing facilities is sometimes forbidden. The second part of this chapter will focus on the chemistry of CSD methods. It will also discuss the capacity to decrease the toxicity of hazardous solvents by the use of two different synthetic approaches, i) an entirely aqueous solution-gel route [13, 14] and ii) a diol-based sol-gel route [15, 16]. In the first case, water is used as innocuous and exclusive solvent of the reactions. This reduces health risks and costs. In the case of the diol-based sol-gel process, mixtures of 1,3-propanediol and water are used as solvent. The toxicity of this diol is considerably lower than that of the compounds widely used in the preparation of ferroelectric thin films by CSD. We will evaluate the main features of both synthetic approaches on the physicochemical characteristics, of the precursor solutions (rheology, "spinnability") and on the properties of the derived films (microstructure, ferroelectricity).

Finally, in the results presented, attention has been paid to the calcium modified lead titanate system, $(Pb_{1-x}Ca_x)TiO_3$ (PCT). This is because films of this lead-based composition are a clear example of multifunctionality with a broad compositional range ($0 \leq x \leq 0.50$). Here, the material results are highly competitive for an extensive deal of ferroelectric and dielectric applications [16, 17, 18, 19, 20, 21]. Compositions with partial substitutions of Pb^{2+} by Ca^{2+} of 24 at% ($x = 0.24$), have been widely studied due to their excellent ferro-, pyro- and piezoelectric properties. For high Ca^{2+} contents ($0.40 \leq x \leq 0.50$), the system is known to behave as a relaxor-ferroelectric, showing diffuse ferro-paraelectric transitions, high values of the dielectric constant at room temperature, and a large non-linearity with voltage. Thus, the properties of the PCT films can be easily tuned depending on the Ca^{2+} content introduced in the perovskite composition. These results are of high technological interest, since a wide spectrum of different microelectronic components of different functionality can be fabricated with PCT thin films by merely modifying the partial substitution of Pb^{2+} by Ca^{2+}.

Table 5.1 Ultraviolet light sources (after [25]).

	Wavelength, nm (photon energy, eV)
Incoherent	
Ultrahigh-pressure mercury lamp (UHM)	270-460
High-pressure mercury lamp (HPM)	250-440
Low-pressure mercury lamp (LPM)	254 (4.9), 185 (6.7)
XeCl excimer lamp	308 (4.0)
KrCl excimer lamp	222 (5.6)
Xe_2 excimer lamp	172 (7.2)
Synchrotron radiation	Soft X-ray ~visible
Coherent	
XeCl excimer laser	308 (4.0)
KrF excimer laser	248 (5.0)
ArF excimer laser	193 (6.4)

5.2 Photochemical Solution Deposition as a Reliable Method to Avoid Lead Volatilisation during Low-Temperature Processing of Ferroelectric Thin Films

5.2.1 The UV Sol-Gel Photoannealing Technique

The ultraviolet (UV) sol-gel photoannealing technique was initially employed in the low-temperature processing of single oxide films, such as Ta_2O_5, ZrO_2 or SiO_2, [22, 23]. The decrease of the leakage current density associated with the capacitors after exposure of the films to UV-light, under oxygen atmosphere conditions [24], was observed. Sol-gel (or solution deposited) materials prepared in the thin-film form were soon considered as the ideal candidates for such studies. This was because the penetration depth of the UV photons is usually limited to the surface region of condensed matter. During the last decade, the effects of the UV-irradiation on sol-gel materials together with the development of new excimer UV lamps of narrow-band (108-354 nm) and high-density radiation (see Table 5.1), have opened novel and interesting ways in the material processing of polycrystalline thin films [25]. The use of the UV-irradiation on solution deposited (e.g., sol-gel) thin films is the main basis of what we have called Photochemical Solution Deposition (PCSD) methods.

The fundamental effects of UV-light on sol-gel films and their applications are summarised in Table 5.2. Energetic photons coming from UV lamps can lead to the thermal or electronic excitation of the substances in the sol-gel film. Thermal excitation is relatively important when the local temperature at the surface is increased by irradiation with intense UV-lights (phonon mode). However, UV-lamps predominantly exhibit electronic rather than thermal excitation (photon mode). In this section, we will deal with the capacity of UV light to produce electronic excitation of chemical bonds, thus enabling its use as a *pre-treatment for crystallisation* of ferroelectric sol-gel films. A broader view of the rest of phenomena produced by UV-light may be had from the comprehensive textbooks and manuscripts as [25, 26, 27, 28, 29, 30, 31, 32].

Table 5.2 Fundamental effects of UV light on sol-gel materials (after [25]).

	UV-IRRADIATION		
	electronic excitation		*thermal excitation*
	Activation of organic components	Activation of inorganic components	
Patterning	O	O	
Pre-treatment for annealing (low-temperature crystallisation)	O	O	
Reduction		O	
Densification	O	O	O
Crystallisation		O	O

Several multioxide ferroelectric compositions [33] have proved their competitiveness in novel electronic micro and nano devices (piezoelectric actuators, infrared sensors and ferroelectric memories) by making use of the "integrated ferroelectric" concept (the integration of an active ferroelectric layer into the silicon circuit). However, all of them share a handicap that prevents their compatibility with the silicon technology. The high processing temperature (over 600ºC) at which the film is normally treated may produce serious damage to the Si-based semiconductor substrate [34]. Because of this concern, different low-temperature CSD methods were born [35, 36, 37, 38, 39, 40, 41], including the UV sol-gel photoannealing technique. At the beginning, this technique was scarcely explored with ferroelectric multioxide films. For these films, UV sol-gel photoannealing was used just as a photo-lithographic method for patterning. As previously mentioned, the development of new excimer lamps together with the capacity of CSD methods for tailoring the solution chemistry (e.g., photosensitivity) have led to large efficiency of the UV sol-gel photoannealing technique. This is in the reduction of the processing temperature (pre-treatment for crystallisation) of multioxide ferroelectric films. The basis of this approach consists in the activation of chemical bonds of compounds in the gel film by the use of UV-light. This enhances the decomposition of the organic species within the film prior to the crystallisation of the material by a further thermal annealing at a relative low temperature. A schematic representation of the process is depicted in Fig. 5.1.

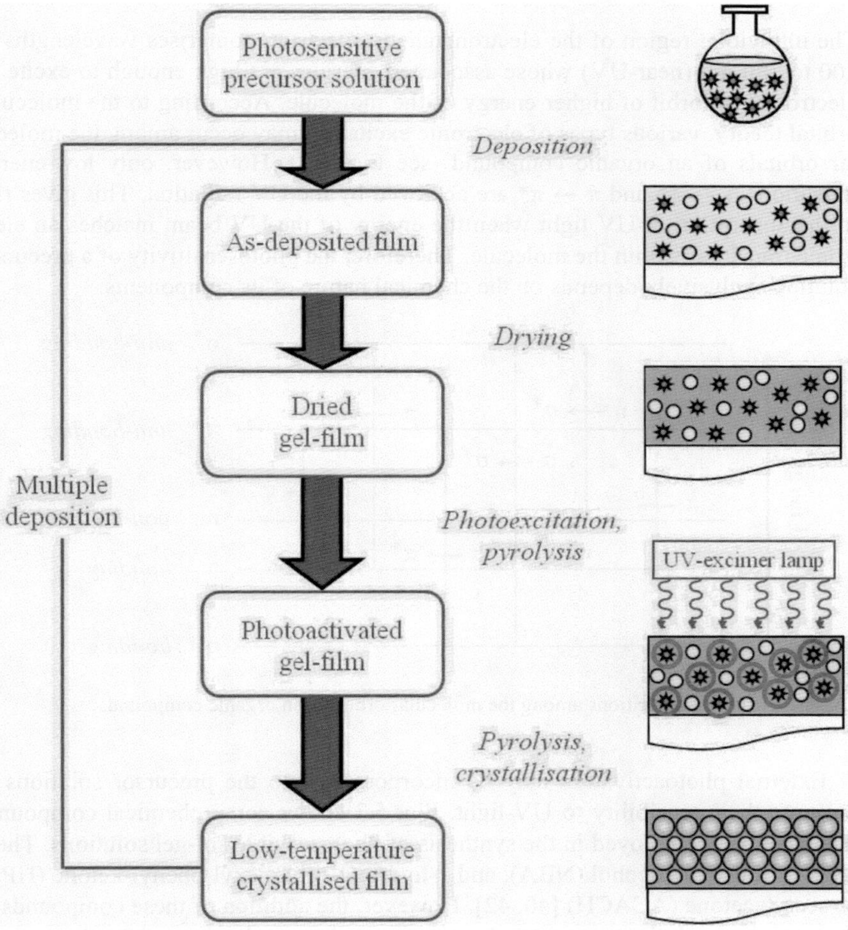

Fig. 5.1 Flow chart of the Photochemical Solution Deposition technique.

Thus, UV sol-gel photoannealing emerged (in 2004) as a successful approach to the optimum fabrication of multioxide ferroelectric thin films. This was at temperatures compatible with those required in the Si technology (450°C, from [10]). After further studies, the authors of this chapter [9, 11] have demonstrated that the processing temperature of the films may be lowered by this technique. This may also be considered an effective way to reduce lead volatilisation (emission of toxic volatiles), and to lessen energy consumption and thermal load of the process.

5.2.2 Photosensitivity of Precursor Solutions

The ultraviolet region of the electromagnetic spectrum comprises wavelengths of 200 to 380 nm (near-UV) whose associated energies are high enough to excite an electron to an orbit of higher energy of the molecule. According to the molecular orbital theory, various types of electronic excitation may occur among the molecular orbitals of an organic compound (see Fig. 5.2). However, only low-energy transitions n $\rightarrow \pi^*$ and $\pi \rightarrow \pi^*$ are achieved by the UV radiation. This gives rise to the absorption of UV light when the energy of the UV beam matches an electronic transition within the molecule. Therefore, the photosensitivity of a precursor solution exclusively depends on the chemical nature of its components.

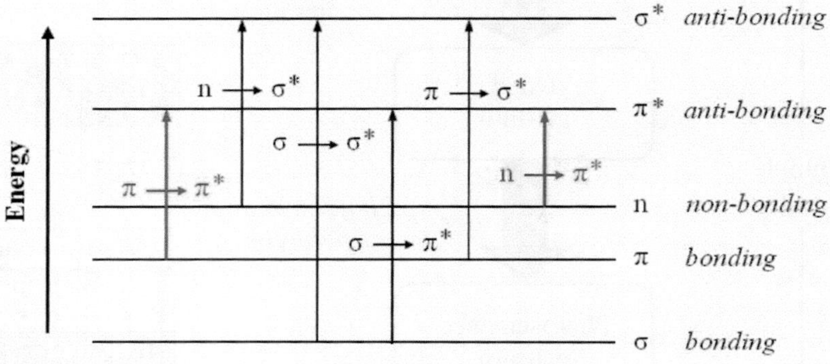

Fig. 5.2 Electronic transitions among the molecular orbits of an organic compound.

External photoactivators may be incorporated into the precursor solutions to promote their sensibility to UV-light. Fig. 5.3 shows some chemical compounds that have been employed in the synthesis of photosensible sol-gel solutions. These are o-nitrobenzyl alcohol (NBA), and 1-hydroxy-cyclohexyl-phenyl-ketone (HPK) or acetylacetone (ACACH) [40, 42]. However, the addition of these compounds to the precursor solutions can alter the decomposition mechanism of the derived gel films. This is either by increasing the organic weight fraction, which must be eliminated during pyrolysis, or by leading to inter- or intra-molecular interactions with the rest of the chemical reagents in the solution (metal precursors, solvents).

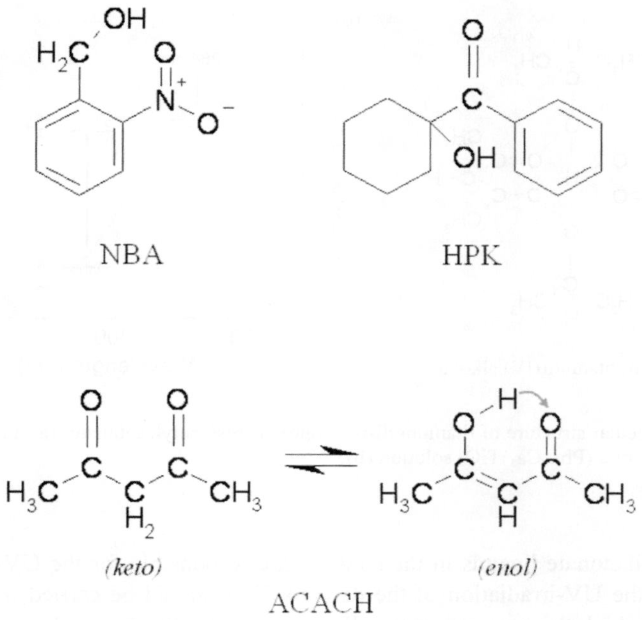

Fig. 5.3 Some examples of external photo-activating molecules.

The use of precursor solutions with compounds, which can be considered as intrinsic photoactivators, is the most effective way to overcome the drawbacks of the aforementioned external photo-reactive generators. Thus, solutions can be synthesised from metal precursors containing ligands that absorb UV-light [40, 43, 44]. In the sol-gel chemistry of most ferroelectric multioxide compositions, acetylacetone is usually employed as a complexing agent that reacts with the transition metal (e.g., Ti^{4+}, Zr^{4+}, Nb^{5+}) alkoxides. This leads to the formation of glycolated compounds, which are more stable towards hydrolysis than the initial metal precursors. These β–diketonate complexes have characteristic absorption bands in the near-ultraviolet region of the electromagnetic spectrum (200-380 nm) due to $\pi \rightarrow \pi^*$ electronic transitions [45]. The photo-excitation of this transition produces the dissociation of the chelating bonds [46]. Thus, an appreciable photosensitivity of these solutions is expected.

For the synthesis of $PbTiO_3$-based compositions, the use of a modified titanium alkoxide [titanium di-isopropoxide bis(acetylacetonate), $Ti(OC_3H_7)_2(CH_3COCHCOCH_3)_2$], which has two acetylacetonate groups, provides solutions that are inherently photosensitive to UV-light. Fig. 5.4 shows the molecular structure of this compound used in the synthesis of $(Pb_{1-x}Ca_x)TiO_3$ precursor solutions. The UV-absorption spectrum of such a solution is also depicted in the figure. A maximum at $\lambda < 275$ nm is observed, which is in the range where the acetylacetonate groups absorb UV-light.

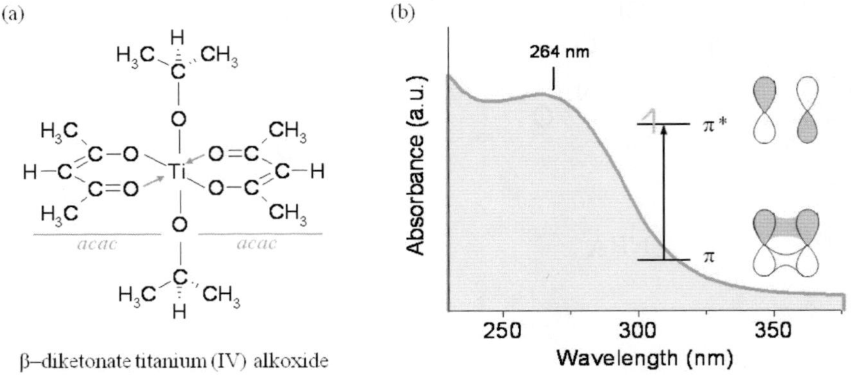

Fig. 5.4 Molecular structure of titanium di-isopropoxide bis(acetylacetonate) (**a**) and UV absorption spectrum of a $(Pb_{1-x}Ca_x)TiO_3$ solution (**b**).

The β–diketonate ligands in the solution are responsible for the UV-absorption. Therefore, the UV-irradiation of the derived films should be carried out at a temperature at which these compounds still remain inside the chemical system. Fig. 5.5 shows the thermal decomposition of gel powders derived from these photosensitive solutions studied by means of Differential Thermal and Thermogravimetric Analysis (DTA-TGA). The weight loss observed below 200°C is ascribed to the evaporation of water and alcohol (solvents) still entrapped within the gel network. The elimination of organic compounds in the system occurs at temperatures between 200°C and 500°C. Thus, 250°C is selected as the temperature at which these films should be irradiated (see solid line in the graph). At this temperature, the acetylacetonate groups responsible for UV-absorption are still present in the gel film. Therefore, their electronic excitation and subsequent activation of the organic compounds of the system will be produced by the UV-light.

The effectiveness of the UV-irradiation on the photoexcitation, and subsequent dissociation of the alkyl group-oxygen bonds of the organic components in the gel film, may be inferred from Fig. 5.6. The data correspond to the infrared spectra of two films deposited onto sapphire substrates and treated in air at 250°C for 240 seconds, with and without UV-irradiation. The spectrum of the non-irradiated gel film shows the stretching vibrations of the CH_2 groups (ν_{asym} ~2930 cm^{-1} and ν_{sym} ~2828 cm^{-1}) and of the CH_3 groups (ν_{asym} ~2967 cm^{-1} and ν_{sym} ~2898 cm^{-1}). This indicates the presence of C-H bonds belonging to residual organics. However, these bands have almost disappeared in the irradiated film. Here, UV-irradiation has produced the cleavage of the C-H bonds enhancing the elimination of the organics from the film at a relatively low temperature.

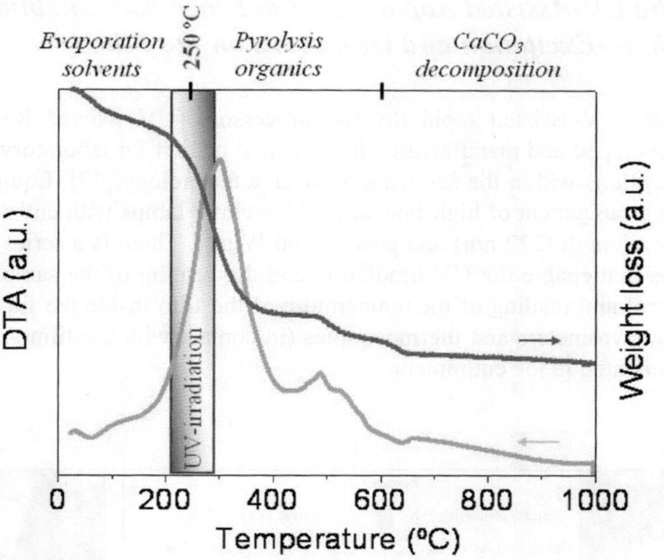

Fig. 5.5 DTA-TGA of $(Pb_{1-x}Ca_x)TiO_3$ solution derived gel powders.

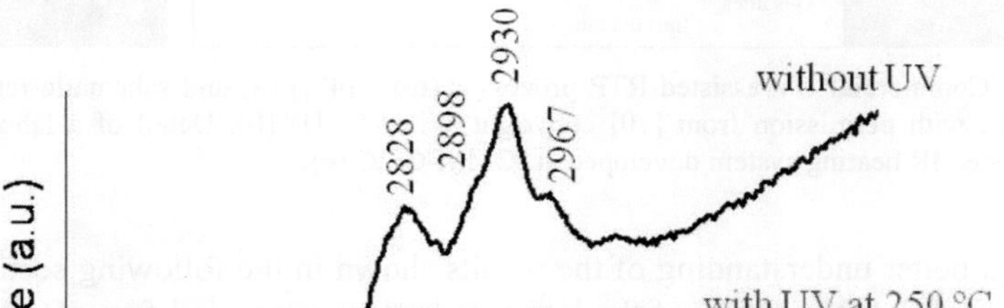

Fig. 5.6 FTIR spectra of the gel layers treated at 250°C with and without UV-irradiation.

5.2.3 *The UV-Assisted Rapid Thermal Processor: Enabling Photo-Excitation and Ozonolysis on the Films*

Commercial UV-assisted rapid thermal processors (UV-assisted RTP) are designed, developed and manufactured in the actual market for laboratory and industrial applications within the semiconductor area technology [47]. Equipment consists in an arrangement of high-intensity UV excimer lamps with emission set at a certain wavelength (222 nm) and power (300 W/m^2). There is a series of infrared (IR) lamps that enable the UV-irradiation and the heating of the sample (see Fig. 5.7). Control and reading of the temperature of the film inside the furnace is carried out by pyrometers and thermocouples (in contact with the film surface) usually implemented in the equipment.

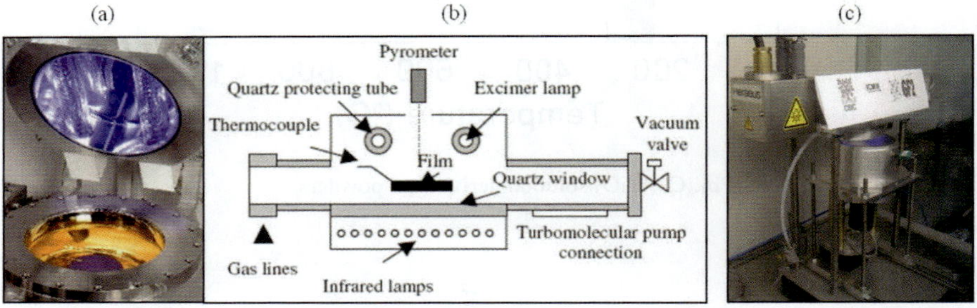

Fig. 5.7 Commercial UV-assisted RTP processor (from [47]) (**a**) and schematic representation (reprinted with permission from [10] copyright Wiley-VCH) (**b**). Detail of a laboratory-scale UV-assisted IR heating system developed at ICMM-CSIC (**c**).

For a better understanding of the results shown in the following sections, a description of the fabrication of the low-temperature processed ferroelectric films is advisable. Thus, films from a precursor solution leading to the $(Pb_{0.76}Ca_{0.24})TiO_3$ nominal composition were deposited onto $Pt/TiO_2/SiO_2/(100)Si$ substrates by spin-coating. This solution contained a 10 mol% of PbO excess. The as-deposited films were dried on a hot-plate in air at 150°C for five minutes prior to the introduction of the samples into the UV-assisted RTP processor. UV-irradiation of the gel films was carried out at 250°C for four minutes. The photo-activated films were subsequently crystallised at 450°C for an hour, both annealing processes in either air or oxygen local atmospheres (pressure of 1 atm) with a heating rate of 30 °C/s. Deposition, drying, irradiation and crystallisation were repeated four times to grow films with an average thickness of 200-250 nm.

By this method, two phenomena are promoted in the films by the use of UV-light. First, the activation of chemical bonds due to $\pi \rightarrow \pi^*$ electronic transitions would lead to the *photo-excitation* of the gel films. This enhances the quick decomposition and easier elimination of organic species from the system, thus advancing the crystallisation of the films. On the other hand, *ozonolysis* would be produced when the films are irradiated under oxygen atmosphere. Oxygen (O_2) is readily dissociated under UV-light, forming active oxygen species $O(^1D)$ and subsequent development of ozone (O_3). Ozone is a strong oxidant agent that produces the rapid combustion of the organic compounds. The active oxygen species can react with suboxides present in the film, thus improving the stoichiometry and decreasing the density of defects and oxygen vacancies [48, 49].

$O_2 + h\nu \rightarrow O + O(^1D)$ (1)

$O + O_2 \rightarrow O_3$ (2)

Ozonolysis effect; formation of active oxygen species (1) and ozone (2)

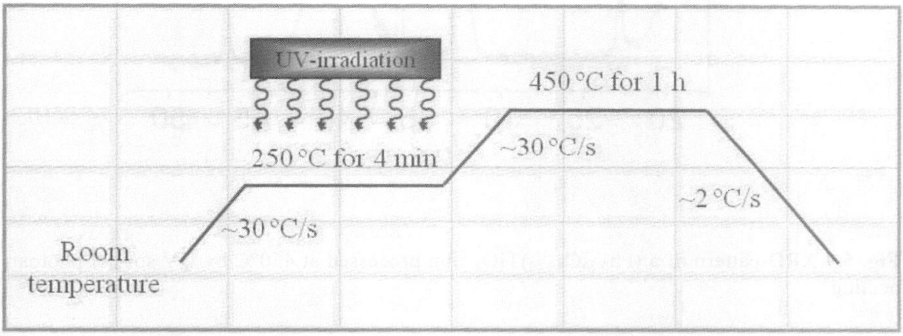

Fig. 5.8 Thermal budget used for the crystallisation of the lead-based ferroelectric films by UV sol-gel photoannealing.

5.2.4 Particular Features of the Low-Temperature Processed Films by UV Sol-Gel Photoannealing

5.2.4.1 Structure, Microstructure and Surface Morphology

Low-temperature processing of ferroelectric thin films has a major drawback. The crystalline oxide phase is usually obtained together with organic residuals and/or secondary non-ferroelectric phases. These lower and even cancel the electrical response of the material. Fig. 5.9 shows the crystal structure of a $(Pb_{0.76}Ca_{0.24})TiO_3$ ferroelectric thin film processed at 450°C by UV sol-gel photoannealing. Reflections corresponding to a single perovskite phase are observed in the XRD pattern, without a secondary phase (e.g., fluorite, pyrochlore).

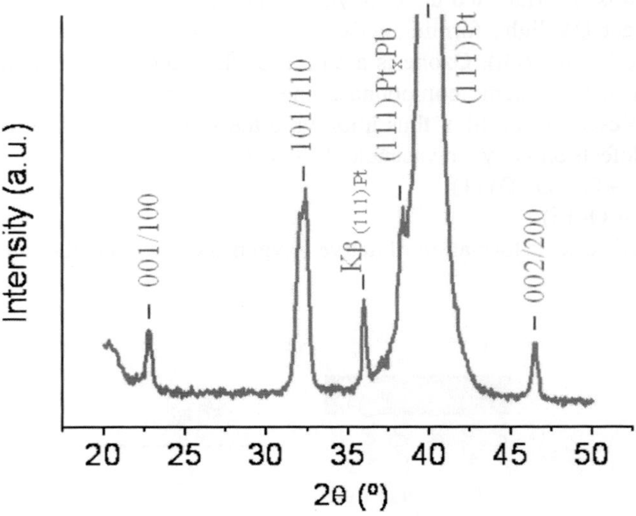

Fig. 5.9 XRD pattern of a $(Pb_{0.76}Ca_{0.24})TiO_3$ film processed at 450°C by UV sol-gel photoannealing.

Despite the strong <111> orientation of the Pt bottom electrode, a clear peak at $2\theta = 38.5°$ may also be inferred from the diffractogram of Fig. 5.9. This must be ascribed to the (111) reflection of an fcc Pt_xPb intermetallic phase [37, 50, 51, 52, 53, 54]. Among the scarce literature on the low-temperature fabrication of ferroelectric thin films, this is probably the most significant result obtained by several authors. The identification of a metastable intermetallic phase at the interface between the ferroelectric layer and the bottom metal electrode has been studied in PZT films deposited onto Pt silicon substrates during the early state of annealing. The composition of this transient phase, which varies depending on the annealing conditions, is referred to Pt_xPb. Its influence on the preferred orientation of the final PZT crystalline films has been thoroughly discussed [51, 52, 53]. The formation of this intermetallic phase arises from specific pyrolysis conditions to which the CSD-derived films are usually subjected. Thus, the presence of organic compounds in the system, which must be pyrolysed prior to the thin film crystallisation, may provide low local values of oxygen partial pressure on the films during this step. Under these circumstances, Pb^{2+} would be reduced to metallic Pb. This could further react with the Pt bottom electrode of the substrate forming a Pt_xPb alloy. The concentration of hydrocarbons in the system decreases as higher annealing temperatures are applied (600°C -700°C). The transient Pt_xPb phase vanishes from the films as a consequence of the converse effect of Pb oxidation. This incorporates the resulting Pb^{2+} ions into the perovskite crystalline structure. However, the use of a low crystallisation temperature (450°C) hampers this last process in the films processed by UV sol-gel photoannealing. Therefore, the Pt_xPb interlayer remains between the perovskite film and the Pt bottom electrode of the substrate. From the XRD pattern of Fig. 5.9, it may also be concluded that photoactivation of the gel film by UV-irradiation leads to the effective decomposition of organic species in the system even at temperatures as low as 450°C. No indications of intermediate decomposition products or rests of amorphous phases (e.g., organic residuals) are detected in the crystalline structure of the film.

The decomposition mechanism of solution precursors are known to be directly dependent on the chemical nature of the reagents (metal precursors, solvents) used in the preparation of multimetal oxide films by CSD methods. Thus, most of the reaction pathways occurring during heat treatment of organic precursors of low-valent metal cations (mainly short-chain carboxylates) lead to the formation of intermediate carbonates in the system prior to the total conversion into the final oxide phase [55, 56]. In the case of alkali and alkaline earth cations, carbonates formed from these elements normally decompose at high temperatures. They thus delay the onset of the crystallisation of the system. This has been, for example, the main cause ascribed to the far lower crystallisation rate (even 300 times slower) of solution-derived $(Ba,Sr)TiO_3$ films as compared to $PbTiO_3$ ones. Regarding the $(Pb_{1-x}Ca_x)TiO_3$ system, weight losses above 600°C have been observed in solution-derived powders that are associated with elimination of carbon dioxide, CO_2, coming from the decomposition of residual calcium carbonate, $CaCO_3$ (see previous Fig. 5.5). The formation of this compound arises from the two-step decomposition (between 200°C - 400°C) of the calcium acetate reagent [57], $Ca(OCOCH_3)_2$, as:

$Ca(OCOCH_3)_2 \rightarrow CaCO_3 + CH_3COCH_3$ (acetone)

$CaCO_3 \rightarrow CaO + CO_2$

Fig. 5.10 shows the binding energy spectra of the C 1s and O 1s core levels measured by X-ray Photoelectron Spectroscopy (XPS). This is during depth profiling of the $(Pb_{0.76}Ca_{0.24})TiO_3$ film processed at 450°C by UV sol-gel photoannealing. At the beginning of the analysis, the experimental signal of the C 1s level shows peak maxima at ~284.7 and ~287.8 eV (Fig. 5.10a). These binding energies correspond to adventitious carbon, and carbon, from carbonate groups respectively [58]. The photoemission peaks at these values of the C 1s core level disappear from the spectrum of the film after a minute of sputtering cleaning with Ar^+. In the case of the O 1s spectrum (Fig. 5.10b), the signal obtained during the first minutes of Ar^+ sputtering reveals two different peaks with maxima at ~529.9 and ~532.5 eV. The former peak is attributed to oxygen atoms of CO_3^{2-} (carbonate) groups and those of the PbO oxide [59] that is preferentially sputtered at the beginning of the Ar^+ bombardment. After a minute of sputtering and cleaning, only the later peak (~532.5 eV) corresponding to oxygen atoms of the $(Pb_{0.76}Ca_{0.24})TiO_3$ perovskite is observed in the O1s spectrum.

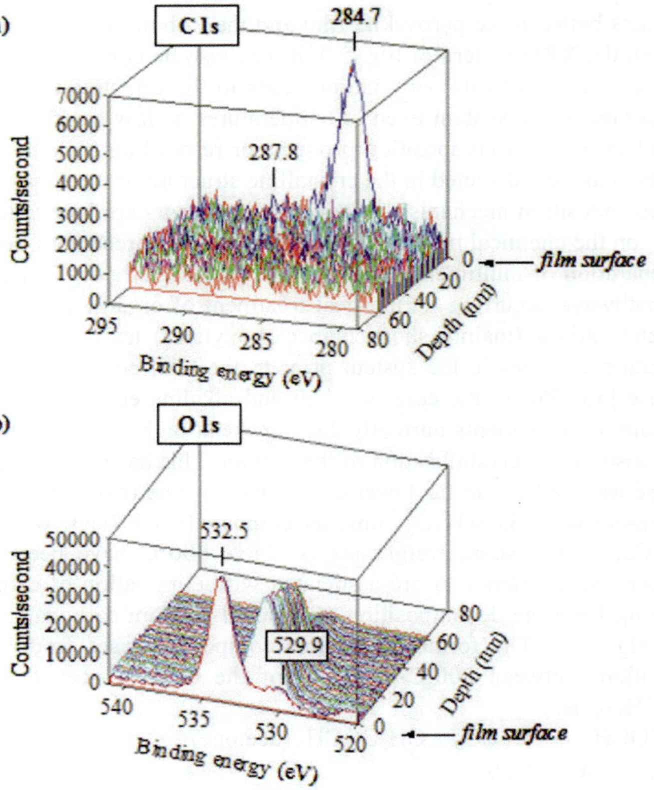

Fig. 5.10 Photoelectron C 1s **a** and O 1s **b** core levels measured in the low-temperature processed film (reprinted with permission from [9] copyright American Chemical Society).

The accumulation of organic contaminants from the atmospheric ambient is responsible for the presence of carbon-containing contaminants of different nature (e.g. graphitic carbon, hydrocarbons, carbonates) on the sample surface prior to the Ar^+ bombardment [60, 61]. After a minute of sputtering and cleaning, these carbonaceous contaminants are removed from the sample. The film surface appears free from carbon residuals. This last result rules out the possibility of having $CaCO_3$ still not decomposed within the bulk film and may not be detectable by XRD. However, it must be taken into account that thermal processes developed in bulk powders may slightly differ from those developed in thin film form. This is mainly due to the specific conformation of the latter and to the different thermal treatments of crystallisation carried out on both type of materials (conventional furnace or RTP). Despite this, films obtained by Photochemical Solution Deposition show no evidence of remaining residual organic species (mainly calcium carbonate). Even at temperatures at which these compounds would do remain in the system under conventional CSD processing.

Among the different driving forces that may govern a crystallisation event, annealing temperature (thermal energy available via heat treatment) is one of the most critical parameters defining the nucleation and crystal growth phenomena in solution-derived thin films. Thus, the microstructure and surface morphology of the films prepared by UV sol-gel photoannealing are strongly affected by the use of such a low processing temperature (450°C). Following the glass crystallisation theory proposed by several authors [62, 63, 64, 65], the transformation from the amorphous film into the crystalline ceramic is thermodynamically favoured at low crystallisation temperatures. This means that a high number of crystalline nuclei are formed at these temperatures to reduce the high free energy associated with the amorphous state. As a consequence, grain growth in the films is limited by the presence of numerous nuclei. On the contrary, differences between free energies of amorphous and crystalline states are reduced at high crystallisation temperatures. Under this situation, only the most energetically favourable crystals will nucleate, thus reducing the number of crystal nuclei in the film. This promotes the grain growth phenomenon at high temperatures resulting in the formation of larger grains. Fig. 5.11 shows SEM images of the $(Pb_{0.76}Ca_{0.24})TiO_3$ film processed at 450°C by UV sol-gel photoannealing. A cross-section with a homogeneous microstructure and grains with an average size of ~30 nm can be observed. As expected, the use of a low annealing temperature leads to the formation of nano-structured films (grain sizes <100 nm).

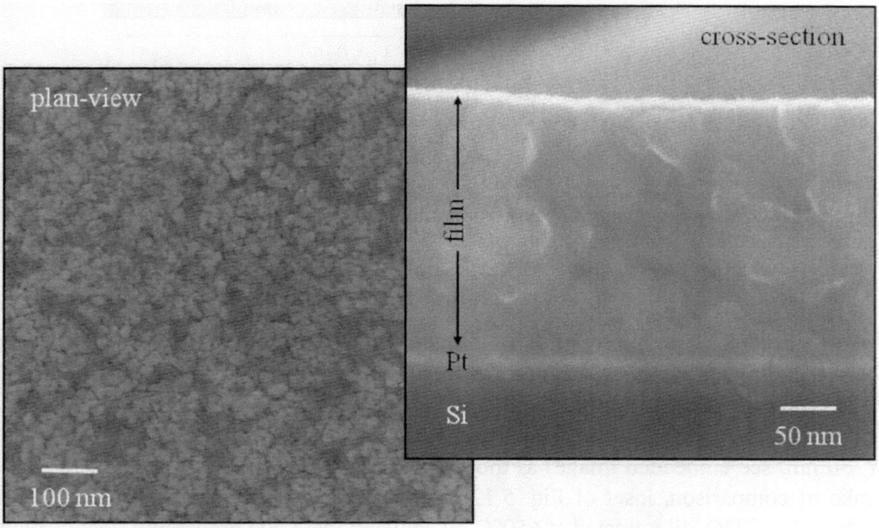

Fig. 5.11 SEM images of the low-temperature processed film.

Regions with different contrast are shown in the micrograph of the film surface, which could be associated with differentiated phases in the film (phases of different grain size, different amount of porosity, different composition, etc.). However, these differences in contrast were not seen in the previous cross-section image. They must be therefore ascribed to the irregular topography of the sample, and not to the presence of a secondary phase. This is also confirmed by observing the surface morphology of the sample by SFM (Fig. 5.12). Areas with variations in height up to around 15 nm are inferred from the line profile, whose aspect is rather similar to the SEM images obtained on the same surface.

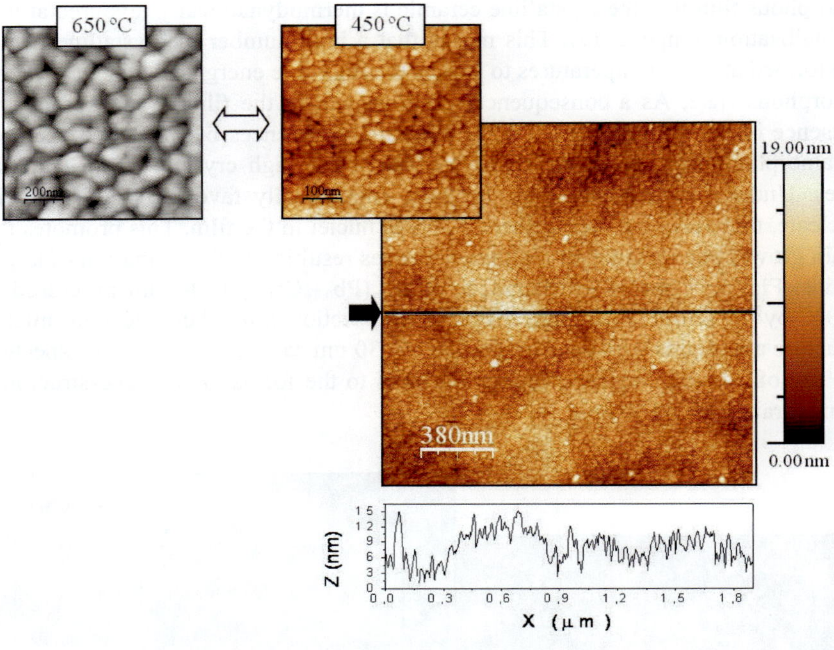

Fig. 5.12 Surface morphology by SFM of the low-temperature processed film. Inset shows an image obtained in a counterpart film processed by conventional CSD at 650°C (note the lower magnification of the photograph).

The irregular topography of these low-temperature processed films may be related to the non-planar nature of Pb-Pt intermetallic layers [52]. The rough surface of this interlayer would be transmitted to the rest of the deposited layers, thus producing the observed effect. On the other hand, similar grain sizes are measured from these images (~30 nm, see embedded image) as those obtained in previous SEM analyses. For the sake of comparison, inset of Fig. 5.12 in grey scale shows an image obtained on a $(Pb_{0.76}Ca_{0.24})TiO_3$ film treated at 650°C by conventional CSD processing (that is, without any UV-irradiation). Here, much larger grains are obtained due to improvement of grain growth mechanism by using high crystallisation temperatures.

5.2.4.2 Heterostructure and Compositional Depth Profile

The influence of the annealing conditions (mainly, the low-processing temperature) associated with the UV sol-gel photoannealing technique is shown to have drastic consequences on the lead volatilisation and composition of the PCSD derived thin films. Thus, information about the particular heterostructure (e.g., film-electrode interface) and compositional depth profile of the low-temperature processed films may be obtained by means of surface. This may also be done in-depth, by analytical techniques such as X-ray Photoelectron Spectroscopy (XPS) and Rutherford Backscattering Spectroscopy (RBS). In the case of the former, preferential loss of Pb is usually observed during the first minutes of Ar^+ sputtering in lead-based compositions. This is due to the high volatility (high vapour pressure) of this element. This prevents the quantitative analysis of the composition of the $(Pb_{0.76}Ca_{0.24})TiO_3$ films shown here. However, after several minutes of Ar^+ bombardment, a steady-state situation is reached in the sample from which a uniform compositional profile representative of the bulk film can be deduced. Fig. 5.13, corresponding to the photoelectron peaks intensity of the constituent elements of this film versus sputter time, reveals also the heterostructure of the substrate; the Pt electrode, the TiO_2 layer and the (100)Si substrate. Note how the signals of the respective Ti 2p and O 1s core levels increase after the maximum of Pt 4f level appears. This is ascribed to the TiO_2 buffer layer present in the substrate ($Pt/TiO_2/SiO_2/(100)Si$). It must be noticed that the atomic concentration of lead seems to slightly increase as the analysis is focused towards the inner layers of the ferroelectric film, that is, before reaching the Pt bottom electrode of the substrate. Inset of Fig. 5.13 depicts a straight line in the graph showing this tendency.

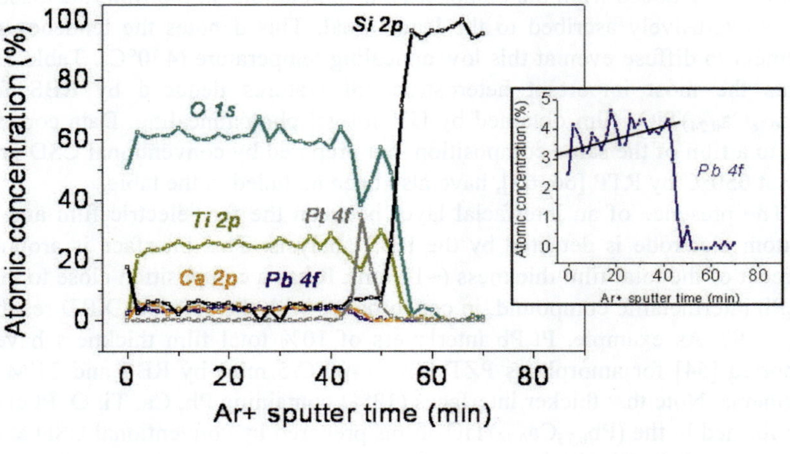

Fig. 5.13 Photoelectron peak intensities obtained by XPS during depth profiling of the low-temperature processed film. Inset shows individual Pb 4f signal.

The presence of chemically different interfacial layers and concentration fluctuations along the low-temperature processed films may also be studied by the RBS technique [52]. The use of light α-particles ($4He^{2+}$ incident ions) allows the detection of the sample elements with a low backscattering energy. The complete heterostructure of the film may be thus obtained (see Fig. 5.14a), denoting the bulk film, the Pt bottom electrode and the TiO_2 buffer layer of the sample.

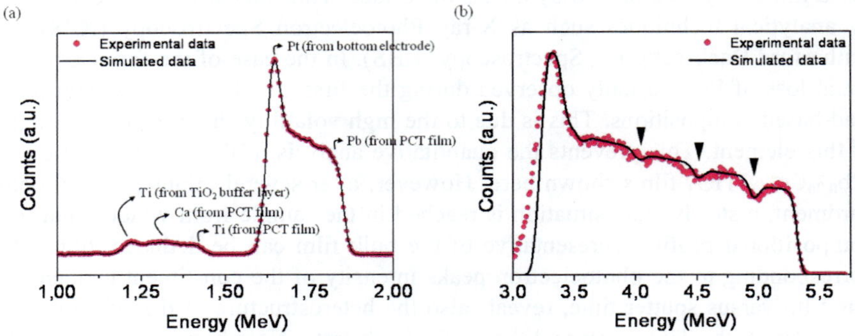

Fig. 5.14 Low-energy **a** and high-energy **b** RBS spectra of the low-temperature processed film by PCSD.

The use of a heavy energetic beam ($14N^{+}$ incident ions) provides measurements with a higher resolution. This makes it possible to infer the four constituent layers of the film. Thus, a clear compositional discontinuity from one layer to another can be observed from the experimental data of Fig. 5.14b, as indicated by arrow heads. As deduced from the compositional simulations in the film, the discontinuity is exclusively ascribed to the lead signal. This denotes the tendency of this element to diffuse even at this low annealing temperature (450°C). Table 5.3 presents the most important heterostructural features deduced by RBS for the $(Pb_{0.76}Ca_{0.24})TiO_3$ film obtained by UV sol-gel photoannealing. Data corresponding to a film of the same composition, but prepared by conventional CSD processing at 650°C by RTP [66, 67], have also been included in the table.

The presence of an interfacial layer between the ferroelectric film and the Pt bottom electrode is detected by the RBS analysis. This interface is around four percent of the total film thickness (~11 nm). It has a composition close to that of a Pt_xPb intermetallic compound, in concordance with the previous XRD results (see Fig. 5.9). As example, Pt_xPb interlayers of 10% total film thickness have been reported [54] for amorphous PZT films (400°C/5 min) by RBS and TEM measurements. Note that thicker interfaces (13%) containing Pb, Ca, Ti, O, Pt elements are formed in the $(Pb_{0.76}Ca_{0.24})TiO_3$ films prepared by conventional CSD at 650°C (see Table 5.3). The higher processing temperature at which these films are annealed is responsible of the larger diffusion of the constituent elements of the perovskite. This leads to the formation of thicker interfaces between the ferroelectric layer and the metal bottom electrode. On the other hand, the lower crystallisation temperatures used by UV sol-gel photoannealing decrease significantly such interdiffusion. Therefore, very fine interfaces are obtained in the films. The reduction in the formation of detrimental interfaces is expected to improve the electrical properties (permittivity, remnant polarisation) of the ferroelectric thin films. This is due to the approximation of an in-series capacitor (ferroelectric bulk film and non-ferroelectric interface), analogous to the "dead layer effect" [68]. One of the major failure mechanisms hindering ferroelectric switching-devices commercialisation is directly related to processes occurring in the ferroelectric layer-bottom metal electrode interface [69].

Table 5.3 Summary of RBS results of the low-temperature processed film by PCSD. Data corresponding to a film prepared at high temperature by conventional CSD are also included.

	Oxide film		Bottom interface	
Nominal composition of the precursor solution	Average elemental composition from RBS (1)	Percentage of film thickness from RBS	Composition from RBS (1)	Percentage of film thickness from RBS
$(Pb_{0.76}Ca_{0.24})TiO_3$ + 10 mol% PbO [0.836 Pb + 0.240 Ca + 1.000 Ti + 3.076 O]	PCSD processed film at 450 °C			
	0.84 Pb + 0.24 Ca + 1.00 Ti + 3.08 O*	96%	Pt_xPb	4%
	CSD processed film at 650 °C [66,67]			
	0.77 Pb + 0.24 Ca + 1.00 Ti + 3.01 O*	87%	Pb-Ca-Ti-O-Pt	13%

(1) Error of ~10 %. The formation of films with perovskite structure is assumed
* The atomic concentration of oxygen has been determined from the atomic concentrations calculated by RBS in the metal cations (Pb, Ca, Ti)

The compositional calculations obtained from the RBS analysis reveal a significant finding with crucial environmental and technological implications. The average elemental concentration simulated for the low-temperature processed film prepared by UV sol-gel photoannealing (0.84 Pb + 0.24 Ca + 1.00 Ti + 3.08 O) presents a 10 mol% of lead excess respect to the stoichiometric composition of the expected perovskite ($Pb_{0.76}Ca_{0.24}TiO_3$). Precursor solutions with a 10 mol% of PbO excess were synthesised for preparation of the films by PCSD described here (see previous section 2.3). This would mean that the aforementioned composition calculated by RBS is exactly the same as the nominal composition of the respective precursor solution (0.836 Pb + 0.240 Ca + 1.000 Ti + 3.076 O). As it would be explained in detail later (section 2.5), some lead excess is usually introduced during the processing of lead based-on ferroelectric compositions, either to the precursor solutions or to the metal targets in CSD and CVD methods, respectively. This is to compensate the loss of this element produced during the thermal annealing of the films at high temperatures. Thus, stoichiometric $(Pb_{0.76}Ca_{0.24})TiO_3$ films are obtained when the initial lead excess of the precursor solution is lost by evaporation during the annealing at 650°C (see data corresponding to conventional CSD films in Table 5.3). However, the low processing temperature applied by Photochemical Solution Deposition (450°C) avoids the volatilisation of this compound. The films therefore retain the 10 mol% of lead excess incorporated to the respective precursor solution. The ultraviolet sol-gel photoannealing technique proves thus to be a feasible method for the soft fabrication of advanced functional materials (ferroelectric ceramics). This reduces hazards volatilisation towards the atmosphere, besides lowering energy consumption and thermal load of the process. Apart from these ecological considerations, the results here shown could be extrapolated not only to other ferroelectric compositions based on the lead titanate perovskite, e.g. $Pb(Zr,Ti)O_3$ (PZT), $Pb(Mg,Nb)O_3$ (PMN), but also to ferroelectric lead-free compositions containing elements with a high vapour pressure such as volatile bismuth, e.g., $(Bi,Na)TiO_3$ (BNT), $SrBi_2Ta_2O_9$ (SBT). This results in a highly interesting technological aspect.

The distribution of lead through the bulk film is not completely homogeneous. A lead gradient with a concentration increasing from the top to the bottom is deduced from both simulated spectra of Fig. 5.14, in agreement with the earlier XPS analysis shown. Although stoichiometric control is a frequent advantage associated with the sol-gel process, compositional gradients are obtained in crystalline films prepared by CSD methods. Contrary to what are shown, lead enrichment is sometimes observed at the top surface of $PbTiO_3$-based films after annealing at conventional processing temperatures (600°C -700°C) [52, 66, 70, 71, 72]. There are several models that try to address the surface enrichment phenomenon of certain elements (mainly lead and bismuth) in multimetal-oxide thin films [73, 74, 75, 76]. One of them [73] postulates that the low local oxygen partial pressures, produced by the pyrolysis of the organic compounds in the gel film, may lead to the formation of elemental Pb^0 and/or Bi^0.) This would diffuse towards the reacting surface (exposed oxide surface) to get oxidised, analogous to the well-known metal corrosion event. Another explanation could be attributed to the evaporation of these high-vapour pressure elements during the annealing. This is known to occur at the more superficial layers of the film. To balance this loss, and therefore to reduce the chemical potential gradient generated, a transfer of material would occur in the condensed phase. Involved cations (Pb^{2+}, Bi^{3+}) would then migrate towards the surface of the film [77]. This would lead to up-graded films, that is, films with a concentration of lead/bismuth increasing from bottom to top. These are usually obtained when conventional annealing temperatures at high temperatures are used. A schematic representation showing this process is depicted in Fig. 5.15.

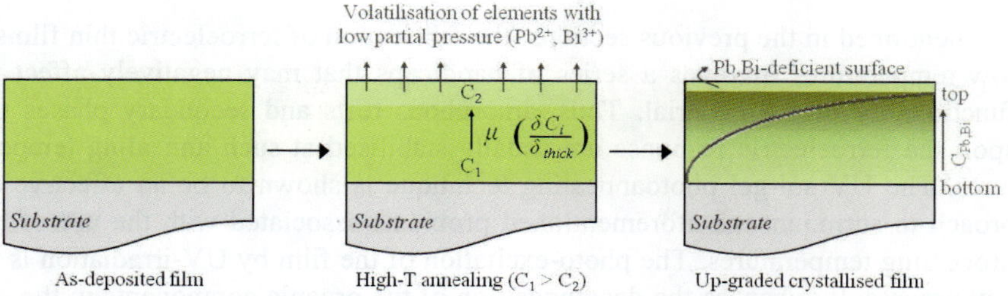

Fig. 5.15 Illustration of cation diffusion in a film driven by a chemical potential gradient.

On the other hand, the absence of lead volatilisation in the low-temperature processed films by PCSD means the diffusion of atoms driven by the loss of lead at the surface would not be expected. In the case of the $(Pb_{0.76}Ca_{0.24})TiO_3$ films here described, lead diffuses to the bottom electrode. This results in a compositional depth profile with a concentration of this element increasing from top to bottom. Fig. 5.16 compares the different heterostructures obtained for these films as a function of the process followed.

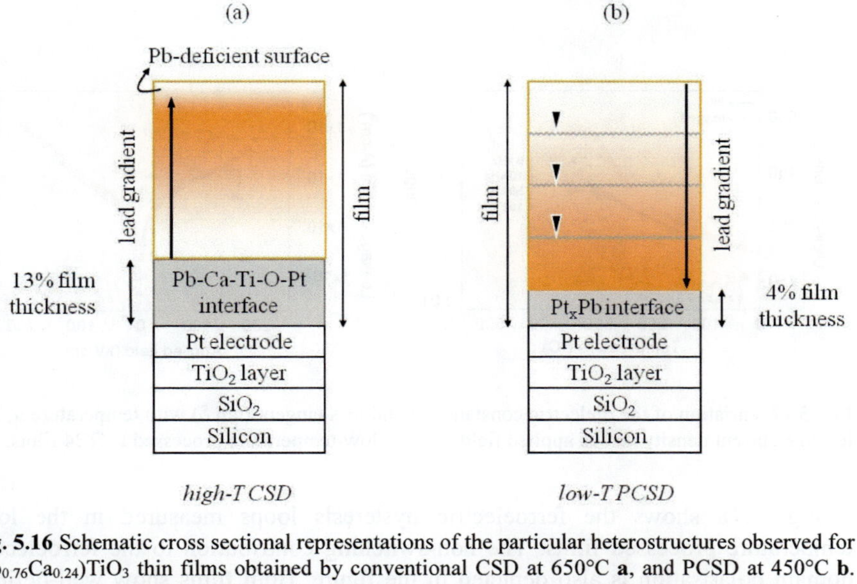

Fig. 5.16 Schematic cross sectional representations of the particular heterostructures observed for $(Pb_{0.76}Ca_{0.24})TiO_3$ thin films obtained by conventional CSD at 650°C **a**, and PCSD at 450°C **b**. Note: precursor solutions with a 10 mol% of lead excess were used for the preparation of the films.

5.2.4.3 Ferro-, Pyro-, Piezo-Electricity

As mentioned in the previous sections, the fabrication of ferroelectric thin films at low temperatures involves a series of handicaps that may negatively affect the functionality of the material. Thus, amorphous rests and secondary phases that spoil the ferroelectric response are usually stabilised at such annealing temperatures. The UV sol-gel photoannealing technique is shown to be an effective approach to surmount the aforementioned problems associated with the use of low processing temperatures. The photo-excitation of the film by UV-irradiation is the key process. It enhances the decomposition of the organic components in the gel-network, leading to the prompt formation of a pure crystalline oxide phase with suitable ferroelectric activity [10, 11].

The dielectric behaviour of the $(Pb_{0.76}Ca_{0.24})TiO_3$ films prepared at 450°C by PCSD is shown in Fig. 5.17. The results corresponding to films processed under air or oxygen local atmospheres are compared in the graphs. Although both films show a clear ferro-paraelectric transition at similar temperatures, higher values of the dielectric constant are obtained for the film processed in oxygen (see Fig. 5.17a). The dielectric loss (tan δ) of the films is close to one percent at room temperature. Their leakage current densities are between 10^{-9}-10^{-7} A/cm^2 for voltages up to ±5 V (see Fig. 5.17b), values which supports the feasibility for their use in capacitive applications.

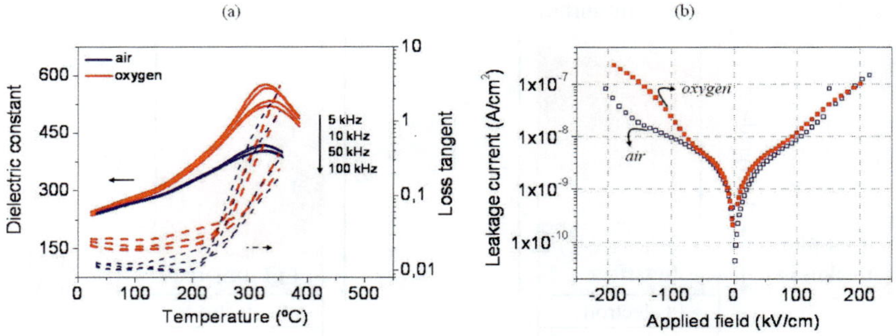

Fig. 5.17 Variation of the dielectric constant (k') and loss tangent (tan δ) with temperature **a**, and leakage current density versus applied field **b** of the low-temperature processed PCT24 films.

Fig. 5.18 shows the ferroelectric hysteresis loops measured in the low-temperature processed films. The non-switching contribution to the ferroelectric domain polarisation is also depicted in the figure. Both films show well-defined ferroelectric hysteresis loops with similar values of remnant polarisation (P_r ~11 μC/cm^2) and spontaneous polarisation (P_s ~14 μC/cm^2). The identical P_s values are obtained for the films, processed in either air or oxygen environment, indicates that a similar amount of active ferroelectric phase is formed in both types of films.

The perovskite crystallisation seems therefore not to be affected by the local atmosphere used in the thermal treatment. However, lower coercive fields are obtained for the film prepared in oxygen with respect to that prepared in air. This is in addition to a larger non-ferroelectric contribution (lower capacity and higher electric displacement, D) observed in this previous film (see Fig. 5.18b).

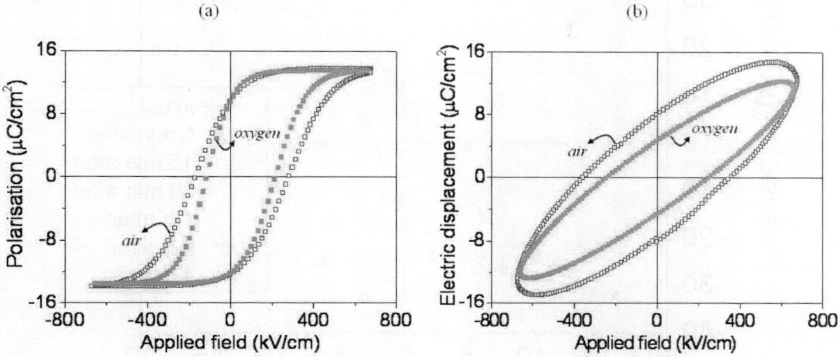

Fig. 5.18 Ferroelectric hysteresis loop **a** and non-switching contribution **b** of the low-temperature processed PCT24 films.

The differences in some of the electrical properties when the UV-irradiation and thermal treatment are carried out, in either air or oxygen, are ascribed to the mechanisms involved during the UV sol-gel photoannealing of the films. Apart from the *photo-excitation* of both films produced by the UV-light, the superior electrical properties observed in the films processed under oxygen in the local atmosphere must be attributed to the *ozonolysis* effect (see previous section 2.3). Thus, an increase in the dielectric constant, a higher capacity and lower conductivity (i.e., the lowering of the non-switching contribution to the ferroelectric domains), or the decrease in the coercive field for the film processed in oxygen, are all consequences of the lower density of defects and oxygen vacancies present in the material due to the ozonolysis of the film [11].

The evolution of the ferroelectric hysteresis loop with time is depicted in Fig. 5.19 for a film processed by UV sol-gel photoannealing at 450°C. After a few writing pulses, an initial square loop is obtained which appears *pinched* after short delay times. This effect may be related to the electrical stabilisation of domain walls by free charge carriers present in the sample (pinning effect). The short time needed for ferroelectric domain pinning in the low-temperature processed film would suggest a fast transport of these charged defects in the crystalline film. Other authors have also reported the possibility of the PbO excess in acting as a pinning centre for the domain walls in multioxide ferroelectric thin films [78, 79, 80, 81]. The low annealing temperature at which the films here described are processed by PCSD means two things: the charged defects (space charges, holes and/or vacancies not redistributed in the material at low temperatures) and PbO excess (present due to the non-volatilisation of this element) may be susceptible in acting as pinning centres for ferroelectric dipoles in the films.

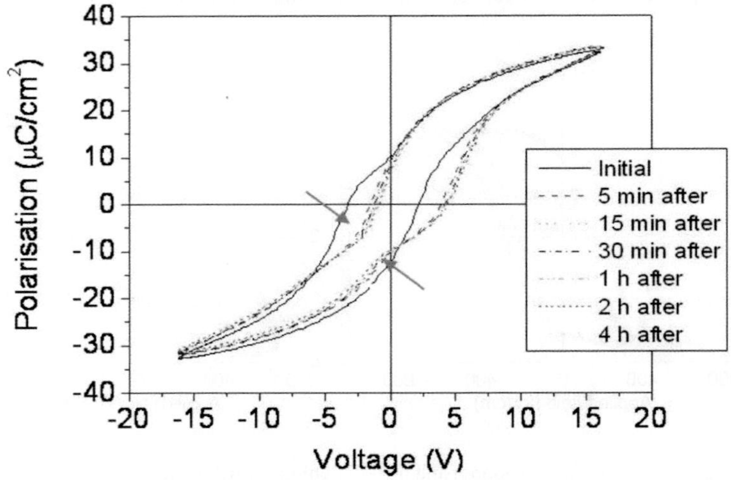

Fig. 5.19 Ferroelectric hysteresis loop with time of a low-temperature processed PCT24 film. Arrows indicate pinning of domain walls.

The presence of PbO excess and free charges in the material leads to the formation of internal fields in the low-temperature processed films. As a result, a large spontaneous pyroelectric coefficient is measured in these films (Fig. 5.20). The positive sign of this coefficient indicates that the polarisation is pointing to the top electrode (since it was configured as the anode in the measurement). This is in agreement with the positive voltage bias observed in the previous hysteresis loops of Fig. 5.18. Poling with negative pulses increases the pyroelectric coefficient without significant aging. The polarisation with positive writing pulses gives rise to a larger γ value but with a drastic aging of this coefficient with time. A stabilisation of the polarisation pointing to the top electrode is therefore obtained in the low-temperature processed $(Pb_{0.76}Ca_{0.24})TiO_3$ films by UV sol-gel photoannealing. The negative self-polarisation and the ferroelectric domain pinning observed in these films may be related to the PbO excess and free charges which remain in the material at this low annealing temperature (450ºC). Unlike the k' and P_r values, the pyroelectric coefficients of these films are comparable to those of films prepared by conventional CSD methods at higher temperatures [82]. The intrinsic microstructural features obtained in these low-temperature processed films (the strong self-polarisation with stabilisation at negative voltages), make them highly suitable non-switching based devices that need only a certain value of polarisation, such as pyroelectric and piezoelectric applications. For a more comprehensive understanding, a schematic representation of the particular microstructure expected for these films according to the former results is included in Fig. 5.20.

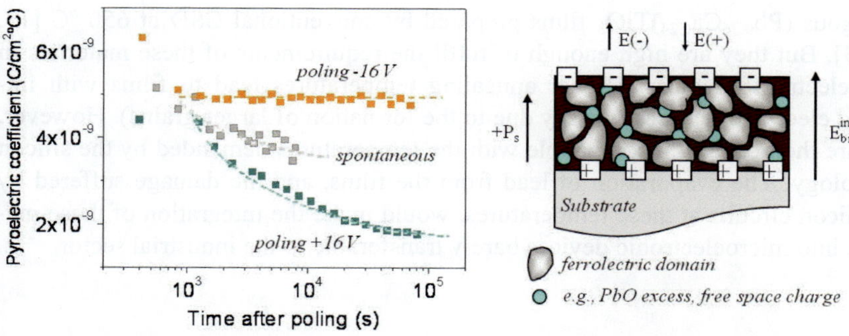

Fig. 5.20 Variation of the pyroelectric coefficient (spontaneous and poled with negative/positive pulses) of a low-temperature processed PCT24 film. A schematic representation is included for the better understanding.

Finally, the piezoelectric response of the films prepared by PCSD is analysed in Fig. 5.21. A local piezoelectric hysteresis loop is shown, denoting the piezoelectric activity of these films.

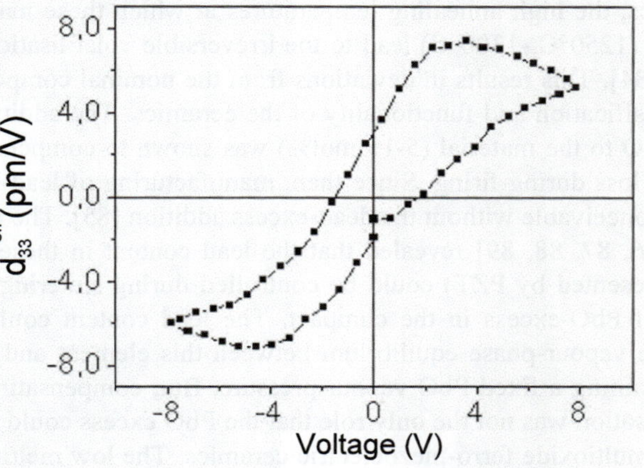

Fig. 5.21 Piezoresponse force microscopy of a low-temperature processed PCT24 film (reprinted with permission from [11] copyright Materials Research Society).

The results shown in this section prove the reliability of the UV sol-gel photoannealing technique for avoiding the volatilisation of hazardous lead during the low-temperature processing of ferroelectric thin films. They also prove the functionality of these materials in a wide range of microelectronic devices (DRAMs, NVFeRAMs, infrared sensors, MEMS). Some electrical properties of the low-temperature processed films shown here are inferior to those reported for analogous ($(Pb_{0.76}Ca_{0.24})TiO_3$ films prepared by conventional CSD at 650 °C [17, 40, 83]. But they are high enough to fulfil the requirements of these materials in microelectronic devices. Higher annealing temperatures lead to films with improved electrical features (mainly due to the formation of larger grains). However, they are theoretically incompatible with the temperatures demanded by the silicon technology. The evaporation of lead from the films, and the damage suffered by the silicon circuits at these temperatures, would make the integration of these materials into microelectronic devices barely transferable to the industrial sector.

5.2.5 Nominally Stoichiometric Solution-Derived Lead-Based Ferroelectric Films: Avoiding the PbO-Excess Addition at Last

5.2.5.1 Historical Background

The high activity (vapour pressure) of the PbO component has historically conditioned the research on lead-based ferro-piezoceramic compositions. Starting with bulk ceramics, the high annealing temperatures at which these materials are usually sintered (1250°C -1300°C) lead to the irreversible volatilisation of PbO from the system [84]. This results in deviations from the nominal composition and detrimental densification and functionality of the ceramics. The addition of a certain excess of PbO to the material (5-15 mol%) was shown to compensate for the undesired lead loss during firing. Since then, manufacturing of lead-based ceramics has been inconceivable without this lead-excess addition [85]. The early studies on this topic [86, 87, 88, 89] revealed that the lead content in these piezoceramics (mainly represented by PZT) could be controlled during sintering under specific conditions of PbO-excess in the compact. The lead content could also be controlled in the vapour-phase equilibrium between this element and an atmosphere powder containing a fixed PbO vapour pressure. But, compensating the lead loss from volatilisation was not the only role that the PbO excess could play during the sintering of multioxide ferro-piezoelectric ceramics. The low melting point of this compound (890°C), as compared to the majority of refractory oxides, allows the formation of a PbO-rich liquid phase at elevated temperatures. This enhances the mobility of the ions during the rearrangement of the solid network [90, 91, 92, 93, 94, 95, 96]. It produces high densification rates at the initial/intermediate stages of the process, and promotes the formation of the final crystalline structure at lower sintering temperatures and times. The addition of PbO excess in lead-based piezoelectric ceramics therefore has two reasons: to compensate the lead loss produced by volatilisation, and to enhance densification kinetics and particle transfer due to the formation of a PbO-rich liquid phase.

Although much lower processing temperatures (600°C -700°C) are employed in the thin film technology, volatilisation of PbO is also observed during the thermal processing of these materials. This may be explained mainly by two reasons. First, the particular thin-film conformation results in a high surface to volume ratio which can alter the physical conditions for the reactions (e.g., evaporation) occurring in the film. Thus, evaporation of lead oxide in films has been observed even at temperatures such low as 400°C [97], although this temperature is dependent on the ceramic composition and annealing conditions. Second, in the case of Chemical Solution Deposition (CSD) methods, residual organic species remain in the gel film after solution deposition. This must be pyrolysed prior to the crystallisation of the material at higher temperatures. The burning of these carbonaceous species requires oxygen consumption from the reaction atmosphere. This could lead to low local values of oxygen partial pressure on the films [98, 99]. The vapour pressure of the PbO component is still low at the usual processing temperatures of the films ($\sim 10^{-8}$ atm). But, elemental Pb can be easily volatilised at slighter reducing conditions ($\sim 10^{-6}$ atm) [100]. Both aforementioned characteristics are partly responsible for the inevitable volatilisation of PbO in thin films at normally annealing conditions. As in the case of bulk ceramics, an amount of PbO excess must be therefore added to the precursor solution to compensate this effect. Lead-deficiency from the nominal composition of most perovskite films usually results in the appearance of secondary crystalline phases (e.g., fluorite, pyrochlore). Their non-ferroelectric character leads to a deteriorated electrical response in the material [81, 101, 102]. The preferential formation of the fluorite and pyrochlore phases over the perovskite rests on the flexibility of the former to accommodate a lead-deficiency on their crystallographic structures. When the lead loss is compensated by incorporating a PbO excess, then the volume of the perovskite phase present in the films is known to increase [103]. However, it is not totally clear how this lead excess influences the crystallisation process of the lead-based ferro-piezoelectric films. It is generally accepted that the PbO excess in the amorphous film also leads to a decrease on the crystallisation temperature of the ceramic material [104, 105]. Until now, the formation of a high-temperature liquid phase (e.g., PbO-rich liquid phase), which would enhance crystal growth and densification in the same way as in equivalent bulk ceramics, has not been observed during the thermal processing of thin films. Some authors have pointed out that the lead excess present in the films may act as a flux for crystal growth [106]. It may also act as a network modifier, improving the mobility of the system [107], and enhancing crystallisation kinetics. It has been suggested that sharper x-ray diffraction peaks, obtained on crystalline films derived from precursor solutions, with a PbO excess, may be a possible indication for improved crystallinity in the material [102, 108, 109]. Anyway, additions of PbO excess to lead-based ferro-piezoelectric compositions (in both bulk ceramic and thin film conformations) represent the state-of-the-art in the preparation of these multifunctional materials. And one critical thought therefore flies over the heads of many researches today. The fabrication of full-crystalline lead-based ceramic films with optimum electrical properties seems to be unattainable without PbO excess addition.

Attending to the extensive literature, the former statement is almost true. To the best of the authors' knowledge, only one work reports on the preparation of homogeneous, single-phase perovskite films from stoichiometric solutions showing optimum ferroelectric properties. Two alternative firing strategies were used by Lefevre *et al.* [110] to control the lead stoichiometry of sol-gel derived PZT films annealed at 650°C. Both approaches are based on the replacement of the usual PbO-excess addition by increasing the PbO activity during firing. This was done either by using a PbO overcoat on the surface of the films [100], or by introducing PbO atmosphere powders in close proximity to the films [91]. The resulting films showed phase-pure perovskite structures with dense and homogeneous microstructures and nearly ideal ferroelectric properties. The detrimental effects of lead deficiency (secondary phases, poor electrical response) were not observed in the films. But, it is worthwhile to mention that compositional studies were not carried out to see if lead volatilisation from the nominal PZT composition was completely avoided.

One may deduce that if lead-based perovskite films were processed at lower annealing temperatures (< 500°C), volatilisation of PbO from the system could be avoided. This is theoretically true, but then additional problems arise that make the task not really straightforward. Thus, the use of low temperatures is known to promote the formation of detrimental secondary phases (e.g., fluorite, pyrochlore) in the films, due to the *kinetically limited crystallisation* concept [107, 111]. The amorphous-to-crystalline phase conversion is favoured at such temperatures. But, the driving force for crystallisation (via thermal treatment) is not large enough to allow the long-range diffusion of the elements within the system. The formation of the disordered fluorite phase over the more ordered structure of the perovskite is therefore encouraged. Generally, amorphous thin films are converted into the perovskite crystalline phase at higher temperatures (over 550ºC). In addition, low-temperature treatments of CSD-derived layers usually lead to films in which an incipient ferroelectric phase is immersed into a matrix containing organic residuals, still not decomposed. Both secondary non-ferroelectric phases and the rest of amorphous compounds strongly deteriorate the electrical response of the films.

In the previous section, the feasibility of the Photochemical Solution Deposition method in overcoming the aforementioned difficulties has been demonstrated. Thus, films processed at 450°C by UV sol-gel photoannealing avoid the volatilisation of lead from the ceramic composition. They also show single-phase crystalline structure and optimum ferro-, pyro-, and piezo-response. As expected, the use of such a low processing temperature prevents the lead evaporation from the system. The photo-excitation of the gel film under a high-intensity UV-irradiation is responsible for the rapid combustion of the organic species, and the prompt formation of the desired crystalline oxide phase.

Based on these results, we will evaluate the fabrication of lead-based ferroelectric films from a nominally stoichiometric solution (that is, a precursor solution without PbO excess) by means of this technique. It is believed that as long as lower annealing temperatures on ferroelectric thin films are to be used, a higher amount of PbO excess must be incorporated to the system. This is to enhance the crystallisation of the ceramic material [112]. Therefore, the effect of lead excess on the nucleation and crystal growth phenomena of the films is expected to be even more dramatic at such low temperatures.

5.2.5.2 Nominally Stoichiometric Solution-Derived Thin Films

The concept of *nominally stoichiometric* solution proposed here refers to a precursor solution leading to the $(Pb_{0.76}Ca_{0.24})TiO_3$ nominal composition. In this, no excess of PbO has been incorporated into the medium (previous results of section 2.4 correspond to films derived from a precursor solution containing a 10 mol% of lead excess). The corresponding films were prepared according to the schematic process shown in Fig. 5.8. In this case, results corresponding to $(Pb_{0.76}Ca_{0.24})TiO_3$ films processed by UV sol-gel photoannealing, using oxygen local atmosphere, would be evaluated. As shown in the previous section, ozonolysis leads to films with superior electrical properties. This is due to the lowering in the density of defects and oxygen vacancies in the material. We would like to clarify that differences in some properties would be shown for these films with respect to those of previous section 2.4. These must be ascribed to the different UV-assisted RTP processors used for their respective fabrication.

The microstructure of the nominally stoichiometric solution-derived films crystallised at 450°C consists of small grains with an average size of ~30 nm (Fig. 5.22a). It thus displays a concept of limited grain growth due to the formation of a high number of crystalline nuclei, promoted at these low annealing temperatures (see section 2.4.1). The XRD pattern of these films (Fig. 5.22b) shows the presence of a single perovskite phase. This is without an indication of secondary phases (fluorite, pyrochlore) or amorphous compounds in the crystalline structure.

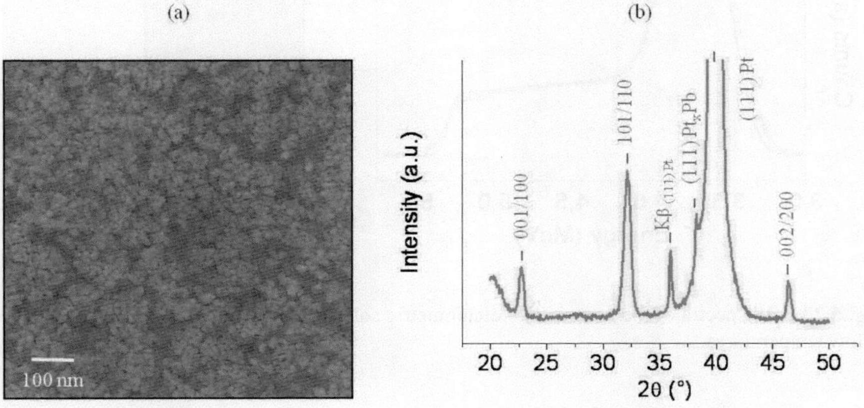

Fig. 5.22 SEM micrograph **a** and XRD pattern **b** of a nominally stoichiometric solution-derived film crystallised at 450 °C by UV sol-gel photoannealing.

The results exposed here demonstrate for the first time that phase-pure, crystalline films of ferroelectric multioxide compositions based on lead can be obtained at low crystallisation temperatures (450°C). This is without using lead excess, neither in the solution nor in the engineering of the process (by using either film overcoats or atmosphere powders of PbO). The photo-excitation, joined to the ozonolysis of the films produced by the use of intense UV-irradiation, is shown to overcome the historical dependence of PbO excess additions. This is for enhancing the crystallisation kinetics of lead-based compositions. Analogous to the PCSD-derived films containing a 10 mol% excess of PbO, a diffraction peak corresponding to the (111) reflection of the Pt_xPb intermetallic phase may be inferred from the XRD pattern of Fig. 5.22b. Note that the intensity of this peak is now lower when compared to the former films with lead excess (see Fig. 5.9). The formation of this Pt_xPb alloy was ascribed to the reduction of Pb^{2+} to metallic Pb and subsequent reaction with the Pt bottom electrode. The low temperature used for the crystallisation of the films prevents the re-oxidation of elemental lead. Therefore, the Pt_xPb interlayer remains in the heterostructure of the films.

The RBS spectra obtained in the nominally stoichiometric solution-derived films can be observed in Fig. 5.23. The heterostructure deduced for these films is also depicted in the figure.

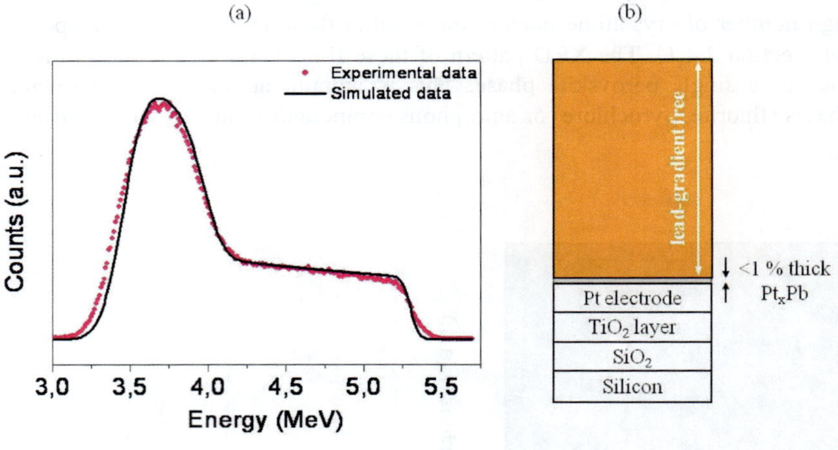

Fig. 5.23 RBS spectra of the nominally stoichiometric solution-derived films prepared by PCSD at low temperatures.

As expected, an average elemental concentration of 0.76 Pb + 0.24 Ca + 1.00 Ti + 3.00 O is calculated from the simulation curve of the films. This concentration, within the range of experimental error, is equivalent to the nominal composition of the respective precursor solution, $(Pb_{0.76}Ca_{0.24})TiO_3$. This denotes that evaporation of lead has not occurred in these films. As in the case of the films with lead excess, a Pt_xPb interlayer between the film and the Pt electrode is required to fit both experimental and simulated curves from Fig. 5.23. However, the thickness of this interface results in now less than the one percent of the total film thickness (~2 nm). This is in contrast to the three percent obtained ion the films with lead excess. This really low value leads us to believe that the formation of the Pt_xPb interlayer in the low-temperature processed films is practically suppressed, when no PbO excesses are present in their microstructure during the fabrication process. The high texture of the compound means the Pt_xPb intermetallic is clearly detected by XRD in the nominally stoichiometric solution-derived film despite its rather low concentration in the sample [53]. The decrease in the intensity of the (111)Pt_xPb peak in these films, with respect to the films with a lead excess, suggests that the formation of this alloy is significantly minimised in the former films. These results would indicate that reduction of lead in these films, and subsequent formation of the Pt_xPb alloy, is mainly produced from the amorphous PbO phase. This comes from the PbO excess addition and not from the $(Pb_{1-x}Ca_x)TiO_3$ perovskite formed. When nominally stoichiometric solution-derived films are used, the lack of a single PbO phase coexisting with the perovskite hinders the formation of this intermetallic layer. This is under such pyrolysis conditions, as shown by the RBS analyses. This phenomenon is analogous to the PbO vapour pressure calculations shown in Fig. 5.24 [113].

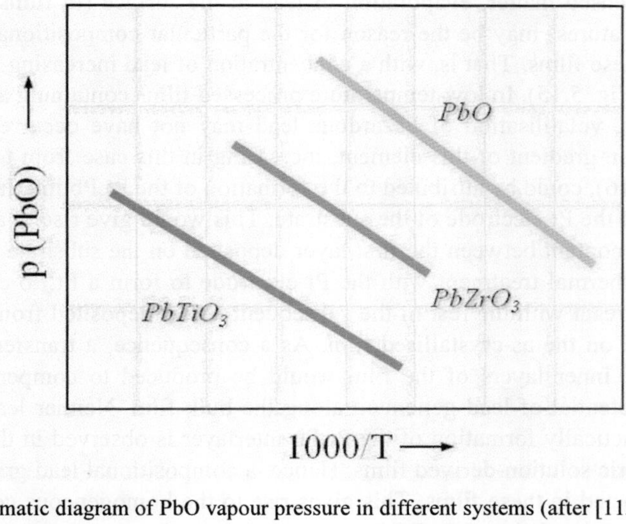

Fig. 5.24 Schematic diagram of PbO vapour pressure in different systems (after [113]).

The equilibrium vapour pressure of PbO over certain lead-based compositions such as $PbTiO_3$ or $Pb(Ti,Zr)O_3$ is smaller than over PbO. This means that when evaporation of lead occurs in ceramic compositions at high annealing temperatures, this firstly comes from the amorphous PbO phase. This is added as a compensating excess for the stoichiometry of the sample and not from the lead-based perovskite. The conditions here imposed by UV sol-gel photoanneling avoid the volatilisation of lead from the films containing a PbO excess. This does not amount to the reduction of this element to form the Pt_xPb intermetallic compound.

If higher annealing temperatures were applied, reoxidation of this elemental lead and subsequent volatilisation would occur in the films. In the nominally stoichiometric solution-derived films, however, the absence of the PbO excess in their microstructure means that reduction of lead should be produced from the $(Pb_{1-x}Ca_x)TiO_3$ perovskite. This is in principle not favoured under such annealing conditions. As a consequence, the formation of the Pt_xPb interlayer is drastically minimised in these films (<1 % total film thickness).

A homogeneous compositional profile is deduced by RBS for the films (comprised of four layers) shown here, in contrast to the compositional discontinuity in the lead signal observed for the low-temperature processed films with a PbO excess. This supports the lack of compositional variations within the bulk material. A lead gradient with a concentration increasing from top to bottom was, on the contrary, found in the PCSD films containing a 10 mol % of PbO excess. The tendency of lead to diffuse is, in most cases, responsible for the strong compositional gradients observed in lead-based ferroelectric films prepared by multilayered CSD methods. This produces variations on the physical properties of the material along the film thickness that can deteriorate the functionality of the device [114]. Following the theory of diffusion of elements driven by chemical potential gradients in the condensed matter, evaporation of lead at the surface (in films annealed at high temperatures) may be the reason for the particular compositional profile observed in these films. That is, with a concentration of lead increasing from bottom to top (see Fig. 5.15). In low-temperature processed films containing a 10 mol% of PbO excess, volatilisation of hazardous lead may not have occurred. Here, the concentration gradient of this element, increasing in this case from top to bottom (see Fig. 5.16), could be attributed to the formation of the Pt_xPb interlayer between the film and the Pt electrode of the substrate. This would give rise to an unbalance on the lead content between the first layer deposited on the substrate and reacting during the thermal treatment with the Pt electrode to form a Pt_xPb compound. It would also react with the rest of the subsequent layers deposited from the precursor solution on the as-crystallised film. As a consequence, a transfer of material towards the inner layers of the film would be produced to compensate for the chemical potential of lead generated along the bulk film. Neither lead volatilisation nor practically formation of the Pt_xPb interlayer is observed in the nominally stoichiometric solution-derived films. Hence, a compositional lead gradient would be not expected in these films. This gives rise to the homogeneous compositional profile previously deduced by RBS.

The effect of the lead excess content on the electrical properties of the low-temperature processed films by UV sol-gel photoannealing may be evaluated from Fig. 5.25. A clear ferro-paraelectric transition is observed for three films displaying different amounts of PbO excess in their composition. The peak maxima of the curves corresponding to films containing 0 and 5 mol% of PbO excess are located at the same temperature. This denotes that the remaining lead has not been incorporated into the $(Pb_{0.76}Ca_{0.24})TiO_3$ perovskite. This would lead to compositionally defective crystals. Shifts in Curie point, due to partial incorporations of PbO into perovskite lattices, are reported for several lead-based ferroelectric ceramics [94,

95]. The slight shift observed in the transition temperature of the film containing a 10 mol% of PbO excess is more likely because of other considerations (e.g., mechanical stresses). Deviations in the nominal composition of the film may not be causing it. The incorporation of PbO into the perovskite lattice of this film would displace its transition temperature toward higher temperatures. This would be in the direction of that of pure $PbTiO_3$, and not to lower values as observed in the graph of Fig. 5.25a. This implies that the PbO excess in the low-temperature processed films must be present as an amorphous phase. This is probably uniformly dispersed in the microstructure of the film with a size small enough not to be apparent to XRD or SEM techniques. The formation of a continuous path of segregated PbO through the film would lead to highly conductive samples. The evidence of which is not supported by the electrical results here shown.

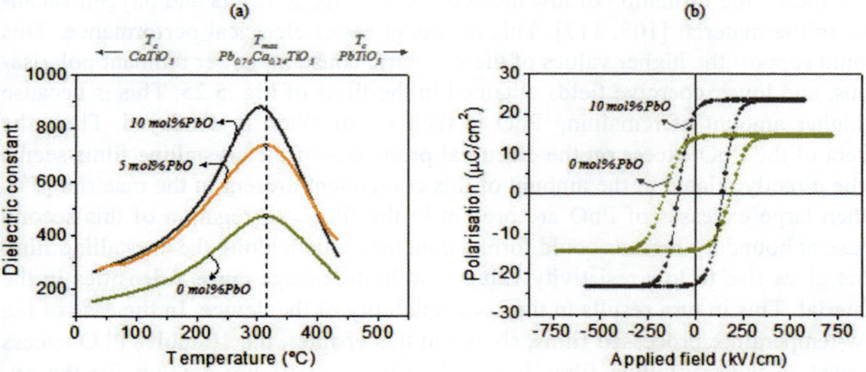

Fig. 5.25 Electrical properties of low-temperature processed PCT24 thin films prepared by UV sol-gel photoannealing. Differences are observed depending on the PbO excess present in the crystalline films.

On the other hand, the compositional gradient obtained in the PCSD films with PbO excess may be related to the internal bias field inferred from the corresponding ferroelectric loop (see different E_c values with voltage sign of Fig. 5.25b). The presence of internal fields in the sample may lead to a faster imprint of the ferroelectric polarisation. There are drastic consequences associated with the use of low annealing temperatures, and null amounts of lead excess (poor crystallinity, formation of fluorite/pyrochlore secondary phases). Despite this, a remarkable ferroelectric response, with no internal bias fields, is obtained in these films (P_r = 14 $\mu C/cm^2$, E_c = 175 kV/cm). This has not been reproducible in the past for any ferroelectric lead-based material processed under similar conditions, like thermal annealing at 450°C and non-incorporation of PbO excess. For example, an addition of 30 mol% of lead excess to the solutions is seen as necessary to crystallise PLZT films at temperatures near 500°C [105]. Conversely, annealing temperatures of 650°C with strategic approaches like PbO atmosphere powders and PbO overcoats are required to obtain single-phase ferroelectric PZT films from stoichiometric solutions with no lead excess [110]. The historical incompatibility between low processing temperatures and an absence of PbO excess has been finally overcome with the successful implementation of the UV sol-gel photoannealing technique.

We can see that improved dielectric and ferroelectric properties are obtained in the low-temperature processed films when a certain amount of lead excess is present within their microstructure. However, controversy arises when the effect of PbO excess on the electrical features of ferroelectric lead-based ceramics is addressed [115, 116]. Some authors claim [78, 79, 81] that the remaining PbO may act as a pinning centre, thus deteriorating the electrical response of the material. Such an effect was observed in the hysteresis loop behaviour with time of the low-temperature processed films containing a 10 mol% of PbO excess (see Fig. 5.18). But, the remnant PbO excess is also believed to improve the interfaces of the films by avoiding the formation of low dielectric space charge layers and oxygen vacancies in the material [105, 117]. This results in better electrical performance. This would support the higher values of the dielectric constant, larger remnant polarisations, and lower coercive fields obtained in the films of Fig. 5.25. This is because a higher amount of remaining PbO in their composition is displayed. Thus, the effect of the PbO excess on the electrical properties of the crystalline films seems to be directly related to the amount of this component present in the material [79]. When large excesses of PbO are present in the films, segregation of this second phase at boundary regions could form a conductive path along the crystalline film. This gives rise to low resistivity values and high leakage current densities in the material. This in turn results in the poor reliability of the device. In the case of the low-temperature processed films, shown in this chapter, the 10 mol% PbO excess present in the crystalline films is found to be an optimum amount for the enhancement of their dielectric and ferroelectric characteristics. This is due to the aforementioned physical effects produced in the films. The volume of this PbO excess, however, is expected to vary as a function of the ceramic composition studied and/or the experimental conditions used.

5.2.6 Remarks

The UV sol-gel photoannealing technique (Photochemical Solution Deposition) is revealed as a reliable method to fabricate lead-based perovskite films. This is with optimum electrical properties at temperatures (450°C) well below to those conventionally used in the thermal annealing of these materials (>600°C). The interest in this technique is two-fold. First, it avoids the emission of hazardous lead into the atmosphere during the film processing. Second, it allows the fabrication of multifunctional ferroelectric ceramics at temperatures compatible with the CMOS process used in the silicon IC (Integrated Circuit) technology.

The use of such low annealing temperatures, however, determines the microstructural features of the resulting films. Since grain growth is not favoured at these temperatures, nano-structured films with an average grain size of 30 nm are obtained. Furthermore, a Pt_xPb interlayer between the crystalline film and the Pt electrode of the substrate is formed when remaining lead excess from the precursor solution is present in the microstructure of the films. The thickness of this interlayer is significantly lower than that of the diffuse interfaces commonly observed in high-temperature processed films. The formation of the Pt_xPb interlayer could act as driving force for the lead diffusion detected in the low-temperature processed films. This is where a lead gradient with a concentration increasing from top to bottom is obtained. Conversely, diffusion of lead towards the more superficial layers of the films is observed when evaporation of this element from the surface of the films occurs at high annealing temperatures.

Dielectric and ferroelectric properties of the low-temperature processed films are in general inferior to those reported for similar films prepared at high temperatures (mainly due to the different grain size). However, they fulfil the requirements of multifunctional electronic devices. These properties are improved when the films are processed in oxygen atmosphere. Here, the photo-excitation of the organics together with their ozonolysis promotes the removing of defects (vacancies, suboxides, etc.) from the films. This gives rise to an enhanced electrical response.

Based on the UV sol-gel photoannealing approach, films derived from a nominally stoichiometric precursor solution may be prepared at these low annealing temperatures. A nominally stoichiometric precursor solution is one without excess lead incorporated into the medium. There is no use of PbO atmosphere powders or PbO overcoats. The absence of excess lead in the microstructure of the crystalline films is shown to minimise the formation of the Pt_xPb alloy with the substrate. As a result, the appearance of chemical potential gradients in the film seems to be avoided. Homogeneous and uniform compositional depth-profiles are thereby observed in the heterostructure of these films. In spite of the drastic conditions for the perovskite crystallisation used here, acceptable dielectric and ferroelectric properties are measured in the resulting thin films. These are, however, slightly inferior to those obtained when 10 mol% of lead excess is present in the microstructure of the low-temperature processed films. The remaining PbO excess is expected to improve the interfaces of the films by avoiding the formation of low dielectric space charge layers. This results in better electrical characteristics.

5.3 Soft Solution Chemistry of Ferroelectric Thin Films

5.3.1 Chemical Solution Deposition Methods

First attempts on solution preparation of oxide-based ferroelectric films were carried out in 1976 by Fukushima *et al.* [118]. They were reporting a partial sol-gel and partial organometallic solution process for the fabrication of $BaTiO_3$ thin films. Subsequent publications by Fukushima *et al.* [119] on Metalloorganic Decomposition (MOD), and Budd *et al.* [120, 121] and Lipeles *et al.* [122] date to the 1980s the developments on sol-gel processing of simple ($PbTiO_3$ and $PbZrO_3$), complex ($Pb(Zr,Ti)O_3$ and $(Pb,La)(Zr,Ti)O_3$) perovskite thin films, and the development of Chemical Solution Deposition (CSD) methods for the preparation of electronic multi-oxide films. Since then, the number of works on CSD processing of ferroelectric thin films has grown exponentially. Several of the most relevant ceramic compositions, including those based on lead (e.g., $PbTiO_3$, $Pb(Zr,Ti)O_3$, $(Pb,Ca)TiO_3$, $xPb(Mg,Nb)O_3$-$(1-x)PbTiO_3$), are now prepared by CSD approaches. The main advantages of these methods are on high compositional control (stoichiometry) and homogeneity of the solutions, the large surface areas deposited, and the low cost associated with the materials and experimental setup.

In general, a typical CSD process involves the initial synthesis of a stable and homogeneous precursor solution. This solution contains multimetal precursors dissolved in an appropriate solvent, or mixture of solvents, with the desired molar ratio. The solution is then deposited onto the substrate by different coating techniques obtaining the as-deposited film. This is then dried, pyrolysed, and crystallised by means of a thermal treatment. Thicker films may be obtained by successive deposition, drying, and annealing of single coatings (layers). Excellent books, reviews, and manuscripts [65, 123, 124, 125, 126, 127, 128] provide a deeper understanding on the basis of the CSD methodology, which is out of the scope of this chapter.

As we said in the Introduction, health effects may arise from occupational exposure to certain chemical reagents and organic solvents commonly used in most CSD methods. For instance, sol-gel and MOD. This chapter deals with ferroelectric ceramic compositions based on lead. Hence, the intrinsic toxicity of lead reagents is assumed and strategies for the soft solution processing of these materials will aim for other components, such as organic solvents. Since the evidence of toxicity in chlorinated solvents in the 1920s, hazards associated with the use of organic solvents have been reported in developed countries ever more. In Table 5.4, several organic solvents are listed that are often employed in the preparation of multioxide ferroelectric ceramic compositions by CSD. Associated risk statements have been included ([129], Risk Phrases and Safety Precautions). Due to the high number of existing solvents, only some of the most representative ones are collected in the table. To this aim, different well-known routes used for the synthesis of precursor solutions have been considered. They may be based on the conventional MOE-route [130] (EGMME [120] and EGMBE [131]), the IMO-route [132], the APP-route [133], diol-based routes [15, 134], the Pechini process [135], the DAAS-technique [136, 137], and some of their slight variations.♣

Table 5.4 Risk statements of some organic solvents often used in CSD methods for the preparation of multioxide ferroelectric ceramic compositions.

Chemical	*Hazard Codes*	*Risk Statements*	*Legend*
2-methoxyethanol	T	60-61-10-20/21/22	T: Toxic Xn: Harmful F: Highly Flammable Xi: Irritant C: Corrosive 10: Flammable 11: Highly Flammable 22: Harmful if swallowed 34: Causes burns 35: Causes severe burns 36: Irritating to the eyes 38: Irritating to the skin 41: Risk of serious damage to eyes 45: May cause cancer 46: May cause heritable genetic damage 60: May impair fertility 61: May cause harm to the unborn child 63: Possible risk of harm to the unborn child 65: Harmful: May cause lung damage if swallowed 67: Vapors may cause drowsiness and dizziness 20/21: Harmful by inhalation and in contact with skin 20/21/22: Harmful by inhalation, in contact with skin and if swallowed 23/24: Toxic by inhalation and in contact with skin 23/24/25: Toxic by inhalation, in contact with skin and if swallowed 36/37/38: Irritating to eyes, respiratory system and skin 36/38: Irritating to eyes and skin 37/38: Irritating to respiratory system and skin 39/23/24/25: Toxic: danger of very serious irreversible effects through inhalation, in contact with skin and if swallowed 48/20: Harmful: danger of serious damage to health by prolonged exposure through inhalation 48/23/24/25: Toxic: danger of serious damage to health by prolonged exposure through inhalation, in contact with skin and if swallowed
2-ethoxyethanol	T	60-61-10-20/21/22	
2-butoxyethanol	Xn	20/21/22-36/38	
1,2-ethanediol	Xn	22	
1,3-propanediol	-	-	
1,4-butanediol	Xn	22	
Methanol	F, T	11-23/24/25-39/23/24/25	
Ethanol	F	11	
1-propanol	F, Xi	11-41-67	
2-propanol	F, Xi	11-36-67	
1-butanol	Xn	10-22-37/38-41-67	
Glacial Acetic Acid	C	10-35	
o/m/p-xylene	Xn	10-20/21-38	
Propionic Acid solution	-	36/37/38	
Butanoic Acid	C	20/21/22-34	
Citric Acid	Xi	41	
2-ethylhexanoic Acid	Xn	63	
Benzene	F, T	45-46-11-36/38-48/23/24/25-65	
Toluene	F, Xn	11-38-48/20-63-65-67	
Ciclohexane	F, Xn, N	11-38-50/53-65-67	
Acetylacetone	Xn	10-22	
Acetone	F, Xi	11-36-66-67	

Historically, the 2-methoxyethanol compound has been the elect solvent for the preparation of perovskite films by sol-gel processing. The reason is the easy solubility of carboxylate precursors together with the stabilisation of alkoxides towards hydrolysis due to the alcohol-exchange reaction:

$$M(OR)_x + x\,R'OH \rightarrow M(OR')_x + x\,ROH$$

Where OR is a reactive alkoxy group and OR′ is the less reactive 2-methoxyethoxy group [123]. In the case of lead-based ferroelectric compositions, lead acetates are extensively used as the lead precursors. This is due to the instability and limited commercial availability of lead alkoxides. Therefore, 2-methoxyethanol has been usually employed as the dissolving solvent. There is developmental and reproductive toxicity related to this compound, including tera-

♣ MOE: methoxyethanol, EGMME: ethylene glycol monomethyl ether, EGMBE: ethylene glycol monobuthyl ether, IMO: inverted mixing order, APP: propionate-in-propionic acid, DAAS: deposition by aqueous acetate solution togenic effects observed in test animals. This has limited its use as a processing solvent and/or an analytical reagent in most manufacturing facilities. It is even forbidden in some developed countries. Since the past decade, replacement of 2-methoxyethanol by other solvents is being reported [13, 15, 138, 139] in solution processes of perovskite thin films. Apart from complying with CSD methods, alternative solvents should present a more benign toxicological profile. This helps avoid the teratogenic character and reproductive hazards associated with 2-methoxyethanol. Thus, replacement by some less-toxic (2-butoxyethanol) or even non-toxic (certain carboxylic acids and alcohols) compounds may fulfil these requirements.

In this part of the chapter, we focus on the substitution of hazardous 2-methoxyethanol by a glycol; the 1,3-propanediol compound. The toxicity of this diol is relatively lower than that of 2-methoxyethanol (see Table 5.4). The use of 1,3-propanediol as the solvent for the preparation of perovskite films by CSD has been reported with the aim of obtaining thicker coatings. This helps avoid the laborious multiple-deposition step [140]. For some piezoelectric actuator applications, film thicknesses in the range of 1-10 μm are required. These cannot be obtained by conventional CSD processes due to associated thickness limitations. Crack-free thick films up to 700 nm can be prepared from single-layer deposition by precursor solutions containing 1,3-propanediol.

Apart from the toxicological viewpoint, differences between 2-methoxyethanol and 1,3-propanediol are inferred with regard to their molecular structure and consequent complex behaviour with the rest of metal cations in the solution (see Fig. 5.26). Thus, 2-methoxyethanol is known to act as *chelating agent*. This is because its respective polar groups, ether and alcohol, are prone to bond to the same cation centre during the alcohol exchange reaction with the metal alkoxide. The resulting steric hindrance around the metal cation prevents the nucleofilic attack by a water molecule. The uncontrolled hydrolysis reaction leads to the precipitation of the metal hydroxide, as mentioned previously. On the other hand, 1,3-propanediol is considered a *non-bridging bidentate* ligand. This ligand's polar groups (alcohols) would bond to different metal precursors, like alkoxides and acetates, giving rise to a cross-linked network of molecules. These molecules are often treated with oligomeric nature. Ferroelectric multioxide compositions based on lead may be prepared by soft solution processing using the 1,3-propanediol compound as solvent. Thus, the particular physicochemical characteristics of diol-based precursor solutions of $(Pb_{1-x}Ca_x)TiO_3$ would be addressed during the last sections of this chapter. Understanding the background chemistry involved in this synthetic approach allows obtaining crystalline films with improved microstructural, dielectric and ferroelectric features.

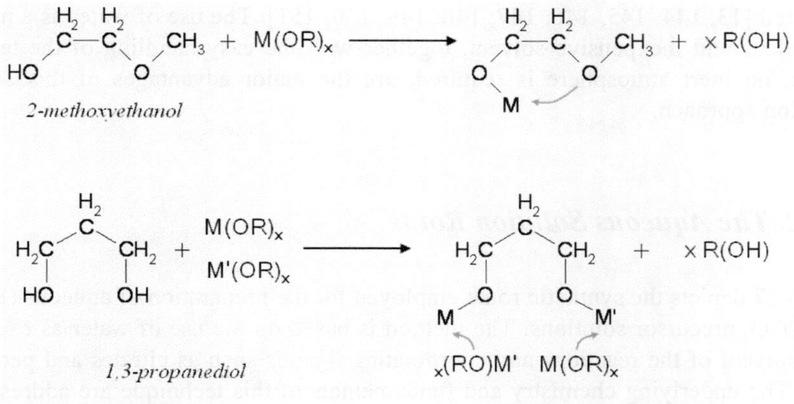

Fig. 5.26 Molecular structure of 2-methoxyethanol and 1,3-propanediol. Expected complex behaviour with metal alkoxides is indicated.

But, let us evaluate before the use of an aqueous solution route for the preparation of lead-based ferroelectric thin films. The substitution of hazardous organic solvents by water would minimise the amount of toxic chemicals during solution processing. This also reduces the costs associated with the technique. This was first attempted by an alternative MOD process using the deposition by aqueous acetate solution (DAAS) technique [136, 137]. Thus, the hazardous metal 2-ethylhexanoates and di-neodecanoates reagents commonly used in this process are replaced by metal carboxylates and β–diketonates compounds. However, alcohols such as methanol [136] or polyvinyl alcohol (PVA)/glycerol [137] must be added to the solutions. This is to obtain crack-free films or to improve the surface wetting adhesion to the substrate. This is analogous to the Pechini process [135, 139]. Despite an alpha hydroxycarboxylic acid, citric acid, being used instead of an organic solvent, a polyhydroxy alcohol, ethylene glycol, must be subsequently incorporated to promote the polyesterification and formation of a spinnable sol-network after heating. Therefore, multimetal oxide precursors may be obtained by aqueous solution methods [135, 136, 137, 141, 142, 143].The preparation of high-quality films from these solutions, however, turns into a difficult task that requires the enhancement of the solution rheology by the addition of organic solvents, normally alcohols. Other solution strategies [141, 142] make use of modified titanium alkoxide reagents that are highly stable towards hydrolysis [titanium bis(ammoniumlacto) dihydroxide], or water-soluble complexes of citrate and ethylendiaminetetraacetic (EDTA) compounds. This could obtain stable aqueous solutions of lead-free ferroelectric compositions like $BaTiO_3$ and $SrBi_2Nb_2O_9$, respectively. Powders of PZT have also been prepared by sol-gel from an aqueous based citratenitrate/oxynitrate process, [143] although NO_x gases are released during the process.

In the next section, the fabrication of $(Pb_{1-x}Ca_x)TiO_3$ films by an *entirely* aqueous method will be shown [14]. The development and successful application of this method to several multioxide ferroelectric compositions has been previously reported [13, 144, 145, 146, 147, 148, 149, 150, 151]. The use of water as a non-hazardous and inexpensive solvent, together with the easy handling of the technique, no inert atmosphere is required, are the major advantages of this soft-solution approach.

5.3.2 The Aqueous Solution Route

Fig. 5.27 depicts the synthetic route employed for the preparation of aqueous $(Pb_{1-x}Ca_x)TiO_3$ precursor solutions. The method is based on the use of water as exclusive solvent of the reactions and coordinating ligands such as citrates and peroxides. The underlying chemistry and fundamentals of this technique are addressed in the original publications from refs. [152, 153, 154, 155]. They may be summarised as follows. A water-stable Ti(IV) precursor is first synthesised from the precipitate (hydroxide) obtained after hydrolysis of the titanium alkoxide reagent (titanium tetra-isopropoxide). This is done through coordination of Ti(IV) by the citric acid and hydrogen peroxide chelating agents. This prevents the solution of the metal cation by the water molecules. The metal-chelate complexes formed are highly stable in the aqueous media. Stock solutions $(Pb_{1-x}Ca_x)TiO_3$ may be obtained after addition of the lead and calcium citrate precursors. The pH of these solutions was adjusted to ~8.5 by the addition of ammonia, since lower pH values lead to precipitation of lead citrate [156].

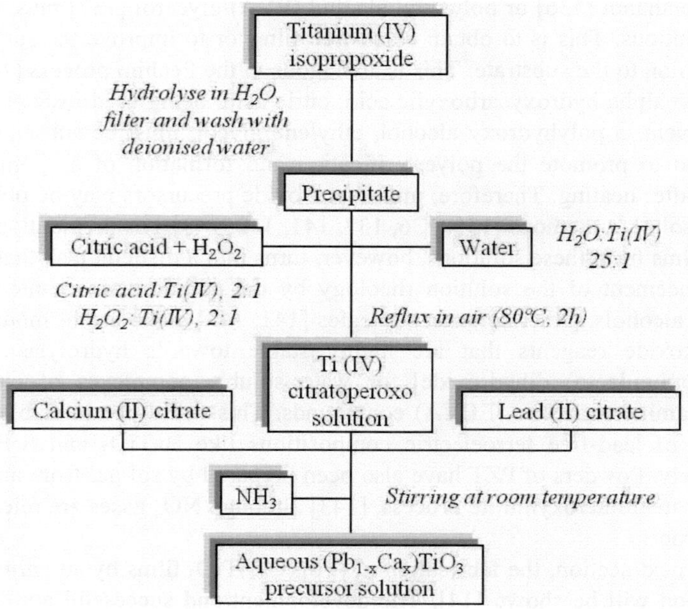

Fig. 5.27 Aqueous solution-gel route.

The physicochemical properties of the $(Pb_{1-x}Ca_x)TiO_3$ solutions synthesised by this aqueous method are shown to be strongly affected by the nominal composition thereby prepared. Thus, we will focus on substitutions of Pb^{2+} by Ca^{2+} of 24 and 50 at percentage, denoted as PCT24 and PCT50, respectively. The interest on these compositions is because of the different dielectric and ferroelectric properties obtained in the derived films (see previous Introduction section).

The low solubility of the calcium citrate reagent in the aqueous media (solubility of 0.9 g/L at room temperature for pH levels above 4.5) means that a volume of additional water is necessary, to add to the PCT50 solution to obtain a transparent and precipitate-free solution. The poor solubility of alkali/alkaline compounds in the organic solvents usually employed in CSD methods seems to be aggravated in the aqueous media. As a result, the stability of the PCT50 aqueous solution is affected in short-term aging. The PCT24 solution remains stable with time, for more than a year. A precipitate, mainly formed by a Ca^{2+} component with a crystalline structure close to calcium citrate, is obtained in the former solution after storing at room conditions in simple glassware for two weeks (see Fig. 5.28). The gel behaviour of these solutions with temperature (60°C) confirms this effect. Glassy and transparent gels obtained from the PCT50 solution become translucent after a critical value of solvent is evaporated (supersaturation conditions). Representative pictures denoting these phenomena are shown in Fig. 5.28.

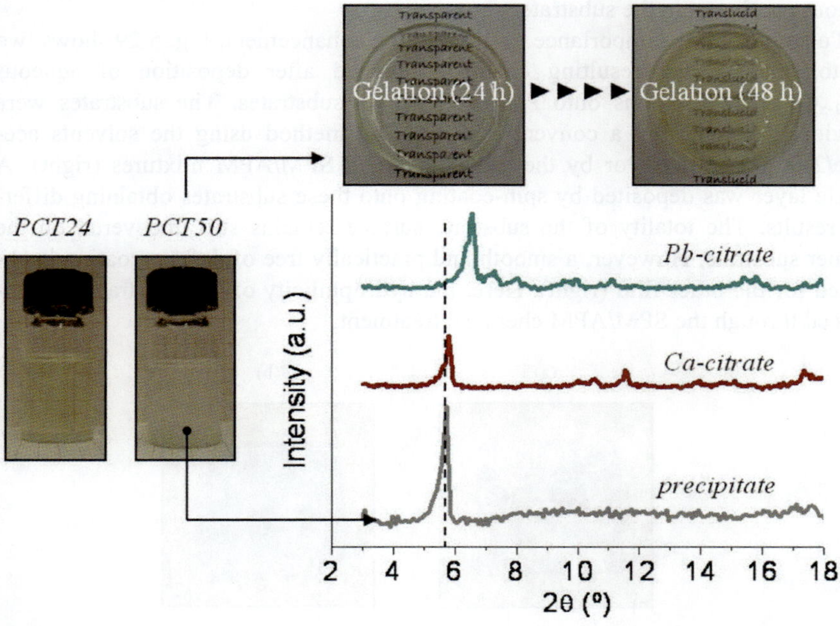

Fig. 5.28 Physicochemical properties of the aqueous PCT precursor solutions (aged) and gels.

The main problem related to aqueous CSD methods concerns the wetting of the substrate (e.g., Pt-coated silicon) by the aqueous solution. The high surface tension of water hinders the wetting of the substrate by the aqueous solution with viscosity values below 1cP. Therefore, defective coatings are usually obtained. This is clearly exposed [146] by means of contact angle measurements on as-deposited films onto different treated substrates. In most cases, this handicap is solved by improving the surface adhesion of the water-based solutions through the addition of surface-wetting reagents to the solution [157]. However, these organic additives may disturb the chemical system and promote the modification of the metallic precursors. Therefore, instead of modifying the characteristics of the precursor solutions, the hydrophilicity of the substrate may be enhanced by a wet chemical or dry treatment [13, 146].

The wet chemical process consists of a cleaning technique generally applied in the field of semiconductor and electronic devices supported on substrate wafers. It is based on the substrate immersion in the chemical mixtures sulphuric acid/hydrogen peroxide (SPM) and ammonia/hydrogen peroxide (APM). The former mixture is believed to strongly oxidise the organic contamination present on the substrate surface. In a second step, further oxidation of remnant organics together with the complexation of metallic traces is promoted by the APM mixture. Consequently, a surface without hydrophobic impurities (organic contaminants) is obtained. The aqueous solution may therefore spread out forming a continuous coating onto the substrate.

To emphasise the importance of this wetting enhancement, Fig. 5.29 shows two photographs of the resulting coatings obtained after deposition of aqueous $(Pb_{1-x}Ca_x)TiO_3$ solutions onto Pt-coated silicon substrates. The substrates were previously cleaned by a conventional cleaning method using the solvents acetone/2-propanol (left), or by the aforementioned SPM/APM mixtures (right). A single layer was deposited by spin-coating onto these substrates obtaining different results. The totality of the substrate surface remains still uncovered on the former substrate. However, a smooth and practically free of defects coating is obtained for the other film (right). Here, the hydrophilicity of the substrate was enhanced through the SPM/APM chemical treatment.

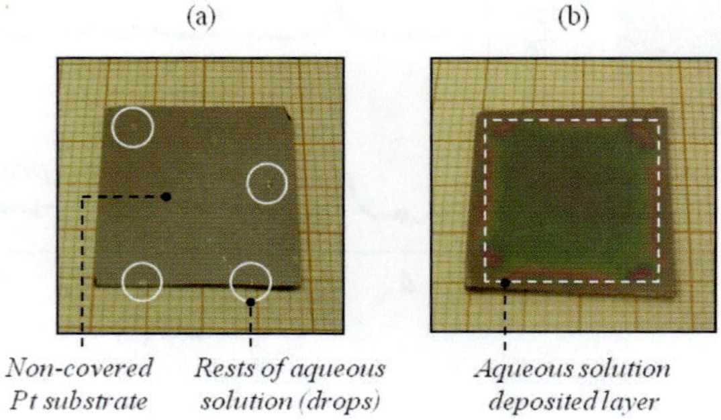

Fig. 5.29 Aqueous solutions derived films deposited onto Pt substrates previously cleaned with acetone/2-propanol **a** and SPM/APM **b** mixtures.

Single-phase polycrystalline films with homogeneous microstructures may be obtained from the aqueous method shown here, after pyrolysis and crystallisation at optimum temperatures. Experimental results from Fig. 5.30 correspond to $(Pb_{1-x}Ca_x)TiO_3$ multilayered films annealed at 650°C by Rapid Thermal Processing (RTP). Characteristic peaks corresponding to the perovskite structure are inferred from the XRD patterns, with no reflections due to secondary phases. In general, the morphology of the films consists of spherical grains whose average size varies from ~100 to ~50 nm. The size depends on the perovskite composition, PCT24 and PCT50, respectively. Differences in grain size within the $(Pb_{1-x}Ca_x)TiO_3$ system are attributed to the effect of the Ca^{2+} content on the nucleation, and growth of crystals in the material [158, 159]. Neither secondary phases nor phase segregation are observed in the micrographs of the films.

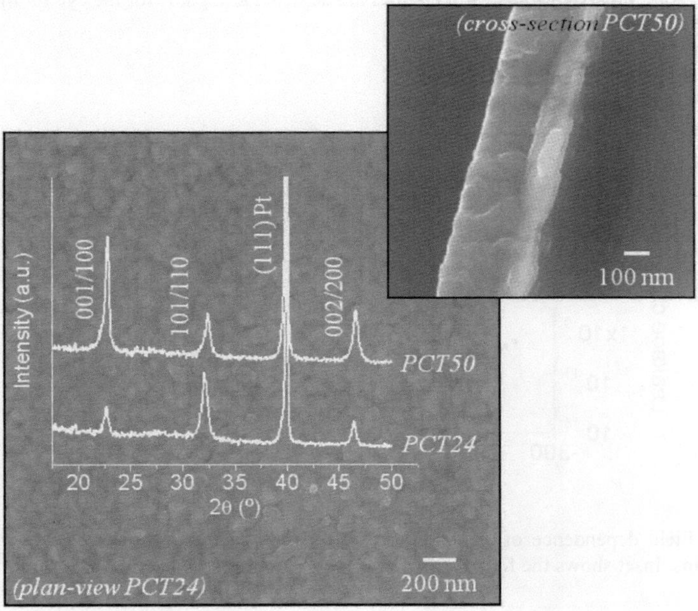

Fig. 5.30 Structure and microstructure of the aqueous solutions derived PCT films.

The functionality of the lead-based ferroelectric thin films here, prepared by an entirely aqueous synthetic route, is evaluated within the technology of microelectronic devices from the dielectric and ferroelectric measurements. This is shown in the next figures. Acceptable values of the dielectric constant are obtained in the films belonging to the PCT24 ferroelectric composition (Fig. 5.31a). But, a strong dielectric relaxation with frequency is observed. This could suggest the presence of extrinsic defects in the film (e.g., low quality interfaces). These would yield an increase on the series resistance of the sample producing the observed phenomenon.

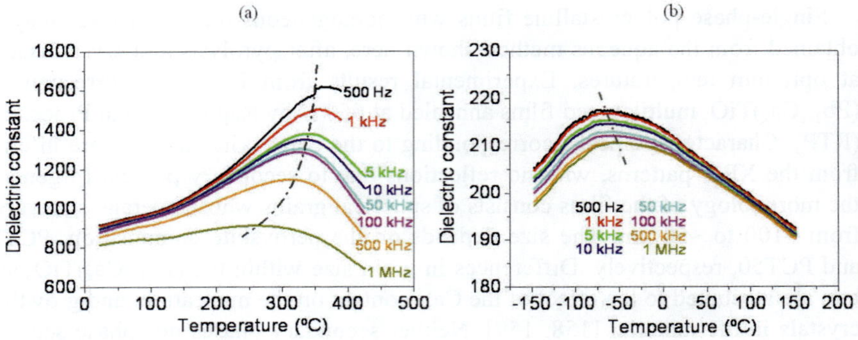

Fig. 5.31 Variation of the dielectric constant with temperature of the aqueous solutions derived PCT24 (**a**) and PCT50 (**b**) films. Broken lines are depicted as a guide for the eye for the temperature maximum shift.

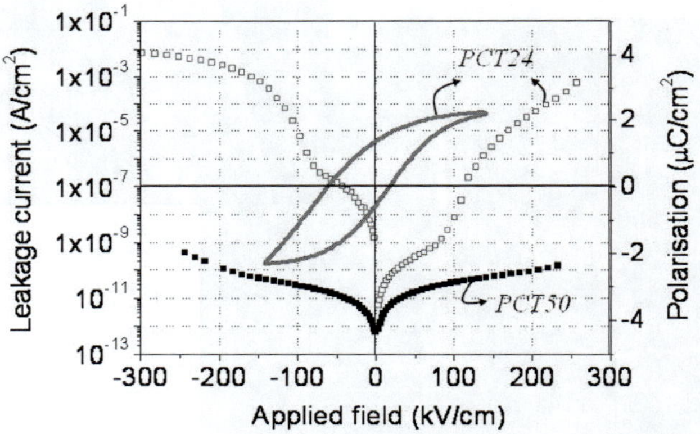

Fig. 5.32 Field dependence of leakage currents for the aqueous solutions derived PCT24 and PCT50 films. Inset shows the ferroelectric hysteresis loop measured in the PCT24 film.

This would support the large and asymmetric leakage current densities also measured in this film (Fig. 5.32). From the experimental setup of the measurement, a higher leakage current density is obtained at the top electrode. This could be an indication of significant roughness present at the surface of the film [160]. Though, when multiple-layer deposition is performed from the aqueous precursor solutions, highly rough surfaces are obtained in the films. This is after a critical thickness is reached by successive deposition and crystallisation steps. The presence of roughness in the films may be inferred from the picture shown in Fig. 5.33. A shining and smooth surface (left) in the crystalline film reflects the tweezers placed in front of it (as a mirror). A rough surface (right) scatters the light and the reflected image appears consequently blurred.

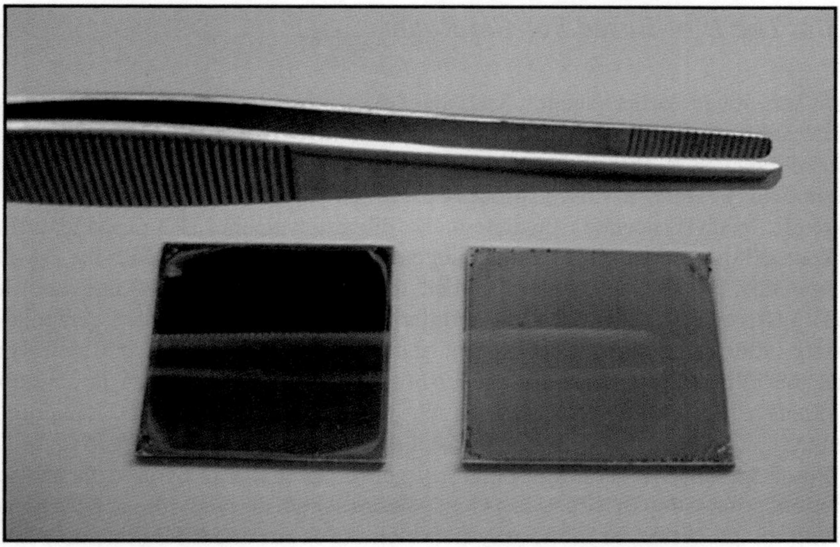

Fig. 5.33 Crystalline films derived from the aqueous solutions. A rough surface can be inferred from the film on the right.

The detrimental interfaces thus formed in the heterostructure of the PCT24 film would be responsible of the dielectric relaxation discussed earlier. They would also be responsible for the large values of leakage current density measured. This prevents the optimum polarisation of this film (see inset of Fig. 5.32), leading to lower values than those that would be obtained if larger voltage bias could be applied.

On the other hand, a clear relaxor-like ferroelectric behaviour is observed for the PCT50 film prepared by this soft aqueous solution processing. The diffuse ferro-paraelectric transition, together with the displacement towards higher temperatures upon the measuring frequency increases, is indicative of this type of electronic multioxide materials [161]. The relatively high k' values and low tanδ measured (< 1%) at room temperature shows the viability of this material for applications in DRAMs and high-frequency components like consumer portable communications and radar systems. Furthermore, the low leakage current density (< 10^{-9} A/cm^2) obtained in this film would prevent the thermal breakdown of the film when the capacitor is successively charge-injected/ released (refreshed). This is associated with the relaxor character of the material and hence the practical absence of domain switching. It makes the role of the system not only feasible but also effective [162].

5.3.3 The Diol-Based Sol-Gel Route

As it was stated earlier (section 3.1), alcohols are usually one of the most appropriate solvents for the dissolving and subsequent polycondensation of the multimetal precursors that are employed within the polymeric systems of sol-gel methodology. Here, a sol-gel process based on the use of a modified titanium alkoxide, lead (II) acetate trihydrate and a dihydroxyalcohol (1,3-propanediol) as solvent will be shown. This avoids the health hazards associated with the 2-MOE compound. The method can be applied for the preparation of A-site modified $PbTiO_3$ ferroelectric compositions and derived films. The glycolated compounds formed after sol-gel synthesis present a higher stability towards the hydrolysis than the initial metal alkoxides precursors [163, 164]. This results in air-stable solutions and crack-free coatings even when the film thickness is over a micrometre [165].

Fig. 5.34 shows several diol-based sol-gel routes that may be used for the soft solution processing of $(Pb_{1-x}Ca_x)TiO_3$ solutions and derived films. Differences among the processes lie in the way in which the A-site modifier (calcium) is incorporated into the chemical system. This can be done:

1. as a water solution of calcium acetate hydrate ($Ca(OCOCH_3)_2{\cdot}xH_2O$) ***Route A***
2. as a diol–water solution of calcium acetate hydrate ($Ca(OCOCH_3)_2{\cdot}xH_2O$) ***Route B***
3. as a diol–water solution of calcium acetylacetonate hydrate ($Ca(CH_3COCHCOCH_3)_2{\cdot}xH_2O$) ***Route C***

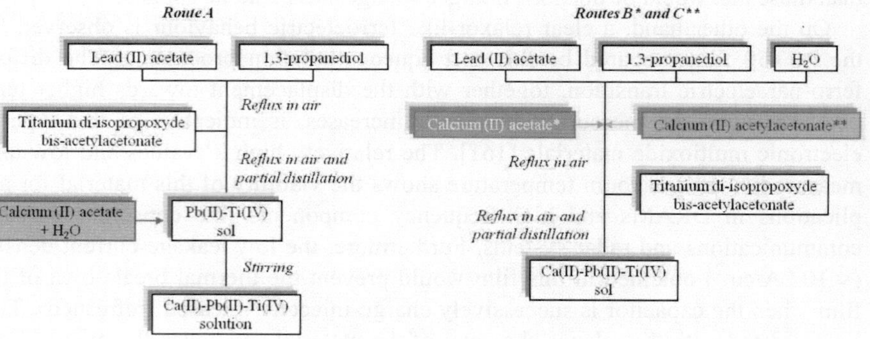

Fig. 5.34 Diol-based sol-gel synthetic routes A, B*, and C**.

Stable *sols* or *solutions* are obtained by the routes A, B and C. The calcium precursor, calcium acetate or calcium acetylacetonate, is introduced at the same time as the other metal reagents and solvents in routes B and C. This leads, after reflux, to the formation of a real Ca(II)-Pb(II)-Ti(IV) *sol*. On the other hand, a stable *solution* is obtained when a water solution of the calcium reagent (calcium acetate) is simply added to the previously synthesised Pb(II)-Ti(IV) sol in synthetic route A. Substitutions of Pb^{2+} by Ca^{2+} up to 50 at percentage may be introduced in the $(Pb_{1-x}Ca_x)TiO_3$ system through the routes A and B. However, only compositions with Ca^{2+} contents below 30 at percentage may be prepared from the synthetic route C. In this section, solutions, gels, and films derived from the synthetic routes A and C, and with a nominal composition of $(Pb_{0.76}Ca_{0.24})TiO_3$, would be denoted as PCT24-A and PCT24-C, respectively. Solutions, gels, and films derived from the synthetic routes A and B, and with a nominal composition of $(Pb_{0.50}Ca_{0.50})TiO_3$, would be named PCT50-A and PCT50-B, respectively. Table 5.5 summarises the main characteristics of the different diol-based sol-gel synthetic routes followed for the preparation of the precursor solutions shown here.

Table 5.5 Physicochemical characteristics of the PCT precursor solutions prepared from the diol-based sol-gel synthetic routes A, B and C.

Nominal composition	Synthetic route	Type of precursor	Incorporation of Ca^{2+}	Concentration, density, viscosity
$(Pb_{0.76}Ca_{0.24})TiO_3$	A	*Solution* (PCT24-A)	Water solution of $Ca(OCOCH_3)_2xH_2O$	~0.80 M, ~1.30 g/mL, ~13 cP
	C	*Sol* (PCT24-C)	Diol-water solution of $Ca(CH_3COCHCOCH_3)_2xH_2O$	~0.80 M, ~1.30 g/mL, ~12 cP
$(Pb_{0.50}Ca_{0.50})TiO_3$	A	*Solution* (PCT50-A)	Water solution of $Ca(OCOCH_3)_2xH_2O$	~0.90 M, ~1.30 g/mL, ~40 cP
	B	*Sol* (PCT50-B)	Diol-water solution of $Ca(OCOCH_3)_2xH_2O$	~0.90 M, ~1.30 g/mL, ~14 cP

The chemical process followed for the solution synthesis by the diol-based sol-gel method is seen to affect the homogeneity and physicochemical features of the resulting $(Pb_{1-x}Ca_x)TiO_3$ sols or solutions (both hereinafter denoted as "solutions"). Measurements obtained by Dynamic Light Scattering (Fig. 5.35) reveal that solutions synthesised from route A are formed by two groups of particles. These have well-differentiated hydrodynamic radii, while an almost symmetrical distribution is obtained for the solutions processed by routes B and C. This shows that solutions derived from the route A consist of a heterogeneous mixture of a Pb(II)-Ti(IV) sol and a calcium acetate water solution. The routes B and C lead to homogeneous Ca(II)-Pb(II)-Ti(IV) sols.

The non-dependence of the viscosity values of the solutions with the shear rate (e.g. shear thinning, shear thickening) observed in Fig. 5.36 indicates the Newtonian nature of these fluids [166]. It confirms that heterogeneity in particle size of solutions derived from route A is not due to particle aggregation. This would have produced a decrease in viscosity as the shear rate increases [167], was it so. Thus, molecular level heterogeneity is the most significant structural feature of the solutions processed by the route A. This route shows two particle size distributions which differ in both nature and size. On the other hand, the larger viscosity of the PCT50-A solution (~40 cP) is related to the presence of particles of much bigger size (~50 nm) in respect to the others.

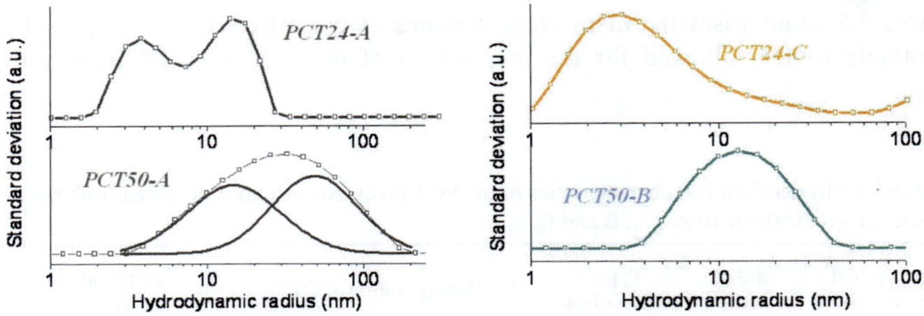

Fig. 5.35 Particle size distributions of the diol-based sol-gel PCT solutions.

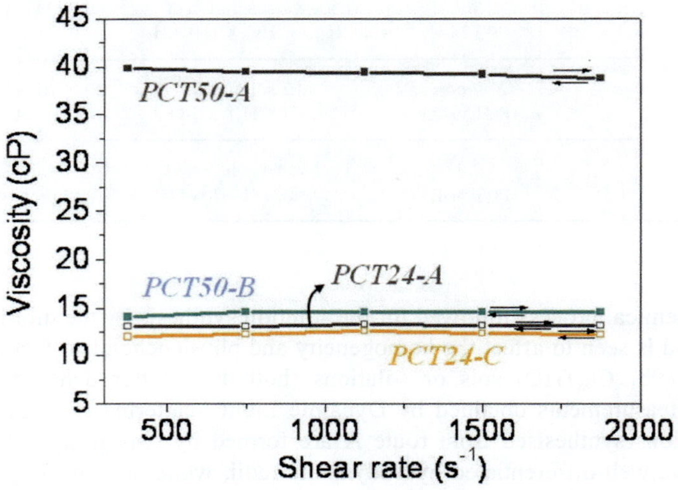

Fig. 5.36 Flow behaviour of the diol-based sol-gel PCT solutions.

From these studies, a proposed molecular structure for the different diol-based sol-gel solutions is depicted in Fig. 5.37. Thus, a cross-linked network of molecules involving the three metal cations would be obtained in the homogeneous $(Pb_{1-x}Ca_x)TiO_3$ solutions processed by routes B and C. In route A, the calcium acetate compound would occupy random positions within the Pb(II)-Ti(IV) sol network. This precludes the formation of a uniform polymeric M-O-M framework. The non-incorporation of A-site precursors into the oligomeric structure of the sol has also been observed in hybrid sol-gel processing of solutions of ABO_3 perovskite compositions, such as $BaTiO_3$ [168] and $Pb(Zr,Ti)O_3$ (PZT) [169]. Instead, the A-site species are proposed to be adsorbed or occluded on the surface of the B-site oligomers [168]. They are hence not intimately involved in the formation of the gel network [170].

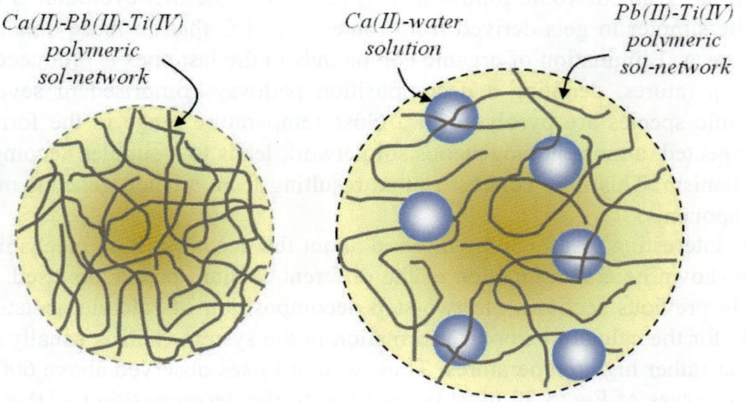

Fig. 5.37 Proposed sol networks for the different solutions obtained through the diol-based sol-gel method.

The stability and aging behaviour of the diol-based sol-gel solutions shown here are directly related to the respective molecular sol network. The evolution of the particle size distributions with time, in the PCT50-A and PCT50-B solutions, is depicted in Fig. 5.38. Results correspond to fresh solutions and after aging times of 6, 14 and 20 months. Viscosity of these solutions, together with thickness values of the derived single-layer crystallised films, is also included in the graphs. As explained before, two different particle distributions with peaks maxima at ~15 and ~50 nm are initially obtained in the solutions derived from route A. These correspond to the mixture of a Pb(II)-Ti(IV) sol and a calcium acetate water solution (see previous Table 5.5). Further aging times promote the growth of the initially bigger particles (~50 nm) up to ~65 and ~120 nm (6 and 14 months, respectively). This is when the formation of a precipitate was observed in the solution. The presence of particles with larger sizes produces an increase in the viscosity of this solution with the aging time. As the solution is formed by bigger particles, their resistance to flow must be consequently higher. Regarding the thickness values of the derived films, the more viscous the solution is, the thicker the film will be. The heterogeneity of solutions processed by route A is therefore observed to increase with time. On the contrary, the PCT50-B solutions remain stable and with a high degree of homogeneity for long times. A narrow and symmetrical distribution with an average particle size of ~15 nm is measured initially and after 6 and 14 months of aging. The solution aged for 20 months has the same average particle size. The distribution, however, shows a slight asymmetry indicating a slight increase in the heterogeneity of this solution with long aging times. The thickness of the derived crystalline films remains practically constant even after 20 months of solution aging.

Differences are also found in the thermal decomposition of the $(Pb_{1-x}Ca_x)TiO_3$ gels obtained from the diol-based sol-gel method evaluated in this section. This depends on the synthetic route followed (Fig. 5.39). The thermal evolution is observed to be simpler in gels derived from routes B and C than in those obtained through route A. Elimination of organic compounds in the last ones is produced at various temperatures, denoting a decomposition pathway comprised of several steps. Organic species are pyrolysed in a close temperature range in the former gels. As expected, a more homogeneous sol network leads to a simpler decomposition mechanism. This is in contrast to that resulting from a heterogeneous mixture of components.

Another interesting result can be inferred about the decomposition mechanism of the gels shown here as a function of the different calcium precursors used. As described in previous sections, the two-step decomposition of calcium acetate is responsible for the calcium carbonate formation in the system. This is usually decomposed at rather high temperatures. Thus, weight losses observed above 600°C in the TGA curves of Fig. 5.39 must be ascribed to the decomposition of this residual carbonate. However, this intermediate compound is not formed in the gels obtained from synthetic route C, where calcium acetylacetonate was used as the Ca^{2+} source. The thermal decomposition of this chemical reagent is shown in the inset of this figure. Weight losses associated with acetone and calcium carbonate decomposition (~700°C) are observed in the respective TGA curve. For the gels derived from route C, however, the near total elimination of the organic species is produced at temperatures below 500°C. This denotes the absence of carbonates formation in this system.

The influence of alkaline earth precursors, consisting in carboxylate compounds of different alkyl chain, on the formation of intermediate complex carbonates phases in the $(Ba,Sr)TiO_3$ system has been previously discussed [171, 172]. As it was concluded, short-chain carboxylates of Ba and Sr compounds (e.g., acetates) exhibit a two-step decomposition process. This involves the formation of intermediate carbonates delaying the onset of the perovskite crystallisation up to high temperatures. On the other hand, long-chain carboxylates (e.g., 2-ethylhexanoates, Sr-propionate) decompose directly into the respective oxides at lower temperatures. This happens without evidence of formation of intermediate carbonate phases. Although being a ketonate, the calcium acetylacetonate reagent used here would belong to this last family of alkaline earth carboxylate precursors. It thus explains the results. For the rest of the systems, using calcium acetates as the Ca^{2+} precursor, the formation of intermediate carbonates delays the total elimination of organics from the gel-network towards higher temperatures.

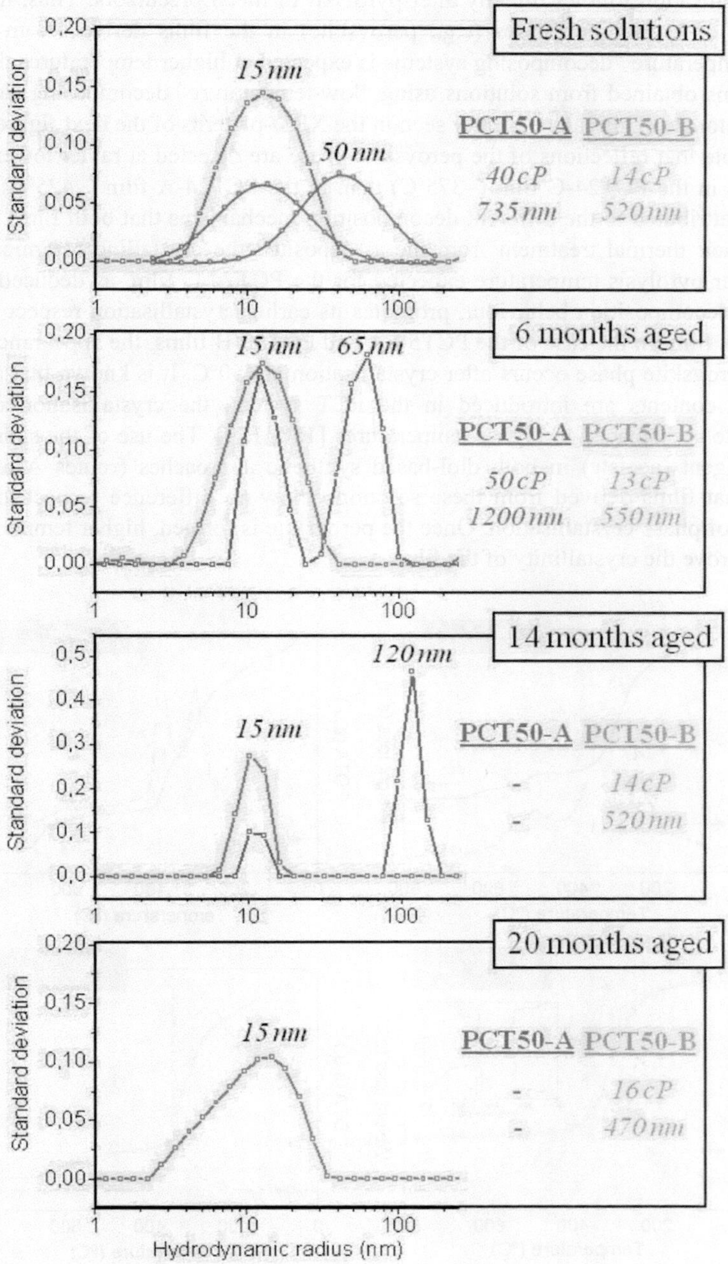

Fig. 5.38 Particle size distributions of diol-based sol-gel PCT solutions at different stages of aging. Viscosity of the solutions (in cP) and thickness of derived monolayers (in nm) are also included.

Crystallisation follows directly after pyrolysis of metal precursors. Thus, the appearance of crystalline phases (e.g., perovskite) in the films derived from these "high-temperature" decomposing systems is expected at higher temperatures than in those films obtained from solutions using "low-temperature" decomposing calcium acetylacetonate. This can be clearly seen in the XRD-patterns of the next figure (Fig. 5.40). Note that reflections of the perovskite phase are detected at rather lower temperatures in the PCT24-C film (~375°C) than in the PCT24-A film (~425°C). This must be attributed to the different decomposition mechanisms that both films suffer during their thermal treatment from the as-deposited the crystalline ceramic state. The lower pyrolysis temperature expected for the PCT24-C film, as deduced from the gels decomposition behaviour, promotes its earlier crystallisation respect to the PCT24-A film. In the case of the PCT50-A and PCT50-B films, the appearance of a single perovskite phase occurs after crystallisation at 450°C. It is known that as larger Ca^{2+} contents are introduced in the PCT system, the crystallisation of the perovskite is displaced to higher temperatures [158, 159]. The use of the same calcium reagent (acetate) in both diol-based synthetic approaches (routes A and B) means that films derived from these solutions show no difference respect to temperature of phase crystallisation. Once the perovskite is formed, higher temperatures only improve the crystallinity of the films.

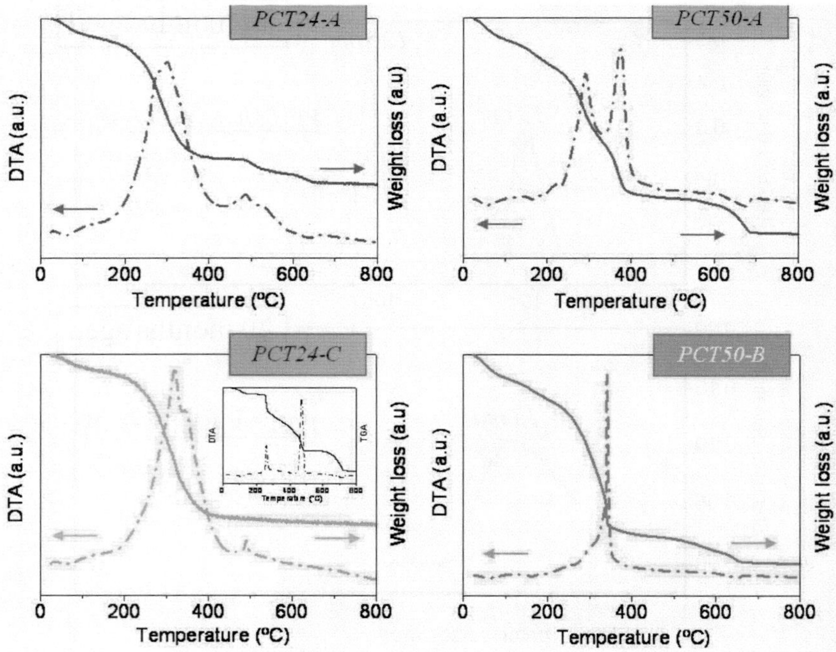

Fig. 5.39 DTA-TGA of the gels powders derived from the diol-based sol-gel PCT solutions. Inset in PCT24-C data shows the thermal decomposition of the calcium acetylacetonate chemical reagent.

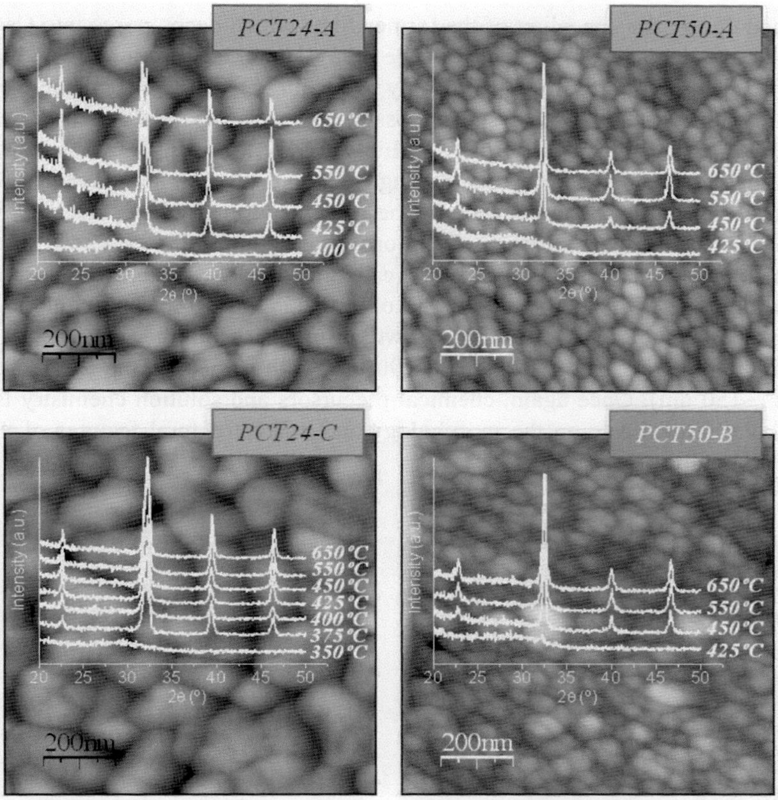

Fig. 5.40 XRD patterns and SFM images of the films obtained from the different diol-based sol-gel PCT solutions.

The different chemical processes based on the diol sol-gel method exposed here are shown to influence not only the temperatures for the perovskite crystallisation in the system, but also the resulting microstructure and morphology of the films. Images of Fig. 5.40 correspond to different $(Pb_{1-x}Ca_x)TiO_3$ films annealed at 650°C by RTP. Homogeneous microstructures without morphological differentiated secondary phases are observed. However, significant differences in grain size can be deduced from the images. Average grain sizes of ~150 and ~190 nm are measured for the PCT24-A and PCT24-C films, respectively, whereas a value of ~50 nm is obtained in both PCT50-A and PCT50-B films. The influence of the Ca^{2+} content of the PCT system on the grain size of the crystalline films was also observed in previous section 3.2. Crystallite growth in this system is known to be hindered in high-Ca compositions [158, 159]. Therefore, resulting films are formed by smaller grains when compared to films of lower Ca^{2+} contents. On the other hand, the different morphology obtained in the PCT24-A and PCT24-C films must be explained on the basis of the decomposition process that both films suffer during their thermal treatment. As discussed before, calcium carbonate is formed during the annealing of the former films which decomposes into CaO and CO_2 at a temperature close to that of the perovskite crystallisation. The release of CO_2 at these high temperatures probably yields the formation of pores in the crystalline films. Fig. 5.41 shows the cross-section micrographs obtained in these films. A larger amount of porosity within the bulk film can be inferred in those films derived from synthetic routes using calcium acetate as initial reagent (routes A and B). This is a consequence of the aforementioned $CaCO_3$ decomposition. However, the expected absence of carbonate formation in the PCT24-C film, due to the "low temperature" decomposing calcium acetylacetonate, would promote a microstructure with a higher density, as observed in the figure. This could improve the phenomena of coalescence and growth of grains within the film, thus explaining the larger size of these grains (~190 nm) with respect to the PCT24-A film ones (~150 nm). Once again, chemical precursors and solution chemistry is seen to play a critical role on the morphology and microstructural features of the derived crystalline films.

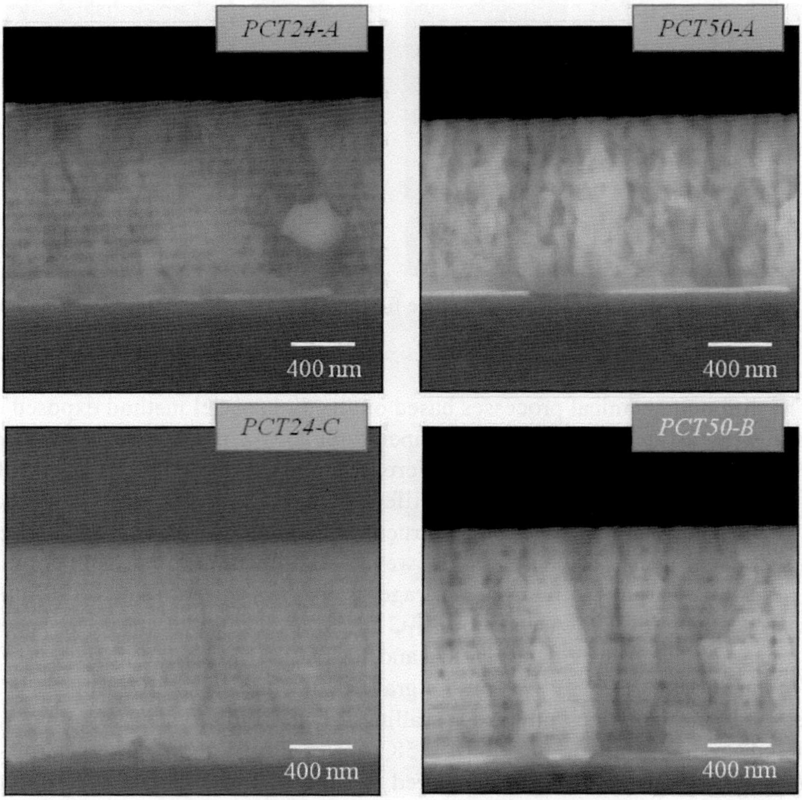

Fig. 5.41 SEM images (cross-section) of the crystalline films (650°C) obtained from the different diol-based sol-gel PCT solutions. A slight lower content of porosity can be inferred in the PCT24-C film.

Homogeneity of precursor solutions is a crucial goal to be achieved in the fabrication of ceramic materials by CSD methods. In the particular case of the sol-gel process, the local atomic arrangement in the sol is expected to be transferred to the final crystalline phase [173, 174, 175]. Therefore, metal atoms in the solution network should be arranged in such a manner that resembles the structure of the desired oxide phase (e.g., perovskite). A relationship between solution homogeneity and film homogeneity may be established for the lead-based ferroelectric films processed by the different diol-based synthetic approaches shown here. A comparison between the heterostructures obtained by RBS measurements in the $(Pb_{1-x}Ca_x)TiO_3$ films, as derived from either homogeneous sols (routes B and C) or heterogeneous solutions (route A), is depicted in Fig. 5.42. The results indicate that the heterostructure of the films is directly related to their respective precursor solutions processing.

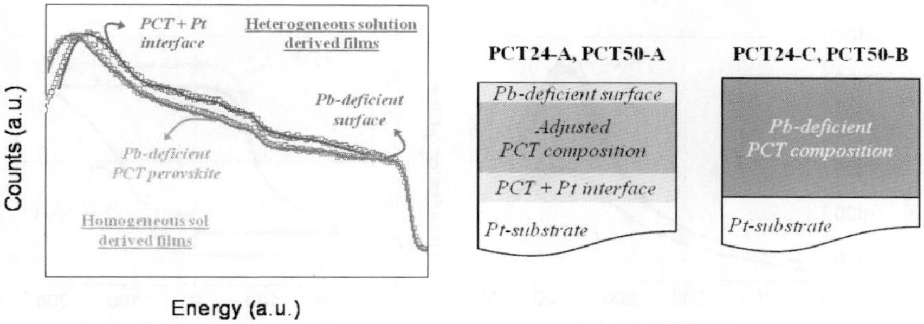

Fig. 5.42 RBS spectra of the crystalline films (650°C) obtained from the different diol-based sol-gel PCT solutions and schematic of their respective heterostructure.

The high homogeneity of solutions obtained through routes B and C (see previous results on DLS measurements) is shown to provide a better compositional control over the derived films. The average elemental concentration calculated in these films denoted a slightly lower content of lead (~10 %) with respect to the nominal perovskite composition. However, significant fluctuations on the compositional depth profile are not deduced from the simulated curves. On the contrary, crystalline films processed by route A show an average composition adjusted to the expected stoichiometric perovskite. They are rather heterogeneously distributed along the film thickness (see scheme of Fig. 5.42). Thus, the more superficial layers of these films present a deficiency of lead with respect to the bulk. This is attributed to the PbO volatilisation during firing. In addition, the formation of a chemically distinct interface region between the film and the substrate (Pt-coated silicon), usually promoted at these high annealing temperatures, is also deduced from the compositional calculations of the films. The thickness of this interlayer is around 13 % of the respective total films thickness. It has a composition resulting from the interdiffusion between the $(Pb_{1-x}Ca_x)TiO_3$ layer and the Pt bottom electrode. The heterogeneous mixture of components proposed for solutions obtained through the route A (a Pb(II)-Ti(IV) sol and a calcium acetate water solution) is believed to yield weaker network structures. This allows the compositional fluctuations and atomic diffusion with the substrate observed in the respective films.

Finally, the feasibility of the diol-based sol-gel method shown here to produce films with optimum dielectric and ferroelectric properties is evaluated in the next figures. Depending on the synthetic route followed, different microstructures were obtained in these films. They show a clear influence on the electrical response of the material. The presence of a small capacitive interface in the films derived from route A leads to a decrease on the effective dielectric constant of the system. This is due to the equivalent approximation of an in-series capacitor with the rest of the bulk film (Fig. 5.43 and Fig. 5.44).

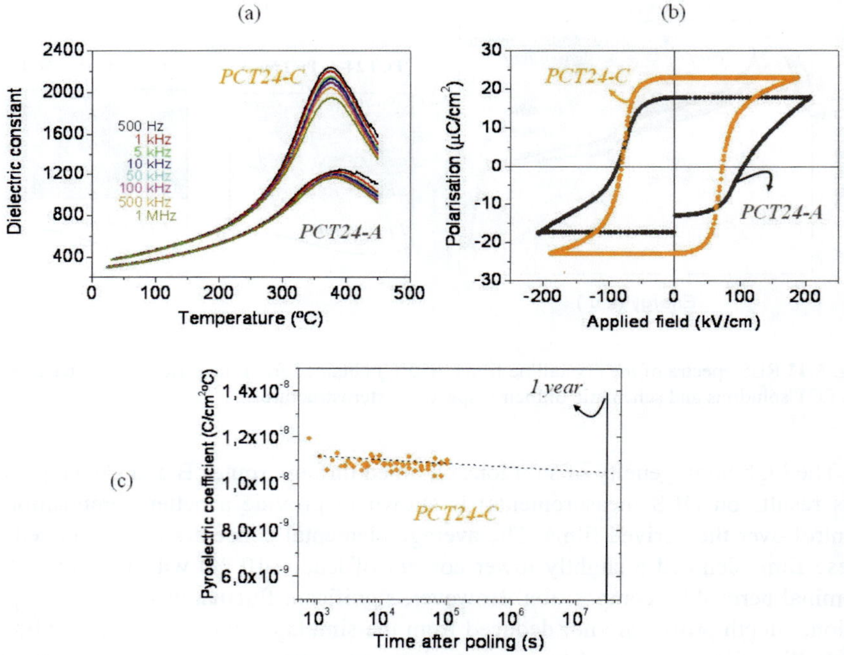

Fig. 5.43 Variation of the dielectric constant with temperature **a,** ferroelectric hysteresis loops **b,** and evolution of the pyroelectric coefficient with time **c** of PCT24 films obtained from the diol-based sol-gel routes A and C.

In the case of the PCT24 ferroelectric composition, gradient terms associated with the heterogeneous heterostructure of these films (lead-deficient surface layer, bulk film and diffuse interface) are consistent with the dielectric peak broadening observed at the ferro-paraelectric transition. The incomplete closing of the compensated P(E) loop (measured by the method of pulses [176], which eliminates the contribution of possible non-ferroelectric charges) in the PCT24-A film, denotes a small effect of leakage contribution. Close and well-defined ferroelectric loops are obtained from the PCT24-C film. The practical absence of detrimental interfaces in this last film, together with a microstructure formed by bigger grains, is responsible for the higher P_r values (23 $\mu C/cm^2$) and lower coercive fields (77 kV/cm) with respect to the PCT24-A film (16.5 $\mu C/cm^2$, 96 kV/cm) [66, 173, 177]. The polarisation-retention characteristics of the PCT24-C film may be illustrated by the evolution of the pyroelectric coefficient with time, included also in Fig. 5.43. A pyroelectric coefficient of $1.2x10^{-8}$ $C/cm^{2o}C$ is obtained after poling, in the range of the values obtained for multiple deposited and crystallised sol-gel $(Pb_{1-x}Ca_x)TiO_3$ films [82]. Near 90 % of this value is retained up to ~1 year, which indicates the potential use of this film for pyroelectric detectors and/or non-switching devices.

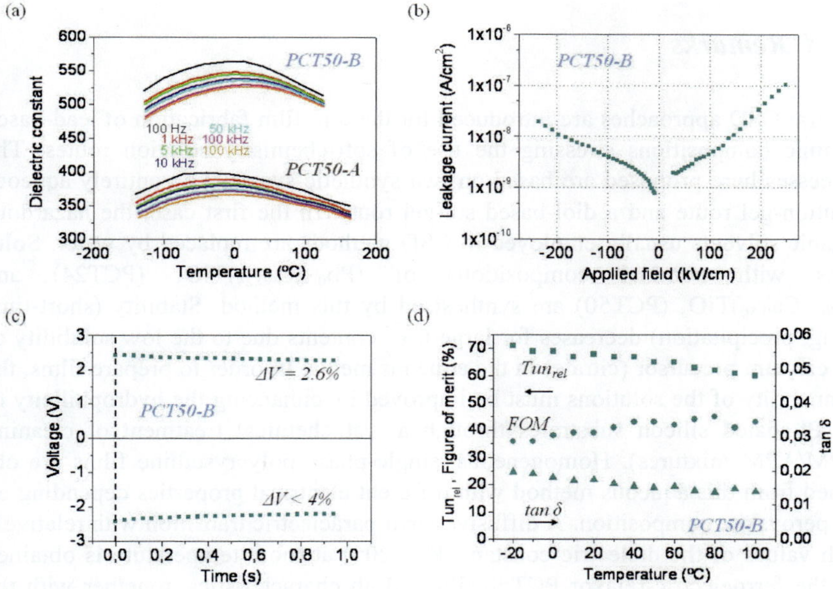

Fig. 5.44 Variation of the dielectric constant with temperature (**a**), leakage current density (**b**), time-voltage drop after writing pulse (**c**), and values of relative tunability, figure of merit and dielectric loss tangent (**d**) of PCT50 films obtained from the diol-based sol-gel routes A and B.

On the other hand, the high values of the dielectric constant at room temperature and the relaxor behaviour shown by the PCT50 films look more promising for capacitor applications (e.g., DRAMS) and voltage-tunable devices (see Fig. 5.44). Leakages currents below 10^{-8} A/cm^2 are obtained in a 100 nm thick PCT50-B film for the range of voltages between ±2.5 V. Values below 10^{-7} A/cm^2 are required for charge storages devices, proving the feasibility of this film for such applications. In accordance with current recommendations [178], a capacitor for DRAMs applications should retain more than the 90 % of its charge after 200 ms, the usual refreshment time in the memory. The voltage drop of this film after a writing pulse is less than five percent.

The variation of the capacitance with voltage of the PCT50-B films allows also us to evaluate their use in the light of voltage-tunable devices (e.g., varactors). From these measurements, the temperature dependence of tunability, figure of merit and tan δ can be obtained.♦ Values of tunability are around 60-70 % and do not change more than 10 % with temperature. Figure of merit is close to 40 % and increases with temperature. This effect is due to the relaxor-like behaviour of these films, in which the dielectric loss tangent is observed to decrease with increasing temperature. Both tunability and figure of merit values of the PCT50-B film are comparable to those reported for the best alternative materials [179, 180].

5.3.4 Remarks

Several CSD approaches are introduced for the thin film fabrication of lead-based ceramic compositions stressing the use of soft-chemistry solution routes. The processes here proposed are based on two synthetic systems; an entirely aqueous solution-gel route and a diol-based sol-gel route. In the first case, the hazardous organic solvents usually employed in CSD methods are replaced by water. Solutions with nominal compositions of $(Pb_{0.76}Ca_{0.24})TiO_3$ (PCT24) and $(Pb_{0.50}Ca_{0.50})TiO_3$ (PCT50) are synthesised by this method. Stability (short-time aging, precipitation) decreases for large Ca^{2+} contents due to the low solubility of the calcium precursor (citrate) in the aqueous media. In order to prepare films, the spinnability of the solutions must be improved by enhancing the hydrophilicity of the Pt-coated silicon substrates through a wet chemical treatment of cleaning (SPM/APM mixtures). Homogeneous, single-phase polycrystalline films are obtained from this aqueous method with different electrical properties depending on the perovskite composition. A diffusive ferro-paraelectric transition with relatively high values of the dielectric constant (k' = 205) at room temperature is obtained for the ferroelectric-relaxor PCT50 films. Both characteristics, together with the low leakage current density measured (below 10^{-9} A/cm^2), indicate the suitability of these films for capacitive applications. On the other hand, the possible presence of detrimental interfaces (rough surfaces) in the ferroelectric PCT24 films means that leakages increase significantly. This hampers their correct polarisation.

Concerning the diol-based solution strategy, the high toxicity of the 2-methoxyethanol solvent, historically employed in the preparation of these materials by sol-gel methods, is avoided by the use of mixtures of 1,3-propanediol and water. Thus, precursor solutions of lead-based $(Pb_{1-x}Ca_x)TiO_3$ compositions with

♦ $Tun_{rel} = (C_{max} - C_{min}) / C_{max}$; $FOM = Tun_{rel} / \tan \delta$

tailored physicochemical properties may be synthesised by several diol-based sol-gel routes using different synthetic pathways (routes A, B and C) and/or calcium precursors (calcium acetate or calcium acetylacetonate). Two differentiated distributions of particle size are observed in solutions derived from synthetic route A. These correspond to the heterogeneous mixture of components (a Pb(II)-Ti(IV) sol and a calcium acetate water solution) that form this solution. On the other hand, homogeneous solutions with an uniform and single distribution of particle size are obtained by the routes B and C (a Ca(II)-Pb(II)-Ti(IV) sol). The heterogeneity of solutions derived from route A increases with time (aging effect), leading to precipitation several months after the synthesis. Solutions from routes B and C remain stable with time (more than a year). The thermal decomposition of the gels derived from these diol-based sol-gel solutions is more homogeneous in those obtained through routes B and C. In route A, the elimination of organics is produced in a wider temperature range. Calcium carbonate rests, which decompose at high temperatures (>600°C), are only present in systems (routes A and B) that use short-chain carboxylates (acetates) for the Ca^{2+} precursor. The two-step decomposition of these reagents via intermediate carbonate formation delays the onset of the crystallisation up to high temperatures. On the other hand, long-chain carboxylates or ketonates (acetylacetonate) of calcium used in route C decompose directly into the single oxide at much lower temperatures (<500°C).

The microstructure of the crystalline films (650°C, RTP) is influenced by both perovskite composition and solution chemistry. PCT24-A films are formed by grains with a mean size of ~150 nm. Average grain sizes of ~190 nm constitute the microstructure of the PCT24-C films. In the case of the PCT50-A and PCT50-B films, microstructures formed by grains of ~50 nm are obtained. The grain growth phenomenon in the films is known to be hindered for high Ca^{2+} concentrations. Porosity is observed in the films in which calcium acetate was used as precursor for the synthesis of the respective solutions (routes A and B). This originates from the release of CO_2 at high temperatures owing to calcium carbonate decomposition. The absence of this intermediate compound in the films derived from solutions using calcium acetylacetonate (route C) leads to dense microstructures formed by grains of a bigger size. Heterogeneous heterostructures are obtained in the films derived from route A, with formation of bottom interfaces with the substrate and lead-deficient surfaces. On the contrary, the presence of detrimental interfaces in the films derived from the homogeneous solution strategies B and C is shown to be minimised. This fairly improves the electrical characteristics of the films. Higher values of remnant polarisation and lower coercive fields are measured in the hysteresis loops of the PCT24-C film (23 $\mu C/cm^2$ and ~77 kV/cm) with respect to the PCT24-A film (16.5 $\mu C/cm^2$ and 96 kV/cm). Furthermore, the values of dielectric constant are much higher in the former than in the latter (2210 and 1190 respectively, at transition temperature). The PCT50 films have a clear relaxor-like ferroelectric character with higher values of the dielectric constant at room temperature for the PCT50-B film (545) than for the PCT50-A film (380). The leakage current density measured in the former films is below 10^{-7} A/cm^2, with optimum values of tunability (60-70 %) and figure of merit (40 %). All these results support the viability of the diol-based sol-gel method to produce functional thin films with similar (even better) properties than those obtained by the hazardous 2-methoxyethanol sol-gel route.

5.4 Summary

Several approaches have been presented in this chapter to show the reader how the toxicity associated with the fabrication of lead-based ferroelectric films may be partially lowered by means of different and effective CSD strategies. Despite the inherent health hazards related to the use of lead reagents, the environmental load may be reduced by avoiding the volatilisation of lead to the atmosphere usually produced at high temperatures. This may also be done by soft solution processing of precursor solutions using organic solvents with significantly decreased toxicity with respect to the historically used 2-methoxyethanol compound.

Lead loss from the system is shown to be successfully avoided by the UV sol-gel photoannealing technique (Photochemical Solution Deposition). This combines the photo-excitation of organic components present in the gel-film by UV-irradiation with the subsequent crystallisation of the photo-activated film at rather low annealing temperatures. Films obtained by this method develop pure perovskite phases and fine-grained microstructures, the consequence of the limited grain growth promoted at such temperatures. Excesses of PbO with respect to the nominal composition of the films yield the formation of thin Pt_xPb interlayers between the film and the metal electrode of Pt silicon substrates. The aforementioned Pt_xPb interface remains in the low-temperature processed films because the annealing temperature is not high enough to produce the volatilisation of this excess lead. The formation of this interlayer is observed to be practically suppressed in nominally stoichiometric solution-derived films. That is, the films without excess PbO addition. Furthermore, compositional gradients of lead, typically obtained in lead-based multilayered films prepared by solution methods, are not detected in these films. The volatilisation of lead from the film surface (at high temperatures) and the presence of Pt_xPb bottom interfaces (at low temperatures) are expected to act as driving mechanisms for the lead diffusion within the film. This leads to the formation of heterogeneous compositional profiles. Films obtained by the UV sol-gel photoannealing technique show optimum dielectric and ferroelectric properties. These are improved when processing under oxygen atmosphere, due to the ozonolysis effect. The functionality of these films, together with the thermal budget of the process (<500°C) and total absence of lead volatilisation (potential source for integrated circuits contamination), makes them superior candidates for integration within the silicon semiconductor technology.

On the other hand, the toxicity of the organic solvents mostly employed in CSD methods can be significantly reduced and/or even suppressed by the use of 1,3-propanediol and water solvents, respectively. Thus, stable precursor solutions of lead-based compositions can be obtained through an entirely aqueous solution-gel route and a diol-based sol-gel synthetic method. In the first case, wetting by the aqueous solutions must be improved by enhancing the hydrophilicity of Pt silicon substrates by a wet chemical treatment of cleaning. The resulting crystallised films show reliable dielectric properties, although the detrimental interfaces (e.g., roughness) obtained in some of them prevent their ferroelectric optimisation. Regarding the diol-based sol-gel process, different synthetic routes are proposed which show to affect the physicochemical characteristics (polymeric network, homogeneity, and aging behaviour) of the respective precursor solutions. The introduction of different metal precursors, calcium acetate or calcium acetylacetonate, also influences the decomposition mechanism of the derived gels and films (via formation of intermediate carbonates). Crystalline films with tailored microstructures, heterostructures and electrical properties may therefore be obtained from the soft solution processing shown here.

Acknowledgments

The authors gratefully acknowledge Ricardo Jiménez (ICMM-CSIC) for his valuable contribution to this chapter, especially within the area of the ultraviolet sol-gel photoannealing technique (Photochemical Solution Deposition). Special thanks are due to Marlies K. Van Bael and co-workers from the University of Hasselt for providing fundamental knowledge about the aqueous solution method.

References

1 United Nations Conference on Environment and Development (UNCED), Rio de Janeiro, 3-14 June 1992
2 Xiao DQ (1998) Environmentally conscious ferroelectric research. J. Korean Phys. Soc. 32:S1798-S1800
3 Xiao DQ (1999) Environmentally conscious ferroelectrics research - present and prospect. Ferroelectrics 231:721-729
4 Yamada K, Halada K, Ijima K (2005) Web-site survey of recent ecomaterial developments in Japan. Mater. Trans. 46:2554-2560
5 Directive 2002/95/EC of the European Parliament and of the Council of 27 January 2003 on the restriction of the use of certain hazardous substances in electrical and electronic equipment
6 Gonnard P, Simon-Seveyrat L (2003) Environmentally conscious electroceramics. In: Kosec M, Kuscer D, Malic B (eds) Processing of Electroceramics, Conference Notes Jozef Stefan Institute, Ljubljana, pp 331-345
7 Saito Y, Takao H, Tani T, Nonoyama T, Takatori K, Homma T, Nagaya T, Nakamura M (2004) Lead-free piezoceramics. Nature 432:84-87

8 MN05-00009-5E Fujitsu Semiconductor Memory Manual FRAM Guide Book, September 2005. FUJITSU LIMITED Electronic Devices
9 Bretos I, Jiménez R, Rodríguez-Castellón E, García-López J, Calzada ML (2008) Heterostructure and compositional depth profile of low-temperature processed lead titanate-based ferroelectric thin films prepared by photochemical solution deposition. Chem. Mater. 20:1443-1450
10 Calzada ML, Bretos I, Jiménez R, Guillon H, Pardo L (2004) Low-temperature processing of ferroelectric thin films compatible with silicon integrated circuit technology. Adv. Mater. 16:1620-1624
11 Calzada ML, Bretos I, Jiménez R, Guillon H, Ricote J, Pardo L (2007) Low-temperature ultraviolet sol-gel photoannealing processing of multifunctional lead-titanate-based thin films. J. Mater. Res. 22:1824-1833
12 Yoshimura M (1998) Importance of soft solution processing for advanced inorganic materials. J. Mater. Res. 13:796-802
13 Van Bael MK, Nelis D, Hardy A, Mondelaers D, Van Werde K, D'Haen J, Vanhoyland G, Van Den Rul H, Mullens J, Van Poucke LC, Frederix F, Wouters DJ (2002) Aqueous chemical solution deposition of ferroelectric thin films. Integr. Ferroelectr. 45:113-122
14 Bretos I, Jiménez R, Calzada ML, Van Bael MK, Hardy A, Van Genechten D, Mullens J (2006) Entirely aqueous solution-gel route for the preparation of $(Pb_{1-x}Ca_x)TiO_3$ thin films. Chem. Mater. 18:6448-6456
15 Phillips NJ, Calzada ML, Milne SJ (1992) Sol-gel derived lead titanate films. J. Non-Cryst. Solids 147-148:285-290
16 Calzada ML, Bretos I, Jiménez R, Ricote J, Mendiola J, García-López J, Respaldiza MA (2005) Chemical solution deposition of $(Pb_{1-x}Ca_x)TiO_3$ thin films with x ~0.5 as new dielectrics for tunable components and dynamic random access memories. J. Am. Ceram. Soc. 88:3388-3396
17 Mendiola J, Calzada ML (2002) Ferroelectric thin films of modified lead titanate. In: Nalwa HS (ed) Handbook of Thin Film Materials. Vol. 3: Ferroelectric and Dielectric Thin Films. Academic Press, London
18 Kholkin A, Seifert A, Setter N (1998) Electromechanical properties of sol-gel derived Ca modified $PbTiO_3$ films. Appl. Phys. Lett. 72:3374-3376
19 Poyato R, Calzada ML, Pardo L (2003) Effects of substrate annealing and post-crystallization thermal treatments on the functional properties of preferentially oriented $(Pb,Ca)TiO_3$ thin films. J. Appl. Phys. 93:4081-4090
20 Mendiola J, Jiménez R, Ramos P, Bretos I, Calzada ML (2005) Dielectric properties of $Pb_{0.5}Ca_{0.5}TiO_3$ thin films. J. Appl. Phys. 98:024106
21 Jiménez R, Bretos I, Ricote J, Alemany C, Calzada ML, Mendiola J (2005) Size effects on $Pb_{0.5}Ca_{0.5}TiO_3$ thin films. J. Eur. Ceram. Soc. 25:2319-2323
22 Yu JJ (2004) Formation of high-quality advanced high-k oxide layers at low temperature by excimer UV lamp-assisted photo-CVD and sol-gel processing. Chem. Res. Chinese U. 20:396-402
23 Maekawa S, Ohishi T (1994) Evaluation of SiO_2 thin-films prepared by sol-gel method using photoirradiation. J. Non-Cryst. Solids 169:207-209
24 Boyd IW, Zhang JY (2001) Photo-induced growth of dielectrics with excimer lamps. Solid State Electron. 45:1413-1431
25 Imai H (2005) Ultraviolet (UV) irradiation. In: Sakka S (ed) Handbook of Sol-Gel Science and Technology: Processing, Characterisation and Applications. Kozuka H (ed) Vol. 1: Sol-gel Processing. Chapter 27, pp 639-650. Kluwer Academic Publishers, Boston

26 Ohishi T, Maekawa S, and Katoh A (1992) Synthesis and properties of tantalum oxide-films prepared by the sol-gel method using photoirradiation. J. Non-Cryst. Solids 147:493-498
27 Imai H, Hirashima H (1994) Novel modification method of sol-gel thin films: densification of silica gel films with electronic excitation. J. Ceram. Soc. Jpn. 102:1094-1096
28 Asakuma N, Hirashima H, Fukui T, Toki M, Awazu K, Imai H (2002) Photoreduction of amorphous and crystalline ZnO films. Jpn. J. Appl. Phys. 41:3909-3915
29 Asakuma N, Fukui T, Toki M, Awazu K, Imai H (2003) Photoinduced hydroxylation at ZnO surface. Thin Solid Films 445:284-287
30 Nakao Y, Nakamura T, Hoshiba K, Sameshima K, Kamisawa A, Abe K, Soyama N, Ogi K (1993) Micro-patterning of $PbZr_xTi_{1-x}O_3$ thin films prepared by photo sensitive sol-gel solution. Jpn. J. Appl. Phys. 32:4141-4143
31 Soyama N, Sasaki G, Atsuki T, Yonezawa T, Ogi K (1994) Preparation of dielectric thin-films from photosensitive sol-gel solution. Jpn. J. Appl. Phys. 33:5268-5271
32 Soppera O, Croutxé-Barghorn C, Carré C, Blanc D (2002) Design of photoinduced relief optical devices with hybrid sol-gel materials. Appl. Surf. Sci. 186:91-94
33 Setter N, Damjanovic D, Eng L, Fox G, Gevorgian S, Hong S, Kingon A, Kohlstedt H, Park NY, Stephenson GB, Stolitchnov I, Taganstev AK, Taylor DV, Yamada TJ (2006) Ferroelectric thin films: Review of materials, properties, and applications. J. Appl. Phys. 100:051606
34 International Technology Roadmap for Semiconductors, http://www.itrs.net, as on 28 February 2008
35 Kwoka CK, Desu SB (1993) Low temperature perovskite formation of lead zirconate titanate thin films by a seeding process. J. Mater. Res. 8:339-344
36 Kato K (1998) Low-temperature synthesis of $SrBi_2Ta_2O_9$ ferroelectric thin films through the complex alkoxide method: Effects of functional group, hydrolysis and water vapour treatment, Jpn. J. Appl. Phys. 37:5178-5184
37 Huang Z, Zang Q, Whatmore RW (1998) The role of an intermetallic phase on the crystallisation of sol-gel processed lead zirconate titanate thin films. J. Mater. Sci. Lett. 17:1157-1159
38 Celinska J, Joshi V, Narayan S, McMillan LD, Paz de Araujo CA (2000) Low temperature process for strontium bismuth tantalate thin films. Integr. Ferroelectr. 30:1-8
39 Kosec M, Malic B, Mandeljc M (2002) Chemical solution deposition of PZT thin films for microelectronics. Mater. Sci. Semicond. Process. 5:97-103
40 Calzada ML, González A, Poyato R, Pardo L (2003) Photo-sensitive sol-gel solutions for the low-temperature UV-assisted processing of $PbTiO_3$ based ferroelectric thin films. J. Mater. Chem. 13:1451-1457
41 Wu A, Vilarinho PM, Reaney I, Salvado IMM (2003) Early stages of crystallisation of sol-gel-derived lead zirconate titanate thin films. Chem. Mater. 15:1147-1155
42 Van de Leest RE (1995) UV photo-annealing of thin sol-gel films. Appl. Surf. Sci. 86:278-285
43 Pardo L, Poyato R, González A, Calzada ML (2002) Low temperature preparation of piezoelectric thin films by ultraviolet-assisted rapid thermal processing. Mat. Sci. Semicon. Proc. 5:77-83
44 González A, Poyato R, Pardo L, Calzada ML (2002) Photo-activated Ca-$PbTiO_3$ solutions for the preparation of films at low temperatures. Ferroelectrics 271:1635-1640
45 Barnum DW (1961) Electronic absorption spectra of acetyl-acetonato complexes – I. J. Inorg. Nucl. Chem. 21:221-237
46 Tohge N, Shinmou K, Minami T (1994) Effects of UV-irradiation on the formation of oxide thin films from chemically modified metal-alkoxides. J. Sol-Gel Sci. Techn. 2:581-585

47 JIPELEC, a division of QUALIFLOW THERM, http://www.jipelec.com, as on 28 February 2008
48 Boyd IW, Zhang JY (2001) Photo-induced growth of dielectrics with excimer lamps. Solid State Electron. 45:1413-1431
49 Zhang Y, Terril RH, Bohn PW (1999) Ultraviolet photochemistry and ex situ ozonolysis of alkanethiol self-assembled monolayers on gold. Chem. Mater. 11:2191-2198
50 Chen SY, Chen IW (1994) Temperature-time texture transition of $Pb(Zr_{1-x}Ti_x)O_3$ thin films. J. Am. Ceram. Soc. 77:2332-2344
51 Chen SY, Chen IW (1998) Texture development, microstructure evolution, and crystallization of chemically derived PZT thin films. J. Am. Ceram. Soc. 81:97-105
52 Impey SA, Huang Z, Patel A, Beanland R, Shorrocks NM, Watton R, Whatmore RW (1998) Microstructural characterization of sol-gel lead-zirconate-titanate thin films. J. Appl. Phys. 83:2202-2208
53 Huang Z, Zang Q, Whatmore RW (1999) Low temperature crystallization of lead zirconate titanate thin films by a sol-gel method. J. Appl. Phys. 85:7355-7361
54 Wu A, Vilarinho PM, Salvado IMM, Baptista JL, Zhou Z, Reaney I, Ramos AR, da Silva MFJ (2002) Effect of lead zirconate titanate seeds on Pt_xPb formation during the pyrolysis of lead zirconate titanate thin films. J. Am. Ceram. Soc. 85:641-646
55 Schwartz RW, Clem PG, Voigt JA, Byhoff ER, Van Stry M, Headley TJ, Nancy A. Missert NA (1999) Control of microstructure and orientation in solution-deposited $BaTiO_3$ and $SrTiO_3$ thin films. J. Am. Ceram. Soc. 82:2359-2367
56 Gust MC, Momoda LA, Evans ND, Mecartney ML (2001) Crystallization of sol-gel-derived barium strontium titanate thin films. J. Am. Ceram. Soc. 84:1087-1092
57 Calzada ML, Malic B, Sirera R, Kosec M (2002) Thermal-decomposition chemistry of modified lead-titanate aquo-diol gels used for the preparation of thin films. J. Sol-Gel Sci. Technol. 23:221-230
58 Gopinath CS, Hegde SG, Ramaswamy AV, Mahapatra S (2002) Photoemission studies of polymorphic $CaCO_3$ materials. Mater. Res. Bull. 37:1323-1332
59 Kim KS, O`Learly TJ, Winograd N (1973) X-ray photoelectron spectra of lead oxides. Anal. Chem. 45:2214-2218
60 Blachard Jr DL, Baer DR (1992) The interactions of Co, Mn and water with calcite surfaces. Surf. Sci. 276:27-39
61 Bukhtiyarov VI, Nizovskii AI, Bluhm H, Hävecker M, Kleimenov E, Knop-Gericke A, Schlögl R (2006) Combined in situ XPS and PTRMS study of ethylene epoxidation over silver. J. Catal. 238:260-269
62 Seifert A., Lange FF, Speck JS (1995) Epitaxial growth of $PbTiO_3$ thin films on $SrTiO_3$ from solution precursors. J. Mat. Res. 10:680-691
63 Lange FF (1996) Chemical solution routes to single-crystal thin films. Science 273:903-909
64 Lefevre MJ, Speck JS, Schwartz RW, Dimos D, Lockwood SL (1996) Microstructural development in sol-gel derived lead zirconate titanate thin films: the role of precursor stoichiometry and processing environment. J. Mat. Res. 11:2076-2084
65 Schwartz RW (1997) Chemical solution deposition of perovskite thin films. Chem. Mat. 9:2325-2340
66 Sirera R, Leinen D, Rodríguez-Castellón E, Calzada ML (1999) Processing effects on the compositional depth profile of ferroelectric sol-gel Ca-$PbTiO_3$ thin films. Chem. Mater. 11:3437-3444
67 Bretos, I (2006) Low-toxic Chemical Solution Deposition Methods for the Preparation of Multifunctional $(Pb_{1-x}Ca_x)TiO_3$ Thin films. PhD Thesis Universidad Autónoma de Madrid

68 Zhou C, Newns DM (1997) Intrinsic dead layer effect and the performance of ferroelectric thin film capacitors. J. Appl. Phys. 82:3081-3088
69 Dawber M, Rabe KM, Scott JF (2005) Physics of thin-film ferroelectric oxides. Rev. Mod. Phys. 77:1083-1130
70 Watts BE, Leccabue F, Bocelli G, Padeletti G, Kaciulis S, Pandolfi L (2005) Lead enrichment at the surface of lead zirconate titanate thin films. J. Eur. Ceram. Soc. 25:2495-2498
71 Dana SS, Etzold KF, Clabes J (1991) Crystallization of sol-gel derived lead zirconate titanate thin-films. J. Appl. Phys. 69:4398-4403
72 Vasco E, Bohme O, Roman E, Zaldo C (2001) Origin and control of the lead-enriched near surface region of (Pb,La)TiO_3. Appl. Phys. Lett. 78:2037-2039
73 Watts BE, Leccabue F, Tallarida G, Ferrari S, Fanciulli M, Padeletti G (2003) Surface segregation mechanisms in ferroelectric thin films. J Electroceram. 11:139-147
74 Martin M, Fromm E (1997) Low-temperature oxidation of metal surfaces. J. Alloy Compd. 258:7-16
75 Wagner C (1933) The theory of the warm-up process. Z. Phys. Chem. B 21:25
76 Cabrera N, Mott NF (1948) Theory of the oxidation of metals. Rep. Prog. Phys. 12:163-184
77 Kingery WD, Bowen HK, Uhlmann DR (1976) Introduction to Ceramics. John Wiley & Sons, New York, pp 217-264
78 Song ZT, Ren W, Zhang LY, Yao X, Lin CL (1999) A study on abnormal electric properties of lead lanthanum titanate thin films caused by excess PbO. Thin Solid Films 353:25-28
79 Aggarwal S, Madhukar S, Nagaraj B, Jenkins IG, Ramesh R, Boyer L, Evans JT (1999) Can lead nonstoichiometry influence ferroelectric properties of Pb(Zr,Ti)O_3 thin films? Appl. Phys. Lett. 75:716-718
80 Amarande L, Miclea C, Tanasoiu, C (2002) Effect of excess PbO on the structure and piezoelectric properties of Bi-modified $PbTiO_3$ ceramics. J. Eur. Ceram. Soc. 22:1269-1275
81 Reaney IM, Brooks K, Klissurska R, Pawlaczyk C, Setter N (1994) Use of transmission electron-microscopy for the characterization of rapid thermally annealed, solution-gel, lead-zirconate titanate films. J. Am. Ceram. Soc. 77:1209-1216
82 Poyato R, González A, Calzada ML, Pardo L (2002) High pyroelectric coefficients of Ca-modified lead titanate sol-gel thin films obtained by multiple deposition and crystallization. Ferroelectrics 271:1975-1980
83 Jiménez R, Calzada ML, Mendiola J (1998) Conditioning effects on RTP (Pb,Ca)TiO_3 thin films. Thin Solid Films 335:292-298
84 Northrop DA (1967) Vaporization of lead zirconate-lead titanate materials. J. Am. Ceram. Soc. 50:441-445
85 Jaffe B, Cook WR, Jaffe H (1971) Piezoelectric Ceramics. Academic Press, New York
86 Atkin RB, Fulrath RM (1971) Point defects and sintering of lead zirconate titanate. J. Am. Ceram. Soc. 54:265-270
87 Holman RL, Fulrath RM (1972) Intrinsic nonstoichiometry in single-phase $Pb(Zr_{0.5},Ti_{0.5})O_3$. J. Am. Ceram. Soc. 55:192-197
88 Holman RL, Fulrath RM (1973) Intrinsic nonstoichiometry in the lead-zirconate lead-titanate system determined by Knudsen effusion. J. Appl. Phys. 44:5227-5236
89 Kingon AI, Clark JB (1983) Sintering of PZT ceramics: 1. Atmosphere control. J. Am. Ceram. Soc. 66:253-256
90 Kingery WD (1959) Densification during sintering in the presence of a liquid phase. 1: Theory. J. Appl. Phys. 30:301-306

91 Snow GS (1973) Improvements in atmosphere sintering of transparent PLZT ceramics. J. Am. Ceram. Soc. 56:479-480
92 Snow GS (1974) Elimination of porosity in $Pb(Zr,Ti)O_3$ ceramics by liquid-phase sintering. J. Am. Ceram. Soc. 57:272-272
93 Kingon AI, Clark JB (1983) Sintering of PZT ceramics: 2. Effect of PbO content on densification kinetics. J. Am. Ceram. Soc. 66:256-260
94 Guha JP, Hong DJ, Anderson HU (1988) Effect of excess PbO on the sintering characteristics and dielectric properties of $Pb(Mg_{1/3}Nb_{2/3})O_3$-$PbTiO_3$-based ceramics. J. Am. Ceram. Soc. 71:C152-C154
95 Song BM, Kim DY, Shirasaki S, Yamamura H (1989) Effect of excess PbO on the densification of PLZT ceramics. J. Am. Ceram. Soc. 72:833-836
96 Zhou L, Zimmermann A, Zeng YP, Aldinger F (2004) Effects of PbO content on the sintering behavior, microstructure, and properties of La-doped PZST antiferroelectric ceramics. J. Mater. Res.: Mat. in Electr. 15:145-151
97 Yang JK, Kim WS, Park HH (2000) The effect of excess Pb content on the crystallization and electrical properties in sol-gel derived $Pb(Zr_{0.4}Ti_{0.6})O_3$ thin films. Thin Solid Films 377:739-744
98 Mantese JV, Micheli AL, Hamdi AH, West RW (1989) Metalorganic deposition (MOD): A nonvacuum, spin-on, liquid-based, thin film method. Mater. Res. Bull. 14:48-53
99 Polli AD, Lange FF, (1995) Pyrolysis of $Pb(Zr_{0.5}Ti_{0.5})O_3$ precursors: Avoiding lead partitioning. J. Am. Ceram. Soc. 78:3401-3404
100 Tani T, Payne DA (1994) Lead-oxide coatings on sol-gel derived lead lanthanum zirconium titanate thin-layers for enhanced crystallization into the perovskite structure. J. Am. Ceram. Soc. 77:1242-1248
101 Tuttle BA, Headley TJ, Bunker BC, Schwartz RW, Zender TJ, Hernandez CL, Goodnow DC, Tissot RJ, Michael J, Carim AH (1992) Microstructural evolution of $Pb(Zr,Ti)O_3$ thin films prepared by hybrid metallo-organic decomposition. J. Mater. Res. 7:1876-1882
102 Tu YL, Milne SJ (1995) A study of the effects of process variables on the properties of PZT films produced by a single-layer sol-gel technique. J. Mater. Sci. 30:2507-2516
103 Tuttle BA, Schwartz RW, Doughty DH, Voigt JA (1990) In: Myers ER, Kingon AI (eds) Ferroelectric Thin Films. Mater. Res. Soc. Symp. Proc. 200, Pittsburgh
104 Bernstein DS, Kisler Y, Wahl JM, Bernacki SE, Collins SR (1991) In: Kingon AI, Myers ER, Tuttle B (eds) Ferroelectric Thin Films II. Mater. Res. Soc. Symp. Proc. 243, Boston
105 Hirano T, Kawai H, Suzuki H, Kaneko S, Wada T (1999) Effects of excess lead addition on processing of sol-gel derived lanthanum-modified lead zirconate titanate thin film. Jpn. J. Appl. Phys. 38:5354-5357
106 Ishida M, Matsunami H, Tanaka T (1977) Preparation and properties of ferroelectric PLZT thin films by rf sputtering. J. Appl. Phys. 48:951-953
107 Polli AD, Lange FF, Levi CG (2000) Metastability of the fluorite, pyrochlore, and perovskite structures in the PbO-ZrO_2-TiO_2 system. J. Am. Ceram. Soc. 83:873-881
108 Sato E, Huang YH, Kosec M, Bell A, Setter N (1994) Lead loss, preferred orientation, and the dielectric properties of sol-gel prepared lead titanate thin films. Appl. Phys. Lett. 65:2678-2680
109 Yang JK, Kim WS, Park HH (2000) The effect of excess Pb content on the crystallization and electrical properties in sol-gel derived $Pb(Zr_{0.4}Ti_{0.6})O_3$ thin films. Thin Solid Films 377:739-744

110 Lefevre MJ, Speck JS, Schwartz RW, Dimos D, Lockwood SJ (1996) Microstructural development in sol-gel derived lead zirconate titanate thin films: The role of precursor stoichiometry and processing environment. J. Mater. Res. 11:2076-2084
111 Seifert A, Lange FF, Speck JS (1995) Epitaxial growth of $PbTiO_3$ thin films on (001) $SrTiO_3$ from solution precursors. J. Mater. Res. 10:680-691
112 Jacobs RN, Salamanca-Riba L (2003) Role of Pb excess in the crystallization of lead zirconate titanate films derived via sol-gel processing. J. Mater. Res. 18:1405-1411
113 Härdtl KH, Rau H (1969) PbO vapour pressure in the $Pb(Ti_{1-x}Zr_x)O_3$ system. Solid State Commun. 7:41-45
114 Calame F, Muralt P (2007) Growth and properties of gradient free sol-gel lead zirconate titanate thin films. Appl. Phys. Lett. 90:062907
115 Villegas M, Caballero AC, Kosec M, Moure C, Durán P, Fernández JF (1999) Effects of PbO excess in $Pb(Mg_{1/3}Nb_{2/3})O_3$-$PbTiO_3$ ceramics: Part I. Sintering and dielectric properties. J. Mater. Res. 14:891-897
116 Villegas M, Fernández JF, Caballero AC, Samardija Z, Drazig G, Kosec M (1999) Effects of PbO excess in $Pb(Mg_{1/3}Nb_{2/3})O_3$-$PbTiO_3$ ceramics: Part II. Microstructure development. J. Mater. Res. 14:898-905
117 Du X, Chen IW (1998) Fatigue of $Pb(Zr_{0.53}Ti_{0.47})O_3$ ferroelectric thin films. J. Appl. Phys. 83:7789-7798
118 Fukushima J (1976) Preparation of $BaTiO_3$ films by hydrolysis of organometallic compounds. Am. Ceram. Soc. Bull. 55:1064-1065
119 Fukushima J, Kodaira K, Marsushita T (1984) Preparation of ferroelectric PZT films by thermal decomposition of organometallic compounds. J. Mater. Sci. 19:595-598
120 Budd KD, Dey SK, Payne DA (1985) Sol-gel processing of $PbTiO_3$, $PbZrO_3$, PZT and PZLT thin films. Brit. Cer. Proc. 36:107-121
121 Dey SK, Budd KD, Payne DA (1988) Thin-film ferroelectrics of PZT by sol-gel processing. IEEE T. Ultrason. Ferr. 35: 80-81
122 Lipeles R, Ives N, Leung M (1985) Sol-gel processing of lead zirconate titanate films. In: 2nd International Conference on Ultrastructure Processing of Ceramics, Glasses and Composites. Palm Coast, Florida
123 Brinker CJ, Sherer GW (1990) Sol-Gel Science: The Physics and Chemistry of Sol-Gel Processing. Academic Press, Boston
124 Sakka S (ed) (2005) Handbook of Sol-Gel Science and Technology: Processing, Characterisation and Applications. Kozuka H (ed) Vol. 1: Sol-gel Processing. Kluwer Academic Publishers, Boston
125 Klein LC (1988) Sol–Gel Technology for Thin Films, Fibers, Preforms, Electronics and Specialty Shapes. Noyes Publishers, New Jersey
126 Schwartz RW, Schneller T, Waser R (2004) Chemical solution deposition of electronic oxide films. C. R. Chimie 7:433-461
127 Schneller T, Waser R (2002) Chemical solution deposition of ferroelectric thin films - State of the art and recent trends. Ferroelectrics 267:293-301
128 Livage J, Sanchez C (1992) Sol-gel chemistry. J. Non-Cryst. Solids 145:11-19
129 Directive 2001/59/EC of the European Union Commission of 6th August 2001 on the approximation of the laws, regulations and administrative provisions relating to the classification, packaging and labelling of dangerous substances
130 Blum JB, Gurkovich SR (1985) Sol-gel derived $PbTiO_3$. J. Mat. Sci. 20:4479-4483
131 Willems GJ, Wouters DJ, Maes HE, Nouwen R (1997) Nucleation and orientation of sol-gel PZT-films on Pt electrodes. Integr. Ferroelectr. 15:19-28
132 Schwartz RW, Bunker BC, Dimos DB, Assink RA, Tuttle BA, Tallant DR, Weinstock IA (1991) Solution chemistry effects in $Pb(Zr,Ti)O_3$ thin film processing. Integr. Ferroelectr. 2:243-254

133 Hasenkox U, Mitze C, Waser R (1997) Metal propionate synthesis of magnetoresistive $La_{1-x}(Ca,Sr)_xMnO_3$ thin films. J. Am. Ceram. Soc. 80:2709-2713
134 Tu YL, Calzada ML, Phillips NJ, Milne SJ (1996) Synthesis and electrical characterisation of thin films of PT and PZT made from a diol-based sol-gel route. J. Am. Ceram. Soc. 79:441-448
135 Pechini MP (1967) US Patent 3 330 697
136 Haertling GH (1991) PLZT thin-films prepared form acetate precursors. Ferroelectrics 116:51-63
137 Lin CT, Li JS, Webb JS, Leung MS, Lipeles RA (1995) Deposition and characterization of photoconductive PZT thin films. J. Electrochem. Soc. 142:1957-1960
138 Yi G, Sayer M (1996) An acetic acid water based sol-gel PZT process I: Modification of Zr and Ti alkoxides with acetic acid. J. Sol-Gel Sci. Tech. 6:65-74
139 Pontes FM, Leite ER, Nunes MSJ, Pontes DSL, Longo E, Magnani R, Pizani PS, Varela JA (2004) Preparation of $Pb(Zr,Ti)O_3$ by soft chemical route. J. Eur. Ceram. Soc. 24:2969-2976
140 Tu YL, Milne SJ (1995) A study of the effects of process variables on the properties of PZT films produced by a single-layer sol-gel technique. J. Mater. Sci. 30:2507-2516
141 Lee B, Zhang J (2001) Preparation, structure evolution and dielectric properties of $BaTiO_3$ thin films and powders by an aqueous sol-gel process. Thin Solid Films 388:107-113
142 Asai T, Camargo ER, Kakihana M, Osada M (2000) A novel aqueous solution route to the low-temperature synthesis of $SrBi_2Nb_2O_9$ by use of water-soluble Bi and Nb complexes. J. Alloy. Compd. 309:113-117
143 Schäfer J, Sigmund W, Roy S, Aldinger F (1997) Low temperature synthesis of ultrafine $Pb(Zr,Ti)O_3$ powder by sol-gel combustion. J. Mater. Res. 12:2518-2521
144 Van Werde K, Vanhoyland G, Nelis D, Mondelaers D, Van Bael MK, Mullens J, Van Poucke LC (2001) Phase formation of ferroelectric perovskite 0.75 $Pb(Zn_{1/3},Nb_{2/3})O_3$-0.25 $BaTiO_3$ prepared by aqueous solution-gel chemistry. J. Mater. Res. 11:1192-1197
145 Mondelaers D, Vanhoyland G, Van den Rul H, D'Haen J, Van Bael MK, Mullens J, Van Poucke LC (2002) Synthesis of ZnO nanopowder via an aqueous acetate-citrate gelation method. Mater. Res. Bull. 37:901-914
146 Nelis D, Van Bael MK, Van den Rul H, Mullens J, Van Poucke LC, Vanhoyland G, D'Haen J, Laureyn W, Wouters DJ (2002) Ferroelectric $SrBi_2Nb_2O_9$ thin films by aqueous chemical solution deposition. Integr. Ferroelectr. 45:205-213
147 Hardy A, Mondelaers D, Van Bael MK, Mullens J, Van Poucke LC, Vanhoyland G, D'Haen J (2004) Synthesis of $(Bi,La)_4Ti_3O_{12}$ by a new aqueous solution-gel route. J Eur. Ceram. Soc. 24:905-909
148 Van den Rul H, Mondelaers D, Van Bael MK, Mullens J (2006) Water-based wet chemical synthesis of (doped) ZnO nanostructures. J. Sol-Gel Sci. Tech. 39:41-47
149 Truijen I, Van Bael MK, Van den Rul H, D'Haen J, Mullens J (2007) Synthesis of thin dense titania films via an aqueous solution-gel method. J. Sol-Gel Sci. Tech. 41:43-48
150 Van Werde K, Vanhoyland G, Mondelaers D, Van den Rul, Van Bael MK, Mullens J, Van Poucke LC (2007) The aqueous solution-gel synthesis of perovskite $Pb(Zr_{1-x}Ti_x)O_3$ (PZT). J. Mater. Sci. 42:624-632
151 Hardy A, D'Haen J, Goux L, Wouters D, Van Bael M, Van den Rul H, Mullens J (2007) Aqueous chemical solution deposition of ferroelectric Ti^{4+} cosubstituted $(Bi,La)_4Ti_3O_{12}$ thin films. Chem. Mater. 19:2994-3001
152 Van Bael MK, Arcon I, Van Werde K, Nelis D, Mullens J, Van Poucke LC (2005) Structure determination by EXAFS of Nb-peroxo-citrato complexes in aqueous solution-gel systems. Phys. Scripta, T115:415-417

153 Hardy A, D'Haen J, Van Bael MK, Mullens J (2007) An aqueous solution-gel citratoperoxo-Ti(IV) precursor: synthesis, gelation, thermo-oxidative decomposition and oxide crystallization. J. Sol-Gel Sci. Tech. 44:65-74
154 Narendar Y, Messing GL (1997) Synthesis, decomposition and crystallization characteristics of peroxo-citrato-nioubium: an aqueous niobium precursor. Chem. Mater. 9:580-587
155 Livage J, Henry M, Sanchez C (1988) Sol-gel chemistry of transition metal oxides. Prog. Solid State Ch. 18:259-341
156 Kourgiantakis M, Matzapetakis M, Raptopoulou CP, Terzis A, Salifoglou A (2000) Lead citrate chemistry. Synthesis, spectroscopic and structural studies of a novel lead(II)-citrate aqueous complex. Inorg. Chim. Acta 297:134-138
157 Javoric S, Kosec M, Malic B (2000) Water-based synthesis of $La_{0.5}Sr_{0.5}CoO_3$ (LSCO) electrodes for ferroelectric thin films. Integr. Ferroelectr. 30:309-316
158 Bretos I, Ricote J, Jiménez R, Mendiola J, Jiménez Riobóo R, Calzada ML (2005) Crystallisation of $(Pb_{1-x}Ca_x)TiO_3$ ferroelectric thin films as a function of the Ca^{2+} content. J. Eur. Ceram. Soc. 25:2325-2329
159 Bretos I, Ricote J, Jiménez Riobóo R, Pardo L, Calzada ML (2007) Elastic instability of the nano-structured state as an intrinsic probe to study the early formation stages of sol–gel derived $(Pb_{1-x}Ca_x)TiO_3$ thin films. Appl. Phys. A-Mater. 89:967-973. J. Eur. Ceram. Soc. 25:2325-2329
160 Zhao YP, Wang GC, Lu TM (1999) Surface-roughness effect on capacitance and leakage current of an insulating film. Phys. Rev. B 60:9157-9164
161 Cross LE (1987) Relaxor ferroelectrics. Ferroelectrics 76:241-267
162 Scott JF (2000) Ferroelectric Memories. Springer, Heilderberg
163 Schubert U (2005) Chemical modification of titanium alkoxides for sol-gel processing. J. Mater. Chem. 15:3701-3715
164 Bradley DC, Mehrotra RC, Rothwell IP, Singh A (2001) Alkoxo and Aryloxo Derivatives of Metals. Academic Press, London
165 Milne SJ, Pyke SH (1991) Modified sol-gel process for the production of lead titanate films. J. Am. Ceram. Soc. 74:1407-1410
166 Sakka S (2005) Viscosity and spinnability of gelling solutions. In: Sakka S (ed) Handbook of Sol-Gel Science and Technology: Processing, Characterisation and Applications. Almeida RM (ed) Vol. 2: Characterization and Properties of Sol-gel Materials and Products. Chapter 12, pp 237-263. Kluwer Academic Publishers, Boston
167 Fredrickson AG (1964) Principles and Applications of Rheology. Prentice-Hall, Englewood Cliffs, New Jersey
168 Hennings D, Rosenstein G, Schreinemacher H (1991) Hydrothermal preparation of barium titanate from barium-titanium acetate gel precursors. J. Eur. Ceram. Soc. 8:107-115
169 Sengupta SS, Ma L, Adler DL, Payne DA (1995) Extended x-ray absorption fine structure determination of local structure in sol-gel-derived lead titanate, lead zirconate, and lead zirconate titanate. J. Mater. Res. 10:1345-1348
170 Lakeman CDE, Payne DA (1992) Processing effects in the sol-gel preparation of PZT dried gels, powders and ferroelectric thin layers. J. Am. Ceram. Soc. 75:3091-3096
171 Hasenkox, U, Hoffmann S, Waser R (1998) Influence of precursor chemistry on the formation of $MTiO_3$ (M=Ba, Sr) ceramic thin films. J. Sol-Gel Sci. Tech. 12:67-79
172 Hoffmann S, Waser R (1999) Control of the morphology of CSD-prepared $(Ba,Sr)TiO_3$ thin films. J. Eur. Ceram. Soc. 19:1339-1343
173 Xu Y, Mackenzie J (1992) Ferroelectric thin films prepared by sol-gel processing. Integr. Ferroelectr. 1:17-42

174 Yi G, Sayer M (1991) Sol-gel processing of complex oxide films. Am. Ceram. Soc. Bull. 70:1173-1179
175 Kato K, Suzuki K, Fu DS, Nishizawa K, Miki T (2002) Chemical approach using tailored liquid sources for traditional and novel ferroelectric thin films. Jpn. J. Appl. Phys. Part 1 41:6829-6835
176 Alemany C, Jiménez R, Revilla J, Mendiola J, Calzada ML (1999) Pulsed hysteresis loops on ferroelectric thin films. J. Phys. D Appl. Phys. 32:L79-L82
177 Moulson AJ, Herbert JM (1990) Electroceramics. Chapman & Hall, London
178 Hoffmann S, Waser R (1997) Dielectric properties, leakage behaviour, and resistance degradation of thin films of the soli solution series $Ba(Ti_{1-y}Zr_y)O_3$. Integr. Ferroelectr. 17:141-152
179 Kingon AI, Maria JP, Streiffer SK (2000) Alternative dielectrics to silicon dioxide for memory and logic devices. Nature 406:1032-1038
180 Dimos D, Mueller CH (1998) Perovskite thin films for high-frequency capacitor applications. Ann. Rev. Mater. Sci. 28:397-419

Chapter 6
Synchrotron Radiation Diffraction and Scattering in Ferroelectrics

Luis E. Fuentes-Cobas

6.1 Synchrotron Radiation

A synchrotron [1, 2, 3, 4] is a cyclical particle accelerator in which the electric (accelerating) and magnetic (deflecting) fields are synchronized to ensure that the particles (generally e^-, e^+) follow a prescribed geometric and energetic path.

Fig. 6.1 shows schematically a synchrotron. It is a large instrument, with a size comparable to that of a stadium. Constructing a synchrotron costs from tens to hundreds millions of dollars and each beamline costs approximately another two or three million dollars. The heart of a synchrotron is the storage ring, where a beam of electrons with speeds approaching that of light describes approximately circular paths. The path curvature is produced by a magnetic field. According to the laws of Electrodynamics, an accelerated charge emits electromagnetic radiation. In a synchrotron, centripetal acceleration generates an extremely intense beam of *synchrotron radiation* that is emitted tangent to the electrons path. Under the operation conditions of a synchrotron, the emitted radiation typically goes from ultraviolet (10^{16} Hz) to x-ray (10^{18} Hz).

With the help of Fig. 6.1, we describe in simplified form the operation of a synchrotron. The elements numbered from 1 to 6 are at high vacuum. The process begins in the source of electrons at the end of the linear accelerator (1: LINAC). In this section the electrons acquire energies on the order of 100 MeV. The second phase of acceleration takes place in the booster (2), where energy rises to some GeV and speed reaches a value of approximately 0.99999995c (c = speed of light).

Centro de Investigación en Materiales Avanzados, S. C.
Complejo Industrial Chihuahua, Miguel de Cervantes 120
31109 Chihuahua, México.
Phone (52 614) 439 11 59
Fax (52 614) 439 48 23
luis.fuentes@cimav.edu.mx

With the acquired speed, electrons are injected to the storage ring (3), where bending magnets (4) guide them along the planned trajectory.

Any curvature in the path generates synchrotron radiation. The electrons lose energy in this process. The radio frequency systems (RFS, 5) perform physical work on the electron beam, restoring the radiated energy.

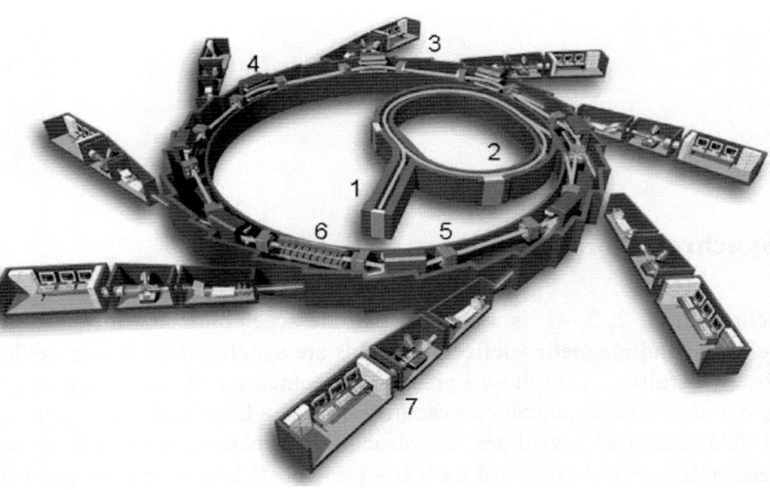

Fig. 6.1 Diagram of a synchrotron. 1: Linac. 2: Booster. 3: Storage ring. 4: Bending magnet. 5: RFSystem. 6: Insertion device (wiggler or undulator). 7: Beamline (optical and experimental hutches, work station). Reproduced from: http://commons.wikimedia.org/wiki/Image:Sch%C3%A9ma_de_principe_du_synchrotron.jpg. Copyright © EPSIM 3D/JF Santarelli, Synchrotron Soleil.

In the storage ring, due to absorption and scattering of electrons by the (few) particles remaining in the high vacuum, the population of circulating electrons decreases over time. The synchrotrons are periodically recharged with electrons, say twice a day.

Historically, the first generation of synchrotrons was created to study elementary particles, for example through electron-positron collisions. The synchrotron radiation was considered "a nuisance" because it represented a loss of energy for the investigated particles. Soon researchers acknowledged that the synchrotron could be regarded as a very important source of x-rays, which can be used to study the structure of matter. Thus came the second generation of synchrotrons during the 1970s and 1980s. These synchrotrons generated x-rays in the bending magnets. The third generation of synchrotrons (1990s) was born with the introduction of the insertion devices (wigglers and undulators – Fig. 6.1, item 6), which significantly increased the flow of photons.

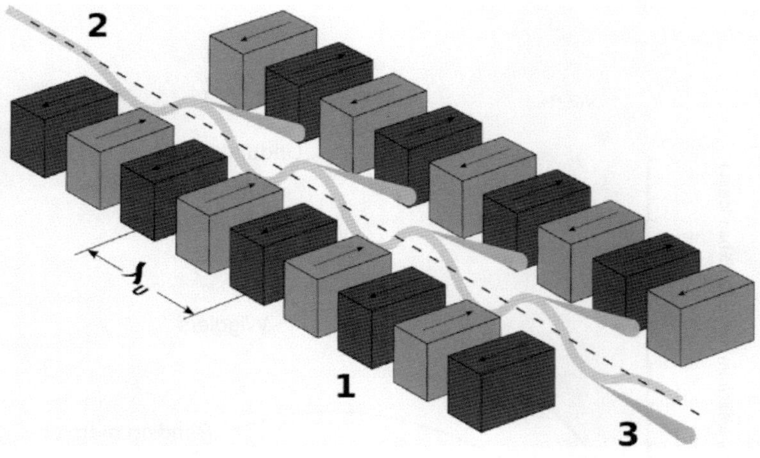

Fig. 6.2 Schematic description of an Undulator. 1: magnets, 2: electron beam, 3: radiation. λ_U: undulator wavelength. Reproduced from: http://commons.wikimedia.org/wiki/Image:Undulator.png Figure by: Bastian Holst.

Insertion devices are collections of electromagnets inserted into the straight sections of the storage ring. These magnets are placed with alternating polarity, so that they oblige the electron beam to move through a swinging trajectory. The oscillation generates synchrotron radiation. The wigglers produce intense beams of photons in a broad energy spectrum. In undulators, the frequency of oscillations is designed so that the constructive interference associated with certain frequencies produce an important strengthening of the intensities. Undulators thus produce ultra-high intensity in a narrow frequency range. Fig. 6.2 shows schematically the operation of an undulator and

Fig. 6.3 compares schematically the brightness of bending magnets, wigglers and undulators.

The intense radiation beams produced by the synchrotron emerge tangentially from the storage ring and are used for research in the beamlines (Fig. 6.1, item 7).

A beamline consists of:

1. Optical hutch, where radiation is suited to the specific needs of the investigation. For example, the left side of Fig. 6.4 shows a focusing and monochromator system.
2. Experimental hutch, for example, a high-resolution diffractometer (Fig. 6.4, to the right of the fixed exit slip).
3. Work station, computer for numerical control of the experiment and data collection.

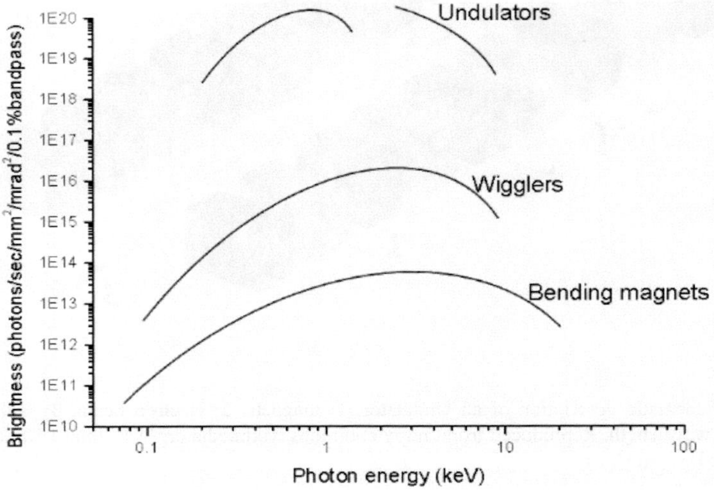

Fig. 6.3 Brightness of different radiation sources as functions of photon energy.

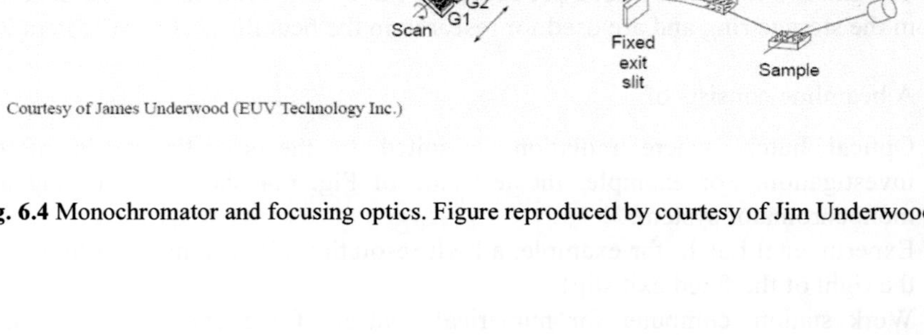

Courtesy of James Underwood (EUV Technology Inc.)

Fig. 6.4 Monochromator and focusing optics. Figure reproduced by courtesy of Jim Underwood.

Fig. 6.5 Some representative synchrotrons around the World

Fig. 6.5 shows aerial views of some synchrotrons representing the third generation. Today (2010), a fourth generation of synchrotrons, with an even greater flow, is being born [5, 6]. Fourth generation synchrotrons use the constructive interference of various bundles of synchrotron radiation, in a manner that resembles a laser, to amplify the radiation intensity.

Characteristics of synchrotron radiation:

- High brightness and high intensity, exceeding X-ray tubes and other artificial and natural light sources by many orders of magnitude. Third generation sources typically have a brightness larger than 10^{18} photons s^{-1} mm^{-2} $mrad^{-2}$ $(0.1\%BW)^{-1}$, where 0.1%BW denotes a bandwidth $10^{-3}\omega$ centred around the frequency ω.
- High collimation (small angle divergence of the beam).
- Low emittance, i.e., the product of source cross section and solid angle of emission is small.
- Widely tunable in energy/wavelength by monochromatization (from eV to MeV).
- High level of polarization (linear or elliptical).
- Pulsed light emission (pulse durations at or below one nanosecond).

A comparison among different types of synchrotron radiation generators is given in Table 6.1.

Table 6.1 Advantages of different types of synchrotron radiation generators.

Bending magnet	Wiggler	Undulator
Broad spectrum	Higher photon energy	Brighter radiation
Good photon flux	More photon flux	Smaller spot size
No heat load	Expensive magnet structure	Partial coherent
Less expensive	Expensive cooled optics	Expensive
Easier access	Less access	Less access

Some uses of synchrotron radiation:

- Structural analysis of crystalline and amorphous materials
- Powder diffraction analysis
- Crystallography of proteins and other macromolecules
- Magnetic scattering
- Small angle X-ray scattering
- X-ray absorption spectroscopy
- Inelastic X-ray scattering
- Tomography
- X-ray imaging in phase contrast mode
- Photolithography for MEMS structures.
- High pressure studies
- Residual stress analysis
- X-Ray Multiple Diffraction

Interested researchers have access to a number of synchrotrons, where they can go with their samples and conduct experiments. The investigator writes a proposal, a scientific panel evaluates its quality and, whether the proposal is approved, beamline time is assigned. The websites of synchrotrons contain assistance on how to apply for requesting beamline time.

6.2 X-Ray Diffraction and Scattering: Fundamentals

6.2.1 Bragg Law, Reciprocal Lattice and Ewald Representation

We review X-ray diffraction (XRD) in bulk samples, with sample size of the order of 10x10x1 mm^3 or smaller and crystallite size of the order $t \sim 1$ μm or smaller. The basic necessary condition for the manifestation of a XRD peak is the Bragg law:

$$2d \sin\theta_B = n\lambda \qquad 2d_h \sin\theta_B = \lambda \tag{1}$$

$d_h = d/n$ is the interplanar distance associated with the considered diffraction maximum (eg $d_{200} = d_{100}/2$), n is the reflection order, θ_B is the Bragg angle and λ is the radiation wavelength.

Let's introduce the description of the Bragg condition in the reciprocal – Fourier-space representation. The reciprocal basis $\{\mathbf{b}_i\}$ (i = 1, 2, 3) corresponding to the direct basis $\{\mathbf{a}_i\}$ is defined by:

$$\mathbf{b}_1 = 2\pi\frac{\mathbf{a}_2\times\mathbf{a}_3}{V_a}; \qquad \mathbf{b}_2 = 2\pi\frac{\mathbf{a}_3\times\mathbf{a}_1}{V_a}; \qquad \mathbf{b}_3 = 2\pi\frac{\mathbf{a}_1\times\mathbf{a}_2}{V_a} \tag{2}$$

$V_a = \mathbf{a}_1\cdot\mathbf{a}_2\times\mathbf{a}_3$ is the direct unit cell volume. The orientation of the $\mathbf{b}_i$ vectors is rigidly linked to that of the $\mathbf{aa}_i$. The scale used to represent the $\mathbf{b}_i$ in a direct space diagram is conventional.

Each reciprocal space node

$$\mathbf{B}_h = \sum_i h_i\mathbf{b}_i \quad (-\infty < h_i < \infty)$$

describes a family of planes in crystal direct space: a) Reciprocal vectors characterize the orientation of crystal planes [vector $\mathbf{B}_h$ is always normal to the family $\mathbf{h}$ = (h_1 h$_2$ h_3)]. b) The interplanar distance for family h is given by the inverse of the module of the associated reciprocal vector

$$\left(d_h = {}^{2\pi}\!\big/\!\left|\mathbf{B}_h\right|\right)$$

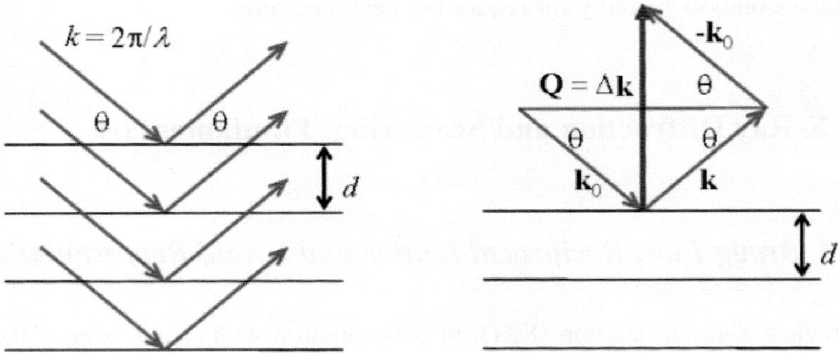

Fig. 6.6 The Bragg condition in terms of wave vectors.

Fig. 6.6 shows a family of crystalline planes oriented according to the Bragg condition. The incident and diffracted beams are characterized respectively by their wave vectors $\mathbf{k}_0$ and $\mathbf{k}$, both of module $2\pi/\lambda$. It is easy to demonstrate that the Bragg law, in terms of the so-called scattering vector $\mathbf{Q} \equiv \Delta\mathbf{k} = \mathbf{k} - \mathbf{k}_0$, may be written as follows:

$$\mathbf{Q} = \mathbf{B}_h \tag{3}$$

Demonstration: Suppose (3) is valid. The modules of involved vectors will fulfil:

$$|\mathbf{Q}| = 2k\sin\theta = \frac{4\pi}{\lambda}\sin\theta = \frac{2\pi}{d_h} \tag{4}$$

$$2d_h \sin\theta = \lambda \quad \text{(Bragg's law).} \tag{4a}$$

To visualize geometrically Equation (3) consider the construction shown in Fig. 6.7. The so-called *Ewald sphere* has a radius equal to the module of $\mathbf{k}_0$. As defined in the picture, this sphere passes through the origin of the reciprocal lattice. If the surface of the sphere intersects a reciprocal lattice node, the way described by vector $\mathbf{B}_h$, then the condition $\mathbf{k} = \mathbf{k}_0 + \mathbf{B}_h$ is fulfilled. This is simply another way to write down the diffraction condition. We come then to an alternative formulation of the Bragg law: If the Ewald sphere intersects a reciprocal lattice node, then there will be a diffraction peak. The diffracted wave vector $\mathbf{k}$ will point from the centre of the sphere toward the node $\mathbf{B}_h$ (the reciprocal origin is trivial, it represents the transmitted direct beam).

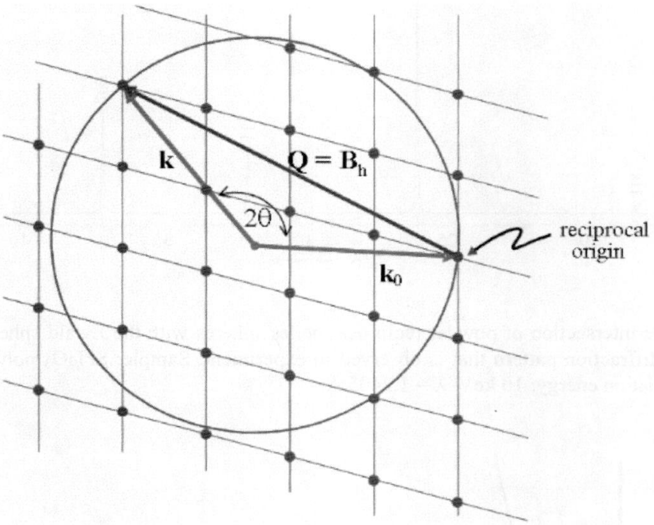

Fig. 6.7 The Ewald construction.

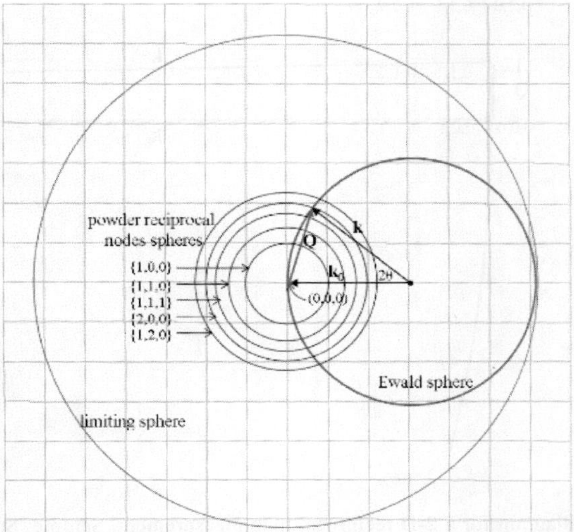

Fig. 6.8a Ewald description of a powder diffraction experiment. The random distribution of crystallites' orientations gives rise to the powder reciprocal nodes spheres.

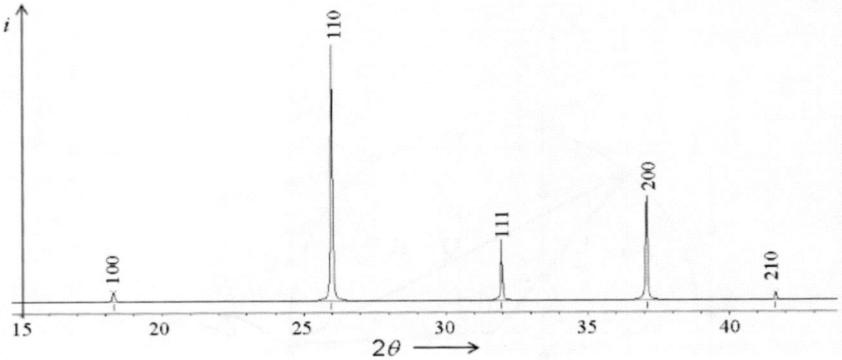

Fig. 6.8b The intersection of powder reciprocal nodes spheres with the Ewald sphere generates the powder diffraction pattern that is observed in experiment. Sample: $SrTiO_3$ polycrystal (a = 3.90 Å). Radiation energy: 10 keV. λ = 1.2405 Å.

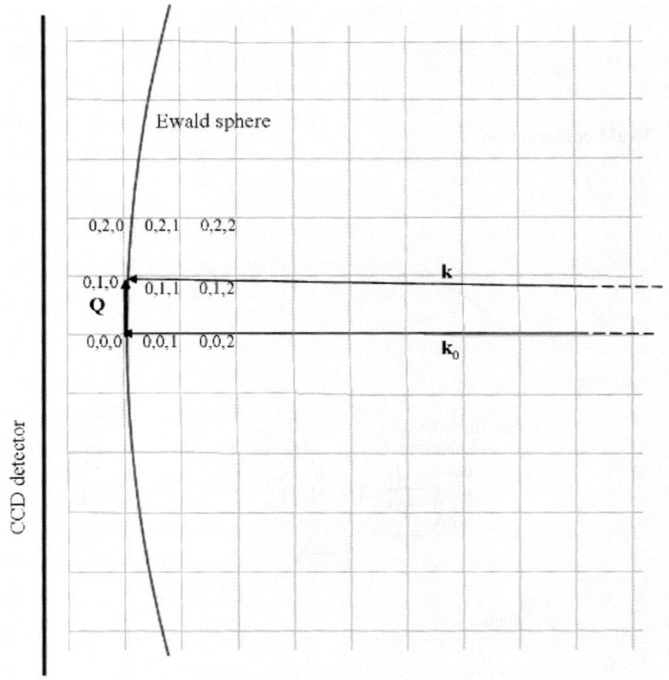

Fig. 6.9a Ewald diagram for a 2-D reciprocal map measurement. Sample: $SrTiO_3$ single crystal. Radiation energy: 67 keV. The Ewald sphere radius, k = 33.9 $Å^{-1}$, is about 21 times the reciprocal lattice parameter. Vector $\mathbf{b}_1$ points entering the paper. The incident beam points anti-parallel to vector b_3.

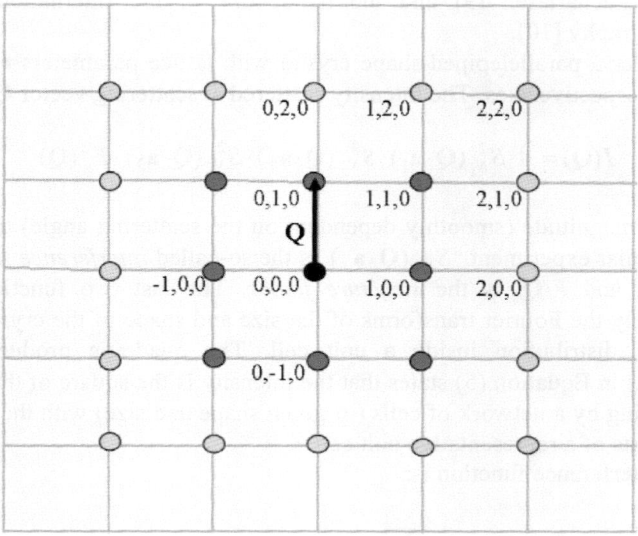

Fig. 6.9b Ewald diagram of Fig. 6.9a, detail of the $l = 0$ reciprocal plane ($\mathbf{b}_3$ points outwards from the drawing). Darker nodes are closer to fulfilment of the $\mathbf{Q} = \mathbf{B}_h$ condition (the Bragg law).

Fig. 6.8 and Fig. 6.9 respectively show simulations, following Ewald's representation, of: a) A conventional powder diffraction experiment at Beamline 2-1 in the Stanford Synchrotron Radiation Laboratory (SSRL), Stanford Linear Accelerator Centre, with 10 keV radiation, $\lambda = 1.2405$ Å. b) An experiment for two-dimensional mapping of reciprocal space in the Beamline X17B1 of the National Synchrotron Light Source (NSLS), Brookhaven National Laboratory, with 67 keV radiation, $\lambda = 0.1852$ Å. In both cases the hypothetical sample is $SrTiO_3$ ($a = 3.9019$ Å).

We recommend the University of Cambridge web page [7] as a friendly introduction to the reciprocal lattice and the Ewald representation.

6.2.2 Diffraction Peaks

6.2.2.1 Interference Function and Structure Factor

We analyze the distribution of intensities, the quantity that is measured, in a XRD pattern. We apply the so-called kinematical diffraction theory, valid as a satisfactory approach in cases of our interest (excluding perfect crystals, which are uncommon). Our exposition follows to a certain extent the excellent review by

Renaud [8]. For detailed deductions and discussions, see for example the classic work of Zachariasen [9] and the handbooks of the International Union of Crystallography [10].

Consider a parallelepiped-shape crystal with lattice parameters $\mathbf{a}_i$ and N_i cells over the respective axes. The intensity scattered at scattering vector $\mathbf{Q}$ is given by:

$$I(\mathbf{Q}) = A \cdot S_{N_1}^2(\mathbf{Q}\cdot\mathbf{a}_1)\cdot S_{N_2}^2(\mathbf{Q}\cdot\mathbf{a}_2)\cdot S_{N_3}^2(\mathbf{Q}\cdot\mathbf{a}_3)\cdot F^2(\mathbf{Q}) \tag{5}$$

A is a magnitude (smoothly dependent on the scattering angle) that represents the particular experiment. $S_{N_j}(\mathbf{Q}\cdot\mathbf{a}_j)$ is the so-called *interference function* for N_j unit cells and $F(\mathbf{Q})$ is the *structure factor*. The last two functions represent respectively the Fourier transforms of the size and shape of the crystal and of the electronic distribution inside a unit cell. The quadratic product of Fourier transforms in Equation (5) states that the intensity is the square of the convolution of scattering by a network of cells (of given shape and size) with the scattering by the contents of a representative unit cell.

The interference function is:

$$S_{N_j}(\mathbf{Q}\cdot\mathbf{a}_j) = \sum_{n=0}^{N-1} \exp(i\mathbf{Q}\cdot\mathbf{a}_j\cdot n), \qquad j = 1,2,3 \tag{6}$$

Its squared module is:

$$S_{N_j}^2(\mathbf{Q}\cdot\mathbf{a}_j) = \frac{\sin^2(N_j\mathbf{Q}\cdot\mathbf{a}_j/2)}{\sin^2(\mathbf{Q}\cdot\mathbf{a}_j/2)}, \qquad j = 1,2,3 \tag{7}$$

If the crystal is large enough, say, $N_j \sim 10^4$, the interference functions tend to a three-dimensional periodic collection of Dirac delta functions. The intensity is different from zero only if the scattering vector satisfies the so-called Laue conditions [equivalent to Equation (3)]: $\mathbf{Q}\cdot\mathbf{a}_1 = 2\pi h$, $\mathbf{Q}\cdot\mathbf{a}_2 = 2\pi k$, $\mathbf{Q}\cdot\mathbf{a}_3 = 2\pi l$. In the diffraction maxima directions the intensity is:

$$I_{hkl} = AN_1^2 N_2^2 N_3^2 F_{hkl}^2 \tag{8}$$

For finite crystals, the diffraction maxima given by (7) show broadenings inversely proportional to N_j.

The structure factor associated with the scattering vector $\mathbf{Q}$ represents the amplitude of the electric field scattered by the considered unit cell. It is the Fourier transform of the scattering matter average density in a cell [11]:

$$F(Q) = \mathcal{F}[\langle\rho(\mathbf{r})\rangle] = \int\langle\rho(\mathbf{r})\rangle e^{i\mathbf{Q}\cdot\mathbf{r}} d^3\mathbf{r} \tag{9}$$

$\rho(\mathbf{r})$ is the matter density at a given point in a given instant. $\langle\rho(\mathbf{r})\rangle$ involves averaging first the time vibrations of atoms around their equilibrium positions and then averaging a representative number of unit cells (each of them averaged over time).

The widely used isolated-atom approximation calculates the average density of each atom in its particular vibration and then applies superposition to find the total average density. This approach is considered satisfactory, but loses some accuracy in cases where coupled motion of atoms is important (for example, covalent bond). In the considered approximation, the average density is as follows:

$$\langle\rho(\mathbf{r})\rangle=\sum_{k=1}^{N} n_k \int \rho(\mathbf{r}-\mathbf{r}_k)\,p_k(\mathbf{r}_k-\mathbf{r}_{k_0})\,d^3\mathbf{r}_k \tag{10}$$

N is the number of atoms in the cell unit, n_k is the occupation factor of atom k, $\rho(\mathbf{r}-\mathbf{r}_k)$ is the electron density (X-ray case) and $p_k(\mathbf{r}_k-\mathbf{r}_{k_0})$ is the *probability density function*, that expresses the probability for atom k to show a displacement $(\mathbf{r}_k-\mathbf{r}_{k_0})$ with respect to its reference position. Equation (10) does not imply that atoms are spherical, but it assumes that they are not deformed under vibration.

Combining (9) and (10) one obtains:

$$\langle\rho(\mathbf{r})\rangle=\sum_{k=1}^{N} n_k F_k(\mathbf{Q}) \tag{11}$$

with:

$$F_k(\mathbf{Q})=\int\left\{\int \rho_k(\mathbf{r}-\mathbf{r}_k)\,p_k(\mathbf{r}_k-\mathbf{r}_{k_0})\,d^3\mathbf{r}\right\}e^{i\mathbf{Q}\cdot\mathbf{r}}\,d^3\mathbf{r} \tag{12}$$

6.2.2.2 Atomic Scattering and Debye-Waller Factors

With the definitions $\mathbf{u}=(\mathbf{r}_k-\mathbf{r}_{k_0})$, $\mathbf{v}=(\mathbf{r}-\mathbf{r}_k)$ and through application of the Fourier convolution theorem, Equation (12) can be transformed into the most familiar expression for the structure factor:

$$F_k(\mathbf{Q})=\sum_{k=1}^{N} n_k f_k(\mathbf{Q})T_k(\mathbf{Q})\,e^{i\mathbf{Q}\cdot\mathbf{r}} \tag{13}$$

with:

$$f_k(\mathbf{Q})=\int \rho_k(\mathbf{v})e^{i\mathbf{Q}\cdot\mathbf{v}}\,d^3\mathbf{v} \tag{14}$$

and:

$$T_k(\mathbf{Q})=\int p_k(\mathbf{u})e^{i\mathbf{Q}\cdot\mathbf{u}}\,d^3\mathbf{u} \tag{15}$$

Equation (14) defines the atomic scattering factor, also called "form factor" and (15) defines the Debye-Waller factor (DWF), traditionally known as "temperature factor" (the latter name is not the best one; see Equation (19) and its explanation).

In the directions of the diffraction maxima, the structure factor adopts the form

$$F_{hkl} = \sum_{j\_unit\_cell} n_j \cdot f_j \cdot T_j \cdot \exp\left[2\pi i\left(hx_j + ky_j + lz_j\right)\right] \qquad (16)$$

The magnitudes x_j, y_j, z_j are the atomic coordinates of the atom j.

In the intensities formula (Equation 5), the structure factor appears squared. The measurement of scattered intensities (and not of electromagnetic fields amplitudes) leads to a loss of information known as *the phase problem* in Crystallography. The well-known Friedel's law

$$\left(\left|F_{hkl}\right|^2 = \left|F_{-h-k-l}\right|^2\right)$$

is a particular case of the phase problem.

In the X-ray case, if atom k shows spherical symmetry, the atomic scattering factor is calculated by:

$$f_j^0(|\mathbf{Q}|) = 4\pi \int_0^\infty \rho_j(u) \frac{\sin(|\mathbf{Q}|u)}{|\mathbf{Q}|u} u^2 du \qquad (17)$$

If the energy of the incident X-rays is close to an irradiated atom absorption edge, then the so-called *anomalous scattering* takes place, absorption increases and the scattered intensities are affected significantly. In these cases, corrections such as (18) are applied:

$$f = f^0 + f' + if'' \qquad (18)$$

The imaginary component correction (if'') characterizes the X-ray atomic absorption.

The anomalous scattering is anisotropic, leading to the breakdown of the Friedel Law. It opens possibilities for solving the phase problem. The technique known as Multiple Anomalous Scattering (MAD) [12, 13] is the representative tool developed for the mentioned objective.

The DWF is the Fourier transform of the probability density function $p_k(\mathbf{u})$ that describes statistically the displacement of atom k relative to its average position. There are various simplified expressions for the DWF. It is often assumed that the probability density function shows a Gaussian tendency with spherical (isotropic) or tri-axial (anisotropic) symmetry. For the anisotropic case, the expression that is obtained for the DWF of an atom is:

$$T(\mathbf{Q}) = e^{-\frac{1}{2}\langle \mathbf{Q}\cdot\mathbf{u}\rangle^2} = \exp\left[-8\pi^2\left\langle u_{\mathbf{Q}}^2\right\rangle\frac{\sin^2\theta}{\lambda^2}\right] \tag{19}$$

The magnitude $u_{\mathbf{Q}}$ is the projection of atomic displacement in the direction of the scattering vector:

$$\left(u_{\mathbf{Q}} = \mathbf{u}\cdot\frac{\mathbf{Q}}{|\mathbf{Q}|}\right)$$

Equation (19) was initially established to characterize *atomic vibrations*, according to the laws of Molecular Dynamics, under the so-called harmonic approximation. Today we know that also applies in the case of *static displacive disorder*.

It is correct, then, to interpret the magnitude

$$\left\langle u_{\mathbf{Q}}^2\right\rangle$$

in Equation (19) as the anisotropic mean square displacement of the considered atom from its equilibrium position, with both static and kinetic contributions.

Under isotropic displacements, the DWF adopts the form:

$$T(|\mathbf{Q}|) = \exp\left[-8\pi^2\left\langle u^2\right\rangle\frac{\sin^2\theta}{\lambda^2}\right] \tag{19a}$$

The usual dimensions of $\left\langle u^2\right\rangle$ are Å^2. In practice the abbreviation $B = 8\pi^2\left\langle u^2\right\rangle$ is also employed. It is as well expressible in Å^2 (though not exactly associated with the same physical interpretation).

The DWF characterizes the intensity decrease of the DRX peaks generated by random variations of the atomic positions. The effect increases with the displacement amplitudes $\left\langle u^2\right\rangle$ and becomes important for large values of the scattering angle (2θ). Where does the intensity lost by the diffraction maxima go? To the diffractogram background, as diffuse scattering. The Debye-Waller factor is essentially linked with diffuse scattering, i.e., signals scattered by the sample in directions different from those given by the Laue conditions. As an initial characterization of the general tendencies of scattering, we have the following:

The radiation that is not contributing to the diffraction maxima will be part of the scattered radiation. It may be isotropically diffuse or exhibit mono-, bi- or three-dimensional structure in the reciprocal space, depending on the dimensionality of the direct-space disorder.

The more important the structural disorder (time oscillations, static periodicity faults or overlap of both) the higher the intensity of scattered radiation.

Each type of departure from periodicity shows its characteristic geometrical pattern of diffuse scattering. The $\langle u^2 \rangle$ dependence of DWF represents a general tendency.

6.2.3 *Diffuse Scattering*

6.2.3.1 Pair Distribution Function

Diffraction analysis leads to long-range space-and time-averaged high-resolution structure determination via information contained in the diffraction peaks. Material information contained in the background is frequently neglected. But every point in a diffraction pattern, including the background, contains structural information. Local (static or dynamic) atomic positions deviations from the reference points decrease the peak intensities, in the amount characterized by the Debye-Waller factor. The intensities not appearing in the Bragg peaks contribute to the so-called diffuse scattering. Careful measurement and data processing of [diffraction + diffuse scattering] data leads to the so-called *pair distribution function* (PDF), containing important information regarding the local environment of atoms in the investigated materials. In the particular case of ferroelectrics, differences between local and long-range ordering may be significant.

The PDF is defined as [14, 15]:

$$G(r) = \frac{2}{\pi} \int_0^\infty Q[S(Q) - 1] \sin Qr \, dQ \qquad (20)$$

S (Q) is the corrected and normalized powder X-ray scattering intensity, which depends on the scattering vector length *Q*. Required corrections include absorption, scattering and multiple inelasticity. In words, *G (r)* is the Fourier transform of the scattering factor *Q*[*S(Q)* -1]. *G (r)* represents the probability (relative to the average electron density ρ_0) of finding pairs of atoms separated by the distance *r*:

$$G(r) = 4\pi[\rho(r) - \rho_0] \qquad (21)$$

G(*r*) is weighted by the respective atomic numbers of the considered atoms.

6.2.3.2 Crystal Dimensions

The diffraction pattern produced by a "large" and practically perfect crystal, consists of extremely sharp spots at the reciprocal space nodes, with intensities given (kinematical theory) by Equation (8). If the three-dimensional crystal is *small*, say with linear dimensions of a few hundred nanometres or smaller, the widths of the spots in reciprocal space acquire significant dimensions. A very well-known and useful manifestation of this effect is the broadening of the diffraction peaks in a powder diffraction experiment. This effect is characterized by the *Scherrer Equation* (22), helpful for estimating the average crystal size from the measurement of peak broadening.

$$t = \frac{K\lambda}{\beta \cos\theta_B} \tag{22}$$

K is a constant with an approximate value of 0.9, λ is the wavelength, t is the average crystal size, β is the so-called *diffraction integral broadening* (over the scattering angle 2θ scale, in radians, corrected for "instrumental" broadening) and θ_B is the Bragg angle.

If the investigated crystal shows the parallelepiped form associated with Equation (8), then the crystal size considered in (22) is:

$$t = \sqrt{\frac{\left(\frac{h}{a_1}\right)^2 + \left(\frac{k}{a_2}\right)^2 + \left(\frac{l}{a_3}\right)^2}{\left(\frac{h}{N_1 a_1^2}\right)^2 + \left(\frac{k}{N_2 a_2^2}\right)^2 + \left(\frac{l}{N_3 a_3^2}\right)^2}} \tag{23}$$

Consider now a two-dimensional quasicrystal, virtually infinite along $\mathbf{a}_1$ and $\mathbf{a}_2$ ($N_1, N_2 \rightarrow \infty$), but formed by a single layer of cells in the $\mathbf{a}_3$ direction ($N_3 = 1$). The scattered intensities remain Dirac delta functions with respect to $\mathbf{a}_1$ and $\mathbf{a}_2$, but no longer on $\mathbf{a}_3$. The dependence of the interference function (6) on ($\mathbf{Q}\cdot\mathbf{a}_3$) is now identically equal to 1. Mathematically, considering as variable the magnitude l in $\mathbf{Q}\cdot\mathbf{a}_3 = 2\pi l$, ·the scattered intensity becomes:

$$I_{hkl}^{2D} = AF_{hkl}^2 N_1^2 N_2^2 \tag{24}$$

different from zero for all l, while the conditions $\mathbf{Q}\cdot\mathbf{a}_1 = 2\pi h$ and $\mathbf{Q}\cdot\mathbf{a}_2 = 2\pi k$ are fulfilled. The intensity modulation is given by the structure factor. The representation of the considered phenomenon on a reciprocal map is illustrative: Intensity is scattered by the 2D-quasicrystal if the scattering vector $\mathbf{Q}$ ends at a point within any of the rods parallel to $\mathbf{b}_3$, defined by the Laue conditions on $\mathbf{a}_1$ and $\mathbf{a}_2$. See Fig. 6.10.

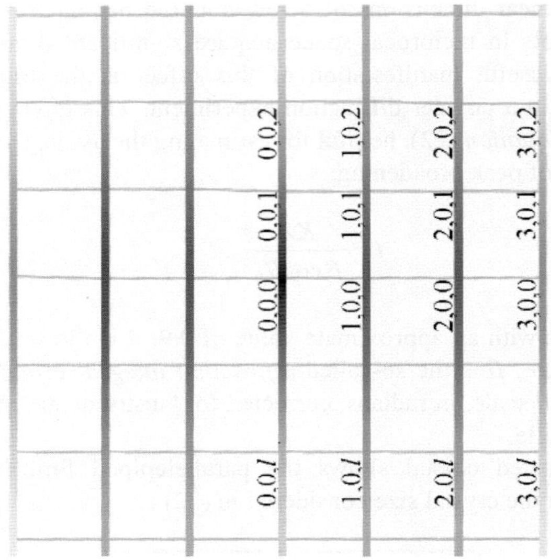

Fig. 6.10 Reciprocal scattering rods associated with a 2-D quasi-crystal. The considered quasi-crystal occupies the ($\mathbf{a}_1$, $\mathbf{a}_2$) direct plane. Vector $\mathbf{b}_2$ points inwards the drawing. The intensity scattered along the rods is determined by the structure factor.

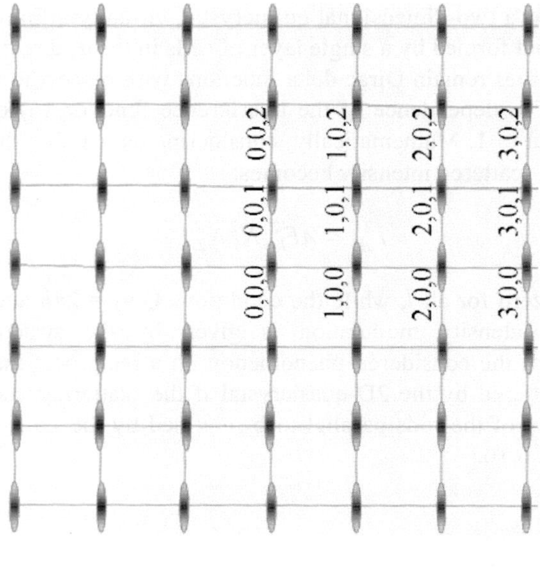

Fig. 6.11 Scattering cigar-shaped reciprocal nodes associated to a pancake-shaped crystal. The direct crystal plane is ($\mathbf{a}_1$, $\mathbf{a}_2$). Vector $\mathbf{b}_2$ points inwards the drawing.

The case of a pancake shaped crystal, intermediate between the infinite and the quasi 2-D crystal, leads to the condition of scattering allowed in elongated nodes of reciprocal space. Fig. 6.11 shows diffracting volumes in reciprocal space for the mentioned case. Eqs. (22) and (23) apply.

How is the diffraction pattern produced by a semi-infinite crystal? This question is important because it characterizes the investigation of a crystal through the incidence of an x-ray beam on its surface. A semi-infinite crystal is the combination of an ideal (infinite) crystal and a step function, representative of the crystal being *truncated*. The intensity scattered by the crystal may be obtained through Fourier convolution of the above mentioned functions. The interference function associated with the direction perpendicular to the sample limiting surface is:

$$S_{N_3}(Q_3a_3) = \sum_{n_3=-\infty}^{0} \exp(iQ_3a_3n_3) \tag{25}$$

$$\left|S_{N_3}(Q_3a_3)\right| = \frac{1}{2\sin(Q_3a_3/2)} \tag{26}$$

If the truncated crystal is *large* in the $\mathbf{a}_1$ and $\mathbf{a}_2$ directions:

$$I(Q_3) = \frac{A\left|F_{hkl}\right|^2 N_1^2 N_2^2}{4\sin^2(Q_3a_3/2)} \tag{27}$$

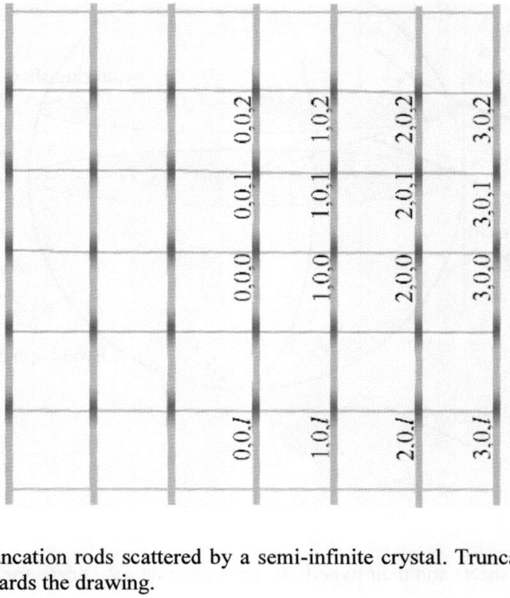

Fig. 6.12 Crystal truncation rods scattered by a semi-infinite crystal. Truncation plane is $z = 0$. Vector $\mathbf{b}_2$ points inwards the drawing.

The rods in reciprocal space represented in Fig. 6.12 are called *crystal truncation rods* (CTR). They represent the directions of measurable scattered intensity. The new effect in the considered case is the dependence on Q_3 of the intensity scattered along a CTR, represented by the distribution of gray shades. The intensity maxima are found at the Bragg points, $Q_3a_3 = 0, 2\pi, 4\pi,...$ Differentiation relative to the 2D quasicrystal case may be easily seen by comparing Equations (24) and (27).

The CTR provide important information on the crystal surface. If the surface is tilted, the CTR are as well inclined. The so-called anti-Bragg points of reciprocal space ($Q_3a_3 = \pi, 3\pi,...$) are interesting. In general, the intensity at these points is not zero, as would be in a large or a pancake-shaped crystal. It has been found that the intensity at these points is highly sensitive to variations in surface roughness.

The case of a one-dimensional regular distribution of matter, specifically along the $\mathbf{a}_3$ axis, is described as follows. For $N_1 = N_2 = 1$, the interference functions (Equation 6.6) associated to directions $\mathbf{a}_1$ and $\mathbf{a}_2$ simplify to unity, whereas the one corresponding to $\mathbf{a}_3$ tends to a collection of Dirac delta functions as $N_3 \rightarrow \infty$. In the reciprocal planes satisfying $\mathbf{Q}\cdot\mathbf{a}_3 = 2\pi l$, the scattered intensity is

$$I_{hkl}^{1D} = AF_{hkl}^2 N_3^2 \tag{28}$$

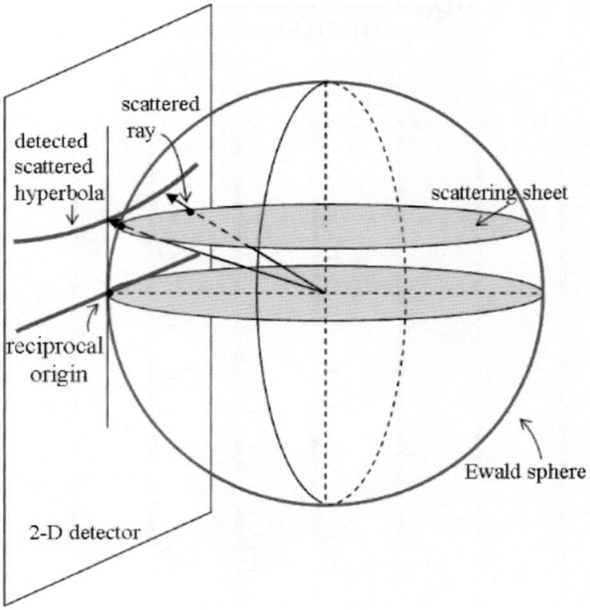

Fig. 6.13 Scattering sheets and their hyperbolic projection on a 2-D detector.

F^2_{hkl} depends on h, k = continuous variables and l = integer. Fig. 6.13 shows the intersection of the Ewald sphere with the scattering sheet associated with a linear atomic periodic array, and the hyperbolic projection of this intersection on a two-dimensional radiation sensor.

6.2.3.3 Static and Kinematical Disorders

Real crystals normally show deviations from ideal periodicity. These imperfections reduce the intensity of the diffraction maxima and generate intensity distributions around the reciprocal space nodes. Depending on the nature, dimensionality and importance of these imperfections, the diffuse scattering distributions can form streaks, planes or rods connecting the reciprocal nodes. The diffuse radiation is produced by static disorder (chemical heterogeneities, micro-deformations) and / or kinematic disorder (vibration, phonons). The following Equation characterizes both cases [16, 17]:

$$I(\mathbf{Q}) = \frac{Q^2}{q^2} X_{\mathbf{q},j} \left| \hat{\mathbf{Q}} \cdot \boldsymbol{\varepsilon}_{\mathbf{q},j} \right|^2 \qquad (29)$$

In (29), $I(\mathbf{Q})$ is the contribution of the harmonic perturbation with vector $\mathbf{q}$ and polarization j to the intensity scattered in the direction $\mathbf{Q}$. Lattice distortions can take up to three polarizations: one longitudinal (L) and two transverse (T).
$\mathbf{q} = \mathbf{Q} - \mathbf{B}_h$, represents the point of observation in reciprocal space, referred to the closest network node $\mathbf{B}_h$. $\hat{\mathbf{Q}}$ is the unit vector in the direction of the scattering vector $\mathbf{Q}$. $\boldsymbol{\varepsilon}_{\mathbf{q},j}$ characterizes atomic or ionic displacement in a distortion (L or T) wave. It is named "polarization" and in ferroelectric crystals (for example in an optical phonon) it is associated with the electric polarization.
Case A) Static disorder:

$$X_{\mathbf{q},j} = \left| U_{\mathbf{q},j} \right|^2 \qquad (30)$$

$U_{\mathbf{q},j}$ is the displacement amplitude for a wave vector $\mathbf{q}$ and a polarization j. For small q, $U_{\mathbf{q},j}$ is virtually independent of q [18, 19] and the $I(Q)$ dependence is proportional to q^{-2}.
Case B) A phonon with wave vector $\mathbf{q}$:

$$X_{\mathbf{q},j} = \frac{2kTq^2}{\rho\omega^2}, \quad I(\mathbf{Q}) = \frac{2kTQ^2}{\rho\omega^2} \left| \hat{\mathbf{Q}} \cdot \boldsymbol{\varepsilon}_{\mathbf{q},j} \right|^2 \qquad (31)$$

T is the temperature, k the Boltzman constant, ρ the density, ω the angular frequency. The dependence on ω^{-2} implies that "soft" (low frequency) phonons generate greater diffuse scattering than "hard" ones. In the case of a linear scattering relation ($\omega = vq$, acoustic modes, v is the speed), the expression (31) may represent a q^{-2} dependence, similar to the static case. This is not a general rule.

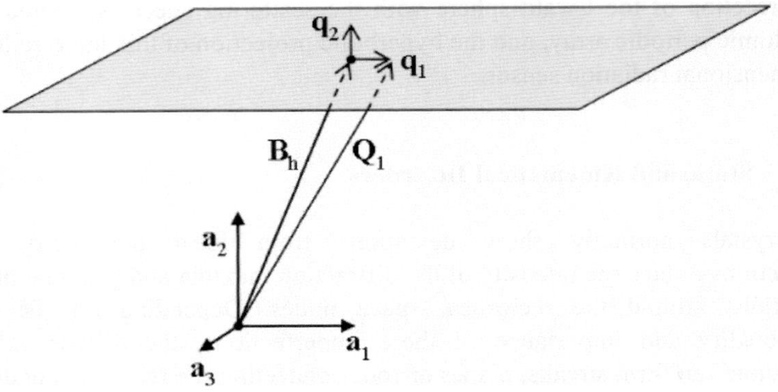

Fig. 6.14 Interpretation of the **Q·ε** factor in Equation (31).

The functional relationship

$$I \propto \left|\hat{\mathbf{Q}} \cdot \boldsymbol{\varepsilon}\right|^2$$

is interesting and important. Its meaning is illustrated with the help of Fig. 6.14. An observation point $\mathbf{Q}_1$ in reciprocal space, close to node $\mathbf{B}_h$, is plotted. $\mathbf{Q}_1 = \mathbf{B}_h + \mathbf{q}_1$ belongs to the reciprocal plane $\mathbf{Q}\cdot\mathbf{a}_2 = 2\pi k$. The other wave vector shown, $\mathbf{q}_2$, is perpendicular to the mentioned plane. For the phonon with wave $\mathbf{q}_1$ as well as for the one with $\mathbf{q}_2$, the dot product $\mathbf{Q}\cdot\boldsymbol{\varepsilon}$ is approximately equal to $\mathbf{B}_h\cdot\boldsymbol{\varepsilon}$. One detail on which to focus attention is the orientation of $\boldsymbol{\varepsilon}$ with respect to vector $\mathbf{B}_h$. Consider the cases associated with phonons of wave vectors $\mathbf{q}_1$ and $\mathbf{q}_2$.

Case A) $\mathbf{B}_h$ belongs to the plane passing through the reciprocal origin ($k = 0$). The longitudinal phonons $\mathbf{q}_1$ generate high-intensity diffuse scattering, since $\boldsymbol{\varepsilon} \parallel \mathbf{B}_h$. Transverse phonons $\mathbf{q}_1$, with $\boldsymbol{\varepsilon} \perp \mathbf{B}_h$, do not produce diffuse scattering. Transverse phonons $\mathbf{q}_2$ produce diffuse scattering, while longitudinal $\mathbf{q}_2$ do not.

Case B) $\mathbf{B}_h \parallel \mathbf{a}_2$. Phonon $\mathbf{q}_1$ generates diffuse scattering only if it is transverse, while $\mathbf{q}_2$ only if it is longitudinal.

6.2.3.4 Linear Disorder

Because of its importance in the study of ferroelectrics, we devote some space to the scattering by linear disorders. We present the model by Comès, Lambert and Guinier (CLG) [20] for this type of imperfection.

Consider a crystal of ferroelectric perovskite ABO_3. For example, it may be $KNbO_3$, in which the cation "B" (Nb) contributes the bulk of the intensity scattered by the unit cell. Ferroelectricity is associated with a small displacement of B cations from the centre of the cell unit. In the CLG model the off-centring equilibrium displacement of the B cation is along a cube diagonal <111 >. In different cells the displacements of the B cations occur in different directions and, according to the distribution of these movements through the crystal, so different average structures, symmetries and polarizations will take place.

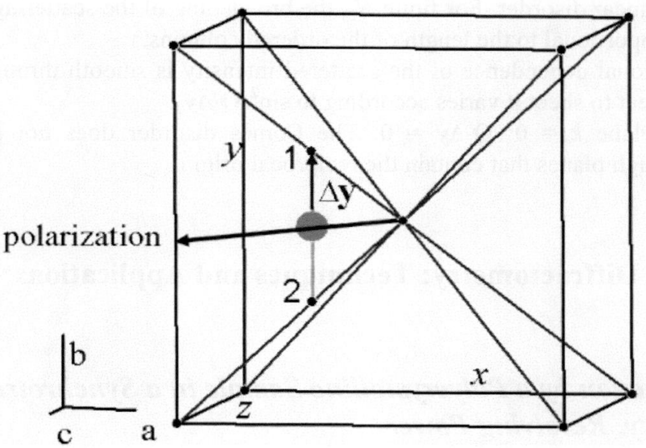

Fig. 6.15 Atomic displacements and polarization in the Comès-Lambert-Guinier linear disorder.

We analyze in some detail the orthorhombic case. Fig. 6.15 shows the positions "1" and "2" which, statistically occupied within different cells, give rise to a mean [-1, 0, 1] polarization.

In the model under consideration, the structure consists of vertical rows of N_2 cells (average estimated value: $N_2 \sim 15$) in which the displacement **Δy** preserves its direction (say, upwards).

Consider a domain consisting of N_1N_3 vertical rows, each containing N_2 cells like the one in Fig. 6.15. In the rows type "1", B atoms are located as in point 1, displaced vertically by +**Δy**, while in type "2" columns, the movement is -**Δy.** The distribution of columns "1" and "2" in the domain is random, so that the average polarization points in the direction [-1, 0, 1]. All the domain cells are equal, except for the **Δy** direction (which varies from column to column). This type of crystalline imperfection is what is known as a *linear disorder*, associated in this example with axis y.

The article by CLG contains the mathematics for the intensity scattered by the considered domain. The representative equation is:

$$I = AN_1N_3 \frac{\sin^2(N_2\mathbf{Q}\cdot\mathbf{a}_2/2)}{\sin^2(\mathbf{Q}\cdot\mathbf{a}_2/2)} \sin^2 \mathbf{Q}\cdot\Delta\mathbf{y} \tag{32}$$

Equation (32) features several interesting effects:

- Take into consideration that Δy is small ($\Delta y << a_2$). This leads to $I(\mathbf{Q}) \propto \sin^2(\mathbf{Q}\cdot\Delta\mathbf{y}) \approx |\mathbf{Q}\cdot\Delta\mathbf{y}|^2$, analogous to the tendency described in Equation (29), with $\boldsymbol{\varepsilon} = \Delta\mathbf{y}$.
- For $N_2 >> 1$, the intensity is zero except at the planes $\mathbf{Q}\cdot\mathbf{a}_2 = 2\pi k$. This is the fingerprint of linear disorder. For finite N_2, the broadening of the scattering sheets is inversely proportional to the length of the ordered columns.
- The functional dependence of the scattered intensity is smooth through each sheet. From sheet to sheet it varies according to $\sin^2(\mathbf{Q}\cdot\Delta\mathbf{y})$.
- For the plane $k = 0$: $\mathbf{Q}\cdot\Delta\mathbf{y} = 0$. The Comès disorder does not produce scattering through planes that contain the reciprocal origin.

6.3 Powder Diffractometry: Techniques and Applications

6.3.1 Diffraction by a Polycrystalline Sample in a Synchrotron Facility. Resolving Power

Bulk technologically important ferroelectric samples usually show polycrystalline nature and are thus investigated by means of powder diffraction experiments. Powder diffractometry projects the three dimensional reciprocal lattice into a one dimensional space. Such projection causes partial /full/ overlapping of peaks with similar /equal/ lattice spacing. Peak overlapping leads to loss of information, so single-crystal methods would be better for the determination of electron-density distributions and other structural data. On the other hand, the one-dimensional nature of powder diffraction practically eliminates the influence of zero-shift and other systematic errors in the detection of minute lattice spacing differences. This way, powder peak overlap becomes a useful factor in investigations of subtle symmetry breakdowns, which are often missed in single crystal measurements. The mentioned advantage of powder diffraction becomes more important when we consider the high resolution of (synchrotron and neutron) present-day powder diffractometers.

As an example, consider the discovery of a monoclinic phase in the PZT system. It was a surprising finding [21]. The measured lattice parameters were a = 5.717 Å, b = 5.703 Å, c = 4.143 Å, β = 90.53°. The experiment was a synchrotron high-resolution *powder* diffraction investigation.

Synchrotron powder diffractometry exhibits the following characteristics:

- High brightness and vertical collimation

- Strong wavelength dependence of the scattering process in the absorption edges vicinities. Resolution of complex structures.
- Magnetic scattering cross-sections dependence on resonance conditions. Solution of magnetic structures.
- Radiation broad-band accessibility. Wide spectrum of resonances available.

The preceding comments support the convenience of careful synchrotron powder diffraction experiments focused on the detection of subtle details and crystal-structure transformations in ferroelectric phases. Mentioned investigations may contribute significant elements to our understanding of the materials properties.

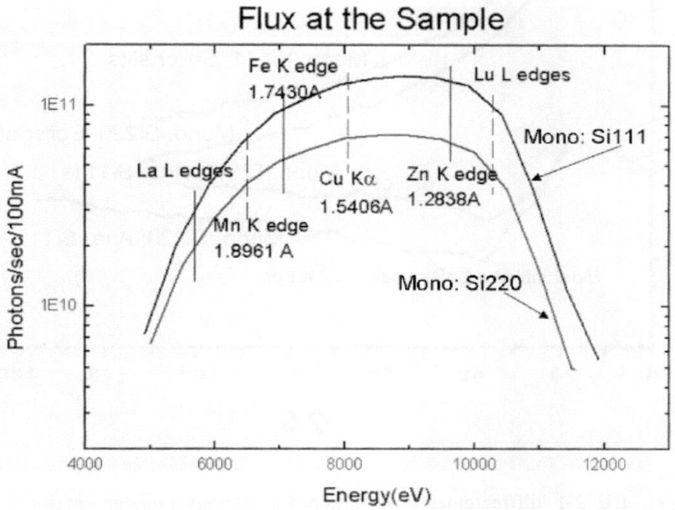

Fig. 6.16 Photon flux at sample for two crystal monochromators. SSRL, beamline 2-1. http://smb.slac.stanford.edu/powder/flux.gif. Courtesy A. Mehta.

As an example of a working synchrotron powder diffractometer, we give some data of the facility at Beamline 2-1, SSRL. This equipment uses the radiation from a focused bending magnet beamline. The size of the focused beam is 2x1 mm^2. For photon energies between 7 and 10 keV, some 10^{11} photons/sec strike the sample. Fig. 6.16 shows the flux dependence on energy for two configurations. The broadband nature of the x-ray beam opens the possibility of choosing the wavelength that best fits the research requirements. The absorption edges indicated in the Figure show the opportunities that exist for performing anomalous scattering experiments (see Section 6.2.2.2). Fig. 6.17 represents the diffractometer resolution for some monochromator-analyzer combinations. The Figure includes, for each configuration, the diffracted intensity (in kilo-counts per second) from the most intense peak of a LaB_6 standard. Ultrahigh resolution, of approximately 0.1% for lattice parameters, is obtained with the Si220 monochromator and the Si111 analyzer. Naturally, gaining resolution implies loosing intensity. The selection of a working configuration, using Fig. 6.16 and Fig. 6.17, leads to a compromise solution.

The considered diffractometer is capable of performing temperature-dependent measurements, from 2 to 1500 K.

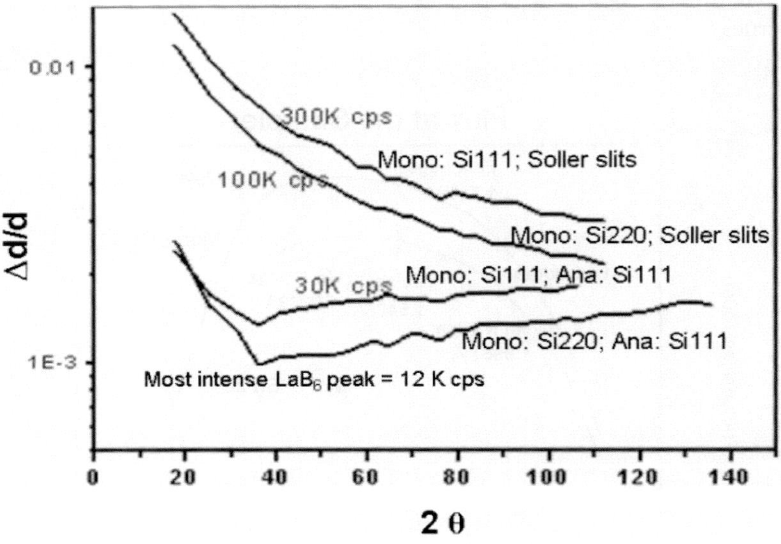

Fig. 6.17 SSRL BL 2-1 diffractometer resolution for various monochromator – analyzer combinations. http://smb.slac.stanford.edu/powder/diffres.gif. Courtesy A. Mehta.

6.3.2 The Rietveld Method: Basic Ideas, Formulae and Software

Synchrotron experiments show significantly higher statistics and resolution (and are considerably more expensive) than measurements in a conventional x-ray diffractometer. Consequently, it frequently happens that the samples to be analyzed in a synchrotron diffractometer are materials for which the qualitative phase identification, as well as an initial structural model, is known. What one looks for are fine details of the crystal structure or other subtle aspects requiring high resolution. The most acknowledged computer-aided technique for processing powder diffractometry data is the so-called Rietveld method [22, 23, 24, 25, 26], to which we devote the present section. We shall frequently refer to software FULLPROF [23] , one of the most popular Rietveld codes.

A single-phase Rietveld refinement starts with the proposition of a model structure for the investigated crystal and the computer simulation of the diffraction pattern that this material produces in a polycrystalline diffraction experiment. After it has been verified that the calculated pattern is qualitatively similar to the experimental one, a meticulous simulation process of systematic variations in experimental conditions and in the sample structure and microstructure is developed. The effects of model variations in the calculated diffraction pattern are evaluated and this process is adjusted until the simulated pattern satisfies a least-squares convergence criterion in relation to the experimental one. At this point it is understood that the structure has been *refined.*

The general scheme for simulating the point-by-point diffraction pattern, starting from a model structure, is as follows.

Knowledge of the crystal space group and cell parameters leads to the generation of the complete collection of possible peak positions, $2\theta_k$. Starting from the structure factors and other known data, the integral intensities of all the peaks are calculated. With this information, a formula $\phi = \phi(2\theta)$ is selected to describe the peak profile. Gauss and Lorentz bell-shaped curves, among others, are frequently used peak shapes. The intensity y_{ci} of the calculated pattern at the observation point $2\theta_i$ is given by the superposition of contributions from all the diffraction peaks to the measurement at the considered angle:

$$y_{ci} = s\sum_k L_k |F_k|^2 \phi(2\theta_i - 2\theta_k) P_k A + y_{bi} \tag{33}$$

Variables in (33) have the following meaning: s is a scale factor; L_k groups the Lorentz-polarization and multiplicity factors; F_k is the structure factor (including the Debye-Waller factors of the different atoms); $\phi(2\theta_i - 2\theta_k)$ is the peak form function, centred at Bragg's $2\theta_k$ angle; P_k describes the texture; A is the absorption correction and finally y_{bi} is the background intensity at the $2\theta_i$ position. Indexes k and i scan diffraction peaks and pattern points respectively. Naturally, the contribution of the peak with a maximum at $2\theta_k$ is only appreciable in positions near this maximum. In the most-used Rietveld programs this contribution is neglected for points with $(2\theta_i - 2\theta_k)$ larger than a few times the peak width.

Equation (33) expresses the dependency of the diffraction pattern on an elevated number of parameters. For example, the average positions and the atomic form factors of all atoms in the unit cell determine the structure factor. The peak-profile function depends on the energy spectrum of the incident beam, on the perfection and dimensions of the crystals, and on a number of instrumental parameters.

The general idea of the least-squares refinement is as follows. The residual S_y is calculated thus:

$$S_y = \sum_i w_i (y_i - y_{ci})^2 \tag{34}$$

In this expression y_i correspond to the experimental intensities of the diffraction pattern and w_i to the respective weights given to these intensities. Generally, criterion w_i = variance inverse (σ_i^{-2}) is adopted.

The amount S_y is a complicated function of all the parameters that form the y_{ci} pattern. The goal is, then, to optimize all these parameters, so that S_y adopts the minimum value.

The mathematics of the minimization of S_y are described by Young [27]. Briefly, it is necessary to solve a so-called *normal equations system*. This is made by the inversion of a (*m* × *m*) *normal matrix*, where *m* is the number of elements to be refined. The residual function is non-linear. Convergence to the real solution of the refinement problem, and not to a false minimum of S_y, strongly depends on the closeness of the initial model to this solution. Also, the order in which the instrumental and structural parameters are refined is important. Stable parameters (scale factor, lattice parameters, background) should be refined first.

In the following sections we give details of the basic factors in the Rietveld working formula (33).

6.3.2.1 Structure Factor Calculations

Starting information for structure factor calculation consists of unit cell geometry and atomic contents. The atoms in the unit cell are characterized by giving the crystal space group and the so-called asymmetric unit.

The calculation of the atomic scattering factors $f(sen\theta/\lambda)$ has been systematized by Cromer and Waber [28], using previous results obtained by Mann [29], for the case of atoms with spherical symmetry. The working Equation (35) requires the knowledge of 9 coefficients, a_i, b_i (i = 1, 2, 3, 4), c, and the wavelength λ.

$$f^0(\sin\theta/\lambda) = \sum_{i=1}^{4} a_i \cdot e^{-b_i(\sin\theta/\lambda)^2} + c \tag{35}$$

The required coefficients are part of the information contained in several Rietveld programs.

Rietveld software generates all the atoms of the unit cell by application of the space group operators to the asymmetric unit. The structure factor is determined by application of Equation (16).

In structure refinement work, plausible initial model and chemically valid final structure are necessary. We mention two tools that may be used as criteria regarding the cited requirements.

Staring structure models may be generated by the so-called *simulated annealing* [30, 31] technique. The method consists of a Montecarlo generation and energy optimization of favourable configurations. The first configuration variations allow large atomic displacements. As optimization proceeds, the sample is virtually "cooled". This means, atomic displacements are gradually restricted as the structure tends to the one with minimum energy.

Final configuration must fulfil, besides diffraction conditions, crystal-chemistry requisites. Minimization of the so-called *combined figure of merit*, R_{comb}, may be used for consideration of both requirements:

$$R_{comb} = w \cdot R_{wp} + (1-w) \cdot R_{energy} \tag{36}$$

where

$$R_{wp} = 100 \left(\sum w_i \left| y_i - y_{ci} \right|^2 \Big/ \sum w_i y_i^2 \right)^{1/2} \tag{36a}$$

is the usual Rietveld disagreement factor and R_{energy} characterizes the deviation of any proposed structure energy with respect to the (estimated) optimized energy. Configurational (considered and optimized) energies are calculated by the methods of quantum chemistry. The weight factors w are given by the researcher.

As for Debye-Waller factors, besides $\langle u^2 \rangle$ and B, Rietveld software uses the dimensionless anisotropic tensors of atomic displacements β_{jk}. The relationships among the three mentioned descriptions are discussed in detail in the recommendations of the International Union of Crystallography [10, 11] . For orthorhombic (or higher) crystal symmetry, the following conversion formula is fulfilled:

$$\langle u_j u_k \rangle = \frac{a_j a_k \beta_{jk}}{2\pi^2} = \frac{B_{jk}}{8\pi^2} \tag{37}$$

As an example, consider a Rietveld refinement in which the lattice parameters are: $a = 4$ Å, $b = 5$ Å, $c = 6$ Å. Isotropic refinement gives $B_{iso} = 0.5$ Å^2. Follows β_{jk} refinement. Which initial values would be good? Application of Equation (37) gives: $\beta_{11} = 0.5/(4 \cdot 4^2) \approx 0.008$, $\beta_{22} = 0.5/(4 \cdot 5^2) = 0.005$, etc.

6.3.2.2 Lorentz-Polarization and Background

The Lorentz-polarization factor is

$$L_k = \frac{1 - K + K\cos^2 2\theta_{monochr} \cos^2 2\theta}{2\sin^2\theta\cos\theta} \tag{38}$$

For synchrotron radiation, with Si monochromator: $\cos^2(2\theta_{111}) = 0.9009$, $\cos^2(2\theta_{220}) = 0.7471$, $K \approx 0.1$.

The pattern background is usually represented as a fifth-degree polynomial:

$$y_{bi} = \sum_{m=0}^{5} B_m \left[(2\theta_i / B_{KPOS}) - 1\right]^m \qquad (39)$$

B_m are coefficients to be refined and B_{KPOS} is the 2θ coordinate of the polynomial origin, selected by the investigator.

6.3.2.3 Peak Shape. Instrumental and Diffraction Broadening. Crystallite Size and Microstrains

An important factor in the Rietveld method is the peak shape. There are about ten functions, representing the formalisms that are used in the most widespread programs. Following are some of the most commonly used peak forms. We first analyze the case of sharp peaks, associated with well-crystallized samples. Peak broadening in this case is considered "instrumental" (no small crystal size or microstrain broadening).

In the formulations that follow, the width of the peaks is characterized by the magnitude H_k, *full width at half maximum* (FWHM). For the instrumental contribution to this magnitude the empirical formula (40) [32], is generally used. U, V and W are to be refined.

$$H_k^2 = U \tan^2\theta + V \tan\theta + W \qquad (40)$$

We identify the shape functions through the parameter "NPROF", as assigned in the programs by Young [22] and Rodríguez-Carvajal [23].

Gauss (NPROF = 0):

$$G = \frac{\sqrt{C_0}}{H_K \sqrt{\pi}} \cdot e^{\left(-C_o (2\theta_i - 2\theta_K)^2 / H_K^2\right)}, \text{ with } C_0 = 4\ln 2 \qquad (41)$$

Lorentz (NPROF = 1):

$$L = \frac{\sqrt{C_1}}{\pi H_K} \cdot \frac{1}{\left[1 + C_1 \dfrac{(2\theta_i - 2\theta_K)^2}{H_K^2}\right]}, \text{ with } C_1 = 4 \qquad (42)$$

Lorentz modified 1 (NPROF = 2):

$$L_1 = \frac{2\sqrt{C_2}}{\pi H_K} \cdot \frac{1}{\left[1 + C_2 \frac{\left(2\theta_i - 2\theta_K\right)^2}{H_K^2}\right]}, \text{ with } C_2 = 4(\sqrt{2} - 1) \quad (43)$$

Lorentz modified 2 (NPROF = 3):

$$L_2 = \frac{\sqrt{C_3}}{2H_K} \cdot \frac{1}{\left[1 + C_3 \frac{\left(2\theta_i - 2\theta_K\right)^2}{H_K^2}\right]^{1.5}}, \text{ with } C_3 = 4\left(2^{2/3} - 1\right) \quad (44)$$

Pseudo-Voigt (pV, NPROF = 5) and Thompson-Cox-Hastings (TCH, NPROF = 7):

$$f_{pV} = \eta L + (1 - \eta) \cdot G \quad (45)$$

The pseudo-Voigt model is given by a linear combination of the Gauss and Lorentz functions. The parameter η defines how Gaussian or Lorentzian is the peak shape. $\eta = 0$ equals NPROF = 0; $\eta = 1$ equals NPROF = 1.

The Thompson-Cox-Hastings model is a pseudo-Voigt variant. The algorithms associated with NPROF = 5 and NPROF = 7 both satisfy the equation (45). The difference between these treatments consists in the selection of which parameters are refined and which are calculated based on the refined magnitudes. In general there are four interrelated parameters: the shape parameter (η), the total peak broadening (H) and the partial broadenings of the Gaussian and Lorentzian components (H_G and H_L). In pV, (η, H) are refined and (H_G, H_L) = f (η, H) are calculated. Conversely, in TCH, (H_G, H_L) are refined and (η, H) = f^{-1}(H_G, H_L) are calculated.

Formulas for the pseudo-Voigt model (NPROF = 5, instrumental broadening):

$$\eta = N_A + N_B \cdot 2\theta \quad (46)$$

In FULLPROF: N_A = Shape1; N_B = X.

$$H^2 = U \tan^2\theta + V \tan\theta + W \quad (47)$$

$$\frac{H_G}{H} = (1 - 0.74417\eta - 0.24781\eta^2 - 0.00810\eta^3)^{1/2} \quad (48)$$

$$\frac{H_L}{H} = 0.72928\eta + 0.19289\eta^2 + 0.07783\eta^3 \quad (49)$$

The shape parameter η is refined through N_A and N_B. The total broadening H is determined through the refinement of the magnitudes U, V, W. The formulas (48) and (49) determine H_L and H_G as functions of η and H.

Next, in parallel with the description of the TCH model, we analyze the case of small crystallites (t <100 nm) and/or microstrains. Peak broadening results from the convolution of *instrumental* and *diffraction* effects associated with these microstructural characteristics of the sample. FULLPROF software has systematized this case through the profile NPROF = 7.

The first requirement for a microstructural study based on peaks profile analysis is to have a quantitative characterization of the employed diffractometer resolution. This is obtained by measuring the diffraction pattern of a well crystallized calibration sample, often LaB_6. The experimental information of the instrumental peak width (FWHM), produced by the standard sample as function of the scattering angle 2θ, conforms the so-called *instrumental resolution function* (IRF). As an example of IRF associated with a synchrotron powder diffractometer, we mention the SSRL reference [33].

Having the instrumental resolution, the sample of interest is measured, the broadening parameters are refined, the instrumental broadening is "subtracted" (more exactly, it is "deconvoluted") and the equations that relate the diffraction broadening with the crystallite size and the microstrains are applied. Small crystallite size, characterized by means of the well-known Scherrer Equation [our Equation (22)], increases with the scattering angle according to $(\cos\theta)^{-1}$. Heterogeneous microstrain (ε) produces a peak broadening β_{strain} proportional to $\tan\theta$. Equation (50) shows the two tendencies:

$$\beta_{size} = \frac{K\lambda}{t\cos\theta_B} \qquad \beta_{strain} = 4\varepsilon\tan\theta_B \tag{50}$$

Next we present the formulas that are applied in the TCH model for the DRX peak profile. Instrumental broadening as well as crystallite size and microstrains are considered. The shown equations refer to isotropic broadening. Later on we give an introductory idea on anisotropic broadening.

$$H_G^2 = U\tan^2\theta + V\tan\theta + W + \frac{I_G}{\cos^2\theta} \tag{51}$$

$$H_L = X\tan\theta + \frac{Y}{\cos\theta} \tag{52}$$

$$H = \left(H_G^5 + AH_G^4H_L + BH_G^3H_L^2 + CH_G^2H_L^3 + DH_GH_L^4 + H_L^5\right)^{1/5} \tag{53}$$

with: $A = 2.69269$ $B = 2.42843$ $C = 4.47163$ $D = 0.07842$.

$$\eta = 1.36603q - 0.47719q^2 + 0.1116q^3\text{, with } q = H_L/H \tag{54}$$

The Gaussian broadening H_G (Equation 51) characterizes the instrumental profile and has also contributions from crystallite size (term in I_G) and microstrains ($U = U_{instr} \oplus U_{strain}$).

The terms in the formula for H_L (Equation 52) represent microstrain and small crystallite size components of the Lorentzian peak broadening.

In the specific case of software FULLPROF, the procedure for microstructural analysis is as follows.

To determine the diffractometer IRF and to supply it to the software. Related instructions are given in the software user's manual.

To plan the refinement of the diffraction broadening parameters, according to the following scheme:

- Gauss Lorentz
- Strain *U X*
- Size I_G (GauSiz) Y

To run FULLPROF. The program, patiently led by the researcher, will: a) Refine the total broadening, b) Optimize the Lorentzian-Gaussian representation of the computerized profile, c) Separate diffraction from instrumental contributions, d) Determine to what extent the correlation "broadening *versus* diffraction angle" tends to $(\cos\theta)^{-1}$ or to $(\tan\theta)$ and e) Estimate the average crystal size and the weighted mean strain of the sample. Create output.

Possible cases of anisotropic broadening are considered in FULLPROF. We illustrate the work in this line with the case of crystals with revolution ellipsoid shape. It includes needles and disks. To investigate this type of samples, additional parameters that describe the crystal-size broadening are refined. The Lorentzian size-broadening width H_L^{size} has two components, the "isotropic" *Y* and the "anisotropic" $F(S_z)$, ec. (55).

$$H_L^{size} = \frac{Y + F(S_Z)}{\cos\theta} \tag{55}$$

For example, in the case of disk-like crystals (Size-Model = 1 in ".pcr"), $F(S_z)$ is as defined in equation (55a).

$$H_L^{size} = \frac{Y + S_z \cos\varphi}{\cos\theta} \tag{55a}$$

φ is the angle between the normal to the family of planes that produce a reflection and the normal to the face of the disk. *Y* and S_z (LorSiz) are refinable.

Crystal size (in Å), in different crystallographic directions, is obtained from *Y* and S_z. Refinement results for crystal sizes are published in the output (".mic") file.

Other anisotropic crystal shapes are discussed in the FULLPROF Manual. Anisotropic strains are also considered.

6.3.2.4 Texture

An effect that produces appreciable changes in the diffracted intensities is the non-random distribution of orientations in a polycrystalline aggregate, that is, the texture. Using texturist jargon, the factor P_k that transforms a random powder diffraction pattern into that of a textured sample is the *sample normal inverse pole figure*. To characterize this effect, Rietveld [34] introduced a Gaussian-like formula:

$$P_k = G_2 + (1 - G_2)\exp(-G_1\alpha_k^2)\text{, Rietveld} \tag{56}$$

α_k is the angle between the normal to the family of planes "k" and the normal to the family with preferred orientation. G_1 and G_2 are refinable parameters. In the Rietveld treatment, the non-existence of texture is expressed by putting $G_1 = 0$.

March and Dollase [35] have proposed the following alternative algorithm:

$$P_k = \left(G_1^2 \cos^2 \alpha_k + \left(\frac{1}{G_1} \right) \sin^2 \alpha_k \right)^{-3/2}\text{, MD} \tag{57}$$

In the March-Dollase formula, the non-texture condition is expressed as $G_1 = 1$. An advantageous characteristic of the MD formalism is that, in it, the preferred peak intensity increases while all the others decrease. In the Rietveld algorithm this requirement is not necessarily fulfilled. In other words, the MD equation preserves the inverse pole figure normalization condition and the Rietveld formula does not. This element favours the MD model in quantitative phase analysis.

More complete information on the texture analysis of polycrystalline ferroelectrics can be found in Chapter 8.

6.3.2.5 Quantitative Phase Analysis

Polycrystalline X-ray diffractometry is a powerful tool for quantitative phase analysis. The Rietveld technique reveals itself as a powerful alternative for this purpose. For Rietveld Quantitative Phase Analysis, a particular algorithm is applied. For a mixture of n phases, scale factors s_j (Equation 33) are adjusted according to the relative abundance of each phase. Scale factor refinement leads to weight fraction (W_j) determination using Equation (58):

$$W_j = \frac{s_j Z_j M_j V_j / t_j}{\sum_i (s_i Z_i M_i V_i / t_i)} \tag{58}$$

Z_j is the number of molecules per elementary cell; M_j is molecular mass; V_j is the volume of the elementary cell and t_j is the so- called *Brindley factor* for absorption contrast. This factor is tabulated in FULLPROF and its numerical value differs significantly from unity if particle sizes are not small enough.

Deviations of Rietveld results from real concentrations in calibrated samples have been measured [36, 37, 38]. Typical errors are in the range from ± 0.5% to ± 3%, depending on experimental conditions and sample characteristics.

6.3.3 *Ferroelectric Applications*

6.3.3.1 The Structure and Microstructure of Aurivillius Phases

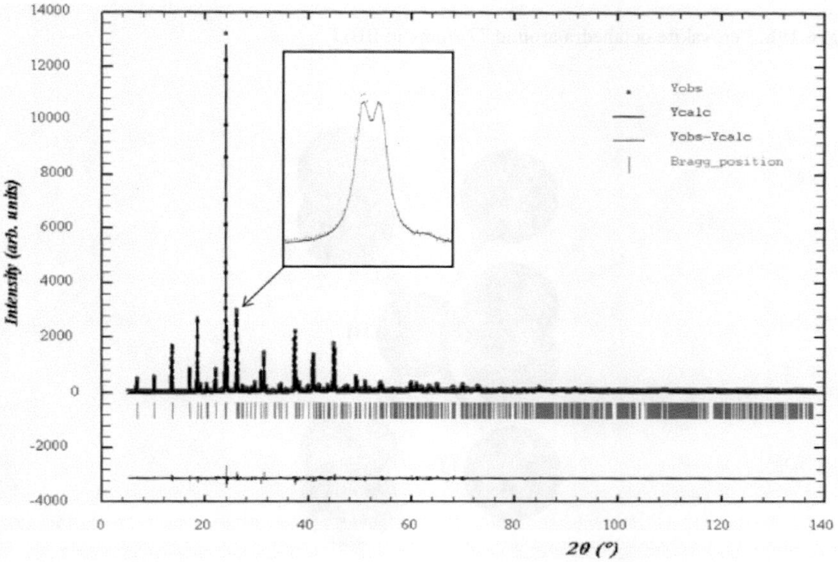

Fig. 6.18 Observed and calculated high-resolution diffraction patterns of BBIT. Powder diffractometer at SSRL beamline 2-1. The inset shows resolution of the (020/002) doublet. Agreement factors Rp = 10.3; $\chi^2 = 1.95$.

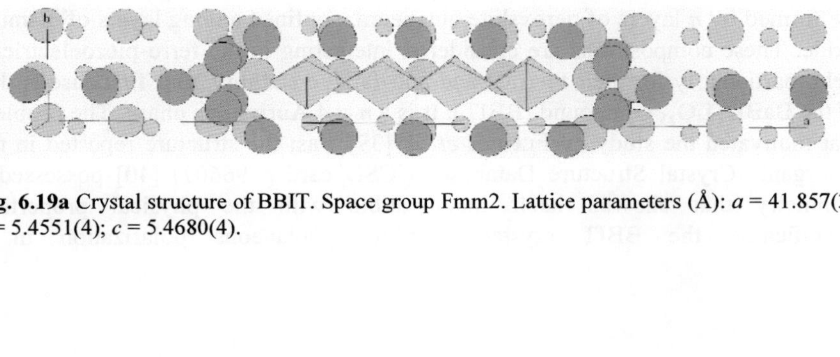

Fig. 6.19a Crystal structure of BBIT. Space group Fmm2. Lattice parameters (Å): $a = 41.857(3)$; $b = 5.4551(4)$; $c = 5.4680(4)$.

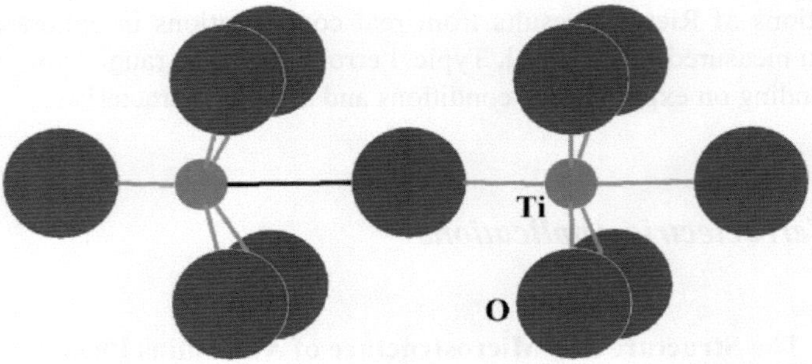

Fig. 6.19b. Perovskite octahedra around Ti atoms in BBIT

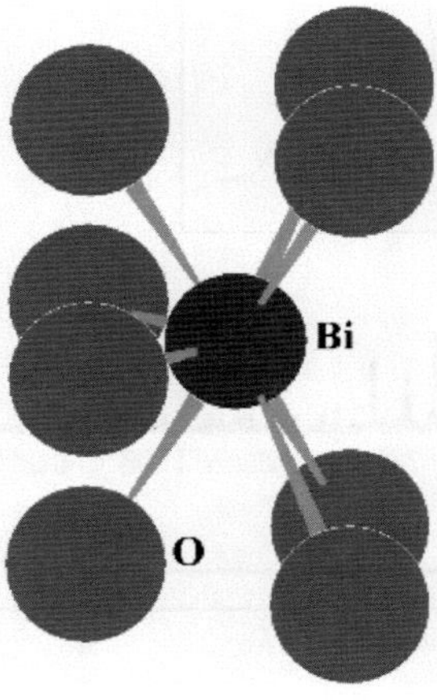

Fig. 6.19c. Environment of Bi atoms in the bismuth oxide layer of BBIT.

We give two examples of Rietveld applications on layered perovskite structures. The studied materials belong to the Aurivillius family. Their structures are formed by *n* layers of perovskite octahedra, confined among layers of bismuth oxide. These compositions are considered interesting in the ferro-piezoelectricity field, particularly for their thermal and anti-fatigue stability. Our first case is that of the $BaBi_4Ti_4O_{15}$ compound (BBIT). It is a $n = 4$ Aurivillius phase. The problem that motivated the study by Fuentes *et al.* [39] was: the structure reported in the "Inorganic Crystal Structure Database" (ICSD card # 96607) [40] possessed a symmetry that showed itself incompatible with the physical properties. Specifically, the BBIT crystals exhibit spontaneous polarization in a crystallographic axis belonging to the plane of bismuth oxide and the structure reported ICSD was centro-symmetrical tetragonal. The reported space group was I4/mmm, with $a = b = 3.8640(1)$ Å, $c = 41.8341(14)$ Å. Inversion symmetry is contradicted by the spontaneous polarization and the tetragonality is contradicted by the observed anisotropy in the bismuth oxide plane. The refinement of the BBIT structure was carried out by means of high resolution DRX experiments at channel 2-1 of the Stanford Synchrotron Radiation Laboratory. The sample was mounted on a zero-background sample holder and data was collected in reflection geometry at 10 keV (1.240Å) from 2° to 138° in 2θ. The scanning step was of 0.01° in 2θ. Experiment was controlled by means of software SUPER [41]. Fig. 6.18 shows the comparison between experimental and calculated diffraction patterns. The proposed space group is Fmm2. The inset presents a zoom of the doublet at $2\theta \approx 25°$. The considered space group explains all the observed diffraction peaks and is also consistent with the observed spontaneous polarization. Fig. 6.19 represents the investigated crystal structure.

Our second example describes a microstructural study of Fe-containing multiferroic Aurivillius phases obtained by a molten salt method [42]. The considered material is $Bi_5Ti_3FeO_{15}$ (BFT), orthorhombic, with lattice parameters (Å): $a = 5.4759(2)$, $b = 5.4420(2)$, $c = 41.157(1)$. Electron micrographs in Fig. 6.20 are representative of crystallite size and shape dependence on reaction temperature. Crystals are plate-shaped and their dimensions increase with increasing reaction temperature. Crystallites large surfaces are parallel to bismuth oxide layers in Aurivillius phases. Diffraction peaks exhibit small crystallite size broadening. Whole pattern observation reveals a somewhat larger broadening for the (0, 0, l) peaks as compared with the (h, 0, 0)/(0, k, 0). Detected anisotropic broadening is caused by the plate-like shape of BFT crystals.

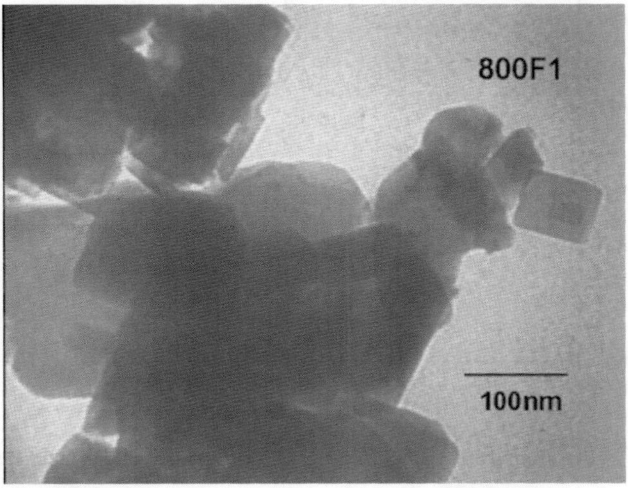

Fig. 6.20a. Transmission electron micrograph of $Bi_5Ti_3FeO_{15}$. Reaction temperature: 1075K.

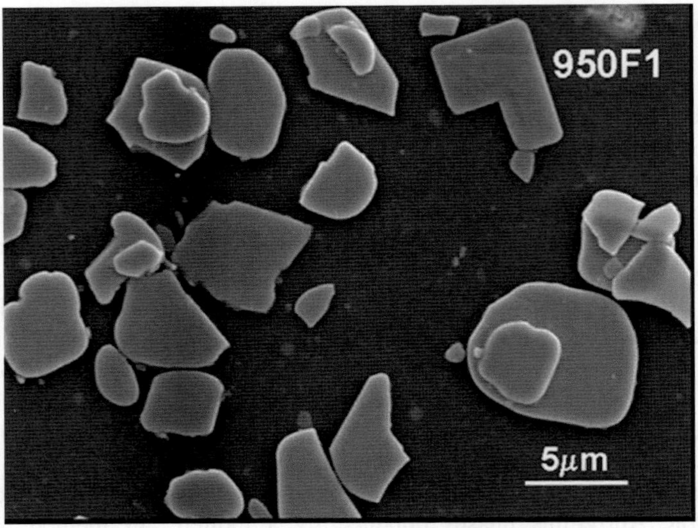

Fig. 6.20b Scanning electron micrograph of $Bi_5Ti_3FeO_{15}$ (BFT). Reaction temperature: 1225K.

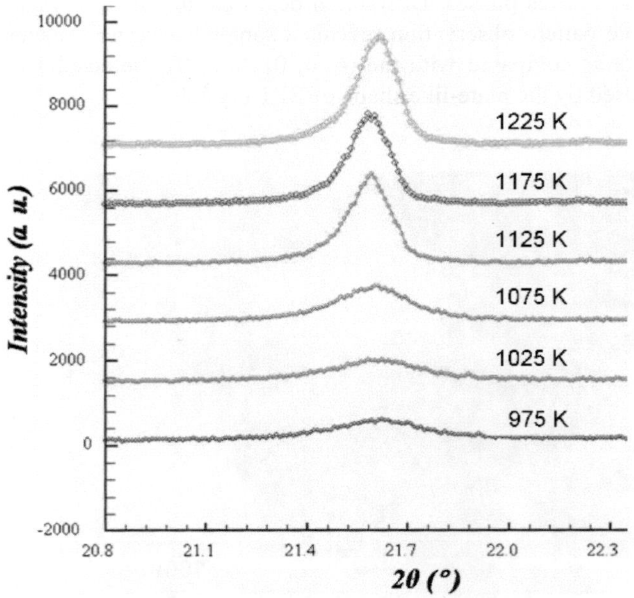

Fig. 6.21a. BFT. Broadening of XRD 0,0,10 peak as function of reaction temperature.

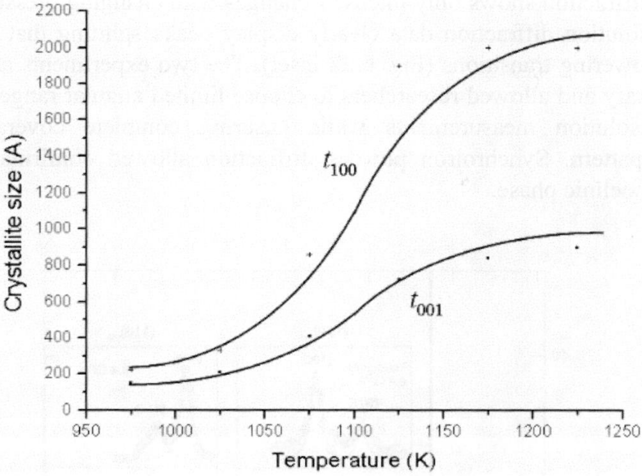

Fig. 6.21b BFT. Diffraction crystal sizes, t, from 100 and 001 peak broadening as a function of reaction temperature.

Rietveld software FULLPROF permits an approximation to crystal size and shape from pattern simulation. Fig. 6.21a shows (0,0,10) peak broadenings for different reaction temperatures and Fig. 6.21b represents the XRD results for crystallite thickness and lateral dimensions as a function of temperature. For all temperatures, broadening of (00*l*) diffraction peaks gives valuable information on crystallite thickness, while (*h*00)/(0*k*0) peak broadenings lead to an estimate of the lateral dimensions of crystallites. This kind of measurement reveals itself useful for $T \leq 1075$K. Above that temperature, diffraction (*h*00)/(0*k*0) peaks are too sharp to allow a quantitative determination of crystallite size. From the XRD analysis it can only be concluded that crystallites are large, at least, in the micrometre range.

6.3.3.2 The Monoclinic Phase of $PbTiO_3$

A recent paper shows interesting findings, via synchrotron powder diffraction, in a classic ferroelectric. $PbTiO_3$ has long been considered a simple ferroelectric, with a single phase transition from the ferroelectric tetragonal structure to cubic perovskite at 766 K and ambient pressure. A Raman study showed two soft modes, both vanishing at a pressure of 12 GPa at room temperature. For the zero temperature theoretical computations to be consistent with the room-temperature data, low temperature symmetry-lowering phase transitions are required. On this basis, Ahart *et al.* [43] performed cryogenic high pressure in situ Raman and synchrotron powder X-ray diffraction experiments, to explore the eventual materialization of theoretical predictions of phase transitions in $PbTiO_3$. Fig. 6.22 shows representative diffraction patterns measured at 10 K, indicating pressure-induced structural transitions. Energy dispersive diffraction shows only intensity changes with changing pressure whereas the high-resolution diffraction data clearly display peaks splitting that result from symmetry-lowering transitions (Fig. 6.22 inset). The two experiments are therefore complementary and allowed researchers to choose limited angular ranges to scan in the high-resolution measurements while ensuring complete coverage of the diffraction pattern. Synchrotron powder diffraction allowed confirmation of the $PbTiO_3$ monoclinic phase.

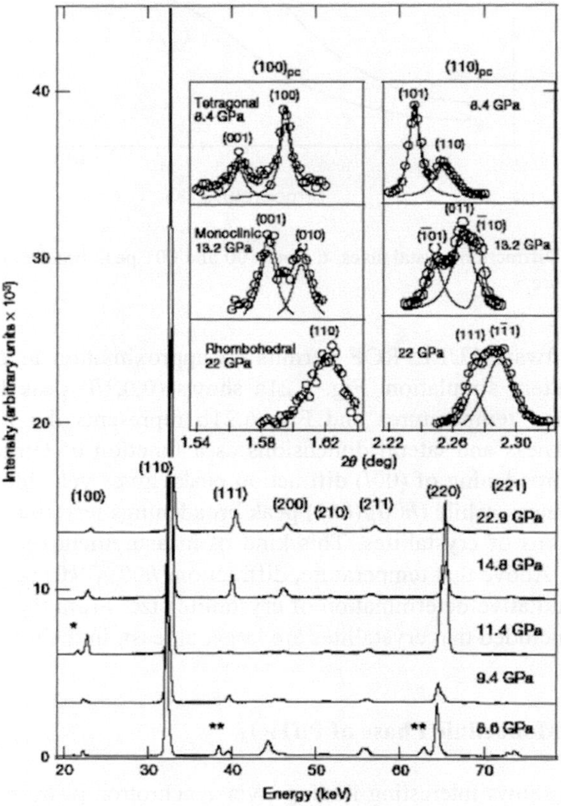

Fig. 6.22. Energy dispersive and high-resolution angle-dispersive X-ray diffraction spectra at different pressures at 10 K of $PbTiO_3$. Major reflection lines indexed with a pseudocubic symmetry (pc). The inset shows high-resolution diffraction data: the left panels show the pseudocubic (100) reflection at 8.4 GPa (tetragonal phase), at 13.2 GPa (monoclinic phase) and at 22 GPa (rhombohedral phase); the right panels show the pseudocubic (110) reflection at 8.4 GPa (tetragonal phase), at 13.2 GPa (monoclinic phase) and at 22 GPa (rhombohedral phase). Pseudocubic (110) reflection splits into a doublet in the tetragonal phase and a quadruplet in the monoclinic phase (the (101) reflection is missing). Experiments by Ahart M, Somayazulu M, Cohen RE, Ganesh P, Dera P, Mao H, Hemley RJ, Ren Y, Liermann P, Wu Z. Reprinted by permission from Macmillan Publishers Ltd: Nature 451: 545. Copyright (2008).

6.3.4 Phase and Texture Identification in Thin Films

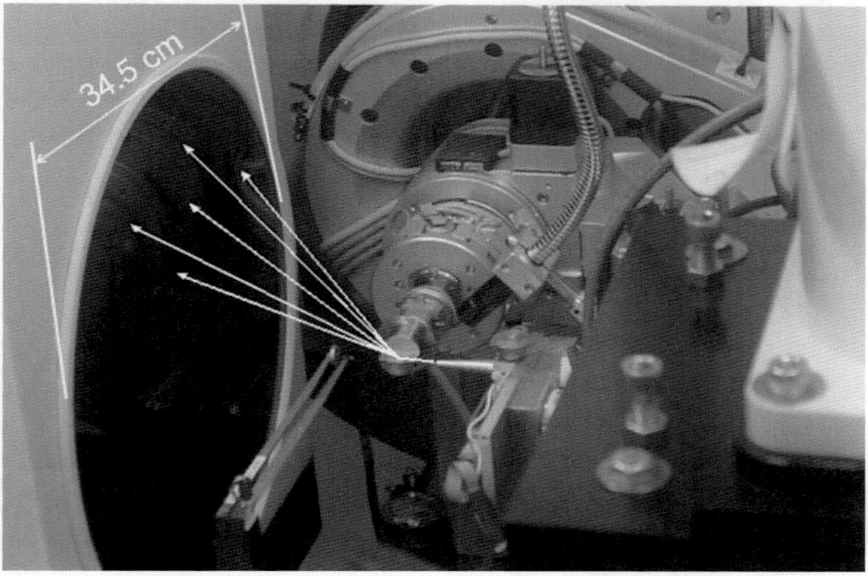

Fig. 6.23. Experimental hutch at SSRL beamline 11-3. A sample is mounted for a grazing incidence experiment. Hypothetical incident and scattered x-rays have been drawn schematically.

We conclude our study of diffraction methods with an introduction to devices and experiments that are also used in diffuse scattering studies.

Fig. 6.23 is a photograph of the two-dimensional (2-D) scattering measurement system installed at the SSRL beamline 11-3. Scattered intensity distribution is collected by means of a two-dimensional charge-coupled device (CCD) detector. The shown setup is the one employed for grazing incident experiments, suitable for the study of thin films and some nanostructures. Irradiation is usually done with incident angles near the critical angle for total reflection. Larger angles are used to obtain information about the substrate or, in general, to explore the sample volume. The sample-detector distance is determined by the investigator, generally seeking for a compromise between resolution and scattering angle interval. Measurement time per scattering pattern varies from a few seconds to several minutes, often significantly smaller than in conventional-source diffractometry. The conversion of information collected in the 2-D CCD into maps of scattered intensity as a function of reciprocal space coordinates is done by specialized software [44].

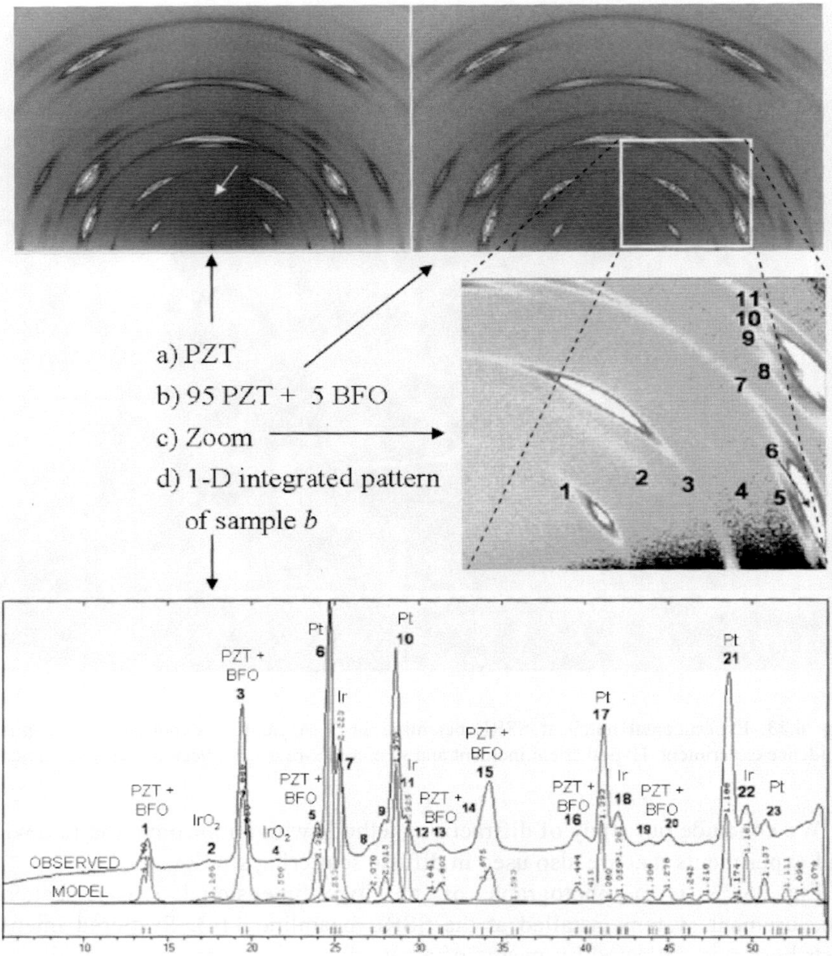

Fig. 6.24 Grazing-incidence investigation of two ferroelectric thin films. (a) 2-D scattering diagram of sample *a*: PZT ($Zr_{0.35}Ti_{0.65}O_3$). (b) 2-D experiment with sample *b*: 95% PZT + 5% BFO ($BiFeO_3$). (c) Zoom of a selected area in the pattern of sample *b*. The diffraction maxima are numbered for identification. (d) Integrated 1-D diffraction pattern. All peaks identified on the basis of the phase composition given in Table 6.2. Samples: courtesy of Ji Young Jo.

Fig. 6.24(a-c) shows the grazing-incidence 2-D scattering patterns of two ferroelectric thin films, crystallized at 875K after sol-gel coating. The composition of sample *a* is $PbZr0_{0.35}Ti_{0.65}O_3$ (PZT) and that of sample *b* is 95% PZT + 5% $BiFeO_3$ (BFO). The substrate is Pt / IrO_2 / Ir / Ti / SiO_2 / Si.

Samples for this study were kindly provided by Ji Young Jo (Seoul National University, Korea).

Data collection is performed by means of software Blue-Ice [45].

We describe the qualitative interpretation process of the investigated materials.

The first stage of the procedure consists of a calibration routine with a LaB_6 standard. This provides an exact relationship among the coordinates of a point on the 2-D detector screen, scattering angle and the associated reciprocal-space position.

Secondly, by means of software FIT2D [46] , the 2-D diagram is integrated, for each polar angle (2θ), along the azimuth. This way, a 1-D diffraction diagram, comparable to conventional $I(2\theta)$ patterns, is obtained. Use of powder diffraction software PowderCell [47] permits identification of present phases by means of model-observed comparison. The result of the identification process is shown in Table 6.2 and in Fig. 6.24(d).

Table 6.2 Phases detected on samples *a* and *b*. In sample *b*, the ferroelectric perovskite is a solid solution of BFO on tetragonal PZT. Subtle peaks displacements due to lattice parameters variations are masked by peak broadening.

Phase	Ir	Pt	PZT (+BFO)	IrO2
Lattice parameters (Å)	3.85	3.94	4.03, 4.14	4.50, 3.16
Comments	Debye rings	Ring sectors		Nanocrystalline "humps"

Integration facilitates phase identification, but does not allow the study of other characteristics of the polycrystalline films, such as texture. To analyze preferred orientation, we revisit the 2-D diagrams and examine the distribution of intensities along the Debye rings.

As a general tendency, the distribution of intensities along the Debye rings is non-uniform. This indicates preferred orientation of crystallites in both samples. Debye rings caused by PZT+BFO (numbered 1, 3 and 5) and Pt (6 and 10) are somewhat more heterogeneous than those produced by Ir (7 and 11). Essentially, the 2-D patterns of the first two phases in both samples are recognized as characteristic of fibre <111> textures. The smoother distribution of intensities in the rings from Ir indicates a weaker preferred orientation for this phase. Sample *a* shows a differentiating PZT $\langle 001 \rangle$ weak texture component. It is indicated by a white arrow in Fig. 6.24a. To compare samples *a* and *b* with respect to texture, use of software ANAELU [48] allows plotting the *intensity versus azimuth* dependence of both samples. The result is shown in Fig. 6.25. The presence of PZT <001> texture component in sample *a* is noticeable.

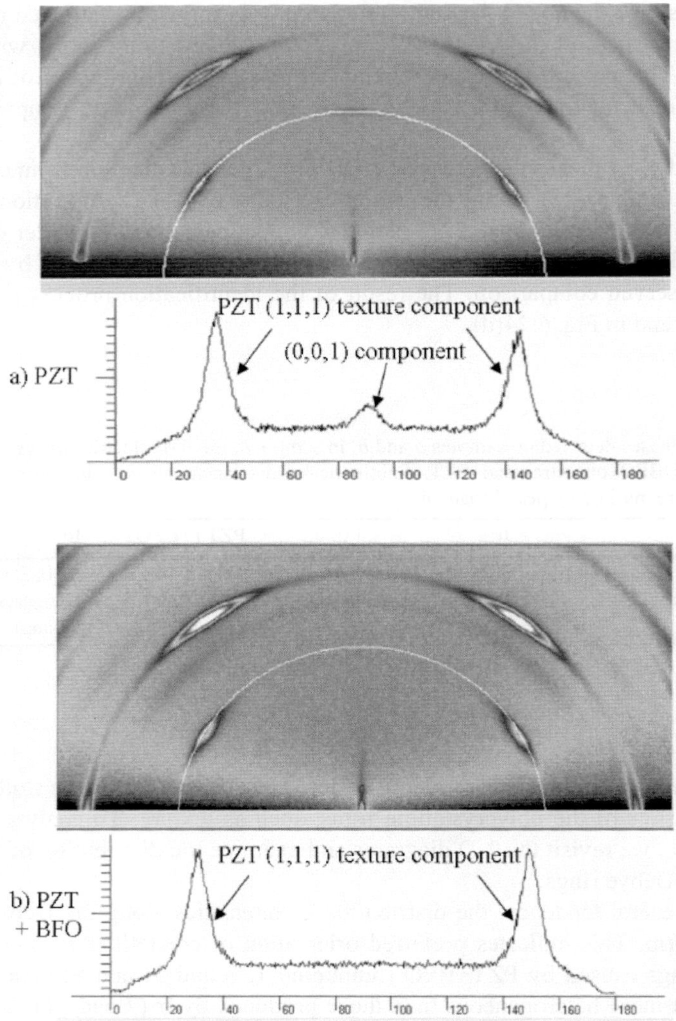

Fig. 6.25. Low angle sectors of 2-D diagrams and intensity *versus* azimuth dependence for samples *a* and *b*. The presence of PZT (001) texture component in sample *a* is observable.

6.4 Diffuse Scattering: Techniques and Applications

6.4.1 Pair Distribution Function

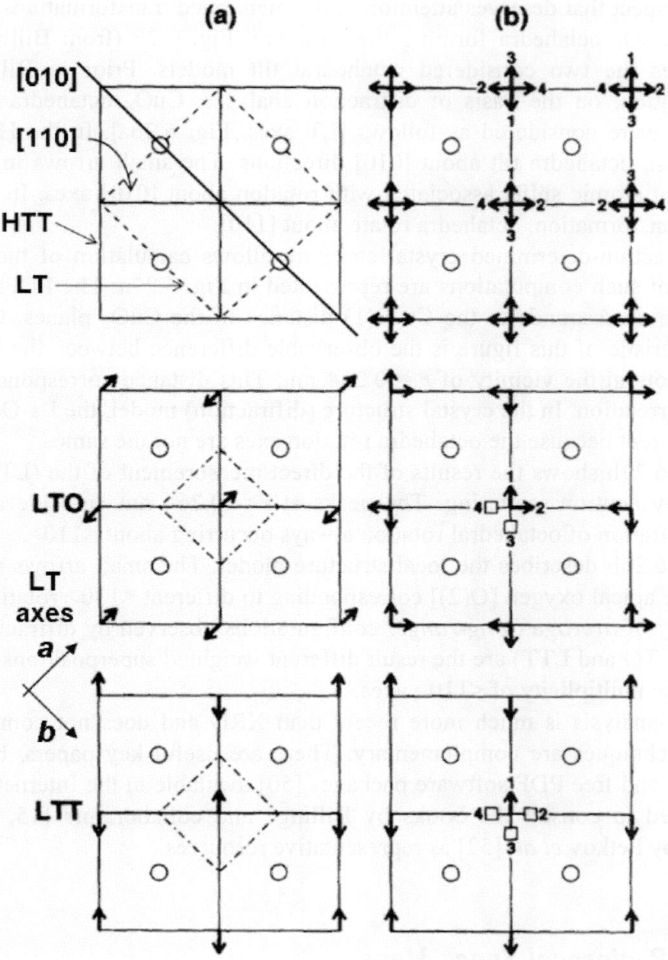

Fig. 6.26 Crystal structure **a** and local structure **b** models for octahedral tilts in La2-xBaxCuO4. The smallest continuous squares represent HTT unit cells. The dashed lines outline the LT (LTO and LTT) cells. Cu sits under the apices, La at the position denoted by open circles. The dot-dashed lines illustrate the orientation of the CuO6 octahedra. The arrows indicate the tilt directions predicted by the considered models. See text for details. Figure reprinted with permission from Billinge S, Kwei GH, Takagi H. Physical Review Letters 72: 2282. Copyright (1994) by the American Physical Society.

A key paper by Billinge et at [49] emphasized the role of PDF analysis in crystallography. The investigated material was the high-T_c superconductor

$La_{2-x}Ba_xCuO_4$. On cooling, this material undergoes the following transformations: high-temperature-tetragonal (HTT, space group I4/mmm, a_{HTT} = 3.779 Å, c_{HTT} = 13.28 Å) → low-temperature orthorhombic (LTO, space group Abma, a_{LTO} = 5.349 Å ≈ $\sqrt{2} \cdot a_{HTT}$, b_{LTO} = 5.398 Å, $c_{LTO} \approx c_{HTT}$) → low-temperature tetragonal (LTT, space group $P4_2$/ncm, a_{LTT} = 5.348 Å ≈ a_{LTO}, $c_{LTT} \approx c_{HTT}$).

An aspect that deserves attention in the mentioned transformations is the tilting of the CuO_6 octahedra forming the structure. Fig. 6.26 (from Billinge's paper) describes the two considered octahedral tilt models. Prior to Billinge's PDF contribution, on the basis of diffraction analyses, CuO_6 octahedra rotations on cooling were considered as follows [LT axes, Fig. 6.26a]. In the HTT → LTO transition, octahedra tilt about [010] directions. The small arrows in Fig. 6.26(a) represent atomic shifts associated with rotation about [010] axes. In the HTO → HTT transformation, octahedra rotate about [110].

Diffraction-determined crystal structure allows calculation of the PDFs. The results of such computations are represented in Fig. 6.27a. The PDFs peak at r = 0.19 nm corresponds to the Cu-O(1) distance in the CuO_2 planes. One relevant characteristic of this figure is the observable difference between the LTT and the LTO plots in the vicinity of r = 0.264 nm. This distance corresponds to the La-O(2) correlation. In the crystal structure (diffraction) model, the La-O(2) distances are different because the octahedra rotation axes are not the same.

Fig. 6.27b shows the results of the direct measurement of the (LTO and LTT) PDFs by neutron scattering. The peaks at r = 0.264 nm coincide and this is a demonstration of octahedral rotation always occurring about <110>.

Fig. 6.26b describes the local structure model. The small arrows represent the shifts of apical oxygen [O(2)] corresponding to different <110> rotation axes. The diversity of *average* (*long-range*) configurations observed by diffraction methods (HTT, LTO and LTT) are the result different weighted superpositions of *local* tilts about the multiplicity of <110> axes.

PDF analysis is much more recent than XRD and does not compete with it. Both techniques are complementary. There are useful key papers, bibliographic reviews and free PDF software packages [50] available in the Internet. The reader is invited to consult the books by Billinge and collaborators [15, 51] and the article by Petkov *et al.* [52] as representative resources.

6.4.2 Reciprocal Space Maps

The current boom in nanoscience entails, naturally, an impressive demand for tools and methods to learn about the structure of matter at the nanometric scale. Nanocrystals and other nanostructures do not produce sharp diffraction peaks, but intensity distributions scattered in regions of diverse dimensionality in reciprocal space. The intensities of these diffuse signals are thousand times weaker than those of conventional diffraction, so that x-ray and other radiation sources should be thousands or millions of times brighter than the ones used conventionally. This clearly explains why one of the lines of development in nanoscience is the study of nanostructures by means of diffuse scattering of synchrotron radiation.

The review papers by Dawber *et al.* [53], Fong and Thompson [54], Renaud [55], Robinson [56] and Vlieg [57] represent excellent summaries on the mentioned subject.

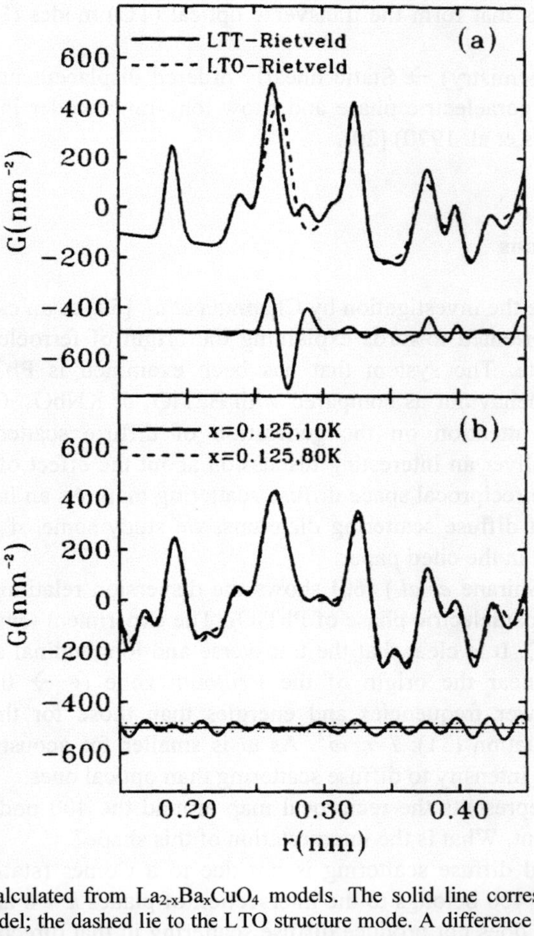

Fig. 6.27a PDFs calculated from $La_{2-x}Ba_xCuO_4$ models. The solid line corresponds to the LTT crystal structure model; the dashed lie to the LTO structure mode. A difference plot is show below them. **b** PDFs data collected from the x = 0.125 sample in the LTT phase (solid line, 10 K) and in the LTO (dashed line, 80 K). A difference is plotted below. The dotted lines indicate the random errors at the 1σ level. Figure reprinted with permission from Billinge S, Kwei GH, Takagi H. Physical Review Letters 72: 2282. Copyright (1994) by the American Physical Society.

6.4.3 *Diffuse Scattering in the Vicinity of Bragg Peaks*

We present three examples of research using ferroelectric diffuse scattering in the three-dimensional vicinity of Bragg peaks.

Work on this research line represents the current tendency aimed at clarifying the static and dynamic characteristics of ferroelectrics above the Curie temperature T_c. The final objective is to elucidate the gestation of ferroelectricity from the paraelectric state.

Two models that represent extremes in the existing diversity of approaches are:

1. (Solid state physics) → Softening of the dynamic chains of correlated displacements that form the transverse optical (TO) modes (Huller 1969) [58] and
2. (Quantum chemistry) → Static linearly ordered displacements that have short range in the paraelectric phase and show long-range order in the ferroelectric phase (Comès et al. 1970) [20].

6.4.3.1 Phonons

We introduce the investigation by Chapman *et al.* [59] as an example of current investigations oriented towards explaining the origin of ferroelectricity from the paraelectric state. The system that has been examined is $PbTiO_3$, showing a differentiating behaviour as compared with $BaTiO_3$ or $KNbO_3$. Chapman and co-workers focus attention on the generation of diffuse scattering by acoustic phonons and deliver an interesting discussion about the effect of different crystal disorders on the reciprocal space diffuse scattering map. As an introduction to the interpretation of diffuse scattering diagrams, we study some of the experimental results analyzed in the cited paper.

Fig. 6.28a (Shirane *et al.*) [60] shows the dispersion relations for low-energy phonons in the paraelectric phase of $PbTiO_3$. The experiment temperature was $T = 783$ K $= 1.04\ T_c$. It is clear that the transverse and longitudinal acoustic phonons (TA and LA) near the origin of the Brillouin zone ($g \rightarrow 0$, $\lambda \rightarrow \infty$) have significantly lower frequencies and energies than those for the optical modes (TO). From equation (31): $I \propto \omega^{-2}$. As ω is smaller for acoustic phonons, they contribute more intensity to diffuse scattering than optical ones.

Fig. 6.28b represents the reciprocal map around the 400 node. The dumbbell contour is evident. What is the interpretation of this shape?

The observed diffuse scattering is *not* due to a Comès (static) disorder. The reciprocal point 400 belongs to the intersection of planes $k = 0$ and $l = 0$ and thus Comès disorder does not produce diffuse scattering in that direction. Neither does Huller disorder give rise to diffuse scattering in planes passing through the reciprocal origin. The origin of the observed diffuse scattering in our Fig. 6.28b is to be found in *thermal phonons*.

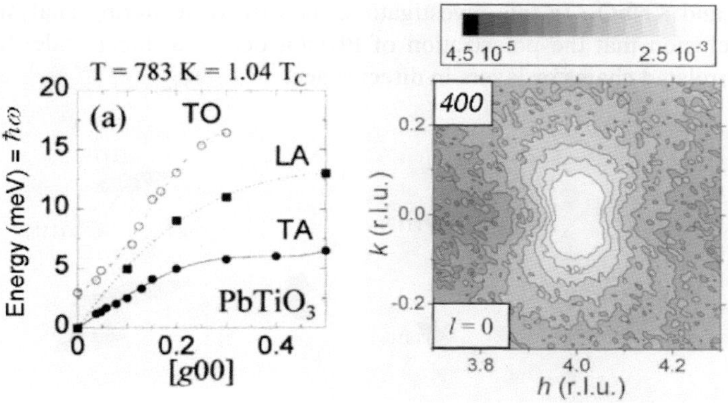

Fig. 6.28 Chapman *et al.* investigation on diffuse scattering in $PbTiO_3$. **a** Dispersion relations for the lowest energy phonon modes along [*g*00] [60]. **b** Diffuse scattering profiles from the cubic phase, on the (*h,k0*) plane, around (400). $T = 800$ K = 1.05 T_C. Figure reprinted with permission from Chapman Bd, Ster Ea, Han Sw, Cross Jo, Seidler Gt, Gavrilyatchenko V, Vedrinskii Rv, Kraizman V. Physical review b 71: 020102(r). Copyright (2005) by the American Physical Society.

The interpretation of how the dumbbell shape appears is derived by applying equation (31).

As already mentioned, acoustic phonons play here the major role. The factor $\left|\hat{\mathbf{Q}}\cdot\boldsymbol{\varepsilon}_{\mathbf{q},j}\right|^2$ in equation (31) can clarify the contributions of LA and TA phonons to the diffuse intensity scattered in the considered neighbourhood. Consider a reciprocal point $\mathbf{Q}$ = [*g*00] slightly to the right of node 400. In the drawing, vectors $\mathbf{Q}$ and $\mathbf{q}$ (= $\mathbf{Q} - \mathbf{B}_h$) are horizontal and the product $\hat{\mathbf{Q}}\cdot\boldsymbol{\varepsilon}_{\mathbf{q},j}$ will be maximum if $\boldsymbol{\varepsilon}_{\mathbf{q},j}$ is also horizontal. Physically, the diffuse scattering intensity is high if the polarization $\boldsymbol{\varepsilon}_{\mathbf{q},j}$ associated with phonons is parallel to the phonon wave vector $\mathbf{q}$. In other words, the phonons that produce scattering along the horizontal line [*g*00] are LA. For a reciprocal point $\mathbf{Q}$ = [4*g*0] shifted vertically from 400, the vector $\mathbf{q}$ is vertical. But, as $\mathbf{Q} \approx 400$, $\hat{\mathbf{Q}}\cdot\boldsymbol{\varepsilon}_{\mathbf{q},j}$ will only have a significant value if $\boldsymbol{\varepsilon}_{\mathbf{q},j}$ is horizontal. Therefore, $\boldsymbol{\varepsilon}_{\mathbf{q},j} \perp \mathbf{q}$. Phonons now are TA.

The diffuse scattered intensity is proportional to the inverse square of the phonon frequency (or energy). In turn, the square of the energy of a LA phonon is proportional to the elastic constant C_{11}. For a TA the proportionality is relative to C_{44}. This leads to $I_{diff\,scatt} \propto C_{ii}^{-1}$. From the literature [60], for $PbTiO_3$: $C_{11} = 234$ GPa, $C_{44} = 62$ GPa $\approx C_{11}/4$. It can be seen that ($1/C_{44} \approx 4/C_{11}$), which can explain in this case why along direction *k* the scattered intensities are approximately four times greater than along axis *h,* producing the dumbbell shape of the diffuse scattering profile shown in Fig. 6.28b.

Chapman *et al.* explores other reciprocal points and goes further in the analysis of substantial differences among $PbTiO_3$, $BaTiO_3$ and $KNbO_3$. $PbTiO_3$ does neither generate diffuse scattering sheets nor diffuse rods. These peculiarities imply that the mechanisms of polarization in $PbTiO_3$ are different from those of

$BaTiO_3$ and $KNbO_3$. In this investigation, the diffuse scattering analysis leads to the conclusion that the polarization of $PbTiO_3$ occurs at local scale. It does not form correlated chains or layers in direct space.

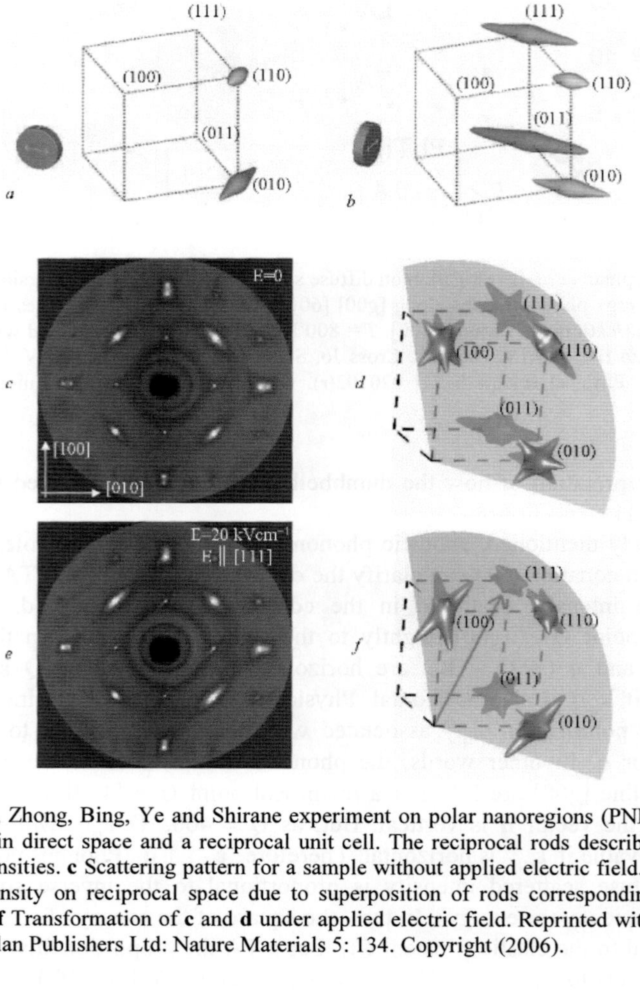

Fig. 6.29 Xu, Zhong, Bing, Ye and Shirane experiment on polar nanoregions (PNR) in PZN. **a** and **b**: PNR in direct space and a reciprocal unit cell. The reciprocal rods describe the diffuse radiation intensities. **c** Scattering pattern for a sample without applied electric field. **d** "Stars" of scattered intensity on reciprocal space due to superposition of rods corresponding to several PNRs. **e** and **f** Transformation of **c** and **d** under applied electric field. Reprinted with permission from Macmillan Publishers Ltd: Nature Materials 5: 134. Copyright (2006).

6.4.3.2 Polarized Nanoregions

We now study the interesting work of Xu and colleagues [61]. The investigated system is the perovskite $PbZn_{1/3}Nb_{2/3}O_3$ (PZN), a known ferroelectric relaxor, recognized for the broad temperature maximum and frequency dependence of its dielectric permittivity. The goal of the article by Xu *et al.* is the characterization, by means of synchrotron diffuse scattering, of polarized nanoregions (PNR) in a

PZN single crystal and the investigation of its dependence on applied electric field. The PNR, present above and below the Curie temperature, play a significant role in the exotic properties of PZN. The experiment was conducted in beamline X17B1 of the National Synchrotron Light Source, with monochromatic x-rays of energy E = 67 keV, energy resolution of 10^{-4} and transmission geometry. The scattered radiation was collected in a two-dimensional CCD detector.

Fig. 6.29 illustrates the results obtained by Xu et al. Parts a and b of the figure represent two orientations for PNR and their corresponding diffuse scattering intensities in the reciprocal space. The small discs show the PNR in direct space and the cubes represent the reciprocal lattice. The analyzed polarizations, of the form ⟨110⟩, are drawn referred to the nodes (100), (010), (110), (011) and (111). The reciprocal rods describe the diffuse radiation intensities corresponding to the polarizations displayed on the selected nodes, according to equation (29). The decisive factor is the dot product

$$\hat{\mathbf{Q}} \cdot \boldsymbol{\varepsilon}_{\mathbf{q},j}$$

Observe that nodes with reciprocal vector **Q** perpendicular to $\boldsymbol{\varepsilon}_{\mathbf{q},j}$ show null intensity. The node **Q** = 011 has the highest intensity of the drawing for $\boldsymbol{\varepsilon}_{\mathbf{q},j} = 011$.

Fig. 6.29c shows the experimental scattering pattern for a sample without external electric field. The x-ray beam is applied in the [001] direction. Structured diffuse scattering, associated to coexisting objects in the reciprocal space, is visible.

The interpretation of Fig. 6.29d is as follows. It is suggested that the investigated sample has PNR distributed according to a given set of orientations, relative to the crystal matrix. The superposition of rods corresponding to several PNRs is represented as "stars", the manner drawn in Fig. 6.29d. The intersection of these reciprocal stars with the CCD detector gives Fig. 6.29c.

The application of an electric field introduces some asymmetry in the stars, which manifests itself in the experiment. This asymmetry is clearly observable in Fig. 6.29e and f.

In the experiment by Xu *et al.*, synchrotron radiation diffuse scattering provides a vivid characterization of the reorientation of the PNR in the presence of an electric field.

6.4.3.3 Ultrathin Ferroelectric Films

The Advanced Photon Source group [62] has achieved a number of interesting research results, by means of synchrotron radiation scattering, on the system of ultra-thin layers of $PbTiO_3$ deposited on a $SrTiO_3$ substrate. With the help of Fig. 6.30 and Fig. 6.31, we illustrate this investigation, which analyzes the possibility of ferroelectric order in a sheet with a thickness of 4 or fewer atomic layers.

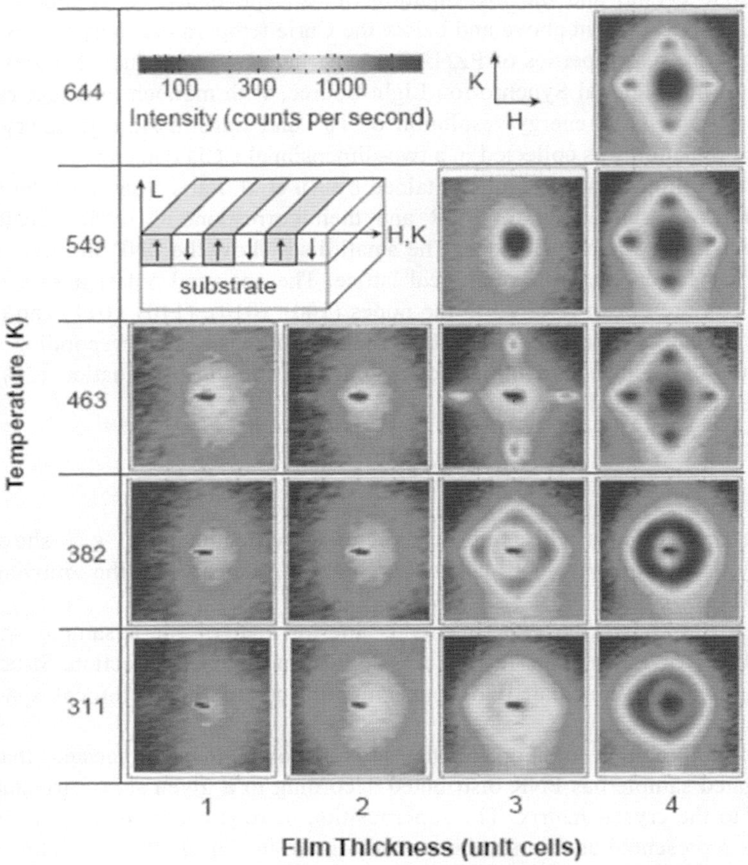

Fig. 6.30 In-plane diffuse x-ray scattering profiles around the $PbTiO_3$ 303 Bragg peak for various thicknesses and temperatures. In each case the region shown extends 0.2 reciprocal lattice units in the in-plane reciprocal space coordinates *H* and *K*. The broadening of the central Bragg peak in the *H* direction is due to instrumental resolution. Inset shows orientation of stripe domains. From Fong DD, Stephenson GB, Streiffer SK, Eastman JA, Auciello O, Fuoss PH, Thompson C (2004). Science 304: 1650. Figure reprinted with permission from AAAS.

The studied system adopts morphology of alternating polarity domains, as represented in the inset of Fig. 6.30. The regular distribution of domains gives rise to satellite diffuse scattering peaks around the $PbTiO_3$ diffraction maxima. The reciprocal distances and intensity distributions of the satellite peaks around the Bragg maxima depend on the periodicity and orientation of the stripe domains relative to the substrate crystal.

The film stripes formed by four layers of cells produce satellite peaks for all the considered temperatures. The distribution of domain orientations tends to be plane-isotropic at low temperatures and approaches a given crystallographic alignment with the substrate for high temperatures. This explains the satellite peaks (circular to 4-fold) symmetry change as temperature increases.

The film with three cell layers gives rise to satellite peaks for temperatures up to 463K, but not for 549K. This difference indicates the temperature range in which the ferro-paraelectric transformation occurs. Samples with one or two cell layers do not give satellite peaks at any temperature. This indicates the absence of ferroelectricity.

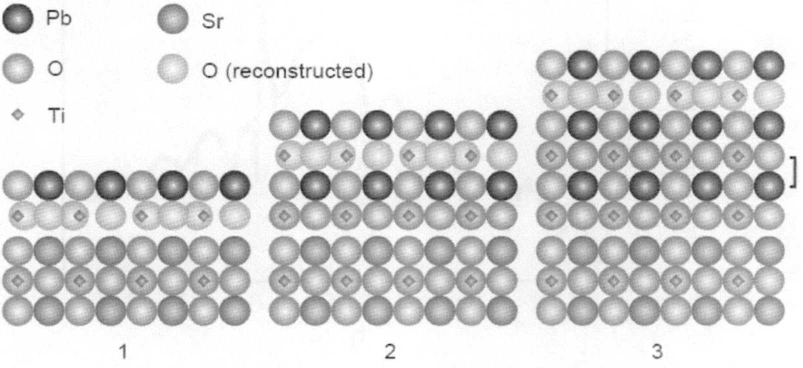

Fig. 6.31 Schematic of $PbTiO_3$ films 1 to 3 unit cells thick with the top unit cell reconstructed, as observed by Fong *et al.* The 3–unit cell film is the thinnest film having PbO and TiO_2 layers with the nearest neighbour environment of bulk $PbTiO_3$ (indicated by bracket). From Fong DD, Stephenson GB, Streiffer SK, Eastman JA, Auciello O, Fuoss PH, Thompson C (2004). Science 304: 1650. Figure reprinted with permission from AAAS.

Why do the three-cell layers reveal themselves as the limit of possible existence of ferroelectricity? The interpretation of Fong *et al.* is that this is the minimum thickness for which $PbTiO_3$ cell is surrounded top-and-bottom by layers of $PbTiO_3$, see Fig. 6.31.

While the positions of Bragg peaks depend on lattice parameters, the locations of satellite maxima depend on ferroelectric domain transverse-size period. Performing proper calculations, it is found that the stripe domain size, along the film surface, is of approximately three unit cells. Resuming the findings associated with different directions, the considered experiment suggests that three unit cells characterize the minimum size required for the appearance of ferroelectricity.

6.4.4 Crystal Truncation Rods

6.4.4.1 Ferroelectric Thin Films

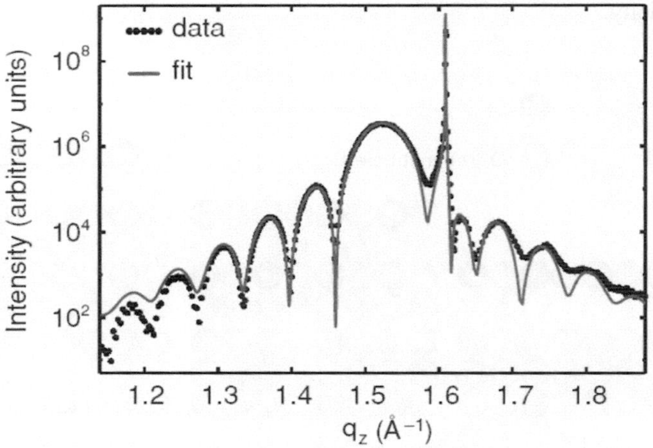

Fig. 6.32. Scattered intensity along the specular rod through the 001 peaks of a 10 nm thick $PbTiO_3$ film on a $SrTiO_3$ substrate. The solid line is the best fit to the data, represented by a dotted line. The results indicate the $PbTiO_3$ film is 24.5 unit cells thick, with c = 0.4128 nm, polarized down and with an interface spacing ξ =0.4830 nm. Figure reprinted with permission from Fong DD, Thompson C, Annual Review of Materials Research 36: 431. Copyright (2006) Annual Reviews.

To illustrate the use of CTR in the characterization of ferroelectric thin films, we present the grazing incidence experiment by Fong and Thompson [49] on a 10 nm thick film of $PbTiO_3$ ($d_{001} = a_3 \approx 4.14$ Å) epitaxially grown on a $SrTiO_3$ substrate ($d_{001} = a_3 \approx 3.90$ Å). Fig. 6.32 shows the 001 scans (also called specular rods) obtained experimentally together with the corresponding best fit to a model that considers the thickness of the film and the width of the interface film-substrate. Data analysis, including conversion of scanned angles into reciprocal space translations, was performed by software SPEC [53] .

In Fig. 6.32, the sharp peak at $q_z \approx 1.6$ Å^{-1} is the reflection 001 of $SrTiO_3$(b_3 = 2π / 3.9 = 1.61 Å^{-1}). The broad peak at the centre of the plot is due to $PbTiO_3$ (b_3 = 2π / 4.14 = 1.52 Å^{-1}). The width of the $PbTiO_3$ layer is estimated easily from the distance between fringes. At first glance it appears that 4 fringes correspond with $\Delta q_z \approx 0.25$ Å^{-1}. Thickness is: $t = (2\pi /$ *fringe spacing*$)$ = 10 nm.

The detailed analysis of the experiment allowed researchers to establish that the polarization of the $PbTiO_3$ film points down to the substrate and lead to the determination of the $SrTiO_3/PbTiO_3$ interface spacing. For details, see the work of Fong *et al.*

6.4.4.2 Atomic Structure

Fong and colleagues have succeeded in characterizing, with sub-Angstrom resolution, ultra-thin films of $PbTiO_3$ on a $SrTiO_3$ substrate [63]. The technique consists of measuring intensities along the Bragg rods and the subsequent data processing by means of the COBRA [64] code to obtain the electronic distribution.

The room temperature x-ray measurements were performed at the MHATT beamline at sector 7 and the PNC beamline at sector 20 of the APS. The experimental setup was as described by Sowwan *et al.* [65]. X-ray energy was 10 keV and Bragg rods scanned *H* and *K* from 0 to 3, over a *L* range typically from 0.5 to 3.1. *H*, *K*, and *L* are Miller indices given in the reciprocal lattice units of cubic $SrTiO_3$.

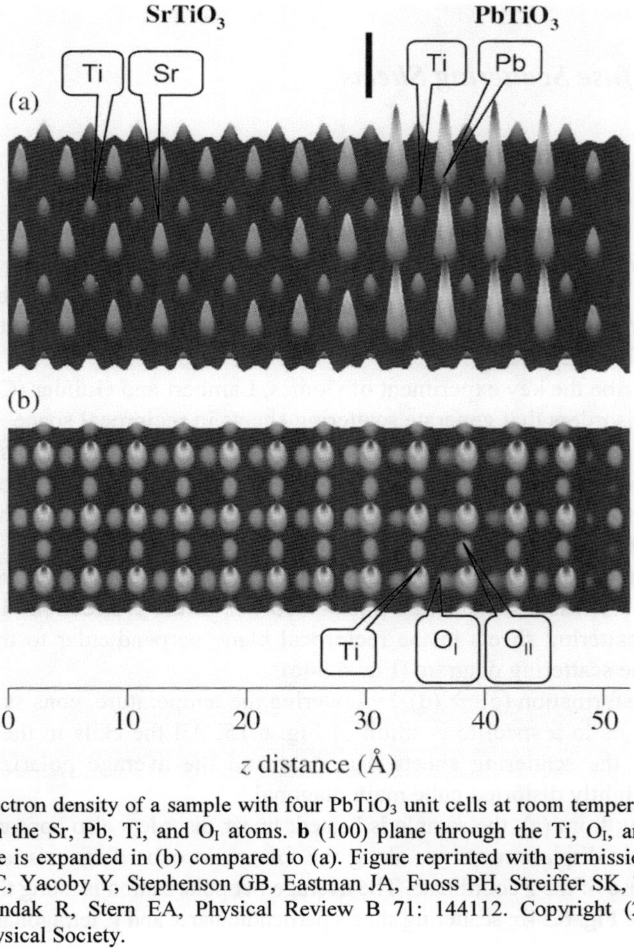

Fig. 6.33. Electron density of a sample with four $PbTiO_3$ unit cells at room temperature. **a** (110) plane through the Sr, Pb, Ti, and O_I atoms. **b** (100) plane through the Ti, O_I, and O_{II} atoms. Intensity scale is expanded in (b) compared to (a). Figure reprinted with permission from Fong DD, Cionca C, Yacoby Y, Stephenson GB, Eastman JA, Fuoss PH, Streiffer SK, Thompson C, Clarke R, Pindak R, Stern EA, Physical Review B 71: 144112. Copyright (2005) by the American Physical Society.

For the considered samples, the intensity distribution was localized on the Bragg rods and no satellites appeared in the diffuse scattering. This result is consistent with the monodomain nature of the investigated film.

The first important calculation by COBRA is to determine, from the intensities distribution in Bragg rods, the complex structure factors. At this stage, to deal with the phase problem, COBRA proposes an original educated guess (ansatz) and performs an iterative process related to those used in other direct methods for surface structure. Finally, it applies the Fourier transform to convert the complex structure factor into electronic distribution in the direct space.

Fig. 6.33 shows representative results of the investigation. A subtle displacement of atoms from their centro-symmetric positions in the $PbTiO_3$ film is observable. The detailed analysis of the positions of all atoms led to the conclusion that $PbTiO_3$ polarization points to the outgoing direction with respect to the $SrTiO_3$ substrate.

6.4.5 Diffuse Scattering Sheets

6.4.5.1 Temperature-Driven Phase Transformations

In this and the following section we briefly present a couple of diffuse scattering experiments that represent the complement to the observations of Chapman *et al.* The idea is to illustrate that careful scattering experiments followed by analytical interpretation allows clarifying the embryonic birth of ferroelectricity in similar but not equivalent cases.

We describe the key experiment of Comès, Lambert and Guinier (CLG) [20] on structural disorders that generate scattering sheets in reciprocal space. In the direct space the disorder is one-dimensional. Fig. 6.34 illustrates this interesting phenomenology. The sequence of two-dimensional scattering diagrams a-b-c-d corresponds respectively with the cubic-tetragonal-orthorhombic-rhombohedral structures that adopt $KNbO_3$ upon cooling and ordering.

The case (c) corresponds to the detailed description that we have studied in section *6.2.3.4*, associated with a *linear disorder* in the ***y*** direction. This disorder produces scattering sheets in the reciprocal plane perpendicular to the ***y*** axis, as shown in the scattering diagram (Fig. 6.34c).

The transformation (c) → (d), by lowering the temperature, consists of ordering all Nb cations to a specific position of Fig. 6.15. All the cells in the domain are now equal, the scattering sheets disappear and the average polarization points along the slightly distorted cube main diagonal.

If, starting from (c), the sample is heated; linear disorders also appear along the ***x*** axis. Cationic displacements are along the four main cube diagonals, always in the growing sense of the *z* coordinate. The average net polarization is along this axis. This is the case of Fig. 6.34b: scattering sheets perpendicular ***x*** and ***y***, tetragonal symmetry.

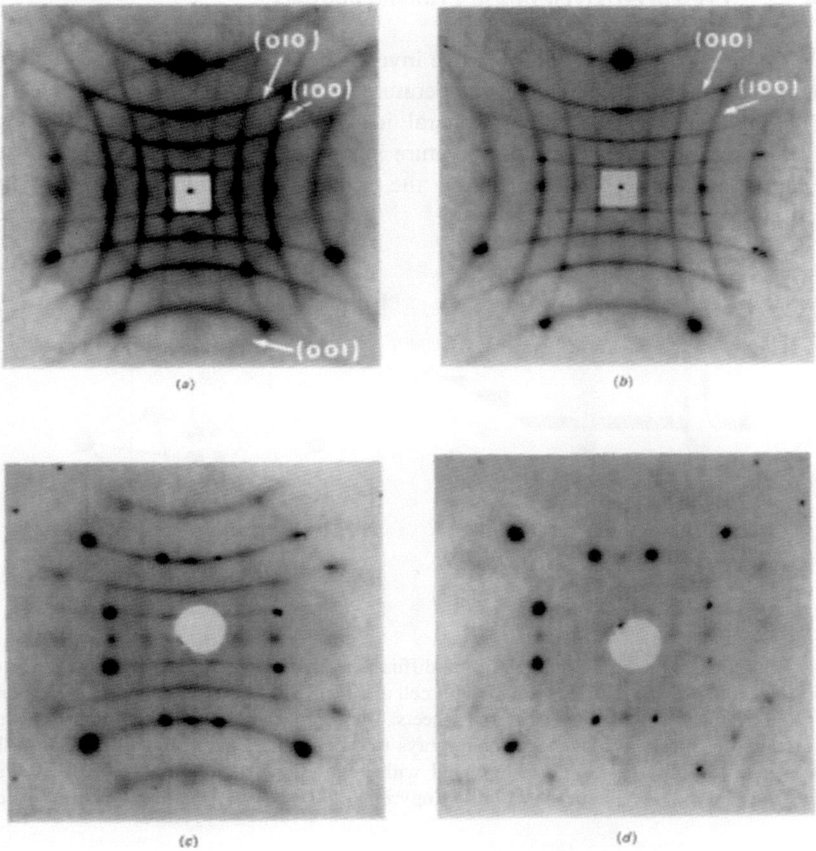

Fig. 6.34. Classical work of Comès, Lambert and Guinier. Scattering sheets produced by linear disorders in a single crystal of KNbo$_3$ at various temperatures. Axis b vertical, axis c parallel to the incident beam, λ = MoKα, exposure time = 2 h. **a** T = 773K: cubic phase. Scattering by the 3 {100} families of planes is observable. **b** T = 523K: tetragonal phase. Scattering by (001) planes has disappeared, signals by (100) and (010) subsist. **c** T = 217K: orthorhombic phase. Only scattering sheets associated with (010) planes survives. T = 217K: Trigonal phase (preserved by thermal hysteresis). No linear disorder sheets. Figure reprinted with permission from Comès R, Lambert M, Guiner A, Acta Cryst. A26: 244. Copyright (1970) by the International Union of Crystallography.

Finally, for high temperature, the disorder is present throughout the three crystallographic axes, Nb cations displace along the four main diagonals in the two directions. Scattering sheets appear perpendicular to $\boldsymbol{x}$, $\boldsymbol{y}$ and z, average symmetry becomes centro-symmetric cubic and ferroelectricity disappears. This is the case (a).

The CLG experiment and its theoretical model provide a simple and convincing explanation for the temperature-symmetry-scattering relationship in $KNbO_3$.

6.4.5.2 Pressure-Driven Phase Transformations

In a recent work, Ravi *et al.* [66] have investigated the possibility of simulating by means of pressure the effect of temperature on the symmetry and ferroelectricity of $BaTiO_3$ and $KNbO_3$. As a general idea, hydrostatic pressure: a) being an isotropic action, resembles a temperature increase because it tends to increase the structure symmetry; b) compacting the sample, exerts an action contrary to thermal expansion.

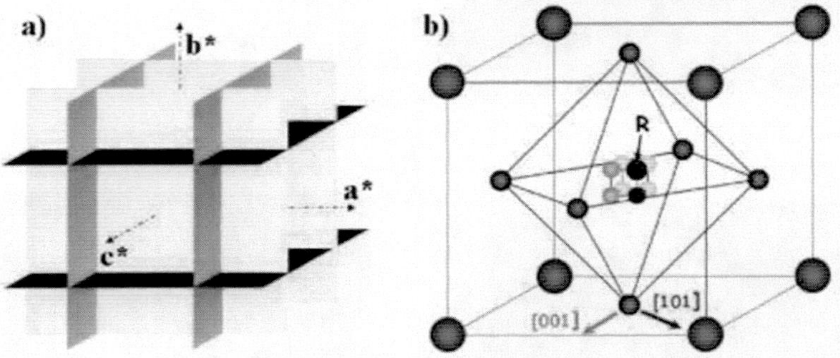

Fig. 6.35a Schematic representation of the diffuse sheets in the reciprocal space of ferroelectric perovskites. **b** Perovskite pseudo-cubic unit cell and the 8 sites of the B atoms. R corresponds to trigonal symmetry, no diffuse scattering sheets. Dark-gray represents orthorhombic occupancy and scattering sheets. Dark- and medium-grays represent tetragonal symmetry. All gray shades correspond to cubic case. Figure reprinted with permission from Ravy S, Itié JP, Polian A, Hanfland M, Phys. Rev. Lett. 99: 117601. Copyright (2007) by the American Physical Society.

Fig. 6.35 recreates the CLG model. For linear, parallel to the ***y*** axis disorder (average orthorhombic symmetry, [101] polarization) diffuse scattering is present at the horizontal reciprocal planes (dark-gray). If the disorder also involves the ***x*** axis, preserving the *z* axis order (average tetragonal symmetry, [001] polarization) then four B perovskite positions, displaced toward the front in Fig. 6.35b, are statistically occupied. In this case the diffuse scattering is also observed in the vertical planes perpendicular to the vector ***a*** (medium-gray). Finally, if the 8 possible positions of the atom B are statistically occupied in columns parallel to vectors ***a***, ***b*** and ***c*** (average cubic symmetry, no polarization), then diffuse scattering can be observed from the three families of mutually perpendicular planes.

Ravy *et al.* experiments have been performed at the ID09A beamline of the European Synchrotron Radiation Facility (ESRF), using a membrane diamond anvil cell. Photon energy was 30 keV. Diffraction patterns were collected by a Mar345 image-plate detector.

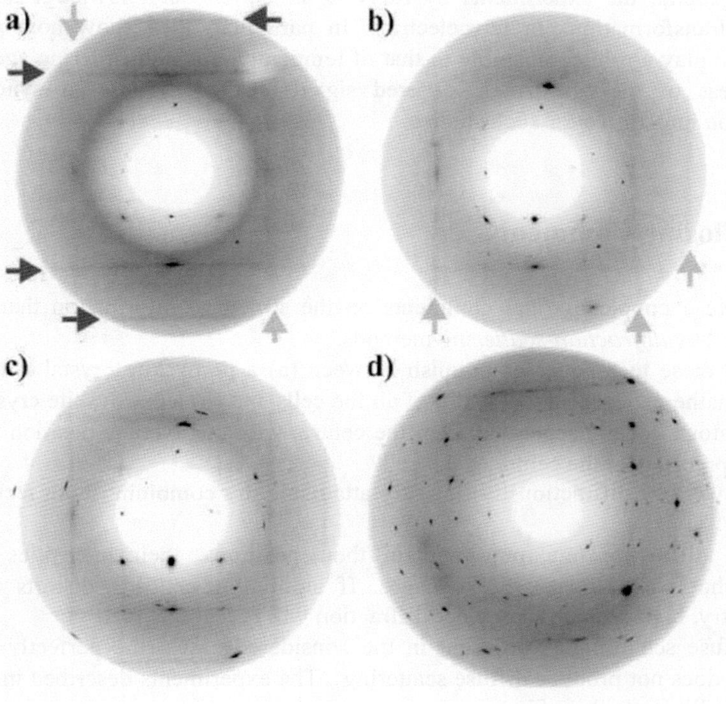

Fig. 6.36. Scattering patterns of $KNbO_3$ at (a) 4.3 GPa, b) 6.3 GPa, c) 12.1 GPa. and (d) 21.8 GPa. Dark-gray (horizontal) and light-gray (vertical) arrows point to (010) and (100) diffuse lines respectively. Figure reprinted with permission from Ravy S, Itié JP, Polian A, Hanfland M, Phys. Rev. Lett. 99: 117601. Copyright (2007) by the American Physical Society.

Fig. 6.36 shows the experimental results of Ravy *et al.* on a $KNbO_3$ crystal. The scan a-b-c represents scattering measurements with sample # 1 under the following pressures (GPa): a) 4.3, b) 6.3 and c) 12.1. Case (d) refers to sample # 2, measured at 21.8 GPa under special high pressure precautions. The orientation of the sample relative to the two-dimensional detector coincides with the one reported in the CLG paper.

The observed diffuse lines correspond, for all the investigated pressures, with planes of the form {001}. For low pressure (Fig. 6.36a, $p < 6$ GPa) the orthorhombic phase is predominant. The (010) diffuse lines are clearly observable (horizontal arrows). Some faint (100) lines are present. For $p = 6.3$ GPa the (100) signal becomes clearly visible, even fairly more intense than the (010). For pressure values higher than 10 GPa, (010) and (100) lines with approximately equal intensities are observed, indicating a stabilization of the tetragonal phase. For high pressure, detection of the cubic phase through the emergence of (001) lines is expected. In Fig. 6.36d ($p = 21.8$ GPa) the intensities of these diffuse lines scrape the detection limit.

In general, the experiments by Ravy *et al.* enrich our knowledge about the phase transformations in ferroelectrics. In particular, they show how pressure tends to play a role comparable to that of temperature as a disordering agent. The weakness of the associated scattered signals makes the use of synchrotron radiation imperative for such studies.

6.5 Closing Comments

We give a couple of final comments on the structural information that can be obtained by *diffraction-scattering* methods.

Can these techniques distinguish between (a) a perovskite crystal with the B atom mathematically at the centre of all the cells and (b) a petrovskite crystal with the B atoms locally separated from the centre, with the average position of these atoms precisely at the centre of the cell?

Answer: by diffraction itself,hardly attainable. by combining both techniques, yes.

The diffraction maxima maintain their positions, their intensities change somewhat when local order changes. If atomic off-centring distorts the cell geometry, high-resolution powder diffraction will reveal this fact.

Diffuse scattering is different in the considered cases (the perfectly ordered crystal does not produce diffuse scattering). The experiments described in our last sections illustrate this effect.

To what extent diffraction-scattering techniques contribute to clarify the microscopic origin of ferroelectricity?

Answer: Diffraction and scattering are of first importance, but –again– the combination of techniques is unavoidable.

At present, two possible phenomena are seen as the main promoters of ferroelectricity at microscopic level. One is the softening of optical phonons at the origin of the Brillouin zone (phonon picture). The other one is the presence of linear chains of correlated local displacements (order-disorder model).

Currat *et al.* [67], by means of inelastic neutron scattering, have found evidence of phonon softening in $KNbO_3$. This evidence points in favour of the phonon model. The diffuse scattering experiments conducted by Chapman *et al.* on $PbTiO_3$, commented in this chapter, are also explained better under the phonon picture. On the other hand, the experiments of Harada *et al.* [68], also with inelastic neutron scattering, support the order-disorder model.

The experiments described here of Comès and Ravy sustain the order-disorder scheme.

The EXAFS technique participates significantly in the current studies about this problem. Ravel's contributions [69] have been significant. According to EXAFS, the Nb cation in $KNbO_3$ is shifted in the direction [111] exactly as in the Comès model.

Investigations about the atomic-scale origin of ferroelectricity are in a decisive moment. Every day, new aspects of this intriguing phenomenon are revealed, new answers are found and new questions arise. Probably, as suggested by Gonzalo and Jiménez [70], separations among different approaches will tend to move closer. Ferroelectricity may be a multifactor phenomenon.

The arsenal of techniques for structural analysis is strong and varied. As part of this arsenal, synchrotron radiation in its several variants will surely play a key role in the next discoveries.

Acknowledgments

Portions of this research were carried out at the Stanford Synchrotron Radiation Laboratory, a national user facility operated by Stanford University on behalf of the U.S. Department of Energy, Office of Basic Energy Sciences. Thanks to A. Metha, B. Johnson, S. Webb, C. Knotts, A. Nilsson and G. Brown Jr. for systematic support at SSRL. Funds from Consejo Nacional de Ciencia y Tecnología, México (CONACYT Projects 25380 and 46515) is gratefully acknowledged. The author appreciates the support given by the Department of Inorganic Chemistry, Universidad Complutense de Madrid, leaded by M. A. Alario-Franco and E. Morán Miguélez. M. E. Montero-Cabrera, M. E. Fuentes-Montero and L. Fuentes-Montero actively participated in the research leading to this chapter.

References

1. http://www-ssrl.slac.stanford.edu/
2. http://www.esrf.eu/
3. http://www.aps.anl.gov/
4. http://www.spring8.or.jp/en/
5. http://www.anl.gov/Media_Center/Frontiers/2001/c2facil.html
6. http://www-ssrl.slac.stanford.edu/lcls/index.html
7. http://www.doitpoms.ac.uk/tlplib/reciprocal_lattice/printall.php
8. Renaud G (1998) Surface Science reports 32 : 1- 90.
9. Zachariasen WH (1945) Theory of x-ray diffraction in crystals. Dover, New York.
10. Internacional Tables for Crystallography http://it.iucr.org/
11. Trueblood KN, Bürgi HB, Burzlaff H, Dunitz JD, Gramaccioli CM, Schulz HH, Shmueli U, Abrahams SC (1996) Atomic displacement parameter nomenclature. Acta Cryst. A52: 770 – 781.
12. Kahn R, Carpenter P, Berthet-Colominas C, Capitan M, Chesne M-L, Fanchon E, Lequien S, Thiaudière D, Vicat J, Zielinski P and Thurman H (2000) Feasibility and

review of anomalous X-ray diffraction at long wavelengths in materials research and protein crystallography. J. Synchrotron Rad. 7: 131-138.
13. Ealick SE (2000) Advances in multiple wavelength anomalous diffraction crystallography. Curr. Opinion Chem. Biol. 4: 495-499.
14. Proffen Th, Billinge S, Egami T, Louca D (2003) Structural analysis of complex materials using the atomic pair distribution function - a practical guide. Z. Kristallogr. 218: 132-143.
15. Egami T, Billinge S (2003) Underneath the Bragg Peaks: Structural analysis of complex materials, Elsevier, Amsterdam.
16. You H, Zhang QM (1997) Diffuse X-ray scattering study of lead magnesium niobate single crystals. Phys. Rev. Lett. 79: 3950-2953.
17. You H (2000) X-ray scattering study of soft-optic-mode freezing in lead magnesium niobate single crystals. J. Phys. and Chem. of Solids 61: 215-220.
18. Ekstein H (1945) Phys. Rev. 68: 120.
19. Huang K (1947) Proc. R. Soc. A 190: 102.
20. Comès R, Lambert M, Guiner A (1970) Désordre linéaire dans les cristaux (cas du silicium, du quartz, et des pérovskites ferroélectriques). Acta Cryst. A26 : 244-254.
21. Noheda B, Cox DE, Shirane G, Gonzalo JA, Cross LE and Park SE (1999) A monoclinic ferroelectric phase in the $Pb(Zr_{1-x}Ti_x)O_3$ solid solution Appl. Phys. Lett. 74: 2059-2061
22. Young R. DBWS Rietveld software: http://www.physics.gatech.edu/downloads/young/DBWS.html
23. Rodriguez-Carvajal J. FULLPROF Rietveld software: http://www.ill.eu/sites/fullprof/
24. von Dreele B. GSAS Rietveld software: http://www.ccp14.ac.uk/solution/gsas/
25. Lutteroti L. RIETQUAN Rietveld software: http://www.ing.unitn.it/~luttero/rietquan/rietquan_20.html
26. Izumi F. RIETAN Rietveld software.: http://homepage.mac.com/fujioizumi/rietan/angle_dispersive/angle_dispersive.html
27. Young RA (1993) The Rietveld Method. Oxford Univ. Press, Oxford.
28. Cromer DT and Waber J T (1965) Acta Cryst 18: 104-109.
29. http://reference.iucr.org/dictionary/Cromer%E2%80%93Mann_coefficients
30. Shankland K (2004) Global Rietveld Refinement. J. Res. Natl. Inst. Stand. Technol. 109: 143-154.
31. Reinaudi L, Leiva EPM and Carbonio RE (2000) Simulated annealing prediction of the crystal structure of ternary inorganic compounds using symmetry restrictions. J. Chem. Soc., Dalton Trans. 2000: 4258-4262.
32. Caglioti C, Paoletti A and Ricci FP (1958) Nucl. Instrum. Methods 3: 223-228.
33. http://smb.slac.stanford.edu/powder/data/list.htm
34. Rietveld HM (1969) A profile refinement method for nuclear and magnetic structures. J. Appl. Cryst. 2: 65-71.
35. Dollase WA (1986) Correction of intensities for preferred orientation in powder diffractometry: application of the March model. J. Appl. Crystalogr. 19: 267-272.
36. Hillier S (2000) Clay Miner. 35: 291.
37. Fullmann T, Neubauer J and Walenta G (1999) Proc. Int. Conf. Cem. Microsc. 21: 103.
38. Neubauer J, Kuzel HJ and Sieber R (1996) Proc. Int. Conf. Cem. Microsc. 18: 100.
39. Fuentes ME, Mehta A, Lascano L, Camacho H, Chianelli R, Fernandez J and Fuentes L (2002) The Crystal Structure of $BaBi_4ti_4O_{15}$. Ferroelectrics 269: 159-164.
40. Nalini G, Guru Row TN (2002) Structure determination at room temperature and phase transition studies above Tc in $ABi_4Ti_4O_{15}$ (A = Ba, Sr or Pb). Bulletin of Materials Science 25(4): 275-281. ICSD card # 96607.
41. SUPER software: http://www-ssrl.slac.stanford.edu/~bart/super.html

42. García-Guaderrama M, Fuentes L, Montero-Cabrera ME, Márquez-Lucero A, Villafuerte-Castrejón ME (2005) Molten salt synthesis and crystal structure of $Bi_5Ti_3FeO_{15}$. Integrated Ferroelectrics 71: 233 – 239; Fuentes L, García M, Bueno D, Fuentes ME, Muñoz A (2006) Magnetoelectric effect in $Bi_5Ti_3FeO_{15}$ ceramics obtained by molten salts synthesis. Ferroelectrics 336: 81-89.
43. Ahart M, Somayazulu M, Cohen RE, Ganesh P, Dera P, Mao H, Hemley RJ, Ren Y, Liermann P, Wu Z (2008) Origin of morphotropic phase boundaries in ferroelectrics. Nature 451: 545-548.
44. SPEC software: http://www.certif.com/index.html
45. BLUE-ICE software: http://smb.slac.stanford.edu/facilities/software/blu-ice/
46. Hammersley A. FIT2D software: http://www.esrf.eu/computing/scientific/FIT2D/
47. Kraus and Nolze. PowderCell software: http://www.ccp14.ac.uk/ccp/web-mirrors/powdcell/a_v/v_1/powder/e_cell.html
48. Fuentes-Montero L, Montero-Cabrera ME and Fuentes-Cobas L: ANAELU software: http://www.cimav.edu.mx/ie/investiga/sw/index.php
49. Billinge S, Kwei GH, Takagi H (1994) Local octahedral tilts in $La_{2-x}Ba_xCuO_4$: evidence for a new structural length scale. Physical Review Letters 72: 2282-2285.
50. Billinge S. PdfGetX and PdfGetN PDF software: http://www.pa.msu.edu/cmp/billinge-group/programs/PDFgetX/, http://pdfgetn.sourceforge.net/. Proffen Th et al. DIFFUSE PDF software: http://www.pa.msu.edu/cmp/billinge-group/programs/discus/index.html
51. Billinge S, Thorpe M (editors, 1998) Local Structure from Diffraction, Plenum.
52. Petkov V, Gateshki M, Choi J, Gillian E, Ren DY (2005) Structure of nanocrystalline GaN from X-ray diffraction, Rietveld and atomic pair distribution function analyses. J. Mater. Chem. 15: 4654.
53. Dawber M, Rabe KM, Scott JF (2005) Physics of thin- ferroelectric oxides. Reviews of modern physics 77: 1083-1130.
54. Fong DD and Thompson C (2006) In situ synchrotron x-ray studies of ferroelectric thin films. Annual Review of Materials Research 36: 431-465.
55. Renaud G (1998) Surface Science reports 32: 1- 90.
56. Robinson IK (1998) X-ray crystallography of surfaces and interfaces. Acta Cryst. A54: 772-778.
57. Vlieg E (2002) Understanding crystal growth in vacuum and beyond. Surface Science 500: 458–474.
58. Huller A (1969) Solid State Commun. 7: 589; Huller A (1969) Z. Phys. 220: 145.
59. Chapman Bd, Ster Ea, Han Sw, Cross Jo, Seidler Gt, Gavrilyatchenko V, Vedrinskii Rv, Kraizman V (2005) Diffuse x-ray scattering in perovskite ferroelectrics. Physical review b 71: 020102(r).
60. Shirane G, Axe JD, Harada J, Remeika JP (1970) Soft Ferroelectric Modes in Lead Titanate. Phys. Rev. B 2: 155-159.
61. Xu G, Zhong Z, Bing Y, Ye ZG, Shirane G (2006) Electric-field-induced redistribution of polar nano-regions in a relaxor ferroelectric. Nature Materials 5: 134-140.
62. Fong DD, Stephenson GB, Streiffer SK, Eastman JA, Auciello O, Fuoss PH, Thompson C (2004) Ferroelectricity in ultrathin perovskite films. Science 304: 1650-1653.
63. Fong DD, Cionca C, Yacoby Y, Stephenson GB, Eastman JA, Fuoss PH, Streiffer SK,Thompson C, Clarke R, Pindak R, Stern EA (2005) Direct structural determination in ultrathin ferroelectric films by analysis of synchrotron x-ray scattering measurements. Physical Review B 71: 144112.
64. Yacoby Y. COBRA software. http://www.aps.anl.gov/Science/Highlights/2003/cobra.htm
65. M. Sowwan, Y. Yacoby, J. Pitney, R. MacHarrie, M. Hong, J. Cross, D. A. Walko, R. Clarke, R. Pindak, and E. A. Stern (2002) Direct atomic structure determination of epitaxially grown films: Gd_2O_3 on GaAs(100). Phys. Rev. B 66: 205311.

66. Ravy S, Itié JP, Polian A, Hanfland M (2007) High-Pressure Study of X-Ray Diffuse Scattering in Ferroelectric Perovskites. Phys. Rev. Lett. 99: 117601.
67. Currat R, Comès R, Dorner B and Wiesendanger E (1974) Inelastic neutron scattering in orthorhombic $KNbO_3$. J. Phys. C: Solid State Phys. 7: 2521-2539; Currat R, Buhay H, Perry CH, and Quittet AM (1989) Phys. Rev. B 40: 10741.
68. Harada J, Axe JD and Shirane G (1971) Neutron-Scattering Study of Soft Modes in Cubic $BaTiO_3$. Phys. Rev. B 4: 155 - 162.
69. Ravel B, Stern EA, Vedrinskii RI, and Kraizman V (1998) Ferroelectrics 206: 407.
70. Gonzalo A, Jiménez *B* (2005) Ferroelectricity: The Fundamentals Collection, WILEY-VCH, Weinheim

Chapter 7
X-Ray Absorption Fine Structure Applied to Ferroelectrics

Maria Elena Montero Cabrera

Abstract

This chapter is devoted to explaining the foundation and merits of the application of synchrotron radiation for studying the X-ray absorption fine structure (XAFS) in ferroelectric materials. XAFS in the local order allows the follows: determining the oxidation states, interatomic distances, Debye-Waller factors and the coordination number of atoms at the first few shells around the absorbing atom, up to 0.5 nm. The text explains the features of the photoelectric effect, as well as its relationship with the absorption edges of each element. Applying the Fermi's Golden Rule, it is explained that the fine structure of absorption edge spectra is the result of the interference of the photoelectron single- and multiple-scattering. This gives rise to the effects of X-ray Absorption Near Edge Structure (XANES) and Extended X-ray Absorption Fine Structure (EXAFS). The content includes significant effects for the study of ferroelectric materials in XANES: energy shift of the edge position, pre-edge transitions and white-lines. Subsequently, the experimental methods and EXAFS spectra processing, by both empirical methods and theoretical models with multiple scattering paths, are presented. Finally, we offer some works representing XAFS applied to ferroelectrics, from the explanation of the displacive or order-disorder nature of the materials $PbTiO_3$ and $BaTiO_3$, to those dedicated to relaxors and Aurivillius oxides.

Centro de Investigación en Materiales Avanzados, S.C.
Miguel de Cervantes 120, Complejo Industrial Chihuahua,
31109 Chihuahua, Chih., Mexico
Phone (52 614) 439 1123
Fax (52 614) 439 1170
elena.montero@cimav.edu.mx

7.1 Introduction: X-Ray Absorption Fine Structure

X-ray Absorption Fine Structure (XAFS) is a structure observed in X-ray Absorption Spectroscopy (XAS). It is associated with an amplitude modulation of the X-ray absorption coefficient of a chemical element, in energies near and above its X-ray absorption edge. Conventionally, the observed effects are divided in two regions, near to the absorption edge (so called X-ray Absorption Near Edge Structure- XANES or Near Edge X-ray Absorption Fine Structure-NEXAFS), and the extended one (Extended X-ray Absorption Fine Structure – EXAFS), from about 50 eV above the absorption edge. These zones contain different information on the chemical states of the absorbing element and neighbouring atoms. Considering XAFS as a modulation effect, it will be seen that its origin is the interference of the wave functions of the photoelectrons. This is while scattering on atoms from molecules or crystalline lattice of the sample, where the absorption is detected.

XANES offers information on the *oxidation state* of the absorbing element and in general on the *electronic structure of the bonds* of the target element. XANES is related to the interaction with the first neighbours of the absorbing element.

EXAFS allows the determination of distances, coordination numbers and degree of thermal and/or structural disorder of the local structure (first few atomic coordination shells) surrounding the absorbing atom. Both spectroscopies offer information on the local atomic coordination to distances normally of 0.5 nm and sometimes until of 0.8 nm. These techniques may be applied to almost any element of the periodic table despite crystallinity state (samples can be amorphous or crystalline) or concentration. That is, XANES and EXAFS limits of detection are 10 and 100 ppm respectively. These properties are particularly useful when the samples are in a liquid state, colloidal, or are adsorbed on surfaces. The sample preparation is usually simple, fulfilling the requirements of homogeneity and uniform thickness.

XAFS experiments require X-ray sources of high intensity and tunable energies offered by Synchrotron Radiation facilities (SR). The general characteristics of synchrotron radiation, such as production, intensity, spectra and management, are explained in Chapter 7, Section 7.1 Synchrotron Radiation, of this handbook.

XAFS has been known since the beginning of the twentieth century [1, 2]. But, the correct interpretation and the experimental implementation were achieved by the work of Stern, Lytle and Sayers [3, 4, 5, 6], who initially performed experiments in conventional X-rays sources. Kincaid and Eisenberger [7] introduced definitive advances when implementing X-rays produced by synchrotron radiation in Stanford Synchrotron Radiation Laboratory SPEAR e^+e^- storage ring. This provided a factor of 5×10^4 improvement in intensity over conventional X-ray sources. Recently, Rigaku Corp. has commercialized a conventional *in-house* XAFS spectrometer, with 3.0 kW X-ray generator with replaceable filaments and target materials to optimize X-ray flux, and a range of optical engineering for tuning the incident energy beam [8].

XAFS provides information, as it has been pointed out above, on the local spatial and electronic structure of materials. That is why it is important for studying ferroelectricity. It allows, for example, the determining of local distortions of crystal lattice not apparent in X-ray diffraction. As it will be seen later, XAFS has been applied to clarify the origin of ferroelectricity in materials such as $BaTiO_3$ and $PbTiO_3$. From these results it has been concluded that there is no fundamental difference between the *displacive* or *order-disorder* behaviour of ferroelectrics [9].

This text will explain the fundamentals of physics, experimental methods, data reduction and analysis, as well as some applications of XAFS to ferroelectric materials. General ideas about XANES spectroscopy would be looked at, and major attention would be paid to the EXAFS region of spectrum.

7.2 X-Rays Absorption in Materials

7.2.1 X-Rays Absorption

X-rays are electromagnetic radiation with very short wavelengths λ. They have high frequency $\nu = c/\lambda$ and hence high energy $E_X = h\nu = hc/\lambda$, where c is the speed of light and h is the Plank constant. In Chapter 7, diffraction and scattering phenomena are discussed. Coherent wave characteristics of X-rays are considered. In the case of XAFS, the predominant mechanism of absorption of X-rays is *corpuscular* instead of *wave-like*, as we see next.

When an incident X-ray beam passes through a material its intensity is reduced. But, it never becomes zero. This phenomenon is called *X-rays attenuation*. The interaction of X-rays with matter is determined according to specific probabilities. That is to say, a photon may travel within the material without undergoing change until an interaction occurs. Once it interacts, disappearing from the beam, each photon independently transfers its energy to an electron in the medium. For that reason, the X-ray beam is attenuated, and the energy of emergent rays is equal to that of the incident rays, although the intensity of the beam is smaller. This probabilistic mechanism is expressed by means of the attenuation law.

Let's consider a monoenergetic and parallel X-rays beam incident perpendicularly to a material of thickness x (Fig. 7.1). Let I_0 be the intensity of the incident beam and I – the intensity of the transmitted beam. The change in X-ray intensity due to attenuation is expressed according to:

$$dI = -\mu(E)\, I\, dx \tag{1}$$

$$I = I_0 e^{-\mu(E)x} \tag{2}$$

where dI is the differential change of intensity of the beam when crossing the differential thickness dx, and the magnitude $\mu(E)$ is called *attenuation or absorption coefficient*. The absorption coefficient depends on the probability of interaction of X-rays with matter and on the density of the absorbing material.

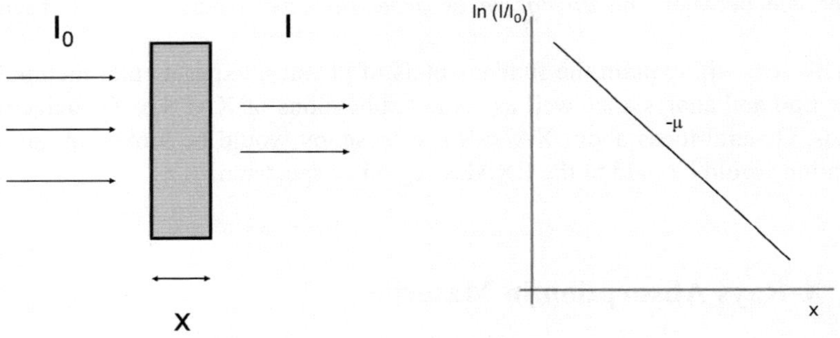

Fig. 7.1 X-ray absorption experiment. I_0: incident beam intensity; I: transmitted beam intensity; x: thickness of material crossed by the beam; μ: X-rays absorption coefficient.

The absorption of X-rays in materials is the consequence of an incoherent interaction of photons with the atomic species (electrons and nuclei). The interaction with the nuclei at characteristic X-rays energies can be neglected [10]. Therefore, we will concentrate in the following section on revising the photon absorption by electrons. The main mechanisms of X-ray absorption in materials are the *photoelectric effect* and *scattering*. Each one of them has an occurrence probability and an absorption coefficient $\mu(E)$ that depends on the X-rays energy, and on the atomic number of the material that is intersected.

In the *photoelectric effect,* a photon of energy $E_X = h\nu$ is absorbed by a system in which the electrons are bound with energy B_i. The incident energy is transferred to the system, giving rise to the release of a *photoelectron* of energy $E_e = h\nu - B_i$. This phenomenon requires the photon energy to be greater, but of the same order, than the binding energy of the electron, $E_X > \approx B_i$. The following is a necessary condition of the photoelectric effect to take place (Fig. 7.2): An X-ray cannot be completely absorbed by a free electron because conservation of energy and linear momentum laws cannot be fulfilled simultaneously [10]. However, if the electron is bound, the residual system moves and both the energy and the linear momentum are conserved.

When the photoelectric effect takes place in the X-ray region (energies of the order of keV), the *released photelectrons* are those belonging to the inner atomic shells. X-rays photoelectric effect is not observed in light elements from H to Be. However, the heavier it is the element, the higher the probability of occurrence. On its way out, the photoelectron interacts with the neighbours of the absorbing atoms and sometimes has enough energy to produce ionization along its path. When this takes place, it loses all its energy and the photoelectron is completely absorbed by the material.

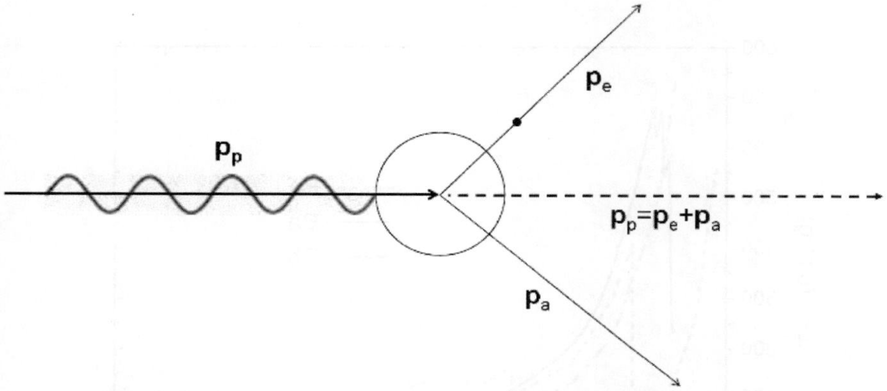

Fig. 7.2 Linear momentum conservation for the photoelectric effect in a system where the electron is bound. $\mathbf{p}_e$ – linear momentum of the photoelectron; $\mathbf{p}_a$ – linear momentum of the ionized atom; $\mathbf{p}_p$ – linear momentum of the photon.

7.2.2 X-Rays Absorption Edges

Fig. 7.3 shows the dependency of the photoelectric absorption coefficient with the X-rays energy for Fe, Ba and Pb in the interval of energies frequently used in synchrotron sources.

There is no general expression for the dependency of the probability of photoelectric absorption for all E_X or Z. An approach of this dependency is

$$\mu \cong const. \times \frac{Z^n}{E_X^3} \tag{3}$$

where n is approximately 4.3 in the energy zone of X-rays. From this relationship, it is seen that the photoeffect in the lightest elements does not take place. The potential law Z^4E^{-3} may be understood qualitatively, according to Teo [11], in the following manner. Linear momentum conservation is required in the photoeffect. The capability of momentum conservation is proportional to the mass of the target atom, and the cross section of photoelectron ejection is then proportional, too. The exponent 4 arises when considering the probability of emission of the photoelectron from the absorbing atom. This probability is proportional to the squared matrix element of the electrical dipole transition. This in turn contains a Z^2 term, and therefore, the probability is approximately proportional to Z^4. For a fixed Z, the transfer of linear momentum increases when X-ray energy increases, which corresponds also to the inverse dependency of the effect probability with the X-rays energy.

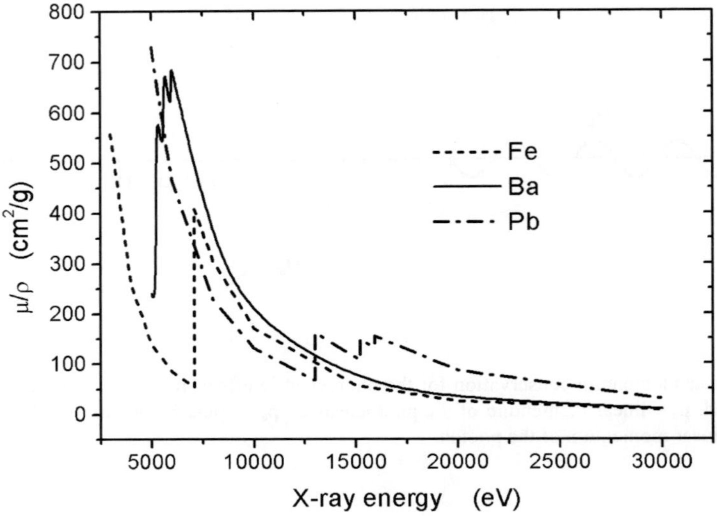

Fig. 7.3 Mass absorption coefficient for Fe, Ba and Pb, at X-ray energies from 5 to 30 keV. Observed absorption edges are the L_{III}-, L_{II}- and L_{I}- edges for Ba and Pb and the K- edge for Fe, respectively. Data were extracted from [12].

The photoelectric absorption coefficient reflects the existing relation between E_X and B_i. As E_X is decreasing and approaching B_i, the interaction probability increases and μ also does. When $E_X = E_0 = B_i$, μ reaches its maximum. For E_X immediately below the binding energy of each shell, the absorption coefficient becomes a minimum, giving the aspect characteristic *saw teeth* shape. This type of behaviour is called *absorption edge*, and it corresponds to each electronic shell of the absorbing element. Another way of describing the absorption edge energy B_i or E_0 is as the minimum energy necessary to create a vacancy in the corresponding internal layer of the absorbing atom.

The absorption coefficient plot appearing in Fig. 7.3 is characteristic of isolated atoms. When absorbing atoms are in binding systems, the absorption coefficient near to the edge presents a *fine structure*. The energy values at the K edges of absorption are proportional to Z^2, according to Moseley's Law. All the elements with $Z \geq 39$ (potassium) have a K or L_{III} edge between 3 and 35 keV that are the energies obtained in synchrotron sources. Fig. 7.4 shows the energy dependency of the K and L_{III} absorption edges with the atomic number for all chemical elements up to Z=94.

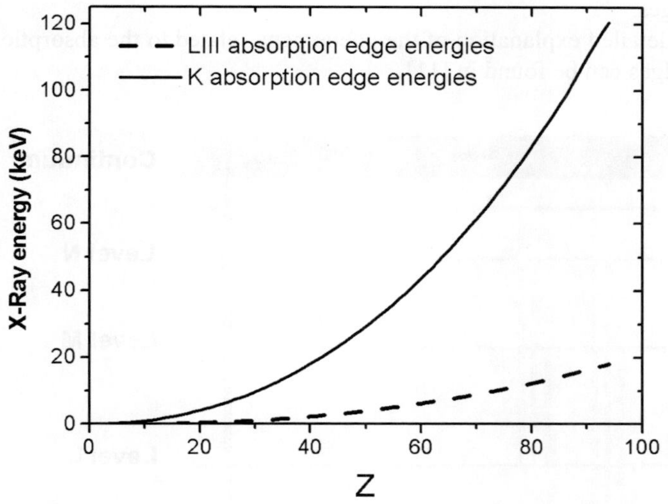

Fig. 7.4 Energy of the of K- and L_{III}- absorption edges, in function of the atomic number. Data were obtained from [13].

Photoelectric effect is also called *true absorption* [11]. Scattering relates to deflection of X-ray photons from their original direction of propagation, leaving the beam. Scattering takes place by collisions with electrons or atoms, either with or without loss of energy. The elastic scattering is called *Rayleigh Effect*, and the inelastic is called *Compton Effect*. The *Compton Effect* consists of the dispersion of the incident photon with a *free* electron. An X-ray with energy much greater than the binding energy of electrons in the system, $E_X >> B_i$, finds those electrons as *free*. As a result of energy and linear momentum conservation laws, the scattered photon has an energy $E_{X'} = h\nu'$ less than the incident photon energy. The Compton electron absorbs the difference in energy, $E_e = h\nu - h\nu'$, which determines the X-ray scattering angle with respect to the incident direction. Scattered X-rays can again undergo some other interactions in material, or escape.

A more careful look at the photoelectric effect allows us to observe that the photoelectron leaves a hole or vacancy in the electronic layer of the absorbing atom. The photoelectron leaves to the energy zone of the unoccupied or continuous spectrum. If the absorbing atom is bonded, its interaction with neighbours accounts for the fine structure in the absorption edge as detailed below.

The hole in the internal electronic layer causes a pronounced imbalance in the atom, and is occupied quickly with electrons from more external layers. These electronic transitions give rise to the emission of *X-rays fluorescence* characteristic of the element of the absorbing atom, or of *Auger electrons.* The process of Auger electron emission can be multiple [11]. Fig. 7.5 schematically presents the production of X-ray fluorescence and Auger effects. The energy of fluorescent X-rays and Auger or secondary electrons is always lesser than the energy of the absorption edges.

A more detailed explanation of the phenomena related to the absorption coefficient and edges can be found in [11].

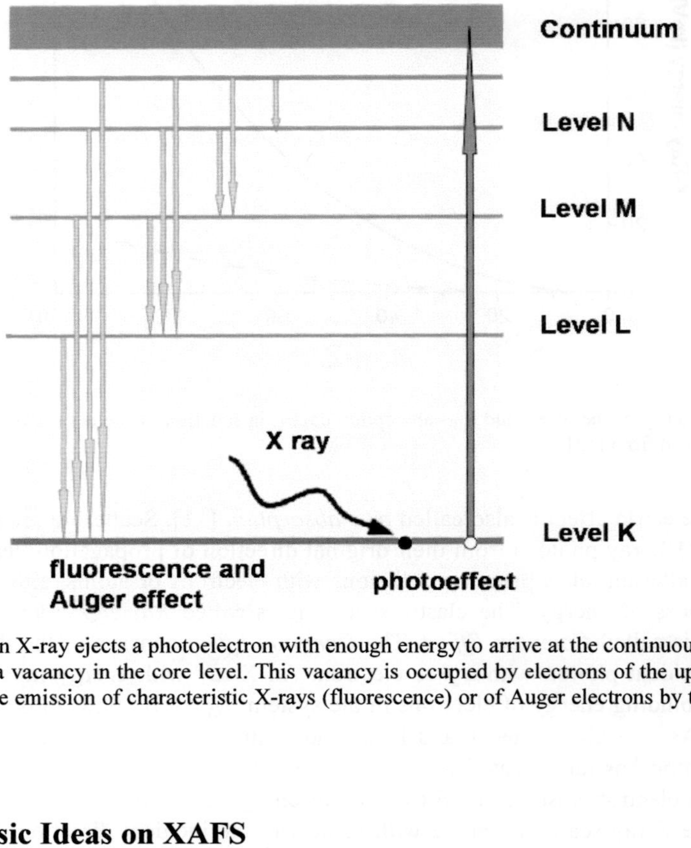

Fig. 7.5 An X-ray ejects a photoelectron with enough energy to arrive at the continuous zone and produces a vacancy in the core level. This vacancy is occupied by electrons of the upper shells, causing the emission of characteristic X-rays (fluorescence) or of Auger electrons by the absorbing atom.

7.3 Basic Ideas on XAFS

7.3.1 The EXAFS Function

This section will briefly introduce the reader to the experimental details of XAFS. An experiment in a synchrotron station requires an X-rays beam of tunable energy through a monochromator. Focused on the sample under study, the initial intensity I_0, the transmitted I_1 and/or the fluorescent I_f must be known. It is advisable to also measure the absorption of a reference sample, with the corresponding intensity I_2. A typical setup appears in Fig. 7.6 that represents an experiment of XAFS in synchrotron DESY at Hamburg [14].

Fig. 7.6 Beamline E4 at storage ring DORIS III. TM: toroidal mirror with Au-coating; PM: plane mirror with C-, Ni- or Au-coating; WBS: white beam slit; RBS: reflected beam slit; Si(111): double crystal monochromator with fixed exit; ES: exit slit; I0: ion chamber for I_0 beam intensity; S: sample; I1: ion chamber for transmission determination through the sample; RS: reference sample; I2: ion chamber for transmission determination through the reference sample

The absorption coefficient $\mu(E)$ may be determined in transmission mode [3, 7]. Transforming expression (2), it becomes:

$$A = \mu(E)x = \ln(I_0 / I_1) \quad (4)$$

The quantity A is known in spectrometry as *absorbance*. The coefficient $\mu(E)$ is also determined in fluorescence mode, introduced by [15], where the intensity of fluorescent beam I_f approximately fulfils

$$\mu(E) \propto I_f / I_0 \quad (5)$$

The oscillations of $\mu(E)$ appearing on the extended region of the XAFS spectra are clearly observed by representing the absorption coefficient in the form of [5, 16]:

$$\chi(E) = \frac{\mu(E) - \mu_0(E)}{\Delta\mu_0(E)} \quad (6)$$

where the smooth function $\mu_0(E)$ – *the background of the bare atom* – has been subtracted from $\mu(E.)$ The result has been divided by $\Delta\mu_0(E)$ – *the height of the absorption edge* – to normalize the absorption event to 1. Equation (6) is the definition of the so-called *EXAFS function*. In Fig. 7.7, the meaning of both parameters is graphically presented.

XAFS is an interference effect and depends on the wave nature of the photoelectron. In this regard, the XAFS function is presented in terms of the photoelectron wave number, k, instead of in the energy of X-rays:

$$k = \sqrt{\frac{2m(E - E_0)}{\hbar^2}} \quad (7)$$

Here $\hbar = h/2\pi$. The resulting spectrum is multiplied by k^2 or k^3, for amplifying the oscillations in the k regions [17], see Fig. 7.8.

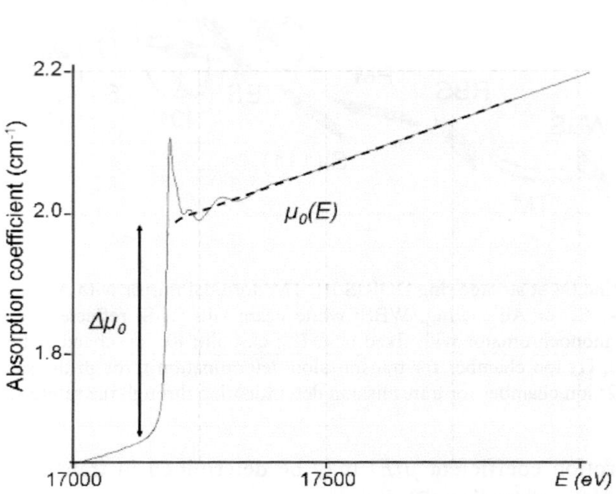

Fig. 7.7 Average raw X-ray fluorescence spectrum of uranium electrodeposited on stainless steel plate. Dashed curve corresponds to the smooth background $\mu_0(E)$of uranium atoms without neighbours (calculated) and $\Delta\mu_0$ is the height of the absorption edge

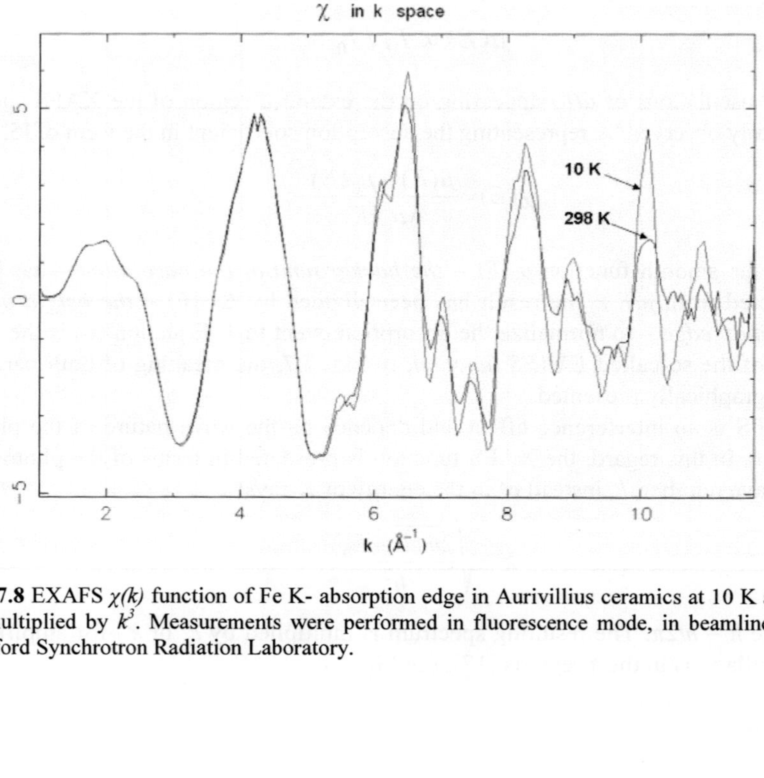

Fig. 7.8 EXAFS $\chi(k)$ function of Fe K- absorption edge in Aurivillius ceramics at 10 K and 298 K, multiplied by k^3. Measurements were performed in fluorescence mode, in beamline 2-3 at Stanford Synchrotron Radiation Laboratory.

In the EXAFS region of the absorption spectra, the oscillations of $\mu(E)$ are an interference effect due to the presence of neighbouring atoms. When an atom absorbs an X-ray of energy E, it emits a photoelectron of energy $(E - E_0)$. For an isolated atom, $\mu(E)$ has the absorption edge for the binding energy of the internal electron, and is a smooth function of the energy above the edge. In the presence of neighbouring atoms, the emitted photoelectron is backscattered and returns to the emitting atom. The wave of the returning photoelectron interferes with itself. This is in a form analogous to the interference of waves in water when a stone is thrown and the wave is reflected by obstacles in a pond. The coefficient $\mu(E)$ depends on the states to which the photoelectron in the absorbing atom accesses. These are altered by interference effects. The amplitude of the backscattered photoelectron wave in the absorbent atom varies with energy, causing oscillations in the coefficient $\mu(E)$, producing the fine structure.

All these processes may be detected through the scattering cross section of X-rays, which provides the probability of interaction with the system. The value of the cross section is related to the scattering of the photoelectron in the neighbouring atoms of the target or absorbent atom. When this scattering takes place, the photoelectron acquires the so called *scattering amplitude F* and *phase shift* δ, which are related to the *photoelectron energy E*. Or, it acquires the *wave number k* with the attributes of the *scattering atom*. The characteristics of these quantities are decisive for the effects of XAFS.

7.4 X-Ray Absorption near Edge Structure – XANES

This section will summarize the general ideas on the X-ray Absorption Near Edge Structure – XANES and its application to ferroelectrics. The phenomenological fundamentals will be provided and the theoretical basics, which are described in more detail in section 7.5, will be outlined. The specific elaboration of XANES spectra through the use of programs, with Green functions algorithms, will not be considered in this text. Readers interested in the theory of XANES are referred to the texts [18,19,20].

7.4.1 The XANES Zone: Photoelectron Multiple Scattering and Allowed Transitions

The XANES area covers the energy zone from the absorption edge up to the extended zone or EXAFS. The division between areas near the edge and extended is arbitrary, since the effects overlap at the border. Fig. 7.9 provides an experimental spectrum obtained by Bianconi *et al.* [21] where the two areas are defined.

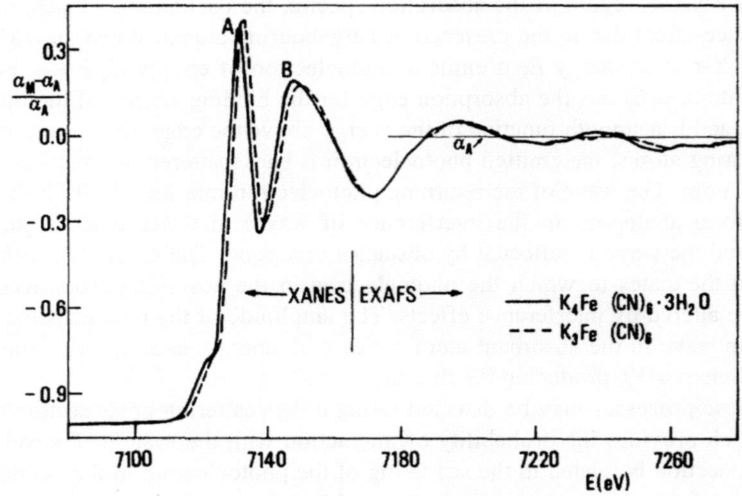

Fig. 7.9 Fe K- edge XAFS spectra of $K_3Fe(CN)_6$ and $K_4Fe(CN)_6$. The relative absorption with respect to the high-energy continuum atomic background α_A is plotted. Reprinted figure with permission from Bianconi A, Dell'Ariccia M, Durham PJ, Pendry JB, Phys Rev B, 26, 6502-6508 (1982). Copyright (1982) by the American Physical Society.

Bianconi [22] takes the concept of wave number of photoelectron defined by the expression (7) to limit the XANES area. This area is limited to that from the absorption edge up to the energy where $k<2\pi/R$, or where the wavelength $\lambda=2\pi/k$ of the excited photoelectron is $\geq R$, the distance between the absorbing atom and its first neighbours.

The absorption edge energy E_0 may be defined in three ways [22]: a) The *absorption threshold* or lowest energy state reached by the core electron excitations; b) the *absorption jump edge* or the energy where the absorption coefficient is at half-height of the bare atom absorption jump, defined above as $\Delta\mu_0(E)$; and c) the *ionization threshold* or the energy where the photoelectron is ejected to the continuum, which is the vacuum in atoms and molecules, the Fermi level in metals and the bottom of the conduction band in insulator.

For practical reasons, in this text it we will adopt E_0 as the absorption jump at half $\Delta\mu_0(E)$. Programs for data reduction use this definition, or that of the inflection point of the absorption coefficient function, for finding E_0.

In the wave number zone characteristic of XANES, the photoelectron backscattering amplitude is high, sharply increasing the probabilities of multiple scattering. That is why the effects of XANES are more intense than in the EXAFS area (see Fig. 7.9). As a consequence, XANES spectroscopy is strongly sensitive to oxidation state and coordination chemistry of the absorbing atom.

XANES spectrum zone is divided in two parts [18, 20]: at low energies, the so-called *edge or threshold region* , extended over 8 to 10 eV; and the region of multiple scattering in the continuum, called the *XANES region.*

The expression for the X-ray absorption cross section is based on the transition probability provided by the Fermi's Golden Rule [18] (see below, section 7.5). At the heart of this calculation is the *Hamiltonian of the multipolar interaction.* This corresponds to the transition allowed by symmetry, and by unoccupied states to which the photoelectron can be excited.

The solutions of the equations, calculated using Green's functions and the T matrix [23, 24, 25], are the eigenvalues of the energy of the electronic bands in the case of crystals. In the case of atoms and molecules, solutions are energy eigenvalues in the unoccupied *p* states of the excited atom.

Each type of absorption edge, namely K, L_I, L_{II} or L_{III}, will have a probability of preferential transition, since the value of the dipolar transition cross section is several orders higher than that of quadrupolar transitions. This fact imposes selection rules on the possible transitions for photoelectrons extracted by the X-ray, and therefore for XAFS. Selection rules offer restrictions to quantum numbers of both states by the relationships: $\Delta m = 0, \pm 1$; $\Delta l = \pm 1$ restricting in this way the final states which allow non-zero intensity transitions. In the K- and L_I-edge cases, this corresponds to $s \rightarrow p$ transition (1s, for K, and 2s, for L_I). For the L_{II}-edge it implies, for $\Delta l = +1$, $p_{1/2} \rightarrow d_{3/2}$, and for $\Delta l = -1$, $p_{1/2} \rightarrow s$ (with low intensity). Similarly, the L_{III} -edge is dominated by the $p_{3/2} \rightarrow d_{3/2}$ and $d_{5/2}$ transitions.

Bianconi [22] has summarized the principal conclusions on XANES multiple-scattering features in condensed systems, which have been derived from experiments. They are observed in the following points:

1. The spectrum dependence in the main edge and in the continuous area of XANES is due to the distribution of neighbouring atoms. There are fingerprints of the valence electrons structure of the occupied levels towards the transitions take place, particularly in the area of the main edge (~ 8 eV).
2. In XANES, the symmetry around the absorbing atom is very important, namely geometry, angles and relative positions of atoms in the environment of the target.
3. The photoelectron multiple scattering is the main physical effect determining the XANES in crystalline structures. This gives multiple-scattering resonances and some features in given systems, as we will see below.
4. The range of the XANES effects is given by the core hole lifetime and inelastic scattering of the photoelectron with the valence electrons, as in EXAFS.

Knowing these characteristics of allowed photoelectron transitions conveniently leads to the interpretation of XANES spectra. The following are several features of the spectra that are useful for the study of materials in general and of ferroelectrics in particular.

7.4.2 Edge Energy Position

The position of the K- or L- absorption edges in a target undergoes a systematic shift toward higher energies when oxidation states increase. In other words, there is a linear correlation between the energy position of the absorption edge and the oxidation state. This effect has been known since this technique was emerging, and is well documented and applied since the 1980s [23, 26]. The effect is verified for cations with two or more different valences. This linear correlation is normally measured at the inflection point of the edge, as defined above, for metals and oxides. This feature is clearly observed in transition metal oxides, such as Ti, V, Cr, Mn, Fe, Mo, and so on. It has been defined an extended concept of coordination charge to substitute oxidation state in order to describe metallic/alloyed or covalent systems [23], and the linear correlation of the edge position conserves also for these cases. This way, by measuring the edge position, it is possible to determine the average oxidation state of an element in a given configuration. Fig. 7.10 shows a number of XANES spectra from K-edge absorption and the calibration curve performed with them for different oxidation states of Mo, from the metal up to MoO_3. This was conducted by Ressler *et al.* [26] in a catalysis study. A feature of the spectrum above the absorption edge that appears on all spectra was employed for calibration. In Fig. 7.10b, it is observed that the absorption edge energy changes in amounts to ~ 6.5 eV with the average Mo valence per increase in oxidation state by one, from metallic Mo to Mo^{6+}. The origin of the effect of the absorption edge shift with the oxidation state is attributed thus. A change in the core-electron binding energy occurs due to the electrostatic potential of the valence electrons that are extracted in the oxidation of the metal to offset the Madelung potential of crystals. A phenomenological discussion of this effect can be found in the review of Fernandez-Garcia [27].

This effect of the absorption edge shift has been applied in the study of oxidation states of ruthenium in perovskite-related materials by Liu *et al.* [28] and Bos *et al.* [29]. Stoupin *et al.* [30] have made use of this effect in studying the oxidation state of Mn in a multifferroic material. A solid solution of Mn replacing Ti in $PbTiO_3$ has been prepared. The solubility limit has been found to be 20 mol % and the material remains tetragonally distorted. The ferroelectric transition temperature decreases with Mn concentration and the transition becomes more diffuse, consistent with the behaviour of solid solutions. XANES confirms the presence of Mn^{3+} and Mn^{4+} oxidation states.

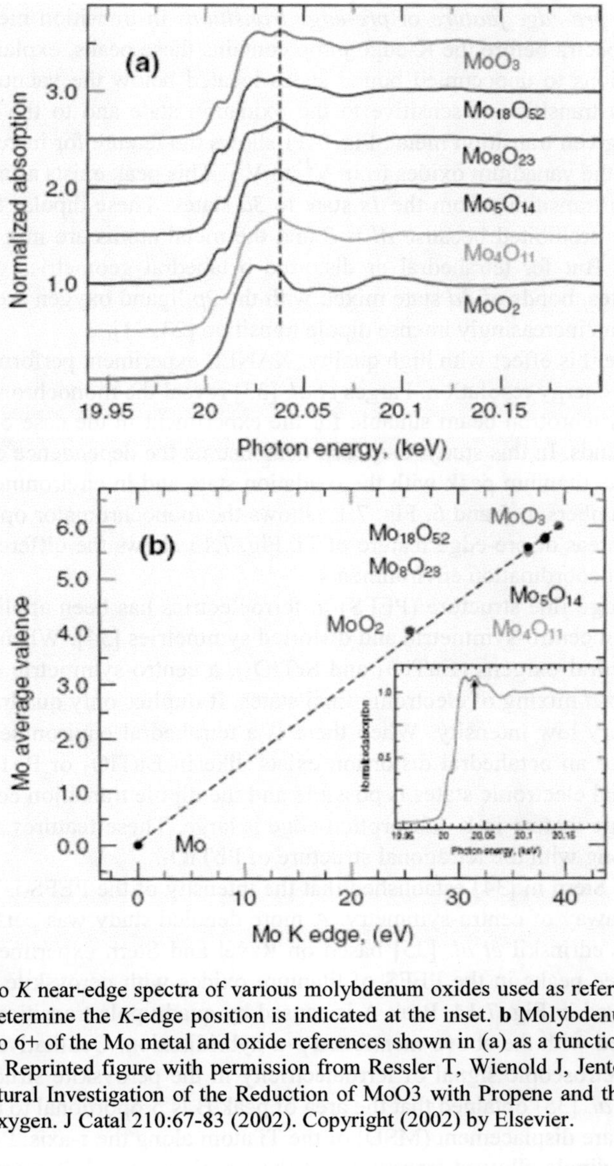

Fig. 7.10 a Mo *K* near-edge spectra of various molybdenum oxides used as references. The feature used to determine the *K*-edge position is indicated at the inset. **b** Molybdenum average valence from 0 to 6+ of the Mo metal and oxide references shown in (a) as a function of the Mo *K*-edge position. Reprinted figure with permission from Ressler T, Wienold J, Jentoft RE, Neisius T, Bulk Structural Investigation of the Reduction of MoO3 with Propene and the Oxidation of MoO2 with Oxygen. J Catal 210:67-83 (2002). Copyright (2002) by Elsevier.

7.4.3 Pre-Edge Transitions

Another interesting spectrum element that is applied for studying ferroelectricity is the so called *pre-edge feature* or *pre-edge transitions* in transition metals. The region of the spectra before the K-edge jump contains these peaks, explained by electronic transitions to unoccupied bound states located below the vacuum level [23, 31, 32]. This transition is sensitive to the oxidation state and to the coordination number of a given transition metal. Fig. 7.11 shows the feature for increasing oxidation states in the vanadium oxides from V^{2+} to V^{5+}. This peak exists as a result of the photoelectron transitions from the *1s* state to *3d* states. These dipolar transitions in the metal are prohibited because $\Delta l = 2$ and the metal atoms are in a symmetrical environment. But for tetrahedral or distorted octahedral geometries of increasing oxidation states, bonds of *3d* state mixed with the *2p* ligand oxygen are formed and it is possible an increasingly intense dipole transition [23, 31].

To observe this effect with high quality, XANES experiment performance should have optimal energy resolution. Farges *et al.* [33] reveal the monochromator optimization of a synchrotron beam suitable for the experiment in the case of model titanium compounds. In this study, they also demonstrate the dependence of height and position of the titanium peak with the oxidation state and in environments with coordination numbers 4, 5 and 6. Fig. 7.12 shows the monochromator optimization to obtain good areas of pre-edge feature of Ti. Fig. 7.13 shows the different spectra of Ti in different coordination environments.

The pre-edge fine structure (PEFS) in ferroelectrics has been applied to distinguish between centro-symmetric and distorted symmetries [34]. When Ti is placed inside octahedral oxigens ($EuTiO_3$ and $SrTiO_3$), a centro-symmetric environment provides no *p-d* mixing of electronic final states. It implies only quadrupole transitions with very low intensity. When there is a tetrahedral environment (in molecules [31]), or an octahedral distortion exists, like in $BaTiO_3$ or $PbTiO_3$, the *p-d* mixing of final electronic states is possible and the dipole transition corresponding to the pre-edge feature in K- absorption edge is large. These features are shown in Fig. 7.14, along with the tetragonal structure of $PbTiO_3$.

Ravel and Stern in [34] established that the intensity of the PEFS is proportional to distortion away of centro-symmetry. A more detailed study was performed theoretically by Vedrinskii *et al.* [35] based on Ravel and Stern experiments. Authors identified three peaks in the PEFS of titanium oxides with perovskite structure, as can be observed in Fig. 7.14. Peak A is caused by quadrupole transitions. Peak B is directly related with the above mentioned *p-d* hybridization. Qualitatively, it is considered a spectroscopic signal of ferroelectricity in the perovskite structure crystal. Vedrinskii *et al.* [35] obtained that the area of peak B is proportional to the square of the mean-square displacement (MSD) of the Ti atom along the *z*-axis. Peak C has its origin in the dipole-allowed transitions of the transition metal *1s*-electron into the transition metal *3d*-originated molecular orbits (MO) of neighbouring octahedra. These are caused by overlapping of these MO with the *1s*-wavefunction of the absorbing atom (band effect). The capability of individualization of these peaks increases the applicability of XANES to the study of ferroelectricity.

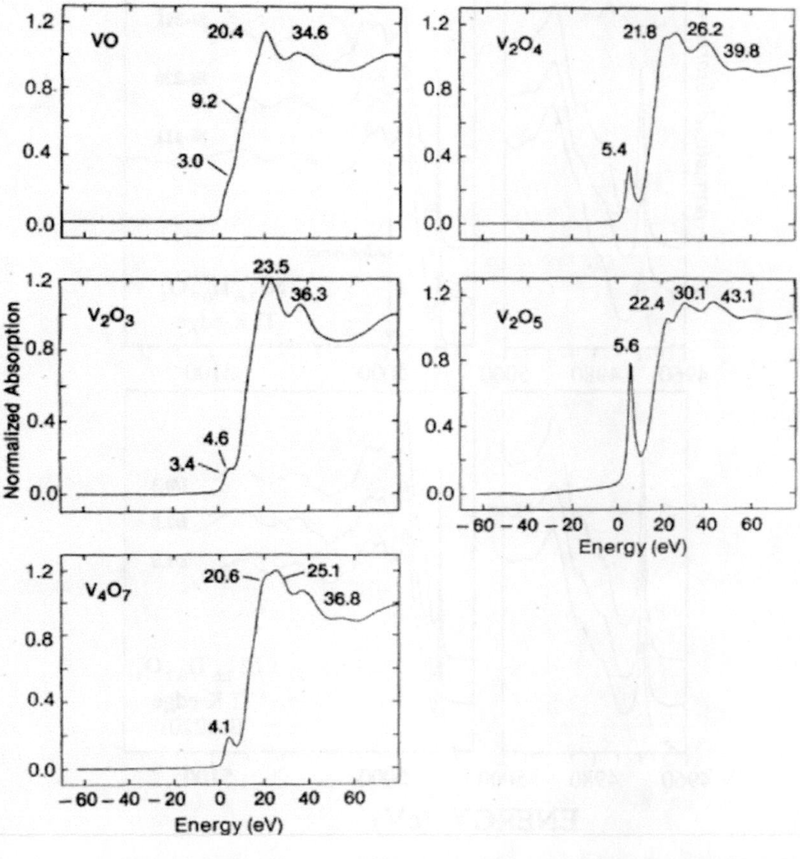

Fig. 7.11 Normalized K-edge XANES spectra of vanadium oxides, the zero of energy taken at 5465 eV. The known crystal structures provide a useful series of materials for the systematic study of valence and coordination geometry on the spectrum of the central metal atom co-ordinated by the same ligand. The formal valence of vanadium is, from top left to bottom right, 2, 3, 3-4, 4 and 5, respectively. Reprinted figure with permission from Wong J, Lytle FW, Messmer RP, Maylotte DH, Phys Rev B, 30, 5596-5610 (1984). Copyright (1984) by the American Physical Society.

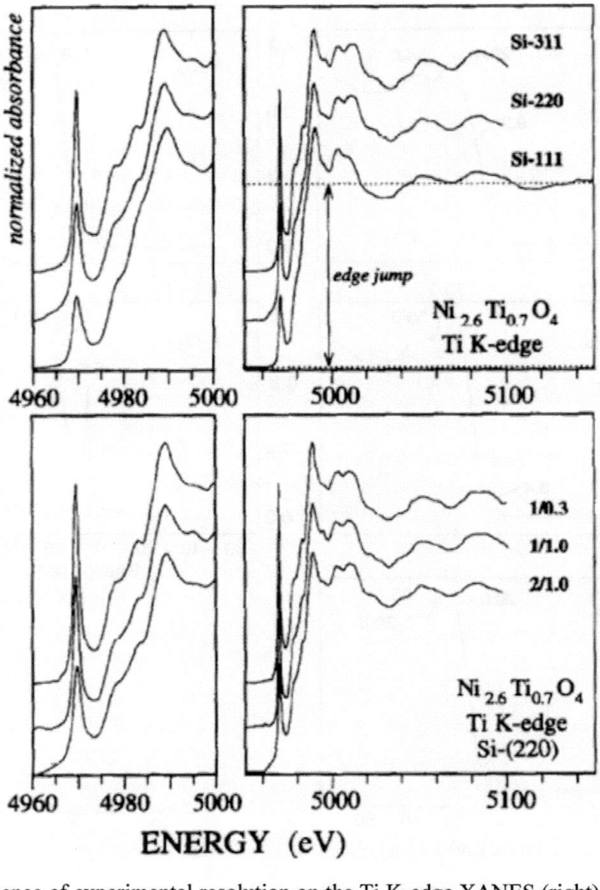

Fig. 7.12 Influence of experimental resolution on the Ti K-edge XANES (right) and its pre-edge region (left) for $Ni_{2.6}Ti_{0.7}O_4$. Top: spectra collected with different Si monochromators: Si- (111), Si-(220), Si-(311) (the slits before and after the monochromator are 2 and 1 mm, respectively). Bottom: spectra collected with (Si-220) monochromator with different slits before (first number, in mm) and after (second number, in mm) the monochromator. The resulting differences in energy resolution strongly affect the height of the Ti pre-edge feature- (not pre-edge position). Note that very high resolution (Si-(220) and 1/0.3 slits) makes the pre-edge height greater than the edge jump. Reprinted figure with permission from Farges F, Brown GE, Rehr JJ, Coordination chemistry of Ti(IV) in silicate glasses and melts: I. XAFS study of titanium coordination in oxide model compounds. Geochim Cosmochim Acta, 60, 3023-3038 (1996). Copyright (1996) by Elsevier.

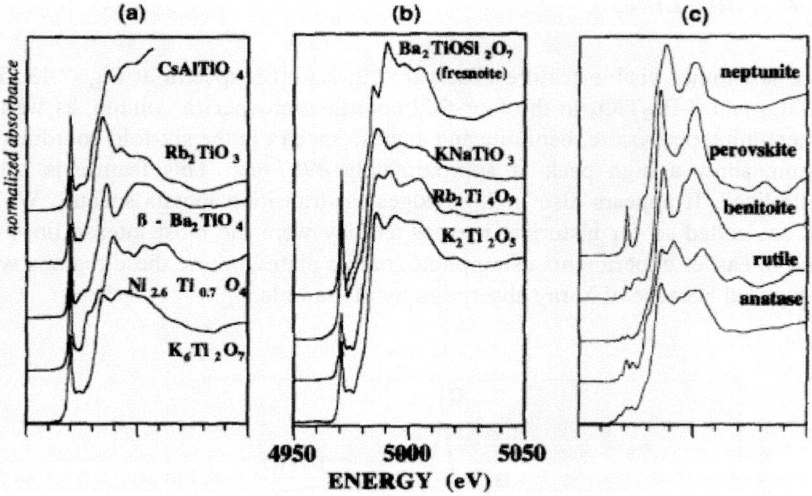

Fig. 7.13 Ti K-edge XANES spectra for Ti-model compounds containing **a** 4-coordinated, **b** 5-coordinated , and **c** 6-coordinated Ti. Note also the *white-lines* in Rb_2TiO_3 and β-Ba_2TiO_4 in the four-fold coordination spectra column, as well as the neptunite, perovskite, benitoite and anatase spectra in the six-fold coordination column. Reprinted figure with permission from Farges F, Brown GE, Rehr JJ, Coordination chemistry of Ti(IV) in silicate glasses and melts: I. XAFS study of titanium coordination in oxide model compounds. Geochim Cosmochim Acta, 60, 3023-3038 (1996). Copyright (1996) by Elsevier.

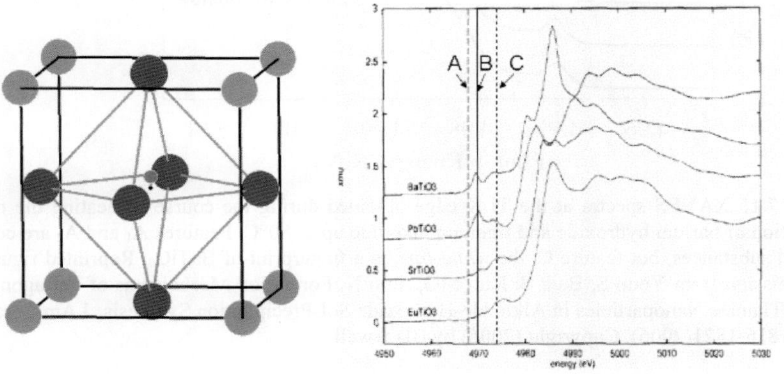

Fig. 7.14 Left: Tetragonal unit cell of $PbTiO_3$, where large black spheres → O; small gray sphere → Ti; large gray spheres → Pb. The displacement of Ti atom from the centre of symmetry is exaggerated and shown by a segment. Right: Pre-edge features of $EuTiO_3$, $SrTiO_3$, $BaTiO_3$, and $PbTiO_3$ at room temperature. Features labelled as A correspond to quadrupole transitions. The 1s→3d dipole transition is the peak B at ~ 4971 eV in the $BaTiO_3$ and $PbTiO_3$ spectra, less visible in the others. Peaks C correspond to the dipole-allowed transitions of the transition metal *1s*-electron into the transition metal *3d*-originated molecular orbits of neighbouring octahedral. Reprinted figure with permission from Ravel B, Stern EA, Local disorder and near edge structure in titanate perovskites, Physica B, 208-209, 316-318 (1995). Copyright (1995) by Elsevier.

7.4.4 White-Lines

There is another visible feature observable in XANES spectra at Fig. 7.13. The Rb_2TiO_3 and β-Ba_2TiO_4 in the four-fold coordination spectra column, as well as the neptunite, perovskite, benitoite and anatase spectra in the six-fold coordination column, show a high peak at approximately 4987 eV. This feature is called "white-line". It appears also in $L_{II,III}$ edges, in transition metals spectra. White-lines are called so for historical reasons as they were the most intense lines detected in earlier experiments using photographic plates, where these regions were not exposed because of X-ray absorption by the sample.

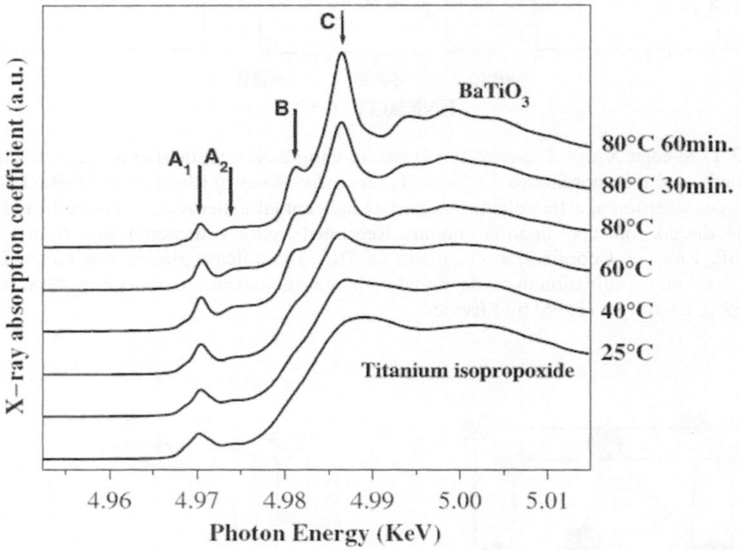

Fig. 7.15 XANES spectra at the Ti K edge obtained during the course of heating the mixture solution of barium hydroxide and titanium alkoxide up to 80°C. Features A_1 and A_2 are common to all substances, but feature C, the *white-line*, is a fingerprint of $BaTiO_3$. Reprinted figure with permission from Yoon S, Baik S, Kim MG, Shin N, Formation Mechanisms of Tetragonal Barium Titanate Nanoparticles in Alkoxide-Hydroxide Sol-Precipitation Synthesis, J Am Ceram Soc 89, 1816-1821(2006). Copyright (2006) by Blackwell.

White-lines have been applied successfully in identification of ferroelectric structures. One example is the work of Yoon *et al.* [36], where the white-line of the Ti K-edge in $BaTiO_3$ has been applied as fingerprint. The experiment consisted on the synthesis of $BaTiO_3$ nanoparticles by dissolution of $Ba(OH)_2$ and the hydrolysis of Ti^{IV} isopropoxide in isopropanol. In Fig. 7.15, it is shown how the $BaTiO_3$ white-line is appearing at 60°C and is completely formed after 60 minutes at 80°C, in particles of 7.5 nm. The advantage of applying XANES technique in this experiment was the possibility of probing the evolution of local structure around Ti atoms as a function of temperature.

Finally, we have to stress some remarks about XANES. The EXAFS equation (next paragraph 7.5) fails at low *k* values, which complicates XANES interpretation. XANES can be described *qualitatively* in terms of coordination chemistry, molecular orbital and band structure. All these features have been described in this paragraph. Quantitative interpretation of XANES using multiple scattering theory and Green functions is in progress [21], and some attempts have been done to describe ferroelectrics. Ravel [37] has applied FEFF8 [38] program with a multiple scattering algorithm and increasing number of coordination shells to describe the $PbTiO_3$ XANES spectrum. Fig. 7.16 shows these results.

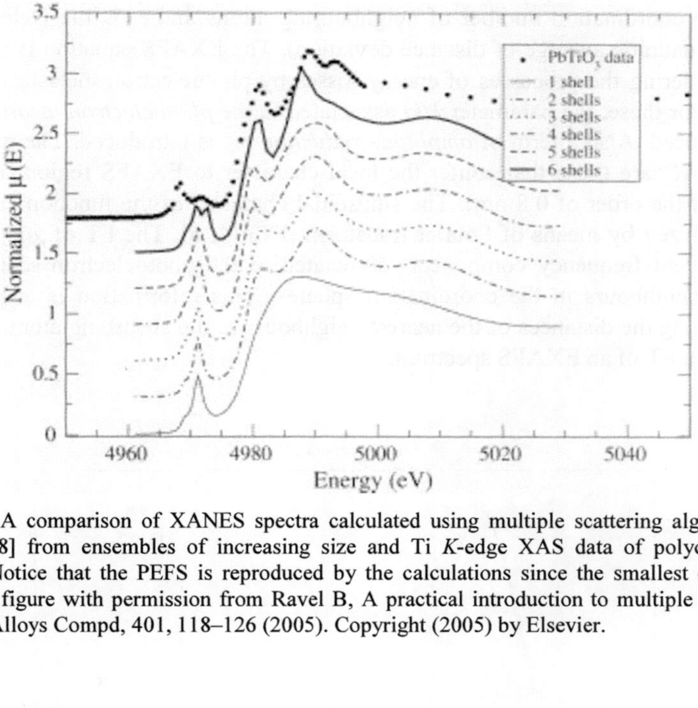

Fig. 7.16 A comparison of XANES spectra calculated using multiple scattering algorithm of FEFF8 [38] from ensembles of increasing size and Ti *K*-edge XAS data of polycrystalline $PbTiO_3$. Notice that the PEFS is reproduced by the calculations since the smallest ensemble. Reprinted figure with permission from Ravel B, A practical introduction to multiple scattering theory, J Alloys Compd, 401, 118–126 (2005). Copyright (2005) by Elsevier.

7.5 Formal Characterization of XAFS

7.5.1 The EXAFS Equation

In this section the bases of theoretical characterization of the extended X-ray absorption fine structure will be given. EXAFS zone begins at energies greater than those of XANES zone. As it has been pointed out above, XANES and EXAFS effects merge in the zone where photoelectron wavelength $\lambda \leq R$, the distance from the absorbing atom to its first neighbours [18].

To extract information from this fine structure, the *EXAFS equation* was proposed by Stern's group [4]:

$$\chi(k) = \sum_j S_0^2 \frac{N_j F_j(k) e^{-2R_j/\lambda(k)} e^{-2k^2\sigma_j^2}}{kR_j^2} \sin\left[2kR_j + \delta_j(k)\right] \quad (8)$$

where $F_j(k)$ and $\delta_j(k)$ are the amplitude and phase shift of the scattered photoelectron (Z-dependent), and hence carry information on the number and species of neighbouring atoms. R_j is the distance of the absorbing atom to neighbouring atoms, N_j is the coordination number of neighbouring atoms and σ_j is the Debye-Waller factor (quadratic average of distance deviation). The EXAFS equation is completed by considering the processes of energy losses by photoelectron inelastic scattering events. For these, the parameter $\lambda(k)$ associated to the *photoelectron mean free path* is introduced. Also a term of *amplitude reduction* S_0^2 is introduced. The parameters $\lambda(k)$ and R^{-2} are those that confer the local character to EXAFS region (up to distances of the order of 0.8 nm). The sinusoidal character of the function can be further analyzed by means of Fourier transform (FT) [4, 6]. The FT of $\chi(k)$ produces the different frequency components associated to the photoelectron scattering by nearest neighbours in the coordination spheres. This information is employed in determining the distances of the nearest neighbours to the absorbing atom. Fig. 7.17 presents a FT of an EXAFS spectrum.

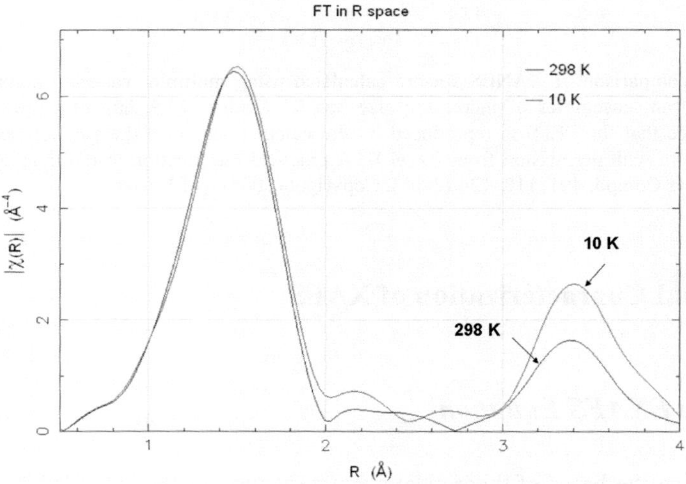

Fig. 7.17 Module of Fourier transform of the EXAFS functions for Fe K- absorption edge in Aurivillius ceramics, obtained at 10 K and 298 K, and presented in Fig. 7.8.

The EXAFS spectra are therefore interpreted by following the sequence indicated above. Resuming, this procedure consists on obtaining the function $\chi(k)$, performing the FT in k space producing R space spectra and finally separating the contributions of $\chi(k)$ correspondent to each coordination sphere. Modelling of chemical structures is typically done using purpose-built computer programs, such as FEFF [20], EXCURV [39] and GNXAS [40, 41]. The models calculate the values of $F(k)$ and $\delta(k)$ associated to the scattering paths from the different coordination spheres. Based on the EXAFS equation (8), the contribution of each scattering path to the model is refined by fitting the values of N, R and σ^2 using programs such as IFEFFIT [42], EXAFSPAK [43], PAXAS [44], WinXAS [45] among others.

In order to understand and apply XAFS, it is necessary to study the origin of these phenomena. Therefore, this section will provide a deeper insight into the physics underlying X-ray absorption spectroscopy. Several reviews and texts on XAFS have been reported. Among others, the work from Teo [11], Stern [46], Rehr and Albers [20] as well as Newville [47] and Filipponi *et al.* [42] have produced extensive material to which the reader is referred to for further studies.

7.5.2 One-Electron Golden Rule Approximation

The basis of XAFS interpretation is the application of Fermi's Golden Rule [11, 20]. That may be expressed as follows.

In the one-electron approach, the probability of interaction of X-rays with the absorbing atom, may be expressed by the interaction cross section as well as by the absorption coefficient $\mu(E)$. The absorption coefficient is proportional to the square of the matrix element of the transition that undergoes the photoelectron from the internal layer of the atom to the continuum:

$$\mu(E) = \left|\langle i|H|f\rangle\right|^2 \qquad (9)$$

Here $\langle i|$ is the wave function of the initial state (the internal or core electron and the photon) and is not altered by the presence of neighbouring atoms. H is the Hamiltonian of the interaction. $|f\rangle$ is the wave function of the final state that describes the photoelectron and the hole in the internal state. This function is indeed altered by the presence of neighbouring atoms and therefore:

$$|\mathrm{f}\rangle = |\,\mathrm{f}_0\rangle + |\Delta \mathrm{f}\rangle \qquad (10)$$

$|f_0\rangle$ represents the approximation of the atom without neighbours and $|\Delta f\rangle$ represents the change in the photoelectron final state due to the backscattering from neighbouring atoms.

The oscillations of $\mu(E)$ results from the interference of both emergent and backscattering waves. A representation of those waves is shown in Fig. 7.18.

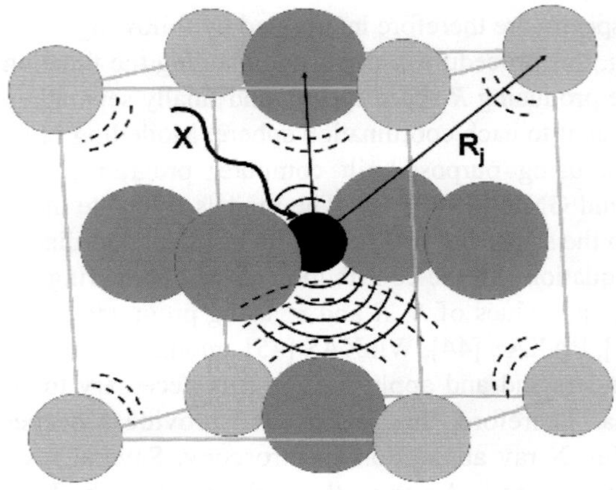

Fig. 7.18 Schematic representation of backscattering waves in a perovskite crystal. The black atom absorbs an X-ray and releases a photoelectron, which interacts and scatters from N_j neighbours at distance R_j. Photoelectron waves interfere and produce the oscillations in the X-ray absorption coefficient. Multiple-scattering effects are possible.

From the definition of the EXAFS function in equation (6), the absorption coefficient can be expressed as:

$$\mu(E) = \mu_0(E)[1 + \chi(E)] \tag{11}$$

This expression together with equation (10) results in:

$$\mu_0(E) = \left|\langle i|H|f_0\rangle\right|^2 \tag{12}$$

$$\chi(E) \propto \langle i|H|\Delta f\rangle \tag{13}$$

Calculations are simplified by considering the outgoing spherical wave (with short wavelength) as simple exponentials and the photoelectron scattering as a plane wave.

In the electric dipole approach, when the X-rays wavelength is much greater than the dimensions of the absorbing atom and in cases of deep-core excitations, the Hamiltonian approaches:

$$\mathrm{H} = \mathrm{e}^{\mathrm{ikr}} \sim 1 \tag{14}$$

This way, equation (13) for the $\chi(E)$ function is associated to the probability of the scattering event:

$$\chi(E) \propto \int dr\, \psi_i^* \,\psi_{scatt}(r) \tag{15}$$

The different interactions of the emergent photoelectron can be envisaged in the following three steps:

- It leaves the absorbent atom as a spherical wave
- It travels a distance R and scatters in a neighbouring atom, acquiring its wave function a scattering amplitude $F(k)$ and a phase shift $\delta(k)$; and
- It returns to the absorbing atom as a spherical wave, travelling the distance R once again.

The EXAFS equation gathers together the characteristics of the wave function of the scattered photoelectron by N neighbouring atoms at the coordination distance R. Considering atoms of different species and at different distances, the EXAFS equation becomes:

$$\chi(k) = \sum_j N_j \left[F_j(k) e^{i\delta_j(k)} \right] \frac{e^{i2kR_j}}{kR_j^2} + \textit{complex conjugate} \quad (16)$$

7.5.3 Fluctuations in Interatomic Distances and the Debye-Waller Factor

Due to temperature effects, atoms have vibrations around their equilibrium positions which must be taken into account. It has been pointed out [11] that the EXAFS phenomenon takes place in a time shorter than the motion of atoms. However, EXAFS experiments require much longer time, and therefore the spectra reflects an average over the atomic distances. The average of the atomic instantaneous relative displacements from the equilibrium position is considered in the exponential term of the EXAFS equation by introducing the following exponential factor:

$$e^{i2kR_j} e^{-2k^2\sigma_j^2} \quad (17)$$

where σ^2 is the Debye-Waller factor (DWF). Only the component which lies *along the equilibrium bond direction* needs to be considered.

Assuming a harmonic oscillation approach, let $\mathbf{u}_0$ and $\mathbf{u}_j$ be the displacements from equilibrium of the absorbing and scattering atoms, respectively (Fig. 7.19). Neglecting terms of order u_j^2, the relative position of the scattering atom $\mathbf{r}_j$ when displaced is

$$\mathbf{r}_j = \mathbf{R}_j + (\mathbf{u}_j - \mathbf{u}_0) \quad (18)$$

The DW factor being the average of the displacements can be written as:

$$\sigma_j^2 = \left\langle (\mathbf{r}_j - \mathbf{R}_j)^2 \right\rangle \tag{19}$$

from equation (18):

$$\sigma_j^2 = \left\langle \left[(\mathbf{u}_j - \mathbf{u}_0) \cdot \hat{\mathbf{r}}_j \right]^2 \right\rangle \tag{20}$$

$$\sigma_j^2 = \left\langle (\hat{\mathbf{r}}_j \cdot \mathbf{u}_j)^2 \right\rangle + \left\langle (\hat{\mathbf{r}}_j \cdot \mathbf{u}_0)^2 \right\rangle - 2\left\langle (\hat{\mathbf{r}}_j \cdot \mathbf{u}_j)(\hat{\mathbf{r}}_j \cdot \mathbf{u}_0) \right\rangle \tag{21}$$

where $\hat{\mathbf{r}}_j$ is the corresponding directing unit vector along the path from the absorbing to the scattering atoms.

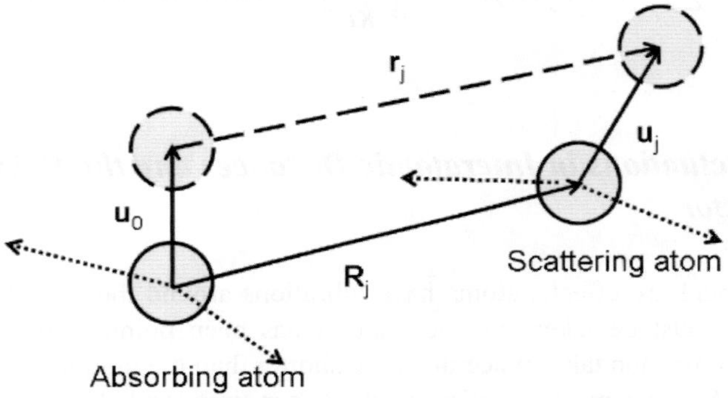

Fig. 7.19 Displacements of absorbing and scattering atoms around equilibrium position. $\mathbf{R}_j$ is the position vector of relative equilibrium of the scattering atom; $\mathbf{u}_0$ and $\mathbf{u}_j$ are vectors of relative displacements from equilibrium of the absorbing and scattering atoms; $\mathbf{r}_j$ is the relative position vector of the scattering atom when moving.

The first and second terms in (21) are the averages of the position vectors of the absorbing and the scattering atoms respectively. They account for any kind of oscillations, but the third one is the correlation term and vanishes only if both atoms move independently. For example, in covalently bonded systems, the nearest neighbours are connected to each other, and this feature introduces correlations in their motion.

DWF in EXAFS differs from that for X-ray diffraction (XRD). In XRD, DWF takes into account the average of the displacement of the atom in the vicinity of the equilibrium position in the unit cell. Mathematically, it contains $<\mathbf{u}_j^2>$ (see equation 7.19 in this Handbook). In words, EXAFS σ^2 informs us about the *disorder in interatomic distances* while XRD u^2 informs us on the *disorder about crystal sites*.

Because DWF is due partly to thermal effects, it increases with temperature T. This effect can be observed in Fig. 7.8 and Fig. 7.17, where the effect of reducing the amplitude of the EXAFS function is evidently larger at high k and R for T=298 K. DWF may even cut off the EXAFS spectra at sufficiently high energy beyond $k \sim 1/\sigma$. For this reason, it is difficult to perform EXAFS analysis beyond k=10 Å^{-1}. In some cases, lowering the temperature during experiments is an alternative option. For this however, it is important to ensure that no phase changes take place in the sample. The temperature dependence of the DWF also implies that it is often negligible in XANES, when $k^2\sigma^2 << 1$ [20].

With the inclusion of the DWF from equation (17) and considering the complex conjugate of equation (16), the function $\chi(k)$ becomes:

$$\chi(k) = \sum_j \frac{N_j F_j(k) e^{-2k^2\sigma_j^2}}{kR_j^2} \sin\left[2kR_j + \delta_j(k)\right] \qquad (22)$$

7.5.4 Curved Waves and Multiple Scattering of Photoelectrons

The short wave approach for the photoelectron emission and the plane wave approach for the scattering process do not completely provide the whole interpretation of experiments. A more rigorous approach takes into consideration that, for the first nearest neighbours, the curved wave is important [48], and for more distant spheres of coordination, photoelectron multiple scattering effects are important [20]. Multiple scattering consists on the propagation of the photoelectron wave from a neighbouring atom to another one before returning to the emitting atom. Multiple scattering is taken into account for accurate calculations of the absorption coefficient in materials. In this particular case, when atoms are in a linear path, the multiple scattering can be greater in magnitude than the backscattering contributions [48]. This case is called *focusing effect.*

The EXAFS equation is based on the Fermi's Golden Rule for a single electron [20, 37]:

$$\mu \propto \sum_f \left| \langle \psi_f | \hat{\varepsilon} \cdot \mathbf{r} | \psi_i \rangle \right|^2 \delta(E_f - E_i) \qquad (23)$$

where $\hat{\varepsilon} \cdot \mathbf{r}$ is the operator of the electrical dipole transition of the incident photon and the sum is for all the energies above the Fermi level. The photoelectron behaves like a quasiparticle that moves in a *complex optical potential*, in which the scattering amplitudes and the phase shifts are calculated. The characteristics of that potential make the emitted photoelectron move in a spherical attractive potential δV. This is when it travels through atoms or ions crosses approximately a free interstitial zone generated by the average of the potentials of all atoms. This potential V_{int} is called *muffin-tin potential*, because it resembles this familiar shape. Fig. 7.20 presents graphically this potential in one and two dimensions.

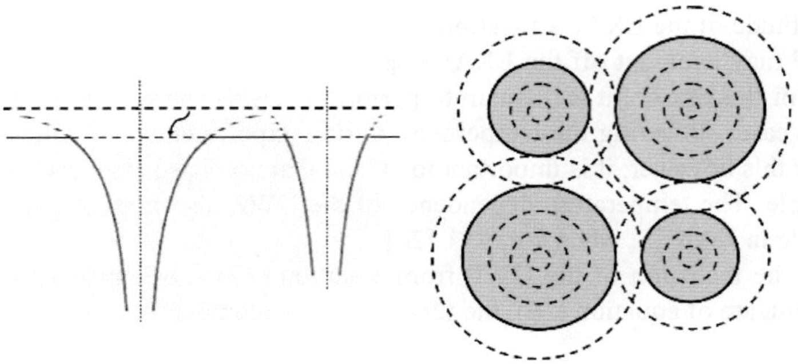

Fig. 7.20 Schematic drawing of one- and two-dimensional muffin-tin potentials. The solid lines between the spherical potentials in the one-dimensional drawing represent the flat interstitial potential for the free movement of the photoelectron. The zero level has moved to this muffin-tin interstitial potential. In the two-dimensional drawing, the gray shadows represent the atom spherical wells, the solid lines represent the interstitial zero level, and the dashed lines over white background, the original true shape of the potentials.

The expression (23) can be solved explicitly by evaluating the corresponding integrals. However, the development of multiple scattering theory has been facilitated by the use of Green functions [20, 37]. The system Hamiltonian is

$$\mathrm{H} = \mathrm{H}_0 + \mathrm{V}_{\mathrm{int}} + \delta\mathrm{V} \tag{24}$$

The Green function is written as

$$\mathbf{G}(E) = 1/(E - H + i\zeta) \tag{25}$$

G*(E)* is interpreted as a one-electron propagator. This describes all possible ways in which the photoelectron can scatter from one or more surrounding atoms or ions before the core-hole is refilled. The term H_0 corresponds to the kinetic energy plus the net Coulomb potential felt by the photoelectron. The real term H corresponds to the photoelectron scattering and the imaginary term ζ corresponds to the photoelectron absorption [20, 37].

Using this algorithm, the X-ray absorption coefficient takes the form

$$\mu(E) \propto -\frac{1}{\pi}\mathrm{Im}\langle\psi_i|\hat{\boldsymbol{\varepsilon}}^* \cdot \mathbf{r}\,\mathbf{G}(E)\,\hat{\boldsymbol{\varepsilon}} \cdot \mathbf{r}|\psi_i\rangle\Theta(E - E_F) \tag{26}$$

Here Θ is a step function that ensures that the scattering cross section is different from zero only when $E > E_F$. Within the theory of curved wave and multiple scattering, the scattering amplitude $F(k)$ of the plane wave is replaced by an *effective scattering amplitude* $F_{eff}(k,R)$.

The effective scattering amplitude $F_{eff}(k,R)$ together with the phase shifts $\delta(k)$ have been determined experimentally for a number of elements. $F_{eff}(k,R)$ and $\delta(k)$ are obtained by first principles (*ab initio*) calculation, with common use EXAFS analysis codes, such as EXCURV [39], GNXAS [40, 42] and FEFF [49]. These programs have been constantly improved over the past two decades. Fig. 7.21 shows the scattering amplitude $F_{eff}(k,R)$ and the phase shifts $\delta(k)$ of the photoelectron emitted by a Fe atom and scattered in O, Bi, Fe and Ti, calculated by program FEFF6 [50].

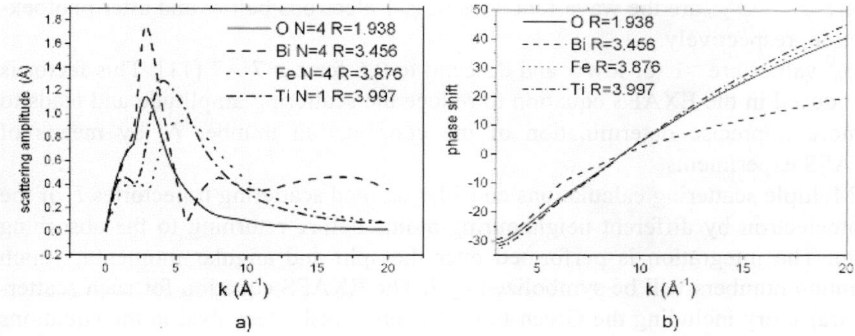

Fig. 7.21 a) Effective scattering amplitude and b) phase shift of the photoelectron emitted by an iron probe when interacting with different elemental neighbours in an Aurivillius ceramic, calculated by FEFF6 [50] code. N represents the coordination number of the given sphere, and R is the coordination radius.

In Fig. 7.21, it is observed that the scattering amplitude depends strongly on the atomic number of the scattering element, such as O, Fe and Bi. Nevertheless, like in other X-ray related phenomena, XAFS badly differentiates close atomic numbers, such as Ti and Fe.

7.5.5 Inelastic Scattering

Let us analyze inelastic scattering phenomena. Up to this moment, if parameters $F_j(k)$ and $\delta_j(k)$ are determined, one must be able to obtain the coordination number of a given sphere, its distance R to the absorbing atom and its uncertainty. However, this vision is oversimplified. Processes of inelastic losses are evident from the experiments and need to be taken into account in the physical model.

First, we will treat the only process in EXAFS that depends significantly on the absorbing atom. This process is called *intrinsic losses* [20]. During the photoionization process, in which the photoelectron acquires energy *for a* direct transition from the core level to the continuum, other processes take place such as excitation of the Z-1 remaining low-binding electrons to superior levels (or to the continuum). The result is equivalent to a process, in which the photoelectron acquires less energy, and where the final state of the absorbing atom is less defined. This is different than the assumed final states in the deduction of the EXAFS equation. Teo [11] explains the convenience of introducing a factor of amplitude reduction $S_0^2(k)$ in the EXAFS equation and proposes the following term:

$$S_0^2(k) = \left|\langle \phi'_{Z-1} | \phi_{Z-1} \rangle\right|^2 < 1 \tag{27}$$

where ϕ_{Z-1} y ϕ'_{Z-1} are the wave functions of Z-1 electrons before and after photoexcitation, respectively.

S_0^2 values are ≈1 for low k and descend to 0.6 for $k \geq 7$ Å^{-1} [11]. This factor is introduced in the EXAFS equation to reduce the scattering amplitude and leads to a more imprecise determination of the coordination number N_j by means of EXAFS experiments.

Multiple scattering calculations consider defined scattering trajectories Γ of the photoelectron by different neighbouring atoms before returning to the absorbing atom. The integration is performed over the spin and angular momenta, which quantum numbers will be symbolized by l. The EXAFS equation for each scattering trajectory including the Green functions previously described in the equations (24-26), is expressed according to [20]:

$$\chi_\Gamma(k) = S_0^2 \,\mathrm{Im}\left[\frac{F_{eff}}{kR^2} e^{2ikR+2i\delta_l} e^{-2\sigma^2 k^2}\right] \tag{28}$$

Processes of extrinsic losses also exist. They are caused by the excitation of plasmons, electron-hole pairs and inelastic scattering in which the photoelectron loses energy. These are well-known processes that electrons undergo when being transported in solids. In EXAFS this lead to a reduction of the scattering amplitude which is considered in the following term:

$$e^{-2R_j/\lambda(k)} \tag{29}$$

where $\lambda(k)$ is the *mean free path* (do not confuse with the wavelength of the photoelectron!). The extrinsic losses introduce systematic effects in the behaviour of the photoelectron, which are reflected in the spectrum. These extrinsic losses may be described by the photoelectron self-energy operator Σ(E) [20]. This operator is complex in nature and energy dependent. The real part reflects the systematic shifts of location of the EXAFS peaks compared to the positions obtained for the ground state. The imaginary part accounts for the mean free path $\lambda(k)$ in (8) and (29).

The mean free path $\lambda(k)$ has been introduced to describe the behaviour of electrons in experiments with X-ray beams. In fact, it has been described for the collisions of electrons in pure elements, inorganic solids, organic solids and adsorbed gases. The dependence of the mean free path with the energy of the photoelectron

E_e or with the wave number k fulfils empirical relations usually called *universal curves* for each type of solid. The dependence with k of this universal curve has the form

$$\lambda(k) = \frac{1}{\eta}\left[\left(\frac{\xi}{k}\right)^4 + k\right] \tag{30}$$

Fig. 7.22 shows the dependence of the mean free path *λ(k)* calculated theoretically by program FEFF6 [50]. In that program, an algorithm based on an analytic approximation to the Green functions GW plasmon-pole self-energy Hedin-Lundqvist potential is used [51]. Fig. 7.22 also shows the evaluation of universal curves for pure elements and inorganic compounds.

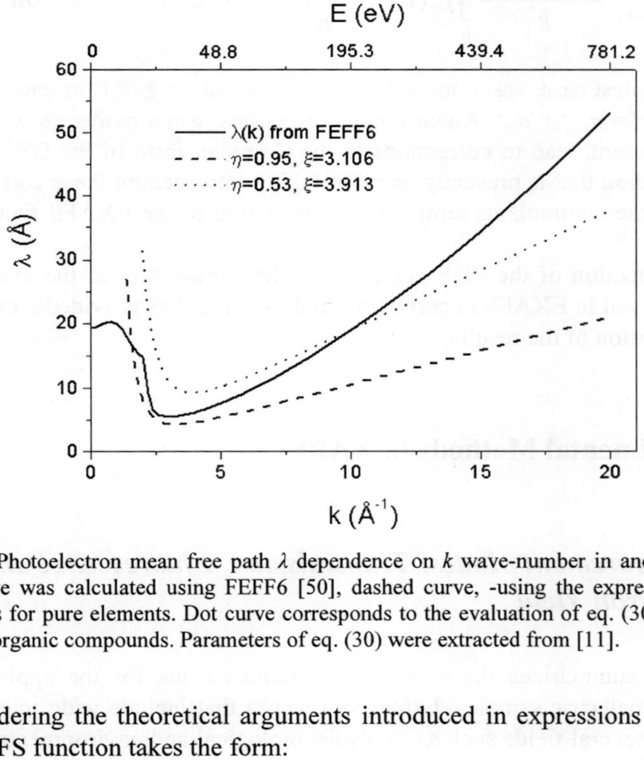

Fig. 7.22 Photoelectron mean free path λ dependence on k wave-number in and on E_e energy. Solid curve was calculated using FEFF6 [50], dashed curve, -using the expression (30) with parameters for pure elements. Dot curve corresponds to the evaluation of eq. (30) with parameters for inorganic compounds. Parameters of eq. (30) were extracted from [11].

Considering the theoretical arguments introduced in expressions (23) to (29), the EXAFS function takes the form:

$$\chi(k) = \sum_j \frac{S_0^2 N_j\, F_j(k) e^{-2k^2\sigma_j^2}\, e^{-2R_j/\lambda(k)}}{kR_j^2} \sin\left[2kR_j + \delta_j(k)\right] \tag{31}$$

In this equation the sum is made on each *photoelectron scattering path j*. In eq. (31) $F_j(k)$ is a simplified notation for the effective scattering amplitude $F_{eff}(k,R)$ introduced in (28) and it corresponds to each *path j*. Equation (31) and (8) are equivalent. As it was pointed out before, it is known as the *EXAFS equation*.

It is important to introduce a statistical vision of the information that offers EXAFS with respect to the atom coordinates and their uncertainties. This vision is provided introducing the *pair distribution function* $g_j(R_j)$ (or PDF) for each path *j* [11, 20]. This way, the DWF may be described by two components, the *static* or *configurational* σ_{stat}^2 and the *thermal* or *vibrational* σ_{th}^2. The configurational and thermal average may be defined in terms of the PDF. For small disorders, the distribution function can be approximated by a Gaussian, with width σ^2. Expression (31) in terms of PDF will be

$$\chi(k) = \sum_j \frac{S_0^2 N_j F_j(k)}{k} \int_0^\infty g_j(R_j) \frac{e^{-2R_j/\lambda(k)}}{R_j^2} \sin\left[2kR_j + \delta_j(k)\right] dR_j \qquad (32)$$

In the simplest case, assuming a Gaussian distribution $g_j(R_j)$ for each contribution implies $\sigma^2 = \sigma_{stat}^2 + \sigma_{th}^2$. Anharmonic corrections, when moderate or large disorders are present, lead to corrections to the Gaussian form of the DWF. An appropriate method that is presently used for taking into account these corrections is to introduce the cummulants terms in the calculation of the EXAFS function [20, 52].

The introduction of the PDF has allowed the explanation of the main effects that are observed in EXAFS experiments, and therefore has provided a more accurate interpretation of the results.

7.6 Experimental Methods in XAFS

7.6.1 Measurement Modes: Transmission, Fluorescence and Total Electron Yield

This section summarizes the experimental requirements for the application of XAFS in ferroelectric samples. Extensive reviews that include wide scientific applications in several fields such as catalysis, biological and environmental studies, may be found in [16, 27, 53, 54].

We are interested in obtaining the XAFS signal by measuring the energy dependence of the X-ray absorption coefficient $\mu(E)$ at and above the absorption edge of a selected element.

Recalling the expression for absorption coefficient:

$$A = \mu(E)x = \ln(I_0 / I_1) \tag{4}$$

In fluorescence mode the absorption coefficient can be expressed as:

$$\mu(E) \propto I_f / I_0 \tag{5}$$

Second and third generation synchrotron facilities provide intense, parallel and tunable X-ray beams that allow these kinds of experiments to be performed in a relatively short time. The energy of the incident beam passes a monochromator with a typical resolution of 1 eV and can scan a range from 3 to 40 keV according to the beamline in use. The energy of synchrotron X-rays depends on the energy of electrons in the storage ring and on the magnetic fields of the insertion devices, (undulators, bending magnets, etc.).

The amplitude of $\chi(k)$ is low in the case of EXAFS and typically $\approx 10^{-2}$ or smaller. Therefore, the signal to noise ratio must be maximized to a factor of 10. This implies that the beam must be intense in the energy zone of interest and that the detectors must be linear. Furthermore, optically well aligned beams and minimization of intensity losses along the beam path is required. That means that beryllium, Kapton or windowless detectors in cases of photon energies below 3 keV, are necessary [53].

After selecting the element and absorption edge to study, the user must design the regimen of energy scanning. Scans are typically taken from 200 eV below the absorption edge to 800 eV above the absorption edge E_0. These scans will cover three spectral zones according to Table 7.1:

Table 7.1 Energy scanning zones in EXAFS experiments

Region	Starting energy (eV)	Ending energy (eV)	Step (eV)
Pre-edge	E_0 - 200	E_0 - 20	5.0 - 10
XANES	E_0 - 20	E_0 + 30 to 50	0.25 – 1.0
EXAFS	E_0 + 30 to 50	E_0 + ~ 800	$\Delta k = 0.05$ Å^{-1}

Energy step size and the number of XANES and EXAFS regions depend on the details of the given experiment. In some cases, the structure of the XANES region is important, for instance, if the determination of the oxidation state or coordination number is desired. On the contrary, the interest may consist in measuring the EXAFS contribution at high *k*. Doing multiple scans and using increased integration time per point at high *k* is common for obtaining good statistics.

A XAFS experiment at room temperature is surprisingly simple. In Fig. 7.23, a scheme of the installation at beamline 2-3 of Stanford Synchrotron Radiation Laboratory (SSRL) is presented. As it has been pointed out above, the beam that produces the synchrotron is incident on the study sample. The detectors allow determining the initial intensity I_0, the transmitted intensity I_1 and/or the fluorescent intensity I_f. If so desired, the I_2 intensity informs us on the reference sample. In the case of beamline 2-3, the collimation and position of the table adjustments are obtained by modifying the relative position between the monochromator and the slits. They are not presented in the scheme. If changes of temperature with a cryostat are required, it is installed in external form.

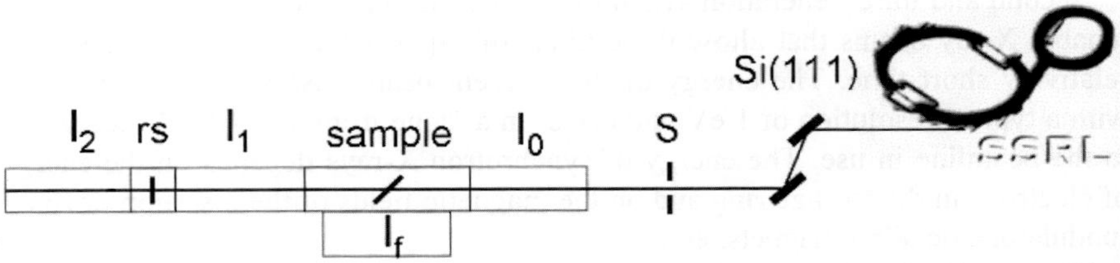

Fig. 7.23 Schematic of beamline 2-3 at Stanford Synchrotron Radiation Laboratory (SSRL). SSRL logo means SPEAR3 storage ring; Si(111)-monochromator; S-slits; I_0-entrance ion chamber; sample-sample chamber; I_f-fluorescence (Lytle) detector; I_1-transmission ion chamber; rs-reference sample chamber; I_2-reference ion chamber.

The detectors more frequently used in XAFS are the ion chambers, to which the X-ray beam is made to cross along, with the purpose of obtaining a high ionization signal. The ion chambers are connected in the so-called *current regime* of radiation detection [55].

When planning XAFS experiments, the knowledge of absorption lengths of the material before and after the measuring absorption edge is needed. If X-ray absorption cross-sections are σ_i (do not confuse it with DWF!), density is ρ, and the mass fractions are m_i/M, to calculate absorbance and absorption length $\tau(E)$ we have

$$\mu(E) = \rho \sum_i \frac{m_i}{M} \sigma_i(E) \tag{33}$$

$$\tau(E) = 1/\mu(E) \tag{34}$$

For these calculations the tables [12, 56] or the PC code HEPHAESTUS from [57] may be used.

For samples with element concentration greater than 10%, XAFS is best measured in transmission regime. If this is the case, enough transmission through the sample should be obtained to get a good signal for the I_1 detector. The sample thickness x should be chosen so that absorbance $A \approx 2.5$ for energies immediately above the absorption edge and/or the edge step should be $\Delta\mu(E)x \approx 1$. The sample has to be uniform, and free of pinholes. For a powder, the grain size should not be greater than an absorption length.

If the sample is dilute, too thin or inhomogeneous, then it is better to use the fluorescence regime for measurements. This is also if the sample is deposited on a substrate. X-rays emitted from the sample will include the *fluorescence line of interest* as well as *scattered* X-rays and other fluorescence lines. Scattered X-rays include both *elastic* (at the same energy as the incident beam) and *inelastic* (e.g., Compton scattering) processes, at lower energy. For dilute or deposited samples, the fluorescence peaks of interest are much weaker than the scatter peak. The isolation of the fluorescence line of interest will be obligatory and the rejection of the remaining radiation should be performed either physically or electronically.

The fluorescent radiation that is excited in the sample is due to the photoelectric effect in the element of interest Z. It is also due to the elements of atomic numbers Z-1, Z-2, Z-n whose absorption edges are neighbouring and where the photoeffect is probable. For the case of ferroelectric synthetic samples under XAFS studies, this effect is evident for example in $(La,Bi)(Fe,Cr)O_3$: if the Fe is excited, the Cr lines must be observed. In the Aurivillius ceramics $Bi_6Ti_3Fe_2O_{18}$: if the Fe is excited, the Ti lines may be observed.

Due to the characteristics of the fluorescent excitation, this way of detection is recommended for samples of low concentration of the element of interest. This should be considered as follows: 1) the element can be considered *diluted*, in the regime of *thick sample*. That is to say, the thickness of sample $x > 10\ \tau$; 2) low weight in very *thin sample*, $x << \tau$.

If the concentration of the probe is very low, the possibility of measuring the fluorescence by means of the electron current that the sample emits from the surface must be considered. That is the regime of *total electron yield* [58, 59]. In this case, the limit of very thin sample is applied because the electrons escape from ~ 100 Å of the surface.

Fluorescence detection may be performed in integral or pulse regime. The integral regime is far more frequently used, with the Stern-Heald ion chamber (also called Lytle detector) [60, 60] or a scintillation detector with electronic photomultiplier tube (PMT) connected in current regime. Stern-Heald-Lytle ion chamber suppresses the high energy photons with a well optimized Z-1 filter [61] and suppresses the filter fluorescence with Soller style slits. Fig. 7.24 shows the image and the scheme of the Stern-Heald-Lytle ion chamber employed in synchrotrons worldwide.

Fluorescence measurement in the pulse-height regime [55] has advantages and disadvantages. The advantages include the possibility of rejecting the elastic and inelastic scattering. Also, after knowing the composition of the sample, it allows the rejection of other fluorescence lines. The drawbacks consist of the count-rate limits (saturation of electronics and non-linearity) as well as more complicated setting up and maintaining of the measurements. Commonly used detectors are scintillator-photomultiplier tubes (PMT) in pulse counting mode and solid-state detectors and arrays, which use Ge or Si as the X-ray absorber [62, 63].

The intensities I_0, I_1 and I_f, based on the X-ray energy E, are generally provided automatically in the management program of the beamline station at the synchrotron. One has to process the data after obtaining them. Explaining this procedure is the purpose of the following paragraph.

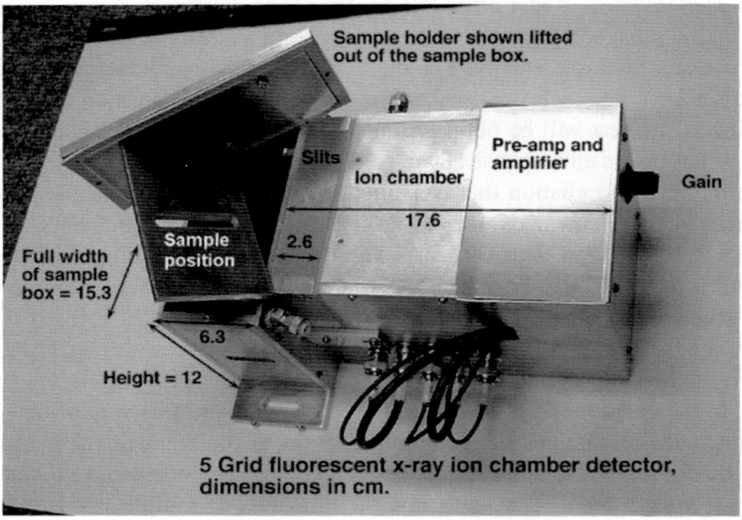

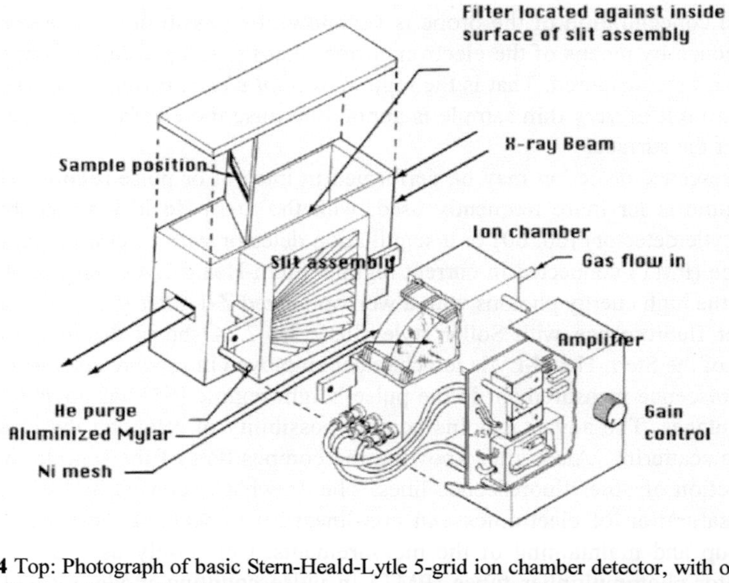

Fig. 7.24 Top: Photograph of basic Stern-Heald-Lytle 5-grid ion chamber detector, with open sample box. Bottom: Schematic illustration of fluorescent X-ray 3-grid ion chamber. By reversing the direction of the slit assembly and the sample holder position, the detector may be positioned for X-ray beams from right to left or from left to right. Courtesy of Dr. Farrel Lytle, 2008.

7.7 Data Reduction

7.7.1 Steps for Obtaining XAFS Experimental Function

Techniques for spectra processing with the information of I_0, I_1 as well as I_f intensities, are reported since the publication of [4] and have been refined with time. They are discussed in detail and are explained in didactic form by Lee *et al.* [53], Teo [11] and Sayers [64]. Currently, it is possible to obtain detailed information through documents and *Tutorials* in internet, by the web page XAFS.ORG [65], closely linked with the International X-ray Absorption Society (IXAS). Synchrotron radiation source organizations usually have sections dedicated to XAFS and its processing. IXAS, the International Union of Crystallography (IUCr) as well as synchrotron source organizations, coordinate periodic schools to explain XAFS by specialists. The texts of these schools are part of the bibliographic collection of XAFS.ORG.

Frequently, at the moment, spectra are elaborated by means of computer programs of the type *graphical user interface* (GUI). These programs provide a platform the user can interact with, for the tasks of gathering XAFS spectra and information for producing the resulting physical parameters. These programs allow the application of algorithms and recommended criteria. They provide the user options to accept or to modify processing and graphics parameters, directories for reading and saving information and other services. In the mentioned site XAFS.ORG, recently updated to the date of writing of this chapter, in the section xafs.org/Software there were 29 programs registered for processing XAFS spectra. Therefore, we will present in this paragraph a panoramic view of the steps of data processing, followed by two figures that illustrate the main steps of the procedure through an example, executed with the GUI SIXPack [66].

In summary, the steps to execute for obtaining $\mu(E)$ and later $\chi(k)$ using measurement data processing programs are:

1. Reading of the intensities I_0, I_1, I_f as well as I_2 to obtain X-ray absorption coefficient $\mu(E)$.
2. Correction of possible systematic errors in spectra due to monochromator, so-called *deglitching*.
3. Determination of E_0 by means of the first derivative from $\mu(E)$ or from a reference.
4. Averaging of the manifold scans to obtain the representative spectrum $\mu(E)$.
5. Self-absorption or dead time corrections in the case of fluorescence spectra from semiconductor detectors.
6. Subtraction of a smooth pre-edge function, for removing instrumental background and absorption from other edges. This step helps finding $\Delta\mu_0(E_0)$.
7. Normalization of the obtained $\mu(E)$ from 0 to 1, by using $\mu_{norm}(E) = \mu(E)/\Delta\mu_0(E_0)$.

8. Obtaining the post-edge background (abstraction of the bare absorbing atom) by using a spline function, and subtracting it from $\mu_{norm}(E)$ to obtain $\chi(E)$.
9. Isolation of the fine-structure oscillations, transforming E to k using eq. (6): *$\chi(E)\rightarrow\chi(k)$, the EXAFS function.*
10. The amplitude of $\chi(k)$ diminishes quickly when increasing k, but effects are present in the function. As have been mentioned above, $\chi(k)$ is multiplied by k^2 or k^3 to better detect the effects at large k.
11. The Fourier Transform (FT) $\chi(k)\rightarrow \chi(R)$ is performed. It must be remembered that the FT is a complex function and that both parts, real and imaginary, will be used later in spectra processing. The FT $\chi(R)$ allows the identification of, in general, the distances to the coordination spheres around the absorbing atom. These are given by maxima positions, corrected by the phase shift δ. The magnitude of the FT is also related to the radial distribution function.
12. EXAFS data $\chi(k)$ is a sum of discrete sine waves. Multiplying the discrete sine wave by a *step function or window $\Omega(k)$* that gradually increases the amplitude of the data smoothes the FT for obtaining $\chi(R)$ [64]. The Fourier transform may be written as follows:

$$\chi(R) = \mathrm{FT}[\chi(k)] = \frac{1}{\sqrt{2\pi}} \int_{-\infty}^{\infty} dk\, e^{i2kR}\, k^{w}\, \chi(k)\Omega(k) \tag{35}$$

13. *Single-shell in R-space* or *Fourier filtering:* Frequently the GUI has the option of performing back FT of a segment, or a so-called *window*, of the maxima of $\chi(R)$. The resulting function is called $\chi(q)$. The procedure consists of choosing the limits of the maximum, multiplying that section of $\chi(R)$ by a window function and performing the FT. This produces another function in the wave number q of regular form. This procedure is more useful if the maxima are well resolved, and if there are no coordination spheres with mixed atoms of different species at approximately the same distance from the absorbing atom.
14. Once all these steps have been performed, it is the moment to apply a model for the interpretation of the experiment, by means of empirical methods or by applying programs that model the EXAFS equation from first principles (*ab initio*).

Fig. 7.25 and Fig. 7.27 present the sequences of images of spectra processing of an EXAFS experiment of an Aurivillius ceramic $Bi_6Ti_3Fe_2O_{18}$ studied in beamline 2-3 at SSRL. The Fe K absorption edge at different temperatures was studied, from T = 10 up to 298 K [67, 68]. The figures show the sequence of processing I_0, I_1, I_f, intensities in function of the energy E up to obtaining $\chi(R)$ from spectra taken at T=10 K.

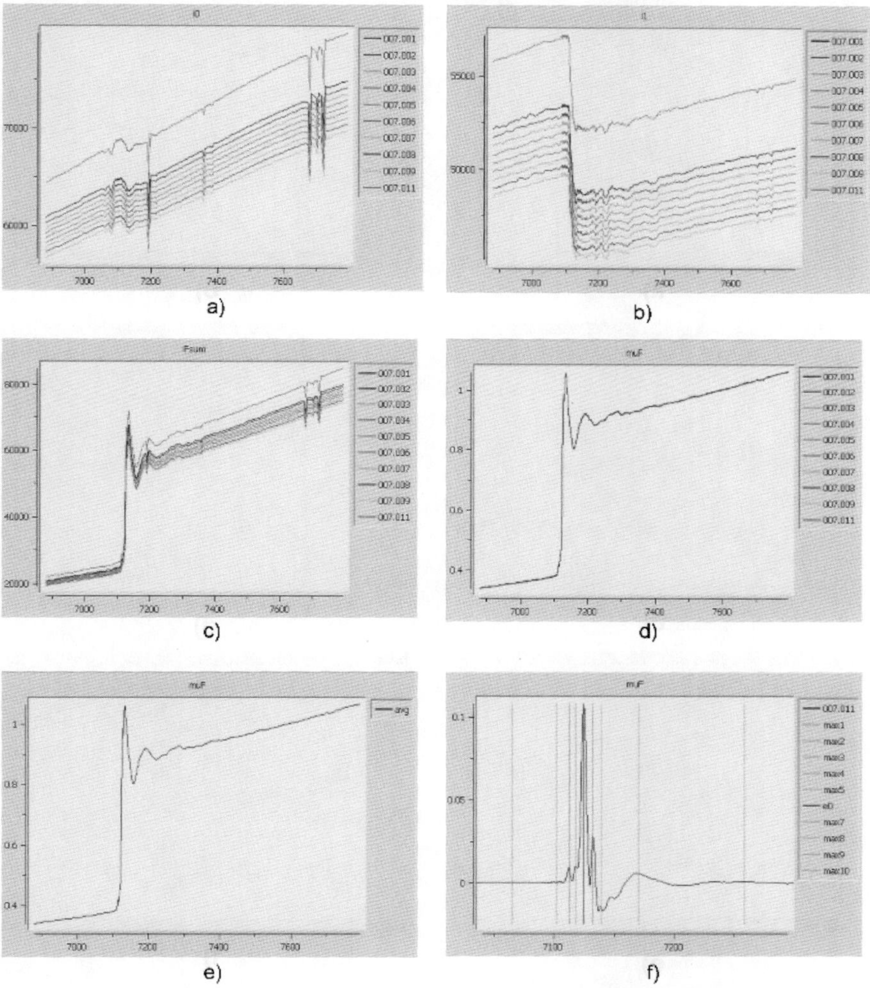

Fig. 7.25 Processing raw I_0, I_1 and I_f intensities for obtaining the X-ray absorption coefficient $\mu(E)$. **a** Ten spectra from entrance ion chamber I_0. **b** Ten spectra from transmission ion chamber I_1. **cs** Ten spectra from fluorescence Lytle detector. **d** Ten individual spectra of x-ray absorption coefficient $\mu(E)$. **e** Average spectrum of X-ray absorption coefficient, so-called *raw* $\mu(E)$. **f** First derivative of spectrum No. 11, with 4 points for smoothing, to define E_0=7124 eV.

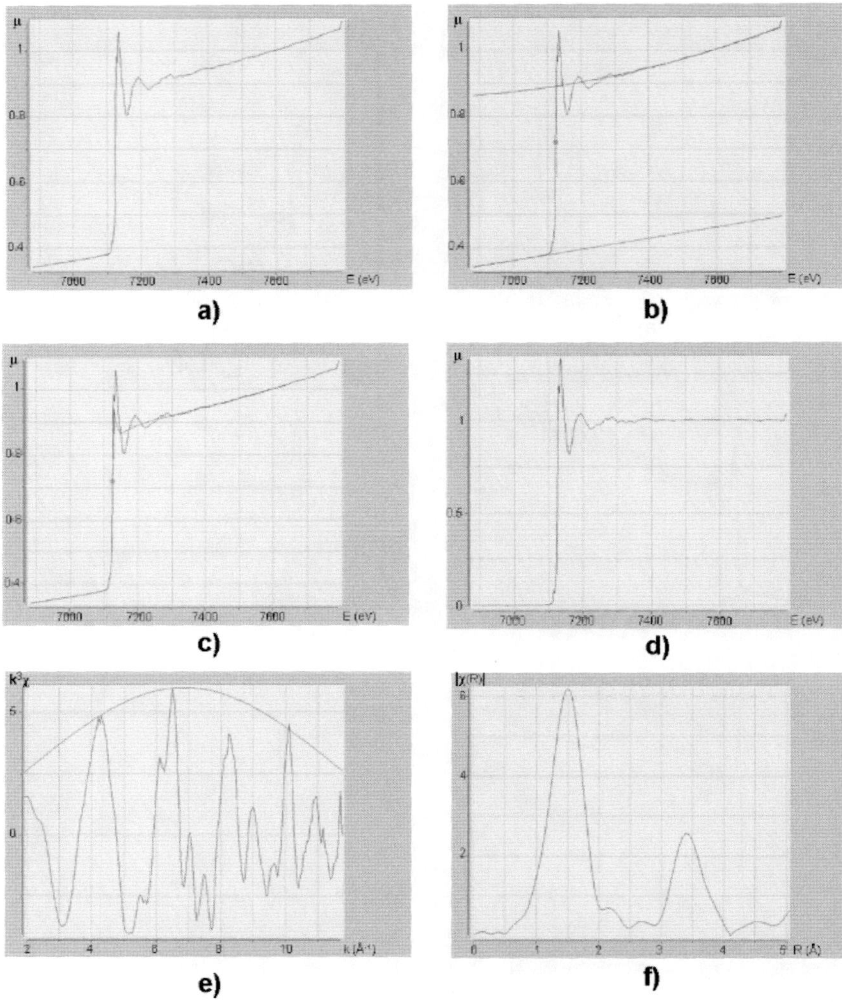

Fig. 7.26 Processing raw $\mu(E)$ for obtaining EXAFS function $\chi(k)$ and $\chi(R)$. **a** Raw $\mu(E)$. **b** Superimposed lines over raw $\mu(E)$ are pre-edge and post-edge functions. **c** Superimposed line over raw $\mu(E)$ is the spline function with seven knots, for subtraction of the bare absorbing atom. **d** Normalized $\mu(E)$. **e** Function $k^3\chi(k)$. Superimposed line is the Kaiser-Baiser window for the Fourier Transform for obtaining $\chi(R)$. **f** Module of the FT up to distances from the absorbing Fe atom equal to 5 Å.

7.8 XAFS Data Analysis

When performing a XAFS experiment, the aim is to determine or confirm physical and chemical properties of the object. In the case of ferroelectrics, the interesting quantities are the structure, angles, charges, local order, thermal properties, and so forth. EXAFS may offer the average values of interatomic distances, coordination numbers and Debye-Waller factors. XANES may provide the oxidation states of the absorbing atom and bonding information. When one has a model to verify, XAFS may also provide arguments of acceptance of that model. We will present in this paragraph the outline of the general methods for determining physical properties from the EXAFS function $\chi(k)$ and its FT $\chi(R)$, obtained from the experiments.

The methods of analysis are classified regarding the use of *empirical spectra of reference* compound or *theoretical spectra of reference.*

The methods that use empirical reference spectra were introduced since the appearance of the technique with publication of [6]. These methods are separated in those that use *curve-fitting* of the unknown spectrum with components of the reference spectrum [11, 69, 70], and those that use the method of *log-ratio/phase-difference* [6, 52] of both the sample and the reference spectra.

The methods that are supported on theoretical reference spectra were previously mentioned in paragraph 7.3.3, based on calculating the functions $\chi(k)$ and $\chi(R)$ from *multiple scattering path expansion.*

From the very beginning of experiment planning, it is necessary to have an idea of the kind of data analysis method to be used. This is because the empirical methods require the measurement of one or several reference samples. In complex cases, even in the case of theoretical spectra of reference, it is necessary to have information about an empirical reference compound. Reference spectra should be measured in the same conditions that the unknown material and processed by theoretical methods [71].

7.8.1 Empirical Methods of Data Analysis

The methods that rely on empirical reference compound spectra consider the possibility of isolating single-shell data in the radial distribution function $\chi(R)$, and then back-transforming or Fourier filtering it (step 13 in paragraph 7.3.5) for obtaining amplitude and phase of isolated shells. The process of isolating a shell may be observed in Fig. 7.27, where both the function $\chi(R)$ and the limits of the selected window to perform the filtering are shown. Fig. 7.28 shows the function $\chi(q)$ obtained for that window and the envelope curve (amplitude), which allows for the application of the *log-ratio/phase-difference* method.

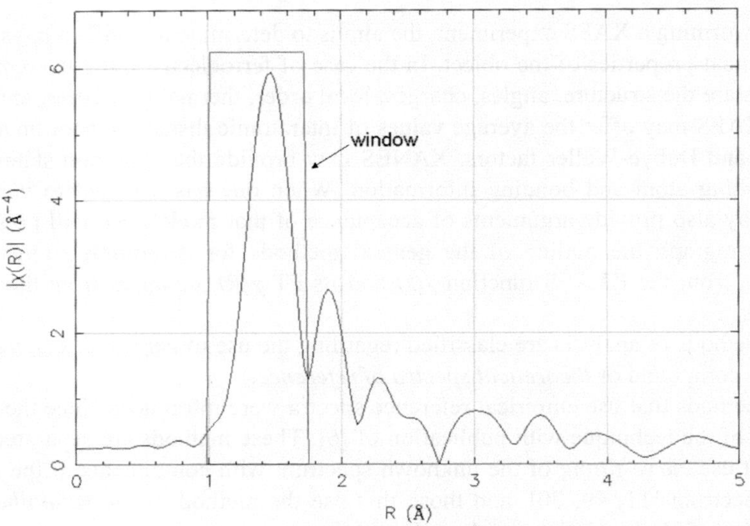

Fig. 7.27 Module of the Fourier transform $\chi(R)$ of the L_{III} absorption edge of uranium, electrode-posited on stainless steel, measured by total electron yield in the beamline 2-3 at SSRL. The superimposed line is the Kaiser-Bessel window for back-transforming the isolated shell.

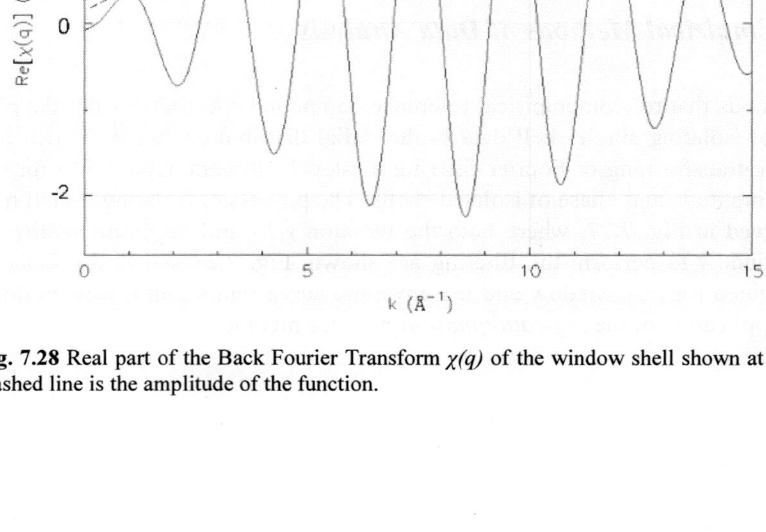

Fig. 7.28 Real part of the Back Fourier Transform $\chi(q)$ of the window shell shown at **Fig. 7.27**. Dashed line is the amplitude of the function.

The determination of Debye-Waller factors (DWF) and coordination numbers (CN) using the log-ratio/phase-difference method is performed by plotting the natural logarithm of the ratio of the sample and standard amplitudes versus k^2 [64]:

If Fourier transform is given in terms of $\chi(q)$ as

$$\chi(q)=A(k)\sin[\varphi(k)] \tag{36}$$

the ratio between amplitudes $A_i(k)$ (the envelope values for a given k) of the unknown and the reference samples fulfils

$$\frac{A_1}{A_2}=\frac{(N_1/R_1^2)S_{01}^2F_1(k)\exp[-2k^2\sigma_1^2-2R_1/\lambda_1]}{(N_2/R_2^2)S_{02}^2F_2(k)\exp[-2k^2\sigma_2^2-2R_2/\lambda_2]} \tag{37}$$

To complete the procedure, the natural logarithm of both members should be extracted. If the reference model is good for the sample, the quantities S_0^2, $F(k)$ and λ can be considered equal. Expression (37) transforms into

$$\ln\left(\frac{A_1}{A_2}\right)=\ln\left(\frac{N_1R_2^2}{N_2R_1^2}\right)+2k^2(\sigma_2^2-\sigma_1^2)-2(R_1-R_2)/\lambda \tag{38}$$

Because of $(R_1\text{-}R_2) << \lambda$, the last term in (37) may be neglected. Plotting several values of the log-ratio $ln(A_1/A_2)$ vs. k^2 enables the slope to provide for the difference in DWF values. The intercept leads to the CN values.

The curve-fitting method consists in a generalization of the method described above. It was conceived on the assumption that the isolated shells of the reference spectra would serve as orthogonal functions to fit them to the spectrum of the study material. It was also assumed that the parameters DWF, CN and R, would be calculated as combinations of the reference's values. Molecular structures of reference compounds produce, in XAFS spectra, the mentioned orthogonal functions. If the unknown compound has a spectrum that may be decomposed in the same orthogonal functions, it is made up of similar molecular structures that those of the reference compound.

The curve-fitting method may be performed by the separation of single-shells in the study of simple and reference spectra [69]. Alternatively, by means of a more advanced method, curve-fitting is performed by calculating the number of principal components of the set of experimental spectra. Later, a least-squares fitting of the experimental spectra is applied with linear combinations of these principal components (model compounds) spectra [70, 72, 73]. These procedures are frequent in the study of environmental and geologic samples, where the analysis of oxidation states, species and even non-destructive microscopic chemical analysis [70] by means of XANES and EXAFS is performed.

For ferroelectric synthetic samples, interest may be focused on the determination of the number and type of probable species and the oxidation state of the absorbing atom by means of the XANES spectra, in experimental series. In that case, it is useful to apply the method of principal component analysis (PCA). The method is implemented in calculation programs like SIXPack [66] that also performs least-squares fitting of the functions of model compounds. These model compounds should be known from libraries or have been specially measured to the effects of determining the oxidation states from well-known substances.

7.8.2 Theoretical Models for Data Analysis

Ferroelectrics have structures that can be relatively simple, such as $PbTiO_3$, or complex, for instance Aurivillius ceramics. In all of them, the distortions of the crystalline cells that cause the electric dipole moments are present. This attribute frequently implies superposition of components in the first coordination shell. In such cases, it will be difficult to apply empirical methods that assume only photoelectron single scattering (SS). Selecting theoretical models for the interpretation of the EXAFS spectra will be unavoidable.

A good strategy for planning a XAFS experiment [74] in ferroelectrics is suggested with the following steps:

- Work with an initial conceptual model that would later be refined. Obtain information about the crystalline structure of the compound to be confirmed. Atomic positions must be known. The geometric relations between the atoms that have relative distances to each other, or with the absorbing atom that are susceptible to be optimized in the model, should be mathematically expressed. Perform experiments with techniques that offer complementary information.
- Design a strategy of analysis by XAFS. Establish which reference compounds must be measured before or simultaneously with the study samples. Accomplish the necessary decisions in order to maximize the number of degrees of freedom in the analysis. Choose the suitable synchrotron beam.
- Finally, use the computational package that offers more advantages.
- Once these steps have been taken, return to refine the initial model by using the information of reference compound and collected complementary data. Finally, process the spectra of the study samples.

The structures of ferroelectric materials lead to considering the effects of photoelectron multiple scattering (MS), described in paragraph 7.5. The spectra processing becomes more complicated, but will simultaneously allow information on angles. The existence of computer programs that provide theoretical reference spectra allows the processing of data without isolating single shells, as well as avoids finding the reference compound adapted for least square fitting.

The strategy of work with these programs will be outlined next. The real work with them is complicated, because programs propose a theoretical spectrum from a well-known structure, but do not solve the inverse problem. So, it is not an exaggeration to recommend attending schools and workshops that synchrotron source organizations, the IXAS, and other organizations offer to train their users.

7.8.2.1 Multiple Scattering Paths

When there is a set of experimental spectra to analyze in the form of $\chi(k)$ and $\chi(R)$, and the series of steps enumerated above have been fulfilled, one of the packages of programs based on theoretical calculation of reference spectra is used. Many programs exist that perform these calculations. The following programs have been already mentioned. Their web pages are:

EXCURV98: http://srs.dl.ac.uk/xrs/computing/Programs/excurv97/intro.html
GNXAS: http://gnxas.unicam.it/XASLABwww/pag_gnxas.html
FEFF: http://leonardo.phys.washington.edu/feff/welcome.html

In the MS platform, for the zone of XANES, another program that has many users is MXAN [75]: http://www.lnf.infn.it/theory/CondensedMatter/

In site XAFS.ORG, there is an extensive list of software with the corresponding description and links. Many of them are free.

The structural parameters of the study sample that appear in the EXAFS equation (31) are extracted when fitting the experimental spectra to the theoretical calculations. For clarity, let's repeat here the EXAFS equation:

$$\chi(k) = \sum_j \frac{S_0^2 N_j \, F_j(k) e^{-2k^2\sigma_j^2} \; e^{-2R_j/\lambda(k)}}{kR_j^2} \sin\left[2kR_j + \delta_j(k)\right] \tag{31}$$

The programs of calculation of theoretical spectra precisely obtain the mean free path $\lambda(k)$, the effective scattering amplitude $F_j(k)$ and the phase shift $\delta_j(k)$ for each trajectory or *path j*.

The beginning of the interpretation procedure consists of introducing the crystalline or molecular configuration of the initial model to the program so that it may generate these functions. Also it is required to find the configuration *paths*, how much contributes each path, and how many paths are important. In other words, determining *how many terms to be added in the equation* is required. The fitting of the experimental spectrum to the sum of the proposed trajectories is performed later, varying the contribution of each *path j* until theoretical and experimental spectra agree satisfactorily. The parameters N_j, R_j and σ^2_j of each fitted *path j* will be the results of the experimental spectrum, and they will introduce modifications to the crystalline or molecular structure of the model originally proposed.

In order to better understand the working algorithm within these programs, let's begin from the foundations of the Green functions formalism that take us to the expression (28). The photoelectron may scatter many times, as shown in Fig. 7.29. Programs based on MS formalism introduce a *Path Formalism* in the *Real Space* calculations [20, 37]:

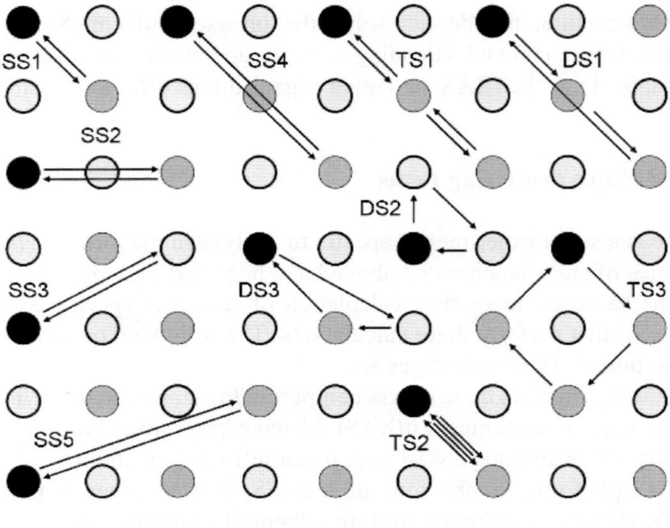

Fig. 7.29 Schematic representation of different paths used for XAFS multiple scattering data analysis. The fcc structure is shown (proper for pure metals like Cu, Pt, etc.): out of plane atoms are denoted by empty circles; in plane atoms are gray. Black circles denote absorbing (central) atoms. Single scattering paths are denoted by SSn; double scattering paths –by DSn; and triple scattering paths –by TSn. This picture is inspired in a figure published by Frenkel *et al.* [71].

$$\mathbf{G} = \mathbf{G}^0 + \mathbf{G}^0\mathbf{t}\mathbf{G}^0 + \mathbf{G}^0\mathbf{t}\mathbf{G}^0\mathbf{t}\mathbf{G}^0 + \mathbf{G}^0\mathbf{t}\mathbf{G}^0\mathbf{t}\mathbf{G}^0\mathbf{t}\mathbf{G}^0 + \ldots \quad (39)$$

where $\mathbf{G}^0$ is a photoelectron propagator and $\mathbf{t}$ is the scattering matrix from the neighbouring atom.

The scattering of photoelectrons is represented in real space in a picture of a *path Γ,* visualizing the absorbing atom as the central atom in the configuration (Fig. 7.29). The propagation of the photoelectron between atoms is referred to as a *path leg*. Path legs always connect two different atoms. Each path begins and ends at the absorbing atom. Paths are always closed.

Various kinds of paths are shown schematically in Fig. 7.29. Paths labelled by SS1, SS2,…SS5 are *single scattering* paths, or paths with *two* legs. Paths labelled by DS1, DS2 and DS3 are examples of *double scattering* paths, or paths with *three* legs. Paths labelled by TS1, TS2 and TS3 are examples of *triple scattering* paths, i.e., paths with *four* legs. Paths DS1 and TS1 are called *collinear or focusing scattering* paths because they involve one or more scattering event in the forward direction.

The initial model data that will be refined should be introduced in the program packages for XAFS data analysis. ATOMS [76] is an example of programs to prepare the data, proper for working with ferroelectric crystalline structures.

The programs of MS analysis help in determining all possible paths, their effective radius R_{eff}, their multiplicity (so-called *degeneracy* in this language; this is not to be confused with scattering multiplicity) and angles between legs.

To properly evaluate equation (31) with the formalism of the equation (39), it is necessary to obtain all possible scattering paths *j* with their geometry and degeneracy, including the nonlinear trajectories and paths which scatter from the absorber itself like TS2 in Fig. 7.29. The number of paths which will be added in (31) will depend, first, on the spectrum zone that is being analyzed. For the XANES zone [20, 37], the *Full Multiple Scattering* approach is adopted. That is to say, *all paths*. For the EXAFS zone, the inclusion of more paths is analyzed gradually, considering the data quality collected in $\chi(k)$ and therefore up to what *R*-value may the FT-$\chi(R)$ be obtained.

Let's keep in mind the objective of this paragraph: to interpret the experimental spectra $\chi(k) \rightarrow \chi(R)$ with a theoretical model. According to the strategy mentioned above, there are a model to refine, several spectra of model compounds, complementary information and a package of programs. In order to support us on a specific algorithm, not being an explicit recommendation of preference over other programs, we will continue with examples based on the use of FEFF [20] and IFEFFIT [42]. To the date of writing this chapter, in XAFS.ORG, there were dozens of programs for processing XAFS spectra related to FEFF. For that reason, this text does not lose its general scope.

FEFF associates paths Γ with the radii of the coordination spheres around the absorbing atom, that is to say, with the effective radii R_{eff} that the photoelectron passes through. When the program is run, it produces for each path Γ a file *feffnnnn.dat* that contains R_{eff}, the mean free path $\lambda(k)$, the effective scattering amplitude $F_j(k)$, the phase shift $\delta_j(k)$ and other data of interest for each trajectory. Program IFEFFIT and its associated GUIs perform the fitting of windows for the experimental spectrum, between limits that the user chooses with an increasing number of the theoretical functions that are generated in those paths, contained in files *feffnnnn.dat*. When the user generates the limits of the windows, he/she is generating the functions $\Omega(k)$ and $\Omega(R)$, to apply operations similar to the equation (35).

When fitting the EXAFS equation using expression (31), the sum is done by scanning index *j* (that is to say, by paths Γ). As the window becomes wider, *j* grows. For *each path* the parameters that appear in Table 7.2 [77] are refined:

Table 7.2. Parameters to be refined for each path while fitting theoretical spectra in EXAFS analysis using FEFF and FEFFIT.

NS_0^2	Amplitude factor: N and S_0^2. If degeneration is equal 1, then N is refined
E_0	Energy shift (it sets where k = 0)
ΔR	Change in path length, $R = \Delta R + R_{eff}$
σ^2	Mean-square-displacement in R: Debye-Waller Factor
C_3	Third cummulant: asymmetric term in Debye-Waller Factor
C_4	Forth cummulant: asymmetric term in Debye-Waller Factor

The meaning of parameters C_3 and C_4 or cummulants is discussed in [20, 52].

The difficulty of the procedure consists in the growing number of variables to fit that are generated with the number of paths included in the fitting. Furthermore, while processing data, the information that may be extracted from an experimental spectrum $\chi(k)$ (or from a given spectrum window) is limited. The maximum number of independent parameters N_{ind} that can be extracted from a periodic signal is [77, 78]:

$$N_{ind} \approx 2\Delta k\,\Delta R / \pi \tag{40}$$

where Δk and ΔR are the widths of $\Omega(k)$ and $\Omega(R)$, respectively. In order to avoid that difficulty, and to validate the model, imposing *equations of constraints* to the fitting variables is required.

A system with constraints and algebraic relations among the parameters of a single path, of different paths or even of different sets of data, is desirable [77]. An example of these constraints, that always have physical sense, is having all paths with equal E_0 and S_0^2. Another example is to assign a thermal behaviour to the parameter σ^2 given by the Debye or Einstein models.

In the procedure of fitting the given window of $\chi(k)$, some variables of the verifying model are adjusted and the fitting program compares the sum of the selected path functions with the functions of the experimental window. The parameters of the model and those of the EXAFS equation that fulfil the convergence criteria are the *fitting results*. The *reduced chi-square* χ^2_v and *R-factor* criteria, as defined in Chapter 7, paragraph 7.3.2, are used for verifying the *Goodness-of-Fit statistics*. The difference here is that χ^2_v is always greater than 10 and is normal to be 100, while R-factor should be kept less than two percent.

In Fig. 7.30, a graphic example of this procedure is presented. The spectrum of the Aurivillius ceramics taken at T=10 K, presented in Figs. 7.8 and 7.17, was compared with the resulting theoretical spectra of single scattering paths, and then with the two first coordination spheres around the absorbing Fe atom, which are the atoms of O octahedron. The purpose of the experiment was to verify the model obtained by XRD of preferential occupation of Fe atoms at the centre of the crystalline structure of the Aurivillius ceramics. The complete fitting implied the analysis of up to the tenth coordination sphere [67].

In order to understand the ideas and procedures for fitting experimental spectra using theoretical reference functions, reading the article [77] is useful. The study of tutorials that appear in XAFS.ORG is very informative.

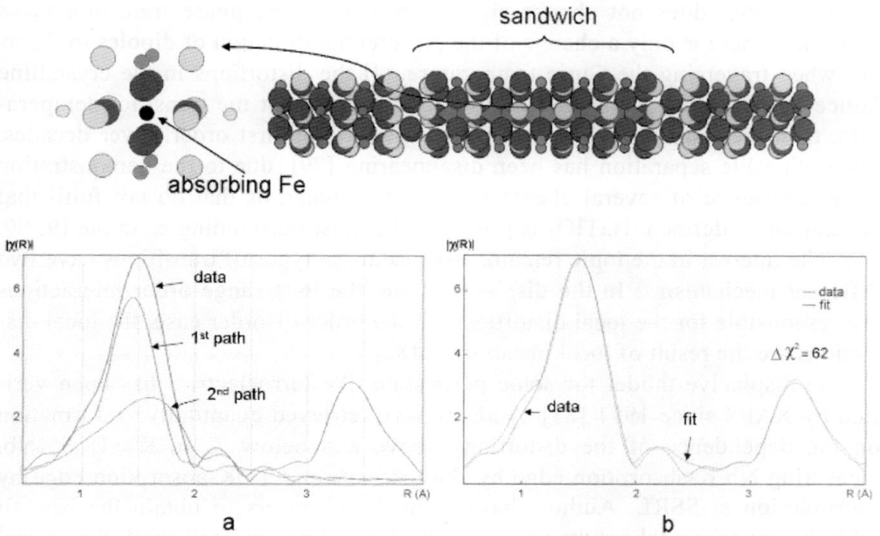

Fig. 7.30 Data analysis of the $\chi(R)$ spectrum corresponding to T= 10 K of the Aurivillius ceramic $Bi_6Ti_3Fe_2O_{18}$. On top right is presented the crystal structure, where large black spheres → O; small gray spheres → Ti/Fe; large light gray spheres → Bi. On top left is presented an insert of the central part of the crystal structure, showing the absorbing Fe atom in black. On bottom: a) Fitting of a window from R=0.5 to R=2.5 Å, showing the module of $\chi(R)$ and the module of the theoretical spectra generated by FEFF6 [50] for the first two paths, for single scattering of O atoms at 1.85 ± 0.04Å; σ^2 = 0.016 ± 0.002 Å^2 and O atoms at 1.97 ± 0.04Å; σ^2 = 0.007 ± 0.002 Å^2. b) Full fitting of the model, for 10 paths, with χ^2_ν= 65 and R-factor = 1.2%.

7.9 XAFS Applied to Ferroelectrics

7.9.1 Pioneering Works on Order-Disorder or Displacive Character of Ferroelectric Materials

The interest in applying XAFS to ferroelectrics is fundamentally due to the possibility of measuring the local order by means of EXAFS, in contrast to long-range average that offers XRD. Also, one takes advantage of the features in XANES to identify materials and determine the oxidation state and coordination of the absorbing atom.

More than one hundred works have been published from 1986 to 2008 on the application of XAFS to ferroelectrics. This interest has been related to the problem of understanding the mechanism of the ferroelectrics phase transition: are they *order-disorder* or are they *displacive*? The type of transition is classified as *order-disorder* if the deformation of the crystalline lattice that gives origin to ferroelectricity does not change significantly when the phase transition takes place, and there is only a change of the preferential direction of dipoles to disorder when traversing the Curie temperature. If the distortions in the crystalline lattice and the order parameters show discontinuities at the transition temperature, then the transformation is *displacive*, and is of first order. Over decades, this noticeable separation has been disappearing [79], due to the demonstration of the existence of several classic ferroelectric materials that do not fulfil that separation so defined. $BaTiO_3$ is probably the most outstanding example [9, 80, 81]. The interest in the topic remains because these types of transitions have two different mechanisms: In the displacive case, the long range order interactions are responsible for the local distortions. In the order-disorder case, the local distortions are the result of local instabilities [82].

The displacive model for some perovskite-like ferroelectrics has been verified by XAFS since 1991 [83]. Authors have retrieved quantitative information on the dependence of the distortions above and below T_c in $KTaO_3$:9%Nb, measuring Nb K-absorption edge by fluorescence and Ta K-absorption edge by transmission at SSRL. Authors have refined parameters to obtain the best fit with the experimental results up to the third coordination shell, with theoretical spectra up to the third-order collinear scattering (TS1 in paragraph 7.8.2.1 and Fig. 7.29).

The work published by Stern, Ravel and collaborators on $PbTiO_3$ [82, 84, 85] is interesting for the scope of this handbook. In this work, they have measured the temperature dependence of the local distortions of $PbTiO_3$ crystals below and above T_c using XAFS. Both Pb L_{III}- and Ti K- absorption edges were measured, providing quantitative determinations of the displacements of the atoms within the unit cell. For determining these distortions, they applied theoretical fitting with multiscattering paths, using FEFF5 code [86]. Fig. 7.31 presents the Fourier transforms of the Pb L_{III} EXAFS spectra $\chi(r)$ for several temperatures, extracted from [82]. Spectra are illustrative of how the fitting up to four coordination shells was important at low temperature (12 K). At higher temperatures, authors have fit up to two shells for Pb EXAFS spectra and one shell for Ti K-edge EXAFS spectra. In the fitting presented at [84], Ravel explains how to determine the bond angles in the chain structure among Ti and O atoms. In [82, 84, 85], authors have found that in the paraelectric phase, the Pb and Ti distortions are about 50% and 70% of the corresponding low temperature values respectively. These results show that an essential element of order-disorder is present even in this ferroelectric crystal, which displays the soft mode behaviour and a dielectric constant typical of displacive ferroelectrics. The article by Sicron *et al.* [82] and the tutorial [84] presented in http://cars9.uchicago.edu/~ravel/talks/course/Welcome.html are strongly recommended for those interested in exercising on EXAFS fitting with theoretical spectras [87].

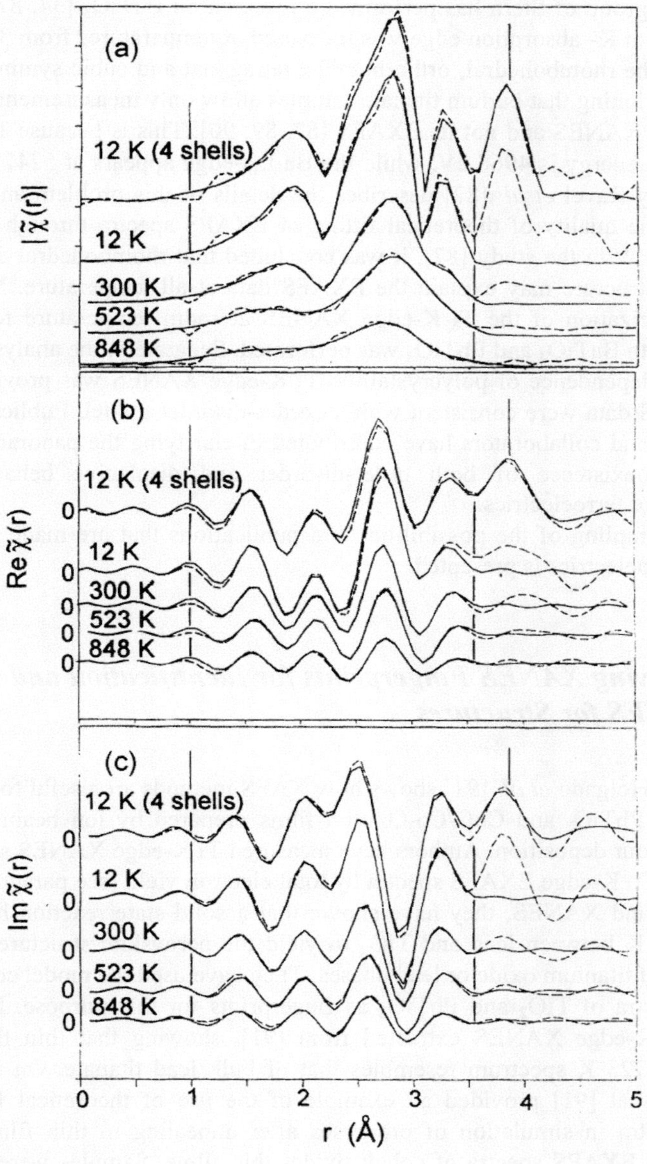

Fig. 7.31 Fourier transforms of the Pb L_{III}-edge EXAFS spectra $\chi(r)$, for several temperatures. Dashed and solid lines represent the experimental results and the two-shell theoretical fits, calculated using FEFF5 code [86] respectively. (a) Absolute value of $\chi(r)$, (b) real part, and (c) imaginary part. Vertical lines indicate the fitting ranges. Note that at 12 K both four- and two-shell fits are shown. Reprinted figure with permission from Sicron N, Ravel B, Yacoby Y, Stern EA, Dogan F, Rehr JJ, Nature of the Ferroelectric Phase-Transition in PbTiO3, Phys Rev B 50,13168-13180 (1994). Copyright (1994) by the American Physical Society.

The same group of Stern has performed a research on $BaTiO_3$ [34, 87, 88]. In [87] the barium K- absorption edge was measured at temperatures from 35 to 750 K, covering the rhombohedral, orthorhombic, tetragonal and cubic symmetries. It is worth mentioning that barium titanate samples allow only measurements of titanium K-edge XANES and not its EXAFS [87, 89, 90]. This is because Ti K- absorption edge energy is 4966 eV, while the Ba L_{III}-edge appears at 5247 eV. The publication by Ravel *et al.* [87] describes the details of this problem and why it has limited the quality of theoretical fitting of EXAFS spectra through the program FEFF6. With the study [87], it was concluded that rhombohedral distortion of the local structure may explain the EXAFS data at all temperature. Measurement of polarization of the Ti K-edge XANES at room temperature for single crystals of both $BaTiO_3$ and $PbTiO_3$ was performed. Separately, the analysis of the temperature dependence of polycrystalline Ti K-edge XANES was provided. All these XANES data were consistent with an order-disorder model. Publications of Stern, Ravel and collaborators have contributed in clarifying the panoramic view about the coexistence of both order-disorder and displacive behaviour of perovskite-like ferroelectrics.

Next, a sampling of the possibilities and publications that are made applying XAFS in ferroelectrics is presented.

7.9.2 Applying XANES Fingerprints for Identification and EXAFS for Structures

The work of Holgado *et al.* [91] shows how XAFS methods are useful for characterization of $PbTiO_3$ and CoO/Co_3O_4 thin films prepared by ion beam induced chemical vapour deposition. Authors have measured Ti K-edge XANES spectra of $PbTiO_3$ and Co K-edge EXAFS spectra by total electron yield (see paragraph 7.6). Using XRD and XANES, they have shown that a solid state reaction has taken place at 723 K between lead and TiO_2 to yield the perovskite structure without segregation of titanium oxide or lead phases. They have used the model compound XANES spectra of TiO_2 and $PbTiO_3$ as fingerprints for this purpose. Fig. 7.32 presents Ti K-edge XANES extracted from [91], showing that thin film after annealing at 723 K spectrum resembles that of bulk lead titanate. On the other hand, authors at [91] provided an example of the use of theoretical fitting of EXAFS spectra in simulation of processes after annealing in thin films. They measured the EXAFS spectra of cobalt oxides thin films. Samples were initially amorphous, and after were annealed at 573 and 873 K. In the Fourier transform spectra, authors recognized the crystalline order resulting from annealing. Then authors, using the FEFF code, fitted theoretical models to the obtained spectra and could interpret the ordering differences in clusters of studied thin films.

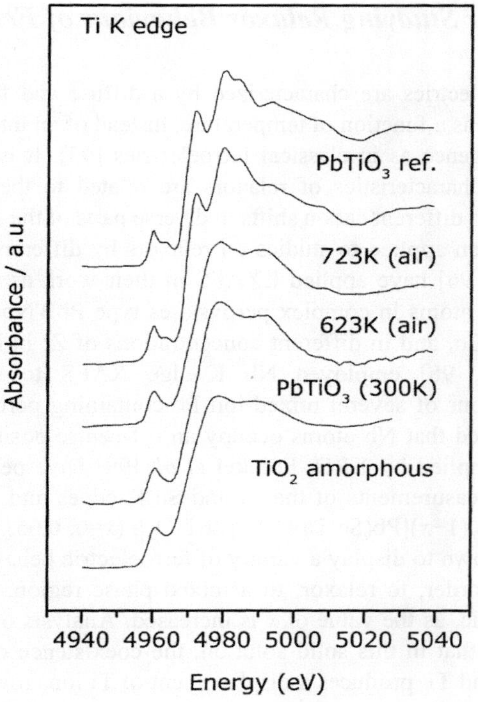

Fig. 7.32 Ti K edge XANES spectra of a lead titanate film after annealing at the indicated temperatures up to 823 K. Included in the figure are the XANES spectra of amorphous TiO_2 and bulk lead titanate, as references. The practically identical shape of the 723 K annealed film and the reference $PbTiO_3$ supports the conclusion that no segregated phase of TiO_2 remains in the film. Reprinted figure with permission from Holgado JP, Caballero A , Espinós JP , Morales J , Jiménez VM , Justo A , González-Elipe AR, Characterisation by X-ray absorption spectroscopy of oxide thin films prepared by ion beam-induced CVD, Thin Solid Films, 377-378, 460-466 (2000). Copyright (2000) by Elsevier.

There is a paper of Neves *et al.* [92] about the short- and long-range order in lanthanum-modified $PbTiO_3$ ceramic materials, ranging from 0 to 30 atom % of La. They have performed both conventional XRD and XAFS. Authors present measurements of the Ti K-edge XAFS in transmission, paying special attention to the pre-edge features of XANES spectra (see paragraph 7.4.3). XRD results show tetragonal structure for all La concentrations except the 30%, for which conventional XRD provided an ideally cubic perovskite structure. The analysis of EXAFS spectra shows a decrease in the local disorder around Ti atoms as the content of La increases. But, according to XANES pre-edge peak intensities, for the sample with 30% of La, a local distortion around Ti atoms persists. Authors conclude that there should really be tetragonal distortions in the 30% La concentration. They relay this argument, in part, on the fact that X-ray diffraction experiment provides structural information about the long-range average structure, whereas XAFS probed the short-range order around the Ti atoms.

7.9.3 XAFS for Studying Relaxor Behaviour of Ferroelectrics

Relaxors in ferroelectrics are characterized by a diffuse and frequency-dependent dielectric anomaly as a function of temperature, instead of an intense and frequency-independent divergence as in classical ferroelectrics [93]. It is generally admitted that the peculiar characteristics of relaxors are related to the existence of polar nanoregions, due to different cation shifts in diverse parts of the structure [94].

EXAFS has been applied to studies on relaxors by different groups. Chen and collaborators [95, 96] have applied EXAFS in their work as an indicator of displacement of lead atoms in complex perovskites type $PbB'NbO_3$, where the atom B' is Ni, Zn and Co, and in different concentrations of Zr in PZT. Shuvaeva and collaborators [97, 98] employed Nb K-edge XAFS to study temperature-dependent behaviour of several mixed-ion Pb-containing perovskite compounds. They have observed that Nb atoms occupy an off-centre position with symmetry lower than that implied by XRD. Frenkel *et al.* [99] have performed room temperature XAFS measurements of the Ti and Sc K-edges and Ta L_{III}-edge on the ferroelectric series $(1-x)[Pb(Sc,Ta)O_3]$-$x[PbTiO_3]$, (x=0, 0.05, 0.1, 0.2, 0.5 and 1). This system is known to display a variety of ferroelectric behaviours ranging from variable order-disorder, to relaxor, to a mixed phase region, and then finally to normal ferroelectric, as the value of x is increased. Analysis of all these data lead to the conclusion that in this solid solution, the coexistence of the decreasing in size ions Ta, Sc and Ti, produces a displacement of Ti ions toward (111) direction for low Ti concentrations and toward (001) direction for higher concentrations.

Shuvaeva *et al.* [100], Laulhé *et al.* [94] and Shanthakumar *et al.* [101] have recently studied local structure of lead-free relaxor titanate-family ferroelectrics by means of XAFS.

Laulhé *et al.* [94] have studied the $BaTi_{1-x}Zr_xO_3$ (BTZ) at the Zr K-edge (17.998 keV). EXAFS experiments were carried out at the European Synchrotron Radiation Facility (ESRF), in transmission mode at room temperature for samples with x =0.25, 0.30, 0.35 and 1. EXAFS spectra for $BaZrO_3$ were collected also at 11 K, and at 11, 90, and150 K for $BaTi_{0.65}Zr_{0.35}O_3$. EXAFS spectra were processed using theoretical models with FEFF8 [38] and taking into account single and multiple scattering. For comparison, authors employed data from a previous EXAFS study [102], showing that the local structure of $BaZrO_3$ can be considered as a perfect cubic perovskite structure. When analyzing the first shell, authors have observed that the length and strength of the bond between the Zr central atom and its first oxygen neighbour are independent of the Zr substitution rate in BTZ samples. When authors analyzed the contributions of photoelectron scattering to reach the second (Ba), third (Zr) and fourth (O) shells, they found that a model of linear position of atoms Zr-O-Zr/Ti did not properly explain spectra. As a result, the authors fit an angle with the apex on first oxygen. They concluded that the EXAFS study of BTZ relaxors at the Zr K-edge may be explained by local structural deviations from the average cubic cell, as well as chemical inhomogeneities. Authors, in the same spirit as the article [99], have assumed a segregation of Zr atoms that consists of small inclusions of $BaZrO_3$ (non-necessarily spherical) in a Ba-

TiO_3 matrix. Taking into account that the cell volume of $BaZrO_3$ is larger than that of $BaTiO_3$, these adjacent TiO_6 octahedra support a tensile strain. Different Ti^{4+} displacements, and thus polarity, are expected in the adjacent regions when compared to the $BaTiO_3$ matrix. These displacements would lead to random electric fields. In such a way, random strains (elastic fields) are considered to lead to local patterns of Polar Regions that cause the relaxor behaviour.

The paper of Shanthakumar *et al.* [101] is another example of using XANES and EXAFS to contribute elucidating the relaxor-like behaviour of $[Ba_{0.6}Sr_{0.4}][(YTa)_{0.03}Ti_{0.94}]O_3$. In this ceramic, there is a charge-balanced substitution of Ti^{4+} by Y^{3+} and Ta^{5+} ions into $Ba_{0.6}Sr_{0.4}TiO_3$ (BST). Potrepka *et al.* [103] previously demonstrated that such substitution can lead to relaxor behaviour. The experiment of Shanthakumar *et al.* [101] planned to explain the difference between two equally Y/Ta doped BST samples: Sample A, sintered at 1550 °C, which is a relaxor; and sample B, sintered at 1600 °C, which is a normal ferroelectric. As a theoretical model for the behaviour of the internal fields, it was assumed that charge compensation occurs at the Ti^{4+} sites where substitution occupancy by Y^{3+} or Ta^{5+} is localized, so that clusters of Y^{3+} and Ta^{5+} impurities form permanent dipoles [104]. For the relaxor behaviour the model of quenched random electric fields after Y/Ta substitution by Westphal *et al.* [105] was accepted.

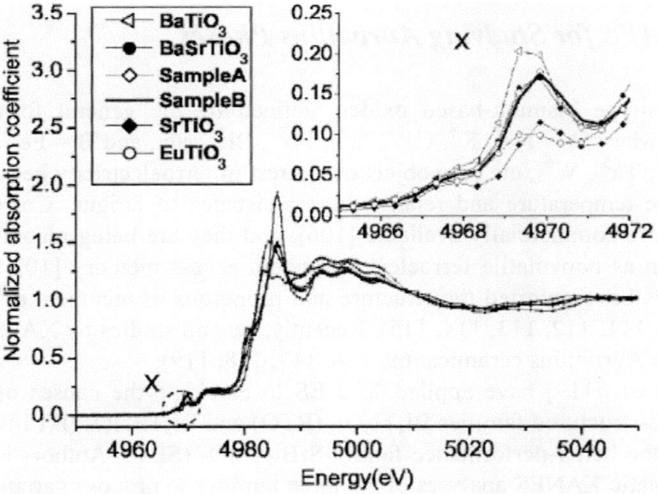

Fig. 7.33 Normalized XANES spectra of Ti K-edge in $BaSrTiO_3$, samples A and B, and $SrTiO_3$. The data for bulk $BaTiO_3$ and a sample of $EuTiO_3$ are given for comparison. The feature *X* denotes the energy region of the PEFS. The insert shows the blow up region of PEFS, where the areas of the peak corresponding to the 1s–3d transition is equal for samples BST, A and B. The vertical scale is in relative units. Reprinted figure with permission from Shanthakumar P, Balasubramanian M, Pease DM, Frenkel AI, Potrepka DM, Kraizman V, Budnick JI, Hines WA, X-ray study of the ferroelectric $[Ba_{0.6}Sr_{0.4}][(YTa)_{0.03}Ti_{0.94}]O_3$, Phys Rev B 74,174103 (2006). Copyright (2006) by the American Physical Society.

After verifying the cubic perovskite character by XRD, authors performed Ti K-edge XANES of $SrTiO_3$, BST, samples A and B by fluorescence at room temperature, at the National Synchrotron Light Source (NSLS) in the US. Ta L_{III}-edge EXAFS spectra were obtained also in NSLS, but Y K-edge EXAFS, which was a more complex measurement (see [101]), was performed at the Advanced Photon Source (APS) at Argonne National Laboratory of the US. Fig. 7.33 presents the XANES spectra obtained in [101], along with $BaTiO_3$ and $EuTiO_3$ given for comparison. Careful application of the pre-edge characterization introduced by Vedrinskii *et al.* [35] (see 7.4.3) allowed concluding in [101] that BST, sample A and sample B have in average the same Ti displacement from the centro-symmetric position and this is not the reason for differences in relaxor behaviour. EXAFS spectra were processed using Artemis and Athena software [57]. By modelling with multiple scattering paths, authors concluded that there are no great differences between the environments of the Ta atoms in sample A vs. sample B, but there are large differences between the Y edges of these two samples. In sample A, authors obtained Ta-O bond lengths shorter than expected from XRD and Y bonds greater than expectations. In sample B, in contrast, authors suspect that some Y has left the perovskite lattice. In conclusion, authors assert that internal strains observed in sample A are an important component in the explanation of its relaxor behaviour, hence the model of Cross [93] cannot be ruled out.

7.9.4 XAFS for Studying Aurivillius Phases

Aurivillius-type bismuth-based oxides, defined by the general formula $Bi_2A_{n-1}B_nO_{3n+3}$, where A = Na^+, K^+, Ca^{2+}, Sr^{2+}, Pb^{2+}, Bi^{3+}, etc., and B = Fe^{3+}, Cr^{3+}, Ti^{4+}, Zr^{4+}, Nb^{5+}, Ta^{5+}, W^{6+}, etc., are object of interest in ferroelectricity because of their high Curie temperature and relative high resistance to fatigue. Currently, these ceramics are commercially available [106], and they are being recommended for application as nonvolatile ferroelectric random access memory [107,108]. Many researchers have reported the structure and properties of members of this family [109, 110, 111, 112, 113, 114, 115]. Recently, several studies by XAFS have been devoted to Aurivillius ceramics [69, 116, 117, 118, 119].

Kim *et al.* [119] have applied XANES to elucidate the causes of fatigue of Aurivillius-structured families $Bi_4Ti_3O_{12}$ (BTO) and Bi_3TaTiO_9 (BTT) in comparison with the better-performance family $SrBi_2Ta_2O_9$ (SBT). Authors have carried out systematic XANES analyses of the three families to propose variations of their chemical bonding nature upon the increase of electronic charge. From these results, they have deduced the effect of bond covalency on lattice stability and fatigue behaviour of the specific family head.

Authors have prepared polycrystalline samples of SBT, BTO and BTT by solid-state reactions. The two chemically reduced derivative members of each family have been produced by Ar-annealing and lithiation [117]. The analysis was done at the Pohang Accelerator Laboratory in Korea by means of XAFS of the Ti K-, Ta L_{III}- and Bi L_{III}-edges [119].

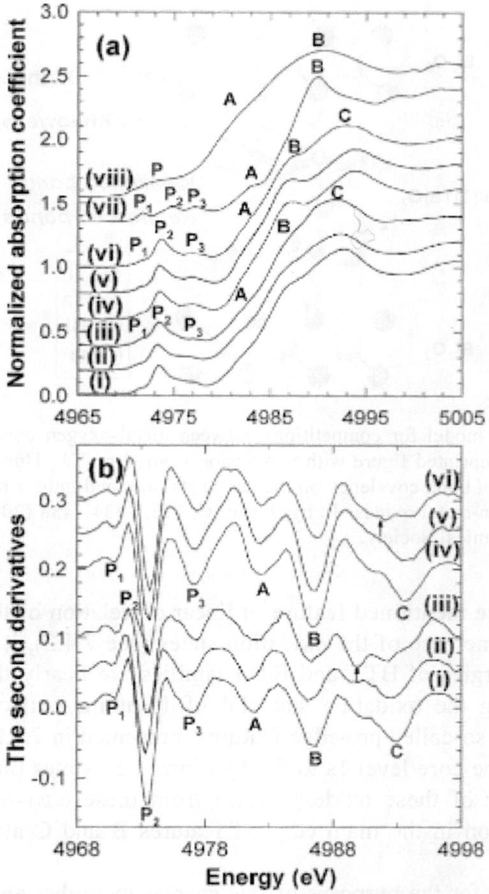

Fig. 7.34 Ti K-edge XANES (a) spline and (b) second derivative spectra for (i) BTT, (ii) Ar-annealed BTT, (iii) lithiated BTT, (iv) BTO, (v) Ar-annealed BTO, and (vi) lithiated BTO, in comparison with those for the reference (vii) anatase TiO_2 and (viii) Ti_2O_3. See text and reference [117] for clarifying the meaning of reduced derivative of BTT and BTO. Arrows put an emphasis on spectral variations after the chemical reduction. Reprinted figure with permission from Kim TW, Hur SG, Han AR, Hwang SJ, Choy JH, Effect of bond covalency on the lattice stability and fatigue behaviour of ferroelectric bismuth transition-metal oxides. J Phys Chem C, 112, 3434-3438 (2008). Copyright (2008) by the American Chemical Society.

In the present text, attention will paid to the Ti K- and Ta L_{III}- XANES results. Fig. 7.34, extracted from [119], shows the spline function and the second derivative of the Ti K-edge XANES spectra for the BTT and BTO families, together with those of the references anatase TiO_2 and Ti_2O_3. It is worth noting the usefulness of using the spectrum-second derivative, because it emphasizes the positions of features that allow discriminating spectra. In Fig. 7.35, the crystal structure of BTT is presented. Bonds and model competitions, as reported in [119], are shown.

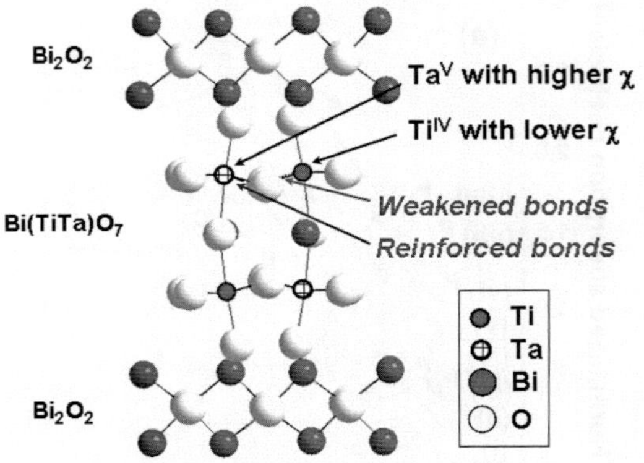

Fig. 7.35 Schematic model for competitions between metal-oxygen bonds of TiO_6 (or TaO_6) octahedra in BTT. Reprinted figure with permission from Kim TW, Hur SG, Han AR, Hwang SJ, Choy JH, Effect of bond covalency on the lattice stability and fatigue behaviour of ferroelectric bismuth transition-metal oxides. J Phys Chem C, 112, 3434-3438 (2008). Copyright (2008) by the American Chemical Society.

Using the above mentioned feature of linear correlation of the absorption edge energy, with the increase of the oxidation states (see 7.4.2), it is easy to observe that the edge energies of BTT and BTO families are nearly the same as that of anatase, indicating the oxidation state IV of Ti atoms. All compounds, except Ti_2O_3, present the so-called pre-edge features (presented in 7.4.3) explained by the transitions from the core level 1s to the hybridized 3d states (and related to ferroelectric behaviour of these oxides). Apart from these *easy-to-observe* features, detailed observation in the main edge of features B and C allows more conclusions.

It is instructive for the purposes of this chapter to further analyze in detail Fig. 7.34. The feature identified as B corresponds to the transition ($1s \rightarrow 4p_z$), while feature C corresponds to the transition ($1s \rightarrow 4p_{x,y}$). Kim *et al.* [119] have explained that the energy position of peak B is lower than that of C because the Z axis is elongated in the Aurivillius structures. In Fig. 7.34, it is observed that the peaks B and C are weaker and broader for the BTT family than the corresponding ones in the BTO family. This attribute is a result of distortions in the BTT octahedral structure, where Ta- and Ti-cations coexist. Distortions create dispersions in the (Ti-O) bond distances and energies of the final Ti 4p orbits. Furthermore, the peak C in the BTT family has lower energy than that of the BTO family. Kim *et al.* [119] have justified this feature on the basis of a competing bond model [120]. In Fig. 7.35, the Ta^V cation shows itself more electronegative than the adjacent Ti^{IV}. As a result, the Ta^V cation attracts towards itself the electron density of the adjacent (Ti-O) bonds, leading to the increased covalency of (Ta-O) bonds and the weakening of the (Ti-O) bonds. Moreover, the broadening and splitting that induces the process of lithiation in the peak C is better observed by the graphics of the second derivative from the

XANES spectra (see Fig 7.34). This is interpreted by Kim *et al.* [119] as evidence of the low stability of TiO_6 octahedra, which may be responsible for the phenomenon of fatigue in the Aurivillius-structured titanates.

Ta L_{III}-edge XANES spectra correspond to BTT and SBT families, and complement the conclusions extracted from Ti K-edge XANES spectra. Several annotations are provided in the article of Kim *et al.* [119] that will not be reproduced in this text. Nevertheless, it is worth to note that in contrast to the Ti K-edge spectra, Ta L_{III}-edge XANES from the two chemically reduced derivative does not show neither broadening nor splitting of peaks. This highlights that the (Ta-O) bonds tested by this probe are more covalent and stable than (Ti-O) bonds.

The analysis of Bi L_{III}-edge XANES spectra in all three families have allowed drawing conclusions about the strong interaction between (Ti-O) and (Ta-O) collinear σ-bonds, and the weak interaction between the perpendicularly aligned (Bi-O) π- and (Ta-O) σ-bonds. All these results explain the better performance for fatigue of SBT family.

This article demonstrate the usefulness of studying different absorbing atoms XANES spectra with detailed analysis of features, leading to interesting conclusions on ferroelectric materials behaviour.

7.9.5 Concluding Remarks: Comparing Information from XAFS and X-Ray Diffraction and Scattering

X-ray absorption fine structure spectroscopy and diffraction/scattering are complementary techniques. They provide a thorough structural characterization, ranging from short- to long-range scales.

The information contained in diffraction peaks gives the long-range average structure. It provides the general structure diagram, precise lattice parameters, average atomic positions and (in synchrotron applications), and is highly sensitive to symmetry breaking. The way random atomic position disorders affect peaks intensities is characterized by the Debye-Waller factor (DWF), a magnitude related to deviations from the ideal crystallographic positions and occupancies.

Scattered intensity, if not in the Bragg peaks, converts into diffuse scattering. Measuring and processing the (structured) diffuse scattering lead to determination of the *pair distribution function* (PDF), a magnitude that provides details of the short-range order not accessible to diffraction methods. Scattering-measured PDF is centred on an "average" atom, representative of the investigated phase.

XAFS investigations in ferroelectrics usually start from a diffraction-obtained structural model. A general characterization of the considered phase is available and XAFS investigation focuses on the determination of short-range structural characteristics invisible for the averaging eyes of diffraction techniques.

EXAFS PDF is similar but not identical to scattering PDF. A XAFS investigation focuses on absorption by a selected element in the sample and consequently the final PDF is centred in an atom of the considered element. This is a clear advantage of XAFS PDF relative to scattering PDF. XANES gives information about absorbing atom electronic transitions; this is another advantage over XRD-scattering.

EXAFS DWF characterizes disorder in *atomic distances*. It is different from diffraction DWF, characterizing disorder in *crystal positions*.

As a closing example, consider a ferroelectric perovskite powder. The small B-position cation is slightly off-centred, so the unit cell is not mathematically cubic. Unit-cells electric dipoles are randomly oriented. How would this structure be characterized by diffraction/scattering and XAFS investigations?

- In a conventional diffractometer experiment, a first important result would be the determination of perovskite crystal structure. The subtle cubic→tetragonal (or other system) symmetry breakdown may be missed. In a synchrotron diffraction experiment, peak splitting (at least anomalous peak broadening) should be observed. Diffraction peaks intensities would lead to a centred non-polar average unit cell. B-site atom DWF would be suspiciously high. Scattering PDF would possibly show central atom off-centring.
- A XANES pre-edge detail examination, with application of the characterization introduced by Vedrinskii *et al.* [35], would allow the quantitative determination of the off-centre displacement of the B-cation, in case of being Ti, and probably in case of V, Cr, Mn, Fe and Mo. Coordination chemistry would be verified by analysis of the pre-edge features. The displacement of the main edge would confirm the oxidation state of the target element.
- The EXAFS short-range PDF would demonstrate that B-cations are locally non-centred. It would be consistent with the perovskite model. A temperature scan of EXAFS spectra would allow researchers to separate dynamic (thermal) from static (structural) interatomic distances disorders. Careful application of theoretical fitting would allow suggesting models of behaviour of ferroelectrics. In case of synchrotron measurement, XRD results would be compatible with the EXAFS B-cation investigation.

Acknowledgments

Portions of this research were carried out at the Stanford Synchrotron Radiation Laboratory, a national user facility operated by Stanford University on behalf of the US Department of Energy, Office of Basic Energy Sciences. The SSRL Structural Molecular Biology Program is supported by the Department of Energy, Office of Biological and Environmental Research, and by the National Institutes of Health, National Centre for Research Resources, Biomedical Technology Program. Funds from Consejo Nacional de Ciencia y Tecnología, CONACYT (Projects 25380, 26040 and 46515) of Mexico is gratefully acknowledged. The author appreciates the support of the Department of Inorganic Chemistry of Universidad

Complutense de Madrid that provided access to *Consortium Madroño* for library cooperation and made possible fruitful discussions within the group led by Miguel Angel Alario Franco, during a scientific visit in 2008. The author wishes to thank Farrel Lytle, who kindly provided key figures; A. Metha, S. Webb and D. Singer, for cooperation during measurements at SSRL, as well as Angela M. Beesley, Maria E. Fuentes Montero and Luis E. Fuentes Cobas for reading and commenting the manuscript.

References

1. Kronig RL (1931) Zur Theorie der Feinstruktur in den Röntgenabsorptionsspektren. Zeitschrift für Physik A Hadrons and Nuclei 70:317-323
2. Kronig RL (1932) Zur Theorie der Feinstruktur in den Röntgenabsorptionsspektren. II. Zeitschrift für Physik A Hadrons and Nuclei 75:191-210
3. Lytle FW, Sayers DE, Stern EA (1975) Extended x-ray-absorption fine-structure technique. II. Experimental practice and selected results. Phys Rev B 11:4825
4. Sayers DE, Stern EA, Lytle FW (1971) New Technique for Investigating Noncrystalline Structures: Fourier Analysis of the Extended X-Ray Absorption Fine Structure. Phys Rev Lett 27:1204
5. Stern EA (1974) Theory of the extended x-ray-absorption fine structure. Phys Rev B 10:3027
6. Stern EA, Sayers DE, Lytle FW (1975) Extended x-ray-absorption fine-structure technique. III. Determination of physical parameters. Phys Rev B 11:4836
7. Kincaid BM, Eisenberger P (1975) Synchrotron Radiation Studies of the K-Edge Photoabsorption Spectra of Kr, Br_2, and $GeCl_4$: A Comparison of Theory and Experiment. Phys Rev Lett 34:1361
8. Rigaku (2006-2007) R-XAS X-ray absorption spectrometer. http://www.rigaku.com/xas/rxas.html.
9. Stern EA (2004) Character of Order-Disorder and Displacive Components in Barium Titanate. Phys Rev Lett 93:037601
10. Burcham WE (1973) Nuclear Physics, an introduction. Longman, London
11. Teo BK (1986) EXAFS: Basics Principles and Data Analysis. Springer-Verlag, Berlin Heidelberg New York Tokyo
12. Hubbell JH, Seltzer SM (1996) Tables of X-Ray Mass Attenuation Coefficients and Mass Energy-Absorption Coefficients. U.S. Secretary of Commerce on behalf of the United States of America
13. Donovan JJ (2000) Table of Absorption Edge Energies. Electron Probe Microanalysis Laboratory, University of California at Berkeley, Berkeley, CA
14. Welter E, Zajac DA, Herrmann M (2008) DESY-Hamburger Synchrotronstrahlungslabor HASYLAB, Beamline and Source parameters E4 (EXAFS II). http://hasylab.desy.de/facilities/doris_iii/beamlines/e4_exafs_ii/beamline_layout/index_eng.html.Feb 8
15. Jaklevic J, Kirby JA, Klein MP, Robertson AS, Brown GS, Eisenberger P (1977) Fluorescence detection of exafs: Sensitivity enhancement for dilute species and thin films. Solid State Commun 23:679-682

16. Koningsberger DC, Prins Re (1988) X-ray absorption: principles, applications, techniques of EXAFS, SEXAFS, and XANES. John Wiley and Sons, New York, NY, United States
17. Eisenberger P, Kincaid BM (1978) EXAFS: New Horizons in Structure Determinatiions. Science 200:1441
18. Durham PJ (1988) Theory of XANES. In: Koningsberger DC, Prins R (eds) X-ray absorption: principles, applications, techniques of EXAFS, SEXAFS, and XANES. John Wiley and Sons, New York, NY
19. Rehr JJ, Albers RC (2000) Theoretical approaches to x-ray absorption fine structure. Rev Mod Phys 72:621-654
20. Rehr JJ, Ankudinov AL (2005) Progress in the theory and interpretation of XANES. Coord Chem Rev 249:131-140
21. Bianconi A, Dell'Ariccia M, Durham PJ, Pendry JB (1982) Multiple-scattering resonances and structural effects in the x-ray-absorption near-edge spectra of Fe II and Fe III hexacyanide complexes. Phys Rev B 26:6502-6508
22. Bianconi A (1988) XANES Spectroscopy. In: Koningsberger DC, Prins R (eds) X-ray absorption: principles, applications, techniques of EXAFS, SEXAFS, and XANES. John Wiley and Sons, New York, NY
23. Wong J, Lytle FW, Messmer RP, Maylotte DH (1984) K-edge absorption spectra of selected vanadium compounds. Phys Rev B 30:5596-5610
24. Lytle FW, Greegor RB, Panson AJ (1988) Discussion of x-ray-absorption near-edge structure: Application to Cu in the high-T_c superconductors $La_{18}Sr_{0.2}CuO_4$ and $YBa_2Cu_3O_7$. Phys Rev B 37:1550
25. Soderholm L, Goodman GL (1989) USE OF X-RAY ABSORPTION-SPECTROSCOPY IN ELECTRONIC-STRUCTURE STUDIES OF HIGH-TC SUPERCONDUCTING OXIDES. Journal of the Optical Society of America B-Optical Physics 6:483-489
26. Ressler T, Wienold J, Jentoft RE, Neisius T (2002) Bulk Structural Investigation of the Reduction of MoO_3 with Propene and the Oxidation of MoO_2 with Oxygen. J Catal 210:67-83
27. Fernández-García M (2002) XANES analysis of catalytic systems under reaction conditions. Catalysis Reviews 44:59-121
28. Liu RS, Jang LY, Hung HH, Tallon JL (2001) Determination of Ru valence from x-ray absorption near-edge structure in $RuSr_2GdCu_2O_8$-type superconductors. Phys Rev B 63:212507
29. Bos JWG, Attfield JP, Chan TS, Liu RS, Jang LY (2005) Orbital ordering and valence states in $(La_{1+x}Ca_{1-x})CoRuO_s$ double perovskites. Phys Rev B 72
30. Stoupin S, Chattopadhyay S, Bolin T, Segre CU (2007) High concentration manganese doping of ferroelectric PbTiO3. Solid State Commun 144:46-49
31. Kutzler FW, Natoli CR, Misemer DK, Doniach S, Hodgson KO (1980) USE OF ONE-ELECTRON THEORY FOR THE INTERPRETATION OF NEAR EDGE STRUCTURE IN K-SHELL X-RAY ABSORPTION-SPECTRA OF TRANSITION-METAL COMPLEXES. J Chem Phys 73:3274-3287
32. Bianconi A, Fritsch E, Calas G, Petiau J (1985) X-ray-absorption near-edge structure of 3d transition elements in tetrahedral coordination: The effect of bond-length variation. Phys Rev B 32:4292-4295
33. Farges F, Brown GE, Rehr JJ (1996) Coordination chemistry of Ti(IV) in silicate glasses and melts: I. XAFS study of titanium coordination in oxide model compounds. Geochim Cosmochim Acta 60:3023-3038
34. Ravel B, Stern EA (1995) Local disorder and near edge structure in titanate perovskites. Physica B 208-209:316-318

35. Vedrinskii RV, Kraizman VL, Novakovich AA, Demekhin PV, Urazhdin SV (1998) Pre-edge fine structure of the 3d atom K x-ray absorption spectra and quantitative atomic structure determinations for ferroelectric perovskite structure crystals. J Phys-Condes Matter 10:9561-9580
36. Yoon S, Baik S, Kim MG, Shin N (2006) Formation Mechanisms of Tetragonal Barium Titanate Nanoparticles in Alkoxide-Hydroxide Sol-Precipitation Synthesis. J Am Ceram Soc 89:1816-1821
37. Ravel B (2005) A practical introduction to multiple scattering theory. J Alloys Compd 401 118–126
38. Ankudinov AL, Ravel B, Rehr JJ, Conradson SD (1998) Real-space multiple-scattering calculation and interpretation of x-ray-absorption near-edge structure. Phys Rev B 58:7565-7576
39. Binsted N, Hasnain SS (1996) State-of-the-art analysis of whole X-ray absorption spectra. Journal of Synchrotron Radiation 3:185-196
40. Filipponi A, DiCicco A (1995) X-ray-absorption spectroscopy and n-body distribution functions in condensed matter .2. Data analysis and applications. Phys Rev B 52:15135-15149
41. Filipponi A, DiCicco A, Natoli CR (1995) X-ray-absorption spectroscopy and n-body distribution functions in condensed matter .1. Theory. Phys Rev B 52:15122-15134
42. Newville M (2001) IFEFFIT: interactive XAFS analysis and FEFF fitting. Journal of Synchrotron Radiation 8:322-324
43. George GN, Pickering IJ (1995) EXAFSPAK, a Suite of Computer Programs for the Analysis of X-ray Absorption Spectra. . Stanford Synchrotron Radiation Laboratory, Stanford, CA
44. Binsted N (1988) PAXAS. An EXAFS Background Subtraction Program. University of Southampton, England
45. Ressler T (2008) WinXAS: A Program for X-ray Absorption Spectroscopy Data Analysis under MS Windows. Hamburg, Germany
46. Stern EA (1988) Theory of EXAFS. In: Koningsberger DC, Prins R (eds) X-ray absorption: principles, applications, techniques of EXAFS, SEXAFS, and XANES. John Wiley and Sons, New York, NY
47. Newville M (2004) Fundamentals of XAFS. Consortium for Advanced Radiation Sources, University of Chicago, Chicago, IL, Chicago, IL, p 41
48. Lee PA, Pendry JB (1975) Theory of the extended x-ray absorption fine structure. Phys Rev B 11:2795
49. Rehr JJ, Mustre De Leon J, Zabinsky SI, Albers RC (1991) Theoretical X-Ray Absorption Fine-Structure Standards. J Am Chem Soc 113:5135-5140
50. Zabinsky SI, Rehr JJ, Ankudinov A, Albers RC, Eller MJ (1995) Multiple-scattering calculations of x-ray-absorption spectra. Phys Rev B 52:2995
51. Mustre de Leon J, Rehr JJ, Zabinsky SI, Albers RC (1991) Ab initio curved-wave x-ray-absorption fine structure. Phys Rev B 44:4146-4156
52. Bunker G (1983) Application of the ratio method of EXAFS analysis to disordered systems. Nuclear Instruments and Methods in Physics Research 207:437-444
53. Lee PA, Citrin PH, Eisenberger P, Kincaid BM (1981) Extended x-ray absorption fine structure; its strengths and limitations as a structural tool. Rev Mod Phys 53:769
54. Newville M (2004) Anatomy of an XAFS Measurement. XAFS.ORG
55. Knoll GF (2000) Radiation Detection and Measurement. Wiley, New York, N.Y.
56. Segre C (2004) X-Ray Cross section data. Center for Synchrotron Radiation Research and Instrumentation, Illinois Institute of Technology
57. Ravel B, Newville M (2005) ATHENA, ARTEMIS, HEPHAESTUS: data analysis for X-ray absorption spectroscopy using IFEFFIT. Journal of Synchrotron Radiation 12:537-541

58. Bouldin CE, Forman RA, Bell MI (1987) Conversion-electron extended x-ray-absorption fine-structure measurements of ion-damaged GaAs. Phys Rev B 35:1429
59. Lytle FW (1998) EXAFS Company-Product List-Basic, 3-grid ion chamber detector and accessories. http://www.exafsco.com/products.html.February 18
60. Stern EA, Heald SM (1979) X-ray filter assembly for fluorescence measurements of x-ray absorption fine structure. Rev Sci Instrum 50:1579-1582
61. Bunker G (1994) Optimizing X-ray Filters. Grant Bunker's XAFS Tutorial documents
62. Canberra I (2003) Germanium array detectors. http://www.canberra.com/products/491.asp.July 24 2008
63. ORTEC (2007) IGLET Germanium (HPGe) Detector for Low-Energy X Rays. http://www.ortec-online.com/detectors/photon/b2_4_2.htm.July 24 2008
64. Sayers DE, Bunker BA (1988) Data Analysis. In: Koningsberger DC, Prins R (eds) X-ray absorption: principles, applications, techniques of EXAFS, SEXAFS, and XANES. John Wiley and Sons, New York, NY
65. Newville M (2007) XAFS, EXAFS, and XANES. http://xafs.org/.
66. Webb S (2007) Sam's Interface for XAS Package SixPACK. Menlo Park, CA
67. Montero-Cabrera ME, García-Guaderrama M, Mehta A, Webb S, Fuentes-Montero L, Duarte-Moller JA, Fuentes-Cobas L (2008) EXAFS determination of cation local order in layered perovskites. Revista Mexicana de Física S54:42-45
68. Fuentes-Montero L, Montero-Cabrera ME, Calzada L, Pérez de la Rosa M, Raymond O, Font R, García M, Torres M, Mehta A, Fuentes L (2008) Synchrotron Techniques Applied to Ferroelectrics: Some Representative Cases. Integrated Ferroelectrics 101:101-113
69. Zhang K, Stern EA, Ellis F, Sanders-Loehr J, Shiemke AK (1988) The active site of hemerythrin as determined by x-ray absorption fine structure. Biochemistry 27:7470-7479
70. Ressler T, Wong J, Roos J, Smith IL (2000) Quantitative Speciation of Mn-Bearing Particulates Emitted from Autos Burning (Methylcyclopentadienyl)manganese Tricarbonyl-Added Gasolines Using XANES Spectroscopy. Environmental Science & Technology 34:950-958
71. Frenkel AI, Hills CW, Nuzzo RG (2001) A View from the Inside: Complexity in the Atomic Scale Ordering of Supported Metal Nanoparticles. J Phys Chem B 105:12689-12703
72. Frenkel AI, Kleifeld O, Wasserman SR, Sagi I (2002) Phase speciation by extended x-ray absorption fine structure spectroscopy. J Chem Phys 116:9449-9456
73. Wasserman SR, Allen PG, Shuh DK, Bucher JJ, M. EN (1999) EXAFS and principal component analysis: a new shell game. Journal of Synchrotron Radiation 6:284-286
74. Frenkel A (2002) First Shell EXAFS Analysis. EXAFS Data Collection and Analysis Workshop, NSLS, September 23-25, 2002. EXAFS Data Collection and Analysis Workshop, NSLS, September 23-25, 2002
75. Benfatto M, D' Angelo P, Della Longa S, Pavel NV (2002) Evidence of distorted fivefold coordination of the Cu^{2+} aqua ion from an x-ray-absorption spectroscopy quantitative analysis. Phys Rev B 65:174205
76. Ravel B (2001) ATOMS: crystallography for the X-ray absorption spectroscopist. Journal of Synchrotron Radiation 8:314-316
77. Newville M (2001) EXAFS analysis using FEFF and FEFFIT. Journal of Synchrotron Radiation 8:96-100
78. Lytle FW, Sayers DE, Stern EA (1989) Report of the International Workshop on Standards and Criteria in X-ray Absorption Spectroscopy. Physica B 158:701-722
79. Gonzalo JA, Jiménez B (2005) Ferroelectricity, The Fundamental Collection. 3-527-40486-4, Berlin

80. Zhong W, Vanderbilt D, Rabe KM (1995) First-Principles Theory of Ferroelectric Phase-Transitions for Perovskites - the Case of $BaTiO_3$. Phys Rev B 52:6301-6312
81. Zalar B, Laguta VV, Blinc R (2003) NMR Evidence for the Coexistence of Order-Disorder and Displacive Components in Barium Titanate. Phys Rev Lett 90:037601
82. Sicron N, Ravel B, Yacoby Y, Stern EA, Dogan F, Rehr JJ (1994) Nature of the Ferroelectric Phase-Transition in PbTiO3. Phys Rev B 50:13168-13180
83. Hanske-Petitpierre O, Yacoby Y, Mustre Deleon J, Stern EA, Rehr JJ (1991) Off-Center Displacement of the Nb Ions below and above the Ferroelectric Phase-Transition of $KTa_{0.91}Nb_{0.09}O_3$. Phys Rev B 44:6700-6707
84. Ravel B (2000) EXAFS Analysis with FEFF and FEFFIT.
85. Ravel B, Sicron N, Yacoby Y, Stern EA, Dogan F, Rehr JJ (1995) Order-disorder behavior in the phase transition of $PbTiO_3$. Ferroelectrics 164:265-277
86. Rehr JJ, Albers RC, Zabinsky SI (1992) High-order multiple-scattering calculations of x-ray-absorption fine structure. Phys Rev Lett 69:3397-3400
87. Ravel B, Stern EA, Vedrinskii RI, Kraizman V (1998) Local structure and the phase transitions of $BaTiO_3$. Ferroelectrics 206:407-430
88. Ravel B (1997) Ferroelectric Phase Transitions in Oxide Perovskites Studied by XAFS. Physics, vol Ph D. University of Washington, Seatle, p 208
89. Ravel B, Bouldin CE, Renevier H, Hodeau JL, Berar JF (1999) X-ray-absorption edge separation using diffraction anomalous fine structure. Phys Rev B 60:778-785
90. Renevier H, Grenier S, Arnaud S, Berar JF, Caillot B, Hodeau JL, Letoublon A, Proietti MG, Ravel B (2003) Diffraction anomalous fine-structure spectroscopy at beamline BM2 at the European Synchrotron Radiation Facility. Journal of Synchrotron Radiation 10:435-444
91. Holgado J.P., Caballero A. , Espinós J.P. , Morales J. , Jiménez V.M. , Justo A. , A.R. G-E (2000) Characterisation by X-ray absorption spectroscopy of oxide thin films prepared by ion beam-induced CVD. Thin Solid Films 377-378:460-466
92. Neves PP, Doriguetto AC, Mastelaro VR, Lopes LP, Mascarenhas YP, Michalowicz A, Eiras JA (2004) XAS and XRD Structural Characterization of Lanthanum-Modified PbTiO3 Ceramic Materials. J Phys Chem B 108:14840-14849
93. Cross LE (1987) Relaxor ferroelectrics. Ferroelectrics 76:241-267
94. Laulhe C, Hippert F, Kreisel J, Maglione M, Simon A, Hazemann JL, Nassif V (2006) EXAFS study of lead-free relaxor ferroelectric $BaTi_{1-x}Zr_xO_3$ at the Zr K edge. Phys Rev B 74:014106
95. Chen IW, Li P, Wang Y (1996) Structural origin of relaxor perovskites. J Phys Chem Solids 57:1525-1536
96. Chen IW (2000) Structural origin of relaxor ferroelectrics—revisited. J Phys Chem Solids 61 197–208
97. Shuvaeva VA, Azuma Y, Raevski IP, Yagi K, Sakaue K, Terauchi H (2004) The local structure of $PbIn_{1/2}Nb_{1/2}O_3$. Ferroelectrics 299:103-108
98. Shuvaeva VA, Pirog I, Azuma Y, Yagi K, Sakaue K, Terauchi H, Raevskii IP, Zhuchkov K, Antipin MY (2003) The local structure of mixed-ion perovskites. J Phys-Condes Matter 15:2413-2421
99. Frenkel AI, Pease DM, Giniewicz J, Stern EA, Brewe DL, Daniel M, Budnick J (2004) Concentration-dependent short-range order in the relaxor ferroelectric $(1-x)Pb(Sc,Ta)O_{3-x}PbTiO_3$. Phys Rev B 70:12
100. Shuvaeva VA, Zekria D, Glazer AM, Jiang Q, Weber SM, Bhattacharya P, Thomas PA (2005) Local structure of the lead-free relaxor ferroelectric $(K_xNa_{1-x})_{0.5}Bi_{0.5}TiO_3$. Physical Review B (Condensed Matter and Materials Physics) 71:174114-174118
101. Shanthakumar P, Balasubramanian M, Pease DM, Frenkel AI, Potrepka DM, Kraizman V, Budnick JI, Hines WA (2006) X-ray study of the ferroelectric $Ba_{0.6}Sr_{0.4}.[(YTa)_{0.03}Ti_{0.94}]O_3$. Phys Rev B 74:174103

102. Petit PE, Guyot F, Farges F (1997) Thermal evolution of $BaZrO_3$ and $SrZrO_3$ perovskites from 4 K to 773 K: An EXAFS study at Zr-K edge. Journal de physique IV 7:2-2
103. Potrepka DM, Tidrow SC, Tauber A (2002) Improved temperature stability of microwave properties in tunable devices using substituted $Ba_{1-x}Sr_xTiO_3$. Integrated Ferroelectrics 42:97-121
104. Crowne FJ, Tidrow SC, Potrepka DM, Tauber A (2002) Microfields induced by random compensated charge pairs in ferroelectric materials. In: Tidrow SC, Horwitz JS, Xi X, Levy J (eds) Materials Issues for Tunable RF and Microwave Devices III. Symposium. Mater. Res. Soc, San Francisco, CA, USA, pp 131-142
105. Westphal V, Kleemann W, Glinchuk MD (1992) Diffuse phase transitions and random-field-induced domain states of the `relaxor' ferroelectric $PbMg_{1/3}Nb_{2/3}O_3$. Phys Rev Lett 68:847-850
106. PiezoTechnologies (2003-2008) PiezoTechnologies, K-12, Ultra High-Temperature Ceramic Provides Piezoelectric Stability in Extreme Environments. http://www.piezotechnologies.com/k12.htm.
107. Paz de Araujo CA, Cuchiaro JD, McMillan LD, Scott MC, Scott JF (1995) Fatigue-free ferroelectric capacitors with platinum electrodes. Nature 374:627-629
108. Park BH, Kang BS, Bu SD, Noh TW, Lee J, Jo W (1999) Lanthanum-substituted bismuth titanate for use in non-volatile memories. Nature 401:682-684
109. Castro A, Millán P, Pardo L, Jiménez B (1999) Synthesis and sintering improvement of Aurivillius type structure ferroelectric ceramics by mechanochemical activation. J Mater Chem 9:1313-1317
110. Durán P, Capel F, Moure C, Villegas M, Fernández JF, Tartaj J, Caballero AC (2001) Processing and dielectric properties of the mixed-layer bismuth titanate niobate $Bi_7Ti_4NbO_{21}$ by the metal-organic precursor synthesis method. J Eur Ceram Soc 21:1-8
111. Ricote J, Pardo L, Moure A, Castro A, Millán P, Chateigner D (2001) Microcharacterisation of grain-oriented ceramics based on Bi_3TiNbO_9 obtained from mechanochemically activated precursors. J Eur Ceram Soc 21:1403-1407
112. Fernandez JF, Caballero AC, Villegas M, Frutos Jd, Lascano L (2002) Relaxor behavior of $Pb_xBi_4Ti_{3+x}O_{12+3x}$ (x = 2,3) Aurivillius ceramics. Appl Phys Lett 81:4811-4813
113. García-Guaderrama M, Fuentes L, Montero-Cabrera ME, Márquez-Lucero A, Villafuerte-Castrejón ME (2005) MOLTEN SALT SYNTHESIS AND CRYSTAL STRUCTURE OF $Bi_5Ti_3FeO_{15}$. Integrated Ferroelectrics 71:233-239
114. Fuentes L, Fernandez J, Fuentes M, Lascano L, Villafuerte M, Montero M, Olivera R, Mehta A, Jardiel T (2006) Synchrotron Radiation Study of Structural Tendencies in Aurivillius Ceramics. Ferroelectrics 339:209-218
115. Villafuerte-Castrejón ME, Camacho-Alanís F, González F, Ibarra-Palos A, González G, Fuentes L, Bucio L (2007) Luminescence and structural study of $Bi_{4-x}Eu_xTi_3O_{12}$ solid solution. J Eur Ceram Soc 27:545-549
116. Rentschler T, Karus M, Wellm A, Reller A (1996) Synthesis and characterization of the Aurivillius phases $Bi_{2-x}Pb_xSr_{1-x}Nd_xNb_2O_9$. Solid State Ionics 90:49-55
117. Choy JH, Kim JY, Chung I (2001) Neutron Diffraction and X-ray Absorption Spectroscopic Analyses for Lithiated Aurivillius-Type Layered Perovskite Oxide, Li2Bi4Ti3O12. J Phys Chem B 105:7908-7912
118. Hur SG, Park DH, Kim TW, Hwang S-J (2004) Evolution of the chemical bonding nature of ferroelectric bismuth titanate upon cation substitution. Appl Phys Lett 85:4130-4132
119. Kim TW, Hur SG, Han AR, Hwang SJ, Choy JH (2008) Effect of bond covalency on the lattice stability and fatigue behavior of ferroelectric bismuth transition-metal oxides. Journal of Physical Chemistry C 112:3434-3438
120. Pauling L (1932) The Nature of the Chemical Bond. IV. The Energy of Single Bonds and the Relative Electronegativity of Atoms. J Am Chem Soc 54:3570-3582

Chapter 8
Quantitative Texture Analysis of Polycrystalline Ferroelectrics

D. Chateigner[1], J. Ricote[2]

8.1 Introduction

A large number of physical properties in most crystals are anisotropic. This is even more important in polar materials, like ferroelectrics, where the polarization determines their behaviour. In polycrystals, researchers have developed techniques to grow crystals along preferential orientations and take advantage of the highest values of the anisotropic properties. The preparation of ferroelectric materials with preferential crystallographic orientations, or textures, is useful in obtaining ferroelectric materials with improved properties for a variety of technological applications, like Non-Volatile Ferroelectric Random Access Memories (FeRAMs) [1, 2], where we use the polarization vector for the 0 and 1 bits, or MicroElectroMechanical Systems (MEMS) [3], where the highest piezoelectric coefficients are associated to specific crystallographic directions. Strong textures along polar axis directions perpendicular to the film surface have been sought in ferroelectric thin films as it improves the final response. But sometimes other orientations are also interesting. For example, in lead titanate-based compositions, while orientations along the <001> polar axis are preferred for pyroelectric applications, <111>-oriented films are more desirable for memory applications, because it provides high remnant polarization and abrupt switching behaviour [4]. Preparation routes have been optimized to obtain highly oriented ferroelectric films [5, 6, 7]. As a consequence of all this, interest on texturing of ferroelectric polycrystalline mate-

[1] Laboratoire de CRIstallographie et Science de MATériaux CRISMAT-ENSICAEN, Institut Universitaire de Technologie (IUT), Université de Caen Basse Normandie, 6 Boulevard du Maréchal Juin, F-14050 Caen, France. Phone: +33(0)231452611
FAX: +33(0)231951600
daniel.chateigner@ensicaen.fr
[2] Instituto de Ciencia de Materiales de Madrid, CSIC, C/ Sor Juana Inés de la Cruz 3, Cantoblanco, E-28049 Madrid, Spain
jricote@icmm.csic.es

rials, the quantitative characterization of texture becomes crucial for polycrystalline ferroelectrics. Different techniques are used, and among them, X-ray diffraction has routinely been used as a non-destructive characterization of texture, strain state, particle size, crystal structure and microstructure in general. However, when applied to anisotropic polycrystalline samples, the conventional diffraction approaches generally fail. For instance, a usual Bragg-Brentano diffraction diagram may not reveal all diffracted lines of a compound when it is strongly textured. This makes its structural determination impossible, which in turn impedes reliable quantitative texture analysis. Also, the presence of strongly overlapping diffraction peaks often makes this difficult analysis. This is usually the case of thin films, where diffraction peaks coming from the substrate may become a serious problem to study the structural characteristics of the film. This may be in some cases poorly understood beforehand. Advanced methods of analysis were therefore needed.

In this chapter, we review the various methodologies to analyze the texture of polycrystalline ferroelectrics by X-ray diffraction. A broad range of techniques are offered to the non-specialist: from the conventional scans with poor quantification in terms of textures, to the most advanced techniques of analysis that allow the simultaneous determination of several structural, textural and microsctructural parameters, the so-called combined analysis. We show that the quantitative characterization of textures is possible in complex ferroelectric multiphase materials in thin film form with a series of examples at the end of the chapter.

8.2 Conventional Texture Analysis

Most natural and artificial solids (rocks, ceramics, metals, alloys...) are constituted by aggregates of grains of different phases, sizes, shapes, stress states and orientations. Grains may be viewed at different scales, depending on the technique used to examine them. Using optical microscopes, grains at the micrometer scale are visible, delimited by regions with different characteristics regarding the reflection or transmission of light. Since visible light is in the micrometer wavelength range, the resolution is of the same order, and, therefore, no direct information about the crystal planes is accessible. Only when there is a relationship between the reflection index and the shape of the crystal with the crystallographic plane, may optical microscopy be used to study crystallographic textures. These methods were used by geologists and metallurgists to study texturing effects before they had access to more sophisticated techniques. With the development of Scanning and Transmission Electron Microscopes, studies down to few nanometres were available. Diffraction of electrons, with lower wavelength, by the material is carried out by examining Kikuchi patterns (Electron Backscattered Scanning Diffraction, EBSD) or Debye-Scherrer electron diffraction patterns in TEM, so the crystallographic orientation of the individual crystals can be analyzed [8].

Diffraction using X-rays or neutrons, which have wavelengths comparable to the distances of crystallographic planes, is used for routine determination of texture. It should be noted that X-ray and neutron diffraction probe crystallites and not grains. This is an important issue to take into account when studying polycrystalline materials, as it may lead to misinterpretation of the results. A crystallite is the largest domain that satisfies the periodic translation of the crystal unit-cell in the three dimensional space. The incident radiation is then coherently scattered in one crystallite, which is then called "coherent scattering domain". But grains can be made of a lot of crystallites, overall in ferroelectrics where regions corresponding to different ferroelectric domains are present. Although both terms, grain and crystallite, are often used indistinctively, it is not correct in most cases.

Before going into detailed description of the textures of ferroelectric polycrystals, we need to introduce the basics of the conventional texture analysis based on diffraction. Firstly, the limitations of the normal θ-2θ diffraction diagrams and rocking curves used for routine texture determination will be described. Secondly, pole figures will be introduced as the best method to obtain a more complete picture of textures of materials.

8.2.1 Qualitative Determination of Texture from Conventional Diffraction Diagrams

8.2.1.1 Bragg-Brentano Diagrams

Fig. 8.1 shows the conventional diffraction arrangement used for powder diffraction. It is generally known as the Bragg-Brentano configuration [9]. The incident and diffracted beams define the incident plane (scattering plane), in which diffraction is measured with a scattering vector $\mathbf{\Delta k}$. The detector is placed at an angle 2θ from the incident beam, itself at θ from the sample surface. Using this geometry, crystallographic planes {hkl} of different d-spacing are successively brought into diffraction for different θ. Diffraction by sets of planes not parallel to the sample surface {hkl}' is not collected. Therefore, the information obtained may represent only a low percentage of the volume of the material.

Fig. 8.2 shows Bragg-Brentano diffraction diagrams corresponding to α-SiO_2, from a randomly oriented powder (Fig. 8.2a), and from a material with a strong orientation with $\{00\ell\}$ planes parallel to the sample surface (Fig. 8.2b). The latter only exhibits the 003 peak, while it is barely visible in the former. As a result, no information can be obtained about other crystal planes, e.g., {h00}, in the oriented sample. Therefore, the texture may only be characterized qualitatively in this case. Even if the diagram had showed diffraction peaks from other planes, it would be only from planes parallel to the sample plane, not giving information about in-plane preferential orientation of the material.

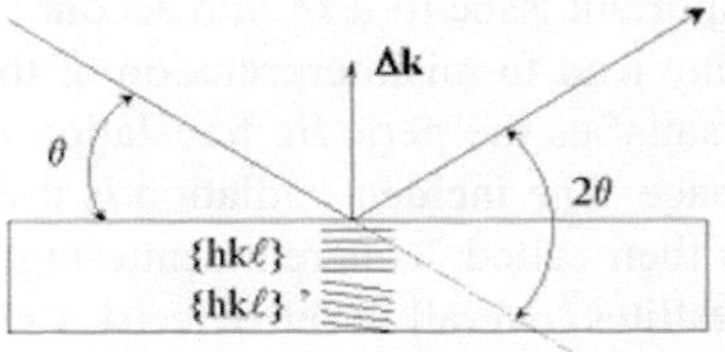

Fig. 8.1 Bragg-Brentano geometry used for diffraction. The plane of the figure is the scattering plane.

Fig. 8.2 Bragg-Brentano diffraction diagrams of α-SiO_2 for **a** a bulk powder without any preferred orientation and **b** an oriented powder showing strong orientation with {00ℓ} planes parallel to the sample surface.

With this geometry, the available range of observed diffraction peaks may be restricted by construction. Normally, ranges of at best 2θ = 165° are measured. In this range, a randomly oriented sample of the diamond structure (Fd3m space group, a = 3.566 Å) using Cr Kα1 radiation (2.2897 Å) will show only two peaks (Fig. 8.3), corresponding to the {111} and {220} planes. If this sample is instead strongly oriented with {100} planes parallel to the surface of the sample, no diffraction peak will be observed in a Bragg-Brentano diffraction diagram.

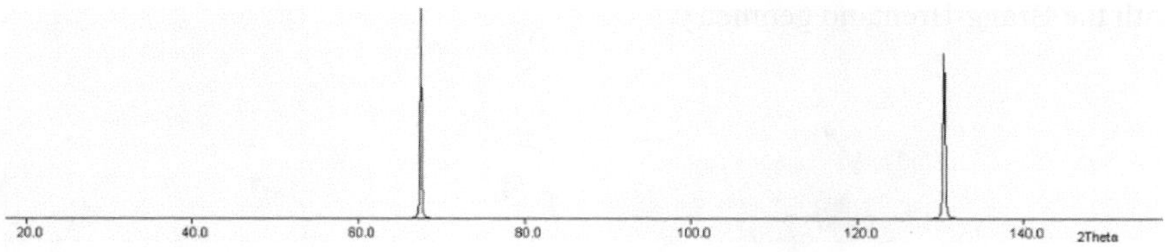

Fig. 8.3 Bragg-Brentano diffraction diagram in the 20-165° 2θ-range, for diamond measured with Cr Kα_1 radiation.

Regarding in-plane orientations in the sample, Fig. 8.4 shows that for a given {hkℓ} plane family, any rotation around the sample normal, **n**, does not change the diffraction diagram. If {hkℓ} are parallel to the surface, the diffracted intensity is kept constant, while for {hkℓ}' planes (Fig. 8.1) no intensity is detected. Both texture components in Fig. 8.4. have the **c** axes of the structure parallel to **n**, but **a** axes are different, rotated around **n**. We may conclude from this that with this configuration it is impossible to check for eventual in-plane alignment of the axes, like the ones involved in epitaxial growth, for example.

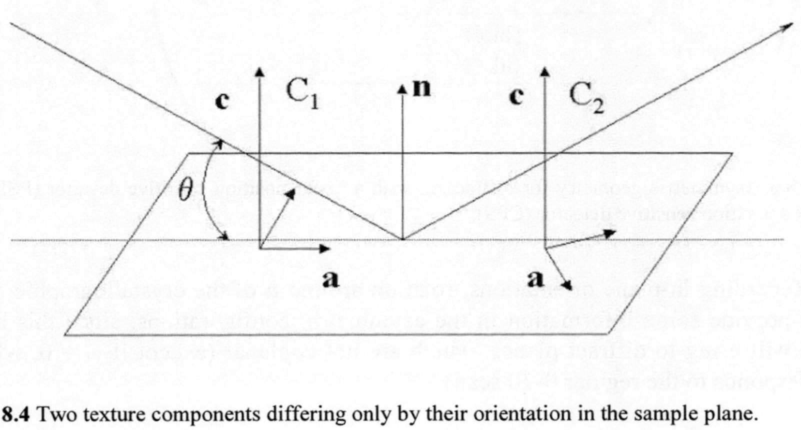

Fig. 8.4 Two texture components differing only by their orientation in the sample plane.

8.2.1.2 Asymmetric Diagrams

The asymmetric configurations for diffraction may be carried out either with a point detector, using an incidence angle ω that is not equal to θ (Fig. 8.5) or with linear and curved position sensitive detectors, PSD [10, 11] (Fig. 8.6). The advantage of PSD is that, while point detectors only probe {hkℓ}' planes that are inclined by $\delta = \omega-\theta$ from the sample plane, they collect information for other sets of planes. In both cases only qualitative characterization of texture is possible, as with the Bragg-Brentano geometry.

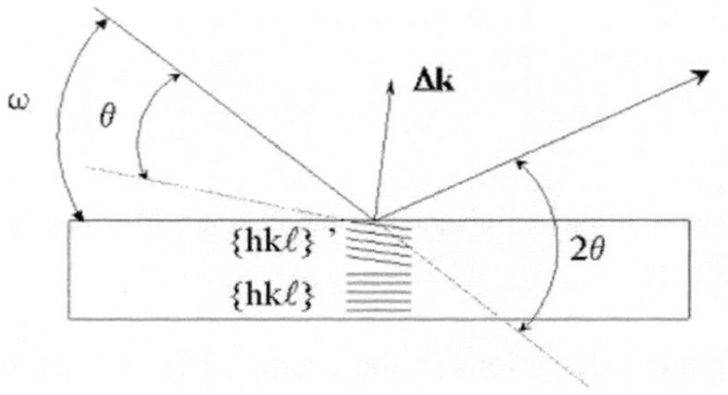

Fig. 8.5 Asymmetric geometry for diffraction with a point detector

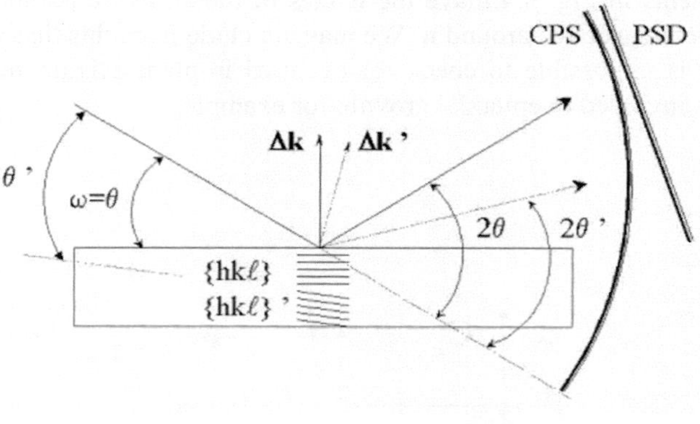

Fig. 8.6 Asymmetric geometry for diffraction with a linear position sensitive detector (PSD) or curved position sensitive detector (CPS).

Regarding in-plane orientations, rotation around **n** of the crystallographic axes will provide some information in the asymmetric configurations, since this rotation will bring to diffract planes, which are not coplanar (except if $\omega = \theta$, which corresponds to the regular θ-2θ scan).

8.2.1.3 ω-Scans: Rocking Curves

The ω–scans, also called rocking curve, are often used to reveal the preferential orientation of specific crystallographic planes with respect to the sample surface. For a fixed θ-2θ position, different sets of planes with the same interplanar distances are brought to diffraction by varying ω (Fig. 8.7). This kind of measurement is used to check the quality of as-grown single crystals, obtaining information about their mosaic spread. This method is also used more to measure the texture of thin structures, like ferroelectric films [12], although the mosaicity is then much larger, and often not suited for such a characterization.

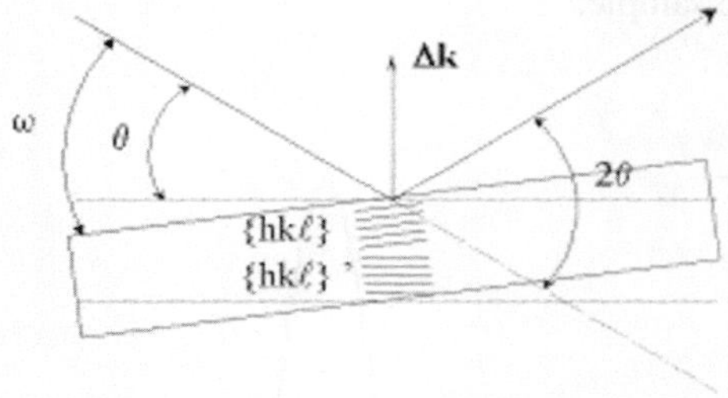

Fig. 8.7 ω-scan (rocking curves) configuration.

A typical rocking curve of a {002} line measured for a single crystal is presented in Fig. 8.8. The FWHM of the curve shown is 0.1° in ω, which means that a certain amount of crystallites are inclined by 0.1° or less with respect to the sample surface. The volume percentage of oriented crystals may also be obtained, and it depends on the curve shape. In this example the curves fits a Gaussian, and approximately 86 % of the intensity is within the FWHM. This means that approximately this is the percentage of oriented crystals in the material.

With this configuration only planes perpendicular to the incident plane can be brought to diffraction. Planes parallel to the scattering plane will never satisfy the Bragg law when rotating around ω (Fig. 8.9).

Rocking curves can produce quantitative information on texture, provided another axis of rotation is added. However, this method suffers from a lack of extent of the available measurements. If we consider a distribution of crystallites well represented by a Gaussian shape with a FWHM of 30°, to bring a plane which makes an angle of ω=-30° with the sample surface into diffraction conditions, one has to rotate the sample by 30°, means ω=θ + 30. In the mean time, if the peak of interest is diffracting at a θ position lower than 30°, the diffracted beam is then fully absorbed in the sample. In this geometry, measurements are limited by the necessary condition ω<± θ. Since Bragg angles for intense peaks (reliable peaks) are located at rather low θ ranges, this limit is far below the necessary ranges of texture measurements. It is of course suited for low angle ranges and, therefore, for single crystals.

Now consider a distribution of crystallite orientations that is not symmetric around **n** (Fig. 8.10). For instance, composed of two orientations with 0.1° FWHMs, one along **n** (with $\mathbf{c}_1$ axes), the other ($\mathbf{c}_2$ axes) at an angle χ from **n** and φ from the intersection of the sample surface and the scattering plane. Measured as positioned in Fig. 8.10, a rocking curve on $\{00\ell\}$ planes will exhibit only one orientation component, C_1, the only one having $\{00\ell\}$ planes perpendicular to the scattering plane (Fig. 8.8). The C_2 component will be revealed on a $\{00\ell\}$ rocking curve if the sample is rotated by φ around **n** before measurement (Fig. 8.11).

We conclude that ω-scans may reveal the texture of a sample, with the absorption limitations discussed above, if several of these scans are measured in different φ orientations of the sample.

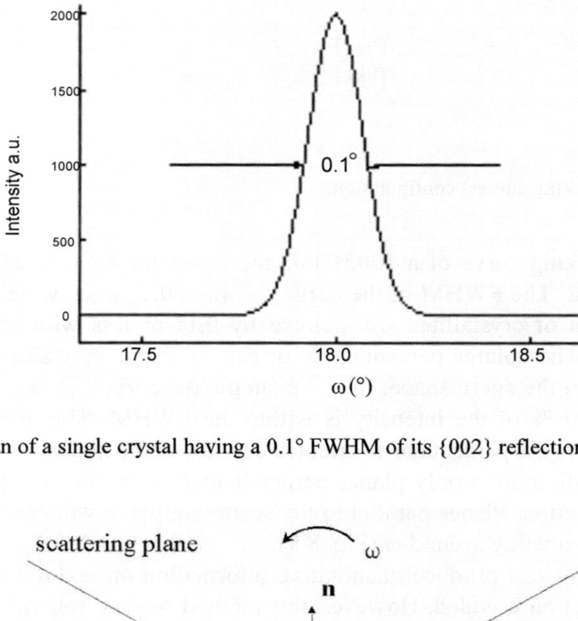

Fig. 8.8 ω-scan of a single crystal having a 0.1° FWHM of its {002} reflection.

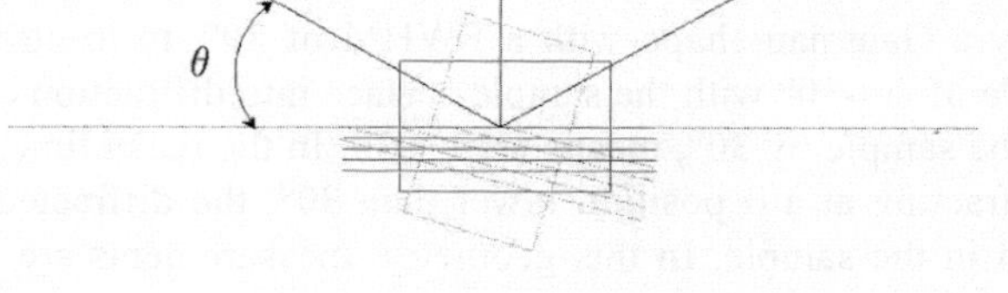

Fig. 8.9 Illustration of crystallographic planes rotation when planes are perpendicular (dotted or continuous lines) or parallel (rectangles) to the diffraction plane.

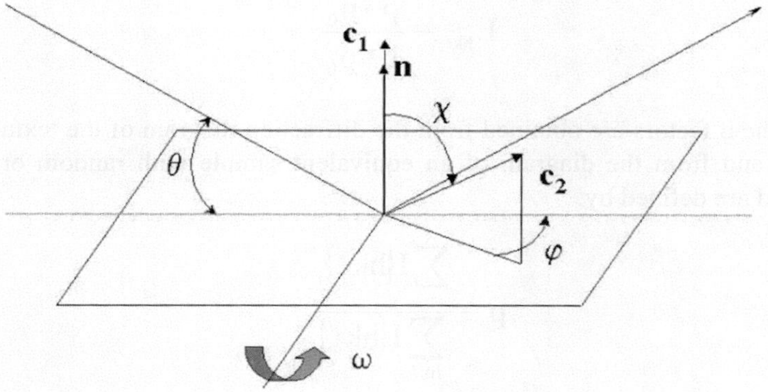

Fig. 8.10 Representation of two orientation components (C_1 and C_2), represented by their c axes.

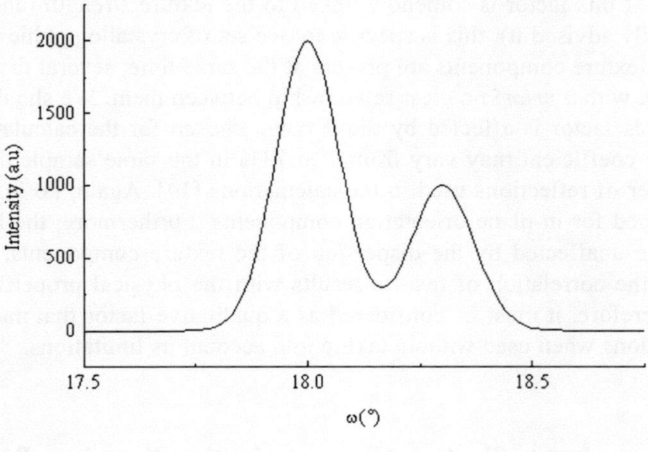

Fig. 8.11 Rocking curves corresponding to the two orientation components as described in Fig. 8.10 after rotation of the sample by φ.

8.2.2 A Quantitative Approach: The Lotgering Factor

Semi-quantitative approaches to determine texture from conventional Bragg-Brentano θ-2θ diffraction diagrams have been developed. The relative ratios of the intensities of the diffraction peaks obtained for a textured material provide information on texture. Lotgering in 1959 [13] derived a quantitative factor, $L_{hk\ell}$, to quantify the degree of orientation of a given material. This factor is defined for a hkℓ orientation by:

$$L_{hk\ell} = \frac{p - p_0}{1 - p_0} \quad (1)$$

where the p factors are obtained from the diffraction diagram of the textured sample (p) and from the diagram of an equivalent sample with random orientation (p_0), and are defined by:

$$p = \frac{\sum_i I\{hk\ell\}_i}{\sum_{hk\ell} I\{hk\ell\}} \quad (2)$$

being $I\{hk\ell\}$ the intensities of the diffraction peaks within a specified 2θ range. p values can vary from p_0 (random oriented sample) to 1 (ideally oriented sample). Therefore, the Lotgering factor may vary from 0 to 1. The higher the value obtained, the more oriented the sample for that direction.

However, if this factor is somehow linked to the texture strength (and Lotgering specifically advised it), this is specific to one set of crystallographic planes. If two or more texture components are present at the same time, several factors must be calculated, with *a priori* no clear relationship between them. We should also be aware that this factor is affected by the θ range chosen for the calculations: the values of the coefficient may vary from 7 to 73% in the same sample depending on the number of reflections used in the calculations [14]. Again, no information can be obtained for in-plane orientation components. Furthermore, the Lotgering factor will be unaffected by the dispersion of the texture components, which is relevant for the correlation of texture results with the physical properties of the material. Therefore, it must be considered as a qualitative factor that may lead to false conclusions when used without taking into account its limitations.

8.2.3 Approaches to Texture Characterization Based on Rietveld Analysis

Preferred orientations have been considered for long in Rietveld analysis as an undesired phenomenon. Many techniques have been described to remove texturing effect prior or during measurements. However, quantitative texture analysis may be included in the Rietveld analysis. To do that, texture can be described using three parameters:

1. The Texturing Direction in the Sample
2. The Crystallographic Direction **h** that is Aligned Preferentially along the Texturing Direction
3. The Texture Strength (the Degree of Orientation and the Angular Dispersion)

The original March approach includes the preferred orientations in the Rietveld equation using the function $P_{\mathbf{h}}$:

$$P_{\mathbf{h}} = G_2 + (1 - G_1)\exp(-G_1 \alpha_{\mathbf{h}}^2) \tag{3}$$

being G_1 and G_2 refinable parameters and $\alpha_{\mathbf{h}}$ the angle between the crystallographic direction **h** and the scattering vector **Δk**.

Later, Dollase modified the function:

$$P_{\mathbf{h}} = \left(G_1^2 \cos^2 \alpha_{\mathbf{h}} + \left(\frac{1}{G_1}\right)\sin^2 \alpha_{\mathbf{h}}\right)^{-3/2} \tag{4}$$

Originally, the correction assumed a Gaussian distribution of the preferred orientation axis of the individual crystallites about the normal to the sample surface. G_1 is the refinable parameter ($G_1 = 1$ means no preferred orientation) that controls the distribution shape. It is an estimate of the preferred orientation strength. The model provides:

- a preferred orientation correction factor that is minimum or maximum at $\alpha = 0°$
- a symmetric and smooth evolution in the [0,90°] $\alpha_{\mathbf{h}}$ range
- a single parameter to be fitted
- the possibility of normalization of the orientation with

$$\int_0^{\frac{\pi}{2}} P_{\mathbf{h}} d\alpha = 1$$

The normalization is important in order to keep constant the total diffracted intensity in a diffraction diagram whatever the distribution shape $P_{\mathbf{h}}$.

More recently, a modification of the March-Dollase approach was proposed [15]:

$$P_{\mathbf{h}} = f\left\{\left(G_1^2 \cos^2 \alpha_{\mathbf{h}} + \left(\frac{1}{G_1}\right)\sin^2 \alpha_{\mathbf{h}}\right)^{-3/2}\right\} + (1 - f) \tag{5}$$

In this formulation, it is expected that the factor f allows the inclusion of the randomly oriented part of crystallites (in volume). However, it must be noted that this factor is linked to the component of orientation that is described by the formula. This latter only describes one orientation component, i.e., a specific **h** distributed around the perpendicular direction to the sample surface. Therefore, f is only the volume fraction of crystallites that do have their **h** direction in this distribution component, (1-f) being the volume fraction of crystallites that are oriented differently, but not necessarily randomly oriented. To illustrate this point, we will give an example: imagine a sample containing tetragonal crystallites with a single orientation component, e.g., with their c-axes distributed around the normal to the sample surface. In this case, if f=0.5, 50% of the crystallites must have [001] directions among those included in that distribution. However, nothing is said for instance about the orientation of a-axes around the orientation component. Therefore, we cannot conclude that the 50% crystallites remaining are necessarily randomly oriented. As a conclusion, the parameter f does not represent the orientation distribution function in equation 5.

In all these calculations, it is always assumed there is only one component in texture and that directions are axially distributed (with a cylindrical symmetry around the scattering vector for a Bragg-Brentano geometry). The approach proved to be efficient for these conditions [16, 17, 18]. It is implemented in many software programs for diffraction analysis, and some of them allow a bi-texture component of this type. For all the other textures, measurements and formalisms to resolve the texture are more complex and have to be adapted for each case. This is unless a destruction of the sample is acceptable, as has been demonstrated for instance by O'Connor *et al.* [19]. Furthermore, the geometry used for the measurements plays an important role. The previous expressions describe the orientations with respect to the sample reference frame, while measurements are carried out in the spectrometer frame. The relationships between the two frames are straightforward when using conventional Bragg-Brentano geometry, but not for other geometries, like rocking curves. It requires, therefore, a localization correction that relates plane normal in the sample frame.

None of these approaches allow the description of textures in terms of distribution densities, due to a lack of normalization of texture results. Therefore, they depend on material characteristics that vary from one to another like porosity, crystalline state...

Another approach is the so-called arbitrary texture correction. This is not a model, and as such cannot be interpreted in terms of physically understandable parameters of texture. It only deserves the fitting possibility of diagrams that show textures, but ones that are not of interest or that cannot be measured (*e.g.* if not enough data have been acquired for this purpose). The correction simply consists in assigning arbitrary intensity values to the peaks in order that it respects observations. Whenever no texture correction would produce a satisfactory fit, one may use this correction to obtain a better fit of the observed intensities. This will allow obtaining more reliable cell parameters to start the analysis.

8.2.4 Representations of Textures: Pole Figures

8.2.4.1 Pole Spheres and Pole Figures

Fig. 8.12 represents a single crystallite orientation with the **c** axis at χ from **n**, and at φ from the macroscopic edges of the sample. The sphere on which all orientations can be distributed is also shown. This sphere of unit radius is called the Pole Sphere, with a $4\pi^2$ surface. The orientations may then be distributed over a solid angle of $4\pi^2$ sr at maximum. The intersection of the crystallographic direction [hkℓ] with the surface of the Pole Sphere is called a "pole", for instance the south or north poles. In these representations, we are interested in the {hkℓ} planes distribution. We choose to locate the orientation of one (hkℓ) plane by its normal [hkℓ]*. Consequently, all {hkℓ} planes will be located by their respective <hkℓ>* directions.

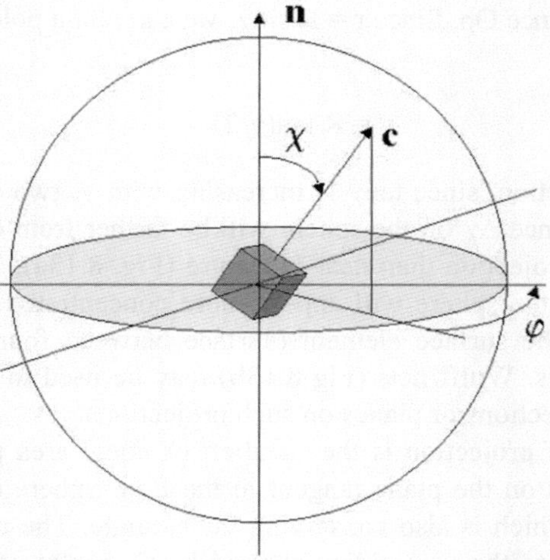

Fig. 8.12 Representation of a crystallite orientation on the Pole Sphere.

A Pole Sphere of a polycrystal will be the representation of all the poles of all the crystallites of the sample. For cubic crystal structures, for which [hkℓ] ⊥ {hkℓ}, the interpretation of the Pole Sphere is then straightforward. But for any other crystal system this relationship is not generally valid, and we may need the help of the reciprocal lattice construction. As evidence, plotting all the poles of all the {hkℓ} planes of all the crystallites of a sample would rapidly tend to a perfect homogeneous coverage of the Pole Sphere. Interesting information will be visible if we represent the poles for only one family of {hkℓ} planes. We will call this a {hkℓ} Pole Sphere. Also experimentally several {hkℓ} plane families may not be separated, and we will have then a {hkℓ/h'k'ℓ'} Multipole Sphere.

One entity that will be used is the unit surface element of the Pole Sphere, dS, the surface drawn for a (dχ, dφ) elementary variation. For the pole sphere of unit radius, this elementary surface is:

$$dS = \sin\chi \, d\chi \, d\varphi \tag{6}$$

It is not easy to represent a three dimensional object like a Pole Sphere and, although some programs can do it, to interpret them this way. Interpreting the Pole Sphere using two-dimensional projections is by far easier. We will call these projections {hkℓ} pole figures. Fig. 8.13a shows the stereographic projection of the pole sphere in the case of the single pole of Fig. 8.12. A pole, P, representing the intersection of [hkℓ]* with the Pole Sphere, is projected on the equatorial plane in p, intersection of SP with the equatorial plane. In this projection, the φ angle is conserved. All points located at the same φ are describing a meridian (large circle) and all points at the same χ are at the same latitude (on a small circle). Hence, a pole P(χ,φ) of the Pole Sphere is represented by the pole p(r',φ) in the {hkℓ} pole figure, where r' is the distance Op. Since r = R sinχ, we can plot a pole figure with any radius R using:

$$r' = R \tan(\chi/2) \tag{7}$$

In such a projection, since tanχ is increasing with χ, two points located at the same angular distance Δχ on the sphere will be farther from one another near the periphery of the projection than near its centre (Fig. 8.13a). Then, randomly distributed points on the sphere will appear more concentrated in the centre of the projection. Also, the surface element (surface between four adjacent points) is larger for larger χ's. Wulff nets (Fig 8.13b) may be used to manually determine angles between directions or planes on such projections.

Another type of projection is the Lambert or equal area projection. The pole P(χ,φ) is projected on the plane tangent to the Pole Sphere containing the north pole (Fig. 8.14), which is also conserving the φ angle. The rotation is around an axis perpendicular to the φ meridian passing by O, centre of the projection. The relation Op = OP is then valid for all points of the projection. For instance, a pole of the equator (χ = 90°) will be at R√2 from O in the projection. The projection p(r',φ) can be obtained whatever R by:

$$r' = 2R \sin(\chi/2) \tag{8}$$

For a given increase in χ, the distance between two points of the projection will decrease slowly, particularly in the high χ-range. This decrease will partially compensate the increase in surface element due to φ. Then this surface element will be similar near the periphery and near the centre of the projection (Fig. 8.15b), which is why this projection type is called "equal-area". Schmidt nets may be used to manually determine angles between directions or planes on this type of projections.

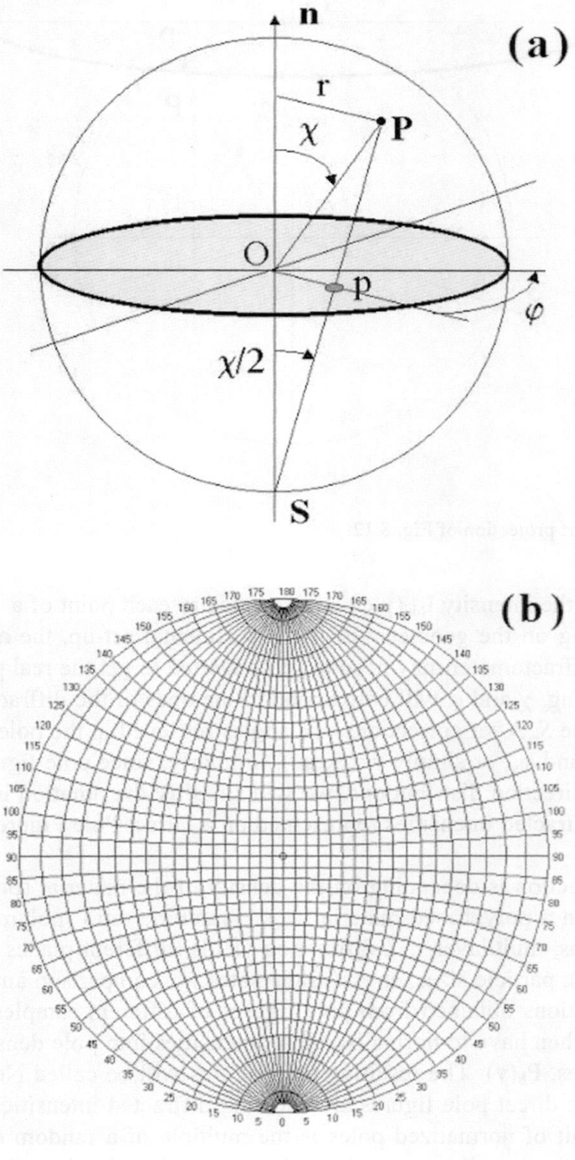

Fig. 8.13a Stereographic projection of figure 12 and **b** Wulff net to manually read angles in the projection

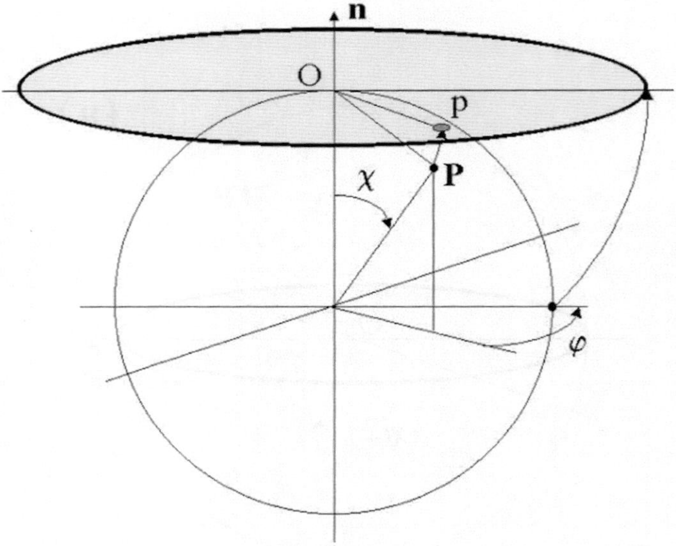

Fig. 8.14 Lambert projection of Fig. 8.12.

A value of the intensity $I_{hkl}(\chi,\varphi)$ is associated to each point of a {hkℓ} pole figure. Depending on the geometry of the experimental set-up, the angles χ and φ used in the diffractometer may need to be converted to get the real pole figures. In all the following, χ and φ will correspond to the ones of the diffractometer (spectrometer) space *S*. After corrections, the angles retrieved in the pole figures space, *Y*, will be ϑ_y and φ_y, or simply $\mathbf{y}=(\vartheta_y,\varphi_y)$. Similarly, since pole figures will represent <hkℓ>* direction distributions, we will simplify the notation into **h**=<hkℓ>*. Hence, the diffracted intensities represented on a Direct Pole Figure will be called $I_{\mathbf{h}}(\mathbf{y})$.

Since diffraction is depending on the density of the material (porosity, density of the phase in a polyphased material …), crystalline ratio (polymers …), thickness (thin films, multilayer …), diffraction yields (different planes diffract different intensities), particle sizes, stress/strain states …, comparison among the orientation distributions obtained from pole figures of different samples becomes impossible. We then have to normalize these intensities into pole densities, or distribution densities, $P_{\mathbf{h}}(\mathbf{y})$. The resulting pole figures will be called Normalized Pole Figures. While direct pole figures are showing diffracted intensities, in diffracted counts, the unit of normalized poles is the multiple of a random distribution, or m.r.d.. Using these units, a sample without a preferred orientation will exhibit normalized pole figures with 1 m.r.d. for all **h** and **y**'s, i.e., $P_{\mathbf{h}}(\mathbf{y})$ = 1 m.r.d.. A textured sample will show minima and maxima of densities, the minimum attainable density being 0.

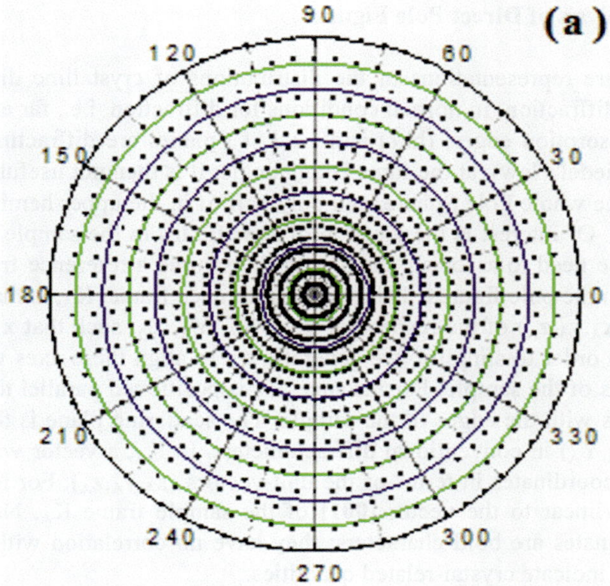

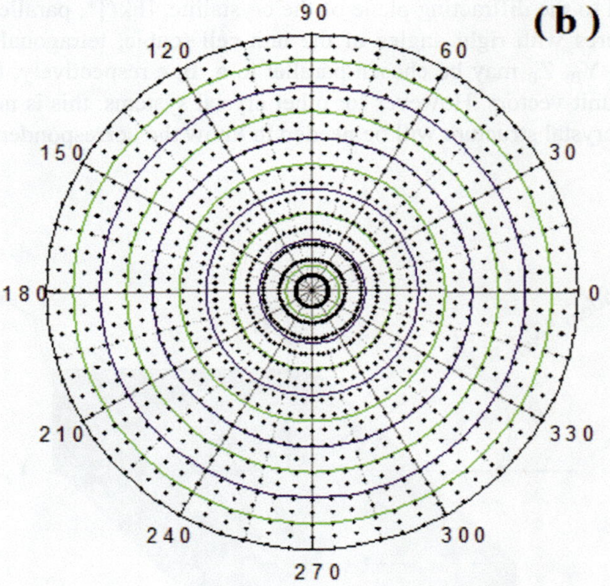

Fig. 8.15a Stereographic and **b** equal-area projections of 1368 points located at every 5° in χ and φ on the Pole Sphere.

8.2.4.2 Analysis of Direct Pole Figures

Pole figures are representations of the distributions of crystalline directions, as measured by diffraction. In normal conditions for diffraction, i.e., far enough from the atomic absorption edges, (hkℓ) and ($\bar{h}\ \bar{k}\ \bar{\ell}$) planes are diffracting the same intensities (Friedel's law), at the same Bragg angle. It is then not useful to measure or represent the whole Pole Sphere, and only half of it, the upper hemisphere, will be considered. Of interest is the location of specific **h**'s in the sample that we are measuring. We need to associate them with the sample a reference frame, which will be used in the pole figures. This sample reference frame, K_A, is made of three unit-vectors ($\mathbf{x_A},\mathbf{y_A},\mathbf{z_A}$) of the respective axes, X_A, Y_A, Z_A, such that $\mathbf{x_A} \wedge \mathbf{y_A} = \mathbf{z_A}$ (Fig. 8.16). In order to simplify, we generally try to align these axes with macroscopic features of the sample. For instance, Z_A is positioned parallel to **n**, and the two other axes with the edges of the sample. The scattering plane is then perpendicular to (X_A,Y_A) in conventional diffractometers. In K_A, a vector will be represented by its coordinates in terms of the unit-vectors ($\mathbf{x_A},\mathbf{y_A},\mathbf{z_A}$). For instance, the Z_A axis is co-linear to the vector **[001]** of the sample frame K_A. Note that the **[XYZ]** coordinates are bold characters; they have no correlation with the Miller indices which indicate crystal-related quantities.

Inside the sample are crystallites, having their own reference frame, K_B, with their own unit-vectors ($\mathbf{x_B},\mathbf{y_B},\mathbf{z_B}$) and axes, X_B, Y_B, Z_B (Fig. 8.17). Similarly, we fix the normal to the diffracting plane of the crystallite, [hkℓ]*, parallel to Z_B. For crystal structures with right angles of the unit-cell (cubic, tetragonal and orthorhombic), X_B, Y_B, Z_B may be chosen parallel to **a**, **b**, **c** respectively, both frames having equal unit-vectors. However for other crystal systems, this is no longer the case, and the crystal structure will be needed to know the correspondence between K_B and **a**, **b**, **c**.

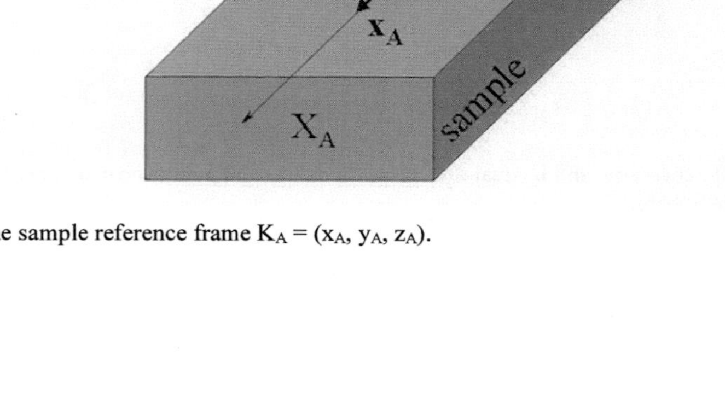

Fig. 8.16 The sample reference frame $K_A = (x_A, y_A, z_A)$.

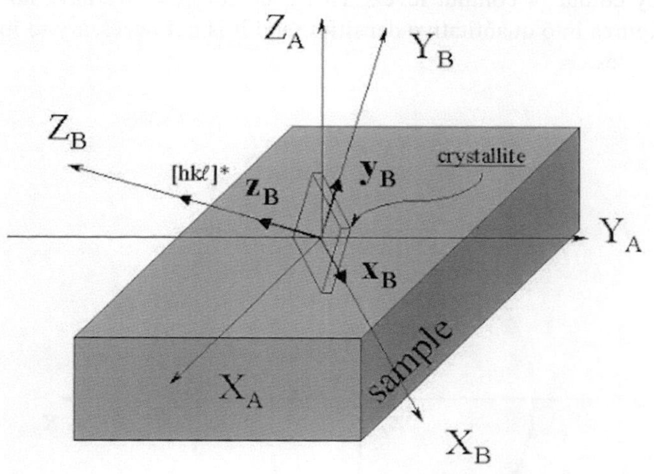

Fig. 8.17 Crystal and sample reference frames $K_B = (\mathbf{x_B}, \mathbf{y_B}, \mathbf{z_B})$ and K_A respectively. Only one crystallite is shown.

In order to represent the textures, we will then use K_B and K_A in the pole figures. The pole figures will be represented in such a way that Z_A is their normal axis, X_A the horizontal axis and Y_A the vertical axis (Fig. 8.18). For one crystallite (Fig. 8.18a), a direction **y** associated to the normal [hkℓ]* of the diffracting plane (hkℓ), is then localized by the polar angle ϑ_y and the azimuth angle φ_y. This latter takes its origin along X_A, and is taken positive for a left-handed rotation when looking down Z_A. It then respects the mathematical definition of K_A for a direct reference frame.

However, in a polycrystal, a large number of crystallites are diffracting, and the pole figure shows regions with large diffracted intensities (or densities, if we deal with normalized pole figures), indicating the tendencies of the crystallites to align around a given direction of the sample. The pole drawn by such an ensemble of crystallites (Fig. 8.18b) is called a texture component. For one component, the pole enlarges around the previous **y** direction, meaning that some of the crystallites are about to align with **y**, but not strictly. It is then useful to describe the component by the same ϑ_y and φ_y angles as **y**, but giving the shape of the distribution and a parameter which quantifies how much the pole spreads, for instance Gaussian with 10° FWHM. But we should never forget that describing a pole by an analytical function imposes that we know the distribution is respecting the imposed shape, which is not easily verifiable. For instance, in thin films, poles may appear very much elongated in one direction, but not in the other [20] due to the substrate interaction with the film. Also, distributions may exhibit shapes that are other than Gaussian, Lorentzian for instance [21]. The diffraction intensity (then density) will depend on how the dispersion is for a given pole. The larger the dispersion, the lower the diffraction will be. The intensity and density values are represented by colour or contour levels. However, for now, we have not normalized the pole figures into quantitative densities, and it is not necessary to indicate these intensity scales.

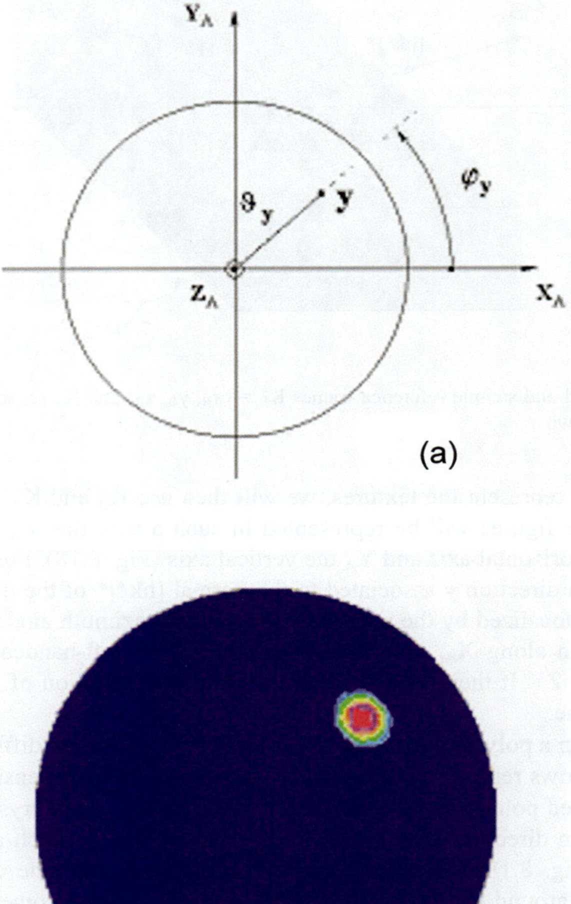

Fig. 8.18a $P_h y$ Diffraction pole figure for one crystallite. The direction **y** is associated to the [hkℓ]* normal. **b** Pole figure of a texture component centred on the previous **y**, having a Gaussian shape of 10° FWHM, for h = <001>* of an orthorhombic crystal structure.

Using diffraction, all the (hkℓ) diffracting at the same Bragg angle will be found on the same $P_{\mathbf{h}}(\mathbf{y})$ pole figure. For instance, (123) and (321) in a cubic crystal system. Then the pole figure has to also show this multiplicity. Taking the C_1 component of Fig. 8.10 in a cubic structure, a pole figure like the one shown in Fig. 8.19 will be obtained. It can be observed a full pole {001} and four halves (corresponding to {100} and {010}) at 90° in ϑ_y from {001}. The multiplicity of this pole figure is three at total, while the crystallographic multiplicity of the {100} reflection of a cubic crystal system is six. This can be explained due to the fact that we are only concerned with the upper hemisphere of the pole sphere in normal diffraction pole figures.

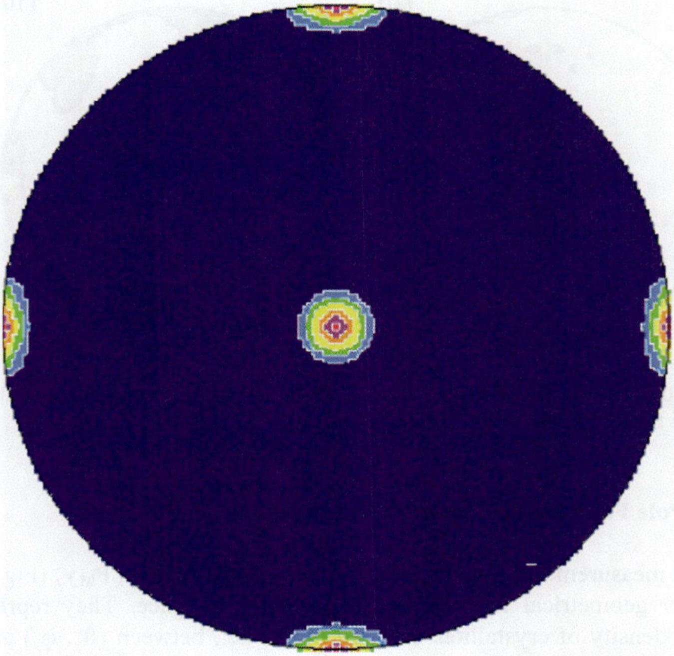

Fig. 8.19 {001} Pole figure for the component C_1 of Figure 8.10, for a cubic crystal structure.

As we have seen in the previous paragraphs, a pole figure can then be a complex object to interpret: it is a combination of a multiplicity, of components, with more or less regular dispersions. Fig. 8.20 shows pole figures that are combining many of these effects. They have been measured on a real sample, an aragonite layer from the sea shell (a cowry) *Cypraea testudinaria.* These pole figures are combining twinning, dispersion and two components, of the orthorhombic crystal system [22]. As it can be seen, pole figures are not easily interpretable, and some tools to relate them to the orientation space are needed.

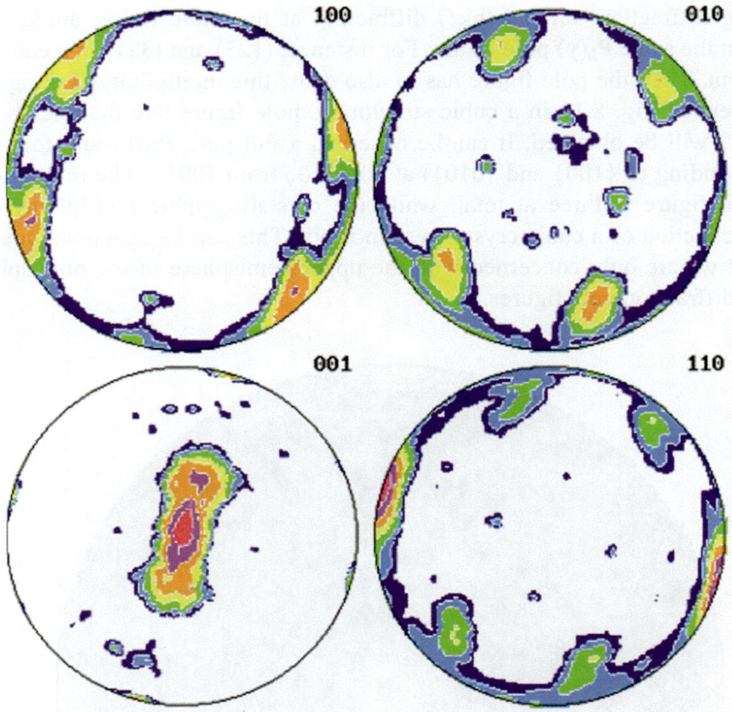

Fig. 8.20 {100}, {010}, {001} and {110} pole figures of an aragonite (orthorhombic crystal system) layer from the sea shell *Cypraea testudinaria.*

8.2.4.3 Pole Figures and Orientation Spaces

Diffraction measurements give in the *Y* space the pole figures, $P_{\mathbf{h}}(\mathbf{y})$ (Fig. 8.18.a), after proper geometrical transformations from the *S* space. They represent the volumetric density of crystallites oriented in **dy**, i.e., between (ϑ_y, φ_y) and (ϑ_y + $d\vartheta_y$, φ_y + φ_y). This can be expressed by:

$$\frac{dV(\mathbf{y})}{V} = \frac{1}{4\pi} P_{\mathbf{h}}(\mathbf{y})\, \mathbf{dy} \tag{9}$$

where V is the irradiated volume of the sample and dV(**y**) the volume of crystallites which orientation is with **h** directions between **y** and **y**+**dy**. Similarly to equation 6 and for consistency we have:

$$\mathbf{dy} = \sin\vartheta_y\, d\vartheta_y\, d\varphi_y \tag{10}$$

and the 1/4π is a normalization factor to distribution densities of pole figures according to:

$$\int_{\varphi_y=0}^{2\pi}\int_{\vartheta_y=0}^{\pi/2} P_{\mathbf{h}}(\vartheta_y,\varphi_y)\sin\vartheta_y \, d\vartheta_y d\varphi_y = 4\pi \tag{11}$$

The orientation of planes is defined by referring them to at least two directions in the real space. We then need another concept, the Orientation Distribution Function (ODF), f(g). This function represents the statistical distribution of the orientations of the crystallites in a polycrystalline aggregate. It is defined similarly by:

$$\frac{dV(g)}{V} = \frac{1}{8\pi^2} f(g)\, dg \tag{12}$$

where dg = sin(β)dβdαdγ is the orientation element (defined in a 3-dimensional space), defined by three Euler angles, g=α,β,γ in the orientation space (or *G*-space) (Fig. 8.21).

These three Euler angles bring the crystal coordinate system K_B collinear with the sample coordinate system K_A. The *G*-space may be constructed from the space groups, taking into account their rotation parts (since orientations are rotations) and the inversion centre (since we are using normal diffraction). The two first angles α and β determine generally the orientation of the <001>* crystallite directions in K_A, they are called azimuth and co-latitude (or pole distance) respectively. The third angle, γ, defines the location of another crystallographic direction, chosen as <010>* (or **b** in the (**a**,**b**) plane of an orthogonal crystal system). V is the irradiated volume of the sample, dV(g) the volume of crystallites which orientation is between g and g+dg.

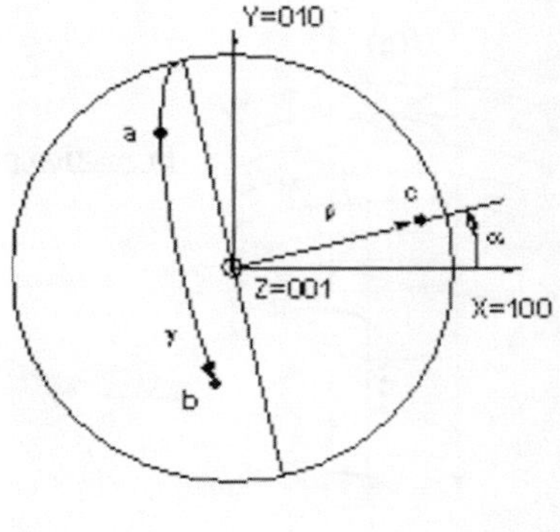

Fig. 8.21 Representation of the three Euler angles that define the position of the crystallite coordinate system K_B of an orthogonal crystal cell in the sample coordinate system K_A. Note, **100**, **010** and **001** are not Miller indices but vectors referring to an ortho-normal frame aligned with K_A.

The function f(g) then represents the volumetric density of crystallites oriented in dg. It is also measured in m.r.d. and normalized to the value $f_r(g)=1$ for a random sample. The normalization condition of f(g) over the whole orientation space is expressed by:

$$\int_{\alpha=0}^{2\pi}\int_{\beta=0}^{\pi/2}\int_{\gamma=0}^{2\pi} f(g)\,dg = 8\pi^2 \tag{13}$$

The function f(g) can take values from 0 (absence of crystallites oriented in dg around g) to infinity (for some of the *G*-space values of single crystals).

From equations (9) and (12) it follows the fundamental equation of texture analysis:

$$P_{\mathbf{h}}(\mathbf{y}) = \frac{1}{2\pi}\int_{\mathbf{h}//\mathbf{y}} f(g)\,d\tilde{\varphi} \tag{14}$$

This equation represents the fact that each pole figure (a 2D object) is a projection along a certain path $\tilde{\varphi}$ of the ODF (a 3D object), which of course depends on the crystal symmetry (Fig. 8.22). Each cell of a given pole figure will then be an average over several cells of the ODF, and each cell of the ODF will be measured by one or more cells from the pole figures. The larger the number of pole figure cells that measure a specific ODF cell, the more statistically reliable is the measurement of this ODF. In practice, one has to measure the largest number as possible of reliable (enough intense) pole figures to define the ODF with the best resolution available.

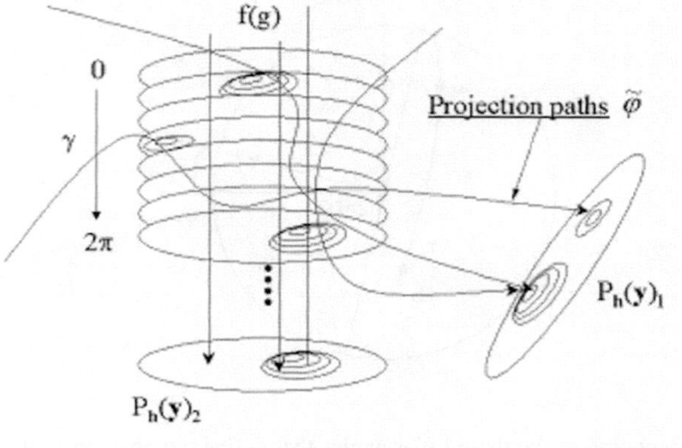

Fig. 8.22 Relationship between the 3D object f(g) and the pole figures $PB_{hB}(y)$. To each pole figure cell correspond several ODF boxes, and each ODF box is linked to several pole figure cells.

Equation (14) was solved several years ago by Bunge, using generalized spherical harmonics formulation [23, 24], but only in the case of high crystal symmetries. An exact solution in an analytical closed form without a series expansion was given [25]. The so-called "vector" [26, 27], entropy maximization [28], component [29] and ADC [30] methods were developed later. The description of these methods is beyond the scope of this chapter. We will describe briefly in the next section how the two main methods used.

The orientation distribution function, which therefore can be calculated from experimental pole figures, provides complete information of all texture components (in-plane and out-of-plane) and allows the calculation of a global degree of orientation. From it we can obtain recalculated direct and normalised pole figures and also the so-called inverse pole figures. They represent the densities of crystalline directions that are parallel to a specific sample direction. Therefore, all texture components along that direction can be visualized directly, and even their relative contributions can be estimated. Several examples of inverse pole figures will be shown in section 5. The objective of the quantitative texture analysis is the calculation of the orientation distribution function.

8.3 Quantitative Texture Analysis

8.3.1 Calculation of the Orientation Distribution Function

To obtain the Orientation Distribution Function is necessary to solve the fundamental equation 14. In the case of the generalized spherical harmonics or the component methods, we obtain a function of the orientation distribution of the crystallites. However, there is no *a priori* need for fitting the experimental data of pole figures to a function, and many “direct” methods (*e.g.* the maximization of entropy, vector or WIMV methods) do not obtain a function. In this case the term "function" can be omitted, and the literature refers to Orientation Distributions (OD). In this respect, since we will only show results from direct methods of f(g) refinements, only "OD" will be used to call f(g). But, even for ODs that have been refined using direct methods, it is somewhat hard to represent ODs on figures without using contour and isolines, which in turn are the result of interpolations of discrete OD points by functions (*e.g.* splines).

8.3.1.1 Williams-Imhof-Matthies-Vinel (WIMV) Method

The WIMV approach [31]- [32 33] for the refinement of the OD is an iterative calculation method based on the numerical refinement of f(g) at step n+1:

$$f^{n+1}(g) = N_n \frac{f^n(g) f^0(g)}{\left(\prod_{\mathbf{h}=1}^{I} \prod_{m=1}^{M_\mathbf{h}} P_\mathbf{h}^n(\mathbf{y}) \right)^{\frac{1}{IM_\mathbf{h}}}} \tag{15}$$

where the product extends over the I experimentally measured pole figures and for all the poles multiplicity $M_\mathbf{h}$. $f^n(g)$ and $P_\mathbf{h}^n(\mathbf{y})$ represent the refined values of $f(g)$ and $P_\mathbf{h}(\mathbf{y})$ at the nth step respectively. The number N_n is a normalizing factor. The $P_\mathbf{h}^n(\mathbf{y})$ values are calculated at each cycle with equation 15. The first step in this procedure is to evaluate $f^0(g)$:

$$f^0(g) = N_0 \left(\prod_{\mathbf{h}=1}^{I} \prod_{m=1}^{M_\mathbf{h}} P_\mathbf{h}^{exp}(\mathbf{y}) \right)^{\frac{1}{IM_\mathbf{h}}} \tag{16}$$

in which $P_\mathbf{h}^{exp}(\mathbf{y})$ stands for the measured pole figures.

The WIMV algorithm maximizes the so-called "phon" (orientation background or minimum value of the OD which represents the randomly oriented fraction of the sample volume) and the texture sharpness.

The regular WIMV method necessitates an OD divided into a finite number of regular cells. Inside each cell a discrete value of the OD is associated. When the WIMV calculation is inserted inside the Rietveld refinement procedure, it requires two additional steps:

- the extraction of the pole figures or texture weights
- the interpolation of these weights to fit the regular grid

This renders non-optimized values of the OD, particularly for sharp textures and coarse irregular coverage of the OD.

The extended WIMV approach (E-WIMV) can be used with irregular coverage of the OD space and includes smoothing based on a concept similar to the tube projection of the ADC method [30]. The extension of the method uses an iterative scheme of the OD refinement, which is close to the maximization of entropy [34]. The E-WIMV method is then often called Entropy-modified WIMV, and it has been applied to ferroelectric materials [35, 36]. The OD cell values are computed through an entropy iteration algorithm that includes the reflection weights:

$$f^{n+1}(g) = f^n(g) \prod_{m=1}^{M_\mathbf{h}} \left(\frac{P_\mathbf{h}(\mathbf{y})}{P_\mathbf{h}^n(\mathbf{y})} \right)^{r_n \frac{w_\mathbf{h}}{M_\mathbf{h}}} \tag{17}$$

in which r_n is a relaxation parameter such that $0 < r_n < 1$, $M_\mathbf{h}$ is the number of division points for the discretisation of the integral of all the orientations around the scattering vector for the pole figure **h**. The reflection weight $w_\mathbf{h}$ is introduced to take into account the different accuracy of the more intense and less overlapped reflections with respect to the smaller ones, and is calculated analogously to the weight factors of the Rietveld analysis. The efficiency of this approach has been proved in a series of works [37, 38, 39, 40]. Figure 8.23 shows the results of two refinements using the WIMV and E-WIMV algorithms for the X-ray diagrams measured at $\chi = 0°$ on a Ca-modified $PbTiO_3$ thin film deposited on a $Pt/TiO_2/SiO_2/Si(100)$ substrate [39]. It can be seen that the E-WIMV approach achieves a better refinement, particularly for the sharp peaks of the highly oriented Pt layer.

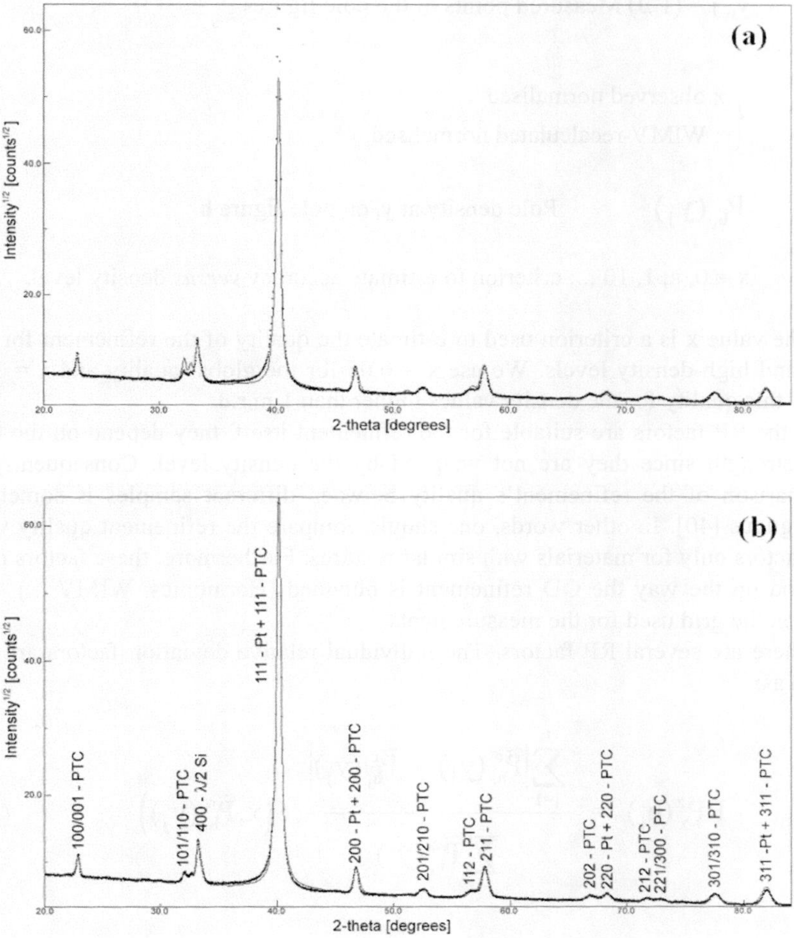

Fig. 8.23 Experimental X-ray diagrams (dotted line) and their corresponding refinement (solid line) using **a** WIMV and **b** E-WIMV approaches.

8.3.1.2 OD Refinement Reliability Factors

The best solution found for f(g) is in most programs given for the minimum averaged reliability factors, RP:

$$\overline{RP_x} = \frac{1}{I}\sum_i \sum_j \frac{\left| P^c_{\mathbf{h}_i}(\mathbf{y}_j) - P^o_{\mathbf{h}_i}(\mathbf{y}_j) \right|}{P^o_{\mathbf{h}_i}(\mathbf{y}_j)} \tag{18}$$

where:

$\mathbf{h}_i$, i = (1..*I*) Measured pole figures
$\mathbf{y}_j$, j = (1..J) Measured points of the pole figures

$$\begin{cases} o\text{: observed normalised} \\ c\text{: WIMV-recalculated normalised} \end{cases}$$

$P_{\mathbf{h}_i}(\mathbf{y}_j)$ Pole density at $\mathbf{y}_j$ on pole figure $\mathbf{h}_i$

x = 0, ε, 1, 10 ...: criterion to estimate accuracy *versus* density level.

The value x is a criterion used to estimate the quality of the refinement for the low and high-density levels. We use x = 0.05 for the global quality and x = 1 to show this quality for the density values higher than 1 m.r.d.

If the RP factors are suitable for the refinement itself, they depend on the texture strength since they are not weighed by the density level. Consequently, a comparison of the refinement's quality between different samples is somehow ambiguous [40]. In other words, one should compare the refinement quality with RP factors only for materials with similar textures. Furthermore, these factors may depend on the way the OD refinement is obtained (Harmonics, WIMV ...), and also on the grid used for the measurements.

There are several RP factors. The individual relative deviation factors are defined as:

$$\mathrm{RP}^z_x(\mathrm{h}_i) = \frac{\sum_{j=1}^{J} \left| \widetilde{\mathrm{P}}^o_{\mathrm{h}_i}(\mathrm{y}_j) - \widetilde{\mathrm{P}}^c_{\mathrm{h}_i}(\mathrm{y}_j) \right|}{\sum_{j=1}^{J} \widetilde{\mathrm{P}}^z_{\mathrm{h}_i}(\mathrm{y}_j)} \, \theta\!\left(\mathrm{x}, \widetilde{\mathrm{P}}^o_{\mathrm{h}_i}(\mathrm{y}_j)\right) \tag{19}$$

$$\text{with } \Theta(\mathrm{x,t}) = \begin{cases} 1 \text{ for } \mathrm{P}_{\mathbf{h}_i}(\mathbf{y}_j) > \mathrm{x} \\ 0 \text{ for } \mathrm{P}_{\mathbf{h}_i}(\mathbf{y}_j) \le \mathrm{x} \end{cases}$$

These individual factors help to detect if some pole figures are particularly badly reproduced after the refinement.

Another type is the averaged relative deviation factors:

$$\overline{RP}_x^z = \frac{1}{I}\sum_{i=1}^{I} RP_x^z(h_i) \qquad (20)$$

These are simply the arithmetic average of the previous ones. They help in comparing results on different samples. The global relative deviation factors are similar, but the averaging scheme differs:

$$RP_x^z = \frac{\sum_{i=1}^{I}\sum_{j=1}^{J}\left|\tilde{P}_{h_i}^{o}(y_j) - \tilde{P}_{h_i}^{c}(y_j)\right|}{\sum_{i=1}^{I}\sum_{j=1}^{J}\tilde{P}_{h_i}^{z}(y_j)}\,\theta\!\left(x, \tilde{P}_{h_i}^{o}(y_j)\right) \qquad (21)$$

The Rietveld-like R-factors or "intensity-weighted", which take into account the normal Gaussian distribution standard deviation for each measured intensity, show less overall variation with the texture strength. It is a better indicator of the OD refinement reliability when comparing different samples. The individual weighted standard deviation factors are defined as:

$$Rw_x^z(h_i) = \frac{\sum_{j=1}^{J}\left[w_{ij}^{o} I_{h_i}^{o}(y_j) - w_{ij}^{c} I_{h_i}^{c}(y_j)\right]^2}{\sum_{j=1}^{J} w_{ij}^{z} I_{h_i}^{z\,2}(y_j)}\,\theta\!\left(x, \tilde{P}_{h_i}^{o}(y_j)\right) \qquad (22)$$

whose averaged factor is:

$$\overline{Rw}_x^z = \frac{1}{I}\sum_{i=1}^{I} Rw_x^z(h_i) \qquad (23)$$

The global weighted standard deviation factors will be:

$$Rw_x^z = \frac{\sum_{i=1}^{I}\sum_{j=1}^{J}\left[w_{ij}^{o} I_{h_i}^{o}(y_j) - w_{ij}^{c} I_{h_i}^{c}(y_j)\right]^2}{\sum_{i=1}^{I}\sum_{j=1}^{J} w_{ij}^{z} I_{h_i}^{z\,2}(y_j)}\,\theta\!\left(x, \tilde{P}_{h_i}^{o}(y_j)\right) \qquad (24)$$

with $I^z_{h_i}(y_j) = \tilde{P}^z_{h_i}(y_j).N_{h_i}$ (Diffracted intensity)

N_{h_i} (Refined normalizing factor)

$$w^z_{ij} = \frac{1}{\sqrt{I^z_{h_i}(y_j)}} \quad \text{(Diffracted intensity weight)}$$

8.3.1.3 Quantitative Texture Analysis Software Programs

Several software programs may be used to do all the calculations necessary for Quantitative Texture Analysis. POFINT [41] is a simple MS-DOS based program developed in Turbo-Pascal used for data reduction and defocusing corrections in the case of conventional texture analysis, prior to the OD refinement. Beartex [42], one of the most used texture software, allows the refinement of the OD from the results obtained with point detectors. It has implemented also the calculation and representation of recalculated and inverse pole figures from the OD, and the deduction of the macroscopic elastic tensor of the textured material from the values of the coefficients of a single crystal, using the OD as a weight factor. The combined analysis methodology is implemented in a user-friendly interface: Materials Analysis Using Diffraction (MAUD) [43]. All the examples shown in this chapter have been obtained using these programs.

8.3.2 OD Texture Strength Factors

Once f(g) is satisfactorily obtained, one can calculate factors that give an estimate of the texture strength. Caution should be taken here when comparing samples on the base of overall texture strength parameters. Samples should have the same crystal symmetry and exhibit similar texture components.

The first texture strength parameter is the so-called 'texture index' [23] (expressed in m.r.d.^{2} units):

$$F^2 = \frac{1}{8\pi^2} \sum_i [f(g_i)]^2 \Delta g_i \tag{25}$$

with $\Delta g_i = \sin\beta_i \,\Delta\beta\, \Delta\alpha\, \Delta\gamma$ is the OD cell volume.

This index varies from 1 (random powder) to infinity (perfect texture or single crystal). It represents the mean square value of the OD. Since this index is expressed in units that are not homogeneous with the distribution density units (m.r.d.), to help comparison and interpretations, it is more convenient to compare the square roots of these values, *i.e.* the Texture Strength:

$$F = \sqrt{F^2} \tag{26}$$

The second overall texture strength parameter is a measure of the texture disorder, evaluated by the calculation of the texture entropy:

$$S = \frac{1}{8\pi^2} \sum_i f(g_i) \ln f(g_i) \Delta g_i \tag{27}$$

8.3.2.1 Characterization of the Randomly Oriented Volume Fraction

A sample exhibiting randomly oriented crystallites has an OD with 1 m.r.d. distribution densities for all g values. However, in some samples only a fraction of the total volume is randomly oriented, V_r, the rest, $V_c(g)$, being the oriented fraction volume having the orientation component or components, $f_c(g)$. The random part produces a "background" level in the OD, sometimes called "FON" or "PHON". The OD can be then expressed as:

$$f(g) = f_r + f_c(g) \tag{28}$$

with the mandatory condition $0 \le f_r \le 1$.

By integrating equation 12, taking into account equation 28:

$$\frac{1}{V} \oint [V_r + V_c(g)]\, dV(g) = \frac{1}{8\pi^2} \oint [f_r + f_c(g)]\, dg \tag{29}$$

which verifies after identification: $V_r/V = f_r$ for the random part, and

$$\frac{dV_c(g)}{V} = \frac{1}{8\pi^2} f_c(g)\, dg$$

for the textured part

Therefore, the minimum value of the OD, f_r, corresponds to the volume fraction of the material that is randomly oriented.

8.3.3 *Estimation of the Elastic Properties of Polycrystals Using the Orientation Distributions*

The OD calculated can be used to estimate the elastic properties of an oriented polycrystal from the values of the single crystal. We will now describe how one may use a geometric average of the tensor quantities to calculate an OD-weighted $C_{ijk\ell}^{M}$, in the case of single-phase materials. Such macroscopic stiffness can be used for example in the calculation of elastic waves propagation in surface acoustic wave devices.

8.3.3.1 Tensor Average

The volume average of a tensor quantity T, which varies inside the volume V can be calculated by:

$$\langle \mathrm{T} \rangle = \frac{1}{V} \int_{\mathrm{V}} \mathrm{T}\, dV \tag{30}$$

In general, tensors are considered constant inside individual grains, and the previous equation can be rewritten as:

$$\langle \mathrm{T} \rangle = \frac{1}{V} \sum_{i} \mathrm{T}\, \mathrm{V_i} \tag{31}$$

for which V_i/V represents the volume fraction or weight associated to each grain in the volume. Since tensor properties of a polycrystal depend on the crystal orientations, this has to be taken into account in the calculations. This can be done through the Orientation Distribution f(g):

$$\langle \mathrm{T} \rangle = \int_{\mathrm{g}} \mathrm{T(g)}\ \mathrm{f(g)}\ dg \tag{32}$$

where g varies in all the orientation space *G*.

It is important to note that using such arithmetic averaging procedures, the average of the inverse of the tensor is in general not equivalent to the inverse of the averaged tensor:

$$\langle T^{-1} \rangle = \frac{1}{V} \sum_{i} T^{-1} V_i \neq \langle T \rangle^{-1} \quad (33)$$

This is a problem if the tensor is aimed at representing a physical property, like the elastic tensor $C_{ijk\ell}$. If the average tensor is $C_{ijk\ell}{}^{M}$, the arithmetic average leads to $(C_{ijk\ell}{}^{M})^{-1} \neq S_{ijk\ell}{}^{M}$ that violates stress-strain equilibrium inside the polycrystal. Therefore, a different averaging procedure must be explored.

The geometric mean of a scalar is:

$$\lceil b \rceil = \prod_{k=1}^{N} b_k{}^{w_k} = \exp(\langle \ln b \rangle) \quad (34)$$

with $\langle \ln b \rangle = \sum_{k=1}^{N} \ln b_k w_k$

However, for a tensor the geometric mean is not straightforward. For the eigenvalues λ_I of a given matrix T, equation 34 can be rewritten as:

$$\lceil \lambda_I \rceil = \prod_{k=1}^{N} \lambda_k{}^{w_{i,k}} \quad (35)$$

which ensures that $\lceil \lambda_I \rceil = 1/\lceil 1/\lambda_I \rceil = \lceil \lambda_I{}^{-1} \rceil^{-1}$. For the matrix T represented in its orthonormal basis of eigenvectors, it can be shown [44] that:

$$\lceil T \rceil_{ij} = \exp(<\ln T>_{i'j'}) = \exp(<\Theta>_{ij,i'j'} \ln T_{i'j'}) \quad (36)$$

In this equation, Θ stands for the transformation applied to the tensor T that represents the property of a given single crystal of orientation g in the single crystal reference frame K_B, in order to bring it coincident to the sample reference frame K_A. Θ then depends on the tensor order, and its average is composed of elements $<\Theta>_{ij}$ given by, similarly as in equation 32:

$$\langle \Theta \rangle_{ij,i'j'} = \int_{g} \Theta_i^{i'}(g) \Theta_j^{j'}(g) f(g)\, dg \quad (37)$$

8.3.3.2 Application to the Estimation of Elastic Properties

The modelling of the mechanical properties has concentrated most of the works of the studies of macroscopic anisotropic properties in geology and metallurgy. We will not describe here the Voigt and Reuss models that produce the lower and upper bounds of the macroscopic elastic coefficients, using arithmetic averaging. Hill developed later another model that derives from the two others [45]. Instead, we will focus on the geometric mean model. Due to the fact that we are dealing with real and symmetric tensors, the eigenvalues of the concerned tensors are real. But neither the $S_{ijk\ell}$ and $C_{ijk\ell}$ nor the s_{ij} and c_{ij} matrices are diagonal, and equation 36 cannot be used. It is necessary first to diagonalize them using an orthonormal basis of eigentensors $\mathbf{b}_{ij}{}^{(\lambda)}$, for instance in the case of the stiffness tensor $C_{ijk\ell}$ of eigenvalues $C^{(\lambda)}$:

$$((\mathbf{b}_{ij}{}^{(\lambda)})^{-1}\, C_{ijk\ell}\, \mathbf{b}_{k\ell}{}^{(\lambda)}) = C^{(\lambda)}\, \delta_{ij} \tag{38}$$

and

$$C_{ijk\ell} = \sum_{\lambda=1}^{6} C^{(\lambda)} \mathbf{b}_{ij}{}^{(\lambda)} \mathbf{b}_{k\ell}{}^{(\lambda)} \tag{39}$$

which extends to

$$\begin{aligned}(\ln C)_{ijk\ell} &= \sum_{\lambda=1}^{6} \ln(C^{(\lambda)}) \mathbf{b}_{ij}{}^{(\lambda)} \mathbf{b}_{k\ell}{}^{(\lambda)} \\ &= \ln\left[\prod_{\lambda=1}^{6} (C^{(\lambda)})^{\mathbf{b}_{ij}{}^{(\lambda)} \mathbf{b}_{k\ell}{}^{(\lambda)}}\right]\end{aligned} \tag{40}$$

Now applying the geometric average over orientations (equation 36), the macroscopic stiffness of the polycrystal can be calculated from:

$$C_{ijk\ell}{}^{M} = \lceil C \rceil_{ijk\ell} = \exp(<\ln C>_{i'j'k'\ell'}) = \exp(<\Theta>_{ijk\ell,i'j'k'\ell'}\, (\ln C)_{i'j'k'\ell'} \tag{41}$$

with

$$\langle \Theta \rangle_{ijk\ell,i'j'k'\ell'} = \int_g \Theta_i^{i'}(g)\Theta_j^{j'}(g)\Theta_k^{k'}(g)\Theta_\ell^{\ell'}(g)\, f(g)\, dg$$

and $(\ln C)B_{i'j'k'\ell'}$ is given by equation 40. However, before calculating the value $\lceil C \rceil B_{ijk\ell}$, one has to first diagonalize $<\ln C>B_{i'j'k'\ell'}$ in order to extract the new eigenvalues and eigentensors for the oriented polycrystal. The four successive tensor transformations relate to the 4th order stiffness tensor character.

The factorial entering the calculation explains the term "geometric mean", in the sense that the oriented polycrystal macroscopic stiffness is obtained by the mean averaging of the single crystal stiffness eigenvalues. Similar expressions may be obtained for the macroscopic compliance tensor $S_{ijk\ell}{}^{M}$, which admits as eigenvalues $S^{(\lambda)} = 1/C^{(\lambda)}$ the reciprocal of the stiffness eigenvalues. This ensures that the same macroscopic elastic properties are calculated when using stiffness or compliances. In other words, the average of the inverse macroscopic property is consistent with the inverse of the average macroscopic property. This may be the reason why this modelling gives rather good estimates of the elastic properties [45], comparable to the ones obtained by more sophisticated models which require larger computation times.

As it is difficult to measure the effective elastic tensor of ferroelectric thin films, its calculation by a volume average of the elastic coefficients of the individual crystals is a good solution that allows, for example, analyzing the anisotropic character of this tensor, and the effect of texture on it. The application of this method to lead titanate-based films has been reported [46].

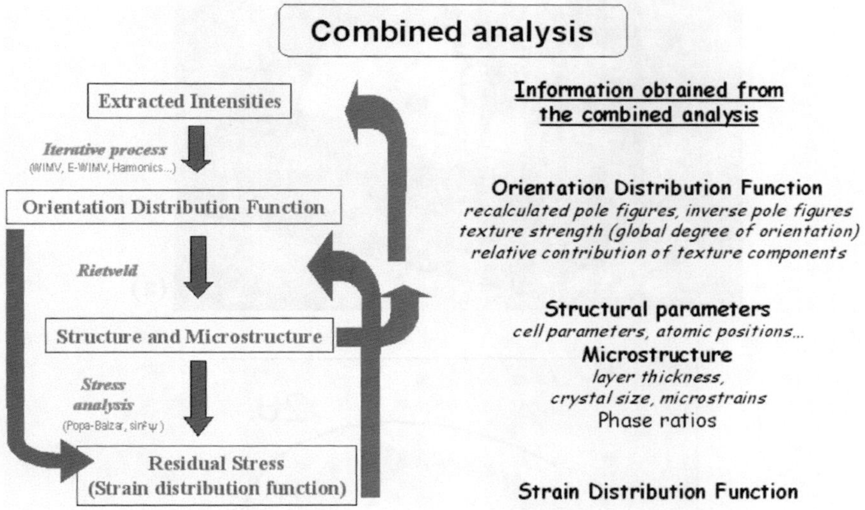

Fig. 8.24 Simplified diagram of the combined method of analysis of diffraction data. A list of the information that can be obtained and some of the algorithms implemented are also included.

8.4 Combined Analysis

Diffraction studies of materials are becoming more complex. Materials present an increasing level of complexity, for which we require as much information about the material characteristics as possible from a non-destructive technique. Heterostructures with several layers, materials with multiple phases, subjected to residual stress and textured are the present challenges for this kind of studies. For the determination of the structure, microsctructural parameters, residual stress and texture exists for each of them a particular method of analysis, which makes the whole characterisation a long and complex process. Furthermore, some parameters are not determined correctly without a precise determination of the others. This requires a global methodology of analysis of the diffraction data.

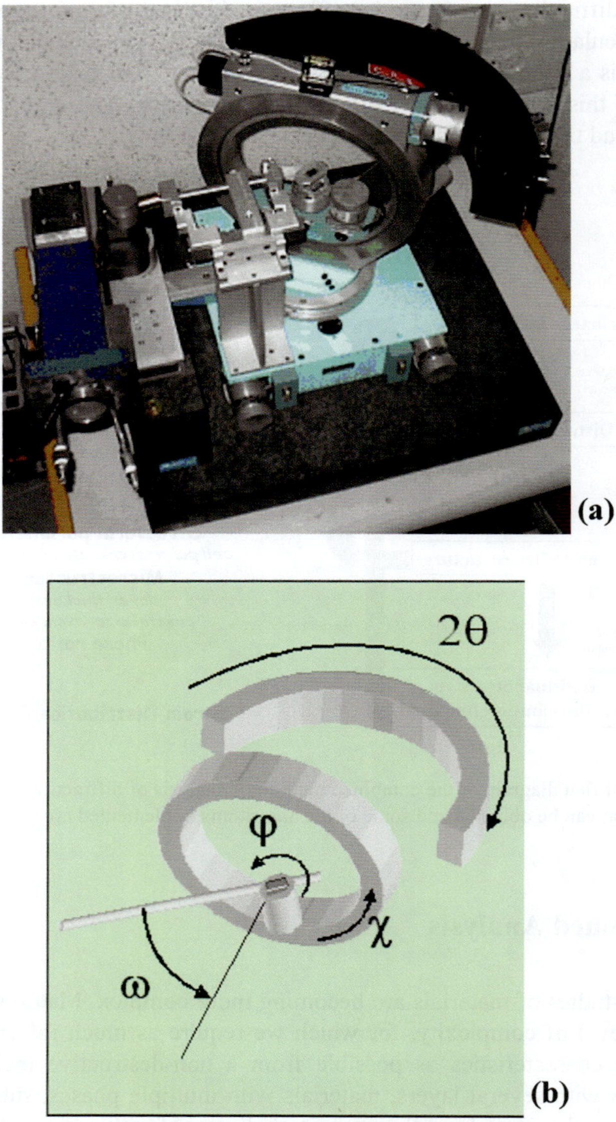

Fig. 8.25a Four-circle goniometer X-ray diffractometer equipped with a curved sensitive position detector. **b.** Schematic diagram showing the angle convention in a four-circle goniometer configuration.

A solution is the so-called combined analysis. This approach takes into account all the previous formalisms for texture, structure, microstructure, residual stresses, macroscopic elastic tensor and layering, by alternatively combining them in a single process. A first Rietveld refinement is operated in a cyclic manner on sets of diffraction diagrams measured at different sample orientations. Then, the extracted intensities are the input data for a Quantitative Texture Analysis cycle, the result of which is used to correct the diffraction intensities in the diagrams for the next Rietveld cycle. The result will be more reliable structural data and orientation distribution function. In between this process, a residual stress calculation of the polycrystal may be operated from the obtained OD. The operation leads to the determination of the parameters that produce the best solution for the whole ensemble of measurements.

Fig. 8.24 shows schematically the process of the combined analysis of diffraction data, the interdependency of the parameters accessible and the corresponding formalisms for refinement.The Materials Analysis Using Diffraction (MAUD) program [43] allows the use of the combined analysis of X-ray (conventional, synchrotron, monochromatic or energy dispersive, symmetric or asymmetric geometries, punctual, linear or planar detectors) and neutron (thermal, TOF) data.

8.4.1 Experimental Requirements for a Combined Analysis of Diffraction Data

The use of the combined analysis requires a large amount of diffraction data from the sample, which needs to be acquired using multiple detectors to avoid long measuring times. The first experiment allowing such an approach was developed using neutron data with a curved position sensitive (CPS) detector [47]. Using X-rays, the first experiment of this kind [35] used a CPS and a 4-circle diffractometer. Other studies follow, among them the characterization of thin ferroelectric structures [36–38, 39, 48].

Firstly, the diffractometer required for combined analysis must be equipped with a four circle goniometer in order to obtain reliable data for texture determination, *i.e.* at least one tilt rotation (χ), one azimuthal rotation (φ), an incidence angle (ω) and a detection circle (2θ), that can be a linear detector (Fig. 8.25).

The use of a curved position sensitive detector accelerates considerably the data acquisition as we can obtain for each (χ,φ) position a complete 2θ diffraction diagram (Fig. 8.26).

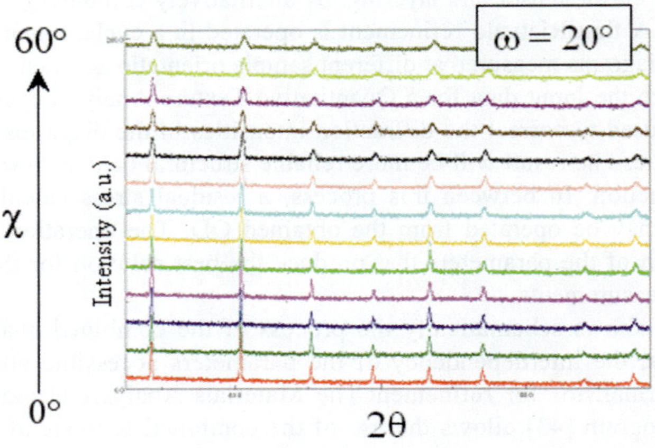

Fig. 8.26 Diffraction diagrams obtained with a diffractometer with a four-circle goniometer and a curve position sensitive detector.

8.4.2 Example of the Application of the Combined Analysis to the Study of a Ferroelectric Thin Film

A ferroelectric thin film of Ca-modified lead titanate of the nominal composition $Pb_{0.76}Ca_{0.24}TiO_3$ (PCT), was obtained by spin-coating deposition of a sol-gel processed solution on a $Pt/TiO_2/SiO_2/Si(100)$ substrate. The Pt layer on which the ferroelectric film is deposited on is a polycrystalline layer with <111> preferred orientation. The deposition of lead titanate based films on substrates with unrelated structures, i.e., without an obvious lattice match, as for this case, leads to a mixed orientation along <001> and <100> perpendicular to the film surface [49], which needs to be precisely characterized as it determines the ferroelectric behaviour of the films. In this tetragonal phase, the polar axis is along the <001> direction, which means that those crystallites oriented along <100> do not contribute to the net polarization of the film. From conventional quantitative texture analysis, we obtain that the films present a fibre-type texture (Fig. 8.27). But, the contributions of the two texture components cannot be determined accurately.

The PCT film texture is difficult to analyze due to the overlap of the diffraction peaks coming from the film and the substrate, and also of the 001 and 100 reflections from the PCT structure (Fig. 8.28). Integration in the conventional Quantitative Texture Analysis (QTA) is carried out over these two overlapped reflections, separating the <001> and <100> texture contributions during the WIMV iterative process [50, 51]. Besides, the information of those peaks overlapped with others from the underlying layers of the substrate is not considered. This reduces considerably the input data used for the OD refinement. In the combined method the 001, 100 peaks are deconvoluted first using the Rietveld refinement. Then the E-WIMV process follows, and an improved evaluation of the texture is achieved. The real separation of the two texture components and the estimation of their contributions have only been possible by the use of this method [38]. This is true for any other material whose texture components are derived from directions whose reflections are close.

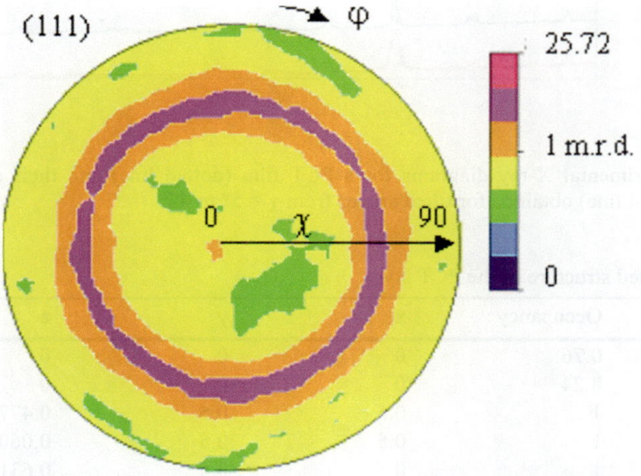

Fig. 8.27 PCT pole figure recalculated from the OD of a PCT/Pt/SiO2/(100)-Si thin structure that shows the fibre-type character of the texture. Equal area projection, logarithmic density scale.

Fig. 8.28 shows a selected series of X-ray diagrams measured at increasing tilt angles (every 5°), with their corresponding refinements using the combined method. It may be seen that all the diagrams are nicely reproduced, with reliability factors R_{Bragg} as low as 5 %. All this show the reliability of the refined values obtained with the combined method, not only for the ferroelectric film, but also for the Pt layer beneath it. The structural, microstructural and texture parameters obtained for both the PCT film and the Pt layer are summarized in Table 8.1 and Table 8.2.

Table 8.1 Structural, microstructural and texture parameters of the PCT and Pt layers obtained from the combined analysis of X-ray diffraction data (R_w = 7%; R_{Bragg} = 5%).

Layer	cell parameters (Å)	thickness (Å)	crystallite size (Å)	μ-strain (rms)	texture index (m.r.d.^{2})	RP_0 (%)
Pt	3.9108(1)	457(3)	458(3)	0.0032(1)	40.8	13.7
PCT	a = 3.9156(1) c = 4.0497(6)	4080(10)	390(7)	0.0067(1)	2.0	11.2

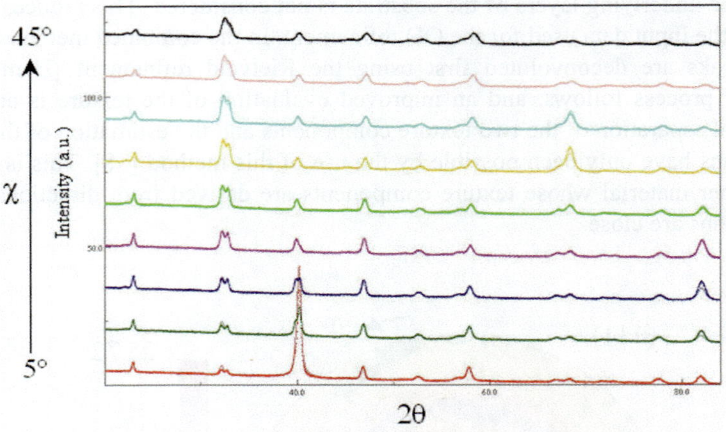

Fig. 8.28 Experimental X-ray diagrams for a PCT film (dotted line) and their corresponding refinement (solid line) obtained for tilted angles from $\chi = 5°$ to 45°.

Table 8.2 Refined structure of the PCT layer.

	Occupancy	x	y	z
Pb	0.76	0	0	0
Ca	0.24	0	0	0
Ti	1	0.5	0.5	0.477(2)
O1	1	0.5	0.5	0.060(2)
O2	1	0	0.5	0.631(1)

The lattice parameter obtained for the Pt is similar to the reported value by Swanson [52]: a = 3.9231 Å. But the ones corresponding to the PCT layer are not close to those reported previously for the same composition [53]: a=3.8939 Å and c=4.0496 Å. Although structural distortions may be present in the polycrystalline thin films, the possible explanation may be the stress state of these films. This difference in the cell parameters supports the idea that in the case of films deposited on a substrate, we should not rely in general on reported values but calculate them by the Rietveld method before starting a reliable texture analysis. The results of the microstructural parameters reveal the presence of larger microstrains in the PCT films than in the Pt layer, which presents the largest mean crystallite size. This is consistent with the fact that crystallites of the PCT film have sizes not larger than a tenth of the total layer thickness, while the Pt layer exhibits an average crystallite size that extends to the full thickness of the layer. The values obtained for the Pt and PCT layers thickness are close to the ones expected from the deposition conditions.

From the refined orientation distribution we can recalculate the pole figures (Fig. 8.29) for the PCT and Pt layers. Calculated texture indices are 2 mrd^2 for PCT and 41 mrd^2 for Pt. The texture of the Pt electrode is strong and characterized by <111> directions perpendicular to the film plane as the only component. The ferroelectric PCT film has a preferential orientation component along <100> perpendicular to the film surface. Interestingly, no significant component along <001> is observed (Fig. 8.29) that contains the polar axis for this tetragonal phase. It must be noted that the conventional quantitative texture analysis gives similar estimated contributions for both directions, which shows the higher accuracy of the combined method.

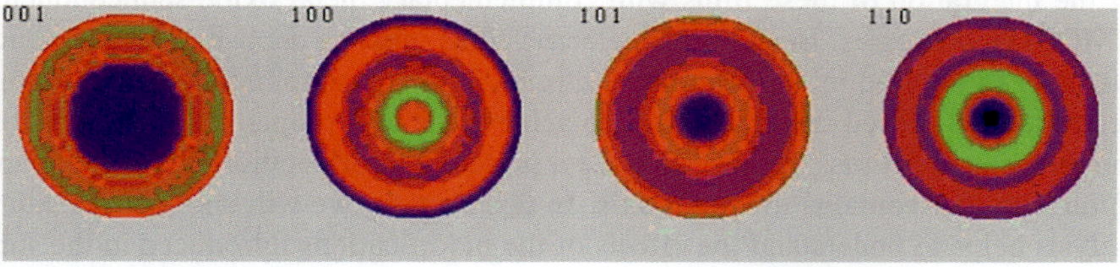

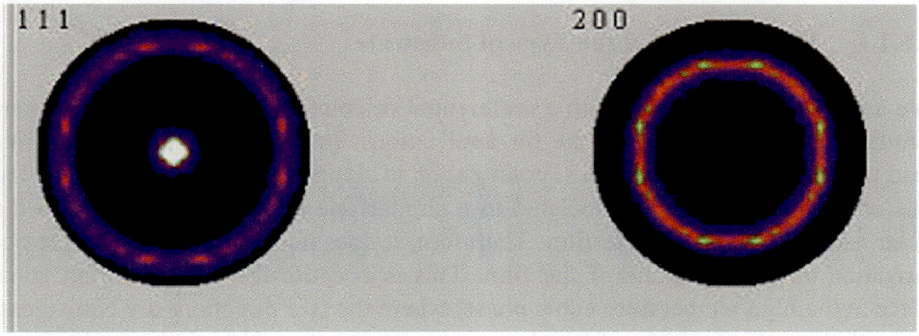

Fig. 8.29a. Recalculated pole figures for the PCT film. Equal area projections, linear density scale: 0.1-3.8 m.r.d.**b** Recalculated pole figures for the Pt layer. Equal area projections, logarithmic density scale: 0-63 m.r.d.

Compared to previous studies [50, 51], the approach appears to be far more powerful in extracting structural, microstructural and texture parameters in complex samples. Parameter divergence is astonishingly low, provided strongly dependent parameters were not released at the same time during the refinement process. This stability is probably due to the high number of experimental pole figures taken into account in the refinements, allowing a decrease of the defocusing effect (large at high χ-ranges) and a reduction of the number of possible OD solutions.

8.5 Texture of Polycrystalline Ferroelectric Films

8.5.1 Substrate Induced Texture Variations

The simplest approach to different textured films is the modification of the substrate on top of which the film is deposited. Silicon based substrates are important for the integration of these films with complementary metal oxide semiconductor (CMOS) technologies. Besides, ferroelectric films must be deposited on a conductive layer to be used in applications. Platinum is the most commonly used bottom electrode on Si-based substrates. But, to achieve the best properties of polycrystalline ferroelectric films, we need to have a preferential orientation out of the plane of the film that contains the polar axis. In this section, we will show how texture analysis helps to understand the effects of the modifications introduced in the substrate used.

8.5.1.1 Modifications of the Type of Substrate

We are interested in a film with a preferential orientation along its polar axis perpendicular to the films substrate for applications. In tetragonal perovskites, like lead titanate based compositions, polarization is along <001>. The problem is that this orientation is always associated to a similar one along <100>, i.e., with the polar axis in the plane of the film. Therefore, it does not contribute to the net polarization out of the plane of the film. This is because the crystallization takes place at the high temperature cubic phase, where the two directions are equivalent. Nevertheless, the presence of tensile or compressive stress during the cooling process may favour one of the two directions [54 55]. Among the possible origins of stress in films, the difference in the thermal expansion coefficients of the film and the substrate is one of them. Therefore, a change of the materials of the substrate may help tailor the contribution of the texture component along <001>. The results obtained by the use of different substrates are shown in Fig. 8.30. While the use of a Si-based substrate (Pt/TiO_2/Si(100)) produces a low textured PCT-Si film, the choice of Pt/MgO(100) or Pt/$SrTiO_3$(100) instead of Si produces a stronger contribution of the <001> texture component, as the larger density values in the centre of the corresponding pole figure show. This is the consequence of the compressive stress developed in these films during crystallization. It is difficult to separate these two texture components, as there is an important overlap of their corresponding diffraction peaks. This is not possible to study with the conventional methods of analysis.

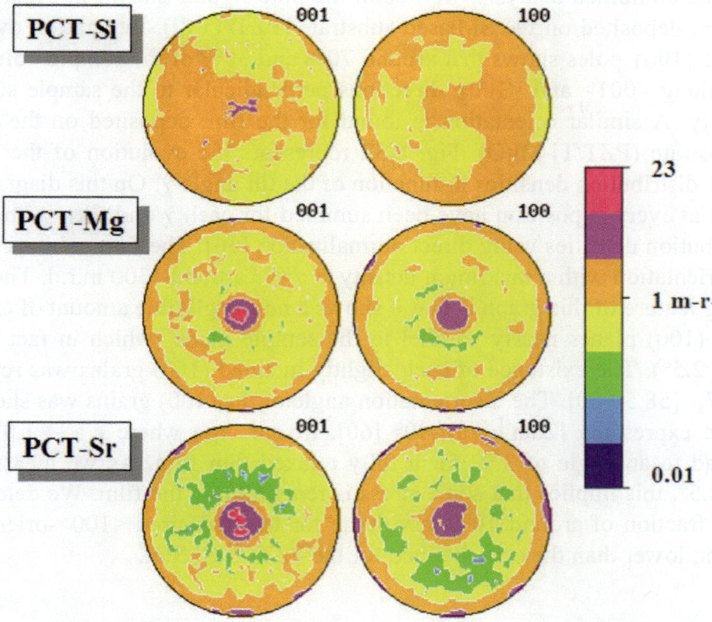

Fig. 8.30 Recalculated normalized pole figures for PCT films deposited on different substrates: $Pt/TiO_2/Si$ (PCT-Si); Pt/MgO (PCT-Mg); $Pt/SrTiO_3$ (PCT-Sr). Equal area projection and logarithmic density scale.

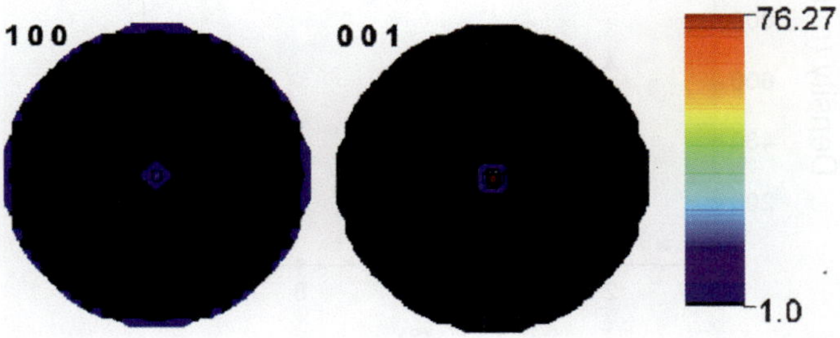

Fig. 8.31 Recalculated {100} and {001} pole figures of the PZT/Ti-Si film. Linear density scale, equal area projection.

It must be noted that the deposition of the ferroelectric film is on a polycrystalline Pt layer. Therefore, any significant change of its texture may be transferred to the film. The deposition of Pt on a Si-based substrate ($TiO_2/SiO_2/Si$) produces a fibre texture along <111>, but if it is on a MgO(100) under the right conditions the Pt film is oriented along <200>. To control the orientation of a $PbZr_{0.53}Ti_{0.47}O_3$ (PZT) film, a thin layer of TiO_x (2 nm) was deposited on top of the Pt before the ferroelectric film is prepared by radio frequency magnetron sputtering.

From the combined analysis, we obtain the pole figures shown in Figure 8.31 for the film deposited on the Si-based substrate (PZT/Ti-Si). Integration over the {001} and {100} poles shows that around 70% and 30% of the sample volume is oriented along <001> and <100> directions perpendicular to the sample surface, respectively. A similar orientation is found for the film deposited on the MgO-based substrate (PZT/Ti-MgO). Fig. 8.32 represents the evolution of the {001} and {100} distribution densities in function of the tilt angle χ. On this diagram, all intensities at every φ position have been summed for each χ and then normalized into distribution densities using direct normalization [56]. The film exhibits a high level of orientation with a maximum density at χ=0° around 1300 m.r.d. The most interesting feature of this graph is that it shows a non-negligible amount of crystallites with (100) planes nearly parallel to the sample plane (which in fact points around χ=2.5°). The existence of such slightly inclined (100) grains was reported earlier [57]- [58 59 60]. The disorientation angle of the (100) grains was shown to follow the expression $[2\tan^{-1}(c/a)]-90°$ [60]. In our case where c/a = 1.07, this should lead to an angle of 3.9° for a fully relaxed thin film. As we measure an angle of 2.5°, this implies that some stress is remaining in the film. We determine a volume fraction of around 10% in volume for the undesired <100>-orientation component, lower than the one obtained for the PZT/Ti-Si film.

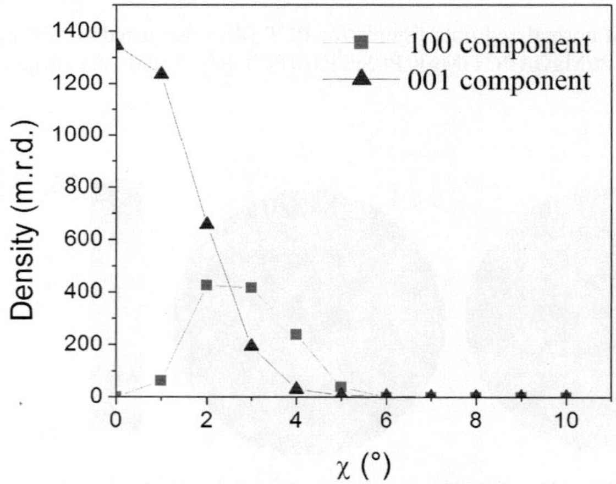

Fig. 8.32 {001} and {100} distribution density plot of the PZT/Ti-MgO film.

A change of the type of substrate may lead to the occurrence of some in-plane orientation contributions. $Pb(Zr_{0.6}Ti_{0.4})O_3$ films were obtained by RF sputtering [61] on a Si-based substrate: TiO_2/Pt(111)/TiO_2/SiO_2/Si(100) (PZT/Si); and on a $LaAlO_3$-based substrate: (TiO_2/Pt(111)/(012)$LaAlO_3$ (PZT-LAO). The TiO_2 layer is used to obtain in-situ crystallization and to promote growth along the <111> directions under previously determined conditions [7, 64, 65]. PZT-Si exhibits strong <111> fibre textures for the PZT and Pt layers, typical of the absence of epitaxial relationship with the substrate. On the contrary, the PZT-LAO sample shows strong hetero-epitaxial relationships between PZT and Pt (Fig. 8.33). The {116}-LAO pole figure allows the determination of epitaxial relationships between LAO, Pt, and PZT. The four poles from the LAO single crystal R-plane allow the location of the $[110]_p$ perovskite-like directions in the equatorial plane using simple crystallography. The {220}-Pt pole figure shows 12 poles up to χ = 60° whereas a perfect single crystal would exhibit only three for this <111> orientation. This means four epitaxial relationships are equivalently present in the Pt film. As for {116}-LAO, intensity variations between the poles are only due to the large scan grid used compared to the pole dispersion. This latter effect is less pronounced for {220}-PZT, indicating that this phase is slightly more distributed than Pt. The four equivalents are explained by the matching possibilities generated between the 6 <110> directions of Pt in the sample plane and the $<110>_p$-LAO directions. The same epitaxial components are stabilized in the PZT layer. Table 8.3 compares the texture found in both films.

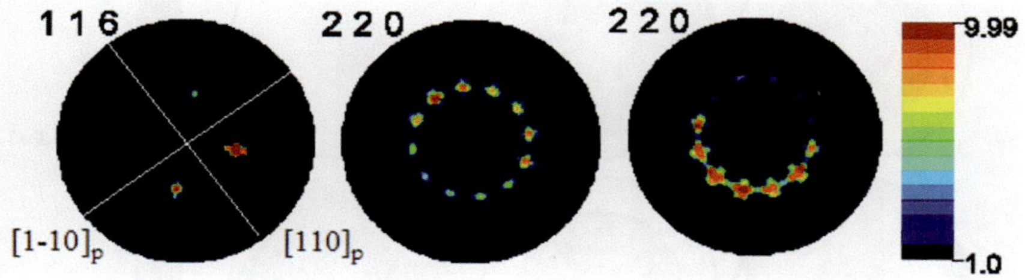

Fig. 8.33a {116}-LAO, **b** {220}-Pt, and **c)** $\{220\}_p$-PZT pole figures of the PZT-LAO film. $[hkl]_p$ state for a pseudo-cubic indexing. Equal-area projections, logarithmic density scale.

Table 8.3 Crystallographic hetero-epitaxial relationships between the substrate and the Pt layer, and between the Pt layer and PZT, out-of-plane and in-plane.

	substrate	Pt layer	PZT film
out-of-plane	Si[100]/SiO_2	Pt<111>	PZT<111>
in-plane	Si[001]/ SiO_2	random	random
out-of-plane	$LaAlO_3$ [012]*	Pt<111>	PZT<111>
in-plane	[110]*	<110>	PZT<110>

The search for stronger orientations along specific crystallographic directions in ferroelectrics leads to the epitaxial growth on specific substrates. Rhombohedral Pb_2ScTaO_6 (PST) films were deposited by RF sputtering [64] onto a magnesium oxide buffer layer that had been deposited on a $(11\bar{2}0)$-Al_2O_3 single-crystal (PST/MgO-AlO) and onto a conventional Pt/Ti/SiO_2/Si(100) (PST/Pt-Si). This sample exhibits strong preferred orientation along <100> perpendicular to the film surface as it may be seen in the pole figures of Fig. 8.34. A good agreement between experimental and recalculated pole figures may be observed, proving the high quality of the quantitative texture analysis. Pole figures show tetragonal symmetry, indicating a moderate in-plane orientation. This is corroborated by the calculated inverse pole figures (Fig. 8.35). The one corresponding to the normal direction Fig. 8.35a shows a preferential orientation along <100>. The strength of the PST in-plane alignment may be revealed by the maximum pole density in the other two inverse pole figures Fig. 8.35b and Fig. 8.35c, corresponding to the sample directions marked in Fig. 8.34 as I (parallel to [10$\bar{1}$0]-Al_2O_3 and [0001]-Al_2O_3) and II (at 45° from I), respectively. They show maxima of 5.5 m.r.d., less than a third of the density observed along the normal to the film surface (18.6 m.r.d.), along <100>-PST and <110>-PST with a broad girdle between the two.

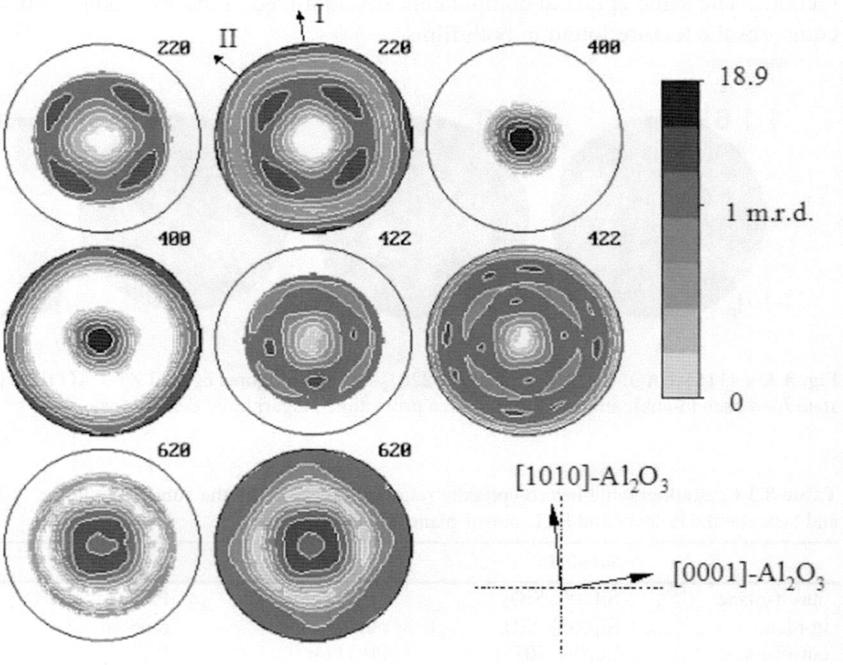

Fig. 8.34 Experimental and recalculated pole figures of film PST/MgO-AlO. Substrate in-plane parameters are indicated to identify the alignment. Equal area projection, logarithmic density scale.

The quantitative texture analysis of film PST/Pt-Si is also reliable as the agreement between recalculated and experimental pole figures of Fig. 8.36a shows. In this case, we obtain a strong preferred orientation along <111> direction at 5° from the normal of the film surface, but randomly distributed around this direction resulting in a fibre texture (Fig. 8.36a). The Pt layer shows also a <111> fibre texture tilted 5° respect to the film normal, which indicates the relation between these two layers. The inverse pole figure corresponding to the perpendicular to the film surface (Fig. 8.36b) corroborates that the only texture component of the PST film in this case is along <111>.

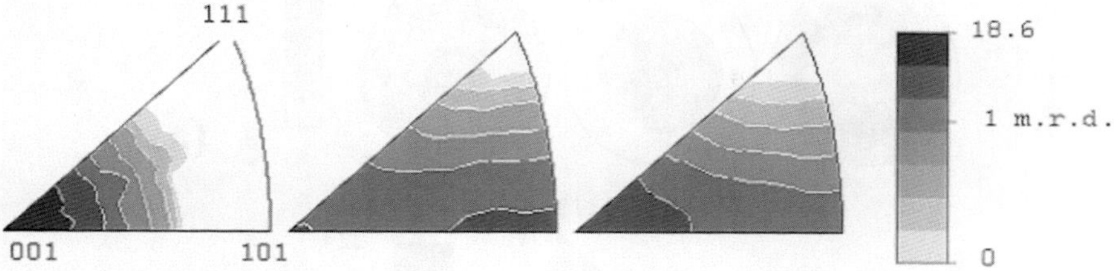

Fig. 8.35 Inverse pole figures of the film PST/MgO-AlO for: **a** direction perpendicular to the film surface, **b** direction I parallel to [100]-Al_2O_3 and **c** direction II at 45° from [100]-Al_2O_3 in the substrate plane. Equal area projection, logarithmic density scale, cubic sector.

The possible epitaxial relationships in film PST/MgO-AlO requires further study. The film was then aligned on the goniometer to have the **a** and **c** axes of the substrate ([$10\bar{1}0$]-Al_2O_3 and [0001]-Al_2O_3 directions respectively) approximately parallel to the vertical and horizontal pole figure axes (inset of Fig. 8.34). A subsequent texture analysis of the MgO layer reveals that the <$\bar{1}01$>-MgO directions align preferentially along the [$10\bar{1}0$]-Al_2O_3 and [0001]-Al_2O_3 axes of the substrate plane. This gives four statistically equivalent orientations with <111>-MgO tilted from the normal. We can deduce from the texture analysis that the major orientation relationships on this heterostructure are:

<100>-PST // ≈<111>-MgO // [$10\bar{2}0$]-Al_2O_3

<110>-PST // <211>-MgO and <110>-MgO // [$10\bar{1}0$] and [000l]-Al_2O_3

The mismatch in d-spacings between d_{110}-PST and two d_{110}-MgO is 3.4%, and 0.4% between three d_{220}-PST and five d_{211}-MgO (Figure 8.37a). The mismatches between MgO and Al_2O_3 are larger and might explain the larger dispersion of the MgO texture. A first orientation (Fig. 8.37b) corresponds to coincidences of four d_{110}-MgO d-spacings with d_{001}-Al_2O_3 (8.3% of mismatch), and seven d_{211}-MgO with three d_{100}-Al_2O_3 and a mismatch of 2.6%. A second orientation (Figure 9.37c) is due to coincidences of three d_{220}-MgO d-spacings with d_{100}-Al_2O_3 (8.5% of mismatch), and eight d_{211}-MgO with d_{001}-Al_2O_3 with 5.9% of mismatch. Due to the weak mismatch differences between these two orientations, MgO crystallites may choose either orientation with an equal probability. This generates, associated with the fibre tilt, the four-fold symmetry of the texture pattern. Even if the texture of MgO is relatively weak, it produces a PST texture strength comparable to the one obtained for the film PST/Pt-Si, and allows the in-plane alignment of low-index crystallographic directions of PST.

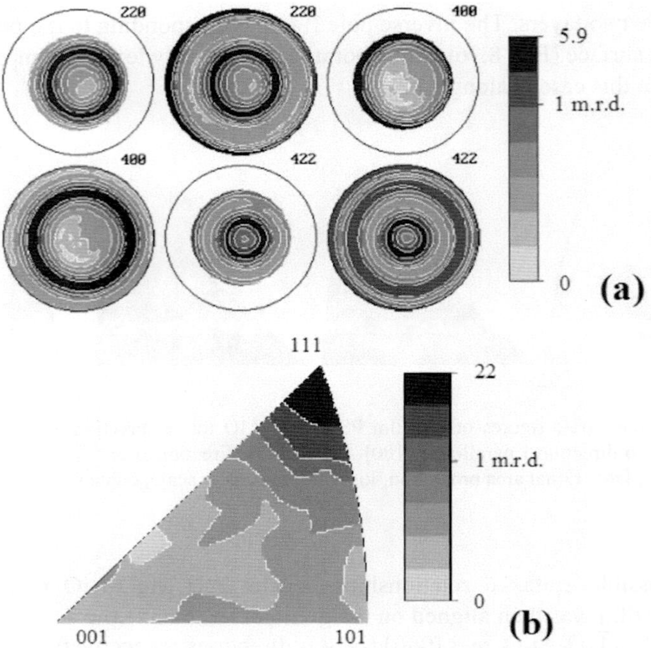

Fig. 8.36a Experimental and recalculated pole figures of the film PST/Pt-Si. **b** Inverse pole figure for the direction perpendicular to the film surface (cubic sector). Equal area projections, logarithmic density scales.

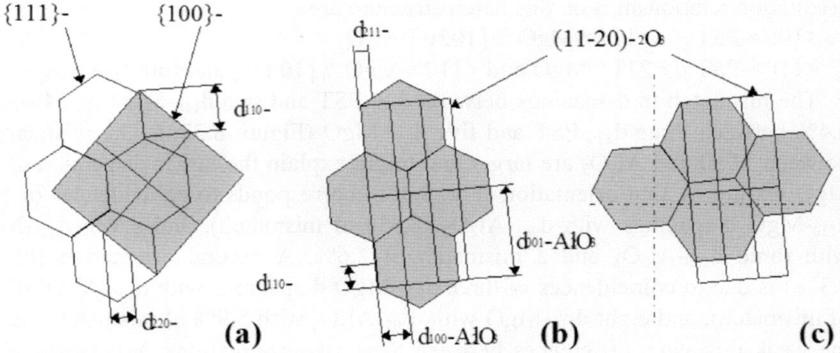

Fig. 8.37 Epitaxial relationships for film PST/MgO-AlO, as deduced from the quantitative texture analysis. Top layers are in gray shade: **a** {100}-PST planes on {111}-MgO, for one orientation of {111}-MgO; **b** first orientation of {111}-MgO plane on (11 $\bar{2}$ 0)-Al_2O_3; **c** second orientation of {111}-MgO planes on (11 $\bar{2}$ 0)-Al_2O_3.

The strategies to obtain the highest preferential orientation along the polarization axis in ferroelectrics sometimes include the choice of the right processing parameters. In the case of $LiNbO_3$ (LN) films, a two-step growth process was used which involves (1) creating a high-nucleation density by radio-frequency sputtering in the early stages of the film growth and (2) enhancing both the crystallinity and the texture by reactive chemical sputtering (also called pyrosol) [65]. The first weakly crystallized sputtered layer acts as a coherent buffer layer for the second pyrosol-layer, thus enhancing the desired texture strength and limiting chemical interfacial reactions. However, the substrate choice is determinant. The analysis of the texture of LN films deposited on Si(111) (LN/Si) and Al_2O_3(001) (LN/AlO) substrates results in a strong <001> preferential orientation [66]. In the LN film deposited on Si(111), the {100} pole figure (Fig. 8.38a) shows also <100> orientation, and the analysis of the {001} pole figure reveals that a ring centred at around 73° from the *c*-axes is a consequence of the occurrence of a <202> texture component. Similarly, in the analysis of the pole figures of the LN films deposited on the Al_2O_3(001) (Fig. 8.38b) we can see a slightly reinforced ring located at around 80° from the *c*-axes. It may be attributed to a texture component along <211>. For both films, the OD minimum is 0 indicating all the crystallites are orientated within the components. The texture index for LN/Si is F^2 = 9.7 m.r.d.^{2}, whereas for LN/AlO is 102 m.r.d.^{2}. This shows the strong hetero-epitaxial character of the texture in LN/AlO film. However, a single component of texture out-of-plane was achieved neither on Si nor on Al_2O_3 substrates. The main difference between the two heterostructures is in their in-plane orientation, as revealed by {100} pole figures (Fig. 8.38).

In the LN/Si film, all texture components are fibre-like (Fig. 8.38a). The film structure depends strongly on the initial growth stages. Crystallite orientation generally arises from either surface free energy or growth rate anisotropy [67]. LN is highly anisotropic and the c-axis is known to be one of the fastest growth directions. Then, after nucleation of the <001>-oriented grains, they grow preferably at the expense of other grain orientations. Besides, (111) planes of the fcc Si-substrate structure are highly dense, thus offering a large number of nucleation sites favourable for a dense and selected oriented growth. However, no coincidence site lattice with reasonable matching of the parameters could be identified between LN and Si. Therefore, there is no reason for ordering in the plane, resulting in an axially symmetric texture.

On the contrary, hetero-epitaxial-like textures are obtained on the <001>-Al_2O_3 substrate, with a d-spacing mismatch of 5 % ($d_{\{110\}}(Al_2O_3)$ = 0.258 nm; $d_{\{200\}}$(Si) = 0.271 nm), resulting in a stronger texture. Assuming a continuous oxygen sublattice at the interface between LN and Al_2O_3, the six-fold symmetry observed on the {100} pole figure (Fig. 8.38b) may be explained by two components of texture coming from differently aligned domains with different cation stacking sequences in each domain [68]. Each domain generates three-fold symmetry in the {100} pole figure, and is separated by 60° around the surface normal (namely $c_{\perp 0}$ being in exact alignment with the substrate and $c_{\perp 60}$). The

The intensities are approximately the same for the two stacking sequences and the film texture consist of two equally-distributed components with the following (1) and (2) hetero-epitaxial relationships:

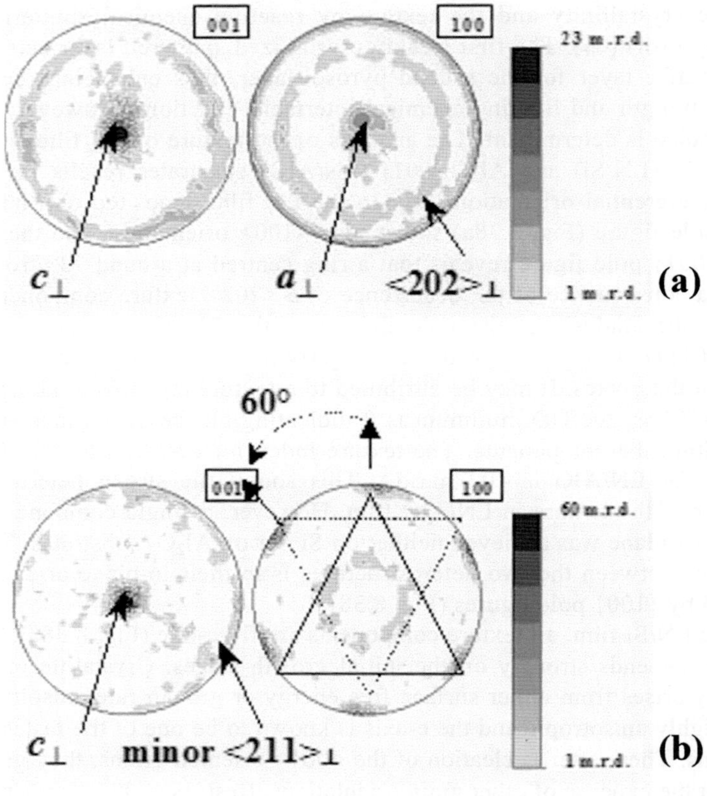

Fig. 8.38a Recalculated {001} and {100} pole figures of $LiNbO_3$ thin films deposited on Si(111) and **b** on $Al_2O_3(001)$ substrates. Equal area projection, logarithmic scale.

($c_{\perp 0}$) <001> $LiNbO_3$ // <001> Al_2O_3 and <100> $LiNbO_3$ // <100> Al_2O_3 (1)
($c_{\perp 60}$) <001> $LiNbO_3$ // <001> Al_2O_3 and <110> $LiNbO_3$ // <100> Al_2O_3 (2)

Both (111)-Si and (001)-Al_2O_3 correspond to close-packed planes. Thus, when adsorbed on the surface of the substrates, the adatoms nucleate and form islands close-packed enough. This is to prevent their in-plane spread during the coalescence step by limiting the lateral growth rate and develop a columnar microstructure while keeping the same out-of-plane orientation (fastest growth direction perpendicular to the film surface). The presence of two in-plane variants in hetero-epitaxial $LiNbO_3$ films deposited onto Al_2O_3 substrates has already been observed and discussed by several authors. In particular, the 60° variant is thought to be partially strain/stress-driven and, therefore, influenced by the lattice mismatch between the two materials

[69]. Even though $LiNbO_3$ and α-Al_2O_3 are commonly indexed using the hexagonal notation, they are really trigonal (R3c and R-3c, respectively) and have only three-fold symmetry. The 60°-rotated grains can easily be shown to have a higher cation alignment ordering energy compared to the 120°-rotated grains because of the reduced distance between Al, Li and Nb at the interface [70]. However, if the adsorbed atoms or adsorbed molecules do not have a sufficient mobility to move far on the growth interface, they may be incorporated into an appropriate low energy site unless they arrive near it. Thus, the growth from these two different nuclei would result in two crystallographic variants in the films. However we explain the 50% observation for each orientation component alternative. It is by a more simple symmetry consideration giving rise to the presence of two texture components, consisting in crystallites having grown with their *c*-axes in opposite direction with respect to one another. In other words, the two crystallographic variants could be compared to parallel and anti-parallel domains, as synthesized.

As a conclusion, different types of substrates may be used to obtain different preferential orientations, sometimes using epitaxial growth. The effects may be studied thanks to the quantitative texture analysis. But, because microelectronic devices are based on Si, special attention must be paid to induce texture on films deposited on Si-based substrates. They may also be modified to obtain different preferential orientations in ferroelectric films.

8.5.1.2 Modifications of Si-Based Substrates

The Si-based substrates with a Pt electrode are normally $Pt/TiO_2/(100)Si$. Pt is the most commonly used bottom electrode due to its resistance to oxidation at the temperatures necessary for the film processing. As the ferroelectric film is deposited on top of this layer, modifications of Pt may lead to variations in texture of the film. One possibility is the increase of the surface roughness of the Pt layer resulting from an annealing process previous to film deposition. This will disrupt the usual nucleation process of the deposited film on the substrate. The inducement of preferential orientation in the lead titanate based films is along the <111> direction on those annealed substrates [71]. Conventional quantitative texture analysis does not give information on the Pt layer, because of the overlaps with the diffraction peaks coming from the ferroelectric film. But the use of the combined method allows the study of the evolution of the structure and texture of Pt before and after the annealing process (Table 8.4). The fibre type <111> orientation of this layer suffers an important increase of the texture index with the annealing process, as a consequence of a preferred growth of the <111> oriented crystals over the others. The $Pb_{0.76}Ca_{0.24}TiO_3$ film deposited on it, PCT-B, shows a decrease of the values of their cell parameters (refined with the combined analysis approach), while a <111> texture component becomes the most important in comparison with a film deposited onto a substrate without any pre-annealing, PCT-A (Table 8.4). This may be a consequence of the increase of the stress during the film formation on the rougher Pt surface.

Table 8.4 Structural and texture data of PCT films and Pt layers on substrates with and without an annealing at 650°C previous to the film deposition.

	lattice parameters (Å)	texture index (m.r.d.)	main texture component
non-treated substrate			
Pt layer	a = 3.9108(1)	129	<111>
PCT-A film	a = 3.9156(1) c = 4.0497(6)	5.2	<100>
annealed substrate			
Pt layer	a = 3.9100(4)	199	<111>
PCT-B film	a = 3.8920(6) c = 4.0187(8)	2.1	<111>

The recalculated pole figures of the two PCT films deposited on untreated and annealed substrates (Fig. 8.39) show that a weak <111> texture component is already present in the PCT-A film deposited on an ordinary substrate. Its small contribution, together with the fact that the 111 reflection of PCT is close to the 111 of Pt, results in an underestimation of its contribution by conventional quantitative texture analysis. This means that this texture component had not been observed in previous studies [71]. Therefore, it seems that <111> becomes the most important texture component as the main <100> texture component does not appear in the annealed substrate. The decrease of the <100>-texture component results also in a decrease of the overall texture index.

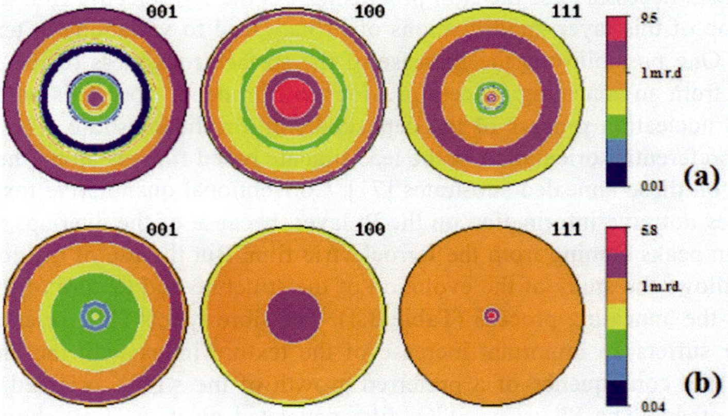

Fig. 8.39 Recalculated pole figures for PCT films deposited on **b** non-treated and **b** annealed at 650°C Pt/TiO_2/SiO_2/Si(100) substrates. Equal area projection and logarithmic density scale.

We can modify the substrate by adding an extra layer on top of the Pt layer to favour the growth of the grains along specific directions. The role of transient intermetallic layers, like Pt_xPb [72, 73], Pt_3Ti [74] and TiO_2 [75] have been reported to explain the occurrence of <111> preferential orientations of lead titanate based materials deposited on Pt layers.

To analyze the role of these intermetallic layers on texture using quantitative texture analysis, $Pb(Zr_{0.30}Ti_{0.70})O_3$ films were deposited by Chemical Solution Deposition methods on a normal Pt/Ti/Si(100) substrate (PZT-A) and on a Au/Pt/Ti/Si(000) (PZT-B). The Au layer will avoid the appearance of Pb, Pt or Ti-based intermetallics. OD calculations with low values of the reliability factors (RP0=20%; RP1=11%) showing the good quality of the refinement, reveal a reduction of the texture index from 32 m.r.d.^{2} (PZT-Pt) to 19 m.r.d.^{2} (PZT-Au). The {111} pole figures of both PZT (Fig. 8.40a) and Pt layers (not shown) present a strong maximum in their centre and random distributions of other axes around, characteristic of the <111> fibre textures. All OD minima for all layers are 0 m.r.d., indicating that all the material volume is textured within the components. The inverse pole figures of the PZT films for a direction perpendicular to the film surface (Fig. 8.40b) corroborate that the main texture component is the <111> fibre, although some minor orientation components along <110> and <100> can also be observed for PZT-Au.

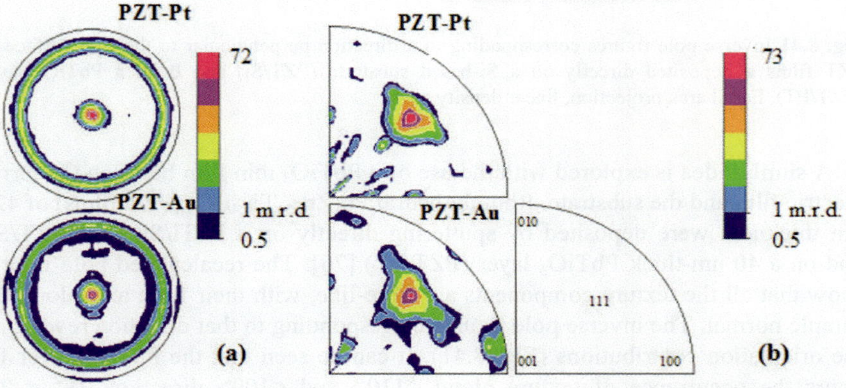

Fig. 8.40a Recalculated {111} pole figures of PZT thin films deposited on Pt/Ti/SiO_2/Si (PZT-Pt) and Au/Pt/Ti/SiO_2/Si (PZT-Au) substrates. **b** Inverse pole figures corresponding to a direction perpendicular to the film surface. Equal area projection, logarithmic density scale.

The use of an extra layer of Au on top of the Pt produces films less textured and containing the intermediate pyrochlore phase. The X-ray diffraction diagram of PZT-Au (not shown) suggests also less crystallinity than in the PZT-Pt film. The Au contributions are difficult to analyze, and reliable results are difficult to obtain. However, since Au crystallizes in an fcc crystal system with a cell parameter around 4.08 Å, we may expect this phase to follow approximately the Pt texture, though with less orientation densities but still accommodating the <111> orientation of PZT. But it seems that Au stops the occurrence of intermetallic layers that promote the preferential orientation of lead titanate based films.

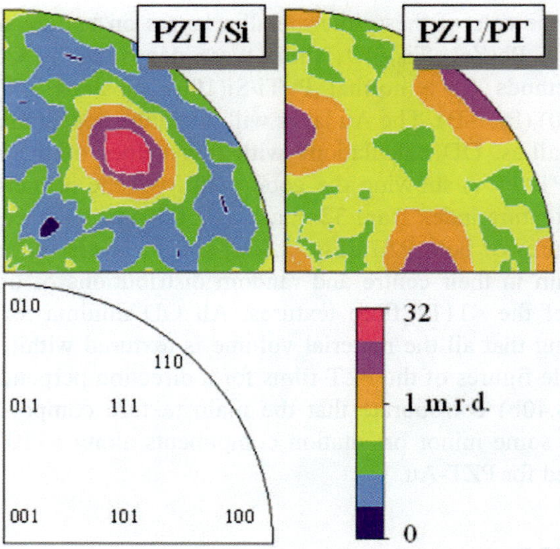

Fig. 8.41 Inverse pole figures corresponding to a direction perpendicular to the film surface of PZT films **a** deposited directly on a Si-based substrate (PZT/Si) and **b** on a $PbTiO_3$ layer (PZT/PT). Equal area projection, linear density scale.

A similar idea is explored with the use of a $PbTiO_3$ thin film between the ferroelectric film and the substrate. Rhombohedral $Pb(Zr_{0.54}Ti_{0.46})O_3$ (PZT films) of 420 nm thickness were deposited by sputtering directly on a Pt/Ti/Si(100) (PZT/Si) and on a 40 nm-thick $PbTiO_3$ layer (PZT/PT) [76]. The recalculated pole figures show that all the texture components are fibre-like, with their fibre axis along the sample normal. The inverse pole figures corresponding to that direction reveals all the orientation contributions (Fig. 8.41). It can be seen that the $PbTiO_3$ layer favours the occurrence of texture along <110> and <100> directions (F^2 = 2.5 m.r.d.^{2}), in contrast to the <111> preferential orientation without that layer (F^2 = 13 m.r.d.^{2}). Similarly, low reliability factors are obtained for the OD refinements (RP0: 15-26%; RP1: 10-13%).

Again, the introduction of a layer that stops the appearance of transient intermetallic layers between film and substrate disrupts the development of <111> textures. Instead, the crystals tend to nucleate on the surface of the $PbTiO_3$ on low energy planes of the high temperature cubic phase, like {001} and {110}. As a consequence, we obtain the texture components along <110> and <100> directions, but none along <111>, when the PZT film is deposited on a $PbTiO_3$ thin film.

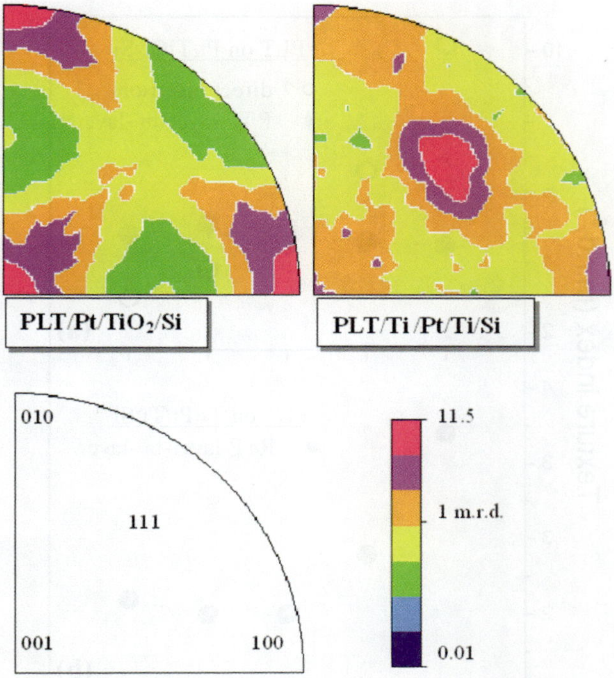

Fig. 8.42 Inverse pole figures corresponding to a direction perpendicular to the film surface for two PLT films deposited on Si-based substrates with and without a Ti layer. Equal area projection and logarithmic density scale.

It seems clear from previous studies [75, 77] that Ti containing layers play an important role in the inducement of <111> texture for lead titanate based thin films. However, none of them was able to clarify whether this orientation, although dominant, is the only one occurring. The results of the texture analysis of tetragonal $Pb_{0.88}La_{0.08}TiO_3$ (PLT) films deposited on a conventional Si-based substrate ($Pt/TiO_2/Si(100)$), and on a similar substrate but with an extra Ti layer (Ti/Pt/Ti/Si(100)), are shown in Fig. 8.42. The presence of the Ti layer produces a strong orientation along the <111> direction, as predicted, but preserving minor texture contributions along <001> and <100>. These contributions that account for 10 and 15 % of the oriented crystals, respectively, appear due to the nucleation of the $PbTiO_3$ perovskites on the low energy {100} planes of the high temperature cubic phase. Crystallization takes place in the cubic phase, and it seems those planes nucleate preferentially in the interface between film and substrate, even when there is no crystallographic relation with the substrate. Therefore, the “natural” orientation along <100>, <001> directions does not disappear when the <111> orientation is induced, but it becomes a small contribution that it is difficult to observe without a detailed analysis of texture.

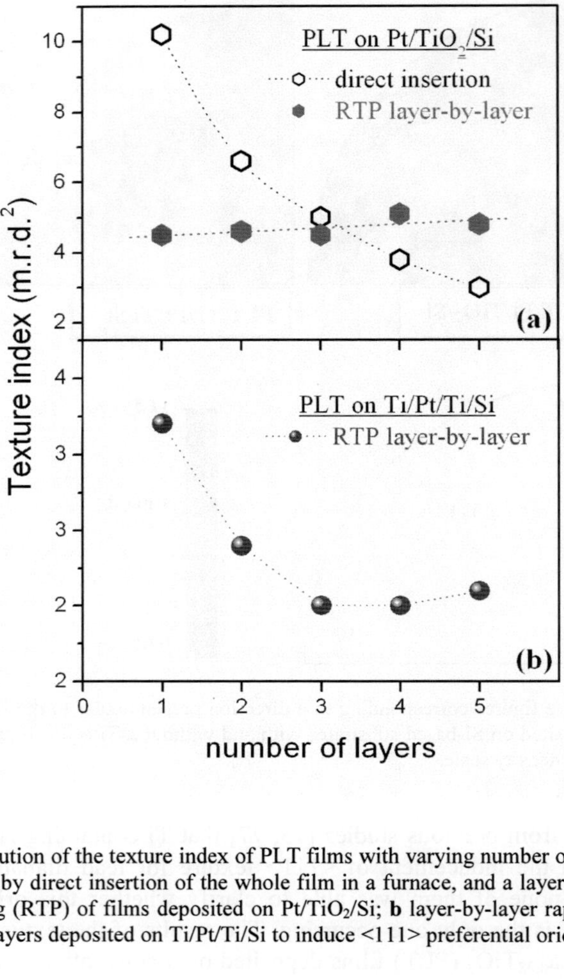

Fig. 8.43 Evolution of the texture index of PLT films with varying number of deposited layers: **a** crystallisation by direct insertion of the whole film in a furnace, and a layer-by-layer rapid thermal processing (RTP) of films deposited on Pt/TiO_2/Si; **b** layer-by-layer rapid thermal processing (RTP) of layers deposited on Ti/Pt/Ti/Si to induce <111> preferential orientation.

8.5.2 *Influence of the Processing Parameters on the Development of Texture in Thin Films*

The quantitative information on the texture of the films may be used to show tendencies that shed light onto the mechanisms involved in the development of preferential orientations, which can be closely related to their preparation process. This is shown in the study of $Pb_{0.88}La_{0.08}TiO_3$ (PLT) films prepared by chemical solution deposition methods with varying thickness [49]. An increasing number of deposited layers results in thicker films. Traditionally, the whole stack of deposited layers is crystallized in one step by direct insertion in a furnace. As we increase the number of layers, we observe a decrease of the degree of orientation, i.e., the texture index. As this means a limitation of the film thickness to obtain highly textured films, we modify the crystallization by the so-called layer-by-layer process. This consists of the rapid thermal processing (RTP) of each layer before the following one is deposited. Variations of the texture index (Fig. 8.42a) of a maximum of 10% are obtained in this case, suppressing the thickness dependence of texture. In both cases, we obtained a mixed <100>, <001> orientation. No significant variations of the contributions of the texture components are observed. We conclude that the relative amount of crystals nucleated at the substrate-film interface with preferential orientation decreases as the film thickness increases. The layer-by-layer crystallization makes possible this kind of nucleation for each layer (this time on the layer-to-layer interface), and as a consequence the relative amount of oriented crystals remains almost constant as thickness increases. Of course, this is valid if the nucleation of oriented crystals takes place only on the interfaces between layers. In the case of induced <111> orientations by the introduction of an extra Ti layer on the substrate (Fig. 8.43b), texture index decreases with increasing number of deposited layers. This is because, in this case, nucleation of <111> oriented crystals is only occurring on the Ti layer. Therefore, the relative amount of these crystals decreases with increasing the thickness of the film. This is also clear looking at the values of <111> contribution to the texture. When the contribution of the <111> component is below ~50%, the texture of the film starts to be dominated by the <001> and <100> components. This means that the thickness effect on the texture disappears, and the value of the texture index reaches a stable value.

Final Remarks

Texture is an important issue for polycrystalline ferroelectrics as it determines their physical properties. The use of advanced methods of analysis of the diffraction data, namely the quantitative texture analysis or the combined method, allows access to quantitative information on the different components of the global texture and to more accurate structural parameters, not available by more conventional approaches. From the results obtained, important conclusions can be drawn regarding the mechanisms of development of preferred orientations and, also, the correlation between them and the ferroelectric behaviour. As these techniques are still evolving, improved and more reliable results are expected that will allow us to solve other problems in the characterization by diffraction of these complex structures.

Acknowledgements

We are indebted to many collaborators and organisations. We would like to thank L. Lutterotti (DIM-Trento), M. Morales (CIMAP-Caen), E. Guilmeau (CRISMAT-Caen) for their assistance in the application of the combined approach on several case studies. Most of this work would not have been possible without the ferroelectric films prepared by several research groups. We wish to thank M.L. Calzada, M. Algueró and R. Poyato (ICMM-CSIC Madrid) for the preparation of the PCT and PLT films; G. Leclerc, R. Bouregba and G. Poullain (CRISMAT-Caen) for the preparation of sputtered PZT films; R. Whatmore and Q. Zhang (Cranfield University) and A. Patel (GEC-Marconi) for providing PZT films obtained by spin-coating; V. Bornand (Univ. Montpellier) for the elaboration of the $LiNbO_3$ and $LiTaO_3$ films; and M. Todd (DRA Malvern) for the elaboration of PST films.

These studies have been funded over years through several projects and contracts. The European Union project ESQUI "X-ray Expert System for electronic films Quality Improvement" within the GROWTH program (G6RD-CT99-00169) deserves special mention as it provided an essential support to develop the tools used in the quantitative analysis of polycrystalline ferroelectrics that we show in this chapter.

References

1. Scott J.F. (2000) Ferroelectric Memories. Springer Series in Advanced Microelectronics 3, Springer-Verlag, Berlin-Heidelberg
2. Arimoto Y., Ishiwara H. (2004) Current status of ferroelectric random-access memory. MRS Bull. 29: 823-828
3. Muralt P., Baborowski J., Ledermann N. (2002) Chapter 12. Piezoelectre microelectromechanical-systems with $PbZr_xTi_{1-x}O_3$ thin films: Integration and application issues. In: N. Setter (Ed.) Piezoelectric Materials in Devices, EPFL Swiss Federal Institute of Technology, Lausanne, pp. 231-260
4. Wouters D.J., Willems G., Lee E.U., Maes H.E. (1997) Elucidation of the switching processes in tetragonal PZT by hysteresis loop and impedance analysis. Integr. Ferroelectr. 15:79-87
5. Jia C.L., Urban K., Hoffmann S., Waser R. (1998) Microstructure of columnar-grained $SrTiO_3$ and $BaTiO_3$ thin film prepared by chemical solution deposition. J. Mater. Res. 13:2206-2217
6. Kim S.-H., Park D.-Y., Woo H.-J., Lee D.-S., Ha J., Hwang C.S., Shim I.-B., Kingon A.I. (2002) Orientation effects in chemical solution derived $Pb(Zr_{0.3},Ti_{0.7})O_3$ thin films on ferroelectric properties. Thin Solid Films 416:264-270
7. Bouregba R., Poullain G., Vilquin B., Murray H. (2000) Orientation control of textured PZT thin films sputtered on silicon substrate with TiO_x seeding. Mater. Res. Bull. 35:1381-1390
8. Randle V., Engler O. (2000) Introduction to texture analysis. Macrotexture, microtexture and orientation mapping. CRC Press, Boca Raton, Florida

9. Brentano J.C.M. (1946) Parafocusing properties of microcrystalline powder layers in x-ray diffraction applied to the design of x-ray goniometers. J. Appl. Phys. 17:420-434
10. Wcislak L., Bunge H.J., Nauer-Gerhardt C.U. (1993) X-ray diffraction texture analysis with a position sensitive detector. Zeitschrift für Metallkunde 84: 479-493
11. Heizmann J.J., Laruelle C. (1986) Simultaneous measurement of several x-ray pole figures. J. App. Cryst. 19:467-472
12. Legrand C., Yi J.H., Thomas P., Guinebretière R., Mercurio J.-P. (1999) Structural characterisation of sol-gel $SrBi_2Nb_2O_9$ thin film deposited on (001) $SrTiO_3$ single crystal. J. Eur. Ceram. Soc. 19: 1379-1381
13. Lotgering F.K. (1959) Topotactical reactions with ferrimagnetic oxides having hexagonal crystal structures-I. J. Inorg. Nucl. Chem. 9:113-123
14. Jones J.L., Slamovich E.B., Bowman K.J. (2004) Critical evaluation of the Lotgering degree of orientation texture indicator. J. Mater. Res. 19: 3414-3422
15. Brosnan K.H., Messing G.L., Meyer Jr. R.J., Vaudin M.D. (2006) Texture measurements in <001> fiber oriented PMN-PT. J. Amer. Ceram. Soc. 89 1965-1971
16. O'Connor B.H., Li D.Y., Sitepu H. (1991) Strategies for preferred orientation corrections in x-ray powder diffraction using line intensity ratios. Advances in X-ray Analysis 34 409-415
17. Capkova P., Peschar R., Schenk H. (1993) Partial multiplicity factors for texture correction of cubic structures in the disc-shaped crystallite model. J. Appl. Cryst. 26:449-452
18. Cerny R, Valvoda V., Cladek M. (1995) Empirical texture corrections for asymmetric diffraction and inclined textures. J. Appl. Cryst. 28:247-253
19. O'Connor B.H., Li D.Y., Sitepu H. (1992) Texture characterization in x-ray powder diffraction using the March formula. /Advances in X-ray Analysis. Advances in X-ray Analysis 35:277-283
20. Pernet M., Chateigner D., Germi P., Dubourdieu C., Thomas O., Sénateur J.-P., Chambonnet D., Belouet C. (1994) Texture influence on critical current density of YBCO films deposited on (100)-MgO substrates. Physica C 235-240:627-628
21. Isaure M.-P., Laboudigue A., Manceau A., Sarret G., Tiffreau C., Trocellier P., Lamble G., Hazemann J.-L., Chateigner D. (2002) Quantitative Zn speciation in a contaminated dredgeg sediment by μ-PIXE, μ-EXAFS spectroscopy and principal component analysis. Geochimica et Cosmochimica Acta 66:1549-1567
22. Chateigner D., Hedegaard C., Wenk H.-R. (1996) Texture analysis of a gastropod shell: Cypraea testudinaria. In Z. Liang, L. Zuo, Y. Chu (eds.) 11th International Conference on Textures of Materials. Vol. 2. Int. Academic Publishers, pp. 1221-1226
23. Bunge H.J., Esling C. (eds) (1982) Quantitative Texture Analysis. DGM, Germany
24. Bunge H.J. (1982) Texture Analysis in Materials Science. P.R. Morris Trans., Butterworths, London
25. Matthies S. (1979) Reproducibility of the orientation distribution function of texture samples from pole figures (ghost phenomena). Physica Status Solidi B 92:K135-K138
26. Ruer D. (1976) Méthode vectorielle d'analyse de la texture. PhD thesis, Université de Metz, France
27. Vadon A. (1981) Généralisation et optimisation de la méthode vectorielle d'analyse de la texture. PhD thesis, Université de Metz, France
28. Schaeben H. (1988) Entropy optimization in quantitative texture analysis. J. Appl. Phys. 64:2236-2237
29. Helming K. (1998) Texture approximations by model components. Materials Structure 5:3-9
30. Pawlik K. (1993) Application of the ADC method for ODF approximation in cases of low crystal and sample symmetries. Mater. Sci. Forum 133-136:151-156
31. Williams R.O. (1968) Analytical methods for representing complex textures by biaxial pole figures. J. Appl. Phys. 39:4329-4335

32. Imhof J. (1982) The resolution of orientation space with reference to pole figure resolution. Textures and Microstructures 4:189-200
33. Matthies S., Vinel G.W. (1982) On the reproduction of the orientation distribution function of texturized samples from reduced pole figures using the conception of a conditional ghost correction. Physica Status Solidi B 112:K111-K114
34. Schaeben H. (1991) Determination of complete ODF using the maximum entropy method. In Bunge H.J., Esling C. (eds) Advances and applications of quantitative texture analysis. DGM, Oberursel, Germany, pp109-118
35. Cont L., Chateigner D., Lutterotti L., Ricote J., Calzada M.L., Mendiola J. (2002) Combined X-ray texture-structure-microstructure analysis applied to ferroelectric ultrastructures: a case study on $Pb_{0.76}Ca_{0.24}TiO_3$. Ferroelectrics 267:323-328
36. Morales M., Chateigner D., Lutterotti L., Ricote J. (2002) X-ray combined QTA using a CPS applied to a ferroelectric ultrastructure. Mater. Sci. Forum 408-412:1055-1060
37. Lutterotti L., Chateigner D., Ferrari S., Ricote J. (2004) Texture, residual stress and structural analysis of thin films using a combined X-ray analysis. Thin Solid Films 450:34-41
38. Ricote J., Chateigner D. (2004) Quantitative microstructural and texture characterization by X-ray diffraction of polycrystalline ferroelectric thin films. J. Appl. Cryst. 37:91-95
39. Ricote J., Chateigner D., Morales M., Calzada M.L., Wiemer C. (2004) Application of the X-ray combined analysis to the study of lead titanate based ferroelectric thin films. Thin Solid Films 450:128-133
40. Chateigner D. (2005) Reliability criteria in Quantitative Texture Analysis with Experimental and Simulated Orientation Distributions. J. Appl. Cryst. 38:603-611
41. Chateigner D. (2002) POFINT: a MS-DOS program for Pole Figure Interpretation. http://www.ecole.ensicaen.fr/~chateign/qta/pofint/
42. Wenk H.R., Matthies S., Donovan J., Chateigner D. (1998) BEARTEX: a Windows-based program system for quantitative texture analysis. J. Appl. Cryst. 31:262-269
43. Lutterotti L., Matthies S., Wenk H.-R. (1999). MAUD (Material Analysis Using Diffraction): a user friendly Java program for Rietveld texture analysis and more. National Research Council of Canada, Ottawa 1999, 1599-1604. http://www.ing.unitn.it/~luttero/maud/
44. Matthies S., Humbert M. (1995) The combination of thermal analysis and time-resolved X-ray techniques: a powerful method for materials characterization. J. Appl. Cryst. 28:31-42
45. Chateigner D. (ed) (2004) Combined analysis: structure-texture-microstructure-phase-stresses-reflectivity analysis by x-ray and neutron scattering. To appear ISTE. http://www.ecole.ensicaen.fr/~chateign/texture/combined.pdf
46. Ricote J., Chateigner D., Algueró M. (2005) Intrinsic effective elastic tensor of ferroelectric polycrystalline lead titanate based thin films with fiber-type texture.Thin Solid Films 491:137-142
47. Chateigner D., Lutterotti L., Hansen T. (1998) Quantitative phase and texture analysis on ceramics-matrix composites using Rietveld texture analysis. ILL Highlights 1997 28-29
48. Lutterotti L., Matthies S., Chateigner D., Ferrari S., Ricote J. (2002) Rietveld texture and stress analysis of thin films by X-ray diffraction. Mater. Sci. Forum 408-412:1603-1608
49. Ricote J., Poyato R., Algueró M., Pardo L., Calzada M.L., Chateigner D. (2003) Texture development in modified lead titanate thin films obtained by chemical solution deposition on silicon-based substrates. J. Am. Ceram. Soc. 86:1571-1577
50. Ricote J., Chateigner D. (1999) Quantitative texture analysis applied to the study of preferential orientations in ferroelectric thin films. Bol. Soc. Esp. Cerám. Vidrio. 38:587-591

51. Ricote J., Chateigner D., Pardo L., Algueró M., Mendiola J., Calzada M.L. (2000) Quantitative analysis of preferential orientation components of ferroelectric thin films. Ferroelectrics 241:167-174
52. Swanson T (1953). Natl. Bur. Stand. (U.S.), Circ. 539, I, 31. (JCPDS file 04-0802)
53. Mendiola J., Jiménez B., Alemany C., Pardo L.,. Del Olmo L. (1989) Influence of calcium on the ferroelectricity of modified lead titanate ceramics. Ferroelectrics 94:183-188
54. Foster C.M., Li Z., Bucckett M., Miller D., Baldo P.M., Rhen L.E. Bai G.R., Guo D., You H., Merkle K.L. (1995) Substrate effects on the structure of epitaxial $PbTiO_3$ films prepared on MgO, $LaAlO_3$ and $SrTiO_3$ by metalorganic chemical-vapor deposition. J. Appl. Phys. 78:2607-2622
55. Ricote J., Chateigner D., Calzada M.L., Mendiola J. (2002) Preferential orientation of ferroelectric calcium modified lead titanate thin films grown on various substrates. Bol. Soc. Esp. Cerám. Vidrio 41:80-84
56. Guilmeau E., Funahashi R., Mikami M., Chong K., Chateigner D. (2004) Thermoelectric properties–texture relationship in highly oriented $Ca_3Co_4O_9$ composites. Appl. Phys. Lett. 85 :1490-1492
57. Foster C.M., Li S., Buckett M., Miller D., Baldo P.M., Rhen L.E., Bai G.R., Guo D., You H., Merkle K.L. (1995) Substrate effects on the structure of epitaxial $PbTiO_3$ thin films prepared on MgO, $LaAlO_3$, and $SrTiO_3$ by metalorganic chemical-vapour deposition. J. Appl. Phys. 78:2607-2622
58. Hsu W.Y., Raj R. (1995) X-ray characterization of the domain structure of epitaxial lead titanate thin films on (001) strontium titanate. Appl. Phys. Lett. 67:792-794
59. Foster C.M., Pompe W., Daykin A.C., Speck J.S. (1995) Relative coherency strain and phase transformation history in epitaxial ferroelectric thin films. J. Appl. Phys. 79:1405-1415
60. Speck J.S., Daykin A.C., Seifert A. (1995) Domain configurations due to multiple misfit relaxation mechanisms in epitaxial ferroelectric thin films. III. Interfacial defects and domain misorientations. J. Appl. Phys. 78:1696-1706
61. Leclerc G., Poullain G., Bouregba R., Chateigner D. (2008) Influence of the substrate on ferroelectric properties of <111> oriented $Pb(Zr_{0.6}T_{0.4})O_3$ thin films. Appl. Surf. Sci. 255:4293
62. Vilquin B., Le Rhun G., Bouregba R., Poullain G., Murray H. (2002) Effect of in situ Pt bottom electrode deposition and of Pt top electrode preparation on PZT thin films properties. Appl. Surf. Sci. 195:63-73
63. Leclerc G., Domenges B., Poullain G., Bouregba R. (2006) Elaboration of (111)-oriented La-doped PZT thin films on platinized silicon substrates. Appl. Surf. Sci. 253:1143-1149
64. Huang Z., Todd M.A., Watton R., Whatmore R.W. (1998) Sputtered lead scandium tantalite thin films: a microstructural study. J. Mater. Sci. 33:363-370
65. Bornand V., Chateigner D., Papet P., Philippot E. (1997). Piezoelectric thin films obtained by pyrosol process. Ann. Chim. Sci. Mat. 22: 683-686
66. Bornand V., Huet I., Chateigner D., Papet Ph. (2002) Oriented Growth of $LiNbO_3$ Thin Films for SAW properties. Mater. Sci. Forum 408-412:1573-1578
67. Smith D.L. (1995) Thin film deposition principles and practice. McGraw Hill, New York, pp. 327-380
68. Fork D.K., Armani-Leplingard F., Kingston J.J., Anderson G.B. (1985) Thin Film Epitaxial. Oxide Optical Waveguides. Mater. Res. Symp. Proc. 392:189-200
69. Derouin T.A., Lakeman C.D.E., Wu X.H., Speck J.S., Lange F.F. (1997) Effect of lattice mismatch on the epitaxy of sol-gel $LiNbO_3$ thin films. J. Mater. Res. 12:1391-1400
70. Guo J., Ellis D.E., Lam D.J. (1992) Electronic structure and energetics of sapphire (0001) and (1-102) surfaces. Phys. Rev. B 45:13647-13656

71. Ricote J., Morales M., Calzada M.L. (2002) Texture analysis of ferroelectric thin films on platinized Si-based substrates with a TiO_2 layer. Mater. Sci. Forum 408-412:1543-1548
72. S.Y. Chen and I.W. Chen (1994) Temperature-time texture transition of $Pb(Zr_{1-x}Ti_x)O_3$ thin films: I. Role of Pb-rich intermediate phases. J. Am. Ceram. Soc. 77:2332-2336
73. Z. Huang, Q. Zhang and R.W. Whatmore (1999) Structural development in the early stages of annealing of sol-gel prepared lead zirconate titanate thin films. J. Appl. Phys. 86:1662-69
74. Y. Liu and P.P. Phulé (1996) Nucleation- or growth-controlled orientation development in chemically derived ferroelectric lead zirconate titanate ($Pb(Zr_xTi_{1-x})O_3$, x=0.4) thin films. J. Am. Ceram. Soc. 79:495-98
75. Muralt P., Maeder T., Sagalowicz L., Hiboux S., Scalese S., Naumovic D., Agostino R.G., Xanthopoulos N., Mathieu H.J., Patthey L., Bullock E.L. (1998). Texture control of $PbTiO_3$ and $Pb(Zr,Ti)O_3$ thin films with TiO_2 seeding. J. Appl. Phys. 83:3835-3841
76. Cattan E., Velu G., Jaber B., Remiens D., Thierry B. (1997) Structure control of $Pb(Zr,Ti)O_3$ films using $PbTiO_3$ buffer layers produced by magnetron sputtering. Appl. Phys. Lett. 70:1718-1720
77. Calzada M.L., Poyato R., García López J., Respaldiza M.A., Ricote J., Pardo L. (2001) Effect of the substrate heterostructure on the texture of lanthanum modified lead titanate thin films. J. Eur. Ceram. Soc. 21:1529-1533

Chapter 9
Nanoscale Investigation of Polycrystalline Ferroelectric Materials via Piezoresponse Force Microscopy

V. V. Shvartsman[1], A. L. Kholkin[2]

9.1 Introduction

Ferroelectrics possess a wide spectrum of functional properties including switchable polarization, piezoelectricity, pyroelectricity, dielectric nonlinearity, and high non-linear optical activity, which make these materials promising for a large number of applications [1]. These include nonvolatile random access memories (FERAM) [2], micro-electromechanical systems (MEMS) [3], infrared detectors, optical modulators and waveguides, and many others [4, 5]. The general trends of miniaturization in modern electronics demand a decrease in the size of the active ferroelectric elements to a submicron scale. This in turn necessitates the development of microscopic techniques allowing for the evaluation of ferroelectric and piezoelectric properties with nanoscale resolution. Several fundamental issues have to be addressed such as the effect of the films thickness and lateral size of the capacitor, or of the single grain on ferroelectric and piezoelectric properties, the relationship between grain/capacitor size and peculiarities of the polarization switching, and mechanisms of degradation effects, such as retention, imprint, and polarization fatigue [2].

To answer these questions both ferroelectric domain structures and their evolution during polarization switching have to be studied at micro- and nanoscales. This can be done using scanning probe microscopy (SPM) techniques, which provide an opportunity for non-destructive visualization of domains in ferroelectric thin films, single crystals and ceramics. SPM has made possible the mapping of the surface potential and charge distribution, evaluation of local

---

[1] Angewandte Physik, University of Duisburg-Essen, Duisburg, Germany
vladimir.shvartsman@uni-due.de
[2] Dept. of Ceramic and Glass Engineering, CICECO, University of Aveiro, Aveiro, Portugal
kholkin@ua.pt

electromechanical properties, and measurements of the non-linear dielectric constants. Since the first SPM imaging of the 180° domain walls in $Gd_2(MoO_4)_3$ [6], a growing number of research papers on nanoscale properties of ferroelectrics studied by SPM have been published (see recent review in [7]). Several novel SPM techniques based on different approaches were adopted or specially developed for these studies [8].

Depending on the type of interaction between the probing tip and the sample – attractive or repulsive – the SPM can operate in non-contact and contact regimes, respectively. In the non-contact regime, the tip is scanned over the surface at a distance of 10-100 nm. The cantilever is mechanically driven to oscillate near its resonance and the feedback loop adjusts the tip-to-sample distance to maintain, for example, the constant amplitude of the oscillation. The tip-sample interaction is dominated by the Van-der-Waals forces and, in the case of polar or charged materials, the electrostatic forces may contribute. In particular, when a small ac-voltage is applied to the tip, the electrostatic interaction between the tip and surface charges results in an oscillation of the cantilever. From the amplitude and the phase of this oscillation, the charge density and polarity of the charges may be estimated [9, 10]. This mode of SPM, called electrostatic force microscopy (EFM), may be used for ferroelectric domain imaging by detecting the sign of the surface polarization charges [6, 11, 12, 13, 14, 15, 16]. In another approach, a small dc-bias is applied to the tip mechanically driven at the resonance frequency. The electrostatic force between the tip and the surface results in a change of the cantilever resonant frequency, which is proportional to the force gradient. The frequency shift is collected as the EFM image [17, 18].

In the Kelvin probe force Microscopy (KPFM), a dc-bias and an ac-voltage are applied simultaneously to the tip $V_{tip}=V_{dc}+V_{ac}\cos\omega t$. The capacitive (Maxwell) force acting between the tip and the surface with a potential V_s is

$$F_{cap}(z) = \frac{1}{2}(V_{tip} - V_s)^2 \frac{\partial C}{\partial z} \tag{1}$$

where z is the distance between the tip and the surface and $C(z)$ is the tip-surface capacitance. The first harmonics of this force is

$$F_{cap}^{1\omega}(z) = (V_{dc} - V_s)V_{ac} \frac{\partial C}{\partial z} \tag{2}$$

The feedback loop is used to nullify this term by adjusting V_{dc}=-V_s. Thus mapping of the nullifying potential, V_{dc}, yields a distribution of the surface potential [19, 20]. This provides important information on the surface electronic properties of ferroelectrics, such as distribution of polarization and screening charges and their evolution during phase transitions [21, 22, 23, 24]. To minimize a possible crosstalk between topography and electrostatic signals, the EFM and KPFM measurements are often done in so-called two-pass technique (Lift Mode) [25]. Each line is scanned twice in this mode. In the first scan the topography of the surface is determined, and during the second scan the tip is lifted to a certain height above the sample surface. This allows for the reconstruction of the distribution of charge or potential on the surface without topographical contribution. General drawbacks of non-contact methods include low-resolution due to large tip-surface separation, sensitivity to sample surface conditions, and susceptibility to screening effects.

Contact modes operate in the repulsive force regime: the tip is in permanent contact with the surface. The feedback loop is adjusted to maintain a constant bending of the cantilever. Ferroelectric domain imaging methods in the contact mode may be divided into static and dynamic ones.

Among the static methods are lateral (friction) force microscopy (LFM) and conventional contact atomic force microscopy (AFM). LFM is based on the detection of the torsion deformation of the cantilever due to frictional forces between the tip and the surface. The structural differences between surfaces of oppositely polarized domains modify the surface potential resulting in two different friction coefficients experienced by the tip [26, 27]. Twinning between domains with in-plane and out-of-plane polarization (*a*- and *c*-domains, respectively) results in surface corrugations at the 90° domain walls. This allows studying ferroelastic domain patterns in single crystals and epitaxial films by topographic imaging of their surfaces [28, 29, 30]. Contact AFM was also used for the visualization of 180° domains in some single crystals via the detection of static thickness change (shrinkage or expansion), piezoelectrically induced by a dc-voltage applied to the tip during scanning [31, 32].

The dynamic methods include scanning non-linear dielectric microscopy (SNDM), atomic force acoustic microscopy (AFAM), and piezoresponse force microscopy (PFM). In SNDM, the sample is a part of a capacitor in a LC resonator circuit. The voltage applied to the tip is modulated in the microwave frequency range. By detecting the voltage-induced changes in the local capacitance SNDM is able to measure point-to-point variations of the non-linear dielectric response of the sample, which translates the distribution of local ferroelectric polarization [33]. This technique may achieve a sub-nanometre lateral resolution [34]. However, the measured non-linear dielectric response is related to a thin surface layer (<10 nm) [35]. SNDM may also be used for the detection of polarization component parallel to the surface [36].

In AFAM the cantilever is mechanically excited at its resonant frequency, while the tip remains in contact with the sample. Variations in elastic properties of the sample result in an apparent shift of the contact resonant frequency and, correspondingly, in a change of the oscillation amplitude. These changes may be monitored either locally in the spectroscopic mode for quantitative evaluation of the local contact stiffness, or in the scanning mode for two-dimensional mapping of the elastic properties [37]. The AFAM method has been applied to reveal the ferroelectric domains at the sub-grain level due to variations in local stiffness and to measure elastic constants in ferroelectric ceramics [37, 38, 39, 40].

9.2 Principle of Piezoresponse Force Microscopy

The most popular and widely used SPM technique for the investigation of ferroelectric materials is piezoresponse force microscopy (PFM) [7, 41, 42, 43, 44]. It is based on the detection of surface deformation due to the converse piezoelectric effect induced by an ac-voltage applied to the conductive SPM tip. Initially, this method was proposed by Güthner and Dransfeld to detect polarized areas in ferroelectric copolymers [45]. During subsequent years it has been intensively developed and now PFM is recognized as an effective tool for the nanoscale study and control of ferroelectric domains in bulk ferroelectric materials and thin films. In particular, in polycrystalline materials along with the domain visualization and evaluation of local piezoelectric coefficients, this technique allows for a direct matching of local properties to the microstructural details, since both piezoresponse image and topography image are acquired simultaneously [46]. In such a way, it is possible to study the interaction between domain wall and grain boundaries, size effect on domain pattern and local piezoelectric properties. The advantages of PFM are high resolution down to several nanometres, a possibility of the three-dimensional reconstruction of the domain structure, and an *in-situ* investigation of effects of local mechanical or electrical impact on domain structure and local piezoelectric properties.

9.2.1 Experimental Setup

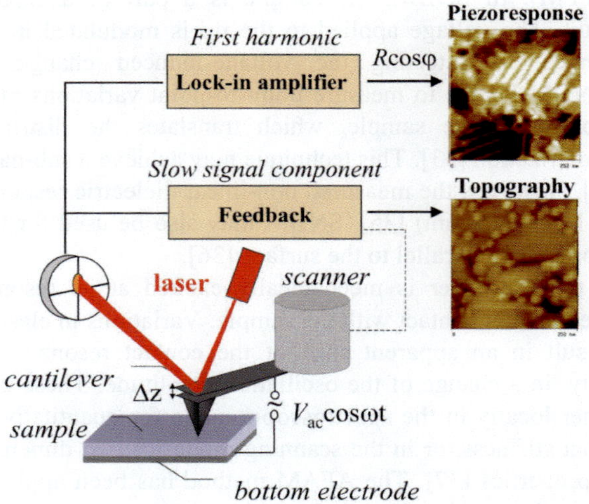

Fig. 9.1 PFM experimental set-up for simultaneous acquisition of topography and piezoresponse. An ac-voltage V_{ac} from a functional generator is applied between the tip and the bottom electrode. The induced cantilever deflection is demodulated with a lock-in amplifier.

A typical experimental PFM set-up consists of a standard scanning probe microscope equipped with a conductive tip/cantilever, a functional generator and one or two lock-in amplifiers (Fig. 9.1). A studied ferroelectric sample is placed between the macroscopic bottom electrode and the conductive tip, which is used as a movable top electrode. The deformation of the sample is transmitted into deflection of the cantilever, which is detected by the optical lever method. In this case the laser beam is reflected from the cantilever on a position sensitive four-quadrant photodetector. Once the cantilever is displaced, the reflected beam moves over the photodiode and changes the amount of light shining on each of the quadrant. Comparing the signals from two upper and two bottom quadrants (top-bottom signal), one may measure a vertical deflection of the cantilever, i.e., vertical displacement of the tip. In the case of PFM, this displacement is caused both by the electromechanical response of the surface due to converse piezoelectric and electrostrictive effects, and by the electrostatic force between tip/cantilever system and the sample.

9.2.2 *Electromechanical Contribution*

For stress-free ferroelectric materials homogeneously polarized in z-direction, the vertical displacement of the surface may be expressed as follows:

$$\Delta z = d_{33}V + \frac{M_{333}}{t}V^2 \tag{3}$$

where V is the applied voltage, t is the sample thickness, d_{33} and M_{333} are the piezoelectric and electrostrictive coefficients, respectively. Under the external voltage $V_{tip}=V_{dc}+V_{ac}\cos\omega t$, the surface displacement will consist of a dc-component and contributions at first and second harmonics [44]:

$$\Delta z_{dc} = d_{33}V_{dc} + \frac{M_{333}}{t}\left(V_{dc}^2 + \frac{1}{2}V_{ac}^2\right) \tag{4}$$

$$\Delta z_{\omega} = d_{33}V_{ac} + 2\frac{M_{333}}{t}V_{dc}V_{ac} \tag{5}$$

$$\Delta z_{2\omega} = \frac{1}{2}\frac{M_{333}}{t}V_{ac}^2 \tag{6}$$

The sign of the piezoelectric effect, i.e., the first term in Equation 3, depends on the mutual orientation of the polarization and the applied electric field. If the polarization vector is pointing at the negative *z*-direction (we assume that *z*-axis is directed from the bottom electrode to the tip), a positive potential applied to the tip induces an expansion of the sample. Displacement Δz has positive sign (Fig. 9.2a).

When the ac-voltage changes the polarity, the sample is contracted and Δz becomes negative. This means that the applied voltage and the first harmonic of surface displacement, $\Delta z_{\omega}(t)$, (hereafter piezoresponse signal) are in phase. If the polarization points at the positive z-direction, the signs of the ac-voltage and the displacement are opposite (Fig. 9.2b), i.e., piezoresponse signal is shifted in phase by 180 degrees relative to the driving voltage. The electrostrictive effect does not depend on polarization direction and causes only a constant background. The contribution related to the electrostriction is typically much smaller than the piezoelectric one in a polarized state [47] and vanishes when no dc-field is applied to the sample. These measurements are referred to as vertical (or out-of-plane) PFM (VPFM). VPFM enables distinguishing domains with different out-of-plane component of the spontaneous polarization. Both the phase, θ, and amplitude, R, of the response are measured using the lock-in technique and recorded together with the regular AFM topography signal. Typically, domains differ by their phase contrast or by the amplitude "mixed" with phase ($x=R\cos\theta$) signal.

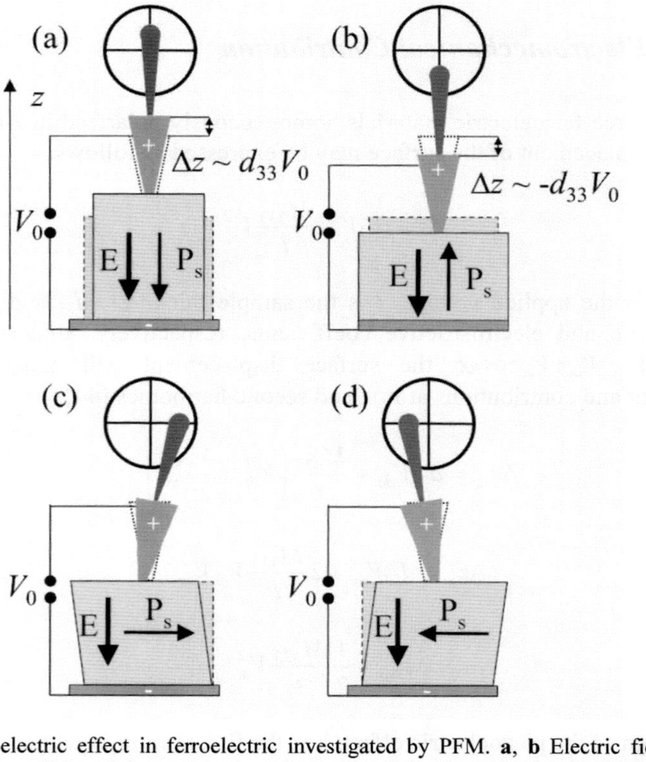

Fig. 9.2 Piezoelectric effect in ferroelectric investigated by PFM. **a**, **b** Electric field aligned parallel or antiparallel to the spontaneous polarization results in a vertical displacement of the cantilever. **c**, **d** Electric field applied perpendicular to the polarization results in a shear deformation related to d_{15} coefficient. This causes a torsional movement of the cantilever forcing a horizontal deflection of laser spot on photo-detector (in-plane signal).

When the polarization is parallel to the sample surface, the voltage applied to the tip results in an in-plane surface deformation. This deformation is transferred via the friction forces to the torsional movement of the cantilever, which is detected by comparison of the signals from two left and two right quadrants of the photodetector (left-right signal). These measurements are usually denoted as lateral (or in-plane) PFM (LPFM) [48]. It has to be mentioned that LPFM is sensitive to the piezoresponse component (polarization) perpendicular to the long cantilever axis. The surface displacement along the cantilever results in buckling of the cantilever and contributes to the VPFM signal.

In the case of a sample with an arbitrary polarization direction the piezoelectric effect will cause both vertical and torsional movement of the cantilever. Thus, combining VPFM and LPFM, it is possible to perform a qualitative reconstruction of the domain pattern. In this case, the second LPFM image has to be taken after rotation of the sample at 90 degrees around the *z*-axis to obtain full information about in-plane polarization distribution. The precise quantitative orientation of polarization can be calculated only if all components of the piezoelectric tensor are known. For the arbitrary oriented sample, the components of the piezoelectric tensor in the laboratory coordinate system, d_{ij}, are linear combinations of the tensor component in the coordinate system related to the principal crystal axis $d^*_{ijk}(\varphi,\theta,\psi) = a_{il}a_{jm}a_{kn}d_{lmn}$ [49], where φ, θ, and ψ are the Euler angles [50] and a_{ij} are the elements of the Euler matrix that describes the rotation defined by the Euler angles. In particular, angle θ describes counter-clockwise rotation from the positive direction of *z*- axis around the *x*-axis. The orientation dependence of the PFM signal has been analyzed by several research groups [7, 51]. As an example, for the materials which have the tetragonal symmetry (point group 4mm), the longitudinal piezoelectric coefficient d_{zz} can be expressed as

$$d_{zz} = (d^0_{31} + d^0_{15})\sin^2\theta\cos\theta + d^0_{33}\cos\theta \qquad (7)$$

where θ is the angle between the normal to the sample surface and polarization direction. As follows from Equation 7, d_{zz} has a maximum at a certain angle, θ_{max}, which depends on the relationship between the components of the piezoelectric tensor. In particular, for $PbTiO_3$ θ_{max}=0 (i.e., d_{zz} maximum corresponds to the polar direction) [52]. But, for $BaTiO_3$ at room temperature θ_{max}=43° [53]. Thus, a precise reconstruction of polarization orientation demands a rigorous analysis of all three PFM images. In principle, the prior knowledge of all components of the piezoelectric tensor is required.

During the reconstruction of the polarization distribution, a possible crosstalk between the VPFM and LPFM signals has to be taken into account. The typical origin of this crosstalk is a possible misalignment, e. g., arbitrary rotation of the four-quadrant photodiode, which thereby may be not in the same plane as the cantilever. As a result, a pure vertical displacement of the cantilever will cause the lateral signal to change and vice-versa. This will result in misinterpretation of the obtained PFM images. This instrumental crosstalk may be corrected, for example, by compensating the circuit as it was proposed by Hoffman *et al.* [54]. Another kind of crosstalk may arise in PFM studies of polycrystalline materials or nanograins. In that case, when the tip is positioned on a slope of a grain, the piezoelectrically induced in-plane deformation may cause buckling or even lifting of the cantilever resulting in some VPFM signal [55]. This crosstalk has relevance for the investigation of size effects in ferroelectric nanostructures, as the observed additional vertical deflection of the laser beam is not necessarily caused by the polarization variation due to nanoscale effects.

9.2.3 Electrostatic Contribution

Besides the pure electromechanical response, there is an additional contribution to the cantilever deflection related to the electrostatic interaction between the tip/cantilever system and the sample. The voltage applied between the tip and counter electrode results in a Maxwell force (Equation 1). If a combination of dc- and ac-voltages ($V=V_{dc}+V_{ac}\cos\omega t$) is applied to the tip, the resulting Maxwell force will have a static component and components oscillating at driving frequency and at a double frequency:

$$F_M = \frac{1}{2}\frac{\partial C}{\partial z}((V_{dc}-V_s)+V_{ac}\cos\omega t)^2 = \frac{1}{2}\frac{\partial C}{\partial z}((V_{dc}-V_s)^2+\frac{1}{2}V_{ac}^2 + 2(V_{dc}-V_s)V_{ac}\cos\omega t+\frac{1}{2}V_{ac}^2\cos 2\omega t) \quad (8)$$

where C is the capacitance of the sample-cantilever system and V_s is the contact potential difference between the tip and the sample. Thus, taking into account the electrostatic contribution, the total 1st harmonic signal measured by VPFM may be presented as [56]

$$PR = \frac{D_{1\omega}}{V_{ac}} = \alpha(h)d_{zz}\frac{k_1}{k_1+k} + \frac{C'_{sphere}+C'_{con}}{k_1+k}(V_{dc}-V_s) + \frac{C'_{cant}}{24k}(V_{dc}-V_s^{av}) \quad (9)$$

The first term in Equation 9 describes the pure electromechanical response, where d_{zz} is the longitudinal piezoelectric coefficient, k_1 is the spring constant of the tip-surface junction, k is the spring constant of the cantilever, and $\alpha(h)$ is an attenuation factor describing the ratio between the nominal applied ac-voltage and the ac-potential on the sample surface. This attenuation can be caused, e.g., by the presence of an adsorbate layer on the sample surface preventing it from direct contact with the tip, by the depletion phenomenon in the doped Si tip, or by the subsurface low-permittivity layer in studied ferroelectrics. The existence of such dielectric "gap", with the thickness $h \sim 0.1\div 1$ nm and the dielectric permittivity $\varepsilon_h \sim 10\div 100$, leads to a significant potential drop between the tip and the surface of a high-permittivity ferroelectric sample. According to Ref. [57], the attenuation factor can be estimated as

$$\alpha(h) \approx \left(1 + \frac{h\varepsilon}{a\varepsilon_h}\right)^{-1}$$

where a is the contact radius, and ε is the dielectric permittivity of the sample.

The last two terms in Equation 9 describe the local and non-local electrostatic contributions due to tip-sample and cantilever-sample interaction, respectively. $C'_{sphere,}$ $C'_{con,}$ and C'_{cant} are the capacitance gradients due to the spherical and conical parts of the tip and due to the cantilever, respectively. Contrary to the electromechanical contribution, the signal related to the Maxwell force is always of the same sign and results in an offset appearing in the VPFM signal.

The electrostatic contribution may be considerable, especially in the vicinity of the phase transition temperature, where the dielectric permittivity of the sample reaches its maximal value. In particular, Luo *et al.* [15] have found that VPFM signal measured on ferroelectric triglycine sulfate does not increase significantly at the phase transition, as expected for the pure electromechanical response. Instead, it follows the same trend as spontaneous polarization, which indicates a large electrostatic contribution.

The effect of the electrostatic interaction may be minimized in so-called regime of strong indentation, when a stiff cantilever (k>>1) is used and high contact forces (10-1000 nN) are applied [57].

9.2.4 Resolution in PFM Experiments

In a typical PFM experiment, the sharp tip plays the role of a movable top electrode. Since usually the thickness of the studied sample is much larger than the tip-sample contact area (5-20 nm), the probing electric field is strongly inhomogeneous and measured PFM response comes from a small volume around the contact point. This provides a high spatial resolution of the PFM method. The natural way to estimate the lateral resolution in the PFM experiment is to measure the width of a domain wall between two antiparallel domains. While the intrinsic width of 180° domain walls in ferroelectrics is expected to be a few unit cells [58, 59, 60], the domain walls measured in a PFM experiment are typically thicker (tens of nm) and, therefore, reflect primarily the spatial resolution of the PFM. Experimentally, the width of the domain wall image, w, is estimated from the profile of the piezoresponse signals across the wall, which is fitted by a suitable function, e.g., by the one used to describe the polarization profile in the mean field theory of ferroelectrics [61].

$$PR(x) = PR_{-} \tanh\left[\frac{(x - x_0)}{w}\right] \tag{10}$$

Fig. 9.3b illustrates the profile of the PFM signal across the 180° domain wall measured on a [001]-oriented $PbTiO_3$ single crystal. The apparent width of the domain wall obtained from the best fit to Equation 10 is about 60 nm.

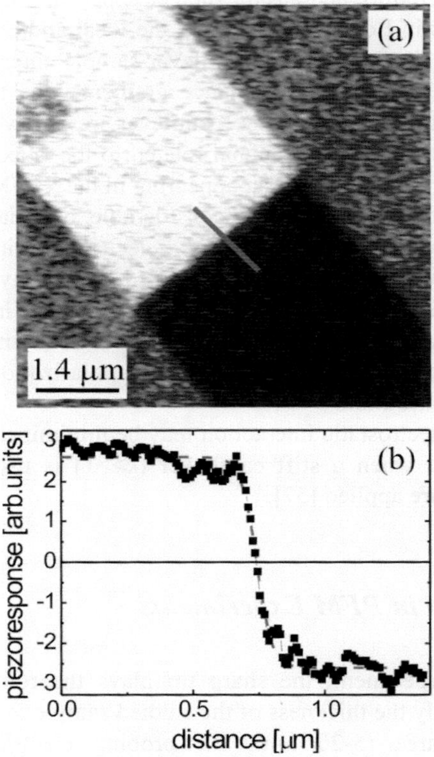

Fig. 9.3 a The LPFM image of [001]-cut of a $PbTiO_3$ single crystal. Bright and dark contrast corresponds to domains with the spontaneous polarization oriented left and right in the figure plane, respectively. **b** The cross-section of the piezoresponse image across the 180° domain wall. The broken line is the best fit to Eq. (10).

The theory of the resolution in PFM was recently considered by Kalinin *et al* [62]. They have shown that for the system with 180° domain walls, the piezoresponse may be presented as a convolution of a function describing the spatial distribution of material properties and a function related to the probe parameters (it is assumed that piezoelectric and dielectric properties are uniform across the sample thickness). In this case, the contrast formation mechanism may be analyzed using the transfer function theory that allows defining both the resolution and information limit. In the linear transfer function theory, the measured image $I(\mathbf{x})$ (where $\mathbf{x}$ is a set of spatial coordinates) is given by the convolution of an ideal image, $I_0(\mathbf{x}-\mathbf{y})$, with the resolution function, $F(\mathrm{y})$ [62]

$$I(\mathbf{x}) = \int I_0(\mathbf{x}-\mathbf{y})F(\mathbf{y})d\mathbf{y} + N(\mathbf{x}) \tag{11}$$

where $N(\mathbf{x})$ is the noise function. In the PFM experiment, the ideal image is the distribution of piezoelectric and stiffness constants that correlate with the domain structure. The resolution function depends on the tip geometry, lock-in amplifier parameters, and scanning conditions. It may be estimated by analyzing an artificial periodical domain pattern created using a template. The Fourier transform of eq. 11 is

$$I(\mathbf{q}) = I_0(\mathbf{q})F(\mathbf{q}) + N(\mathbf{q}) \tag{12}$$

where $F(q)$ is called the object transform function. It may be defined from ratio of the intensities of fast Fourier transformation of the experimental images to the ideal images.

One of the traditional resolution criteria used in optics is the Rayleigh two-point resolution limit – two Gaussian shaped image features of similar intensity can be resolved, if the intensity at the midpoint between them is less than 81% of the maximum [63]. If the object transfer function has a Gaussian shape, the Rayleigh two-point resolution criterion may be defined as $w_r=1/q_r$ for which $F(q_r)=0.58F(0)$. Kalinin *et al.* [62] showed that in the PFM experiment, the Rayleigh resolution correlates with the measured width of the domain wall. Moreover, the quantitative determination of material properties from the PFM experiment requires that typical domain size exceed w_r. Nevertheless, the features with smaller size may still be resolved by PFM. The minimal feature size detectable against the noise corresponds to the information limit defined from the condition $N(q)=F(q)$. However, the intensity of the PFM signal in that case starts to scale with the feature size and no reliable information about material properties can be obtained. For PFM, the information limit may be considerably smaller than the Rayleigh resolution.

The dependencies of the resolution on the tip size, as well as on the sample parameters (thickness, material), were studied experimentally by Jungk *et al.* [64]. They found that for the metal coated tips, the width of the measured walls scales linearly with the tip radius. For the uncoated Si tip, the broader domain walls were measured. It was explained as an effect of the dielectric SiO_2 layer formed on the tip surface. As a result, the probe is electrically separated from the sample surface and the electric field is less localized leading to reduced spatial resolution. No effect of the material parameters (dielectric permittivity, elastic and piezoelectric constant) on the resolution was found. The thinnest domain wall width (measured for the tip with the radius 15 nm) was only 17 nm.

Recently, Rodriguez *et al.* [65] reported that the measured width of the domain wall can be as small as 3 nm when the measurements are done not in the ambient condition (air) but in a liquid environment. They suggested that the mobile ions present in the solution screen the long range electrostatic interactions from the conical parts of a tip, at distances greater than the Debye length enhancing localization of the probing field in the tip-surface junction.

Due to the strong inhomogeneity of the probing electric field, the signal in the typical PFM experiment is collected mainly from a surface layer, whose thickness is a function of the dielectric permittivity and contact conditions, and is typically unknown. To overcome this limitation, the domain structure may be visualized through the top electrode of a ferroelectric capacitor [66]. In this case, the electric voltage may be applied either via the tip or using an external wire attached to the top electrode. In the latter case, the tip is used only for the detection of the piezoelectric displacement. In this configuration, the probing electric field is uniform and a measured response is generated by the entire sample thickness. This method allows the quantitative study of the dynamics of domain walls and polarization reversal mechanisms in ferroelectric capacitors. The drawbacks of this approach is a substantially smaller lateral resolution and inability to measure lateral (LPFM) signal.

9.3 PFM in Polycrystalline Materials. Effect of Microstructure, Texture, Composition

One of the advantages of PFM is the opportunity to correlate peculiarities of observed domain patterns directly with the microstructure of samples (polycrystalline thin films, ceramics). Large grains (> 1 μm) in conventional bulk ceramics are usually polydomain. The shape of the observed domain pattern depends on the symmetry of the crystalline structure and on the crystallographic orientation of the individual grains. Fig 9.4 shows the PFM images taken on $BiFeO_3$ ceramics [67]. Colours ranging between black and white indicate different directions and orientations of $\boldsymbol{P}_s$ with respect to the normal to the sample. Details of the domain structure are analyzed on two large grains selected in Fig. 9.4 a and b. The patterns differ in their overall contrast and depend strongly on which component (VPFM or LPFM) is measured. While the black/white VPFM contrast at the upper left corner transforms into a nearly unstructured brownish colour in LPFM image, the brown and yellow vertical contrast at the lower right corner is essentially unchanged in its LPFM image. Obviously, two different habit planes are encountered. In the grain at the lower right corner, 6 - 15 μm wide domains have straight boundaries. These are parallel and diagonal (*i e.* intersecting at angles of 45°) to each other, respectively. The observation of nearly identical vertical and lateral patterns complies with ferroelastic domains (twins), in which both P_z and P_x (observed by VPFM and LPFM, respectively) change sign simultaneously from one domain to the other. $BiFeO_3$ has rhombohedral symmetry *R3c* [68]. In this case, ferroelastic domains are separated either by 109° or 71° domain walls, corresponding to $\{110\}_p$ and $\{100\}_p$ planes (pseudo-cubic unit cell indices). The variation of the VPFM and LPFM contrast indicates that the crystallographic orientation $(a\bar{a}b)$ of this grain ($b >> a$) is tilted with respect to (001)-plane around the [110] direction (Fig. 9.4c). Peculiarly, however, more irregular stripe patterns on a sub-μm scale are observed within the elastic twin domains. They reflect mere FE twinning, either by 180° or $\pm(P_z + P_x)$

while conserving P_y. Being unrestricted by strain compatibility rules, they nevertheless form stripe-like patterns in order to minimize electric stray field energy. The extreme VPFM contrast observed in the upper left corner of panel (a) hints at domains, which are viewed in the (001) habit plane. Indeed, in the rhombohedral phase of perovskite materials, the longitudinal piezoresponse is known to attain extreme values not in the polar, but approximately along the $[001]_p$ direction [69]. The virtually vanishing lateral contrast (Fig. 9.4b) indicates that the x-component of the polarization is small, *i.e.* the polarization vector is parallel to domain walls. The observed domain walls may be either 180° ferroelectric domain walls or 109° ferroelastic ones (Fig. 9.4d). The electrostatically stabilized parallel walls between $\pm\boldsymbol{P}_s$ domains are slightly irregular, as expected, in the absence of strict strain compatibility rules. The diagonal wall observed in the left part of the grain corresponds most probably to 71° twins (see Fig. 9.4d) whose in-plane polarization component gives rise to a sizable contrast in the lateral PFM image (Fig. 9.4b).

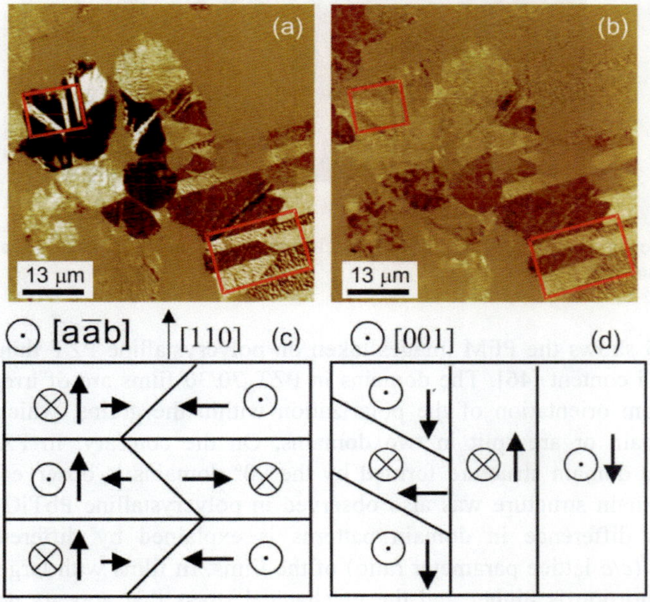

Fig. 9.4 VPFM **a** and LPFM **b** images of $BiFeO_3$ ceramics. Schematic domain configurations in **c** and **d** refer to the grains in bottom right and upper left corners, respectively (indicated by rectangles in **a** and **b**).

The mechanical boundary condition in polycrystalline samples may result in the existence of domain walls, which are forbidden for free-standing single crystals. For instance, Muñoz-Saldaña *et al.* [70]. found that in $Pb(Zr_{1-x}Ti_x)O_3$ (PZT) ceramics, another array of domain walls parallel to {210} planes exists besides the conventional {110}-oriented 90° and 180° domain walls. The reason for this unconventional domain configuration was explained by the clamping of the crystallites in the polycrystalline material.

Fig. 9.5 The topography **a,c** and VPFM **b,d** images taken on PZT 70/30 a,b and 80/20 c,d polycrystalline thin films

Fig. 9.5 shows the PFM images taken on polycrystalline PZT thin films with different Ti content [46]. The domains in PZT 70/30 films are of irregular shape with random orientation of the polarization within the grains, which are either single-domain or are split in two domains. On the contrary, in PZT 20/80, a regular *a-c* domain structure formed by the 90° domains is observed. A similar regular domain structure was also observed in polycrystalline $PbTiO_3$ thin films [71]. This difference in domain patterns is explained by different unit cell distortions (*c/a* lattice parameter ratio) of the films. In films with large amount of Ti, the distortion is higher and the mechanical stress that appears upon cooling through the phase transition temperature has to be relieved by the formation of the ferroelastic (90-degree in this case) domains. These domains form regular patterns because of the minimization of elastic energy [72]. On the other hand, in PZT70/30 films having smaller distortion of the unit cell, the domain walls are likely to separate 180° domains. The orientation of these walls is not restricted by strain compatibility rules. Their "random" structure reflects different local electrical conditions and inhomogeneity of defects, rather than stress relief.

It is known that with the decrease of the grain size in ceramics, the periodicity of 90° domains changes and finally the grain becomes single domain [73]. In the

PZT20/80 thin films, it was found that the relative area occupied by domains with in-plane and out-of-plane polarization depends on grain size. Namely *a*- to *c*-domain surface ratio increases and then drops down with grain size [46]. Such a complex behaviour was attributed to the stress relief originating not only from the substrate but also from neighbouring grains.

Generally, for non-textured polycrystalline materials, both the piezoresponse contrast and the domain patterns vary among individual grains. Nevertheless, important information on distribution of the local polarization may be obtained from the analysis of piezoresponse histograms, i.e., the number of the pixels on the PFM image corresponding to a given piezoresponse signal [74, 75, 76]. The deconvolution of the piezohistograms in several peaks may provide valuable information on relative population of different domain states. From the peak position, an effective piezoresponse value can be estimated. An important parameter is the half-width of the peak; its broadening may indicate coexistence of various polarization directions, which can be the case for polycrystalline films without texture, or even due to existence of oblique domain walls, which will result in a diffuse piezoresponse contrast.

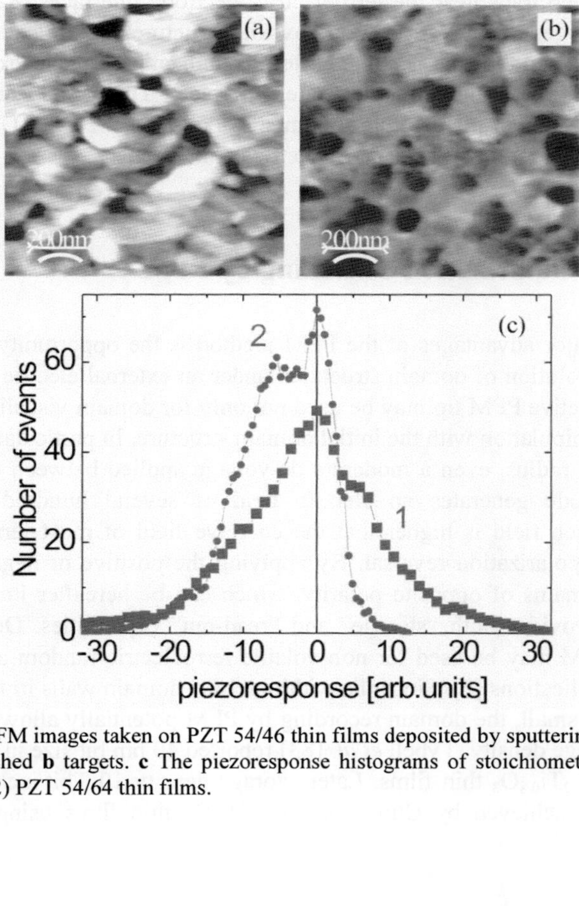

Fig. 9.6 The VPFM images taken on PZT 54/46 thin films deposited by sputtering stoichiometric **a** and lead-enriched **b** targets. **c** The piezoresponse histograms of stoichiometric (1) and non-stoichiometric (2) PZT 54/64 thin films.

Fig. 9.13 shows the histograms of PFM images taken on two PZT 54/46 thin films prepared by RF magnetron sputtering from a stoichiometric target and a target containing an excess of lead oxide [77]. For stoichiometric films, the piezoresponse distributions are approximately symmetric relative to domains of both up and down polarity. On the contrary, in non-stoichiometric films a "negative" shoulder exists on the piezoresponse histograms indicating that the regions exhibiting the negative piezosignal (in this case, domains with the polarization oriented towards the bottom electrode) occupy essentially larger area than those of the positive piezoresponse (domains with the polarization oriented towards the free surface). Thus, from analysis of the piezohistograms, it may be concluded that the PZT films obtained from lead-enriched targets have excess of negative polarization, i.e., are self-polarized. This conclusion agrees well with the macroscopic properties of these films [78]. Self-polarization is often observed in films and is characterized, for instance, by a shift of the polarization hysteresis loops or strong polarization imprint. This phenomenon occurs due to the presence of an internal electric field, which is at least as large as the coercive field at the Curie temperature. This field may have different origins [78, 79, 80, 81]. It was suggested that in the PZT films prepared with excess of lead, the built-in electric field arises due to the negative charges captured by deep traps near the ferroelectric–electrode interfaces [82]. These films have many oxygen vacancies in the perovskite structure, which leads to the n-type conductivity. When the film is cooled after crystallization, the electrons occupy the localized states near the film-electrode interface. The disappearance of self-polarization after high-temperature treatment (above T_c) [82] and UV-illumination [77] confirms the dominance of such "electrical" mechanism.

9.4 Local Polarization Switching by PFM

One of the major advantages of the PFM method is the opportunity to investigate directly the evolution of domain structures under an external electric or mechanical field. A conductive PFM tip may be used not only for domain visualization but also for a local manipulation with the initial domain structure. In particular, due to a very small tip apex radius, even a moderate dc-voltage applied between the tip and the bottom electrode generates an electric field of several hundred kilovolts per centimetre. Such field is higher that the coercive field of most ferroelectrics and induces local polarization reversal. By applying the positive or negative bias, one can create domains of opposite polarity, which can be hereafter imaged by PFM. Thus, PFM provides both "storage" and "read-out" capabilities. Domain patterns written by PFM may be used for non-volatile ferroelectric random access memory (FERAM) applications [2]. Since the width of 180° domain walls in ferroelectrics is typically very small, the domain recording by PFM potentially allows an extremely high data storage density. Tybell *et al.* [83] reported 40 nm bit size in epitaxial [001] oriented $PbZr_{0.2}Ti_{0.8}O_3$ thin films. Later, storage density 10 Tbit/inch^2 (bit size ~ 8 nm) has been achieved by Cho *et al.* in $LiTaO_3$ thin films using the scanning nonlinear dielectric microscopy [84]. Another interesting application is the fabrication of domain gratings with submicron period by PFM-nanodomain engineering. It may be used in optical nonlinear frequency conversion devices, as an example for backward-propagating quasi-phase matched conversion [85, 86]. The domain patterning by PFM may be applied in ferroelectric lithography, a method that explores the relation between surface chemical reactivity of ferroelectric materials and local polarization direction. It allows the allocation of multiple nanostructures of different materials in pre-defined positions [87, 88].

Applications of ferroelectric domain patterning for data storage, electro-optic devices, and ferroelectric lithography necessitate fundamental studies of the domain switching process, including thermodynamics and kinetics of domain nucleation, growth, and relaxation.

9.4.1 Thermodynamics of PFM Tip-Induced Polarization Reversal

Several approaches have been developed to describe the thermodynamics and kinetics of domain switching in PFM.

The switching in the PFM experiment starts from the nucleation of a new domain underneath the tip. The direction of the polarization in this domain coincides with that of the normal component of the applied electric field. The newly-formed domain expands by motion of the domain walls. So far, the electric field is larger than the coercive one, the process of growth of the domain is non-activated, and the size of the domain increases rapidly. At larger distances from the contact point, where the electric field decreases below the coercive one, the movement of domain walls becomes thermally activated and is slowed down. The domain walls continue to move until the inverted domain reaches an equilibrium state.

In the first approximation, the electric field of the tip may be considered as the field of a metallic sphere, the radius of which is equal to the tip apex radius R [89, 90, 91]. In the frame of this model, Molotskii obtained the closed form solution for the equilibrium domain shape [91]. The change of the free energy related to the nucleating domain is

$$\Delta W = W_d + W_s + W_t \tag{13}$$

where W_d is the depolarization energy contribution, W_s is the domain wall surface energy, and W_t describes the electrostatic interaction between the domain and electric field of the tip [91]. W_t term favours the enlargement of the domain, while W_d and W_s contributions hinder domain growth. The shape of the created domain is assumed within the Landauer model [92]: to be a half ellipsoid with the small and large axis r_d and l_d, respectively (Fig. 9.7). In this geometry

$$W_s = b r_d l_d \tag{14}$$

$$W_d = \frac{cr_d^{\ 4}}{l} \tag{15}$$

where

$$b = \sigma_{wall}\pi^2 / 2 \tag{16}$$

$$c = \frac{16\pi^2 P_s^2}{3\varepsilon_a}\left[\ln\left(\frac{2l_d}{r_d}\sqrt{\frac{\varepsilon_a}{\varepsilon_c}}\right) - 1\right] \tag{17}$$

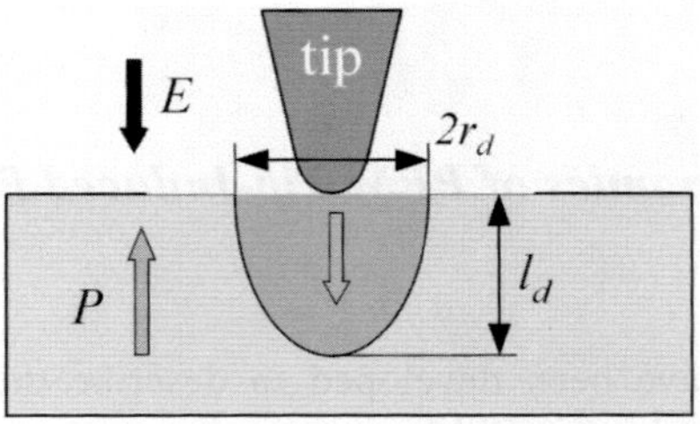

Fig. 9.7 Domain geometry by PFM tip-induced polarization reversal.

Here P_s is the spontaneous polarization, σ_{wall} is the domain wall surface energy density, and ε_c and ε_a are the values of dielectric permittivity in the directions parallel and perpendicular to the polar axis, respectively. When the domain is formed, the polarization value is changed by $2P_s$. Therefore, the energy of the interaction between the domain and the electric field may be presented as

$$W_t = -4\pi \int_0^{l_d} dz \int_0^{r(z)} P_s E_n(r,z) r dr \tag{18}$$

where $E_n(r,z)$ is the normal component of the electric field (parallel to the c- axis),

$$r(z) = r_d\sqrt{1 - z^2 / l^2}$$

By minimization of the free energy, Molotskii found parameters of the equilibrium domain shape as functions of the applied voltage [91]:

$$r_{eq} \sim V^{2/3} \tag{19}$$

$$l_{eq} \sim V \tag{20}$$

for $s \ll r_d$, and

$$r_{eq} \sim V^2 \tag{21}$$

$$l_{eq} \sim V^3 \tag{22}$$

for $s >> r_d$, where s is the distance between the centre of the curvature of the tip and the sample surface. In particular, it was found that the parameter

$$\frac{r_{eq}^{3/2}}{l_{eq}}$$

is an invariant of the equilibrium domain shape. It practically does not depend on PFM experimental parameters and is defined only by the properties of ferroelectrics themselves.

This model describes well the experimental results obtained in $C(NH_2)_3Al(SO_4)_2{\cdot}6H_2O$ (GASH) [93], triglycine sulfate $(NH_2CH_2OOH)_3H_2SO_4$ (TGS) [94] and $BaTiO_3$ [94] single crystals. Interesting results were obtained in $LiNbO_3$ where very long ($l_{eq} > 200$ μm) and relatively thin ($r_{eq} \sim$ 05-0.8 μm) domains are formed at large applied voltage ($V_{dc} > 3.5$ kV) in a good agreement with the aforementioned model [90, 95]. These results are surprising at first sight, since the electric field of the PFM tip rapidly decays from the contact point into the sample and cannot influence directly the elongation of the domain far from the surface. The propagation of the domains in this case is due to decreasing of the depolarization field energy. This process continues until the forces associated with the increase of the domain surface area compensate the driving forces caused by the depolarizing field [95]. The effect of the field created by the tip is indirect. It reveals itself through an increase of the domain radius due to the AFM tip field and a corresponding change of the domain length to satisfy the minimum free energy conditions. Such an effect was called "domain breakdown" since the created domains are similar to the electric breakdown channels [95].

While the model proposed by Molotskii describes well the domains with large size, it is not applicable for description of the polarization switching on length scale comparable to the tip apex radius [96]. In particular, according to the Molotskii model, the field in the vicinity of the tip is infinite and the domain nucleation is induced at arbitrary small bias voltages, which is in contradiction to many experimental observations.

At small length scales, the thermodynamics of switching process requires exact electroelastic field structure to be taken into account. The model that uses the rigorously derived electroelastic field was proposed by Kalinin *et al.* [96]. In particular, it was found that the domain nucleation requires a certain threshold bias 0.1-1 V corresponding to non-zero activation energy for nucleation. Further, this problem was elaborated by Morozovska *et al.* [97, 98] in the case of semi-infinite materials by taking into account the realistic tip geometry, the effects of surface depolarization energy, the surface screening charges, and the finite Debye screening length of domain nucleation.

The process of the polarization reversal in thin films was considered by Emelyanov [99] by taking into account tip geometry and interaction of the nucleated domain with the bottom electrode. He defined four stages of the PFM-induced switching. (a) Nucleation: at threshold voltage, V_{th}, the stable nucleus of hemisphere shape is formed by the polarization reversal in finite volume. (b) Bulk growth: the forward domain growth with a minor lateral expansion $l/r >> 1$ at $V_{th}<V<V_j$. (c) Contact instability: the "jump" to the bottom electrode at $V \sim V_j$, transformation of domain shape into a cylindrical one, and an abrupt increase of the switched volume. (d) Sideways expansion: a thermally activate process impeded by local depolarizing fields and pinning barriers. If the bias is turned off at stages (a) and (b), an unpinned domain will disappear.

9.4.2 Domain Dynamics Studied by PFM

While the high spatial resolution of PFM provides a unique opportunity for investigation of domain dynamics, the time resolution of this method is limited by the time required for image acquisition (approximately 10-20 min at the standard conditions). This means that PFM may be readily used for the investigation of processes with characteristic times of the order of minutes and above. But, the *in situ* study of domain dynamics during fast switching processes is difficult. Several approaches have been developed to circumvent this problem and to investigate domain dynamics in sub-minute and even sub-second time range.

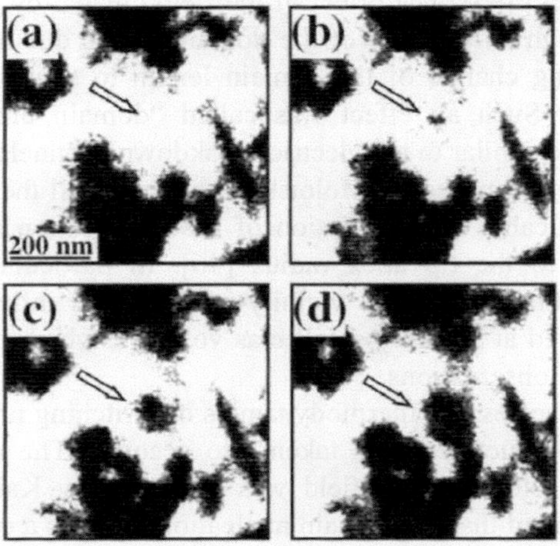

Fig. 9.8 Sequence of PFM images of PZT film showing a sidewise growth of a domain in the centre of the image. The growth is induced by application of 5 V voltage pulses of increasing duration to the same spot by a fixed probing tip. a Original domain structure; (b-d) domain images after application of voltage pulses. Pulse duration: **b** 0.25 s, **c** 0.5 s, d 1 s. Reprinted with permission from Ref. [44]

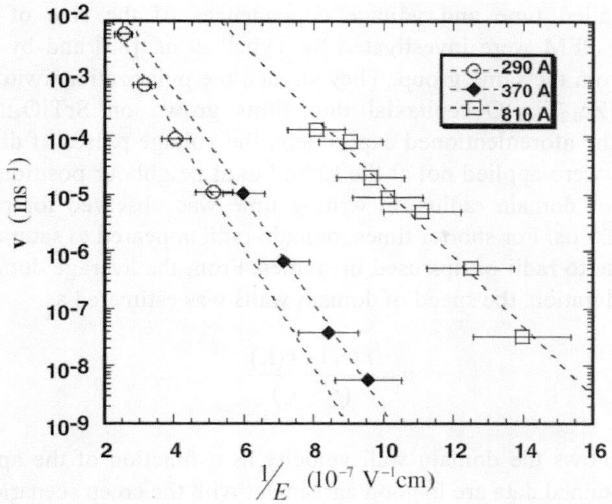

Fig. 9.9 Domain wall speed as a function of the inverse applied electric field for 290, 370, and 810 A thick $Pb(Zr_{0.2}Ti_{0.8})O_3$ films. Straight lines show the best fit of the experimental data to

$$v \sim \exp\left[-\frac{U_c}{kT}\left(\frac{E_0}{E}\right)^{\mu}\right]$$

Reprinted with permission from [83]. Copyright 2002 by the American Physical Society.

In particular, the so-called "step-by-step switching" approach was proposed by Gruverman *et al.* [100] (see Fig. 9.13a). In this case, partial reversal of the polarization is achieved by applying a voltage pulse shorter than the time required for full switching but with the pulse amplitude fixed above a threshold voltage. A consecutive picture of domain dynamics may be obtained by applying a sequence of such short pulses of incrementally increasing duration ($\tau_1 < \tau_2 < \ldots < t_{sw}$), and PFM imaging after each pulse. This approach may provide information on the domain wall velocity, its spatial anisotropy and its field dependence. Fig. 9.8 shows VPFM images of a growing domain in a PZT thin film illustrating this approach. Since the intensity of the PFM contrast for the reversed domain is approximately the same as for the initial state, the authors assumed that the domain fully extends through the film thickness already after the first pulse, and that the observed growth occurs via the sidewise expansion only. The diameter of the smallest stable domain produced by the first voltage pulse of 5 V varied from 20 nm to 40 nm. In the step-by-step approach, time resolution is determined by the pulse length increment, which can be in the nanosecond range. This is particularly important for studies of domain switching dynamics in thin film ferroelectric capacitors used in memory devices, where the switching time is typically below 100 ns. In this case, the time resolution will also depend on the rise time of the input pulse generated by the voltage source and on the size of the capacitor, i.e., its RC-time constant [101].

More detailed time and voltage dependences of the size of the domains generated by PFM were investigated by Tybell *et al.* [83] and by Paruch *et al.* [102, 103] from the same group. They studied the polarization switching in [001] oriented $Pb(Zr_{0.2}Ti_{0.8})O_3$ epitaxial thin films grown on $SrTiO_3$:Nb substrate. Contrary to the aforementioned experiment, the voltage pulses of different height and duration were applied not at the same but at neighbour positions. The strong dependence of domain radius on writing time was observed for pulse duration greater than 20 μs. For shorter times, domain radii appeared to saturate at ~20 nm, which is close to radii of tips used in studies. From the average domain size for a given pulse duration, the speed of domain walls was estimated as

$$v = \frac{r(t_2) - r(t_1)}{(t_2 - t_1)}$$

Fig. 9.9 shows the domain wall velocity as a function of the applied electric field. The obtained data are in good agreement with the creep scenario

$$v \sim \exp\left[-\frac{U_c}{kT} \left(\frac{E_0}{E} \right)^{\mu} \right]$$

over 12 decades of velocity [83]. The value of the dynamical exponent μ was found to be 0.7-1. These results indicate that in PFM experiments, the lateral expansion of the domains is a thermoactivated motion across randomly distributed pinning centres: vacancies and other defects in the lattice structure. Calculations show that the transition from the creep to the slide motion regime occurs at electric field > 180 MV/m [103]. In samples with artificially introduced macroscopic defects, for instance, regions of an amorphous material in highly irradiated samples, or domains with in-plane polarization, a decrease of dynamical exponent was found [103]. It was related to a relaxation of strains caused by macroscopic "defects", which changes the density or pinning force of inherent point defects presented in the film. This affects the disorder potential experienced by the domain wall [103].

In polycrystalline materials, the grain boundaries may strongly affect the switching dynamics. In particular, it was shown that the switched area of polycrystalline PZT films is confined by grain boundaries [104]. Kim *et al.* [105] studied the relation between the size of the switched domain and pulse duration in polycrystalline (111) oriented PZT25/75 thin films. They found that this behaviour is, in principle, similar to that in epitaxial thin films (domain size is proportional to the logarithm of pulse duration [83]). However, some peculiarities were observed related to the interaction between domain walls and grain boundaries. So, for small pulse duration (10 μs), the newly created domain could not pass through a grain boundary. But, this was not the case for domains created by long pulses (5 ms). This effect indicates that grain boundaries act as potential barriers for the movement of domain walls, and that longer "exposition" time is required for domain walls to overcome such a barrier.

The PFM step-by-step switching approach may be also used for the investigation of domain dynamics in a plane capacitor configuration [66, 106, 107, 108]. Contrary to the conventional PFM method, the probing tip in this case is in contact with the thin top electrode deposited on the sample surface. Such configurations have several advantages. The applied field may be considered homogeneous and is well defined even for micron-size thin film capacitors. Also, the electric contact between tip and ferroelectric sample is more reliable. The PFM results acquired in this configuration may be directly compared with the switching current data obtained by conventional measurements. The main disadvantage is the inherently lower spatial resolution due to imaging through the top electrode.

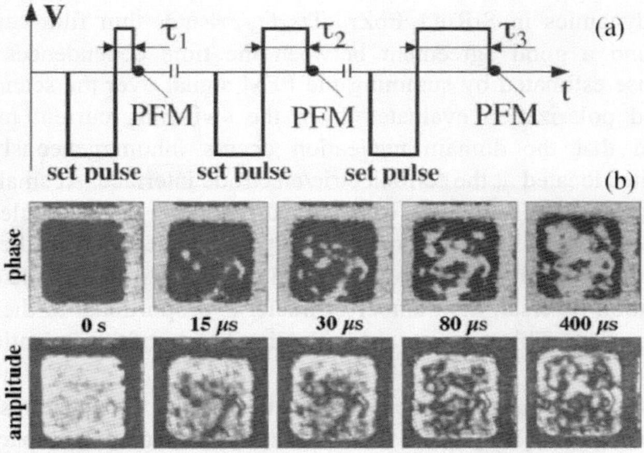

Fig. 9.10 a Voltage pulse sequence applied to study domain kinetics in thin film capacitors using PFM step-by-step approach. Reused with permission from [106]. Copyright 2008 by the American Physical Society. **b** PFM phase and amplitude images of instantaneous domain configurations developing in the 3×3 μm^2 PZT capacitor at different stages of polarization reversal under the 1.1 V bias. Reused with permission from [66]. Copyright 2005, American Institute of Physics.

Using this approach, Gruverman *et al.* performed a direct study of domain dynamics during polarization reversal in PZT memory capacitors [66]. Fig 9.10b shows PFM images obtained at different switching stages. The forward velocity for domain growth (about 0.3–0.4 m/s for the voltage pulse of 1.4 V) was estimated from the amplitude changes of the PFM signal that accompany domain growth across the film thickness. The lateral domain wall velocity has been obtained from the time dependence of domain radius in a way similar to that described in Ref. [83]. It was found that lateral velocity fluctuates during domain growth by three orders of magnitude, most likely due to interaction of domains with grain boundaries or/and microstructural defects. Another interesting effect is the slowing down of domains before coalescence with neighbouring ones. Analysis of data revealed that in the low-field range (just above the threshold field), the polarization reversal process consists of a fast initial part dominated by nucleation of new domains and slow switching part due to lateral domain expansion. The nucleation-dominated stage may be well approximated by the Kolmogorov-Avrami-Ishibashi model [109, 110, 111]. At the slow switching stage, a broad distribution of the domain wall mobility should be taken into account. As the size of capacitors decreases below 1–2 μm^2, the contribution of domain nucleation to polarization reversal is significantly reduced and the switching proceeds mostly via lateral growth of just a few domains [106]. Therefore, in the high-field regime, the smaller capacitors will switch faster than the larger capacitors. In the low-field regime, switching rate is limited by domain wall speed. For these reasons, larger capacitors characterized by considerable contribution of nucleation mechanism to the polarization reversal will switch faster.

The same approach was used by Kim *et al.* [107] for the investigation of switching dynamics in $SrRuO_3/PbZr_{0.4}Ti_{0.6}O_3/SrRuO_3$ thin film capacitors. The authors found a good agreement between the time dependences of the total piezoresponse estimated by summing the PFM signal over the scanned area, and the reversed polarization evaluated from the switching current measurements. They found that the domain nucleation occurs inhomogeneously, mostly at particular sites located at the ferroelectric/electrode interface. At small $t<10^{-5}$ s, the switched polarization correlates well with the number of the nucleus. At larger times, the polarization increased more rapidly thus indicating that the domain wall motion starts to contribute more to the domain switching. During the early stage of the polarization reversal, the number of nuclei is proportional to the logarithm of the switching time. This points to the wide distribution of the activation energy for nucleation.

Recently, Gruverman *et al.* reported on the effect of capacitor shapes on the polarization reversal [108]. It was found that in small circular capacitors (< 1 μm in diameter), domains nucleate at the electrode edge and propagate around the circumference forming a doughnut-type structure (with unswitched central part) in a few microseconds. This structure remains for > 1 s, after which the unswitched central domain collapses to leave a single domain state. The authors have attributed such particular configuration to a vortex domain structure developed during domain switching. This conclusion was supported by numerical simulation of switching process.

9.4.3 Local Piezoelectric Hysteresis Loops

Another way to study the processes of the polarization switching in ferroelectrics is to use PFM in a spectroscopy mode, where the measurements are done on a fixed position of the tip under the dc-voltage swept in a cyclic manner. The obtained dependence of the local piezoresponse on the applied bias is referred to as a local piezoelectric hysteresis loop [100]. On a macroscopic scale, these loops correspond to a weak-field piezoelectric coefficient tuned by a varying bias voltage. According to a linearized electrostriction equation, in a ferroelectric with centrosymmetric paraelectric phase, the longitudinal piezoelectric coefficient can be expressed as [112]

$$d_{33} = 2Q\varepsilon_0\varepsilon_{33}P_3$$

where Q is the longitudinal electrostriction coefficient, ε_0 is the permittivity of vacuum and ε_{33} and P_{33} are the corresponding dielectric permittivity and spontaneous polarization values. Therefore, the variation of piezoelectric coefficient reflects the polarization switching with corresponding tuning of the dielectric permittivity $\varepsilon(E)$, which affects the d_{33} value [113]. The analysis of the piezoelectric hysteresis loops may be used for the identification of the switching mechanism and for study of the effects of ferroelectric degradation. Measurements of local hysteresis loops are of great importance in polycrystalline ferroelectrics because they are able to quantify polarization switching on a scale significantly smaller than the grain size (typically few tens of nm).

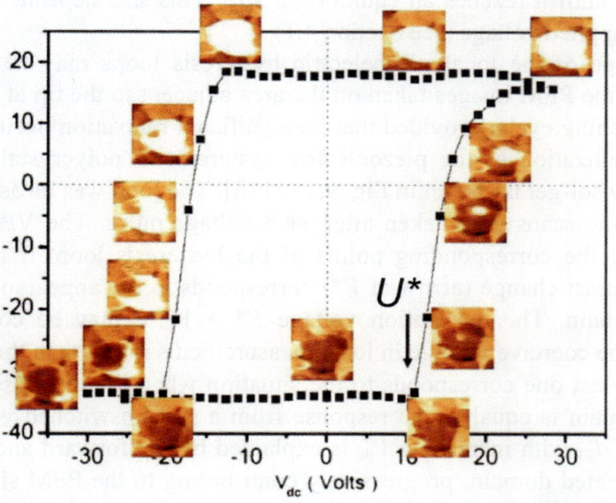

Fig. 9.11 Detailed local hysteresis loop of PZT70/30 film with intermediate scanning. The insets (scan size 1 μm) show the static domain structures obtained by scanning with the ac voltage of the amplitude of 1 V in the corresponding points of the hysteresis [116].

There are two main approaches to measure the local hysteresis in the PFM experiment. In the first one, the probing ac-voltage with moderate amplitude (0.1-1 V) and high frequency (typically 10-100 kHz) is superimposed on a slowly varying (10^{-2} – 1 Hz) switching voltage with high amplitude. Then, the local piezoresponse is plotted as a function of the current value of the switching voltage. The drawback of this method is the large electrostatic contribution into the measured response, which may overshadow the true piezoresponse signal [115]. In the second approach, the loops are measured in so-called step mode. In this case, a sequence of voltage pulses with the same duration and with the amplitude cycled from $-V_{\min}$ to $V_{\max}$ is applied and the piezoresponse is measured at the time interval between the pulses. Since no bias voltage is applied during the measurements, the contribution related to the electrostatic interaction is minimized. However, a partial polarization backswitching after dc-voltage removal has to be taken into account.

Usually, the shape of local hysteresis loop differs strongly from that measured by the macroscopic methods reflecting different physical mechanisms involved. Macroscopically, the switching occurs via nucleation and growth of a large number of reverse domains in a situation where the applied electric field is uniform. Therefore, the d_{33} hysteresis corresponds to the switching averaged over an entire sample under the electrode. In typical PFM experimental conditions (tip is on the electrode-free surface), the electric field is strongly localized and inhomogeneous. Therefore, the polarization switching starts with the nucleation of a single domain just under the tip [116]. Upon increasing the applied dc-bias, this newly formed domain elongates to the bottom electrode, simultaneously expanding in lateral dimensions until it reaches an equilibrium size. This size depends on the value of maximum applied voltage (see Section 4.1).

The shape of the local piezoelectric hysteresis loops may be understood by comparing the PFM images taken on the area adjacent to the tip at different stages of the switching cycle, provided that no significant relaxation occurs. An example of the visualization of the piezoelectric hysteresis in polycrystalline PZT films prepared by sol-gel is shown in Fig. 9.11 [116]. The loop was measured in the step mode and the scans were taken after each voltage pulse. The VPFM images are shown near the corresponding points of the hysteresis loop. It is seen that the sudden contrast change (arrow at V^*) corresponds to the appearance of the stable inverse domain. This nucleation voltage $V^* \approx 11$ V may be considered as an analog of the coercive voltage in local measurements rather than the voltage where d_{33}=0. The last one corresponds to the situation when the response from a newly created domain is equal to the response from a still unswitched volume. A rapid increase of d_{33} with increasing V_{dc} is explained by the forward and lateral growth of the nucleated domain, progressively contributing to the PFM signal. When the size of the inverse domain is much greater than the penetration depth of the probing field, the tip senses only a fully polarized area with aligned polarization and d_{33} is well saturated. One can see from Fig. 9.11 that the shape of the switched area is distorted: the propagation of domain wall in polycrystalline ferroelectrics is strongly influenced by local inhomogeneities (e.g., grain boundaries) and stresses, which result in strong irregularity of the domain boundary. A correlation between the nucleation voltage, estimated from local hysteresis loops, and the composition was reported for PZT thin films [76]. As the Zr/Ti ratio changed from 20/80 to 60/40, the value of V^* is decreased from 4.64 V to 1.43 V indicating that domain nucleation is more difficult in tetragonal PZT films with low Zr content, than in rhombohedral PZT films with high Zr concentration.

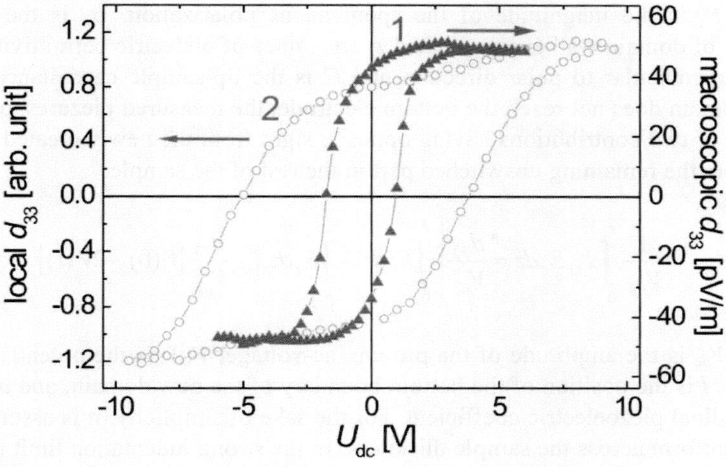

Fig. 9.12 Comparison of macroscopic (interferometry) (1) and local piezoelectric hysteresis loops measured by PFM (2).

Fig. 9.12 represents a comparison between the local and macroscopic piezoelectric hysteresis measured in the polycrystalline $PbZr_{0.45}Ti_{0.55}O_3$ thin films under the same combination of dc- and ac- voltages [117]. The macroscopic piezoelectric coefficient was measured by the laser interferometry method [118] as a macroscopic displacement of metal electrode in a homogenous electric field. One can see that the shape of the loop depends strongly on measurement conditions. A notable decrease of piezoresponse on the reverse branch of the hysteresis curve ($0 < V_{dc} < V_{max}$), in the case of the local measurements, is caused by partial backswitching induced by depolarizing field. This field is much stronger in PFM experiments due to the influence of an interface layer with low dielectric permittivity. Also, the "coercive" voltage (nullifying piezoresponse) is much larger for the local loop, illustrating its different physical meaning.

The one-dimensional model for the local hysteresis loop formation was developed in thermodynamic [76, 114] and kinetic limits [119]. In the first approach, the size of domain is considered to be equal to that at the thermodynamic equilibrium. In the kinetic limit, the velocity of forward and lateral growth of the domain is taken into account.

The polarization switching in the thermodynamic limit was considered in the frame of the model developed by Molotskii [91]. In this case, the length of a newly created domain, l, (in the direction perpendicular to the sample surface) is proportional to the applied bias, V_{dc}.

$$l = \frac{16CP_s}{5\pi(\sqrt{\varepsilon_c \varepsilon_a}+1)\sigma_w} V_{dc} \tag{23}$$

where P_s is the magnitude of the spontaneous polarization, σ_w is the surface density of domain wall energy, ε_c and ε_a are values of dielectric permittivity along and perpendicular to polar direction and C is the tip-sample capacitance. If the new domain does not reach the bottom electrode, the measured piezoresponse will consist of two contributions having opposite sign: from the newly created domain and from the remaining unswitched part in the rest of the sample:

$$d_{eff} = \frac{1}{V_{tip}} \int_0^t d_{33} E_z dz = \frac{d_{33}}{V_{ac}} \left[\int_0^l E_z dz - \int_l^t E_z dz \right] = \frac{d_{33}}{V_{tip}} [V(0) - 2V(l)] \qquad (24)$$

where V_{ac} is the amplitude of the probing ac-voltage, $V(0)$ is the potential on the surface, l is the position of the bottom boundary of the new domain, and d_{33} is the longitudinal piezoelectric coefficient. For the sake of simplicity, it is assumed that d_{33} is uniform across the sample thickness. In the strong indentation limit [57], the potential at distance l from the surface may be presented as $V(l) \sim V(0)R/(R+l)$, where R is the tip radius, and the capacitance of tip-sample junction is

$$C \approx 4\pi R \ln\left(\frac{\sqrt{\varepsilon_c \varepsilon_a}}{2} \right)$$

Thus, for the measured local piezoelectric coefficient, we obtain

$$d_{eff} \propto d_{zz} \left[1 - \frac{2R}{R+l} \right] \propto d_{zz} \left[1 - \frac{2}{1 + \dfrac{64 P_s V_{dc} \ln\sqrt{\varepsilon_c \varepsilon_a}}{5\sigma_w \sqrt{\varepsilon_c \varepsilon_a}}} \right] \qquad (25)$$

It can be seen from Equation. 25 that in thermodynamic limit, the shape of the hysteresis loop is determined both by the dielectric permittivity and the spontaneous polarization of the film. In particular, a tilt of the local hysteresis loop is roughly proportional to the value of spontaneous polarization P_s and inversely proportional to the dielectric permittivity, $\sqrt{\varepsilon_c \varepsilon_a}$. Indeed, the measurements of piezoelectric hysteresis in PZT films of different compositions [76] confirmed such a tendency.

In the absence of pinning, the thermodynamic model suggests that the switching process is reversible and the domain should grow and shrink instantaneously in response to the bias (Fig. 9.13c). In such a situation, no hysteresis loops should be observed, and the piezoresponse versus bias voltage will proceed along the solid line in Fig. 9.13a. If the weak pinning is included [99, 119], the new domain will not shrink when bias voltage decreases (Fig. 9.13d). After the applied bias is reversed, inside this pinned domain another domain of opposite polarity will nucleate at V_c^+. With further decrease in voltage, the size of the inverted domain increases. At voltage $-V_{max}$, the walls of the two domains annihilate and the system returns to the initial state.

In the kinetically limited scenario, the effects related to the finite velocity of the domain wall are taken into account. In this case, the hysteresis loop broadens as compared to thermodynamic case (Fig. 9.13b) [119]. Experimentally, the radius of the switched domain has been shown to follow the phenomenological dependence $r=k_s V \ln t$. Domain wall velocity is thus $v=k_s V/t$, where k_s is the kinetic constant depending on the pinning strength of the material [120]. Note that, while the kinetic data was obtained for the lateral domain size, the time dependences for vertical and lateral domain sizes may be expected to be commensurate. If the switching voltage ramps linearly with time, $V=bt$, then $v=k_s b$ and the domain size, $r=k_s bt=k_s V$. Thus, the domain size depends only on current voltage value and on pinning strength of the material, i.e., the functional form of Equation 25 may be recovered for a robust description of hysteresis loop shape. Both the tilt of the hysteresis loop and coercive voltage are controlled now by the pinning strength of the material.

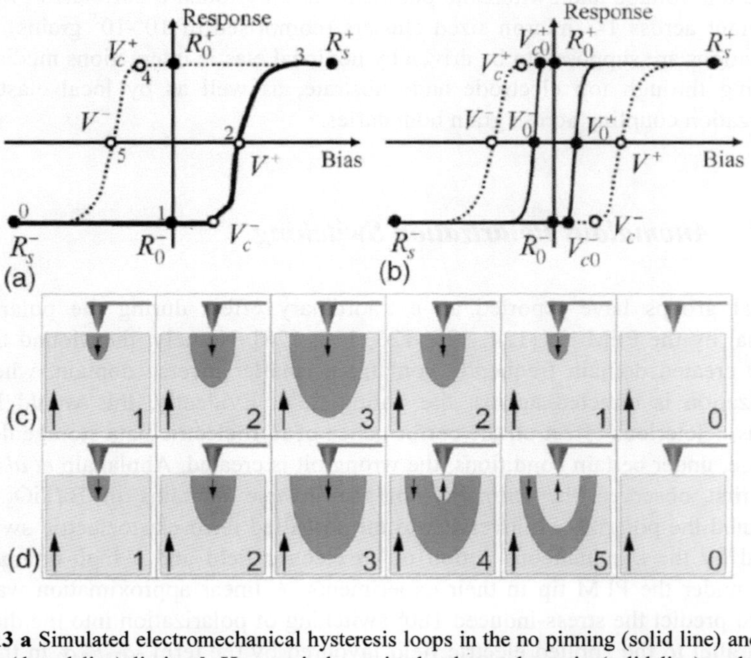

Fig. 9.13 a Simulated electromechanical hysteresis loops in the no pinning (solid line) and weak pinning (dotted line) limits. **b** Hysteresis loops in the thermodynamic (solid line) and kinetic (dotted line) limits. **c** A schematic representation of domain growth for the thermodynamic model without pinning and **d** with pinning. The numbers correspond to points on the hysteresis curve in **a**. Arrows indicate domain orientation. Reused with permission from [119]. Copyright 2006. American Institute of Physics.

Recently, a new technique, Switching Spectroscopy PFM (SS-PFM), was developed for a real-space mapping of switching behaviour in ferroelectric materials [119]. In SS-PFM, the local hysteresis loop is acquired at each point of the scanned area. Then, relevant characteristic describing the switching, like positive and negative voltages corresponding to domain nucleation, switchable piezoresponse, imprint, hysteresis area, etc. are extracted and plotted as two-dimensional maps. Recently, Jesse *et al.* have shown that SS-PFM may be applied to visualize nucleation centres of polarization reversal, determine their spatial and energy distribution, and correlate them to local microstructure [121]. In particular, they found that in epitaxial $PbZr_{0.2}Ti_{0.8}O_3$, enhanced nucleation activity of thin films is observed at the ferroelastic (90°) domain walls and their intersections. In Ref. 122, SS-PFM has been applied to study the polarization switching within single $PbZr_{0.52}Ti_{0.48}O_3$ nanocrystals. An analysis of the spatial variation of the local hysteresis loops allows the estimation of the depth profile of a layer with the non-switchable ("frozen") polarization and mapping the intensity of the built-in electric field in the regions with the switchable polarization [122]. Bintachitt *et al.* applied SS-PFM to investigate polarization switching in polycrystalline $PbZr_{0.52}Ti_{0.48}O_3$ capacitor structures [123]. By analyzing spatial variability of nucleation voltage and switchable polarization, they found a correlated switching behaviour across 1-3 micron sized clusters (comprised of 10^2-10^3 grains). These correlations are supposed to be driven by nonlocal elastic interactions mediated by coupling through top electrode and substrate, as well as by local elastic and polarization coupling across grain boundaries.

9.4.4 Anomalous Polarization Switching

Several groups have reported an extraordinary effect during the polarization reversal by the PFM tip [124, 125, 126, 127, 128]. Namely, they found that the newly created domain frequently contains a smaller inverse domain, where the polarization is directed against the poling field. Evidently, this would have a serious deleterious effect on the performance of ferroelectric data storage devices, because, under certain conditions, the wrong bit is created. Abplanalp *et al.* [124], who first observed the formation of the inverse domains in $BaTiO_3$ films, attributed the polarization inversion to the so-called ferro-elastoelectric switching caused by the simultaneous action of an electric field and a high compressive stress under the PFM tip in their experiments. A linear approximation was then used to predict the stress-induced 180° switching of polarization into the direction antiparallel to the applied electric field favoured by the term $-d_{33}E_3\sigma_3$ in the free-energy expansion. Here, d_{33} is the longitudinal piezoelectric coefficient, E_3 is the electric field intensity, and σ_3 is the mechanical stress [124]. This prediction, however, was not confirmed by the rigorous nonlinear thermodynamic calculations, which showed that the antiparallel polarization orientation never appears to be energetically more favourable than that along the field [129].

Further studies of anomalous polarization switching (polarization inversion), performed by Morita and Cho [125] on lithium tantalate and lead zirconate titanate thin films, demonstrated that the inverse domains arise only after the external electric field is removed. This result rules out the ferro-elastoelectric switching as the origin of the discussed phenomenon. Alternatively, the polarization inversion under the PFM tip may be attributed to the appearance of a local internal electric field in the subsurface region, which is directed against the poling one [126, 127, 128]. Such field may be created by screening charges [127] or by a space charge formed beneath the tip due to the injection of charge carriers and their subsequent drift and trapping [126, 128]. Fig. 9.14 shows the appearance of the "inverted" domain in [111]-oriented $PbZn_{1/3}Nb_{2/3}O_3$-0.045$PbTiO_3$ (PZN-PT) single crystal [128]. At a moderate voltage (+30 V) applied for a short time $\tau = 2$ s, only the 'bright' domain appears inside the initial 'dark' background (Fig. 9.14 d). i.e., the normal polarization switching takes place and the newly created domain remains stable after the removal of the external electric field. However, when the applied voltage exceeds some threshold value V_{th} (at $\tau = 2$ s, $V_{th} \sim 35$ V), an additional 'dark' domain forms in the central part of the written 'bright' one, as demonstrated by the piezoresponse images in Figs. 9.14(e-f). The diameters of both domains (normal and inverse) are functions of the voltage pulse amplitude and duration (Fig. 9.15). It is important that no influence of the PFM set point was found, which indicates that the domain formation is insensitive to the mechanical stress exerted by the tip. On the other hand, it was found that the polarization inversion is highly sensitive to the sign of the applied voltage, being observed only when the potential difference between the tip and the bottom electrode is negative.

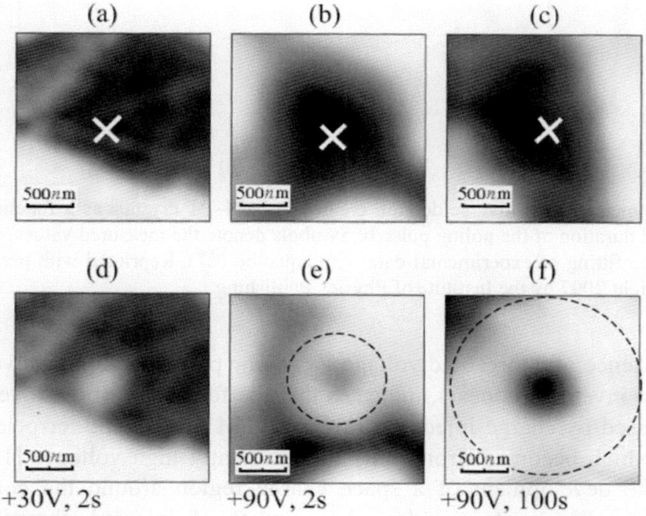

Fig. 9.14 Piezoresponse images of PZN–PT crystals showing three initial domain patterns (**a–c**) and their changes after the application of voltage pulses of different heights and durations (**d–e**). The PFM tip was positioned at the points marked by crosses. Reprinted with permission from [128]. Copyright 2007 by the Institute of Physics Publishing.

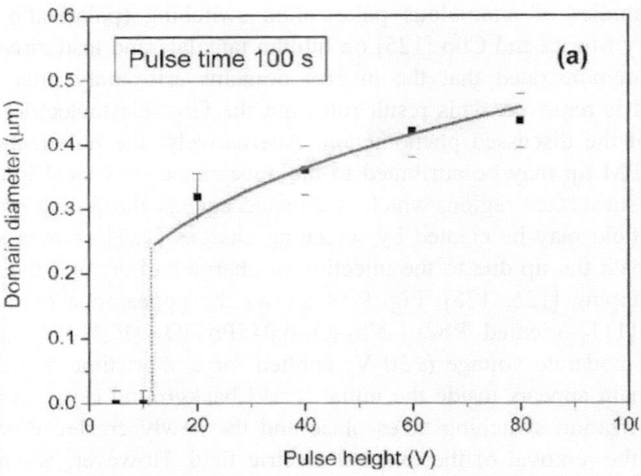

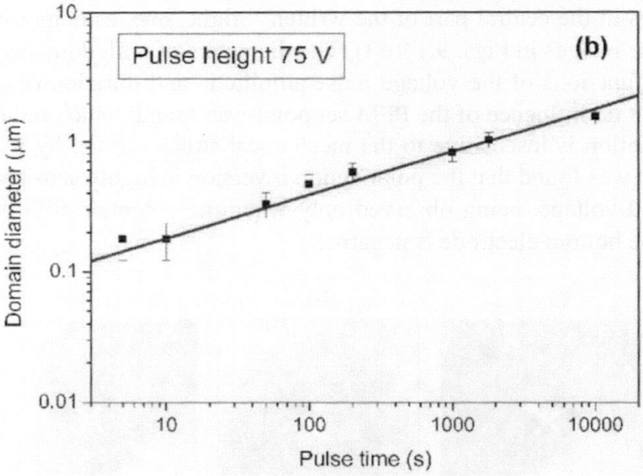

Fig. 9.15 Diameter of the inverse domain created in PZN–PT crystals as **a** function of applied voltage **a** and duration of the poling pulse **b**. Symbols denote the measured values, whereas solid lines show the fitting of experimental data with equation (27). Reprinted with permission from [128]. Copyright 2007 by the Institute of Physics Publishing.

The existence of a threshold voltage V_{th} of one particular sign, above which the polarization inversion appears, indicates that the formation of the inverse domain may be caused by charge injection from the PFM tip into the ferroelectric. Such injection, which becomes strong only at sufficiently high voltages [130], should result in the development of a space charge region around the contact point. During the poling voltage pulse, the number of injected charges increases gradually, and the charge carriers drift away from the tip, filling traps existing in the ferroelectric material. The forming space charge creates an internal electric field $E_i(\boldsymbol{r})$ inside the crystal, which is directed against the applied field $E(\boldsymbol{r})$ just below the tip (but along this field outside the space-charge region). If the charge density is high enough to create $E_i(\mathbf{r})$ exceeding the coercive field in some volume, polarization reversal under the tip will occur after removal of the applied voltage. Once a nucleus of the inverse domain appears, it will grow inside the space-charge region until the field $E_i(\mathbf{r})$ drops down to a level at which the electric driving force acting on the domain boundary becomes too small to overcome the pinning forces caused by defects. Accordingly, the diameter of the inverse domain, D, is expected to correlate with that of the region of trapped charge, $D(\tau)=2\eta R(\tau)$, where $R(\tau)$ is the position of the space-charge front in the subsurface layer, and $\eta < 1$. Since the size of the inverted domain is much larger than the tip radius (10 nm), the electric field of the tip may be estimated in the point-charge approximation [91]. In this case, the in-plane component of the electric field may be written as

$$E_\rho(\rho,z) \approx \frac{CV}{2\pi\varepsilon_0\sqrt{\varepsilon_a\varepsilon_c}} \frac{\rho}{\left[\rho^2 + \frac{\varepsilon_a}{\varepsilon_c} z^2\right]^{\frac{3}{2}}} \tag{26}$$

where ρ and z are the cylindrical coordinates. Assuming that velocity of the propagation of the space-charge front is proportional to the intensity of the electric field: $dR/dt=\mu_{\text{eff}}E_\rho(\rho=R)$. Kholkin *et al.* [128] obtained for the diameter of the inverse domain near the surface ($z << \rho$)

$$D(V,\tau) \approx 2\eta \left(\frac{3C\mu_{eff}}{2\pi\varepsilon_0\sqrt{\varepsilon_a\varepsilon_c}} V\tau \right)^{\frac{1}{3}} \tag{27}$$

The experimentally observed time and field dependences $D(V,\tau)$ follow well the behaviour predicted by Equation 27 (Fig. 9.15). The charge carries mobility μ_{eff} evaluated from the best fit is about 10^{-14} $m^2V^{-1}s^{-1}$ indicates that the charge transport is due to hopping of injected charges between localized states under action of applied electric field [128].

The injection-mediated mechanism of polarization inversion is consistent with the asymmetry of the effect relative to the sign of applied voltage. Most probably, the threshold for hole injection is much larger than the threshold for electron injection. Also, the inverse domain disappears after illumination with ultraviolet light [128]. Indeed, the bandgap illumination of the ferroelectric creates free charge carriers of both signs [131], which neutralize the trapped space charge and so eliminate the internal field $E_i(r)$.

The inverse switching phenomenon has important implications for the functioning of ferroelectric memories. On the one hand, if the voltage applied to a ferroelectric exceeds the threshold value V_{th}, an erroneous bit of information will be written owing to the formation of the inverse domain inside the 'normal' one. Since the trapped charge is stable at room temperature, illumination with UV light or heating will be necessary to refresh the memory cell. On the other hand, the polarization inversion may be employed to increase the density of data storage, if the information has to be stored in a permanent form. Indeed, the inverse domain is always smaller than the 'normal' one and it is stable. According to equation (27), its size D could be diminished further via the reduction of threshold voltage V_{th} and effective mobility μ_{eff} of the injected charges. These parameters may be tailored by choosing an appropriate tip material and doping and by increasing the density of traps in the ferroelectric (e.g., using irradiation with high-energy particles to create radiation induced defects).

9.4.5 Polarization Retention Loss (Aging) in PFM Experiments

The stability of the polarization state created by the external voltage is a prerequisite for the realization of SPM-based ferroelectric memories. One of the general requirements for the development of FERAM is a long-time retention ability. However, a spontaneous reversal of the written polarization state, which is often observed in ferroelectric capacitors, leads to a progressive loss of stored data with time. It has been proposed that this backswitching process (retention loss) is largely determined by depolarizing fields. They always exist in thin ferroelectric films due to a finite separation between polarization charge and compensating free charges on the electrodes [132, 133]. Besides, depoling may be accelerated by an internal bias field, which is typical, e.g., for PZT thin films [134]. This field always points in one direction and is related either to the influence of electrodes (different work functions for both interfaces) or to the non-uniform space-charge distribution due to diffusion of oxygen vacancies under the high-temperature processing of the films. Typically, for macroscopic retention loss, which is an average characteristic on an ensemble of grains, the logarithmic time dependence have been reported, pointing out a broad distribution of relaxation times [133]. PFM, as compared to macroscopic measurements, allows the understanding of this effect locally rather than by averaging over many backswitching events. Fig. 9.16 shows an example of the local aging measurements in PZT45/55 thin films [135]. For these experiments, two different regions were chosen. First, the rectangular area 300x300 nm^2 was polarized with a negative bias (-10 V) to create a uniformly polarized state. Further, the tip was positioned in the centre of a large grain (region I). A +10 V pulse of 1 s duration was applied to switch the polarization mainly inside this grain. In the second case (region II), the tip was located in the centre of a small grain, and dc-pulse with same parameters initiates switching in several neighbouring grains. It may be seen that in both regions the domain size decreases with time. When switching is done in a big grain, this process is rapid, so that the domain almost disappears within an hour. It is interesting that during the backswitching process, the initial large domain splits into a number of small domains in this case. This indicates that the grain is inhomogeneous and increasing the domain wall energy is compensated by the energy gain from the pinning. It is confirmed by the survived domain (see Fig. 9.16c) near the grain boundary. The evolution of the domain pattern in a fine-grain location is quite different, i.e., local piezoelectric aging is much slower in this case. It indicates that the grain boundaries serve as additional pinning sites impeding the domain wall motion.

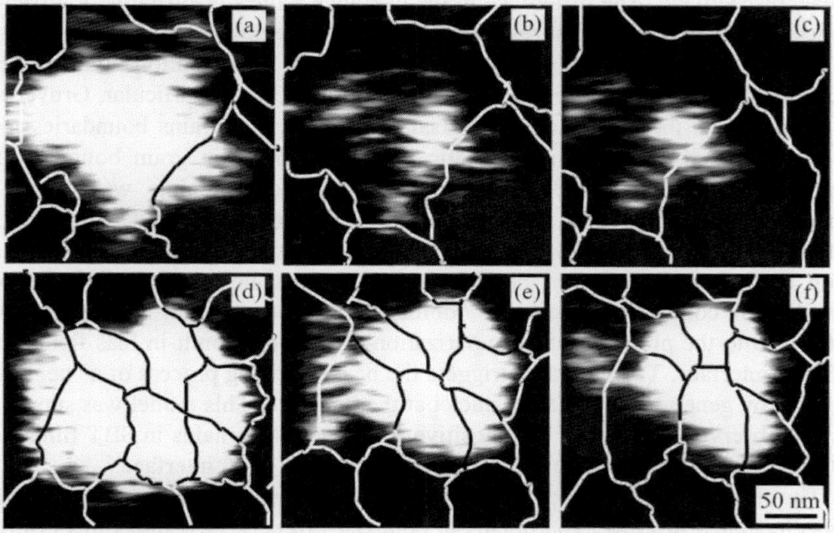

Fig. 9.16 Time evolution of written domain pattern in PZT45/55 films. Switching is done in two different locations (upper row: region I, bottom row: region II). The scans correspond to the following times after poling: **a**, **d** –5 min, **b**, **e** –20 min, **c**, **f** –90 min.

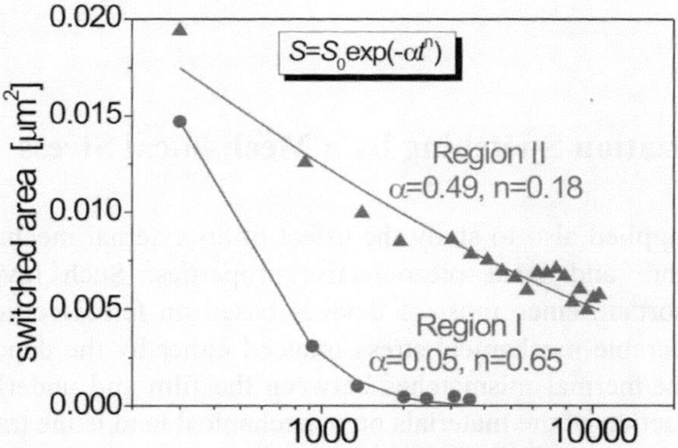

Fig. 9.17 Time evolution of the poled area in coarse-grain (circles) and fine-grain (triangles) locations in PZT45/55 films.

Fig. 9.17 quantifies the grain size effect on the relaxation of the written domain area, *S*, after poling. The obtained dependencies *S(t)* follows the so-called stretched exponential law $S=S_o exp(-\alpha t^n)$, which is frequently used to describe the stochastic processes in physics [136]. Thus, the sidewise shrinkage of the domains after poling may be discussed in terms of the random jumps of segments of domain walls. The best fit yields n=0.65 and 0.18 for coarse-grain and fine-grain areas, respectively. The difference in the exponent values is attributed to different concentration of pinning centres for the two studied regions.

Similar stretched exponential time dependence of retention loss in PFM experiment was also reported by other groups [137, 138]. In particular, Gruverman *et al.* observed that polarization reversal starts along the grains boundaries most probably due to effect of electrostatic interaction between grain boundary and charged domain wall [137]. An asymmetric backswitching was found in polycrystalline $SrBi_2Ta_2O_9$ (SBT) thin films on Pt electrodes [139]. Namely, retention loss was dependent on the sign of the polarization. "Positive" domains with the polarization vector pointing to the bottom electrode exhibit excellent stability in contrast to "negative" domains oriented upward. This effect was attributed to the presence of the polarization-independent built-in bias field at the SBT/Pt interface. This interface triggers the backswitching process of a "negative" domain by generating "positive" nuclei at the interface. This model was supported by the observed stability of both positive and negative domains in SBT films with IrO_2 electrodes, due to inhibited nucleation at the SBT/IrO_2 interface.

The role of 90°-domain (twins) boundaries in the polarization backswitching was elucidated in PFM experiments in epitaxial $PbZr_{0.2}Ti_{0.8}O_3$ thin films [140]. It was found that the backswitching mainly starts close to the existing twin boundaries and propagates with the velocity depending on the curvature of domain wall. In the final stage of domain relaxation, the faceting of the domain wall is seen as result of slowing down the relaxation along certain crystallographic directions.

9.5 Polarization Switching by a Mechanical Stress

PFM may be applied also to study the effect of an external mechanical stress on domain structure and local piezoelectric properties. Such investigations are especially important, since most of devices based on ferroelectric films operate under a considerable mechanical stress induced either by the deposition process, or by the lattice/thermal mismatches between the film and underlying substrate. One kind of reaction of the materials on a mechanical load is the transformation of a domain pattern to minimize the free energy contribution related to a mechanical deformation. Nucleation and propagation of 90°domains under a compressive unidirectional mechanical stress was investigated using PFM in $BaTiO_3$ coarse-grained ceramics [28]. In grains with the initial in-plane polarization (*a*-domains), nucleation of needle-like domains with the out of plane polarization (*c*-domains) was observed at the stress above a critical value 6 MPa. The long axis of these domains was perpendicular to the applied force. An increase in the load resulted in their sideways expansion. An abrupt broadening of the newly created domains was found at σ about 33 MPa, which corresponds to the coercive stress for the studied material.

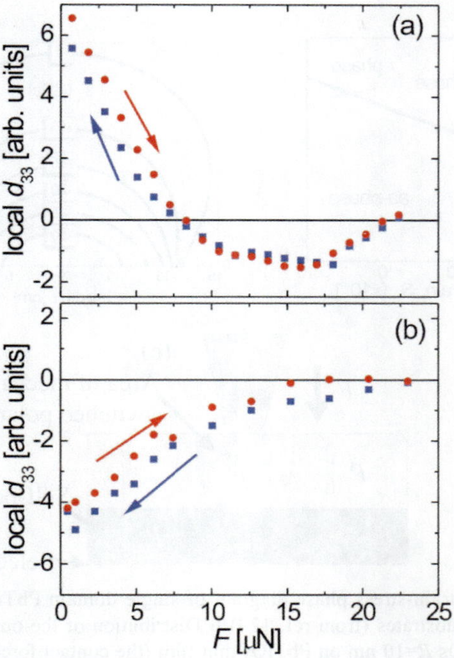

Fig. 9.18 The local piezoresponse measured in 001-oriented $PbTiO_3$: 0.08La thin films as a function of the mechanical force, F, exerted by the tip in domains with the polarization oriented up **a** and down **b** relative to film surface.

It is possible to control the local stress exerted by the tip on the investigated surface by varying the set point in the contact AFM mode. This opens up the possibility of studying *in-situ* effect of mechanical stress on the piezoelectric properties inside a single domain [141, 142]. In particular, Zavala *et al.* [141] have reported a gradual decrease of the VPFM signal (~ 60%) under increasing mechanical stress exerted by the tip in PZT thin films. Kholkin *et al.* investigated the influence of local mechanical force on the piezoresponse in polycrystalline $PbTiO_3$ films doped with 8 mol% La (PTL) [142]. PTL films were chosen because of their simple tetragonal structure, high piezoelectric coefficients, and preferred (100)/(001) orientation. The representative dependencies of d_{33} on the mechanical force, applied in the interior of domains with the polarization oriented up ("bright") and down ("dark") relative to the film plane, are demonstrated in Figs.

9.18a and 9.18b, respectively. It may be seen that the piezoelectric signal strongly decreases with force and finally vanishes for both polarization directions. However, the piezoelectric suppression is qualitatively different for opposite domains. For bright domains, d_{33} varies non-monotonically and switches its sign (phase change of ≈180°). The effect is reversible when the piezoresponse returns to the initial value immediately upon unloading without a substantial hysteresis.

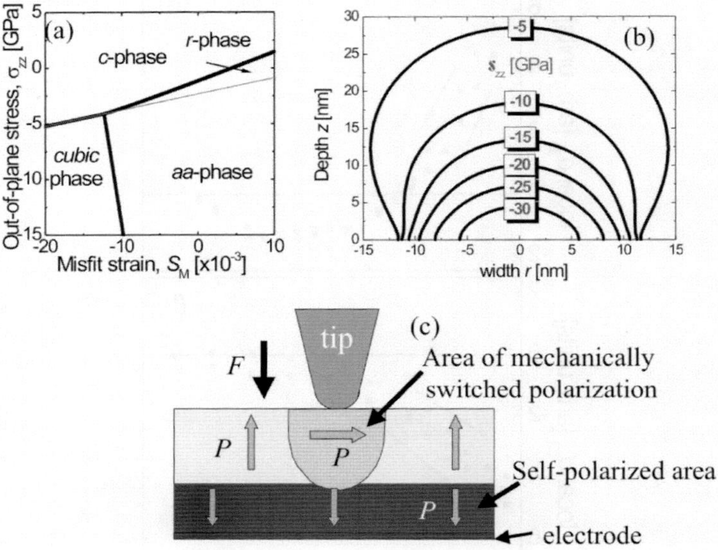

Fig. 9.19 **a** A misfit strain-stress phase diagram of single domain $PbTiO_3$ thin films grown on (001)-oriented cubic substrates (from ref. 129) **b** Distribution of the out-of plane stress exerted by the Si-tip with radius R=10 nm on $PbTiO_3$ thin film (the contact force is 10 μN) **c** Schematic of polarization switching mechanically induced by PFM tip.

The estimation of contact stress using the Hertz elastic solution offers maximum value of the out-of-plane compressive stress of the order of several tens of GPa. This stress is highly inhomogeneous and decays rapidly onto the film. Obviously, such a high stress may significantly modify the polarization states in the vicinity of the contact area. An apparent explanation of the observed effect is based on the phase diagrams for ferroelectric thin films. This predicts the existence of various polarization states with the polarization vector out-of-plane (c-phase), in-plane (aa-phase), and at some angle to film plane (r-phase) depending on the misfit strain [143]. Furthermore, it has been shown that the external compressive stress applied to the surface of epitaxial $PbTiO_3$ film deposited on Si substrate may result in polarization switching from c-phase to aa- or r-phase [129]. This is equivalent to the polarization rotation of the previously *c*-oriented domain. Thus, the experimental results may be interpreted as follows. Under an increasing pressure from the tip, the portion of the film experiencing 90°

polarization rotation propagates gradually from the top to the bottom electrode (Fig. 9.19). The acquired VPFM signal that represents the average piezoelectric coefficient collected from both rotated and still not affected c-oriented volumes (Equation 25) is, therefore, decreasing with force and vanishes when most of the film under the tip is switched into the aa-domain state. The difference observed for the "bright" (signal changes the sign) and "dark" (signal remains negative) domains may be due to the existence of built-in polarization or self-polarization near the bottom electrode interface [81]. Obviously, when the initial polarization state is antiparallel to the direction of self-polarization, which is pointing towards the bottom electrode, the total signal may change the sign under sufficiently high stress where only the contribution from the nearby-electrode region plays a role.

9.6 Investigation of Polarization Fatigue by PFM

Fatigue in ferroelectrics is defined as a loss of switchable polarization during continuous domain switching. Fatigue is a general phenomenon that manifests itself in a similar way in thin films, polycrystalline ceramics, and single crystals [144]. However, the microscopic mechanisms are drastically different. Besides microcracking (which is *a priori* a purely mechanical source of fatigue [145]), the pinning of domains in preferred or random orientations has been assumed to be the major origin of fatigue. This pinning may occur at domain walls via their interaction with defects or through the inhibition of the creation of new domains. The latter mechanism seems to dominate in ferroelectric thin films, where nucleation of opposite domains is inhibited by the injection of charge carriers from the electrode [144]. In ceramics, the pinning of domain walls seems to prevail and may be due to agglomerates of point defects (e.g., oxygen vacancies), defect dipoles, or space charges trapped near grain boundaries [146].

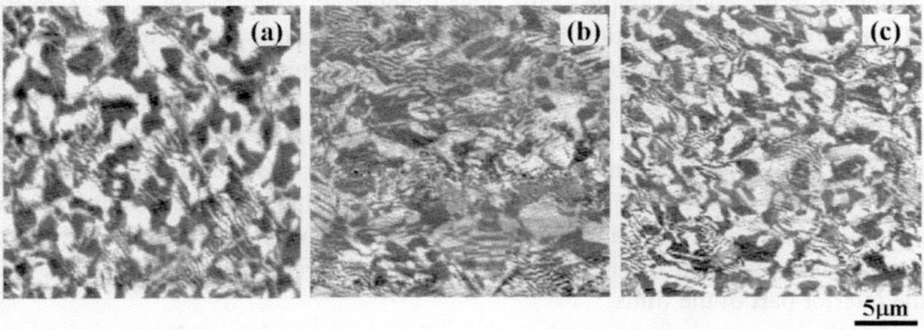

Fig. 9.20 The PFM images of PZT ceramics: **a** - initial state, **b** - fatigued (after $5*10^7$ switching cycles), **c** - annealed after fatigue. Reused with permission from [147]. Copyright 2005, American Institute of Physics.

Fig. 9.20 illustrates domain changes induced by bipolar fatigue in lead zirconate-titanate ceramics [147]. The density of domains in fatigued samples is significantly increased when compared to virgin samples. Many of the new domain walls have well-defined orientations and may hence be attributed to the ferroelastic domain boundaries (90° domain walls for tetragonal symmetry). It was suggested that for polycrystalline samples, fatigue results in different switching abilities in different grains. If the polarization in certain grains can no longer switch, but others are fully switchable, mechanical stresses develop near the boundaries between switchable and unswitchable grains. These stresses may be high and even produce macrocracks [148]. In order to relieve the elastic energy related to these stresses, a splitting of single domains into multiple ferroelastic domains may be energetically favourable. The PFM measurements done on the cross-sections of the fatigued ceramics at different distances from the electrode reveal a strong distortion of ferroelastic domains ("wavy" domains) near the electrode (Fig. 9.21). Under normal circumstances, ferroelastic domains have to be oriented in certain crystallographic directions and be of a regular shape. In the vicinity of both macro- and microscopic defects, some deviation from such regular shapes may take place. In fatigued samples, wavy domain walls are observed far from the meso- and macro- defects and may thus be related to an interaction with microscopic defects: point defects or their agglomerates. The scale of the observed wall deformation (hundreds of nanometres) indicates that domains interact with defect clusters rather than with single point defects. The clustering of point defects is typical for bipolar fatigue [149]. These clusters may form high barriers for the domain-wall motion inside a grain, because stable domain states are induced in their immediate proximity. As the number and size of the agglomerates increase with cycling, more grains become affected. In accordance with aforementioned mechanism, the distribution of distorted domains across the sample thickness (Fig. 9.21d) indicates that the pinning of domains occurs mainly in grains located near the electrodes. In other words, these grains are more affected by fatigue compared to those in the interior part of the sample. The immobilization of domains in grains adjacent to the electrodes seems to be strong enough to block the polarization switching in the entire sample [150]. This is consistent with the reported recovery of the switchable polarization in heavily fatigued samples, after the mechanical removal of a layer underneath the electrodes several grains thick [151].

The importance of the sample-electrode interface for fatigue behaviour was also shown for thin films [152, 153]. After the removal of the top electrodes, the frozen domains were examined in fatigued PZT films under different ac-voltages [152]. Increasing the amplitude of an applied ac-voltage allows probing of the piezoresponse from the interior part of samples. It was concluded that pinned domains are localized in the vicinity of the top electrode and do not extend into the interior part of the films.

Another approach is to emulate the fatigue process by applying a high ac-voltage (enough to switch the polarization in a virgin sample) to the tip that scans the sample surface, and thus induces continuous switching at each point of the scanned region. After such treatment, an attempt to pole the fatigued region is performed by scanning it with the tip under a dc-bias. These studies done in electrode-free films [154] and in thin film capacitors [155] have shown that fatigue proceeds via grain-by-grain suppression of switchable polarization. It is always accompanied by strong domain wall pinning, so that certain domains become frozen and do not participate in switching anymore. The results have been complemented by local piezoelectric hysteresis measurement, which showed a progressive shift of the d_{33} loops in vertical and horizontal directions (also observed by the macroscopic technique [156]). This shift corresponds to a significant imprint of the polarization state (one polarization state is preferred over the other). The orientation of the frozen domains may either have a strong preferential direction (terminated at one of the electrodes), or be randomly distributed in both directions. The size of the frozen regions varies between 100 nm to several microns implying critical consequences when the capacitor size is reduced to similar dimensions.

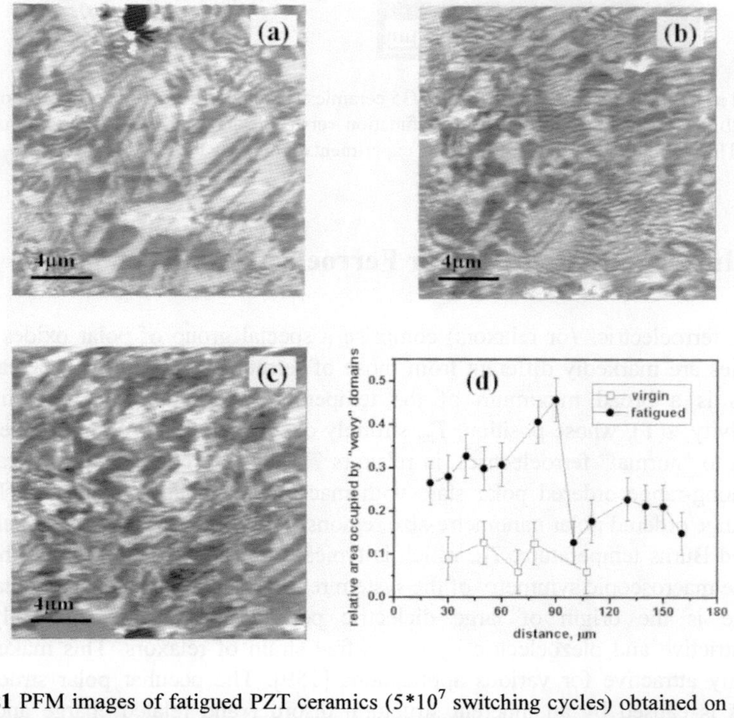

Fig. 9.21 PFM images of fatigued PZT ceramics ($5*10^7$ switching cycles) obtained on different distance from the electrode: 20 μm - **a**; 70 μm –**b**; 110 μm –**c**. Dependence of relative area occupied by distorted domains on distance from electrode –**d**. Reused with permission from [147]. Copyright 2005, American Institute of Physics.

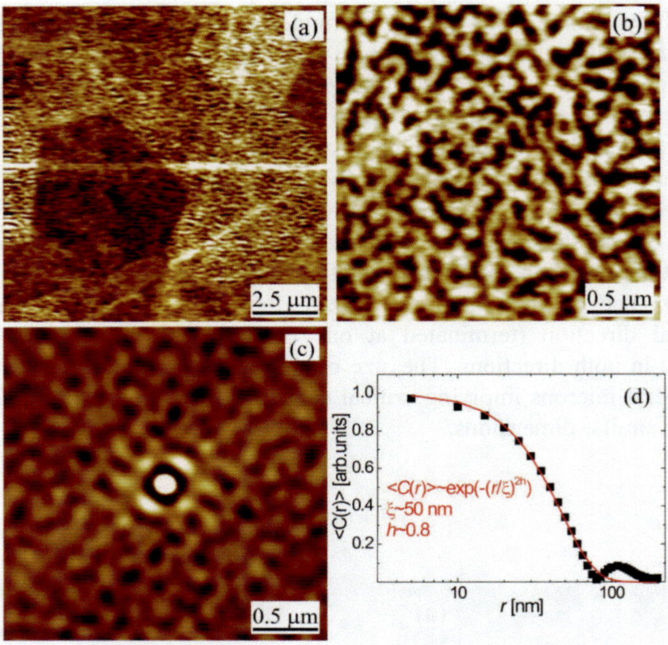

Fig. 9.22 a, b PFM images of PLZT 9.75/65/35 ceramics. **c** Autocorrelation image corresponding to Fig. 9.22b. **d** The averaged autocorrelation function versus distance. The values are normalized to $C(r=0)$. The red line shows the best fit of the experimental points by the Eq. 29.

9.7 Investigation of Relaxor Ferroelectrics by PFM

Relaxor ferroelectrics (or relaxors) comprise a special group of polar oxides whose properties are markedly different from those of ferroelectrics. The typical feature of relaxors is a broad maximum of the temperature dependence of the dielectric permittivity, $\varepsilon(T)$, whose position, T_m, strongly depends on the probing frequency. In contrast to "normal" ferroelectrics, in relaxors T_m does not correspond to a transition into a long-range ordered polar state with macroscopic polarization [157]. Instead, short-range ordered polar nanometre-size regions (PNRs) appear in relaxors below the so-called Burns temperature, T_D, which is typically far above T_m [158]. At the same time, the macroscopic symmetry of the system remains unbroken. This particular polar structure is the origin of large dielectric permittivity, and exceptionally high electrostrictive and piezoelectric hysteresis-free strain of relaxors. This makes them especially attractive for various applications [159]. The peculiar polar structure of relaxors is caused by an inherent structural disorder and related charge and strain inhomogeneities typical for these materials. The charge disorder is a source of quenched electric random fields, whose fluctuations promote nucleation of PNRs below T_D [160]. At high temperatures, PNRs have a size of several nanometres and are highly dynamic. This state is called the ergodic relaxor state, since interactions between PNRs are supposed to be weak and the system quickly goes back to the initial state after some excitation [161]. The evolution of a high-temperature ergodic relaxor state upon cooling may occur in two ways. In some relaxors, the ergodic phase transforms into ferroelectric phase at a Curie temperature, which is usually below T_m. In some other relaxors, e.g., in canonical relaxor $PbMg_{1/3}Nb_{2/3}O_3$ (PMN), a transition into a glassy nonergodic relaxor phase takes place. The ergodicity becomes broken because of dipolar interactions, which leads to a cluster glass state where the dipole moments of PNRs are frozen in random directions [161].

Even below the transition temperature, the size of polar regions in relaxors remains on a nanometre scale. PFM, due to its high spatial resolution, is ideally suited for studying them. Recently, PFM has been successfully applied to investigate polar structures both in relaxor single crystals, such as $Sr_{1-x}Ba_xNb_2O_6$ (SBN) [162, 163, 164, 165], $PbMg_{1/3}Nb_{2/3}O_3$-$PbTiO_3$ (PMN-PT) [166, 167, 168, 169] and $PbZn_{1/3}Nb_{2/3}O_3$-$PbTiO_3$ (PZN-PT) [170, 171], and relaxor ceramics, such as $Pb_{1-x}La_xZr_{1-y}Ti_yO_3$ (PLZT) [172, 173] and $BaTiO_3$-based [174, 175] compounds. An example of polar structure observed by VPFM in PLZT ceramics with the concentration of La x=9.75% is shown in Fig. 9.22 [172]. Piezoactive regions form a complex labyrinth structure, which reflects spatial inhomogeneities determined by random fields and random strains introduced with La doping. In order to quantitatively analyze the PFM images, the autocorrelation function technique was applied. Autocorrelation images were obtained from the original PFM images by the following transformation:

$$C(r_1, r_2) = \sum_{x,y} D(x,y)D(x+r_1, y+r_2) \tag{28}$$

where $D(x,y)$ is the value of the piezoresponse signal. Positive or negative values of the autocorrelation function, $C(r_1, r_2)$, correspond to probabilities to find a region with parallel or antiparallel direction of the polarization after a shift of (r_1,r_2) from an arbitrary point in the original image, respectively. It was found that for relaxor PLZT ceramics, the images of the autocorrelation function (Fig. 9.22c) are regular in two different directions revealing the local rhombohedral symmetry. In order to estimate the mean size of the polar regions, the autocorrelation image is averaged over all in-plane directions and then approximated by Gaussian-like function [169].

$$< C(r) > \sim \exp\left[-\left(\frac{r}{\xi}\right)^{2h}\right] \tag{29}$$

Here ξ has a meaning of the average autocorrelation radius. The best fit gives the value, $\xi \approx 50$ nm. In this case, the measurements were done at room temperature, which is below the temperature of the ergodic relaxor – ferroelectric transition for this PLZT composition.

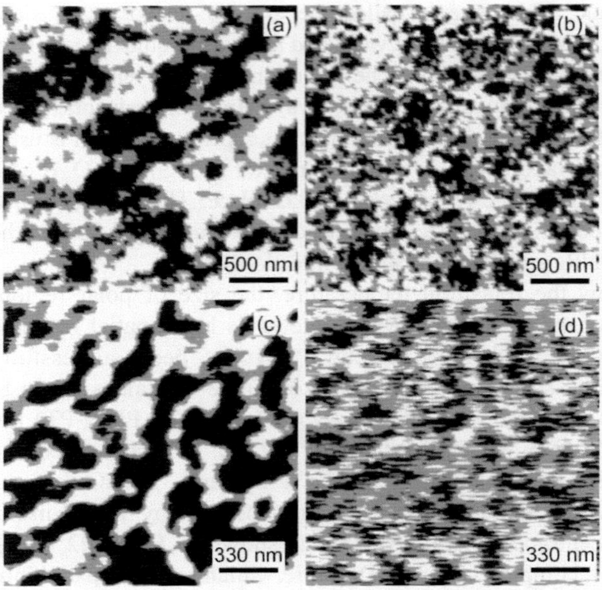

Fig. 9.23 VPFM images of a SBN61 single crystal (T_C = 346 K) at 295 K **a** and 354 K b, and of a PMN-PT10 single crystal (T_C = 280 K) at 295 K **c** and 335 K **d**.

The evolution of polar structures at the phase transition from the ergodic relaxor state into ferroelectric state was investigated by PFM in SBN [163, 165] and PMN-PT [166, 169] single crystalline relaxors. In the ergodic state, only dynamic PNRs with a size less than 10 nm were expected. Nevertheless, both in PMN-PT and in SBN, large quasistatic regions of correlated piezoresponse (i.e., correlated polarization) with size 30-100 nm were found already above the macroscopic T_C (Fig. 9.23). These regions were attributed to the "mesoscopic" PNRs, which grow in the vicinity of the phase transition and become immobilized on the time scale of the PFM experiment due to their large size. The "frozen" PNRs are surrounded by interfaces with negligible piezoresponse, which most probably consist of smaller PNRs being spatially unresolved. These results point out that the transformation from the ergodic relaxor state with dynamic PNRs to the ferroelectric state takes place not at a fixed temperature, but in a relatively broad temperature range. In this range, the system contains both small dynamic and large quasistatic PNRs. The latter ones are precursors of domains persisting in the low-temperature ferroelectric state. Such a system is not an ergodic anymore, as was verified by observation of aging of the dielectric susceptibility in SBN in the same temperature range above T_C [163]. This aging is related to isothermal growth of quasistatic PNRs at the expense of the dynamic ones. Furthermore, the existence of "frozen" PNRs results in a deviation of the critical behaviour at the phase transition observed in experiments from that predicted by theory [163].

In “cubic” relaxors such as PMN-PT, “frozen” PNRs were observed in a much broader temperature range above T_C than in the uniaxial SBN relaxors. It was suggested that this is related to an effect of mechanical strains resulting from the rhombohedral distortion of unit cell in PMN-PT [169]. There are three main factors influencing the polar structure in relaxors: (i) minimization of the depolarization field energy, which drives the formation of 180° domains; (ii) the influence of random fields, which break domains into smaller entities; and (iii) minimization of mechanical stress, which is achieved by formation of ferroelastic domains (twins). The third factor is important for “cubic” relaxors but is not active in the uniaxial ones. In particular, it results in a preferential orientation of boundaries of the polar regions parallel to {110} habit planes, which are invariant for systems with rhombohedral symmetry [169, 171]. It is supposed that the domains observed in PMN-PT with high Ti^{4+} content below T_C are self-assembled agglomerates of smaller nanosized domains [167]. The driving force for such self-organization is the minimization of the free energy related to the mechanical deformation. Furthermore, this tendency to self-organize may stimulate the formation of relatively large agglomerates of PNRs already at high enough temperatures above the nominal T_C. The observed regions of correlated piezoresponse may correspond to such agglomerates. Besides, it should be taken into account that (1-x)PMN-xPT relaxors are spatially inhomogeneous materials. In compositions with $0.05 < x < 0.25$, a transition into a state with macroscopic rhombohedral symmetry was found in a surface layer, while the interior regions of the sample remain macroscopically cubic [176]. The thickness of the outer layer is several tens of microns. The nature of this phenomenon is unclear. In particular, a large compressive strain was found near the surface in PMN single crystals [177]. Such strain may stabilize the ferroelectric state and shift the phase transition to higher temperatures. If there is a gradient of the strain across the outer layer, then the shift of the Curie temperature will also be non-uniform. Then T_C in a thin subsurface layer, which gives the main contribution to the PFM signal, may be considerably higher than an average (nominal) Curie temperature estimated from X-ray measurements. Nevertheless, at least in SBN, the existence of “frozen” PNRs above T_C is not just a surface effect. Indeed, dielectric spectroscopy data reveal a particular relaxation mode related to the boundaries of “frozen” PNRs [178]. Since in this case the response from the entire sample is measured, it may be concluded that the appearance of quasistatic mesoscale PNRs above the nominal Curie temperature takes place in the whole sample.

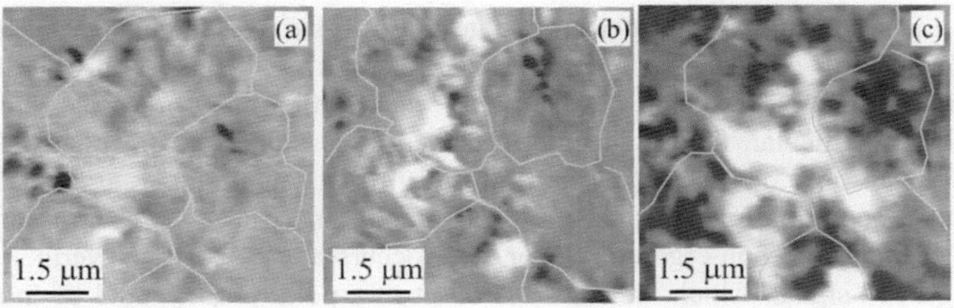

Fig. 9.24 VPFM images of a PMN-PT14 ceramics ($T_C = 330$ K) at 360 K **a**, 315 K **b**, 295 K **c**

The results obtained for single crystals match the experimental data for ceramic samples. Fig. 9.24 shows the VPFM images taken on PMN-PT ceramics with 14% of titanium on cooling through the Curie temperature (T_C~330 K). Like in single crystals, quasistatic piezoactive regions of several hundred nanometres in size appear already above the nominal T_C. It is important that they are concentrated near the grain boundaries, especially near the "triple points" where several grains are joined. Obviously, in these places, mechanical stresses are accumulated, which promote coarsening of PNRs and stimulate their agglomeration in bigger entities, as discussed above.

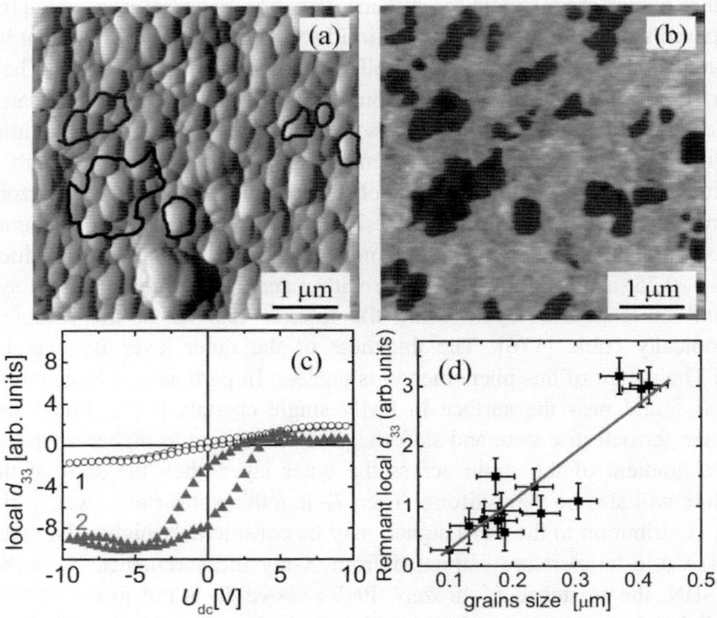

Fig. 9.25 Topographic **a** and VPFM **b** images of 0.9PMN–0.1PT thin film. Black regions on piezoresponse image correspond to grains with self-polarization. Typical local piezoresponse hysteresis loops **c** for initially non-piezoactive **1** and self-polarized **2** grains. The remnant d_{33} value as a function of the grain size **d**.

A transition into the ferroelectric state may be induced in relaxors after the application of a strong enough external electric field [161]. In the PFM experiment, the corresponding value of the electric field may be easily achieved locally. In particular, field-induced effects were investigated by Shvartsman *et al.* in epitaxial (001) oriented PMN thin films [179]. At room temperature, the films exhibit a negligible piezoresponse typical for this material, where the transition from ergodic into non-ergodic relaxor state occurs at T_f=220 K. Nevertheless, a strong PFM signal, which indicates the onset of a polar phase, could be induced after applying a dc-voltage exceeding a threshold value of 1.5–2 V. The induced piezoresponse was unstable and relaxed within tens of minutes after switching off the dc-bias. The time dependence of local d_{33} exhibited an initial fast decay followed by a slow decrease in a longer time. The slow stage of the d_{33} decay may be approximated well with a Kohlrausch-Williams-Watt type dependence: $d_{33}(t) \sim \exp[-(t/\tau)^{\beta}]$, where t is the time, and τ and β are the fitting parameters. Both, the KWW type of the slow relaxation stage, and the increase of the parameter β observed with increasing field, indicate a narrowing of relaxation time spectrum of PMN under a higher dc-bias. The characteristic time τ was in the range of 20-30 min as compared to the seconds reported for the macroscopic experiments [180]. To explain the obtained results, the following model was proposed. Applying short voltage pulses of moderate magnitude results in the reorientation of the polar clusters and their subsequent coarsening into bigger entities, nanodomains. The slow relaxation kinetics reflects the thermally activated breakup of these nanodomains. The longer relaxation times, as compared to macroscopic experiments, were attributed to the stabilization of the polar clusters near the surface that is expected in PFM experimental conditions. If the voltage is higher than the critical value, a macroscopic single domain ferroelectric state could be created. In this case, the relaxation is governed by the depolarization process as in normal ferroelectrics [181]. The decay is much faster because of the existence of strong macroscopic depolarizing field.

In polycrystalline relaxor films, it was found that the transition into ferroelectric state is influenced by the properties of the individual grains (due to their different orientation, chemical composition, mechanical stress, etc.). As an example, Fig. 9.25 shows the typical topography and piezoelectric images of polycrystalline PMN-0.1PT thin films [182]. The majority of the grains exhibit a weak piezoelectric activity (an intermediate contrast on the PFM image), while others have relatively high d_{33}. It means they are strongly polarized without any external bias, i.e., self-polarized. Two conclusions were made: (i) self-polarization can be found at nanoscale in relaxor state, while it is barely found in macroscopic measurements, and (ii) since the piezoelectricity in relaxors is possible only under the bias field, the distribution of self-polarization represents also the distribution of internal field in these materials. Thus, relaxor thin film on the microscopic level behaves as a nanocomposite, the macroscopic response of which can be derived by the averaging of local responses of the individual grains. The induced piezoresponse inside individual grains was found to depend on the grain size: the bigger the grain, the higher the remnant d_{zz} (Fig. 9.25d). There are several possible explanations for the observed phenomena. First, the mechanical clamping of a grain caused by surrounding grains is apparently dependent on its size. It is likely that the mechanical interaction between the grains leads to a high level of the compressive pressure inside small grains, while a mean stress value (or the stress in the centre of the grain) remains relatively low for large grains. The growth and reorientation of polar clusters under an external bias may be considerably restricted in smaller grains via the stronger coupling between local strains and polarization inherent for cubic relaxors. Therefore, the piezoelectric response is substantially reduced in those grains. In very small grains, clusters may be completely frozen due to high mechanical clamping and the response is due to paraelectric matrix. Another reason for the observed size effect may be related to the possible grain size dependence of the average size of a cluster inside a single grain, possibly through the mechanical clamping effect again. Indeed, studies on PMN ceramics [183] have revealed a pronounced decrease in the size and corresponding weakening of the dielectric response of an average cluster with decreasing grain size. The chemical inhomogeneity and local variations of the compositions ($PbTiO_3$ content) could also be partly responsible for the variation of the physical properties from grain to grain.

9.8 Size Effect and Search for the Ferroelectricity Limit

The development of technology allows the fabrication of ferroelectric nanostructures with length scales of the order of or less than several hundred nm [5]. As device dimensions are scaled down, their size reaches the limit where ferroelectric materials show a significant deviation from bulk properties. The following parameters were found to change in ferroelectrics as their size decreases: (i) reduction of the remnant polarization, dielectric permittivity and, consequently, piezocoefficient; (ii) increasing coercive field; (iii) change in the domain structure; and (iv) lower phase transition temperature and smearing of the phase transition. The size effects are a consequence of a size-driven instability in the polar phase followed by the suppression of ferroelectricity below a critical size – a super-paraelectric limit. A number of theoretical approaches have been used to predict the critical size or thickness for the disappearance of ferroelectricity [184, 185, 186, 187, 188]. PFM is ideally suited for polarization detection in small objects such as ultrathin films and ferroelectric nanostructures because the effective tip diameter may be as small as few nm.

The results reported for ferroelectric structures prepared by top-down approaches, which are described in Chapter 1, are contradictory. For instance, Ganpule *et al.* investigated Pt/LSCO/PZT/LSCO/Pt capacitors of 70 x 70 nm^2 [189] and Pt/SBT/Pt capacitors of 100 x 100 nm^2 [190] by using focused ion beam etching. For capacitors milled down to the bottom electrode, a higher piezoelectric response was measured by PFM. This was in comparison to capacitors where only the top-electrode was structured. This result was correlated to the in-plane constraints in the continuous film, rather than to the intrinsic size effects. The hysteresis loops obtained by PFM in the PZT capacitors of different dimensions show no change in the intrinsic properties except that associated with the clamping effect of the substrate [189]. Even for the smallest capacitors, no intrinsic size effects were detected by PFM. Bühlmann *et al.* studied PZT ferroelectric islands of lateral size 100-200 nm fabricated from an epitaxial thin film using e-beam lithography [191]. They found a strong increase of piezoresponse with a decrease in the size of nanocapacitors. It was explained by a decreasing fraction of domains with in-plane polarization (*a*-domains) for the PZT islands as compared to the continuous films. This is again correlated to different mechanical strains. A direct electron beam-writing technique was applied by Alexe *et al.* to fabricate SBT and PZT structures as small as 100 x 100 nm^2 [192]. While the coercive field was found to be size-independent, a decrease of the piezoelectric response with decreasing size was reported. Thus, the majority of reported size effects in the devices structured from thin films via top-bottom approach may be attributed to a reduction of the in-plane strain. The influence of the structuring method has to be taken into account, like damage from etching, doping with sputter ions, etc.

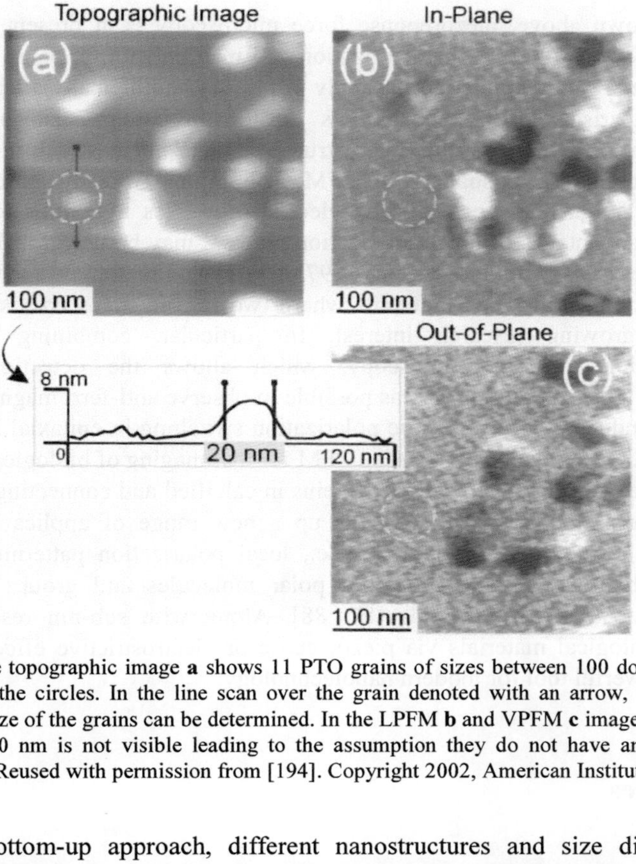

Fig. 9.26 The topographic image **a** shows 11 PTO grains of sizes between 100 down to 20 nm indicated by the circles. In the line scan over the grain denoted with an arrow, shown at the bottom, the size of the grains can be determined. In the LPFM **b** and VPFM **c** images the grain of the size of 20 nm is not visible leading to the assumption they do not have any permanent polarization. Reused with permission from [194]. Copyright 2002, American Institute of Physics.

By a bottom-up approach, different nanostructures and size distributions, ranging from 200 nm to about 10 nm in diameter and 20 nm down to 2 nm in height were obtained [193, 194, 195]. Fig. 9.26 shows examples of the VPFM measurements on such structures. It was found that the separated grains mostly exhibit 180° domains, contrary to the continuous films where the 90° domain pattern was observed [194]. The grains with size smaller than 40 nm become monodomain, and, finally, the grains smaller than 20 nm did not show piezoresponse. It was supposed that this could be the limiting size for the ferroelectric phase. Recently, Clemens *et al.* reported on ferroelectricity in $PbTiO_3$ grains as small as 15 nm [195]. In this case, the authors applied LPFM mode, which generally exhibits a higher resolution and signal/noise ratio than the VPFM mode [196]. This value is closer to the calculated critical size for a lead titanate nanograin, 7.4 nm [186].

Conclusions

As was shown above, piezoreponse force microscopy is at present a powerful technique for quantitative characterization and manipulation in polar materials at the nanoscale. Being applied to the study of polycrystalline materials it allows the direct comparison of the peculiarities of domain patterns and of the local polarization switching to the microstructure. The polar structure of relaxor ferroelectrics may be visualized by PFM. The domain wall dynamics, effect of aging, and fatigue on the local piezoelectric properties is readily addressed by PFM. The advantages of local polarization reversal may be used for development of ferroelectric data storage systems [197]. PFM may be applied for the study of multiferroic materials (i.e., materials where two ferroic states coexist), which is attracting growing research interest. In particular, combining PFM with photoemission electron microscopy, which allows the detection of anti-ferromagnetic configuration, it was possible to observe anti-ferromagnetic domain switching induced by ferroelectric polarization switching in epitaxial $BiFeO_3$ thin films [198]. Recently, a usefulness of PFM for the imaging of biological materials, such as electromechanically active proteins in calcified and connecting issues, has been demonstrated [199]. This opens up a new range of applications of this technique. Ferroelectric lithography, i.e., local polarization patterning, may be used for the recognition of specific polar molecules and groups of various biologically important materials [87, 88]. Along with sub-nm resolution, for probing biological materials via piezoelectric or electrostrictive effects [199], it offers a powerful tool for modern nanotechnology.

References

1. Lines ME, Glass AM (1979) Principles and applications of ferroelectrics and related materials. Clarendon Press. Oxford.
2. Auciello O, Ramesh R (eds) (July 1996) Electroceramic thin films: Device applications. MRS Bulletin. Materials Research Society.
3. Muralt P (2008) Recent progress in materials issues for piezoelectric MEMS, J. Am. Ceram. Soc. 91: 1385-1396.
4. Uchino K (2000) Ferroelectric devices. Decker. New York.

5. Waser R (ed.) (2005) Nanoelectronics and information technology: advanced electronic materials and novel devices. Wiley-VCH. Berlin.
6. Saurenbach F, Terris BD (1990) Imaging of ferroelectric domain walls by force microscopy. Appl. Phys. Lett. 56: 1703-1705.
7. Kalinin SV, Rar A, Jesse S (2006) A decade of piezoresponse force microscopy: Progress, challenges, and opportunities. IEEE Trans. UFFC 53: 2226-2252.
8. Kalinin SV, Gruverman A, Scanning Probe Microscopy: Electrical and Electromechanical Phenomena at the Nanoscale, Springer, New York (2007).
9. Stern JE, Terris BD, Mamin HJ, Rugar D (1988) Deposition and imaging of localized charges on insulator surfaces using a force microscopy. Appl. Phys. Lett. 53: 2717-2719.
10. Martin Y, Abraham DW, Wickramasinghe H (1987) High-resolution capacitance measurement and potentiometry by force microscopy. Appl. Phys. Lett. 52, 1103-1105.
11. Hong JW, Kahng DS, Shin JC, Kim HJ, Khim ZG (1998) Detection and control of ferroelectric domains by an electrostatic force microscope. J. Vac. Sci. Technol. B 16: 2942-2946.
12. Hong JW, Noh KH, Park S, Kwun SI, Khim ZG (1998) Surface charge density and evolution of domain structure in triglycine sulfate determined by electrostatic force microscopy. Phys. Rev. B 58: 5078-5084
13. Hong JW, Park S, Khim ZG (1999) Measurement of hardness, surface potential, and charge distribution with dynamic contact mode electrostatic force microscope. Rev. Sci. Instr. 70: 1735-1739.
14. Lehnen P, Dec J, Kleemann W (2000) Ferroelectric domain structures of $PbTiO_3$ studied by scanning force microscopy. J. Phys. D 33: 1932-1936.
15. Luo EZ, Xie Z, Xu JB, Wilson IH, Zhao LH (2000) *In situ* observation of the ferroelectric-paraelectric phase transition in a triglycine sulfate single crystal by variable-temperature electrostatic force microscopy. Phys. Rev. B 61: 203-206.
16. Soergel E, Pankrath P, Buse K (2003) Investigation of photorefractive SBN crystals with atomic force microscopy. Ferroelectrics 296: 19-27.
17. Lüthi R, Haefke H, Meyer KP, Meyer E, Howald L, Güntherodt HJ (1993) Surface and domain structures of ferroelectric crystals studied with scanning force microscopy. J. Appl. Phys. 74: 7461-7471.
18. Eng LM, Bammerlin M, Loppacher Ch, Guggisberg M, Bennewitz R, Lüthi R, Meyer E, Güntherodt HJ (1999) Surface morphology, chemical contrast, and ferroelectric domains in TGS bulk single crystals differentiated with UHV non-contact force microscopy, Appl. Surf. Sci. 140: 253-258.
19. Nonnenmacher M, O'Boyle MP, Wickramasinghe HK (1991) Kelvin probe force microscopy. Appl. Phys. Lett. 58: 2921-2923.
20. Jacobs HO, Knapp HF, Stemmer A (1999) Practical aspects of Kelvin probe force microscopy, Rev. Sci. Instr. 70: 1756-1760.
21. Kalinin SV and Bonnell DA (2001) Local potential and polarization screening on ferroelectric surfaces. Phys. Rev. B 63: 125411 (1-13).
22. Son JY, Bang SH, Cho JH (2003) Kelvin probe force microscopy study of $SrBi_2Ta_2O_9$ and $PbZr_{0.53}Ti_{0.47}O_3$ thin films for high-density nonvolatile storage devices. Appl. Phys. Lett. 82: 3505-3507.
23. Lu XM, Schlaphof F, Grafstrom S, Loppacher C, Eng LM, Suchaneck G, Gerlach G (2002) Scanning force microscopy investigation of the $Pb(Zr_{0.25}Ti_{0.75})O_3$/Pt interface. Appl. Phys. Lett. 81: 3215-3217.
24. Shishkin EI, Shur VY, Schlaphof F, Eng LM (2006) Observation and manipulation of the as-grown maze domain structure in lead germanate by scanning force microscopy. Appl. Phys. Lett. 88: 252902.

25. Bluhm H, Wadas A, Wiesendanger R, Roshko A, Aust JA, Nam D (1997) Imaging of domain-inverted gratings in $LiNbO_3$ by electrostatic force microscopy. Appl. Phys. Lett. 71: 146-148.
26. Eng LM, Friedrich M, Fousek J, Guenter P (1996) Deconvolution of topographic and ferroelectric contrast by noncontact and friction force microscopy. J. Vac. Sci. Technol. B 14: 1191-1196.
27. Bluhm H, Schwarz UD, Wiesendanger R (1998) Origin of the ferroelectric domain contrast observed in lateral force microscopy. Phys. Rev. B 57: 161-169.
28. Munoz-Saldana J, Schneider GA, Eng LM (2001) Stress induced movement of ferroelastic domain walls in $BaTiO_3$ single crystals evaluated by scanning force microscopy. Surf. Sci. Lett. 480: L402-L410.
29. Zybill CE, Boubekeur H, Li B, Koch F, Schindler G, Dehm C (2001) Domain structure of (100) strontium bismuth tantalate (SBT) $SrBi_2Ta_2O_2$ films. Thin Solid Films 386: 59-67.
30. Ganpule CS, Nagarajan V, Hill BK, Roytburd AL, Williams ED, Alpay SP, Roelofs A, Waser R, Eng LM (2002) Imaging three-dimensional polarization in epitaxial polydomain ferroelectric thin films. J. Appl. Phys. 91: 1477-1481.
31. Wang YG, Kleemann W, Woike T, Pankrath R (2000) Atomic force microscopy of domains and volume holograms in $Sr_{0.61}Ba_{0.39}Nb_2O_6:Ce^{3+}$. Phys. Rev. B 61: 3333-3336.
32. Jungk T, Soergel E (2005) Depth-resolved analysis of ferroelectric domain structures in bulk $LiNbO_3$ crystals by scanning force microscopy. Appl. Phys. Lett. 86: 242901.
33. Cho Y, Kirihara A, Saeki T (1996) Scanning nonlinear dielectric microscope. Rev. Sci. Instr. 67: 2297-2303.
34. Odagawa H, Cho Y (2000) Theoretical and experimental study on nanoscale ferroelectric domain measurement using scanning nonlinear dielectric microscopy. Japan. J. Appl. Phys. 39: 5719-5722.
35. Cho Y, Ohara K (2001) Higher-order nonlinear dielectric microscopy. Appl. Phys. Lett. 79: 3842-3844.
36. Odagawa H, Cho Y (2002) Measuring ferroelectric polarization component parallel to the surface by scanning nonlinear dielectric microscopy. Appl. Phys. Lett. 80: 2159-2161.
37. Rabe U, Kopycinska M, Hirsekorn S, Munoz-Saldana J, Schneider GA, Arnold W (2002) High-resolution characterization of piezoelectric ceramics by ultrasonic scanning force microscopy techniques. J. Phys. D: Appl.Phys. 35: 2621-2635.
38. Rabe U, Scherer V, Hirsekorn S, Arnold W (1997) Nanomechanical surface characterization by atomic force acoustic microscopy. J. Vac. Sci. Technol. B 15: 1506-1511.
39. Liu XX, Heiderhoff R, Abicht HP, Balk LJ (2002) Scanning near-field acoustic study of ferroelectric $BaTiO_3$ ceramics. J. Phys. D: Appl. Phys. 35: 74-87.
40. Tsuji T, Saito S, Fukuda K, Yamanaka K, Ogiso H, Akedo J, Kawakami Y. (2005) Significant stiffness reduction at ferroelectric domain boundary evaluated by ultrasonic atomic force microscopy. Appl. Phys. Lett. 87: 071909.
41. Hong S. (ed.) (2003) Nanoscale phenomena in ferroelectric thin films. Kluwer. Dordrecht.
42. Alexe M, Gruverman A (eds.) (2004) Nanoscale characterization of ferroelectrics: scanning probe microscopy approach. Springer. Berlin.
43. Gruverman A, Kholkin A (2006) Nanoscale ferroelectrics: processing, characterization and future trends. Rep. Prog. Phys. 69: 2443-2474.
44. Kholkin AL, Kalinin SV, Roelofs A, Gruverman A (2007) Review of ferroelectric domain imaging by piezoresponse force microscopy. In Kalinin S, Gruverman A (eds) Electrical and electromechanical phenomena at the nanoscale. Springer, New York, vol 1, pp 173-214.

45. Güthner P, Dransfeld K (1992) Local poling of ferroelectric polymers by scanning force microscopy. Appl. Phys. Lett. 61: 1137-1139.
46. Kholkin AL, Shvartsman VV, Kiselev, DA Bdikin IK (2006) Nanoscale characterization of ferroelectric materials for piezoelectric applications. Ferroelectrics 341: 3-19.
47. Jaffe B, Cook WR, Jaffe H (1971) Piezoelectric Ceramics. Academic. London.
48. Abplanalp M, Eng LM, Cünter P (1998) Mapping the domain distribution at ferroelectric surfaces by scanning force microscopy. Appl. Phys. A 66: S231-S234.
49. Nye JF (1985) Physical Properties of Crystals. Oxford University, Oxford.
50. Goldstein H (1978) Classical Mechanics. Addison-Wesley, Reading.
51. Harnagea C, Pignolet A, Alexe M, Hesse D (2001) Piezoresponse scanning force microscopy: What quantitative information can we really get out of piezoresponse measurements on ferroelectric thin films. Integr. Ferroelectr. 38:667-673.
52. Budimir M, Damjanovic D, Setter N (2005) Enhancement of the piezoelectric response of tetragonal perovskite single crystals by uniaxial stress applied along the polar axis: A free-energy approach. Phys. Rev. B 72: 064107.
53. Budimir M, Damjanovic D, Setter N (2003) Piezoelectric anisotropy–phase transition relations in perovskite single crystals. J. Appl. Phys. 94: 6753-6761.
54. Hoffman A, Jungk T, Soergel E (2007) Crosstalk correction in atomic force microscopy. Rev. Sci. Instr. 78: 016101.
55. Peter F, Rüdiger A, Waser R (2006) Mechanical crosstalk between vertical and lateral piezoresponse force microscopy. Rev. Sci. Instrum. 77: 036103.
56. Kalinin SV, Rodriguez BJ, Jesse S, Karapetian E, Mirman B, Eliseev EA, Morozovska AN (2007) Nanoscale electromechanics of ferroelectric and biological systems: A new dimension in scanning probe microscopy. Annu. Rev. Matter. Res. 37 189-238.
57. Kalinin SV and Bonnell DA (2002) Imaging mechanism of piezoresponse force microscopy of ferroelectric surfaces. Phys. Rev. B 65: 125408.
58. Zhirnov VA (1959) A contribution to the theory of domain walls in ferroelectrics. Sov. Phys. JETP 8: 822-825.
59. Padilla J, Zhong W, Vanderbilt D (1996) First-principles investigation of 180° domain walls in $BaTiO_3$. Phys. Rev. B 53: R5969-R5973.
60. Meyer B, Vanderbilt and D (2005) *Ab initio* study of ferroelectric domain walls in $PbTiO_3$. Phys. Rev. B 65: 104111.
61. Strukov BA, Levanyuk AP (1998) Ferroelectric phenomena in crystals: physical foundations. Spinger. Berlin.
62. Kalinin SV, Jesse S, Rodriguez BJ, Shin J, Baddorf AP, Lee HN, Borisevich A, Pennycook SJ (2006) Spatial resolution, information limit, and contrast transfer in piezoresponse force microscopy. Nanotechnology 17: 3400-3411.
63. Born M, Wolf E (1975) Principles of Optics. Pergamon Press, Oxford.
64. Jungk T, Hoffmann A, Soergel E (2008) Impact of the tip radius on the lateral resolution in piezoresponse force microscopy, New J. Phys. 10: 013019.
65. Rodriguez BJ, Jesse S, Baddorf AP, Kalinin SV (2006) High resolution electromechanical imaging of ferroelectric materials in a liquid environment by piezoresponse force microscopy, Phys. Rev. Lett. 96: 237602.
66. Gruverman A, Rodriguez BJ, Dehoff C, Waldrep D, Kingon AI, Nemanich RJ, Cross JS (2005) Direct studies of domain switching dynamics in thin film ferroelectric capacitors. Appl. Phys. Lett 87: 082902.
67. Shvartsman VV, Kleemann W, Haumont R, Kreisel J (2007) Large bulk polarization and regular domain structure in ceramic $BiFeO_3$. Appl. Phys. Lett. 90: 172115.
68. Teague JR, Gerson R, James WJ (1970) Dielectric hysteresis in single crystal $BiFeO_3$. Solid State Commun. 8: 1073-1074.

69. Damjanovic D, Budimir M, Davis M, Setter N (2006) , Piezoelectric anisotropy: Enhanced piezoelectric response along nonpolar directions in perovskite crystals. J. Mater. Sci. 41: 65-76.
70. Munoz-Saldana J, Hoffmann MJ, Schneider GA (2003) Ferroelectric domains in coarse-grained lead zirconate titanate ceramics characterized by scanning force microscopy. J. Mat. Res. 18: 1777-1786.
71. Roelofs A, Schneller T, Szot K, Waser R (2002) Piezoresponse force microscopy of lead titanate nanograins possibly reaching the limit of ferroelectricity. Appl. Phys. Lett. 81: 5231-5233.
72. Roytburd AL (1998) Thermodynamics of polydomain heterostructures. I. Effect of macrostresses. J. Appl. Phys. 83: 228-238.
73. Arlt G, Sasko P (1980) Domain configuration and equilibrium size of domains in $BaTiO_3$ ceramics. J. Appl. Phys. 51: 4956-4960.
74. Gruverman A (1999) Scaling effect on statistical behaviour of switching parameters of ferroelectric capacitors. Appl. Phys. Lett. 75: 152-1454.
75. Bdikin IK, Shvartsman VV, Kholkin AL, Kim SH (2004). Ferroelectric domain structure and local piezoelectric properties of sol-gel derived $Pb(Zr_{1-x}Ti_x)O_3$ films. In Hoffmann-Eifert S, Funakubo H, Kingon AI, Koutsaroff I, Joshi V (eds.) Ferroelectric thin films XII, pp 55-60.
76. Wu A, Vilarinho PM, Shvartsman VV, Suchanek G, Kholkin AL (2005) Domain populations in lead zirconate titanate thin films of different compositions via piezoresponse force microscopy. Nanotechnology 16: 2587-2595.
77. Shvartsman VV, Pankrashkin AV, Afanasjev VP, Kaptelov EY, Pronin IP, Kholkin AL (2005) Piezoelectric properties of self-polarized $Pb(Zr_{1-x}Ti_x)O_3$ thin films probed by scanning force microscopy. Integrated Ferroelectrics 69: 103-111.
78. Afanasjev VP, Petrov AA, Pronin IP, Tarakanov EA, Kaptelov EJ, Graul J (2001) Polarization and self-polarization in thin $PbZr_{1-x}Ti_xO_3$ (PZT) films. J. Phys.: Condens. Matter 13: 8755-8763.
79. Warren WL, Tuttle BA, Dimos D, Pike GE, Al-Shareef HN, Ramesh R, Evans JT (1996) Imprint in ferroelectric capacitors. Jpn. J. Appl. Phys. 35: 1521-1524.
80. Spierings GACM, Dormans GJM, Moors WGJ, Ulenaers MJE, Larsen PK (1995) Stresses in $PT/Pb(Zr,Ti)O_3/PT$ thin-film stacks for integrated ferroelectric capacitors. J. Appl. Phys. 78: 1926-1933.
81. Kholkin AL, Brooks KG, Taylor DV, Hiboux S, Setter N (1998) Self-polarization effect in $Pb(Zr,Ti)O_3$ thin films. Integr. Ferroelectr. 22: 525-533.
82. Pronin IP, Kaptelov EY, Tarakanov EA, Sorokin LM, Afanasjiev VP, Pankrashkin AV (2002) Self-polarization and migratory polarization in thin-film ferroelectric capacitor. Integr. Ferroelectr. 49: 285-294.
83. Tybell T, Paruch P, Giamarchi T, Trsicone J-M (2002) Domain wall creep in epitaxial ferroelectric $Pb(Zr_{0.2}Ti_{0.8})O_3$ thin films. Phys. Rev. Lett. 89: 097601.
84. Cho Y, Hashimoto S, Odagawa N, Tanaka K, Hiranaga Y (2005) Realization of 10 $Tbit/in.^2$ memory density and subnanosecond domain switching time in ferroelectric data storage. Appl. Phys. Lett. 87: 232907.
85. Rosenman G, Urenski P, Argonin A, Arie A, Rosenwaks Y (2003) Nanodomain engineering in $RbTiOPO_4$ ferroelectric crystals. Appl. Phys. Lett. 82: 3934-3936.
86. Rosenwaks Y, Dahan D, Molotski M, Rosenman G (2005) Ferroelectric domain engineering using atomic force microscopy tip arrays in the domain breakdown regime. Appl. Phys. Lett. 86: 012909.
87. Kalinin SV, Bonnell DA, Alvarez T, Lei X, Hu Z, Ferris JH, Zhang Q, Dunn S (2002) Atomic polarization and local reactivity on ferroelectric surfaces: A new route toward complex nanostructures. Nano Letters 2: 589-593.

88. Kalinin SV, Bonnell DA, Alvarez T, Lei X, Hu Z, Shao R, Ferris JH (2004) Ferroelectric lithography of multicomponent nanostructures. Adv. Mat. 16: 795-799.
89. Durkan C, Welland ME, Chu DP, Migliorato P (1999) Probing domains at the nanometre scale in piezoelectric thin films. Phys. Rev. B 60: 16198-16204.
90. Molotskii M, Agronin A, Uresnki P, Shvebelman M, Rosenman G, Rosenwaks Y (2003) Ferroelectric Domain Breakdown. Phys. Rev. Lett. 90: 107601.
91. Molotskii M (2003) Generation of ferroelectric domains in atomic force microscope. J. Appl. Phys. 93: 6234-6237.
92. Landauer R (1957) Electrostatic considerations in $BaTiO_3$ domain formation during polarization reversal. J. Appl. Phys. 28: 227-234.
93. Gruverman A, Kolosov O, Hatano J, Takahashi K, Takumoto H (1995) Domain-structure and polarization reversal in ferroelectrics studied by atomic-force microscopy. J. Vac. Sci. Technol. B 13: 1095-1099.
94. Eng LM, Abplanalp M, Günter P (1998) Ferroelectric domain switching in tri-glycine sulphate and barium-titanate bulk single crystals by scanning force microscopy. Appl. Phys. A 66: S679-S683.
95. Rosenman G, Urenski P, Argonin A, Rosenwaks Y, Molotskii M (2003) Submicron ferroelectric domain structures tailored by high-voltage scanning probe microscopy. Appl. Phys. Lett. 82: 103-105.
96. Kalinin SV, Gruverman A, Rodriguez BJ, Shin J, Baddorf AP, Karapetian E, Kachanov M (2005) Nanoelectromechanics of polarization switching in piezoresponse force microscopy. J. Appl. Phys. 97: 074305.
97. Morozovska AN, Eliseev EA (2006) Screening and size effects on the nanodomain tailoring in ferroelectrics semiconductors. Phys. Rev. B 73: 104440.
98. Morozovska AN, Eliseev EA, Kalinin SV (2006) Domain nucleation and hysteresis loop shape in piezoresponse force spectroscopy. Appl. Phys. Lett. 89: 192901.
99. Emelyanov AY (2005) Coherent ferroelectric switching by atomic force microscopy. Phys. Rev. B 71: 132102.
100. Gruverman A, Auciello O, Tokumoto H (1998) Imaging and control of domain structures in ferroelectric thin films via scanning force microscopy. Annu. Rev. Mat. Sci. 28: 101-123.
101. Li J, Nagaraj B, Liang H, Cao W, Lee CH, Ramesh R (2004) Ultrafast polarization switching in thin-film ferroelectrics. Appl. Phys. Lett. 84: 1174-1176.
102. Paruch P, Tybell T, Triscone JM (2001) Nanoscale control of ferroelectric polarization and domain size in epitaxial $Pb(Zr_{0.2}Ti_{0.8})O_3$ thin films. Appl. Phys. Lett. 79: 530-532.
103. Paruch P, Giamarchi T, Tybell T, Triscone JM (2006) Nanoscale studies of domain wall motion in epitaxial ferroelectric thin films. J. Appl. Phys. 100: 051608.
104. Gruverman A, Kholkin A, Kingon A, Tokumoto H (2001) Asymmetric nanoscale switching in ferroelectric thin films by scanning force microscopy. Appl. Phys. Lett. 78: 2751-2753.
105. Kim Y, Cho Y, Hong S, Bühlmann S, Park H, Min DK, Kim SH (2006) Tip travelling and grain boundary effects in domain formation using piezoelectric force microscopy for probe storage applications. Appl. Phys. Lett. 89: 172909.
106. Gruverman A, Wu D, and Scott JF (2008) Piezoresponse force microscopy studies of switching behaviour of ferroelectric capacitors on a 100-ns time scale. Phys Rev. Lett. 100: 097601
107. Kim DJ, Jo JY, Kim TH, Yang SM, Chen B, Kim YS, Noh TW (2007) Observation of inhomogeneous domain nucleation in epitaxial $Pb(Zr,Ti)O_3$ capacitors. Appl. Phys. Lett. 91: 132903.
108. Gruverman A, Wu D, Fan HJ, Vrejoiu I, Alexe M, Harrison RJ, and Scott JF (2008) Vortex ferroelectric domains. J. Phys.: Condens. Matter 20: 342201.

109. Kolmogorov AN (1937) Statistical theory of crystallization of metals. (in Russian) Bull. Acad. Sci. USSR. Ser. Math. 1: 355-359.
110. Avrami M (1939) Kinetics of Phase Change. I General Theory. J. Chem Phys. 7: 1103-1112.
111. Ishibashi Y and Takagi Y (1971) Note on ferroelectric domain switching. J. Phys. Soc. Japan 31: 506-510.
112. Iona F, Shirane G (1962) Ferroelectric crystals. Pergamon Press. Oxford.
113. Kholkin AL, Akdogan EK, Safari A, Chauvy PF, Setter N (2001) Characterization of the effective electrostriction coefficients in ferroelectric thin films. J.Appl. Phys. 89: 8066-8073.
114. Kalinin SV, Gruverman A, Bonnell DA (2004) Quantitative analysis of nanoscale switching in $SrBi_2Ta_2O_9$ thin films by piezoresponse force microscopy. Appl. Phys. Lett. 85: 795-797.
115. Hong S, Woo J, Shin H, Jeon JU, Pak YE (2001) Principle of ferroelectric domain imaging using atomic force microscope. J. Appl. Phys. 89: 1377-1386.
116. Kholkin AL, Bdikin IK, Shvartsman VV, Orlova A, Kiselev DA, Bogomolov AA, Kim SH (2005) Local Electromechanical properties of ferroelectric materials for piezoelectric applications. In Kalinin SV, Goldberg B, Eng LM, Huey D (eds) *Scanning-Probe and Other Novel Microscopies of Local Phenomena in Nanostructured Materials (*Mat. Res. Soc. Proc. 838E) O7.6.
117. Shvartsman VV, Pertsev NA, Herrero JM, Zaldo C, Kholkin AL (2005) Nonlinear local piezoelectric deformation in ferroelectric thin films studied by scanning force microscopy. J. Appl. Phys. 97: 104105.
118. Kholkin AL, Wütchrich C, Taylor DV, Setter N (1996) Interferometric measurements of electric field-induced displacements in piezoelectric thin films. Rev. Sci. Instr. 67: 1935-1941.
119. Jesse S, Baddorf AP, Kalinin SV (2006) Switching spectroscopy piezoresponse force microscopy of ferroelectric materials. Appl. Phys. Lett. 88: 062908.
120. Rodriguez BJ, Nemanich RJ, Kingon A, Gruverman A, Kalinin SV, Terabe K, Liu XY, Kitamura K (2005) Domain growth kinetics in lithium niobate single crystals studied by piezoresponse force microscopy. Appl. Phys. Lett. 86: 012906.
121. Jesse S, Rodriguez BJ, Choudhury S, Baddorf AP, Vrejoiu I, Hesse D, Alexe M, Eliseev EA, Morozovska AN, Zhang J, Chen LQ, Kalinin SV (2008) Direct imaging of the spatial and energy distribution of nucleation centres in ferroelectric materials. Nature Mater. 7: 209.
122. Jesse S, Rodriguez BJ, Alexe M, Kalinin SV (2008) Spatially resolved mapping of polarization switching behaviour in nanoscale ferroelectrics. Adv. Mater. 20:109-114.
123. Bintachitt P, Trolier-McKinstry S, Seal K, Jesse S, and Kalinin SV (2009) Switching spectroscopy piezoresponse force microscopy of polycrystalline capacitor structure. Appl. Phys. Lett. 94: 042906.
124. Ablanalp M, Fousek J, Günter P (2001) Higher Order Ferroic Switching Induced by Scanning Force Microscopy. Phys. Rev. Lett. 86: 5799-5802.
125. Morita T, Cho Y (2003) Observation of antiparallel polarization reversal using a scanning nonlinear dielectric microscope. Japan. J. Appl. Phys. 42: 6214-6217.
126. Bühlmann S, Colla E, Muralt P (2005) Polarization reversal due to charge injection in ferroelectric films. Phys. Rev. B 72: 214120.
127. Shishkin EI, Shur VI, Mieth O, Eng LM, Galambos LL, Miles RO (2006) Kinetics of the local polarization switching in stoichiometric $LiTaO_3$ under electric field applied using the tip of scanning probe microscope. Ferroelectrics 340: 129-136.
128. Kholkin AL, Bdikin IK, Shvartsman VV, Pertsev NA (2007) Anomalous polarization inversion in ferroelectrics via scanning force microscopy. Nanotechnology 18: 095502.

129. Emelyanov AY, Pertsev NA, Kholkin AL (2002) Effect of external stress on ferroelectricity in epitaxial thin films. Phys. Rev. B 66: 214108.
130. Tagantsev AK, Stolichnov IA (1999) Injection-controlled size effect on switching of ferroelectric thin films. Appl. Phys. Lett. 74: 1326-1328.
131. Fridkin V M (1979) Photoferroelectrics. Springer. New York.
132. Würfel P, Batra IP (1976) Depolarization effects in thin ferroelectric films. Ferroelectrics 12: 55-61.
133. Benedetto JM, Moore RA, McLean FB (1994) Effects of operating conditions on the fast-decay component of the retained polarization in lead zirconate titanate thin films. J. Appl. Phys. 75: 460-466.
134. Kholkin AL, Tagantsev AK, Colla EL, Taylor DV, Setter N (1997) Piezoelectric and dielectric aging in $Pb(Zr,Ti)O_3$ thin films and bulk ceramics. Integr. Ferroelectr. 15: 317-324.
135. Shvartsman VV, Kholkin AL (2003) Investigation of switching behaviour in $PbZr_{0.55}Ti_{0.45}O_3$ thin films by means of scanning probe microscopy. Ferroelectrics 286: 1013-1021.
136. Kakalios J, Street RA, Jackson WB (1987) Stretched-exponential relaxation arising from dispersive diffusion of hydrogen in amorphous silicon. Phys. Rev. Lett. 59: 1037-1040.
137. Gruverman A, Tokumoto H, Prakash AS, Aggarwal S, Yang B, Wuttig M, Ramesh R, Auciello O (1997) Nanoscale imaging of domain dynamics and retention in ferroelectric thin films. Appl. Phys. Lett. 71: 3492-3494.
138. Jo W, Kim DC, Hong JW (2000) Reverse-poling effects on charge retention in $Pb(Zr,Ti)O_3(001)/LaNiO_3(001)$ heterostructures. Appl. Phys. Lett. 76: 390-392.
139. Gruverman A, Tanaka M (2001) Polarization retention in $SrBi_2Ta_2O_9$ thin films investigated at nanoscale. J. Appl. Phys. 89:1836-1843.
140. Ganpule CS, Roytburd AL, Nagarajan V, Hill BK, Ogale SB, Williams ED, Ramesh R, Scott JF (2002) Polarization relaxation kinetics and 180° domain wall dynamics in ferroelectric thin films. Phys. Rev. B 65: 014101.
141. Zavala G, Fendler JH, Trolier-McKinstry S (1997) Characterization of ferroelectric lead zirconate titanate films by scanning force microscopy. J. Appl. Phys. 81: 7480-7491.
142. Kholkin AL, Shvartsman VV, Emelyanov AY, Poyato R, Calzada ML, Pardo L (2003) Stress-induced suppression of piezoelectric properties in $PbTiO_3$:La thin films via scanning force microscopy. Appl. Phys. Lett. 82: 2127-2129.
143. Pertsev NA, Zembiglotov AG, Tagantsev AK (1998) Effect of mechanical boundary conditions on phase diagrams of epitaxial ferroelectric thin films. Phys. Rev. Lett. 80: 1988-1991.
144. Tagantsev AK, Stolichnov I, Colla EL, Setter N (2001) Polarization fatigue in ferroelectric films: Basic experimental findings, phenomenological scenarios, and microscopic features. J. Appl. Phys. 90: 1387-1402.
145. Kim SJ, Jiang Q (1996) Microcracking and electric fatigue in polycrystalline ferroelectric ceramics. Smart. Mater. Struct. 5: 321-326.
146. Lupascu DC (2004) Fatigue in ferroelectric ceramics and related issues. Springer Heidelberg.
147. Shvartsman VV, Kholkin AL, Verdier C, Lupascu DC (2005) Fatigue-induced evolution of domain structure in ferroelectric lead zirconate titanate ceramics investigated by piezoresponse force microscopy. J. Appl. Phys. 98: 094109.
148. Nuffer J, Lupascu DC, Glazounov A, Kleebe HJ, Rödel J (2002) Microstructural modifications of ferroelectric lead zirconate titanate ceramics due to bipolar electric fatigue. J. Eur. Ceram. Soc. 22: 2133-2142.
149. Lupascu DC, Rabe U (2002) Cyclic cluster growth in ferroelectric perovskites. Phys. Rev. Lett. 89: 187601.

150. Zhang Y, Lupascu DC, Aulbach E, Baturin I, Bell A, Rödel J (2005) Heterogeneity of fatigue in bulk lead zirconate titanate. Acta Mater. 53: 2203-2213.
151. Zhang Y, Lupascu DC, Balke N, Rödel J (2005) Near electrode fatigue in lead zirconate titanate ceramics. J. Phys. IV 128: 97-103.
152. Colla EL, Stolichnov I, Bradely PE, Setter N (2003) Direct observation of inversely polarized frozen nanodomains in fatigued ferroelectric memory capacitors. Appl. Phys. Lett. 82: 1604-1606.
153. Liu JS, Zhang SR, Dai LS, Yuan Y (2005) Domain evolution in ferroelectric thin films during fatigue process. J. Appl. Phys. 97: 104102.
154. Gruverman A, Auciello O, Tokumoto H (1996) Nanoscale investigation of fatigue effects in $Pb(Zr,Ti)O_3$ films. Appl. Phys. Lett. 69: 3191-3193.
155. Colla EL, Hong S, Taylor DV, Tagantsev AK, Setter N, No K (1998) Direct observation of region by region suppression of the switchable polarization (fatigue) in $Pb(Zr,Ti)O_3$ thin film capacitors with Pt electrodes. Appl. Phys. Lett. 72: 2763-2765.
156. Kholkin AL, Colla EL, Tagantsev AK, Taylor DV, Setter N (1996) Fatigue of piezoelectric properties in $Pb(Zr,Ti)O_3$ films. Appl. Phys. Lett. 68: 2577-2579.
157. Cross LE (1987) Relaxor ferroelectrics. Ferroelectrics 76: 241-267.
158. Burns G, Dacol FH (1983) Glassy polarization behaviour in ferroelectric compounds $PbMg_{1/3}Nb_{2/3}O3$ and $PbZn_{1/3}Nb_{2/3}O_3$. Solid State Commun. 48: 853-586.
159. Park SE, Shrout TR (1997) Ultrahigh strain and piezoelectric behaviour in relaxor-based ferroelectric single crystals. J. Appl. Phys. 82: 1804-1811.
160. Kleemann W, Samara GA, Dec J (2005) Relaxor ferroelectrics – from random field models to glassy relaxation and domain states. In Waser R, Boettger U, Tiedke S (eds) Polar oxides: properties, characterization and imaging. Wiley, Weinheim, pp 129-206.
161. Bokov AA, Ye ZG (2006) Recent progress in relaxor ferroelectrics with perovskite structure. J. Mater. Sci. 41: 31-52.
162. Lehnen P, Kleemann W, Woike T, Pankrath P (2001) Ferroelectric nanodomains in the uniaxial relaxor system $Sr_{0.61-x}Ba_{0.39}Nb_2O_6:Ce_x^{3+}$: Phys. Rev. B 64: 224109.
163. Kleemann W, Dec J, Shvartsman VV, Kutnjak S, Braun T (2006) Two-dimensional Ising model criticality in a three-dimensional uniaxial relaxor ferroelectric with frozen polar nanoregions. Phys. Rev. Lett. 97: 065702.
164. Shvartsman VV, Kleemann W (2006) Evolution of nanodomains in the uniaxial relaxor $Sr_{0.61}Ba_{0.39}Nb_2O_6$:Ce. IEEE Transactions UFFC 53: 2275-2279.
165. Shvartsman VV, Kleemann W, Łukaciewicz T, Dec J (2008) Nanopolar structure in $Sr_xBa_{1-x}Nb_2O_6$ single crystals tuned by the Sr/Ba ratio and investigated by piezoelectric force microscopy. Phys. Rev. B 77: 054105.
166. Shvartsman VV, Kholkin AL (2004) Domain structure of $0.8Pb(Mg_{1/3}Nb_{2/3})O_3$-$0.2PbTiO_3$ studied by piezoresponse force microscopy. Phys. Rev. B 69: 014102.
167. Bai F, Li J, Viehland D (2005) Domain engineered states over various length scales in (001) oriented $PbMg_{1/3}Nb_{2/3}O_3$-x%$PbTiO_3$ crystals: Electrical history dependence of hierarchal domains. J. Appl. Phys.; 97: 054103.
168. Zhao X, Dai JY, Wang J, Chen HLW, Chou CL, Wan XM, Luo HS (2005) Relaxor ferroelectric characteristics and temperature-dependent domain structure in a (110)-cut $(PbMg_{1/3}Nb_{2/3}O_3)_{0.75}(PbTiO_3)_{0.25}$ single crystal. Phys. Rev. B 72: 064114.
169. Shvartsman VV and Kholkin AL (2007) Evolution of nanodomains in $0.9PbMg_{1/3}Nb_{2/3}O_3$-$0.1PbTiO_3$ single crystals. J. Appl. Phys. 101: 064108.
170. Abplanalp M, Barosova D, Bridenbaugh P, Erhart J, Fousek J, Günter P (2002) Scanning force microscopy of domain structures in $Pb(Zn_{1/3}Nb_{2/3})O_3$-8% $PbTiO_3$ and $Pb(Mg_{1/3}Nb_{2/3})O_3$-29% $PbTiO_3$ J. Appl. Phys. 91: 3797-3805.
171. Bdikin IK, Shvartsman VV, Kholkin AL (2003) Nanoscale domains and local piezoelectric hysteresis in $Pb(Zn_{1/3}Nb_{2/3})O_3$-4.5%$PbTiO_3$ single crystals. Appl. Phys. Lett. 83: 4232-4234.

172. Shvartsman VV, Kholkin AL, Orlova A, Kiselev D, Bogomolov AA, Sternberg A (2005) Polar nanodomains and local ferroelectric phenomena in relaxor lead lanthanum zirconate titanate ceramics. Appl. Phys. Lett. 86: 202907.
173. Kiselev DA, Bdikin IK, Selezneva EK, Bormanis K, Sternberg A, Kholkin AL (2007) Grain size effect and local disorder in polycrystalline relaxors via scanning probe microscopy. J. Phys. D: Appl. Phys. 40: 7109-7112.
174. Salak AN, Shvartsman VV, Seabra MP, Kholkin AL, Ferreira VM (2004) Ferroelectric-to-relaxor transition behaviour of $BaTiO_3$ ceramics doped with $La(Mg_{1/2}Ti_{1/2})O_3$. J. Phys.: Condens. Matter. 16: 2785-2794.
175. Shvartsman VV, Kleemann W, Dec J, Xu ZK, Lu SG (2006) Diffuse phase transition in $BaTi_{1-x}Sn_xO_3$ ceramics: An intermediate state between ferroelectric and relaxor behaviour. J. Appl. Phys. 99: 124111.
176. Xu G, Viehland D, Li JF, Gehring PM, Shirane G (2003) Evidence of decoupled lattice distortion and ferroelectric polarization in the relaxor system PMN-xPT. Phys. Rev. B 68: 212410.
177. Conlon KH, Luo H, Viehland D, Li JF, Whan T, Fox JH, Stock C, Shirane G (2004) Direct observation of the near-surface layer in $Pb(Mg_{1/3}Nb_{2/3})O_3$ using neutron diffraction. Phys. Rev. B 70: 172204.
178. Dec J, Shvartsman VV, Kleemann W (2006) Domain-like precursor clusters in the paraelectric phase of the uniaxial relaxor $Sr_{0.61}Ba_{0.39}Nb_2O_6$. Appl. Phys. Lett. 89: 212901.
179. Shvartsman VV, Kholkin AL, Tyunina M, Levoska J (2005) Relaxation of induced polar state in relaxor $PbMg_{1/3}Nb_{2/3}O_3$ thin films studied by piezoresponse force microscopy. Appl. Phys. Lett. 86: 222907.
180. Gladkii VV, Kirikov VA, Pronina EV (2003) The slow polarization kinetics of the ferroelectric relaxor lead magnesium niobate. Phys. Solid State 45: 1238-1244.
181. Boikov YA, Goltsman BM, Yarmarkin VK, Lemanov VV (2001) Slow capacitance relaxation in $(BaSr)TiO_3$ thin films due to the oxygen vacancy redistribution. Appl. Phys. Lett. 78: 3866-3868.
182. Shvartsman VV, Emelyanov AY, Kholkin AL, Safari A (2002) Local hysteresis and grain size effect in PMN-PT thin films. Appl. Phys. Lett., 81: 117-119.
183. Mishima T, Fujioka H, Nagakari S, Kamigaki K, Nambu S (1997) Lattice Image Observations of Nanoscale Ordered Regions in $Pb(Mg_{1/3}Nb_{2/3})O_3$. Jpn. J. Appl. Phys. 36: 6141-6144.
184. Scott JF, Diuker HM, Beale PD, Pouligny B, Dimmler K, Parris M, Butler D, Eaton S (1988) Properties of ceramic KNO_3 thin-film memories. Physica B+C 150: 160-167.
185. Zhong WL, Wang YG, Zhang PL, Qu BD (1994) Phenomenological study of the size effect on phase transitions in ferroelectric particles. Phys. Rev B 50: 698-703.
186. Wang YG, Zhong WL, Zhang PL (1995) Lateral size effects on cells in ferroelectric films. Phys. Rev. B 51: 17235-17238.
187. Wang SL, Smith SRP (1995) Landau theory of the size-driven phase transition in ferroelectrics. J. Phys.: Condens. Matter 7: 7163-7171.
188. Li S, Eastman JA, Vetrone JM, Foster CM, Newnham RE, Cross LE (1997) Dimension and size effects in ferroelectrics. Jpn. J. Appl. Phys. 36: 5169-5174.
189. Ganpule C, Stanishevsky A, Su Q, Aggarwal S, Melngailis J, Williams E, Ramesh R (1999) Scaling of ferroelectric properties in thin films. Appl. Phys. Lett. 75: 409-411
190. Ganpule C, Stanishevsky A, Aggarwal S, Melngailis J, Williams E, Ramesh R, Joshi V, Araujo CP (1999) Scaling of ferroelectric and piezoelectric properties in $Pt/SrBi_2Ta_2O_9/Pt$ thin films. Appl. Phys. Lett. 75: 3874-3876.
191. Bühlmann S, Dwir B, Baborowski J, Muralt P (2002) Size effect in mesoscopic epitaxial ferroelectric structures: Increase of piezoelectric response with decreasing feature size. Appl. Phys. Lett. 80: 3195-3197.

192. Alexe M. Harnagea C, Erfurth W, Hesse D, Gösele U (2000) 100-nm lateral size ferroelectric memory cells fabricated by electron-beam direct writing. Appl. Phys. A 70: 247-251.
193. Waser R, Schneller T, Hoffmann-Eifert S, Ehrhart P (2001) Advanced chemical deposition techniques - from research to production. Integr. Ferroelectr. 36: 3-20.
194. Roelofs A, Schneller T, Szot K, Waser R (2002) Piezoresponse force microscopy of lead titanate nanograins possibly reaching the limit of ferroelectricity. Appl. Phys. Lett. 81:5231-5233.
195. Clemens S, Röhrig S, Rüdiger A, Schneller T, Waser R (2006) Variable size and shape distribution of ferroelectric nanoislands by chemical mechanical polishing. Small 2: 500-502.
196. Peter F, Rüdiger A, Waser A, Szot K, Reichenberg B (2005) Comparison of in-plane and out-of-plane optical amplification in AFM measurements. Rev. Sci. Instr. 76: 046101.
197. Hiranaga I, Cho Y (2005) Ultrahigh-density ferroelectric data storage with low bit error rate. Jap. J. Appl. Phys. 44: 6960-6963.
198. Zhao T, Scholl A, Zavaliche F, Lee K, Barry M, Doran A, Cruz MP, Chu YH, Ederer C, Spaldin NA, Das RR, Kim DM, Baek SH, Eom CB, Ramesh R (2006) Electrical control of anti-ferromagnetic domains in multiferroic $BiFeO_3$ films at room temperature. Nature Mater. 5: 823-829.
199. Kalinin SV, Rodriguez BJ, Jesse S, Thundat T, Gruverman A (2005) Electromechanical imaging of biological systems with sub-10 nm resolution. Appl. Phys. Lett. 87: 053901.

Chapter 10
Mechanical Properties of Ferro-Piezoceramics

Doru C. Lupascu[1], Jörg Schröder[2], Christopher S. Lynch[3], Wolfgang Kreher[4], Ilona Westram[5]

Ferroelectric materials provide the best electromechanical coupling factor of known materials, making them the material most often used in piezoelectric devices in a constantly growing market. They are used in a range of applications that includes classical ultrasonic devices, fuel injection systems, smart structures, and high precision positioning equipment for electronic manufacturing and advanced scanning probe microscopy. Each of these applications is dependent not only on the electromechanical coupling, but also on the mechanical properties of the ferroelectric material. This chapter provides an overview of mechanical properties that tend to be overlooked when a linear analysis is used in device design. These properties include nonlinear stress strain relations, bipolar and unipolar electric field strain relations, dynamic response and creep, as well as fracture and failure mechanisms. Approaches to modelling constitutive behaviour, which includes non-linearity and hysteresis associated with polarization reorientation, are presented as this provides considerable insight into such behaviour. The emphasis of this chapter lies on recent advances and the state of the art understanding of ferroelectric material properties.

10.1 Introduction

The piezoelectric effect in single crystals, namely generation of electric field upon application of mechanical stress (direct effect), or generation of strain for an applied electric field (converse effect), has been known for more than a century since the Curie

---

[1] Institut für Materialwissenschaft, Universität Duisburg-Essen, Essen, Germany
[2] Institute of Mechanics, Universität Duisburg-Essen, Germany
[3] Department of Mechanical and Aerospace Engineering, University of California, Los Angeles, USA
[4] Institute for Materials Science, Technische Universität Dresden, Germany
[5] Institute for Materials Science, Darmstadt University of Technology, Darmstadt, Germany

brothers first reported on the effect in 1880 [1]. Rochelle-salt, discovered by Valasek [2] in the 1920s was the first known ferroelectric, but it was mostly synthesized as a monodomain single crystal for which a large piezoelectric effect is found without a prior poling step. Its solubility in water proved to be a technological drawback and necessitated encapsulation of all ultrasonic devices developed and used until the mid 1940s. During the Second World War, several independent groups discovered piezoelectricity in barium titanate. But, it took some time before it was recognized that a poling step was required for macroscopic piezoelectricity in a polycrystalline material (for details see the historical introduction in [3]). The best performance of piezoelectric ceramics is found in the lead zirconate titanate (PZT) family of solid solution compositions, first thoroughly investigated by Jaffe and others [3]. It is used in nearly all piezoelectric devices today. Only single crystals, particularly in the lead magnesium niobate family and recently lead-free ceramics from morphotropic phase boundary niobates, have piezoelectric coefficients exceeding the ceramic values found in PZT [4, 5]. A thorough historical introduction to ferroelectric materials is found in the magnificent book by Lines and Glass [66].

A fairly large number of reviews that address ferroelectrics have appeared recently [6, 7, 8, 9, 10, 11, 12, 13, 14]. It will be the focus of this chapter to emphasize mechanical properties as these are often only given secondary consideration, yet are critical to reliable device design.

Little will be said about the important topics of ferroelectricity, thin film devices, multiferroicity, or other aspects of research in the oxide-based ferroelectrics, which have been topics of other work [9, 10, 11, 12, 13].

10.2 Electromechanical Hysteresis, Experiment

10.2.1 Introduction to Hysteresis

Constitutive behaviour of materials is often modelled as linear. This approach is sufficiently accurate to capture the essential mechanical, electrical, and magnetic properties of many materials. Nonlinearity and hysteretic behaviour, on the other hand, are often observed. Strong non-linearity is rare in pure form without hysteresis but does occur with little hysteresis in certain quadratic electrostrictors [15]. A crystal (or polymer) exhibiting a first order phase transition from a higher to a lower symmetry class displays characteristic intrinsically hysteretic material behaviour. Improper hysteresis may arise due to time-dependent transport of charge carriers or chemical species inducing a delay in material response. This may yield apparent hysteretic behaviour, but this will not be an issue here. A first order transition requires that a finite amount of heat is dissipated in a process that involves a change of at least one extensive material property: strain, polarization, magnetization or even heat itself [16]. Most ferroelectric crystals in use today exhibit one or more first order phase transitions.

For ferroelectrics, polarization is the extensive variable. It may undergo small changes associated with deformation of the crystal lattice in response to an externally applied electric field or addition of heat. These small polarization and strain changes are often referred to as an intrinsic effect. Intrinsic changes of polarization and strain are reversible and do not contribute to hysteresis. Additional changes of polarization are associated with domain wall motion. A domain is a region or an entire crystal in which the polarization of each of the unit cells has the same direction. Under an external electric field the motion of a domain wall results in the volume which is swept by the wall to undergo a transformation from polarization in one direction to polarization in another direction. Work is done during this process. If the domain wall motion takes place at constant electric field and zero stress, the work done per unit volume is given by the dot product of the electric field vector with the polarization change vector. The total work done is given by this dot product multiplied by the volume of material swept by the domain wall. Domain wall motion may be viewed as a phase transition between two crystal variants with equivalent energy states at zero field. The electric field may be considered a driving force that pushes the change past an energy barrier, or it may be considered to lower the energy state of one variant relative to the other at applied field. These are different ways of describing the same effect.

Landau theory was initially developed as a polynomial expansion of a thermodynamic potential to describe critical behaviour of matter near a phase transition. In ferroelectrics, it proved to be valid over a much wider temperature range than initially anticipated [17, 18]. Also, polycrystalline material behaviour may be approximated to a certain degree by a polynomial expansion of energy in polarization. Strain coupling is atomically provided by intrinsic piezoelectricity or electrostriction. In piezoelectrics that are not ferroelectric, like aluminium nitride or quartz, the piezoelectric effect dominates the electrostrictive effect. Ferroelectrics mostly exhibit such strong electrostriction on the atomic scale that intrinsic piezoelectricity becomes similar in magnitude, or even a secondary effect. Strain is related to the atomic scale electric field by electrostriction. The local field in turn is given by the local positions of the ions within the unit cell and thus by the polarization value itself. This is why, in a hysteretic material macroscopically a quadratic relation is found between the absolute value of polarization and strain rather than electric field [6]. If the change of polarization is not too large, relative changes of polarization are approximated by the slope of the electrostrictive curve. This linearization is the basis for most mechanical models developed for describing the material behaviour of macroscopic polycrystalline actuator or sensor materials. Models for this will be compared in Section 10.1. It is this linear part of the complicated hysterestic electrostrictive material response, which results in the linearized piezoelectric behaviour that is taken advantage of in the design of macroscopic devices.

There is a multitude of other ways of dealing with hysteresis which have been summarized in multiple thorough works, the most encompassing treatise being the multivolume book by Bertotti and Mayergoyz [19] to which we refer the interested reader.

10.2.2 Electromechanical Coupling in Single Crystals

The electro-mechanical coupling behaviour in ferroelectric ceramics is the result of phenomena occurring at multiple length scales. Fig. 10.1 describes the scales involved, showing the phenomena at each scale and the associated modelling approaches. The ferroelectric device is typically part of a larger structure in which the piezoelectric is involved with vibration, acoustic, or position control. Actuators are most often multilayer devices with alternating electrodes and ceramic or single crystal on the 1 to 10 mm length scale. At this length scale, the ferroelectric ceramic behaviour can be homogenized and described using phenomenological models even though the underlying mechanisms originate at smaller length scales. At the length scale of the single crystal grains, typically 5 to 10 micrometers, each grain has some average polarization associated with its domain structure. Micromechanical models are used to volume average the behaviour of the grains to obtain the behaviour of the ceramic (see section 10.3.2.). At the next smaller length scale, the material is made up of domains and domain walls. Domain formation and domain wall motion are typically modelled using phase field approaches. Finally, at the length of the unit cell, it is the arrangement of the individual ions in the crystal structure that give rise to the electro-mechanical coupling. Behaviour at this length scale is modelled using *ab-initio* and molecular dynamics simulations.

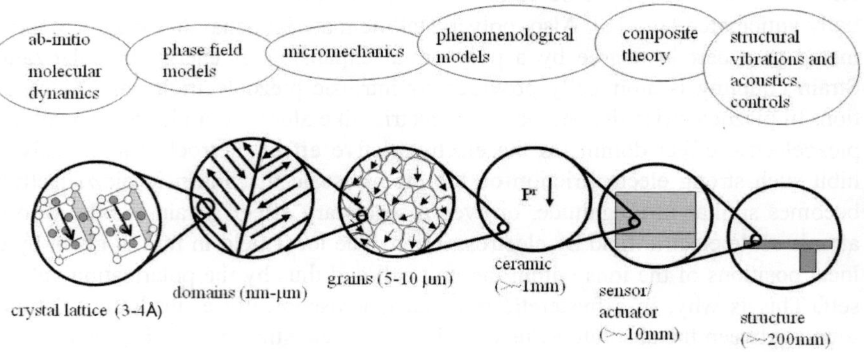

Fig. 10.1 Schematic illustration of the multiscale nature of applications of ferroelectric materials along with the modelling approaches that have been applied at each of the length scales.

Fig. 10.2 is an electron back scatter diffraction (EBSD) micrograph of PZT that shows some of the sub-structure of the ceramic. The large blocks (approximately seven microns across) are the grains. On the grain boundaries there are regions of porosity. The striations on each of the grains are caused by domains, regions of like polarization. It is clear from this micrograph that for certain grain boundaries, likely the low angle boundaries, the domain patterns pass from one grain to the next. Grain boundary electromechanical compatibility conditions affect the domain pattern.

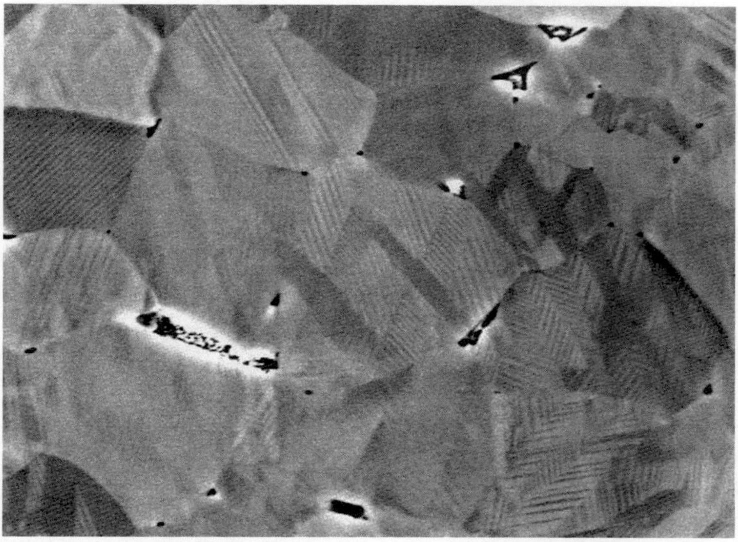

Fig. 10.2 EBSD micrograph of polished PZT (Micrograph obtained in collaboration with Dr. E. Kennik at Oak Ridge National Laboratory).

The crystal structure is the source of the electro-mechanical coupling and the ferroelectric behaviour. Fig. 10.3 is a schematic of the ABO_3 perovskite type structure of lead titanate. The cell shown is comprised of lead ions at each corner, oxygen ions at the centre of each face, and a titanium ion at the body centre. This cell is charge neutral. The eight corner ions each carry a charge of +2. Each is shared with eight other neighbour cells, so the total charge contributed is +2. The six oxygen ions each have a charge of -2 and each is shared with one neighbouring cell. Thus, the oxygen ions contribute a net charge of -6. The body centre titanium ion contributes a net charge of +4. The sum of the charges is zero.

Above the Curie temperature the structure is cubic and centrosymmetric. As the crystal is cooled below the Curie temperature, it undergoes a phase transformation that involves a small relative displacement of each of the ions. The result is shown schematically in Fig. 10.3 (right) where the titanium ion has shifted upward relative to the oxygen ions (in reality, all of the ions shift). The result is that there is a net shift of positive charge upward and of negative charge downward. This separation of charge creates a dipole moment. The dipole moment per unit volume is the polarization of the cell. This is referred to as the spontaneous polarization, because it forms as a result of the temperature-dependent crystal structure change rather than by application of an external field. There is also a change of shape of the cell associated with the phase transformation, with an elongation in the direction of the polarization and contraction perpendicular to the polarization direction. This produces a strain relative to the cubic shape. This is referred to as spontaneous strain.

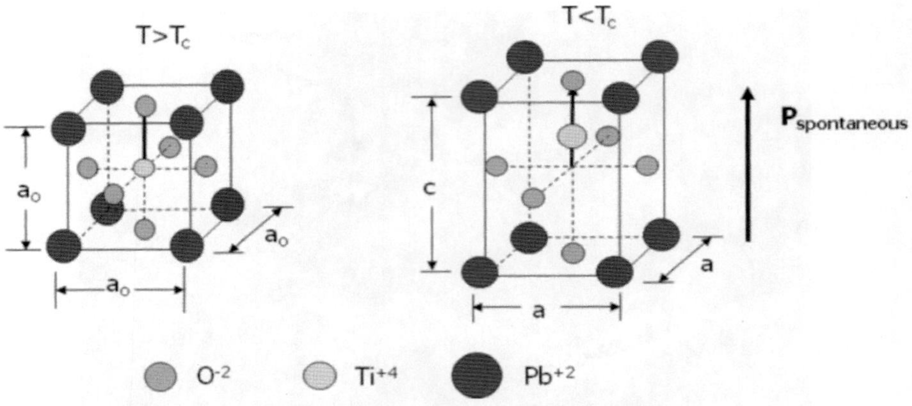

Fig. 10.3 Schematic representation of the high temperature cubic phase (left) and the low temperature tetragonal phase (right) of lead titanate.

Many chemical compositions with interesting physical properties may be found in the ABO_3 perovskite crystal structure. For certain (still many) chemical compositions, a ferroelectric displacement may yield so-called perovskite-like crystal structures (mostly only called perovskites), named so because they only differ slightly from the cubic structure of the mineral Perovskite ($CaTiO_3$). In each of these, the crystal structure is elongated in the polarization direction and contracted in the transverse directions. As the phase transformations occurring by change of temperature generate external electric potentials, these materials are pyroelectrics. In the tetragonal structure, the polarization may point toward any of the six pseudocubic (again because the difference to cubic is small) faces. These are referred to as the six equivalent variants of the tetragonal structure. In the rhombohedral structure, the polarization is toward one of the corners with an extension on the body diagonal and contraction perpendicular to the polarization. This yields eight possible polarization directions. Other structures include orthorhombic and monoclinic. The properties of a variant are typically defined with the introduction of a coordinate system in which the x_3 direction by definition is aligned with the polarization vector. The crystals tend to be elastically anisotropic with a lower stiffness in the polar direction and higher stiffness in the transverse directions (for constant electric field boundary condition). It is reverse for constant electric displacement boundary condition e.g., in $BaTiO_3$ [20]. These polar crystals are also piezoelectric and anisotropic dielectric.

A ferroelectric material has the property that the polarization direction can be changed by application of a sufficiently strong electric field in a direction that will induce the change. A ferroelastic material has the property that the principle values of the strain tensor can be switched by application of a sufficiently strong stress in a direction that will induce the change. These processes are described in Fig. 10.4. The process is dissipative and results in hysteresis.

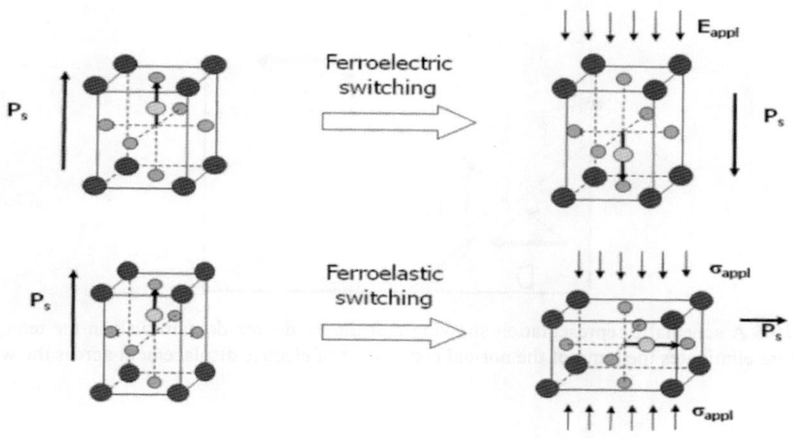

Fig. 10.4 The polarization of the unit cell can be switched to another direction by a sufficiently strong electric field (ferroelectric) or by a sufficiently strong stress (ferroelastic).

Domains are regions of like polarization within a single crystal, where here the word *single crystal* is not to be understood in a crystallographer's sense where each domain would be a single crystal, but in the sense that all variants stem from the same cubic high temperature crystallite, namely the grain. The formation of domains minimizes the free energy of the crystal. This may be understood through the equations of electrical and mechanical compatibility across boundaries. The divergence of the electric displacement vector $\underline{D}$ is equal to the free charge at a point ρ_q.

$$\nabla \cdot \underline{D} = \rho_q$$

This leads to the boundary condition that the jump in the normal component of electric displacement is equal to the charge per unit area on the boundary [21].

$$\|D_n\| = Q / A$$

If the mechanical displacement changes across a boundary, the material is physically stretched and thus there is stress. The result is strain energy associated with the boundary. Domain walls occupy those orientations that minimize the free energy of the system. In the tetragonal crystal structure, these walls are 90 degree or 180 degree. Fig. 10.5 shows how the 90 degree wall eliminates jumps in the normal component of electric displacement across the wall. Fig. 10.6 shows schematic examples of the 90 and 180 degree walls.

$$\|u_t\| = 0$$

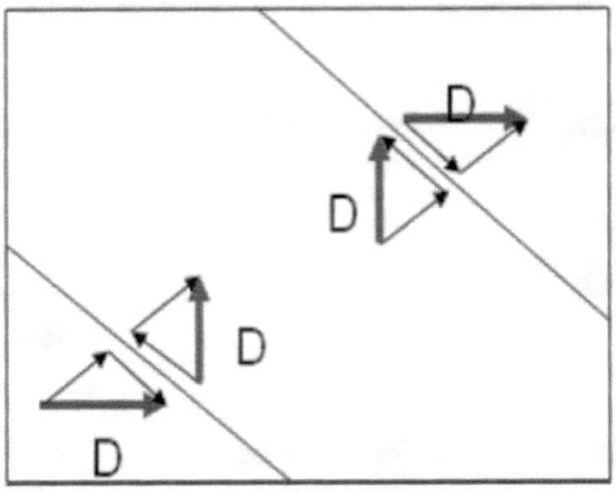

Fig. 10.5 A schematic representation showing that the 90 degree domain wall in the tetragonal structure eliminates the jump of the normal component of electric displacement across the wall.

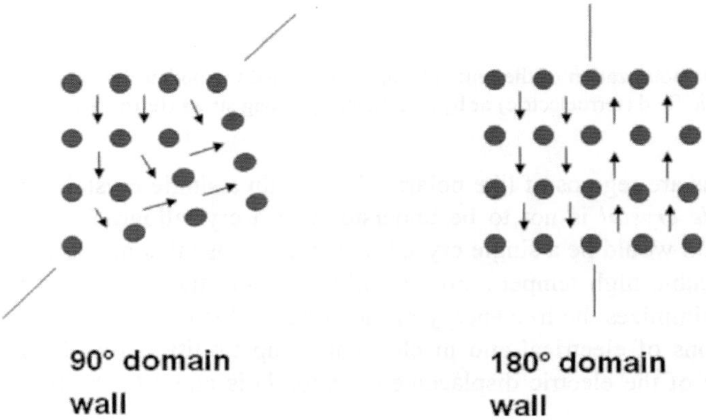

Fig. 10.6 Schematic representation of the structures of 90 and 180 degree domain walls in the tetragonal structure. In each case there is no jump of the tangential component u_t of mechanical displacement across the wall and there is no jump of the normal component of electric displacement across the wall.

Early work on single crystal ferroelectrics focused on barium titanate [22, 23, 24, 25, 26]. From the early 1980s to the end of the 1990s, there has been a growing body of literature on relaxor ferroelectric single crystal PMN-xPT, (1-x)Pb($Mg_{1/3}Nb_{2/3}$)O_3-$x$$PbTiO_3$, and PZN-xPT (1-$x$)Pb($Zn_{1/3}Nb_{2/3}$)$O_3$-$x$$PbTiO_3$ growth and characterization [27, 28, 29, 30, 31, 32, 33, 34, 35, 36, 37, 38, 39, 40, 41, 42, 43, 44, 45, 46, 47, 48, 49, 50, 51, 52, 53, 54].

Recent work on relaxor ferroelectric single crystals has provided considerable insight into ferroelectric material behaviour. Relaxor ferroelectrics display a diffuse phase transformation with a broad Curie peak rather than a distinct Curie point [55].

It is important to note that not only can an electric field drive polarization reorientation, it can also drive a phase transformation. This behaviour is well known from the anti-ferroelectric to ferroelectric field-driven phase transformations and its propensity to result in damage to actuators driven through this transformation [56, 57, 58]. More recently, work on two different relaxor single crystal material systems has resulted in the demonstration of several different field-induced phase transformations. These include rhombohedral to tetragonal and rhombohedral to orthorhombic [59, 60, 61, 62, 63].

Fig. 10.7 shows the effects of the electric field driven step-like rhombohedral to orthorhombic phase transformation in PZN-4.5%PT, and the diffuse rhombohedral to orthorhombic phase transformation that occurs in PMN-32%PT. In each case, the electric field was applied in the [011] crystallographic direction and the strain measured in the [001] crystallographic direction. This is the equivalent of a d_{32} type piezoelectric coupling. The transformation in PZN-4.5%PT is accompanied by a discontinuous change in strain and strong hysteresis that is characteristic of a first-order phase change [62, 63]. The transformation in PMN-32%PT is nearly continuous and spread over a range of field values.

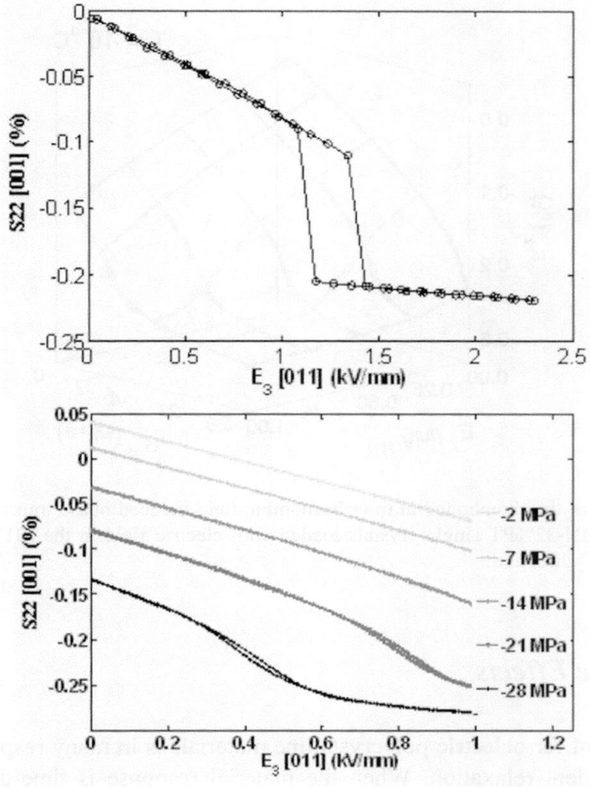

Fig. 10.7 a PZN-0.045PT [59] and **b** PMN-0.32PT [64] strain vs. electric field behaviour during field induced phase transformations. The constant uniaxial compressive stress in the PMN-0.32PT was in the [001] direction.

The field induced phase transformation behaviour results in a dramatic shift in the elastic, dielactric, and piezoelectric constants in these materials. This behaviour has been mapped out experimentally [59, 60, 61, 65]. Some of the results are presented in Fig. 10.8. where the strain is plotted as a function of stress and electric field. The upper flat region is in the rhombohedral phase and the lower flat region is in the orthorhombic phase.

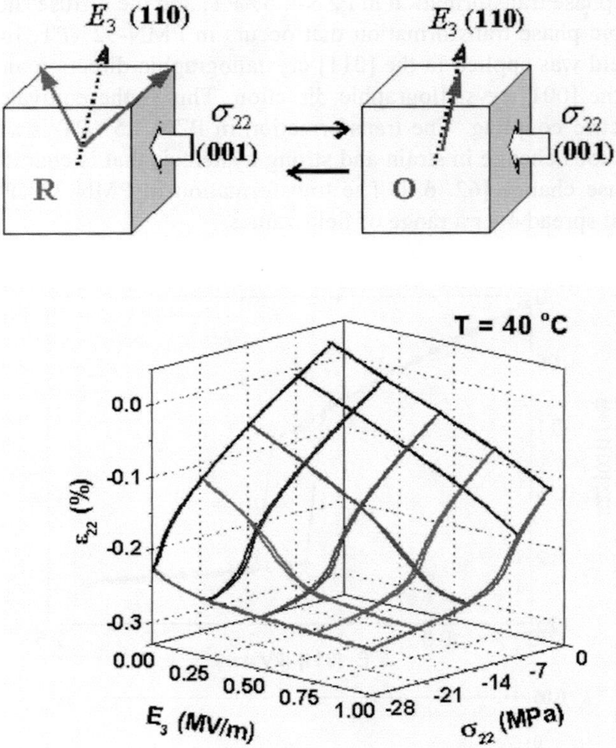

Fig. 10.8 Effect of the rhombohedral to orthorhombic field induced phase transformation on the behaviour of PMN-32%PT single crystals loaded with electric field in the [011] direction and stress in the [001] direction.

10.2.3 Time Effects

The response of ferroelectric polycrystalline materials is in many respects the result of time-dependent relaxation. When the material response is time-dependent, the energy dissipation rate will be a function of the temporal profile of the loads. If the loading function is a triangle wave, the crucial components are the rate of loading the peak load applied, and the bias load. While devices exhibit a component of time-dependent behaviour due to inertia, another component of time-dependent behaviour is a fundamental part of the ferroelectric properties. The elementary or intrinsic piezoelectric effect cannot be faster than the phonon frequency involved in the displacement of the ions in the unit cell that provide the strain coupling. These frequencies lie in the terahertz range and mostly affect the optical or microwave properties. This frequency range is too high to be relevant for most piezoelectric devices [66]. The extrinsic contribution to strain, stemming from domain wall motion, has a much lower frequency limit. Domain wall motion itself consists of two time-dependent components: nucleation and growth [22]. The time dependence of nucleation is still not fully understood. Domain growth in ideal crystals is limited by the macroscopic speed of sound. In a real system defects retard domain wall motion. Therefore, dopants, dislocations, and grain boundary defects may fundamentally alter material properties for device operation.

The fatigue and aging effects are associated with changes of defect structure and result in a degradation of the material properties. Overall, the different types of defects, the grain size distribution in a polycrystalline system, and the orientation distribution of the crystallites result in a material time response spanning many orders of magnitude in time [67, 68]. Electrical and mechanical properties are equally affected by rate-dependent mechanisms. Ferroelectric materials are polarized under unipolar electrical loading. A large degree of poling will be reached shortly, but a fully polarized state may be reached after a long period in time only. Under a reverse unipolar electric field, which drives polarization reversal, the time constants involved depend on the external driving force. Near the coercive field the relaxation is slowest [69, 70], while nearly complete polarization reversal may be reached in much less than a microsecond for twice the coercive field in soft lead zirconate titanate [71].

Fig. 10.9 shows time relaxation of polarization and strain for an initially non-poled material. The characteristic is a power law dependence of polarization and strain on time [72]. Creep under mechanical load has been observed in similar form [67].

Some applications of ferroelectrics require fast response, particularly in the high frequency ultrasonic range. In an ultrasonic device, the material is driven by bipolar fields or sometimes under electrical bias. Rate dependence in the material occurs mostly as a result of switching. In thin films it appears natural that short switching times may be achieved. The reversal time in thick specimens is similarly fast so that it exceeds the speed of sound in the material. It nevertheless stays well below the speed of light in the material. Fig. 10.10 displays the reversal time for different degrees of electric fatigue of a soft ferroelectric (PIC 151 [68]). Switching the non-fatigued material is limited by the electric circuitry in the data shown. Intrinsic time constants are observed in the tens of nanoseconds (not shown). Reverse domain nucleation is thus not limited to the external electrodes as often observed in single crystal studies where domain nucleation is directly observable. It also arises in the bulk of the sample. In order not to be limited by the response of the domain system in the material, practically all ultrasonic devices are manufactured from "hard" doped ferroelectrics. Hard doping is understood to yield domain clamping at the unit cell level via defect dipoles. The reverse effect namely "soft" doping facilitating domain wall motion still remains unexplained [73]. Hard doping reduces the losses in the material and the hysteresis to a narrow minor loop [74]. The response of devices is practically not limited in frequency at the expense of much lower piezoelectric coefficients [75, 76, 77]. Electrical as well as mechanical properties are equally affected.

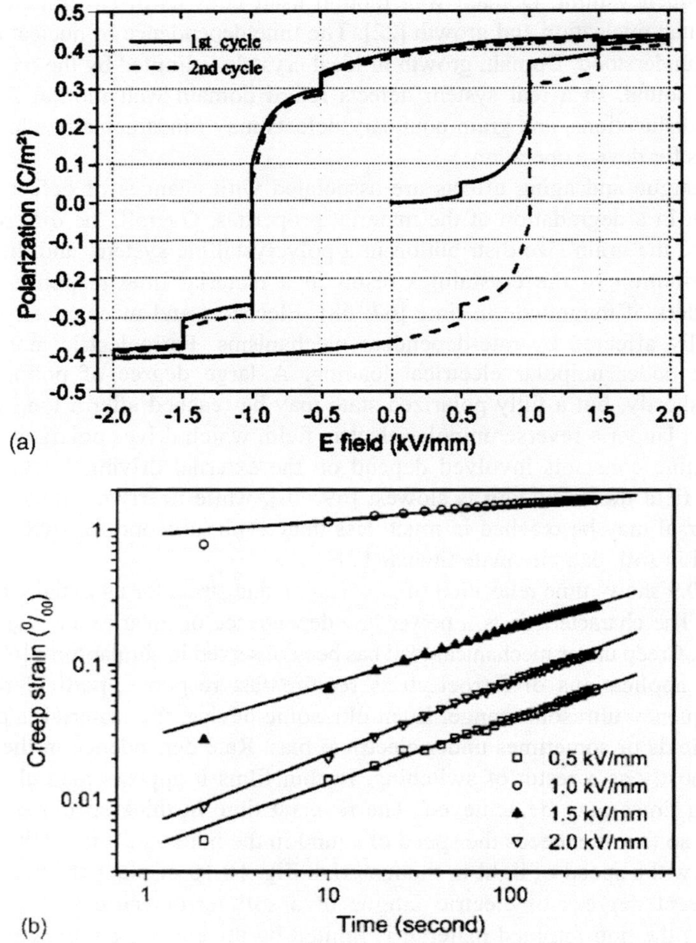

Fig. 10.9 Creep strain measurement scheme **a** and logarithmic representation of creep strain **b** as a function of time for a soft PZT (PIC 151, PI Ceramic, Lederhose, Germany). The electric field was held constant at different levels during the poling process of an initially unpoled sample [72]. The 20 mHz value of E_c is 1 kV/mm. ‰ denotes millistrain. The sequential conduct of the experiment leads to the lower rates at higher fields. A step like application of field would lead to a similar behaviour as in Fig. 10.10 also for strain. ©*Inst. of Phys.*

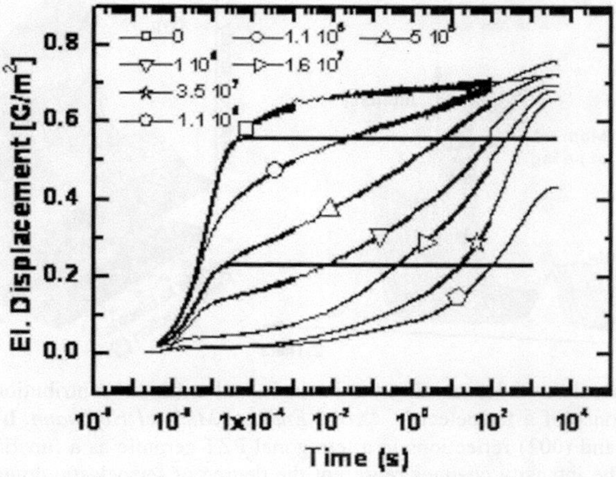

Fig. 10.10 Semi-logarithmic representation of polarization versus time for initially reverse poled samples of PIC 151. The field applied was twice the coercive field. The different curves show the influence of bipolar fatigue on switching dynamics in a polycrystalline soft PZT material. The initial rise-time is due to the external circuitry. A non-fatigued sample reaches a characteristic intrinsic relaxation time of around 30 ns. Fits of exponential relaxation functions are shown for comparison exemplifying the stretched exponential behaviour of the material [68] ©*IOP*.

10.2.4 Electromechanical Coupling in Polycrystalline Materials

10.2.4.1 X-Ray Diffraction-Data on Local and Global Strain

As mentioned above, the total strain output of a ferroelectric is shared between an intrinsic contribution, due to the unit cell extension, and an extrinsic contribution, due to domain wall motion. The latter changes the relative amount of domains of one orientation in favour of an orientation better adopted to the local stress or electric field state. For many years the relative contributions were not very well determined experimentally. A first thorough analysis was performed by M. Hoffman's group in Karlsruhe [78] using X-ray diffraction (XRD). This has since been extended to texture in ferroelectrics due to poling as well as in the non-poled polycrystalline microstructure [79, 80]. Fig. 10.11a displays data on the change of intensity due to poling of a polycrystalline soft lead zirconate titanate composition using a synchrotron source. The intensities of the (200) and (002) XRD peaks change after the moment of poling, the (200) intensity is disfavoured after poling and decreases slightly while (002) increases. A number of sequential measurements are shown in order to also capture potential relaxation effects. It is obvious from the data that a quantitative analysis is a difficult task. Fig. 10.11b displays the orientation dependence of the relative intensity of the (002) and (200) peaks of a fully poled sample of a soft PZT.

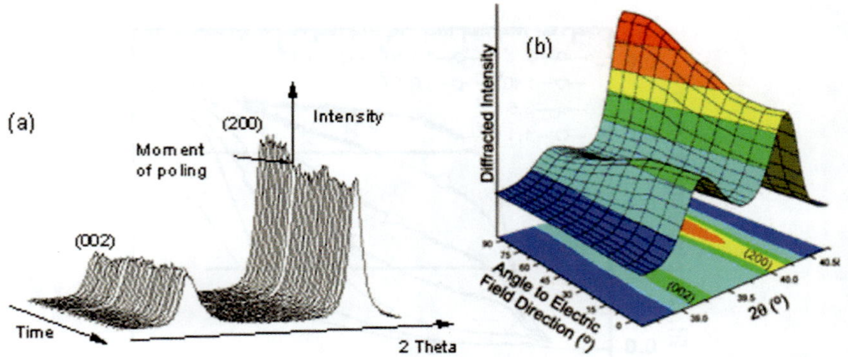

Fig. 10.11 **a** Illustration of determining the intrinsic and extrinsic contributions to the macroscopic strain response of a ferroelectric. ©*Axel Endriss, Michael Hoffmann*. **b** Diffraction profiles of the (200) and (002) reflections in a tetragonal PZT ceramic as a function of angle from the poling axis. The intensity changes represent the degree of ferroelastic domain switching attributed to the electrical poling process. Data obtained using neutron diffraction at the Australian Nuclear Science and Technology Organisation (ANSTO) [79] ©*Springer*.

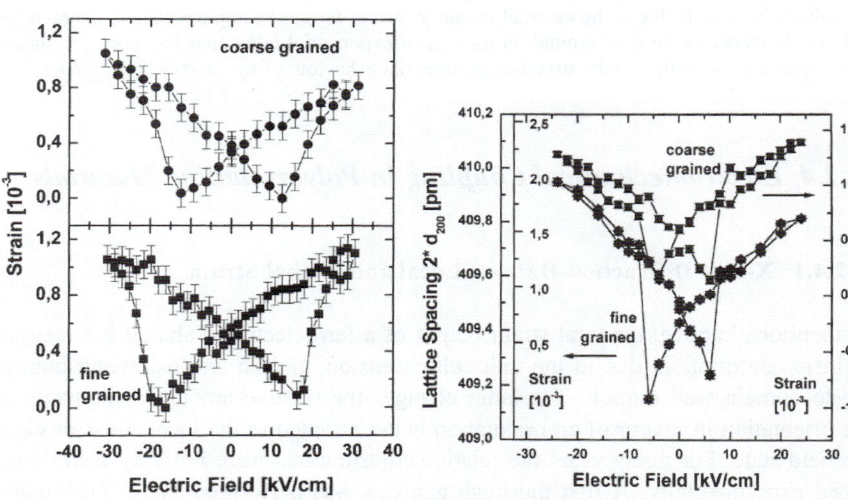

Fig. 10.12 Intrinsic contribution to strain in coarse and fine grained 2% La-doped PZT along maximum strain directions in tetragonal PZT 45/55 [111] (left) and rhombohedral PZT 60/40 (right). The mean grain sizes are 1.6 and 3.0 μm [78]. ©*Elsevier*

The peak position is the second information included in the data. It is easier fitted and is representative of the unit cell strain. Fig. 10.12 displays the resultant values for tetragonal as well as rhombohedral PZT (2% La). Lattice strain contributes roughly half the total strain of a bipolar hysteresis cycle with slight differences for different grain sizes. In a similar manner, the relative intensity differences may be monitored yielding the respective contribution of domain wall motion to strain.

10.2.4.2 Macroscopic Hysteresis in PZT

Figure 10.13 displays typical macroscopic hysteresis curves for three types of ferroelectric lead zirconate titanate (PZT) near the morphotropic phase boundary (MPB). Compositions near the MPB are the technologically most employed range of compositions. The MPB denotes a transition line in the phase diagram, where the crystal structure almost exclusively depends on chemical composition rather than temperature [3]. As with a temperature-dependent transition, the effective material coefficients, which are gradients to the thermodynamic potential, become large near the transition. The near vertical MPB in the phase composition diagram enables the use of compositions near to the boundary exhibiting enhanced electromechanical coupling. The near-vertical MPB renders these properties relatively stable with temperature.

Figure 10.13 displays the classical electrical and mechanical hysteresis loops including their poling cycle from the non-poled material yielding the origin of the graphs. The macroscopic relation between strain and polarization is displayed in the bottom row. It is clear that Landau-Devonshire theory for single crystals is not sufficient to capture macroscopic material behaviour. Landau-Devonshire theory predicts a hysteresis-free parabolic relationship between polarization and strain. Much of the complications in material description stem from the insufficiency of the microscopic theory to describe macroscopic material behaviour, which is significantly influenced by the orientation distribution of the microstructure, defects, grain boundaries, and the macroscopic boundary conditions. It will be the objective of section 10.3.1 to identify pathways to overcome this problem.

The technologically used piezoelectric device is a poled ferroelectric (see section 10.2.4.4 for poling). The resultant macroscopic piezoelectric effect is a superposition of the intrinsic piezoelectric effect and the linearization of electrostriction. Fig. 10.14 displays how the unipolar hysteresis cycle used for actuation (a) is embedded into the overall hysteresis behaviour of the material (b)-(e) at growing partly or entirely negative field amplitude.

A different perspective arises when considering classical ferroelastic hysteresis, namely hysteretic stress strain curves. Fig. 10.15 displays compressive stress strain curves under different electric bias (E_B) with the pure ferroelastic curve for $E_B = 0$. The electric field in polarization direction holds back the switching under compressive loading along the polarization direction. The different starting values in the experiment for rising compressive stress are due to the already present electric field. It reflects the different strain states visible in Fig 10.14 II (e) under applied field with shifted origin (s = 0 for the non-poled state) in Fig. 10.15.

Under tensile loading, the samples are prone to brittle failure from the inherent defects in the ceramic [81, 82]. Therefore, fully saturated tensile mechanical hysteresis data are not available. Nevertheless, a significant difference in tensile and compressive loading has been reported based on bending experiments. Fett *et al.* [83] showed that tensile loading can in principle lead to a much higher inelastic strain than compression. This is easily understood if a random initial domain state is assumed. In a thought experiment, one may imagine a hypothetical material with equal volume fractions of three domain types with polarization along each of three Cartesian coordinate axes. For tensile loading along one axis, two out of three domain types may switch to the type aligned with the tensile direction. The associated change of spontaneous strain contributes to the elongation. If instead a compressive stress is applied, only the domain with polarization along the compressive stress direction will switch. In reality, all crystallite orientations exist (see section 10.3.2) and only projections of the different variants contribute to the macroscopic behaviour. In small volumes, the number of natural initial defects leading to fracture may be negligibly small. Thus, locally tensile yielding at locations of stress concentrations in a mechanical device, e.g., a process zone ahead of a crack or the tip of an electrode in multilayer actuator, may result in a full ferroelastic elongation of the local volume elements even under tension. This macroscopically inaccessible strain state may thus be realized locally under certain circumstances.

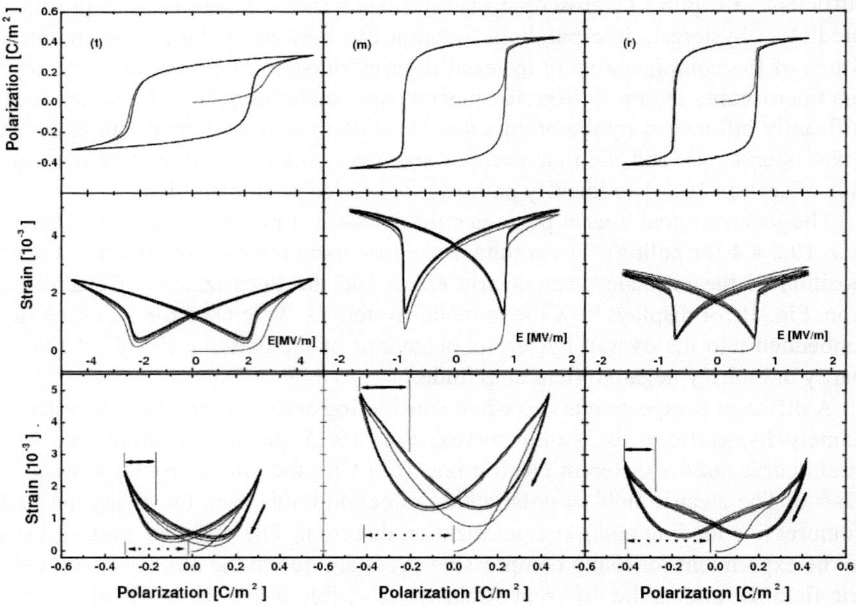

Fig. 10.13 The classical hysteresis loops of ferroelectric lead zirconate titanate (PZT, here $Pb_{1-3x/2}La_x(Zr_{1-y}Ti_y)O_3$) at room temperature for three different compositions 1-y/y around the morphotropic phase boundary: **t** tetragonal 45/55, **m** morphotropic 54/46, and **r** rhombohedral 60/40 [71] ©*Springer*.

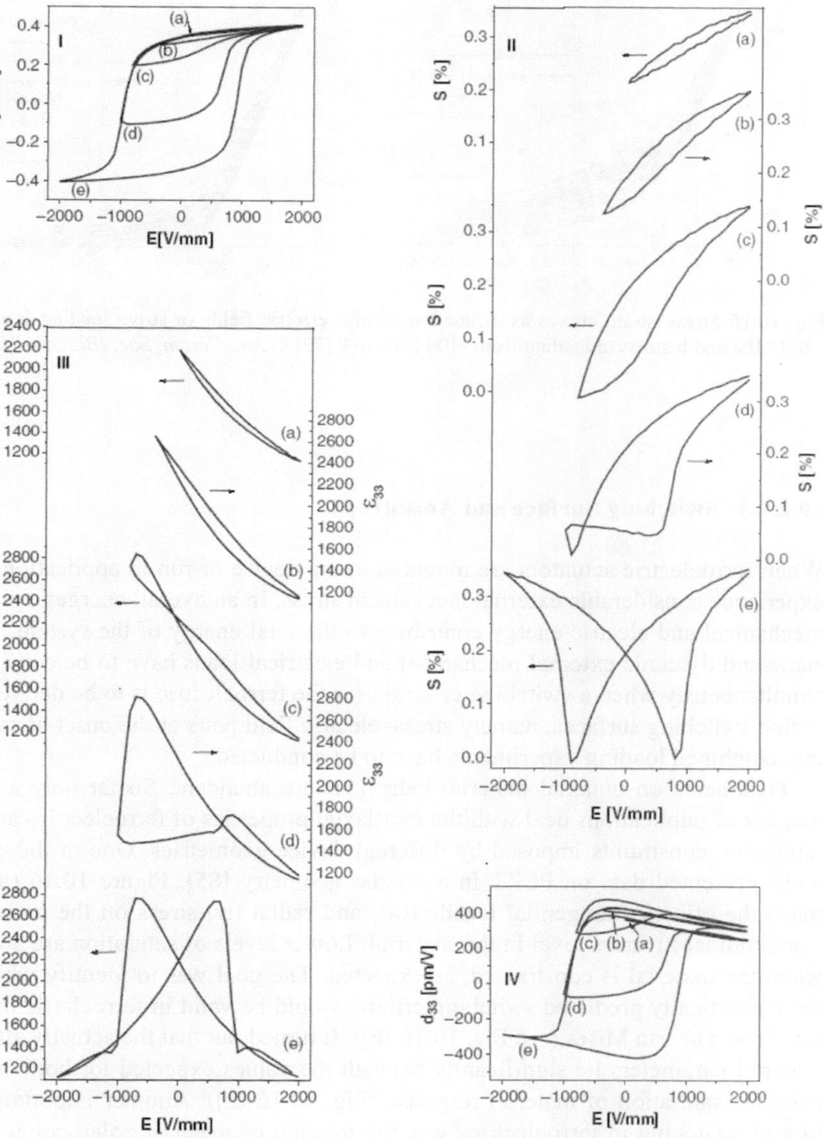

Fig. 10.14 Hystereses of polarization P (I), Strain S (II), dielectric coefficient ε_{33} (III), and piezoelectric coefficient d_{33} (IV) for a unipolar cycle (a), sesquipolar cycles with E_{min} = –400 V/mm (b), –800 V/mm (c), –1000 V/mm (d), and a bipolar cycle (e), in PIC 151 [84] © *Am. Ceram. Soc. (Blackwell)*.

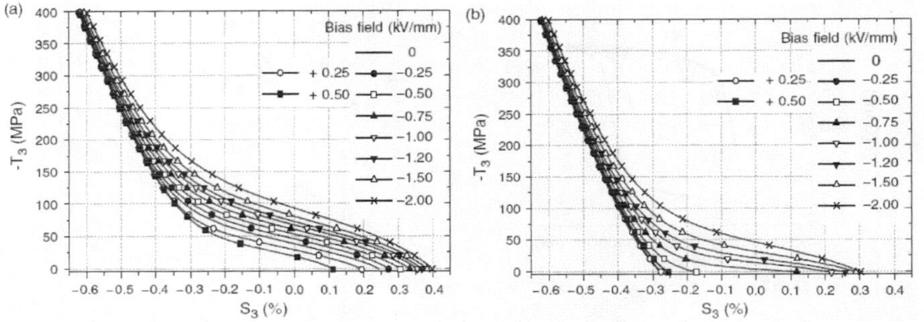

Fig. 10.15 Stress–strain curves as a function of bias electric fields (**a** stress loading from 0 to –400 MPa and **b** stress unloading from -400 MPa to 0 [72] © *Am. Ceram. Soc. (Blackwell)*.

10.2.4.3 Switching Surface and Anisotropy

When ferroelectric actuators are mounted into a device or run in application, they experience considerable external mechanical stress. In an overall energetic picture, mechanical and electric energy contribute to the total energy of the system. Thus, static and dynamic external mechanical and electrical loads have to be considered simultaneously when a switching criterion for the ferroelectric is to be derived. To define switching surfaces, namely stress–electric field pairs at the onset of switching, combined loading experiments have to be conducted.

Treatments on uniaxial material behaviour are abundant. So far only a small number of publications deal with the multiaxial properties of ferroelectrics and the multiaxial constraints imposed by different device geometries. One of the earlier woks presented data on PLZT in a tubular geometry [85]. Figure 10.16 (a) displays the effect of tangential tensile (θθ) and radial (rr) stress on the transverse (longitudinal z) stress level in the material. Lower levels of saturation are attained when the material is constrained, as expected. The goal was to identify which of the theoretically predicted switching criteria would be valid in ferroelastic materials, Tresca or van Mises (see Fig. 10.16 (b)). It turned out that the actually attained material parameters lie significantly beneath the values expected for both criteria even for saturation of material response (Fig. 10.16 (b)). Another important feature of switching in ferroelectrics was the rotation of material polarization under an off-axial electric field [86]. For this end, a pre-poled sample was cut at distinct angles with respect to the initial poling direction given in Fig. 10.16 (e). As expected, intermediate degrees of electric displacement changes are observed for finite angles with respect to the initial poling direction (Fig. 10.16 (c,d)).

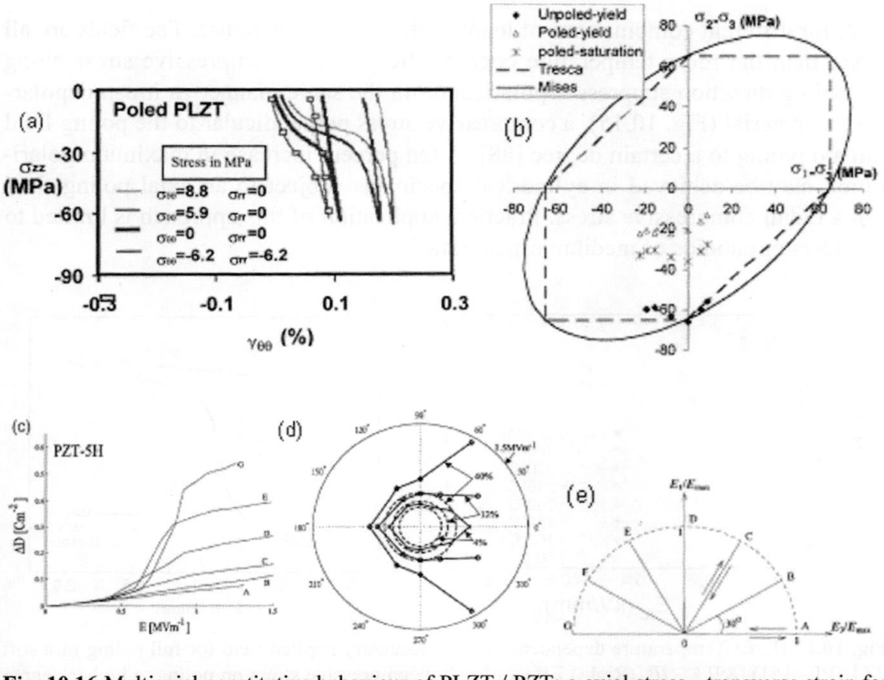

Fig. 10.16 Multiaxial constitutive behaviour of PLZT / PZT: **a** axial stress - transverse strain for various tensile transverse stress levels in a tubular PLZT sample, **b** the resulting experimental switching criteria compared to theoretical criteria: Tresca $\sigma_0 > MAX(|\sigma_1-\sigma_2|, |\sigma_{\square}-\sigma_3|, |\sigma_3-\sigma_1|)$ or van Mises $\sigma_0 > 1/\sqrt{2}\cdot\sqrt{(\sigma_1-\sigma_2)^2+(\sigma_2-\sigma_3)^2+(\sigma_3-\sigma_1)^2}$ [85]. © 2001 *ASME*. **c** shows the change in electric displacement ΔD versus electric field E response of PZT-5H specimens loaded along paths OA to OG shown in e (i.e. inclined at various angles θ to the initial poling direction OA). For this end samples were cut at the respective angles from pre-poled samples as indicated in e. OA is unipolar loading, OG is bipolar loading, OD rotation at 90 degrees. In d the corresponding offset switching surfaces (solid curves) for three levels of offset corresponding to 4%, 12% and 40% of the remnant polarization after cold poling are displayed, with the radial axis showing electric field in MV m-1 and the polar axis showing angle θ. In contrast, the dashed circles are the unpoled (isotropic) switching surfaces for each material corresponding to the same levels of remnant polarization offset [86] © *Elsevier*.

10.2.4.4 Poling and De-Poling

Poling of polycrystalline ferroelectrics has been a fundamental process in device production since the very beginning of piezoelectric technology. Recently, it has found renewed interest because certain properties of the material itself may be derived from the poling process. For poling, an external applied electric field is necessary to induce the breaking of symmetry. The poling field magnitude depends on temperature and on externally applied stress. A minimum value seems to exist even when the poling process is undertaken while cooling through the Curie point of the material [87]. Fig. 10.17 displays the obtainable piezoelectric coefficient for different combinations of loading field and temperature. The fields are all lower than the room temperature coercive field. While compressive stress along the poling direction suppresses polarization in the same manner as it can depolarize the material (Fig. 10.15), a compressive stress perpendicular to the poling field can aid poling to a certain degree [88]. A ten percent increase in maximum polarization may be achieved in cylindrical specimens subject to an axial poling field and a radial compressive stress. Practical application of this approach is limited to cylinders or cuboids of medium aspect ratio.

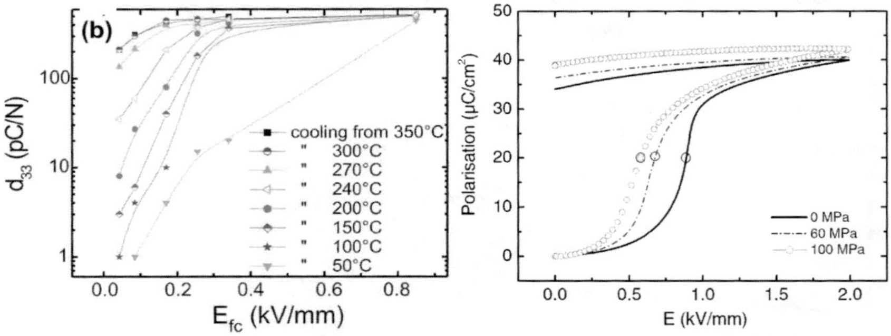

Fig. 10.17 (Left) Temperature dependence of the necessary applied field for full poling of a soft PZT (PIC 151) [89] ©*AIP*. (Right) Effect of radial compressive stress on poling: a higher degree of poling is acquired along with a lower coercive field [88]. ©*Elsevier*.

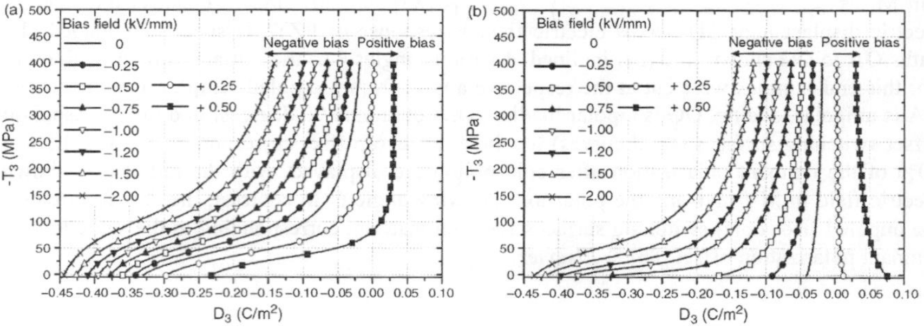

Fig. 10.18 Stress–depolarization curves as a function of bias electric fields in a commercial soft PZT (PIC 151): **a** stress loading from 0 to -400 MPa and **b** stress unloading from -400 MPa to 0 [72] ©*Am. Ceram. Soc. (Blackwell)*.

Depolarization may occur thermally or under applied compressive stress for short circuit electric boundary conditions. Short circuiting may arise in the material itself at elevated temperature when a partial conduction sets in. For open circuit conditions, the uncompensated charges on the electrodes re-polarize the material once the mechanical load is relieved. This behaviour is apparent in Fig. 10.18b. The material is initially poled in the negative direction. Increase of the negative bias field further increases the negative polarization. Compressive stress results in depolarization. A full set of different electrical bias conditions and their influence on the depolarization behaviour was measured by Zhou *et al.* [72] and is given in Fig. 10.18. It is particularly clear from the unloading experiment that the positive bias will completely alter the macroscopic and microscopic state of the sample.

10.3 Electromechanical Hysteresis, Modelling

10.3.1 Models of Hysteresis

Modelling the non-linear material behaviour of ferroelectric ceramics provides a description of the observed material behaviour that is useful for the design of devices. It also provides a deeper understanding of the contributions of the underlying phenomena to the observed behaviour. Some of the design challenges include configurations where the material is subjected to inhomogeneous fields such as at the edges of electrodes in multilayer structures. This may affect the reliability of the devices.

At low applied electric or mechanical field levels, the material behaviour of ferroelectric ceramics may be described by linear constitutive equations. These equations (under isothermal and quasi-static conditions) relate the total strain S_{ij} and the electric displacement D_i to the applied stress T_{ij} and electric field E_i by

$$\begin{aligned} S_{ij} &= s^E_{ijkl} T_{kl} + d_{kij} E_k + S^r_{ij} \\ D_i &= d_{ikl} T_{kl} + \varepsilon^T_{ik} E_k + P^r_i \end{aligned} \qquad (1)$$

$\boldsymbol{s}^E$, $\boldsymbol{d}$ and $\boldsymbol{\varepsilon}^T$ denote the elastic compliance, the piezoelectric moduli, and the permittivity, respectively.[2] The remnant parts of strain and polarization, $\boldsymbol{S}^r$ and $\boldsymbol{P}^r$, are constant in the low field range.

At intermediate field levels, the linear constitutive equations may be amended by second order nonlinear terms. Hysteretic effects which are connected with energy dissipation are taken into account by imaginary parts in the material constants. Here, the empirical Rayleigh relations can be used (for a good overview see [90]).

When a ferroelectric ceramic is subject to large electrical and/or mechanical loading, complete reconfiguration of the domain structure occurs including do-

---

2 Equations (1) and all the following assume the Einstein summation convention summing over double indices. The superscripts denote the thermodynamic variable kept constant during determination of the respective constants.

main wall motion, domain reorientation / switching, and formation / disappearance of domains. Accordingly, non-linear and hysteretic material behavior is observed, which must be described by history-dependent remnant quantities $\boldsymbol{S}^r$ and $\boldsymbol{P}^r$. Also, the material constants in (1) become state-dependent. An overview may be found again in Hall [90] and in the paper by Bhattacharya & Ravichandran [91] who also consider lattice dynamics effects. Huber [92] has reviewed micro-electromechanical models of non-linear hysteretic behaviour and Landis [93] phenomenological constitutive theories. A comparison of the micromechanical and phenomenological models contrasted with experimental results was given by Shieh *et al.* [86].

In this section, we discuss novel developments in the theory of non-linear material behaviour. The modelling schemes may be distinguished by the degree of details of the microstructural processes which are taken into account as discussed earlier (see Fig. 10.1). The phenomenological approaches (subsection 10.3.1.1) commonly use the macroscopic remnant strain and polarization as internal state variables. Micromechanical models apply internal parameters describing the statistical weight fraction of differently oriented domains, often in combination with the orientation distribution function of crystallites (grains) forming the polycrystalline ceramic (see section 10.3.2). These domain orientation models may be separated into the switching models where single-domain grains (or elements in a finite element numerical approach) can only switch between crystallographically fixed orientations (10.3.1.2), the dissipative transformation models which allow for a gradual change of domain volume fractions (10.3.1.3), and models which include dissipative rate effects (10.3.1.4).

10.3.1.1 Phenomenological Models

For a simple and efficient macroscopic description, phenomenological approaches are important. Such an approach is especially useful when devices under complex boundary conditions have to be analyzed using numerical schemes like the finite element method. Models based on microstructural considerations remain too computational time consuming.

A physical approach coming from the description of magnetic behaviour is provided by the Preisach method, which may be considered an extension of the empirical Rayleigh formulas. A good overview of its application to ferroelectric ceramics is given by Robert *et al.* [94]. The Preisach method offers the possibility to relate empirical constants to microstructural features.

Another approach which takes into account the notion of domain orientation has been presented by Zhou & Chattopadhyay [95]. They obtain relatively simple explicit expressions for the coupled electromechanical hysteresis under uniaxial loading.

Real material behaviour, however, requires a full tensorial description of the material law. The model presented by Kamlah & Tsakmakis [96] is based on the theory of plasticity with switching surfaces (the generalization of the yield surface in stress space) and a flow rule. The coupled hardening and saturation behaviour is described by carefully devised rules and conditional statements. Also, Huber & Fleck [97] presented such a phenomenological model in addition to their crystal plasticity model and the viscoplastic formulation.

A more general thermodynamic framework may provide a description which is free from the need of special assumptions on the details of hardening and saturation of ferroelectric transitions. Such models have been developed, e.g., by McMeeking & Landis [98] and Landis [99] (see the review by Landis [93] for more examples). Recently, these general mechanics approaches have been further extended by Schröder & Romanowski [100] (general invariant formulation, application to single-crystal behaviour), Klinkel [101] (irreversible electric field instead of the remnant polarization as internal variable) and Mehling *et al.* [102]. In the latter theory, the models of Kamlah & Tsakmakis [96] and Landis [99] have been expanded by using two fully three-dimensional internal state variables. One is a texture tensor, describing a three-dimensional orientation distribution function (ODF), and the second variable is vector-valued and describes the state of macroscopic irreversible polarization.

In order to apply these phenomenological models, parameter adjustment must be done. Though there are efficient methods to accomplish this task, it is often complicated to extrapolate to other material systems because of the lack of physical interpretation of the model constants.

10.3.1.2 Switching Grain Models

Switching models have in common that a certain element (an individual domain, a single-domain grain, a lattice cell in a Potts model) is completely transformed to a new crystallographic orientation, if a certain energetic condition is fulfilled. In the field of ferroelectrics, this approach originated from Hwang *et al.* [103], who considered statistically oriented single-domain grains. In their theory they neglected the interaction between grains by assuming that every grain experiences the applied stress and electric field. Obviously this is a strong simplification. Often this assumption is masked by calling it a "Reuss approximation" according to a famous approach in the theory of effective elastic constants of polycrystals[3]. Therefore, better schemes have been developed in which one considers a grain embedded in an effective surrounding (see the review by Huber [92]).

Since these homogenization schemes are computational time consuming, one may also try to overcome the assumption of equal load to every grain by empirically adding local field fluctuations to the macroscopically applied fields [104] or to the switching energy barrier [105, 106]. The magnitude of the statistical fluctuation is then adjusted to fit the experimental data.

3 Drawing an analogy to polycrystal plasticity theory, the assumption should be better called a Sachs model, which is the complement to the more famous Taylor model.

There are also some new interesting approaches within the assumption of homogeneous driving fields which try to improve the predictions for the combined action of mechanical and electrical load [107, 108], and for the influence of the sequence of loadings for non-symmetric electro-mechanical behaviour [109].

Another problem arises in connection with the motion of free charge carriers inside the material. It is sometimes argued that poling charges are completely neutralized by those free charges. This has the consequence that only the mechanical part of interaction constraint between a certain poled or switching grain with its surroundings should be considered in the theory [110]. Both possible assumptions (immediate complete neutralization or no free charge motion at all) do not really match the actual material behaviour. Seemingly, the effect of free charge motion must be described within a rate-dependent approach (see 10.3.1.4).

Finally, models inspired by the Ising or Potts models of ferromagnetism have been developed. These approaches describe the material by a Hamiltonian which takes into account the interactions of polarizations and strains of special cells arranged in a two-dimensional lattice. By adopting a suitable algorithm, one may find the equilibrium state as the minimum of total energy [111, 112]. Due to the restriction to a two-dimensional arrangement, these models are especially appropriate for thin film applications. The same applies to the so-called phase field simulations, where the restriction of having predefined phase boundaries can be revoked. Instead, the equilibrium state including domain walls with a finite thickness is obtained by applying the framework of the time dependent Ginzburg–Landau equations (see, e.g., [113, 114]).

10.3.1.3 Dissipatively Transforming Grains (Rate-Independent Models)

The domain microstructure inside a given single crystal of a polycrystalline aggregate is formed by a multi-rank laminate consisting of fine twins. This structure follows from the condition of minimum energy (see the fundamental investigations by Arlt [115]). Under external loadings, the domain structure does not completely switch to another variant. Instead, a gradual change of domain volume fractions is observed according to a local (and global) equilibrium. If this picture is taken, one arrives at a model which resembles features of classical rate-independent plasticity theory.

The simplest approach is provided by a lamellar stack of two types of domains which for tetragonal crystals form 90° domain walls. The domain stack behaviour can be homogenized by analytical methods whereas the interaction of grains may be taken into account by the effective medium approximation [116, 117] (a similar model has been developed by [118]). Domain wall motion is allowed to take place by considering the total potential energy variation of the polycrystalline material in comparison with microscopic retarding forces. These retarding forces are due to the interaction of domain walls with pinning defects. Fig. 10.19 shows a typical outcome of such a model. In the beginning, an optimum poling state is set by switching all grains as near as possible to the poling direction under the condition that the two types of domains still have equal volume fraction inside all crystals. The poling process itself is not well covered by the model since a complete restructuring of the domain configuration is involved. During subsequent unipolar loading cycles, the structure "shakes down" by adjusting the domain volume fractions. The predicted unipolar hysteresis is directly related to the assumption about an energy barrier for domain wall motion.

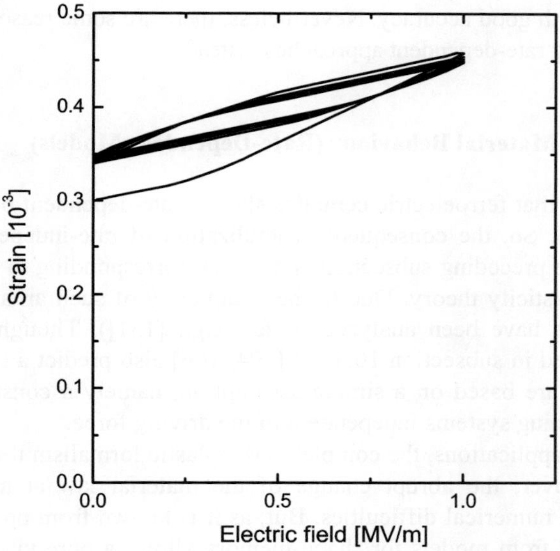

Fig. 10.19 Longitudinal strain under the application of an electric field (macroscopically stress-free sample). One poling cycle followed by four cycles of loading and unloading. The energy barrier for complete switch between 90° domains was set to 100 kJ/m³ (details see [119]).

Obviously the assumption of only two types of domains in a certain single crystal imposes restrictions on the free local electrical and mechanical response. A more general model has been developed by Huber *et al.* [120] following an analogue to the operation of a crystal glide system. All transformations between crystallographically allowed domain orientations are possible and take place if a certain yield criterion is fulfilled where also assumptions on hardening behaviour may be incorporated. The main drawback of this model lies in its computational complexity and in the fact that constraints due to the geometrical arrangement of domains within the crystal are neglected. Consequently, field fluctuations inside a grain disappear from the theory. Nevertheless, the model has been successfully applied to multiaxial and coupled electro-mechanical loadings [97, 121, 122, 123].

Even less detailed models are obtained, if the correlation between crystallographic grain properties and domain orientations is neglected. Here, the ceramic is simply considered as a mixture of domains whose statistical weights (which usually but not necessarily are interpreted as volume fractions) may change if a certain driving force reaches a threshold. The driving force is obtained from energetic considerations similar to the switching models. So, one arrives at a rate-independent behaviour. The models may be distinguished by the number of domain variants which are considered and by the method how the influence of surrounding material on the driving force of a particular domain variant is taken into account [124, 125, 126, 127, 128]. The model parameters may be adjusted so to describe real ceramic material behaviour with good accuracy. Nevertheless, there are some reasons to consider the corresponding rate-dependent approaches instead.

10.3.1.4 Creep Material Behaviour (Rate-Dependent Models)

It is well known that ferroelectric ceramics show a rate-dependent behaviour (see, e.g., [129, 130]). So, the consequent generalization of rate-independent models considered in the preceding subsection is a theory corresponding to the viscoplastic models in plasticity theory. Due to the complexity of such an approach, only special problems have been analyzed so far (e.g., [131]). Though some of the models mentioned in subsection 10.3.1.2 [104, 106] also predict a rate-dependent behaviour, they are based on a simple assumption, namely a constant transition rate for all switching systems independent of the driving force.

For practical applications, the complete viscoplastic formalism thus appears too complex. Moreover, the abrupt change of the material law at a certain field threshold causes numerical difficulties. But, as it is known from polycrystal plasticity theory and from models for shape-memory alloys, a pure viscous approach without a sharp threshold may be successfully utilized. This has been proposed by Huber and Fleck [97] in the context of their crystal transformation model [120].

In a viscous approach, N different domain orientations (variants) are considered. The number N depends on the problem considered (uniaxial or multiaxial loading) and on crystal symmetry (tetragonal, rhombohedral); typical values used are $N = 2,6,8,20,42$. Let ξ_I denote the statistical weight of variant I. These weights are subject to the conditions:

$$0 \leq \xi_I \leq 1 \sum_{I=1}^{N} \xi_I = 1 \tag{2}$$

The transition rate from domain orientation variant I to variant J is approximated by

$$v_{IJ} = v_0 \left(\frac{w_J - w_I}{w_c} \right)^m (N\xi_I)^a \tag{3}$$

for $w_J - w_I \geq 0$ (otherwise $v_{IJ} = 0$)

Here w_J and w_I are generalized driving forces for a transition into variant J or I, respectively, and w_c is a characteristic energy above which the transition rates become large (w_c may depend on the transition type I-J).

The second power term in (3) takes into account the saturation of the transition, if the weight ξ_I of the starting variant becomes zero. v_0 is a characteristic frequency which corresponds to the conditions $w_J - w_I = w_c$ and $\xi_I = 1/N$ (this is the isotropic state where all variants have equal weight). This frequency and the two empirical constants m and α are to be adjusted to experimental data.

By means of the power law (3), a smooth transition between the piezoelectric regime (below the characteristic value w_c) and the ferroelectric/ferroelastic behaviour above w_c can be described. For $m \gg 1$, the characteristic energy w_c becomes a real threshold (i.e., the viscous material law turns into a viscoplastic constitutive model[4]). Pathak and McMeeking (2008) have evaluated the crystal transformation model in that way by assuming $m = 50$. Otherwise, it is possible to obtain equation (3) from a model based on thermally activated domain wall motion. Then, temperature-dependent hysteresis curves may also be modelled and the creep exponent can be expressed by $m = \Delta H_0/kT$, where ΔH_0 is the activation enthalpy [132, 133].

Besides the application of the transition rate model at the single crystal level, it is often sufficient to combine this model with the notion that the polycrystalline-polydomain material can be described by a sufficient number of independent domain orientations as already discussed at the end of the preceding subsection. This methodology was proposed by Huber & Fleck [97]. By comparing different models with experimental results, they have shown that this simplified model is a good compromise between the numerically cumbersome crystal plasticity model and the problematic phenomenological approaches.

Because of these advantages, the viscous approach seems most interesting for practical applications. In its simplest form, we may assume that each domain variant is subjected to the externally applied average fields $\overline{T}_{ij}$ and $\overline{E}_i$. Then the driving force w_I is given by the potential energy $w_p^{(I)}$ (i.e. the free enthalpy) of the domain variant I

$$w_I = w_p^{(I)} = -\frac{1}{2}\overline{T}_{ij} S_{ijkl}^{E(I)} \overline{T}_{kl} - \overline{E}_i d_{ikl}^{(I)} \overline{T}_{kl} - \frac{1}{2}\overline{E}_i \varepsilon_{ik}^{T(I)} \overline{E}_k - \overline{T}_{ij} S_{ij}^{r(I)} - \overline{E}_i P_i^{r(I)} \tag{4}$$

This corresponds to a total potential energy of the material which is simply obtained by the statistical average

4 Nevertheless the viscous approach is most often called "viscoplastic".

$$w_p\left(\overline{T}_{ij},\overline{E}_i,\xi_1,..,\xi_N\right)=\sum_{I=1}^{N}\xi_I w_p^{(I)}\left(\overline{T}_{ij},\overline{E}_i\right) \tag{5}$$

Accordingly, the effective material law has the form of (1) where the effective remnant quantities are easily calculated by

$$S_{ij}^r=\sum_{I=1}^{N}\xi_I S_{ij}^{r(I)} \quad P_i^r=\sum_{I=1}^{N}\xi_I P_i^{r(I)} \tag{6}$$

and similar equations for the effective piezoelectric constants. For more details see [97] or [132].

Due to the assumption that each domain variant is subjected to the same external average fields, one may again call this scheme a Reuss approximation. But, since no definite microstructural arrangement of domains is presupposed, it does not make much sense to allow for a variation of the specifically acting fields. Instead, one should consider this model as being a semi-phenomenological approach with a certain number of physically motivated internal state variables ξ_I.

It is important to note that the right-hand side of (3) is a unique function of state, so that in numerical simulations no iteration is necessary when a certain load increment is prescribed. Instead, one directly obtains the increments of internal state variables by summing all transition rates:

$$\xi_I=\sum_{J=1}^{N}\left(+v_{JI}-v_{IJ}\right) \tag{7}$$

In this way, the above system of equations allows for a history-dependent calculation of the statistical weights ξ_I from which the complete hysteresis curves follow. Fig. 10.20 displays an example for such a simulation. More details of the approach as well as comparison with other models and experiments for non-proportional loading paths may be found in Huber & Fleck [97] and Belov & Kreher [134, 135].

10.3.2 Homogenization

In this section we will develop a macroscopic model based on the homogenization at the mesolevel, in more detail, and also display some of the mathematics to arrive at a complete model. The characteristic feature of ferroelectric ceramics is that the spontaneous polarization may be reversed by an applied electric field. This phenomenon occurs in the ferroelectric phase within a certain range of temperature. In the framework of phenomenological phase transition theory (Landau and Ginzburg) [17], one may study the specific behaviour by expanding the thermodynamic potential in terms of a suitable order parameter. A stability analysis of the response function leads to a classification into different kinds of hysteresis loops in dielectrics, see e.g. Maugin *et al.* [136] and the references therein. In a ferroelectric polycrystalline ceramic, one does not observe piezoelectric property before poling due to the random orientation of the ferroelectric domains, Fig. 10.21, state A.

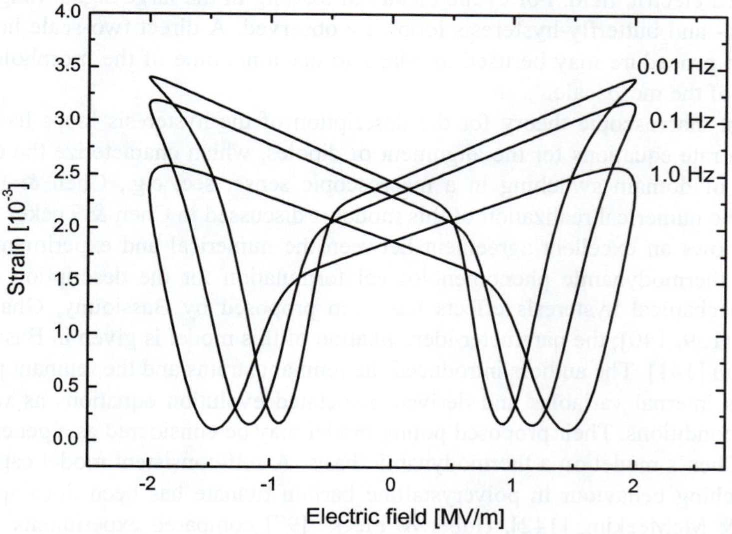

Fig. 10.20 Simulated strain hysteresis curves under sinusoidal loading with different frequency for soft PZT (details of calculation parameters are given in [135]).

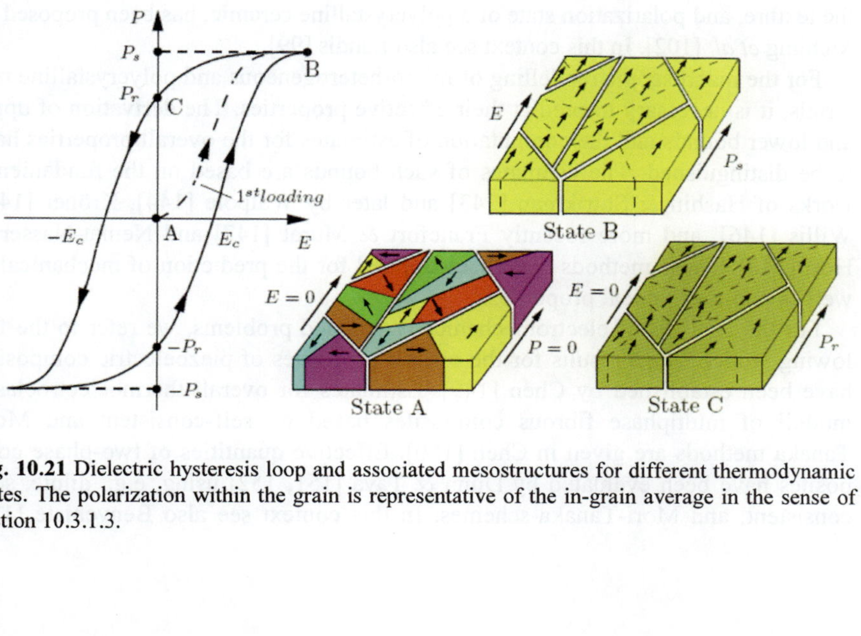

Fig. 10.21 Dielectric hysteresis loop and associated mesostructures for different thermodynamic states. The polarization within the grain is representative of the in-grain average in the sense of section 10.3.1.3.

In our simplified approach, the interdependence of domain orientations within one grain as visible, e.g., in Fig. 10.2 and its relation to the crystallite orientation are neglected. An applied electric field yields a reorientation of the domains in the direction of the applied field (Fig. 10.21, state B). When the electric loading is removed, a remnant polarization and remnant strains remain (Fig. 10.21, state C). The specimen now exhibits piezoelectricity. The polarization may be reversed by an applied electric field. For cyclic electrical loading in the large signal range, the dielectric- and butterfly-hysteresis loops are observed. A direct two-scale homogenization procedure may be used to take into account some of the morphological features of the mesoscale.

A first macroscopic theory for the description of the hysteresis loops has been based on rate equations for the alignment of dipoles, which characterize the consequences of domain switching in a macroscopic sense, see, e.g., Chen & Peercy [137]. The numerical realization of this model is discussed in Chen & Tucker [138], which shows an excellent agreement between the numerical and experimental results. A thermodynamic phenomenological formulation for the description of the electromechanical hysteresis effects has been proposed by Bassiouny, Ghaleb & Maugin [139, 140]; the parameter identification of this model is given in Bassiouny & Maugin [141]. The authors introduced the remnant strains and the remnant polarization as internal variables and derived associated evolution equations as well as loading conditions. Their proposed poling model may be considered as a generalization of Chen's model on a thermodynamic basis. A self-consistent model capturing the switching behaviour in polycrystalline barium titanate has been developed by Landis & McMeeking [142]. Huber & Fleck [97] compared experiments with a self-consistent polycrystalline formulation, a simplified crystal viscoplasticity model, and a classical phenomenological model based on a rate-independent flow theory for multiaxial electrical switching. A thermodynamically consistent phenomenological model using orientation distribution functions for the approximation of the texture, and polarization state of a polycrystalline ceramic, has been proposed by Mehling *et al.* [102]. In this context see also Landis [99].

For the macroscopic modelling of micro-heterogeneous and polycrystalline materials, it is necessary to predict their effective properties. The derivation of upper and lower bounds and the computation of estimates for the overall properties have to be distinguished. The estimates of such bounds are based on the fundamental works of Hashin & Shtrikman [143] and later by Walpole [144], Kröner [145], Willis [146], and more recently Francfort & Murat [147] and Nemat-Nasser & Hori [148]. These methods have been applied for the prediction of mechanical as well as non-mechanical properties.

For the analysis of electromechanically coupled problems, we refer to the following works. Exact results for the overall properties of piezoelectric composites have been established by Chen [149]. Estimates for overall thermoelectroelastic moduli of multiphase fibrous composites based on self-consistent and Mori-Tanaka methods are given in Chen [150]. Effective quantities of two-phase composites have been evaluated by Dunn & Taya [151, 152] using, e.g., dilute, self-consistent, and Mori-Tanaka-schemes. In this context see also Benveniste [153,

154, 155, 156]. Universal bounds for effective piezoelectric properties of heterogeneous materials have been derived by Hori & Nemat-Nasser [157] by using generalized Hashin-Shtrikman variational principles.

The overall properties of (periodic) composites depend on the morphology of their (unit cells) mesostructure and the properties of their individual constituents. Therefore, it is possible to improve the performance characteristics of piezoelectric materials using topology optimization and homogenization techniques, see, e.g., Silva *et al.* [158] and [159]. Utilizing a unit-cell method, Li *et al.* [160] investigated the relation between effective properties and different geometries of microvoids based on a 3D finite element analysis. A multi-scale finite element modelling procedure for the macroscopic description of polycrystalline ferroelectrics has been proposed by Uetsuji *et al.* [161]. A homogenization procedure based on an asymptotic expansion of the displacements and the electric potential was utilized, for the mathematical background see, e.g., Sanchez-Palencia [162]. General works on the homogenization theory are Hill [163, 164], Suquet [165], and Krawietz [166].

A general direct homogenization procedure which couples the macroscopic to the mesoscopic scale, in this context see also Miehe *et al.* [167], Schröder [168], is as follows:

1. At each macroscopic point: localize suitable macroscopic quantities (e.g., the strains and the electric field) to the mesoscale. To be more specific, apply constraint conditions or boundary conditions, e.g., driven by the macroscopic strains and the electric field, on a representative volume element.
2. Next, solve the equations of balance of linear momentum and Gauss' law under the applied macroscopic loading to obtain the dual quantities (e.g., the stresses and the electric displacements) on the mesoscale.
3. Next, perform a homogenization step, i.e., compute the average values of the dual quantities on the mesoscale. These macroscopic variables have to be transferred to the associated points of the macroscale.
4. Finally, solve the electromechanically coupled boundary value problem on the macroscale and proceed with step 1 until converged solutions are obtained on both scales. The numerical solution is based on separate finite element analyses on each scale. The overall algorithmic moduli needed for the Newton-Raphson iteration scheme on the macroscale may be computed in an efficient way during the standard solution procedure on the mesoscale, see, e.g., Schröder [168], Schröder & Keip [169, 170].

Focus now shifts to a simplified approach in which each domain is modelled within a coordinate invariant formulation for an assumed transversely isotropic material as presented in Romanowski & Schröder [171], Romanowski [172], and Schröder & Gross [173]. Furthermore, the individual domains are represented by individual preferred directions and subjected to mechanical and electrical constraints. This means that the macroscopic polycrystals may be approximated by discrete orientation distribution functions, see Schröder *et al.* [174, 175], Kurzhöfer [176]. This procedure is associated with the well known Reuss- and Voigt-bounds.

10.3.2.1 Electromechanically Coupled Boundary Value Problems (BVP)

Macroscopic Electromechanically Coupled BVP: Let $\mathcal{B} \subset \mathbf{R}^3$ be the body of interest on the macroscopic scale which is parameterized in $\bar{x}$. Furthermore, $\bar{u}$ denotes the macroscopic displacement field. The basic kinematic and electric variables are the linear strain tensor $\bar{S}$, which is defined by the symmetric part of the displacement gradient,

$$\bar{S}(\bar{x}) := \operatorname{sym}\left[\bar{\nabla}\bar{u}(\bar{x})\right], \tag{8}$$

and the electric field vector $\bar{E}$, which is given by the negative gradient of the macroscopic scalar potential $\bar{\phi}$,

$$\bar{E}(\bar{x}) := -\bar{\nabla}\bar{\phi}(\bar{x}). \tag{9}$$

The governing field equations for the quasi-static case are the equation of equilibrium and the Gauss' law

$$\operatorname{div}_{\bar{x}}\left[\bar{T}\right] + \bar{f} = 0 \qquad \text{and} \qquad \operatorname{div}_{\bar{x}}\left[\bar{D}\right] = \bar{q} \qquad \text{in } \mathcal{B}. \tag{10}$$

Here $\bar{\nabla}$ denotes the gradient operator and $\operatorname{div}_{\bar{x}}$ the divergence operator with respect to $\bar{x}$. $\bar{T}$ represents the symmetric Cauchy stress tensor, $\bar{f}$ is the given body force, $\bar{D}$ denotes the vector of electric displacements and $\bar{q}$ is the given density of free charge carriers.

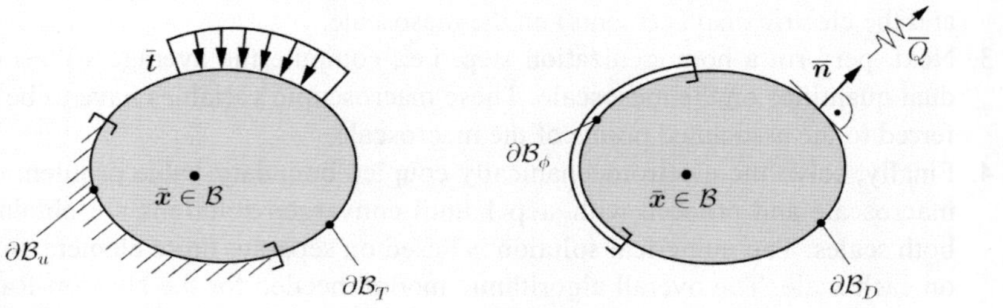

Fig. 10.22 Decomposition of the boundary of the considered body $\mathcal{B}$ into mechanical and electrical parts.

To treat the electromechanical BVP, the surface of the considered body is decomposed in mechanical parts, i.e.,

$$\partial B_u \cup \partial B_T = \partial B \qquad \text{with} \qquad \partial B_u \cap \partial B_T = \phi \tag{11}$$

and electrical parts, i.e.,

$$\partial B_\phi \cup \partial B_D = \partial B \qquad \text{with} \qquad \partial B_\phi \cap \partial B_D = \phi \tag{12}$$

The boundary conditions for the displacements and the surface tractions $\bar{t}$ are

$$\bar{u} = \bar{u}_b \quad \text{on } \partial B_u \quad \text{and} \quad \bar{t} = \bar{T} \cdot \bar{n} \quad \text{on } \partial B_T \,. \tag{13}$$

The electric potential and the electric surface charge Q are

$$\bar{\phi} = \bar{\phi}_b \quad \text{on } \partial B_\phi \quad \text{and} \quad -\bar{Q} = \bar{D} \cdot \bar{n} \quad \text{on } \partial B_D \,, \tag{14}$$

where $\bar{n}$ is a unit vector normal to the surface directed outwards from the volume. In a pure phenomenological approach we assume the existence of a thermodynamical potential and compute the associated thermodynamical quantities by the partial derivatives of the potential with respect to the basic variables. Here, we do not postulate a set of phenomenological constitutive equations. We attach a representative volume element ($\mathcal{RVE}$) to each macroscopic point $\bar{x}$ see Fig. 10.23.

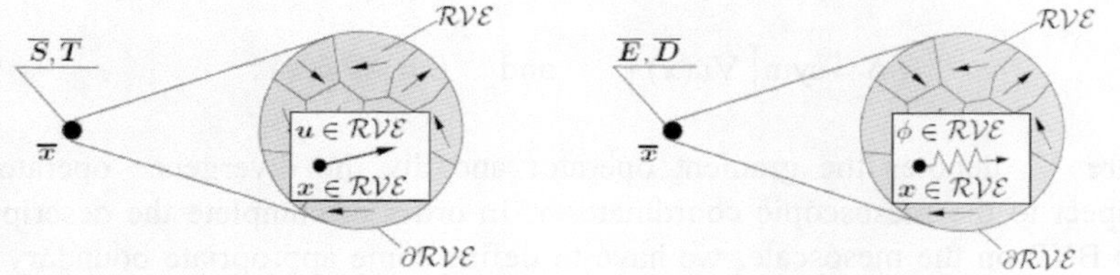

Fig. 10.23 Mesoscopic mechanic and electric variables are defined considering a representative volume element $\mathcal{RVE}$.

In order to link the macroscopic variables $\{\bar{S}, \bar{T}, \bar{E}, \bar{D}\}$ with their microscopic counterparts $\{S, T, E, D\}$ we define the macroscopic variables in this two-scale approach in terms of some suitable surface integrals over the boundary of the representative volume element with volume V. The macroscopic strains and stresses are given by

$$\bar{S} := \frac{1}{V} \int_{\partial\mathcal{RVE}} \text{sym}\left[u \otimes n\right] da \qquad \text{and} \qquad \bar{T} := \frac{1}{V} \int_{\partial\mathcal{RVE}} \text{sym}\left[t \otimes x\right] da \,, \tag{15}$$

where u and t are the displacement and traction vectors at the boundary of the $\mathcal{RVE}$, respectively. Furthermore, the macroscopic electric field and electric displacements are defined by the surface integrals

$$\overline{E} := \frac{1}{V}\int_{\partial\mathcal{RVE}} -\phi\, n\, da \quad \text{and} \quad \overline{D} := \frac{1}{V}\int_{\partial\mathcal{RVE}} -Q\, x\, da\,, \tag{16}$$

which are governed by the electric potential ϕ and the electric charge density Q on $\partial\mathcal{RVE}$.

Mesoscopic Electromechanically Coupled BVP: For the setup of the mesoscopic BVP we consider a representative volume element $\mathcal{RVE} \subset \mathbf{R}^3$ parameterized in the mesoscopic cartesian coordinates x. The governing balance equations are the balance of linear momentum

$$\operatorname{div}[T] = 0 \quad \text{in} \quad \mathcal{RVE}, \tag{17}$$

where we have neglected body forces, and Gauss' law

$$\operatorname{div}[D] = 0 \quad \text{in} \quad \mathcal{RVE}, \tag{18}$$

neglecting the density of free charge carriers. In analogy to (8) and (9) we define the mesoscopic strains and electric field vector as

$$S = \operatorname{sym}\left[\nabla u(x)\right] \quad \text{and} \quad E = -\nabla\phi(x)\,. \tag{19}$$

Here ∇ denotes the gradient operator and div the divergence operator with respect to the mesoscopic coordinates x. In order to complete the description of the BVP on the mesoscale, we have to define some appropriate boundary conditions on the boundary of the representative volume element $\partial\mathcal{RVE}$ or some constraint conditions in the whole $\mathcal{RVE}$. For this, we apply a generalized macro-homogeneity condition, which equates the macroscopic and mesoscopic power, i.e.,

$$\overline{T} : \dot{\overline{S}} + \overline{D}\cdot\dot{\overline{E}} = \frac{1}{V}\int_{\mathcal{RVE}} T : \dot{S}\, dv + \frac{1}{V}\int_{\mathcal{RVE}} D\cdot\dot{E}\, dv\,; \tag{20}$$

in this context see Hill [177]. In the following we assume a decoupling of the mechanical and electrical contribution and define

$$\mathcal{P}_1 := \frac{1}{V}\int_{\mathcal{RVE}} T : \dot{S}\, dv - \overline{T} : \dot{\overline{S}} \quad \text{and} \quad \mathcal{P}_2 := \frac{1}{V}\int_{\mathcal{RVE}} D\cdot\dot{E}\, dv - \overline{D}\cdot\dot{\overline{E}}\,. \tag{21}$$

The simplest conditions which fulfil the condition $\mathcal{P}_1 = 0$ are obtained by setting

$$T = \overline{T} = \text{const.} \qquad \text{or} \qquad \dot{S} = \dot{\overline{S}} = \text{const.} \tag{22}$$

for all points of the mesoscale; these are the well-known Reuss- and Voigt-bounds, respectively. Analogously, we obtain for $\mathcal{P}_2 = 0$ the associated conditions

$$D = \overline{D} = \text{const.} \qquad \text{or} \qquad \dot{E} = \dot{\overline{E}} = \text{const.} \tag{23}$$

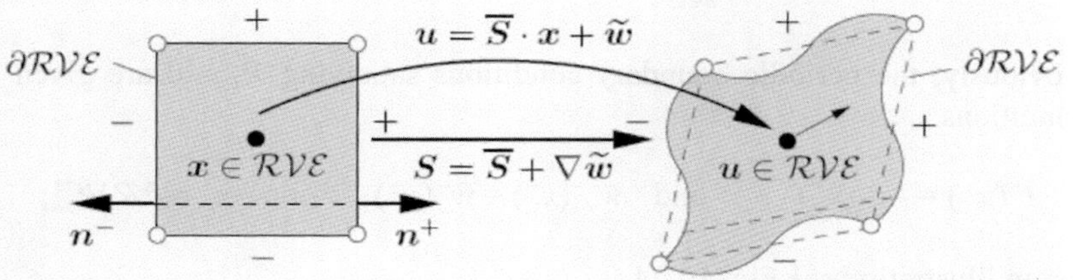

Fig. 10.24 Periodic boundary conditions on the representative volume element.

In the following we denote (22) and (23) as constraint conditions. More sophisticated expressions for the mechanical boundary conditions may be derived from the equivalent expression to $(21)_1$

$$\mathcal{P}_1 := \frac{1}{V} \int_{\partial\mathcal{RVE}} (t - \overline{T} \cdot n) \cdot (\dot{u} - \dot{\overline{S}} \cdot x)\, da\,, \tag{24}$$

where we used the Gauss theorem, the balance of linear momentum (17) and the Cauchy theorem $t = T \cdot n$. Evaluation of the latter equation leads to the Neumann- or Dirichlet-boundary conditions

$$t = \overline{T} \cdot n \quad \text{on} \quad \partial\mathcal{RVE} \qquad \text{or} \qquad \dot{u} = \dot{\overline{S}} \cdot x \quad \text{on} \quad \partial\mathcal{RVE}\,. \tag{25}$$

In order to derive periodic boundary conditions, we decompose the boundary of the representative volume element into $\partial\mathcal{RVE}^+$ and $\partial\mathcal{RVE}^-$ with the corresponding points $x^+ \in \partial\mathcal{RVE}^+$ and $x^- \in \partial\mathcal{RVE}^-$ and the condition on the outward unit normal $n^+(x^+) = -n^-(x^-)$. Furthermore, we assume that the deformation on the mesoscale consists of a constant part $\overline{S} \cdot \text{x}$ and a periodic fluctuation field $\tilde{w}$, i.e.,

$$\dot{u} := \dot{\overline{S}} \cdot x + \dot{\tilde{w}} \qquad \text{with} \qquad \tilde{w}^+(x^+) = \tilde{w}^-(x^-), \tag{26}$$

which implies

$$S = \text{sym}[\nabla u] = \overline{S} + \text{sym}[\nabla \tilde{w}]. \tag{27}$$

Inserting this relation into (24) leads, with $t^+ = T \cdot n^+$ and $t^- = T \cdot n^-$, to the expression

$$\begin{aligned} \mathcal{P}_1 &= \frac{1}{V}\left[\int_{\partial\mathcal{RVE}^+} t^+ \cdot \dot{\tilde{w}}^+ \, da + \int_{\partial\mathcal{RVE}-} t^- \cdot \dot{\tilde{w}}^- \, da\right] \\ &= \frac{1}{V}\int_{\partial\mathcal{RVE}+} (t^+(x^+) + t^-(x^-)) \cdot \dot{\tilde{w}}^+ \, da . \end{aligned} \tag{28}$$

Obviously, the periodic boundary conditions satisfying $\mathcal{P}_1 = 0$ are given by the conditions

$$t^+(x^+) = -t^-(x^-) \qquad \text{and} \quad \tilde{w}^+(x^+) = \tilde{w}^-(x^-) \qquad \text{on} \quad x^\pm \in \partial\mathcal{RVE}^\pm, \tag{29}$$

for an illustration see Fig. 10.24.

In analogy to the procedure mentioned above, possible boundary conditions are obtained for the electrical part of the boundary value problem on the mesoscale by evaluating the expression

$$\mathcal{P}_2 := \frac{1}{V}\int_{\partial\mathcal{RVE}} (Q + \overline{D} \cdot n)(\phi + \dot{\overline{E}} \cdot x)\, da, \tag{30}$$

which is equivalent to $(21)_2$. Equation (30) is derived from $(21)_2$ by using (18), the Gauss theorem and $Q = -\mathrm{D} \cdot \mathrm{n}$.

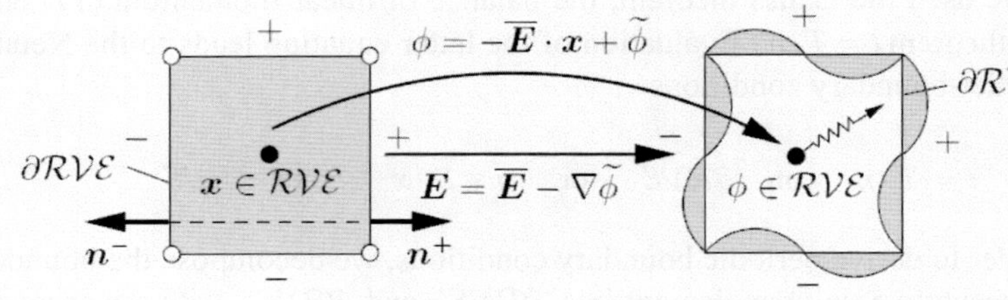

Fig. 10.25 Periodic fields for the electrical variables on the boundary of the representative volume element.

Possible Neumann- or Dirichlet-boundary conditions, obtained from (30), are

$$Q = -\overline{D} \cdot n \qquad \text{on} \quad \partial\mathcal{RVE} \qquad \text{or} \quad \phi = -\dot{\overline{E}} \cdot x \qquad \text{on} \quad \partial\mathcal{RVE}. \tag{31}$$

Now, we will use the same geometrical definitions as for the derivation of the periodic mechanical boundary conditions and assume that the electrical potential consists of a constant part $-\overline{E}\cdot x$ and a superimposed fluctuation field $\dot{\tilde{\phi}}$, i.e.

$$\dot{\phi} := -\dot{\overline{E}}\cdot x + \dot{\tilde{\phi}} \qquad \text{with} \qquad \tilde{\phi}^{+}(x^{+}) = \tilde{\phi}^{-}(x^{-}), \tag{32}$$

which implies

$$E = -\nabla\phi = \overline{E} - \nabla\tilde{\phi}. \tag{33}$$

Inserting (32) into (30) leads, with $Q^{+} = -D\cdot n^{+}$ and $Q^{-} = -D\cdot n^{-}$, to the expression

$$\mathcal{P}_2 = \frac{1}{V}\left[\int_{\partial\mathcal{RVE}^{+}} Q^{+}\cdot\dot{\tilde{\phi}}^{+}da + \int_{\partial\mathcal{RVE}-} Q^{-}\cdot\dot{\tilde{\phi}}^{-}da\right] = \frac{1}{V}\int_{\partial\mathcal{RVE}+}(Q^{+}(x^{+}) + Q^{-}(x^{-}))\cdot\dot{\tilde{\phi}}^{+}da. \tag{34}$$

Periodic boundary conditions, satisfying $\mathcal{P}_2 = 0$ are given by the conditions

$$Q^{+}(x^{+}) = -Q^{-}(x^{-}) \qquad \text{and} \qquad \tilde{\phi}^{+}(x^{+}) = \tilde{\phi}^{-}(x^{-}) \qquad \text{on } x^{\pm}\in\partial\mathcal{RVE}^{\pm}, \tag{35}$$

for an illustration see Fig. 10.25. The basic relations are summarized in Table 10.1.

10.3.2.2 Thermodynamically Consistent Framework

The key assumption of the proposed model is an additive decomposition of the strains and the electric displacements into reversible and remnant (irreversible) parts. Let us now assume the existence of a thermodynamic potential $H = \hat{H}$ (S, S^r, E, P^r), in terms of the total strains S, the remnant strains S^r, the electric field E and the polarization P^r. The evaluation of the second law of thermodynamic yields the constitutive expressions for the stresses and electric displacements and a reduced dissipation inequality remains. In order to model the dissipation process, we have to construct evolution equations for the remnant quantities. For this purpose, the existence of a switching surface Φ is assumed, where the optimization conditions $\partial_{(\bullet)}\mathcal{L}$ of the Lagrangian functional $\mathcal{L}(\tilde{T},\tilde{E},\lambda)$ lead to associated flow rules for the remnant quantities and related loading and unloading conditions (postulate of maximum remnant dissipation). The complete set of equations for the proposed model is summarized in Table 10.2.

Table 10.1 Basic equations of the two-scale homogenization procedure.

Coupling between Macro- and Microscale

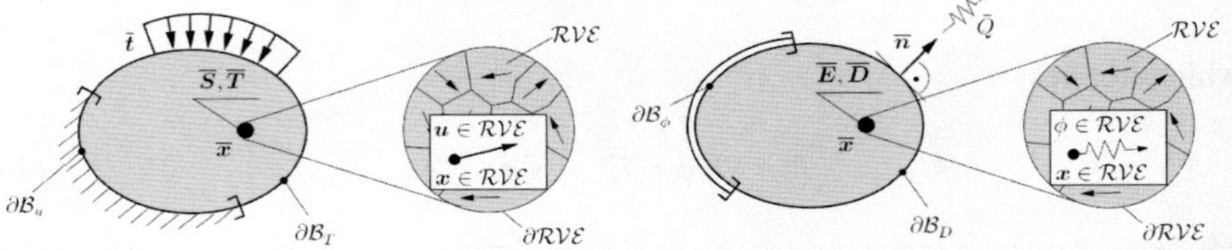

Macroscale: Electro-mechanically coupled boundary value problem

Mechanical field equations & bc's		Electrical field equations & bc's	
$\overline{S} = \mathrm{sym}[\overline{\nabla}\overline{u}]$		$\overline{E} = -\overline{\nabla}\,\overline{\phi}$	
$\mathrm{div}_{\overline{x}}\overline{T} + \bar{f} = 0$	in $\mathcal{B}$	$\mathrm{div}_{\overline{x}}\overline{D} = \overline{q}$	in $\mathcal{B}$
$\overline{u} = \overline{u}_b$	on $\partial\mathcal{B}_u$	$\overline{\phi} = \overline{\phi}_b$	on $\partial\mathcal{B}_\phi$
$\bar{t} = \overline{T}\cdot\overline{n}$	on $\partial\mathcal{B}_T$	$-\overline{Q} = \overline{D}\cdot\overline{n}$	on $\partial\mathcal{B}_D$

Definition of macroscopic field variables ($V = vol(\mathcal{RVE})$)

$$\overline{T} = \frac{1}{V}\int_{\partial\mathcal{RVE}} \mathrm{sym}[t \otimes x]\, da \qquad \overline{D} = \frac{1}{V}\int_{\partial\mathcal{RVE}} -Q\, x\, da$$

$$\overline{S} = \frac{1}{V}\int_{\partial\mathcal{RVE}} \mathrm{sym}[u \otimes n]\, da \qquad \overline{\phi} = \frac{1}{V}\int_{\partial\mathcal{RVE}} -\phi\, n\, da$$

Microscale: Electro-mechanically coupled boundary value problem

Mechanical field equations		Electrical field equations	
$S = \mathrm{sym}[\nabla u]$		$E = -\nabla\phi$	
$\mathrm{div}T = 0$	in $\mathcal{RVE}$	$\mathrm{div}D = 0$	in $\mathcal{RVE}$

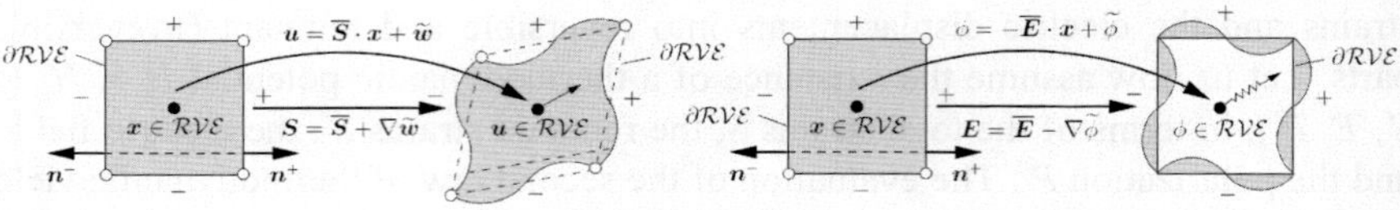

Mechanical bc's		Electrical bc's
$t = \overline{T}\cdot n$	Neumann bc's	$Q = -\overline{D}\cdot n$
$u = \overline{S}\cdot x$	Dirichlet bc's	$\phi = -\overline{E}\cdot x$
$u = \overline{S}\cdot x + \widetilde{w}$; $\widetilde{w}^+ = \widetilde{w}^-,\quad t^+ = -t^-$	periodic bc's	$\phi = -\overline{E}\cdot x + \widetilde{\phi}$; $\widetilde{\phi}^+ = \widetilde{\phi}^-,\quad Q^+ = -Q^-$

Table 10.2. Set of equations for the proposed model.

additive decomposition	$\boldsymbol{S} = \boldsymbol{S}^e + \boldsymbol{S}^r$ and $\boldsymbol{D} = \boldsymbol{D}^e + \boldsymbol{P}^r$
thermodynamic potential	$H = \hat{H}(\boldsymbol{S}, \boldsymbol{S}^r, \boldsymbol{E}, \boldsymbol{P}^r, \boldsymbol{a})$
stresses and electric displacements	$\boldsymbol{T} = \partial_{\boldsymbol{S}} H$ and $\boldsymbol{D} = -\partial_{\boldsymbol{E}} H$
conjugated stresses and electric field	$\tilde{\boldsymbol{T}} = -\partial_{\boldsymbol{S}^r} H$ and $\tilde{\boldsymbol{E}} = -\partial_{\boldsymbol{P}^r} H$
reduced dissipation inequality	$\mathcal{D} = \tilde{\boldsymbol{T}} : \dot{\boldsymbol{S}}^r + \tilde{\boldsymbol{E}} \cdot \dot{\boldsymbol{P}}^r \geq 0$
switching surface	$\Phi = \hat{\Phi}(\tilde{\boldsymbol{T}}, \tilde{\boldsymbol{E}}) \leq 0$
Lagrangian functional	$\mathcal{L}(\tilde{\boldsymbol{T}}, \tilde{\boldsymbol{E}}, \lambda) = -\mathcal{D}(\tilde{\boldsymbol{T}}, \tilde{\boldsymbol{E}}) + \lambda\phi(\tilde{\boldsymbol{T}}, \tilde{\boldsymbol{E}})$
flow rules for remanent quantities	$\dot{\boldsymbol{S}}^r = \lambda \partial_{\tilde{\boldsymbol{T}}} \Phi = \lambda\, \tilde{\boldsymbol{m}}$ and $\dot{\boldsymbol{P}}^r = \lambda \partial_{\tilde{\boldsymbol{E}}} \Phi = \lambda\, \tilde{\boldsymbol{n}}$
loading/unloading conditions	$\lambda \geq 0;\ \ \Phi \leq 0$ and $\lambda\Phi = 0$

In order to regard the anisotropic material behaviour, we use the representation theorems of isotropic tensor functions for the explicit formulation of the invariant constitutive equations. For a general introduction, see, e.g., Boehler [178] and for electromechanically coupled problems we refer to Schröder and Gross [173]. In the following we will focus on transversely isotropic materials and therefore introduce the preferred direction a, with $||a|| = 1$, as an additional argument within the set of variables of the thermodynamic potential, whereas the associated symmetry group for the considered polar material is

$$\mathcal{G} = \{Q \in \mathcal{O}(3),\ Qa = a\}. \tag{36}$$

This yields the following representation of the electric enthalpy function

$$H(S, S^r, E, P^r, a) = H(QSQ^T, QS^rQ^T, QE, QP^r, Qa)\ \forall\ Q \in \mathcal{O}(3), \tag{37}$$

which is the definition of an isotropic tensor function with respect to the whole set of arguments $\{S, S^r, E, P^r, a\}$, see Romanowski and Schröder [171], Schröder and Romanowski [174].

In order to set up a specific model problem for a transversely isotropic material, the thermodynamic potential is defined by the following five terms

$$H = H_1(S, S^r, a) + H_2(E, a) + H_3(S, S^r, E, P^r, a) + H_4(E, P^r, a) + H_5(P^r, a), \tag{38}$$

governed by the set of invariants of interest

$$\begin{gathered} I_1 := \text{trace}\left[S - S^r\right],\quad I_2 := \text{trace}\left[(S - S^r)^2\right],\quad I_4 := \text{trace}\left[(S - S^r)(a \otimes a)\right], \\ I_5 := \text{trace}\left[(S - S^r)^2(a \otimes a)\right],\quad J_1 := \text{trace}\left[(E \otimes E)\right],\quad J_2 := \text{trace}\left[(E \otimes a)\right], \\ K_1 := \text{trace}\left[(S - S^r)(E \otimes a)\right],\quad \bar{N}^P := P^r \cdot a. \end{gathered} \tag{39}$$

The first term in (38) is a purely mechanical part and is chosen as a quadratic function in $S - S^r$

$$H_1 = \frac{1}{2}\lambda I_1^2 + \mu I_2 + \alpha_1 I_5 + \alpha_2 I_4^2 + \alpha_3 I_1 I_4 \ , \tag{40}$$

the set of coefficients $\{\lambda, \mu, \alpha_1, \alpha_2, \alpha_3\}$ specifies the mechanical material properties. For the purely electrical part with the corresponding material properties $\{\gamma_1, \gamma_2\}$ we consider a quadratic function in E, given by

$$H_2 = \gamma_1 J_1 + \gamma_2 J_2^2 \ , \tag{41}$$

whereas the coupling between the mechanical and electrical part, denoted as the piezoelectricity part, is set to

$$H_3 = \left[\beta_1 I_1 J_2 + \beta_2 I_4 J_2 + \beta_3 K_1\right] \frac{1}{P_s} \bar{N}^P \tag{42}$$

with the piezoelectric material properties $\{\beta_1, \beta_2, \beta_3\}$. It becomes apparent that this coupling effect is assumed to increase with an increasing remnant polarization until a saturation value, which is given by the maximum achievable polarization P_s, is reached. The terms H_4 and H_5 with

$$H_4 = -J_2 \bar{N}^P \ \text{and} \ H_5 = f(\bar{N}^P) \tag{43}$$

take the remnant polarization of the material into account, where the function $f(\bar{N}^P)$ governs the form of the dielectric hysteresis curve:

$$f(\bar{N}^P) = \frac{1}{c}\left[\bar{N}^P \ \text{Artanh}(\frac{\bar{N}^P}{P_S}) + \frac{1}{2} P_S \ln(1 - (\frac{\bar{N}^P}{P_S})^2)\right]. \tag{44}$$

For simplicity we express the enthalpy function H in terms of the elements of the polynomial basis

$$H(S, S^r, E, P^r, a) = H(I_1, I_2, I_4, I_5, J_1, J_2, K_1, \bar{N}^P) =: H(L_i | i = 1, \ldots 8) \ , \tag{45}$$

which is invariant under all transformations $Q \in \mathcal{O}(3)$. The explicit form of the stresses and electric displacements appear as

$$T = \frac{\partial H}{\partial S} = \sum_{i=1}^{8} \frac{\partial H}{\partial L_i} \frac{\partial L_i}{\partial S} \quad \text{and} \quad D = -\frac{\partial H}{\partial E} = -\sum_{i=1}^{8} \frac{\partial H}{\partial L_i} \frac{\partial L_i}{\partial E}, \tag{46}$$

in detail we obtain

$$
\begin{aligned}
T &= (\lambda I_1 + \alpha_3 I_4)1 + 2\mu(S - S^r) + \alpha_1[a \otimes (S - S^r)a \\
&+ a(S - S^r) \otimes a] + (2\alpha_2 I_4 + \alpha_3 I_1)a \otimes a \\
&+ [\beta_1 J_2 1 + \beta_2 J_2 a \otimes a \frac{1}{2}\beta_3(E \otimes a + a \otimes E)]\frac{1}{P_S}\bar{N}^P,
\end{aligned} \tag{47}
$$

$$
D = -2\gamma_1 E - 2\gamma_2 J_2 a - [(\beta_1 I_1 + \beta_2 I_4)a + \beta_3 a(S - S^r)]\frac{1}{P_s}\bar{N}^P + P^r.
$$

Here $P^r = -\partial H_4 / \partial E = \bar{N}^P a$ is the remnant polarization with respect to the polarization axis.

In order to describe the evolution of the remnant variables, the existence of a dissipation potential is assumed. This is expressed as a continuous, convex scalar-valued function of the flux variables $\dot{S}^r$ and $\dot{P}^r$. In this context see, e.g., Landis [99] and McMeeking and Landis [98]. Applying a Legendre-Fenchel transformation, leads to a corresponding potential that may be formulated in terms of the dual quantities. Let us now introduce a switching surface Φ in terms of the dual variables $\tilde{T}$ and $\tilde{E}$, with

$$
\Phi(\tilde{T}, \tilde{E}) \le 0 . \tag{48}
$$

By applying the principle of maximum remnant dissipation, a generalization of the principle of maximum dissipation, we construct the Lagrangian functional

$$
\mathcal{L}(\tilde{T}, \tilde{E}, \lambda) = -\mathcal{D}(\tilde{T}, \tilde{E}) + \lambda\Phi(\tilde{T}, \tilde{E}) \tag{49}
$$

with the Lagrange multiplier λ. The optimization conditions

$$
\partial_{\tilde{T}}\mathcal{L} = 0, \quad \partial_{\tilde{E}}\mathcal{L} = 0, \quad \partial_{\lambda}\mathcal{L} = 0 \tag{50}
$$

lead to the associated flow rules of the remnant variables

$$
\dot{S}^r = \lambda\partial_{\tilde{T}}\Phi(\tilde{T}, \tilde{E}) \text{ and } \dot{P}^r = \lambda\partial_{\tilde{E}}\Phi(\tilde{T}, \tilde{E}) \tag{51}
$$

and the loading/unloading conditions $\lambda \ge 0, \Phi(\tilde{T}, \tilde{E}) \le 0$ and $\lambda\Phi(\tilde{T}, \tilde{E}) = 0$. It should be noted that the normality rule is sufficient to satisfy the second law of thermodynamics. Following McMeeking and Landis [98], we simplify our set of equations by introducing the constitutive relation

$$
S^r = \frac{3S_a^r}{2P_S^2}\operatorname{dev}(P^r \otimes P^r) \tag{52}
$$

for the remnant strains, where S_a^r is associated with the maximum achievable remnant strain due to polarization in direction of the polarization axis. This quadratic relationship between the remnant polarization and strains is a commonly reasonable assumption, when the electric fields are strong and the stresses are small, see, e.g., Jaffe *et al.* [3]. In this simple model we choose

$$\Phi = (J_2 - E^B)^2 - E_C^2 \leq 0 \quad \text{with} \quad E^B := \partial_{\bar{N}^P} H_3 + \partial_{\bar{N}^P} H_5, \tag{53}$$

as a switching criterion, where E_c denotes the coercive field strength.

10.3.2.3 Two-Scale Transition Procedure Based on Discrete Orientation Distribution Functions

The simplest two-scale transition procedure of the electromechanically coupled BVP is based on applying the constraint conditions (22) and (23). In the following we set

$$S = \overline{S} = \text{const.} \quad \text{and} \quad E = \overline{E} = \text{const.} \tag{54}$$

for all points of the mesoscale, see Schröder *et al.* [174, 175].

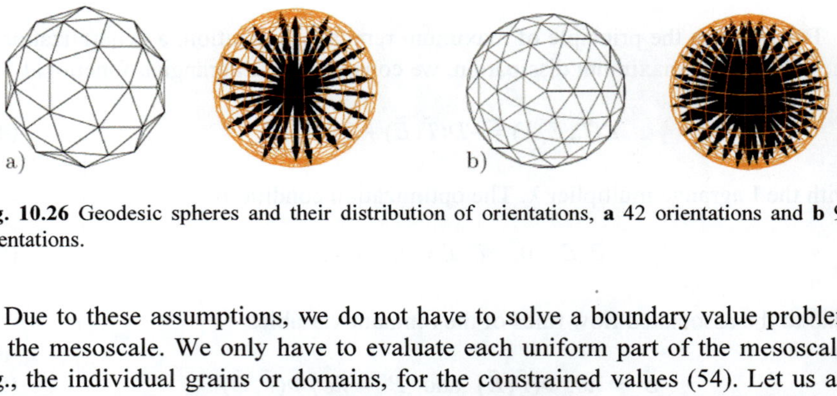

Fig. 10.26 Geodesic spheres and their distribution of orientations, **a** 42 orientations and **b** 92 orientations.

Due to these assumptions, we do not have to solve a boundary value problem on the mesoscale. We only have to evaluate each uniform part of the mesoscale, e.g., the individual grains or domains, for the constrained values (54). Let us assume that the mesoscale is represented by a distribution of preferred directions $a_i|i=1,\ldots,n$. This seems to be a suitable assumption, because the chosen model is transversely isotropic. In order to start with a discrete approximation of a uniform spatial distribution of the preferred directions, we use a partitioning of the sphere surface into parts of equal areas. An innovative treatment of this problem was given by Richard Buckminster Fuller, who separated the surface of the sphere into equilateral congruent triangles; constructions of this type are known as geodesic spheres or geodesic domes. We use this consistent segmentation of the unit sphere and assign each node of the triangles one preferred direction. Fig. 10.26 shows two geodesic spheres with n = 42 and n = 92 nodes and the corresponding distribution of preferred directions.

Based on these assumptions we may now compute the macroscopic stresses and electric displacements by

$$\bar{T} = \frac{1}{n}\sum_{i=1}^{n} T(\bar{S}, S_i^r, \bar{E}, P_i^r, a_i) \quad \text{and} \quad \bar{D} = \frac{1}{n}\sum_{i=1}^{n} D(\bar{S}, S_i^r, \bar{E}, P_i^r, a_i) \tag{55}$$

where we assume an identical volume fraction associated to each orientation. Of course, the evaluation of the constitutive laws for the stresses and electric displacements for the n orientations induces a huge set of history variables as indicated in the bracketed terms in (55). Furthermore, we need the consistent linearization of the governing equations for the treatment of the coupled boundary value problems within the Finite-Element-Method. Let the linear increment of the stresses and electric displacements on the mesoscale for an individual orientation be given by

$$\begin{aligned} \Delta T &= \mathbb{C}_{algo} : \Delta S - e_{algo}^{T} \cdot \Delta E, \\ \Delta D &= e_{algo} : \Delta S + \varepsilon_{algo} \cdot \Delta E. \end{aligned} \tag{56}$$

For the numerical solution of the macroscopic boundary value problem, we need the associated algorithmic expressions

$$\begin{aligned} \Delta \bar{T} &= \bar{\mathbb{C}}_{algo} : \Delta \bar{S} - \bar{e}_{algo}^{T} \cdot \Delta \bar{E}, \\ \Delta \bar{D} &= \bar{e}_{algo} : \Delta \bar{S} + \bar{\varepsilon}_{algo} \cdot \Delta \bar{E}, \end{aligned} \tag{57}$$

with the overall consistent moduli

$$\bar{\mathbb{C}}_{algo} = \frac{1}{n}\sum_{i=1}^{n} \mathbb{C}_{algo}(\ldots, a_i),\ \bar{e}_{algo} = \frac{1}{n}\sum_{i=1}^{n} e_{algo}(\ldots, a_i),\ \bar{\varepsilon}_{algo} = \frac{1}{n}\sum_{i=1}^{n} \varepsilon_{algo}(\ldots, a_i).$$

Details for the numerical treatment, especially of the consistent linearization, are given in Schröder and Romanowski [174].

10.3.2.4 Numerical Examples

The previous sections are concerned with the meso-macro transition of ferroelectric materials. To clarify the macroscopic procedures, a homogeneous and an inhomogeneous electromechanical boundary value problem are discussed. The material parameters used in these examples are chosen in accordance to the one found in Jaffe *et al.* [3] for single crystal barium titanate. The material parameters for the elastic stiffness tensor are set to

$$\mathbb{C}_{11} = 166, \qquad \mathbb{C}_{12} = 76.6, \qquad \mathbb{C}_{13} = 77.5, \qquad \mathbb{C}_{33} = 162 \qquad \mathbb{C}_{44} = 42.9 \tag{58}$$

in units of 10^3N/mm^2. The components of the piezoelectric tensor are chosen to

$$e_{31} = -4.4\ ,\ e_{33} = 18.6\ ,\ e_{15} = 11.6 \tag{59}$$

in units of 10^{-3}N/Vmm. The parameters for the dielectric tensor are set to

$$\varepsilon_{11} = 1260 \cdot \varepsilon_0 = 1.12\ ,\ \ \varepsilon_{33} = 1420 \cdot \varepsilon_0 = 1.26 \tag{60}$$

in units of 10^{-11}C/Vmm, where $\varepsilon_0 = 8.854 \cdot 10^{-15}$C/Vmm is the permittivity of free space. At a temperature of approximate 25°C one can find the following corresponding values for the maximum achievable polarization $P_s = 26 \cdot 10^{-8}$C/mm^2 and the coercive field E_c = 1000 V/mm. For the maximum remnant strain along the polarization direction that is introduced in Equation (52) we choose $S_a^r = 0.001$. For the use of these material parameters in the coordinate-invariant formulation, we have to convert the elastic, piezoelectric and dielectric parameters. A comparison of the parameters used for the invariant formulation (47) with the coordinate-dependent formulation, see Schröder and Gross [173], leads, for the mechanical properties, to the following expressions

$$\begin{aligned} &\lambda = \mathbb{C}_{12},\quad \mu = \frac{1}{2}(\mathbb{C}_{11} - \mathbb{C}_{12}),\ \alpha_1 = 2\mathbb{C}_{44} + \mathbb{C}_{12} - \mathbb{C}_{11}, \\ &\alpha_2 = \frac{1}{2}(\mathbb{C}_{11} + \mathbb{C}_{33}) - 2\mathbb{C}_{44} - \mathbb{C}_{13},\quad \alpha_3 = \mathbb{C}_{13} - \mathbb{C}_{12}, \end{aligned} \tag{61}$$

for the electromechanical coupling properties we obtain

$$\beta_1 = -e_{31},\ \beta_2 = -e_{33} + 2e_{15} + e_{31},\ \ \beta_3 = -2e_{15} \tag{62}$$

and the dielectric properties which describe the purely electrical material behaviour are given by

$$\gamma_1 = -\varepsilon_{11}/2, \qquad \gamma_2 = \frac{1}{2}(\varepsilon_{11} - \varepsilon_{33}). \tag{63}$$

Homogeneous Boundary Value Problem: The first example shows an electromechanical boundary value problem consisting of a piezoelectric specimen that is loaded with an alternating electric field as depicted in Fig. 10.27. As indicated, two distinct initial mesoscopic states are considered. On the one hand, we have a not pre-polarized sample and on the other hand, we consider a material that is pre-polarized in two opposite directions. Both are – and that is an essential condition in this context – initially not polarized in a macroscopic manner, which is a characteristic property of ferroelectric ceramics after sintering. However, as can be seen in Fig. 10.27 the two macroscopically equivalent samples behave differently.

In cases in which we have a not pre-polarized material, a hysteretic behaviour, consistent to the case with one preferred direction, may be observed. The first loading path is linear until the coercive field strength is reached, which is a typical observation in connection to phenomenological models which only consider one preferred direction for the whole specimen. On the other side, where we have a material with two oppositional preferred directions that are pre-polarized, the material response is different. Here, we observe a characteristic first loading path, both for the dielectric and the butterfly hysteresis curve. This is an essential characteristic for virgin ferroelectric samples, which are macroscopically not piezoelectric and which are polarized within a certain range of electric field. The reason for that behaviour lies in the mesoscopic setup of the boundary value problem. There we have two distinct mesoscopic configurations with opposite polarizations, from which one switches due to the outer electric field and the other one does not. The one that is pointing oppositional to the applied electric field is switching in the range of the coercive field strength.

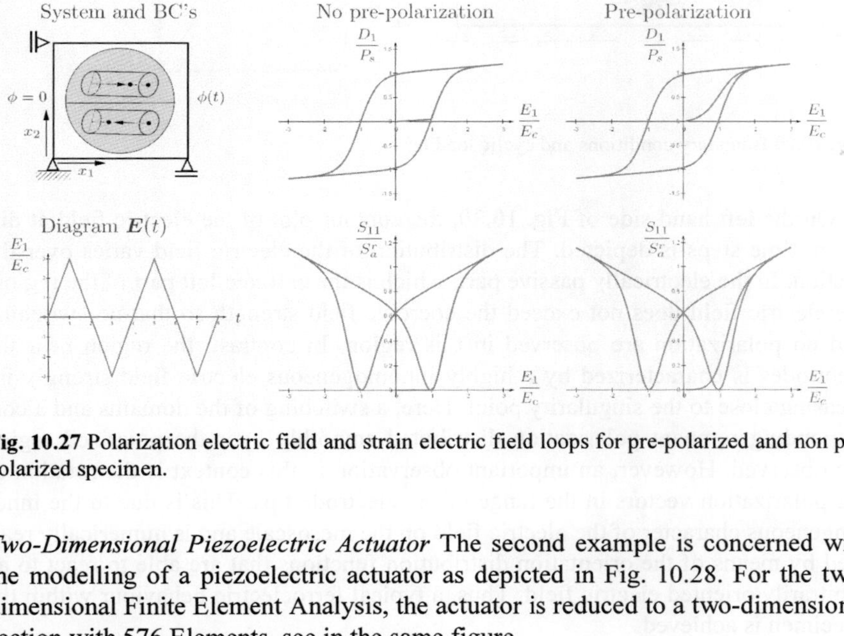

Fig. 10.27 Polarization electric field and strain electric field loops for pre-polarized and non pre-polarized specimen.

Two-Dimensional Piezoelectric Actuator The second example is concerned with the modelling of a piezoelectric actuator as depicted in Fig. 10.28. For the two-dimensional Finite Element Analysis, the actuator is reduced to a two-dimensional section with 576 Elements, see in the same figure.

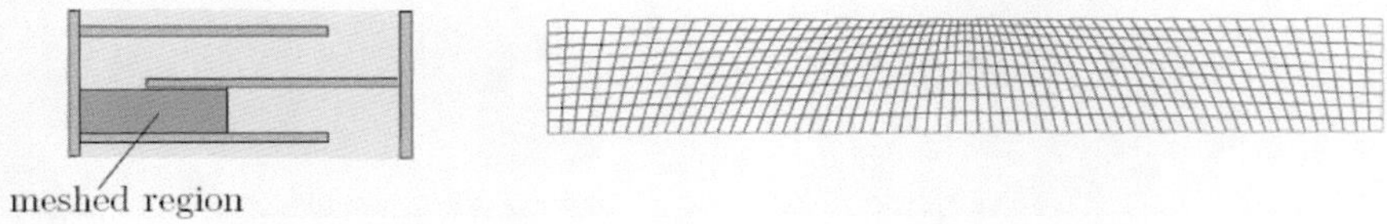

Fig. 10.28 Symmetric section of an actuator and Finite-Element discretization.

The dimensions of the meshed region were taken in adaption to Kamlah and Böhle [179]. To be precise, this means that a region with a height of $h = 57.5\mu m$ and lengths $l_1 = l_2$ given by $l_1 = l_2 = 205\mu m$ is discretized. The poling process is modelled by application of a cyclic electric potential with a maximum value of $\phi_{max} = \pm 333V$, which is sufficient for the polarization of the material in the electromechanically active part. The boundary conditions in terms of load factor over time are depicted in Fig. 10.29. A spatial distribution of 42 orientations is assigned to each Gauss Point of the elements, which is indicated in the same figure.

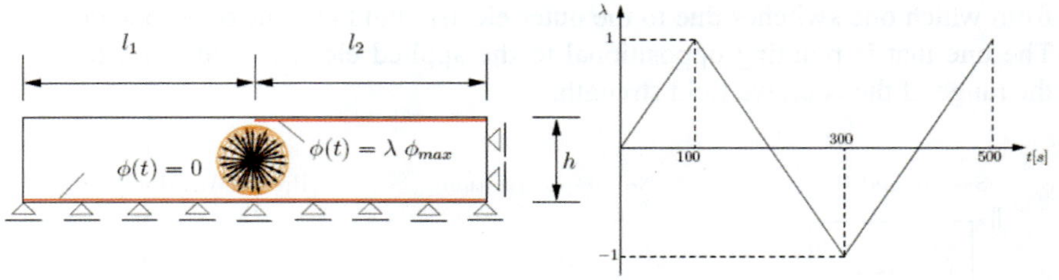

Fig. 10.29 Boundary conditions and cyclic loading.

On the left hand side of Fig. 10.30, the contour plot of the electric field at different time steps is depicted. The distribution of the electric field varies over the section. In the electrically passive part, which is the extreme left part of the region, the electric field does not exceed the coercive field strength so that no switching and no polarization are observed in this region. In contrast, the region near the electrodes is characterized by a highly inhomogeneous electric field strongly increasing close to the singularity point. Here, a switching of the domains and a corresponding remnant polarization distributed smoothly over the polarized region are observed. However, an important observation in this context is the rotation of the polarization vectors in the range of the electrode tips. This is due to the inhomogeneous character of the electric field on the mesoscale and is numerically realized by means of the orientation distribution functions that are able to react to an arbitrarily oriented electric field. Thus, a typical ferroelectric behaviour within the specimen is achieved.

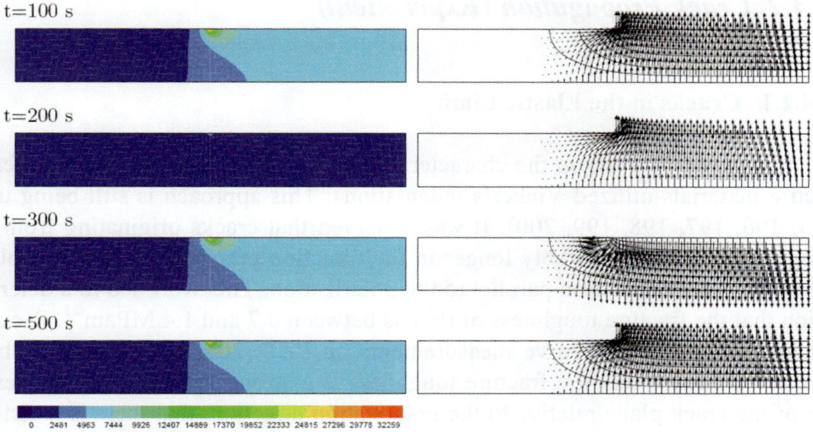

Fig. 10.30 $|E|$ and $\overline{P}^r$ at different time steps of the computation.

10.4 Mechanical Failure

10.4.1 Crack Origins in Devices

Stress and electric field concentrators in ferroelectric materials may lead to premature failure [180, 181]. Field concentrators are regions of inhomogeneities in the electrical or mechanical properties, and may take several forms. They may be intentional parts of the structure such as electrode edges, or unwanted defects like cracks, voids, porosity, surface finish, and conducting paths. A common structure for co-fired multilayer piezoelectric actuators has the electrodes of alternate layers terminate within the device. This design results in field concentrations at the electrode edges and carries with it the potential for various types of processing flaws. Since the work of Winzer *et al.* [182], many researchers have addressed cracking associated with internal electrode edges [183, 184, 185, 57, 186, 187, 188, 189, 61, 190, 191, 192, 193, 194]. Scratches associated with surface finish from machining ferroelectric parts may act as nucleation sites for cracks. Porosity or voids within the material may result in large electric field concentrations within the void. This leads to large stresses in the ferroelectric material around the pores and may lead to dielectric breakdown within the pores. Surface electrode material or other charge carriers (like OH^-) may migrate into the material and develop short circuit paths between the electrodes. Each of these phenomena has been described within a fracture mechanics framework.

10.4.2 Crack Propagation (Experiment)

10.4.2.1 Cracks in the Elastic Limit

Much of the early work on the characterization of the fracture behaviour of ferroelectric materials utilized Vicker's indentations. This approach is still being used [195, 196, 197, 198, 199, 200]. It was observed that cracks originating from the indentation were considerably longer in the direction perpendicular to the polarization than in the direction parallel to the polarization. This work led to a determination that the fracture toughness of PZT is between 0.7 and 1.4 $MPam^{0.5}$. R-curve measurements and v-K curve measurements on PZT [201, 202, 203, 204] have provided evidence that the fracture toughness is a strong function of the orientation of the crack plane relative to the polarization direction, that there is a distinct toughening effect associated with polarization reorientation in the crack tip field, and that sub-critical crack growth is enhanced by the presence of moisture in the surrounding environment (Fig. 10.31. right).

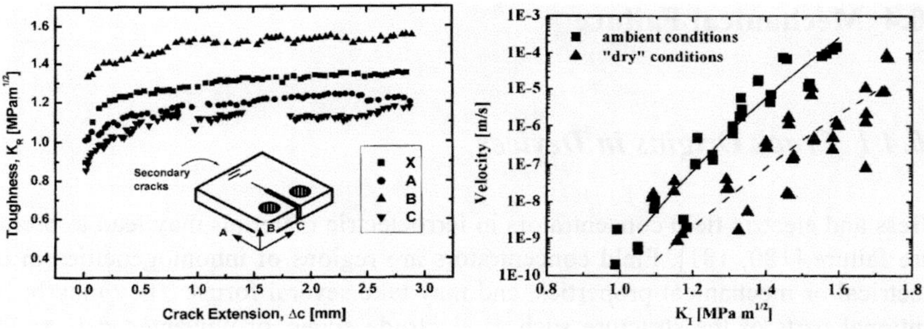

Fig. 10.31 (left) R-curves for PZT specimens in three different poling directions are contrasted to an unpoled specimen in a compact tension test geometry. The secondary cracks formed for the A-direction poled specimens are indicated, X denotes unpoled. The C-direction is weakest, because the electrical poling has already oriented the domains in the direction of the tensile stresses at the crack tip prior to mechanical loading itself. B yields a better toughening effect than A, because the flat sample face permits plastic deformation up to the open upper surface of the sample (plane stress boundary condition). For A-poled samples this ferroelastic toughening is constrained by the sample extension (long edge 50 mm) and the effective process zone is smaller. This geometry is closer to plane strain boundary conditions [206]. For details of the B-poled case see section 10.4.2.2. (right) Crack velocity curves (v-K) for unpoled PZT CT specimens exposed to ambient laboratory conditions (T = 23°C and RH = 35%) and dry-air conditions (T = 23°C and RH ≈ 0.02%). The dry-air specimens and one ambient specimen were heat-treated at 400°C before they were tested. Regression curves were fitted to the data [205] © *Am. Ceram. Soc. (Blackwell).*

R-curve behaviour (increase of fracture resistance if the energy approach is used, or equivalently increase of toughness with crack propagation) in ferroelectric materials is the result of energy dissipation in a process zone. The process zone is the region near the tip of the crack where the stress and electric field are large enough to drive hysteretic processes. In a metal this is typically dislocation motion. In a ferroelectric material it is polarization reorientation. The difference between these two mechanisms is that dislocations can nucleate and move in a manner that generates large scale deformation. Polarization reorientation on the other hand may only facilitate deformation as large as the change in spontaneous strain. This is hindered by induced internal electric fields if there is no charge conduction in the material. The hysteretic strain taking place in the process zone reduces the stress intensity factor at the crack tip, shielding it from some of the applied stress intensity. An increasing load must be applied to obtain the next increment of crack growth. This results in an increase of the hysteretic strain and a further increase of the required load to further propagate the crack. This process saturates after a short amount of crack growth. Although toughening in metals is of great technological importance, it does not yet appear to be of much use in ferroelectric materials. The toughening effect is a strong function of the polarization direction, see Fig. 10.31 left [206], and the material tends to be reset to the bottom of the R-curve each time the load is removed [14]. One must therefore design devices for the intrinsic toughness (around 0.6 $MPam^{0.5}$).

Creep crack growth is another important phenomenon that takes place in ferroelectric materials. This may be described in terms of crack velocity versus applied stress intensity factor curves (v-K curves) [202, 205, 203]. Experimental measurements of v-K behaviour indicate that creep crack growth is enhanced by the presence of moisture. Furthermore, the moisture may be readily absorbed from the atmosphere by PZT over a period of a day or two. This is apparent as a gradual darkening of the colour over time. Oates, *et. al.* [205] found that baking the material just above the boiling point of water drives off this moisture and reduces the creep crack growth behaviour.

Recent work has focused on crack propagation in relaxor ferroelectric single crystals. The initial approach of placing a Vicker's indentation in the side of a crystal and cycling a bipolar electric field at twice the coercive field, while watching the indentation induced crack, gave a surprising result. There was little or no growth of the crack, but crack systems formed at the edges of the crystal. The reason for this was not entirely clear, but there was apparently some kind of mechanical incompatibility that led to the cracking. Fig. 10.33 shows an example of this kind of cracking.

In an attempt to get to the reason for the cracking observed at the right in figure 10.32, several specimens of PZN-0.045PT were polished to an optical finish and an electric field cycled. Once some cracking was observed at the corners of the specimen, the electric field was stopped and the specimen viewed in an optical microscope. When looking in through the sides, the cracks were apparent, but no domain structure could be seen. Next, the electrodes were polished and the specimens viewed from the top. The results are shown in Fig. 10.33. There is an interesting interaction between the observed domain structure and the observed cracks.

Single edge v-notched beams (SEVNB) have been used to measure R-curves in single crystals with the notch oriented in the [001] direction. In these tests the crack at the head of the notch tended to either propagate straight ahead, or at 45 degrees to the plane of maximum tensile stress. The lower stress intensity necessary to drive the cracks at 45 degrees seen in Fig. 10.34 suggests that this represents a cleavage plane in the single crystal.

Fig. 10.32 Cyclic field induced damage in a single crystal of 0.9PZN-0.1PT. Left image shows an indentation with no response to a bipolar cyclic electric field of frequency=20 Hz, amplitude=1.5 MV/m. Right image taken at the bottom right corner of the specimen showing electric field induced damage (picture height = 1mm).

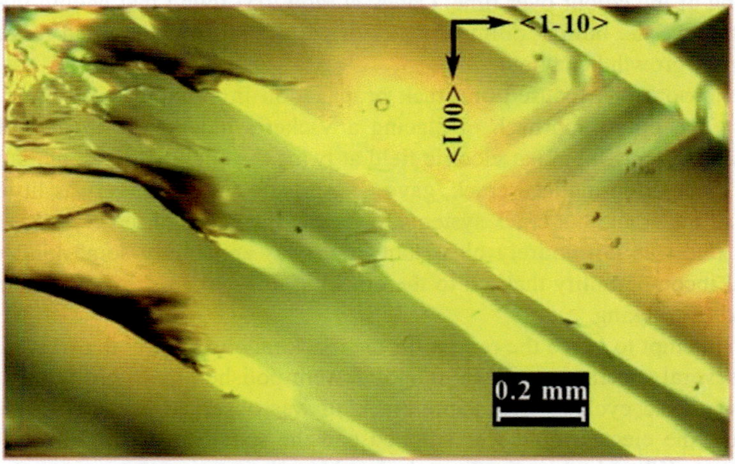

Fig. 10.33 Domains interacting with cracks at the corner of a PZN-0.045PT specimen. The electrodes were polished off and the domains are being viewed with transmitted light in the poling direction.

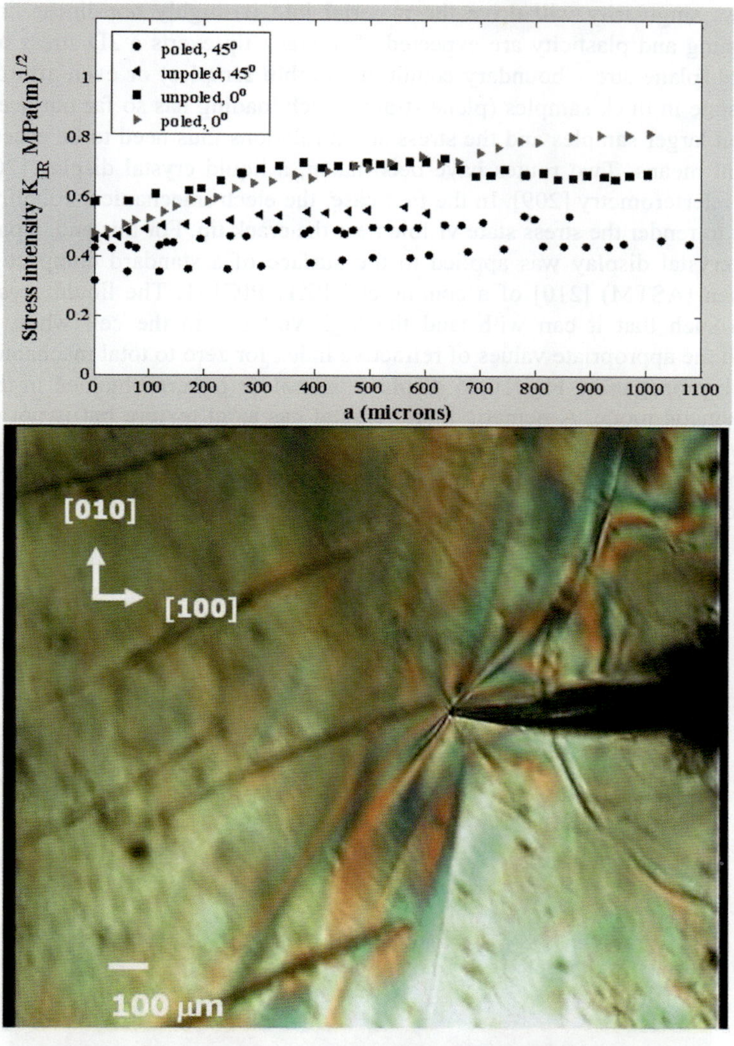

Fig. 10.34 R-curves measured for SEVNB PZN-0.045PT specimens shown (top). Domains are visible at the tip of the zero degree crack (bottom). Figure and image from [207] © *Am. Ceram. Soc. (Blackwell)*.

10.4.2.2 Process Zones at Crack Tips

A ferroelectric ceramic is a highly non-linear material. The crack tip as a location of stress singularity will drive the material into its highly non-linear limit and toughening and plasticity are expected. The crack tip exerts a 2D stress onto the material (plane stress boundary conditions in thin samples) or even affords a 3D stress state in thick samples (plane strain). Such loading has so far not been exercised on larger samples and the stress strain relations thus need to be accessed by different means. Two routes have been taken, a liquid crystal display [208] and Moiré interferometry [209]. In the first case, the electromechanical coupling itself is used to render the stress state visible near the crack tip. For this end, a particular liquid crystal display was applied to the surface of a standard compact tension specimen (ASTM) [210] of a commercial PZT: PIC151. The liquid crystal was chosen such that it can withstand the high voltages in the cell while passing through the appropriate values of refractive index for zero to total mechanical load at crack propagation. Fig. 10.35 displays the colour pattern obtained in the standard nematic mode. A nematic liquid crystal has axial texture but is not ordered otherwise. The texture is induced by an ordered surface morphology here parallel scratches in a thin layer of wax. For more details refer to Waser [211] or other textbooks on LCDs.

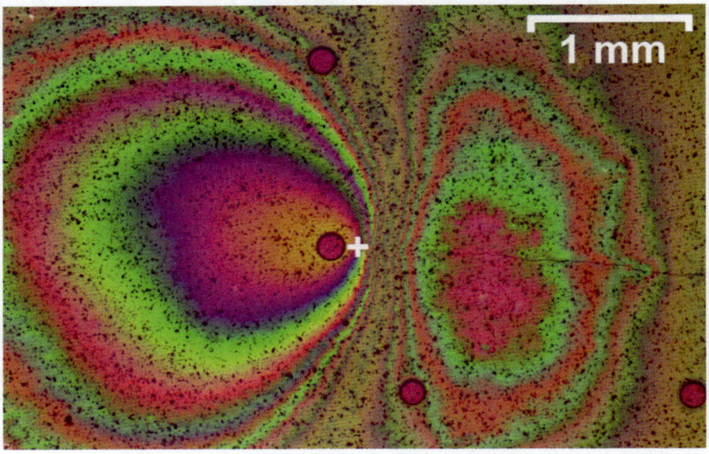

Fig. 10.35 A liquid crystal displaying lines of equal surface potential on a mechanically loaded compact tension specimen of a thickness-poled soft ferroelectric lead zirconate titanate (PIC 151). The surface potential is the thickness integral of all local piezoelectric effects or depolarization switching. The crack tip is marked by the white cross (Lupascu) ©*Am. Ceram. Soc. (Blackwell)*.

Using another LCD mode permits the assignation of actual numerical values to the surface potential. The twisted nematic mode rotates the polarization plane of the optical wave. For crossed optical polarizer orientations, the reflected light is blocked. For low fields across the LCD the image is black. Once the potential difference across the LCD exceeds a critical value the twist mode breaks down, and intermediate polarization rotation values are obtained, again as colour fringes. The edge determines a well reproducible potential value. If the transparent counter electrode at the opposite side of the LCD is addressed by an external voltage, a true surface potential value may be scanned across the entire surface of the sample. This technique was used to determine the edge of the ferroelastic process zone as the line of deviation from the $1/\sqrt{r}$ -type singularity of the elastic K-field at the crack tip. Form and size of the process zone as well as a toughening exponent became accessible through this technique [87].

The shape and effect of the process zones under monotonic loading have thus been clarified. Figure 10.36 displays the growth of the process zone for increasing applied K_I.

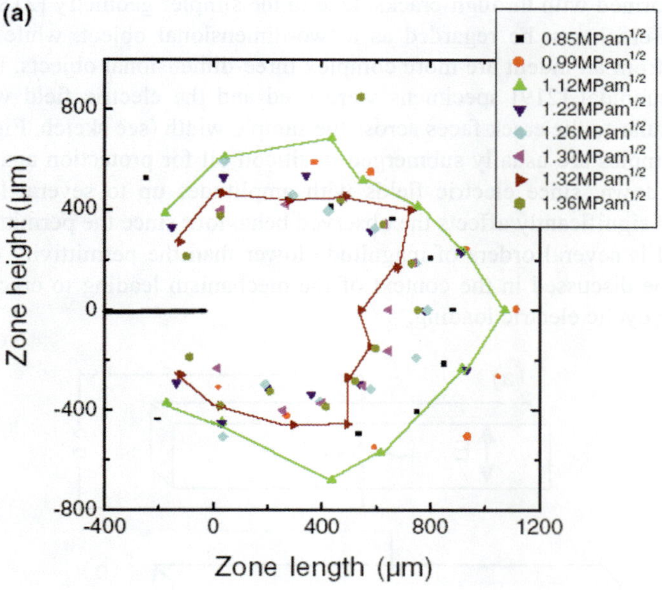

Fig. 10.36 Contours of the boundary to the switching zone as determined from the liquid crystal technique. The points mark the onset of nonlinearity when approaching the crack tip in a compact tension sample. In order to achieve a surface potential value from the LCD image the twisted liquid crystal technique has to be employed. A certain colour line can be assigned to a breakdown of the optical twist mode across the LCD. The size and shape of the process zone as well as the hardening exponent became accessible by this technique [87] ©*Elsevier*.

10.4.2.3 Cyclic Crack Growth

A cyclic mechanical or electrical load may cause cyclic crack propagation in a ferroelectric ceramic under certain conditions. For applications like actuators, it is important to know and understand these conditions to avoid them and thereby avoid failure of devices. Most work in this field has been conducted on crack growth under cyclic electrical loading which will therefore be reviewed in the first and largest part of this section. Work done on mechanical and combined electro-mechanical loading will be described thereafter.

Specimen Geometry, Materials and Setup

In the first works conducted on crack growth under cyclic electric loading [212, 197, 213, 214, 215, 216, 217], beam-like specimens with Vickers indents on one polished face were used. The width of the samples was usually of the order of several millimetres and the cyclic electric field was applied between the two faces perpendicular to the indented one (see sketch, Fig. 10.37 a). Later, experiments were performed with through-cracks. Due to the simpler geometry [218, 219, 220] a through-crack can be regarded as a two-dimensional object, while the cracks resulting from an indent are more complex three-dimensional objects. Beam [218, 220] or cuboidal [219] specimens were used and the electric field was applied perpendicular to the crack faces across the sample width (see sketch, Fig. 10.37 b)

The samples are usually submerged in silicon oil for protection against dielectric breakdown, since electric fields with amplitudes up to several kV/mm are used. This significantly affects the observed behaviour since the permittivity of the silicon oil is several orders of magnitude lower than the permittivity of the PZT and will be discussed in the context of the mechanism leading to crack propagation under cyclic electric loading.

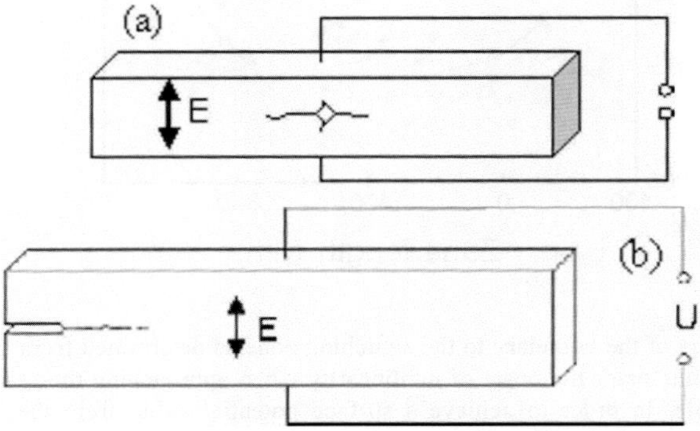

Fig. 10.37 a Geometric sketch of specimens and the applied field used in the early works on electric cycling. **b** Sketch of the specimens with through-cracks that were cycled.

The materials employed in different studies comprise compositions of PZT, both on the tetragonal and rhombohedral side of the morphotropic phase boundary, as well as PLZT (8/65/35 typically). Both poled and unpoled specimens have been investigated.

Typically, bipolar electric fields are applied to the specimens with amplitudes ranging from values below the coercive field of the respective material to values significantly above E_c, i.e., 2 or even 3 E_c. The limit depends on the width of the specimen, on the power-supplying equipment one can provide, and on the breakdown strength of the silicon oil surrounding the specimen. The observation of the crack extension per cycle is performed optically with standard light microscopes.

When low frequencies of the order of 1-20 Hz are employed, there is no significant effect of a frequency change on the crack propagation behaviour [197]. However, Weitzing *et al.* [213] used frequencies between 50 and 500 Hz and observed short intervals of abrupt crack growth when the frequency was changed during cycling. To date, no systematic study has been conducted, but since crack propagation is related to domain switching, as will be shown below, one may suppose that a frequency-dependence exists, because domain switching is also a time-dependent process.

General Observations

Common to all observations is the dominant direction of crack propagation: it always occurs in the direction perpendicular to the electric field. Only a minor amount of crack propagation is observed parallel to the electric field (e.g., [212]) or none at all. Even when a Vickers indent is introduced at an angle to the electric field, and therefore the starting crack is not perpendicular to it, once electric cycling starts, the crack changes direction and propagates perpendicular to the electric field [217]. This observation correlates to the observed anisotropy of fracture toughness, which is lower in the direction perpendicular to the field than in the direction parallel to it.

Furthermore, crack propagation takes place in different regimes. After an initial period of rapid crack growth or a pop-in from a notch, the growth rate settles into a steady state with a constant crack extension per cycle. If cycling is continued, the crack growth rate will typically decrease and eventually crack arrest will occur. These three regimes are differently pronounced in different materials and under different loading conditions.

Influence of electric field, E, on incremental crack extension, da/dN

Generally, a threshold electric field strength needs to be exceeded to cause significant crack propagation. Depending on the material, the threshold may be as low as 0.8 E_c [218], but in most observations it corresponds approximately to the coercive field strength of the material. In that case, poling of a sample only affects the first half-cycle. In a poled sample, crack extension will start in the half-cycle where the electric field is antiparallel to the poling direction. Thereafter, the polarization is switched during every half-cycle and the crack extends in every half-cycle. In an unpoled sample, less domains are available for switching. Therefore, crack extension in the first half-cycle is less pronounced than in a poled sample. Thereafter, the sample is poled and will show the same behaviour. For increasing field amplitude above the threshold, the crack growth rate increases. This is observed in all regimes. During the initial propagation period, values between 10^{-5} to 10^{-4} m/cycle have been observed [197, 215].

When a notch is present in the sample, a pop-in occurs during the first half-cycle of electric loading. In the case of a poled specimen, this is during the first half-cycle applied opposite to the poling direction. This pop-in phenomenon is explained in detail in [220].

An example of the increasing crack growth rate with field for the steady state growth is shown in Fig. 10.38 [221]. Here, the crack extension starts around 10^{-5}m/cycle for E=1.1E_c and increases up to 1.4×10^{-4}m/cycle for E=1.7E_c. These values were obtained during cycle number 2-10 in a soft PZT composition since this was the observed regime of steady-state in this material. Lower rates were presented by Shieh *et al.* [219] for PZT-5H, namely 10^{-10} up to 10^{-7} m/cycle for E approximately between 1.1 to 3E_c. Weitzing *et al.* [213] observed da/dN to be approximately 10^{-8} to 10^{-7} m/cycle for an applied field amplitude of 1.5E_c.

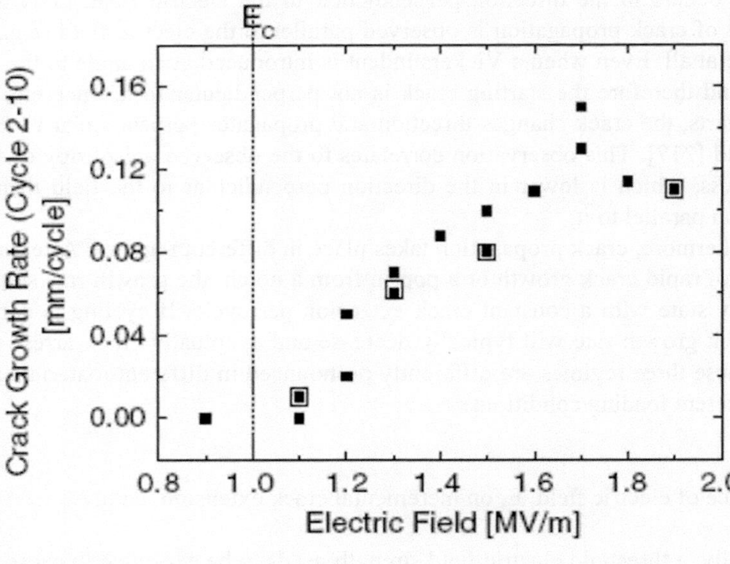

Fig. 10.38 Crack growth rate in mm/cycle vs. applied electric field strength during cycles 2-10. The framed symbols represent two data points lying at the same positions. © *Elsevier* [220].

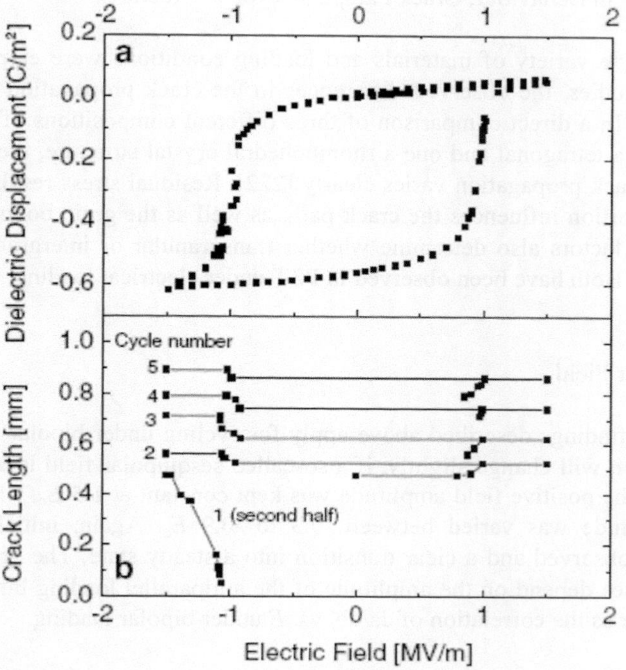

Fig. 10.39 a Dielectric hysteresis during the first five bipolar hysteresis cycles and **b** corresponding crack lengths as a function of applied electric field. © *Elsevier* [220].

In PLZT, crack extension rates ranging from 10^{-8} m/cycle up to 10^{-4}m/cycle have been observed when E was increased from 1 to 3 E_c [212]. Lynch *et al.* [197] observed values of 5×10^{-6} up to 3.5×10^{-5} with electric field amplitudes between 2 and 3 E_c. A much lower rate in PLZT of only 2.4×10^{-10} is reported by Shang & Tan (2001).

The duration of the steady state period during crack advance varies strongly and the transition to a decreasing crack growth rate may be smooth. Even when the growth-rate decreases, crack propagation may still last for up to 10^5 cycles [213, 217, 219].

The exact onset of crack propagation in each half-cycle may be studied if a low frequency of only 0.01 Hz is used. The results for cycling with E=1.5 E_c are displayed in Fig. 10.39. Along with crack propagation, the dielectric hysteresis was measured and is displayed in the top part of the Figure [221]. One clearly observes that crack extension sets in close to the coercive field strength but stops before the maximum amplitude is reached. In the second half of the first cycle, the pop-in of the crack from a notch occurs. It is apparently much more pronounced than subsequent crack growth.

Differences in Behaviour, Crack Path, and Type of Fracture

Since a wide variety of materials and loading conditions were employed in the different studies, the observed differences in the crack propagation rates are not surprising. In a direct comparison of three different compositions of PZT, two of which had a tetragonal and one a rhombohedral crystal structure, the crack initiation and crack propagation varies clearly [222]. Residual stress resulting from the lattice distortion influences the crack path, as well as the grain boundary chemistry. These factors also determine whether transgranular or intergranular fracture will occur. Both have been observed in PZT under electrical loading.

Sesquipolar Field

While the findings described above apply for cycling under bipolar electric field, the situation will change slightly, if a so-called sesquipolar field is used [223]. In this case, the positive field amplitude was kept constant at 1.5 E_c, while the negative amplitude was varied between -0.3 to -0.9 E_c. Again, initial rapid crack growth is observed and a clear transition into a steady state. The growth rates in both regimes depend on the amplitude of the antiparallel loading but the effect is not as clear as the correlation of da/dN vs. E under bipolar loading.

Conducting Cracks

Only few works have been conducted on cracks filled with a conducting medium like NaCl or water, one of them by Lynch *et al.* [197]. In this case, the crack was initiated from one of the electroded faces of the sample by Vickers indentation, and the sample was cycled at 2 E_c with a frequency of 0.2 Hz. Contrary to the insulating crack, the conducting crack only propagated during the positive half of the field cycle. A tree-like structure of the crack was observed to form and grow into the sample.

Mechanism for Cyclic Crack Propagation

Opening and closing of cracks during electrical cycling was observed in several works [212, 197, 220]. This observation inspired Cao and Evans to propose a wedging model in which contacting asperities of the crack cause strain mismatch during cycling. This in turn results in stress that drives the crack tip forward. However, this model lacked an explanation in the case of a pop-in from a notch where no contacting asperities exist.

The role of strain also comes into play in the observations of Weitzing *et al.* [213] who cycled three different PZT compositions. The composition with the highest peak-to-peak strain values of the strain hysteresis also displayed the highest crack propagation rates. Furthermore, a collapse of one wing of the butterfly loop was observed after 10^6 cycles. Since da/dN decreases with increasing cycle numbers, this observation indicates the relevance of strain in the mechanism of crack propagation.

Taking all of the above described observations into account, the following explanation of the crack propagation mechanism is the most precise and also most recent. It is described in detail in [194]. It will only be briefly introduced here. As shown in Fig. 10.40, the material around a notch or crack may be divided into different regions in terms of its respective strain state. Due to the much lower permittivity of a notch or crack, the field intensity at its tip (region 2) will be larger than the applied field. Above and below the notch/crack (regions 1), it will be lower. Therefore, domain switching in those regions takes place at different times during one half-cycle which is indicated in the strain hysteresis in the figure. This yields a strain mismatch and, therefore, stress that drives a crack cyclically forward. This model can also account for differences in the driving force observed for different materials, since all of them have slightly different permittivities. The mechanism was verified by finite element modelling, see [194].

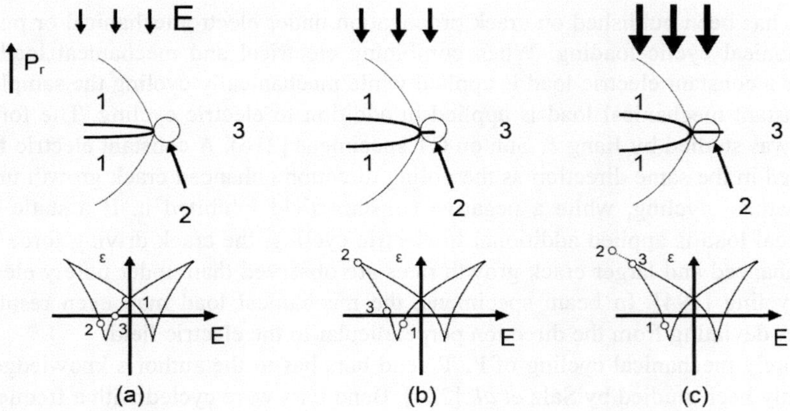

Fig. 10.40 Schematic of the regions of ferroelectric switching near a crack tip with poling direction given by P_r. The butterfly hysteresis loop (strain, ε, vs. electric field, **E**) is used to illustrate three distinct regions of inhomogeneous switching. **a** Small-scale switching in region 2 for applied fields below the coercive field, **b** onset of crack propagation at applied fields near the coercive field, **c** crack extension driven by large mismatch in strain (regions 1 and 2) as the field increases above the coercive field. Contraction in region 1 drives an increment of growth into region 3 followed by arrest. This process is reset for the subsequent half-cycle © *Elsevier* [220].

The observed decrease of crack growth rate with increasing cycle number may also be attributed to the difference in permittivities of the PZT material and the notch. The cyclic opening and closing of the crack results in wear of the crack surfaces and debris mixing with the silicon oil. Hence, the crack is more permeable for the electric field and less strain mismatch results. Further cycling leads to microcracking, and thereby an even more permeable region reducing the driving force further. A schematic of these different regimes is shown in Fig. 10.41 [222].

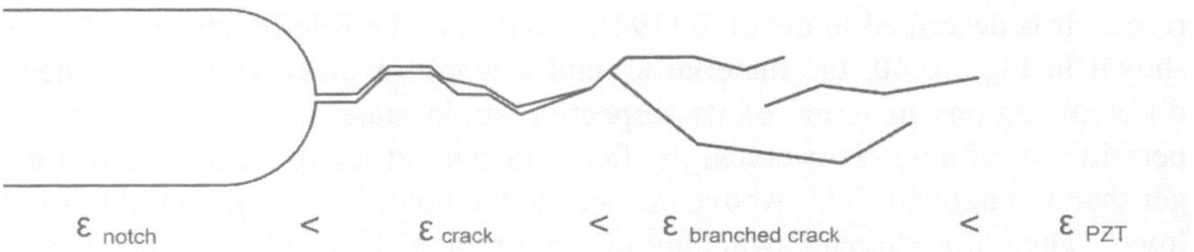

Fig. 10.41 Sketch of notch and different stages of crack propagation together with a ranking of the local relative permittivities © *Elsevier* [220].

Combined Electromechanical and Purely Mechanical Loading

Little has been published on crack propagation under electromechanical or purely mechanical cyclic loading. When combining electrical and mechanical loading, either a constant electric load is applied while mechanically cycling the sample or a constant mechanical load is applied in addition to electric cycling. The former case was studied by Jiang & Sun on CT specimens [216]. A constant electric field applied in the same direction as the poling direction enhanced crack growth under mechanical cycling, while a negative constant field inhibited it. If a static mechanical load is applied additional to electric cycling, the crack driving force will be enhanced and larger crack growth rates are observed than under purely electrical cycling [194]. In beam specimens, the mechanical load may even result in cracks deviating from the direction perpendicular to the electric field.

Purely mechanical cycling of PZT bend bars has to the author's knowledge so far only been studied by Salz *et al.* [224]. Bend bars were cycled with a frequency of 10 Hz for several thousand cycles. The initial minimum stress intensity factor was 0.2 $MPam^{1/2}$ while the maximum stress intensity factor and therefore the amplitude ΔK varied. The resulting crack extension rates were described by a Paris power-law relationship with $A=9.27\times10^{-6}$ and $n=10.8$.

10.4.3 Models for Cracking in Ferroelectrics

One of the significant contributions of fracture mechanics to our understanding of failure in structural materials was the introduction of a single parameter, the fracture toughness, governing failure. The failure criterion becomes a statement that if the applied stress intensity exceeds the fracture toughness, then a crack will propagate. This concept was modified to encompass rising fracture resistance curves that account for a fracture toughness that increases with crack growth, Paris' law behaviour governing fatigue crack growth in which the crack growth per cycle is fit to a power law in cyclic stress intensity, and creep crack growth in which the crack propagation velocity is expressed in terms of the applied stress intensity factor for different environments. These concepts have more recently been extended to ferroelectric materials with the introduction of the concept of electric field intensity and an electric toughness.

Cracks and voids in ferroelectric ceramics are typically filled with air or vacuum. Vacuum has the dielectric permittivity of free space where the surrounding ferroelectric material has a relative permittivity of near 1000. This results in a large electric field concentration in the ferroelectric material just outside of the void or at the tip of the crack. This local field concentration induces a large local piezoelectric response that is constrained by the surrounding material. The result is a large local stress that can lead to electric field induced fracture even in the absence of applied stress. These observations of the interaction of electric field with cracks in ferroelectric materials have led to the development of asymptotic solutions for electro-mechanical crack tip fields with certain simplifying assumptions. One such assumption was that the crack is impermeable to the electric field. This assumption results in an over-estimation of the electric field concentration. Another simplifying assumption was that the interior of the crack is a perfect insulator, even in the presence of electric fields within the crack that are many times the breakdown strength of air. Even though the simplifying assumptions have proven to be only rough approximations to the actual crack tip fields, the concept of an electric field intensity factor [225, 226, 186, 227] has proven to be useful.

10.4.3.1 The Impermeable Crack in a Piezoelectric

The impermeable crack solution for the linear dielectric results in expressions for the components of electric field

$$E_1 = \frac{K_E}{\sqrt{2\pi r}} \cos(\vartheta/2)$$

$$E_2 = \frac{K_E}{\sqrt{2\pi r}} \sin(\vartheta/2)$$

and for the electric potential

$$U = -K_E \sqrt{\frac{2r}{\pi}} \cos(\vartheta/2)$$

Where the electric field intensity factor K_E is a function of geometry. These equations provide a first approximation of the crack tip fields.

More detailed solutions may be obtained by including elastically anisotropic material behaviour and piezoelectric coupling. There are a number of discussions of this in the literature, but the initial assumption of an impermeable crack may introduce an inaccuracy that overshadows the benefits of the more sophisticated analysis. A thorough description of all facets of this problem may be found in the book by Yang [228].

10.4.3.2 The Permeable Crack

The permeable crack addresses the issue of electric field being concentrated within the crack interior. For the linear dielectric with an elliptical through hole, the field concentration solution is a well known solution of Laplace's equation and may be found in many textbooks. The introduction of the permeable boundary conditions in the asymptotic electrical crack solution is problematic, however. The problem is that the field concentration tends to zero as the crack opening goes to zero. If the crack is mechanically wedged open by a finite amount, then there will be a strong interaction between the electrical and mechanical fields.

10.4.3.3 Crack Propagation

Due to the complex constitutive behaviour of PZT, the modelling of this behaviour is challenging, which has been described in more detail in section 10.4.2.3. The theoretical description of crack propagation is even more challenging, and only few attempts have been made.

Lynch *et al.* [197] proposed the first model using step-like electrostriction in a circle around the crack tip. If the electric field is larger than the coercive field strength, the strain will have a certain constant value. In the other case, the strain will be zero. Thereby, a stress intensity factor is found that increases with $a^{1/2}$. The constrained switching zone around the crack tip stops to grow when the crack length becomes comparable to the sample thickness. Therefore, a decreasing crack growth rate with increasing crack length can be explained. However, the cyclic nature of crack growth could not be accounted for.

A more realistic approach of Zhu & Yang [215] included the switching strain caused by 90° domain switching in a switching zone around the crack tip. A body force is induced onto the boundary of the switching zone and integrated along the zone boundary. Only fields below E_c are considered and small-scale switching. In a poled material, different crack-tip stress intensity factors result for positive and negative fields. During crack propagation, the form of the domain switching zone changes, which leads to crack arrest. In the next half-cycle the crack is re-initiated. Both these effects also resulted in the analytical calculations. However, the model did not include large-scale switching.

A similar model was proposed by Mao & Fang [229] who also considered small-scale switching around an impermeable crack. They included effects of electromechanical coupling and derived a crack-tip stress intensity factor for some combinations of cyclic electrical and static mechanical loading. With finite element modelling, all stages of crack propagation may be described, and the effect of an additional mechanical load may also be included realistically [194].

10.5 Summary

This chapter tried to give an overview of the many aspects of mechanical properties of ferroelectrics. It has become apparent that many scales are relevant to material properties and device design. Basically, all aspects known to classical failure in ceramics are amended by the subtleties of a finite size plastic deformation, which yields nonlinear hysteretic material behaviour, its saturation, creep, a crack tip decorated with a dynamic process zone, and everything coupled to the electric field. We have not touched on the influence of temperature, which for certain applications will be highly relevant with respect to external temperature as well a heat generation within the device [230]. We have also only briefly touched the vast body of literature on the theory of impermeable cracks as well as the electric crack intensity factor for cracks along the field direction treated in detail by Schneider [14].

We hope that we were able to introduce the interested newcomer to the field and, at the same time, offer some novelties even for the advanced readership, e.g., in the section on homogenization techniques.

Acknowledgements

The authors are grateful to the ferroelectrics research group at TU Darmstadt and particularly the benevolent guidance of Jürgen Rödel, who has brought together the authors involved in this paper. Furthermore, discussions during the last years with Sergio Luis dos Santos e Lucto, Alain Brice Kounga-Njiwa, Emil Aulbach, Nina Balke, Marc Kamlah, Herbert Balke, Meinhard Kuna, Johannes Rödel, and many others are highly appreciated. Work was partly conducted at TU Dresden.

References

1 Curie P, Curie J (1881) Contractions et dilatations produites par des tensions électriques dans les cristaux hémièdres à faces inclinées. Comptes rendus de l'Acad. Sci., séance du 26 décembre 1881, XCIII, p. 1137
2 Valasek J (1921) Piezoelectric Activity of Rochelle Salt under Various Conditions. Phys Rev. 17: 422-423
3 Jaffe B, Cook Jr. WR, Jaffe H (1971) Piezoelectric Ceramics. Academic Press, Marietta OH
4 Zhang S, Lee S-M, Kim D-H, Lee H-Y, Shrout TR (2007) Electromechanical Properties of PMN–PZT Piezoelectric Single Crystals Near Morphotropic Phase Boundary Compositions. J. Am. Ceram. Soc. 90: 3859 - 3862
5 Zhang S-T, Kounga AB, Aulbach E, Granzow T, Jo W, Kleebe H-J, Rödel J (2008) Lead-free piezoceramics with giant strain in the system $Bi_{0.5}Na_{0.5}TiO_3$–$BaTiO_3$–$K_{0.5}Na_{0.5}NbO_3$. I. Structure and room temperature properties. J. Appl. Phys. 103: 034107
6 Lupascu DC, Rödel J (2005) Fatigue in Bulk Lead Zirconate Titante Actuator Materials, Adv. Eng. Mater. 7: 882-897
7 Morozov M, Lupascu DC (2009) Ferroelectric Properties, in Ceramics Science and Technology. Vol. 2, Chen I-W, Riedel R, Eds, Wiley VCH, Weinheim
8 Muralt P (2008) Recent Progress in Materials Issues for Piezoelectric MEMS, J. Am. Ceram. Soc. 91: 1385–1396
9 Scott JF (2007) Applications of modern ferroelectrics, Science 315: 954-959
10 Nan C-W, Bichurin MI, Dong S, Viehland D, Srinivasan G (2008) Multiferroic magnetoelectric composites: Historical perspective, status, and future directions, J. Appl. Phys. 103: 031101
11 Eerenstein W, Mathur ND, Scott JF (2006) Multiferroic and magnetoelectric materials, Nature 442: 759-765
12 Fiebig M (2005) Revival of the magnetoelectric effect. J. Phys. D: Appl. Phys. 38: R123–R152
13 Damjanovic D (2005) Hysteresis in piezoelectric and ferroelectric materials. Chapter 4 in Vol. III of "Science of hysteresis" ed. G. Bertotti and I. Mayergoyz, Eds., Elsevier, Amsterdam: 337-465
14 Schneider G (2007) Influence of Electric Field and Mechanical Stresses on the Fracture of Ferroelectrics, Annu. Rev. Mater. Res. 37: 491–538
15 Furuta A, Uchino K (1986) Field induced strain in (Pb,La)(Zr,Ti)O_3 Ceramics. Adv. Ceram. Mater. 1: 61-63
16 Strukov BA, Levanyuk AP (1998) Ferroelectric Phenomena in Crystals, Springer, Berlin
17 Devonshire AF (1954) Theory of Ferroelectrics, Adv. Phys. 3: 85-129
18 Strukov BA, Levanyuk AP (1998) Ferroelectric Phenomena in Crystals, Springer, Berlin
19 Bertotti G, Mayergoyz I (2005) Science of hysteresis, Elsevier, Amsterdam
20 Zgonik M, Bernasconi P, Duelli M, Schlesser R, Günter P, Garrett MH, Rytz D, Zhu Y, Wu X (1994) Dielectric, elastic, piezoelectric, electro-optic, and elasto-optic tensors of $BaTiO_3$ crystals. Phys. Rev. B 50: 5941-5949
21 Jackson JD (1975) Classical Electrodynamics. John Wiley & Sons, New York
22 Merz WJ (1954) Domain Formation and Domain Wall Motions in Ferroelectric $BaTiO_3$ Single Crystals. Phys. Rev. 95: 690-698
23 Hooton JA, Merz WJ (1955) Etch patterns and ferroelectric domains in $BaTiO_3$ single crystals. Phys. Rev. 98: 409-413

24 Berlincourt D, Jaffe H (1958) Elastic and piezoelectric coefficients of single-crystal barium titanate." Phys. Rev. 111: 143-148
25 Fatuzzo E, Merz WJ (1961) Surface layer in $BaTiO_3$ single crystals. J. Appl. Phys. 32: 1685-1687
26 Merz WJ (1964) Polarization processes in barium titanate, Ueber Polarisationsvorgaenge in $BaTiO_3$. Nachrichtentechnische Fachberichte 29: 37-40
27 Kuwata J, Uchino K, Nomura S (1981) Phase transitions in the $Pb(Zn_{1/3}Nb_{2/3})O_3$ - $PbTiO_3$ system. Ferroelectrics 37: 579-582
28 Kuwata, J, Uchino K, Nomura S (1982). Dielectric and piezoelectric properties of 0.91 $Pb(Zn_{1/3}Nb_{2/3})O_3$ - 0.09 $PbTiO_3$ single crystals. Jpn. J. Appl. Phys., Part 1: 21: 1298-1302
29 Kato K, Suzuki K, Uchino K (1990) Observation of the domain motion in ferroelectric single crystals with a diffuse phase transition. Nippon Seramikkusu Kyokai Gakujutsu Ronbunshi/J. Ceram. Soc. Jpn. 98: 840-845
30 Waser R, Baiatu T, Härdtl K-H (1990) DC electrical degradation of perovskite-type titanates. II. Single crystals. J. Am. Ceram. Soc. 73: 1654-1662
31 Mulvihill ML, Cross LE, Uchino K (1995) Low-temperature observation of relaxor ferroelectric domains in lead zinc niobate. J. Am. Ceram. Soc. 78: 3345-3351
32 Mulvihill ML, Cross LE, Uchino K (1996) Dynamic motion of the domain configuration in relaxor ferroelectric single crystals as a function of temperature and electric field. Ferroelectrics 186: 325-328
33 Mulvihill ML, Park SE, Risch G, Li Z, Uchino K, Shrout TR (1996) Role of processing variables in the flux growth of lead zinc niobate-lead titanate relaxor ferroelectric single crystals. Jpn. J. Appl. Phys., Part 1: 35: 3984-3990
34 Ye ZG, Schmid H (1996) Growth from high temperature solution and characterization of $Pb(Fe_{2/3}W_{1/3})O_3$ single crystals. J. Crystal Growth 167: 628-637
35 Mulvihill ML, Cross LE, Cao W, Uchino K (1997) Domain-related phase transition like behaviour in lead zinc niobate relaxor ferroelectric single crystals. J. Am. Ceram. Soc. 80: 1462-1468
36 Park SE, Shrout TR (1997) Characteristics of relaxor-based piezoelectric single crystals for ultrasonic transducers. IEEE Trans. Ultrason., Ferroelectrics Freq. Control 44: 1140-1147
37 Park SE, Shrout TR (1997) Ultrahigh strain and piezoelectric behaviour in relaxor based ferroelectric single crystals. J. Appl. Phys. 82: 1804-1811.
38 Wakiya N, Shinozaki K, Mizutani N, Ishizawa N (1997) Estimation of phase stability in $Pb(Mg_{1/3}Nb_{2/3})O_3$ and $Pb(Zn_{1/3}Nb_{2/3})O_3$using the bond valence approach. J. Am. Ceram. Soc. 80: 3217-3220
39 Belegundu, U, Pastor M, Du XH, Cross LE, Uchino K (1998) Domain formation in 0.90 $Pb(Zn_{1/3}Nb_{2/3})O_3$ - 0.10 $PbTiO_3$ single crystals under electric field along [111] direction. Proc. IEEE Ultrasonics Symp. 1: 625-628
40 Harada K, Shimanuki S, Kobayashi T, Saitoh S, Yamashita Y (1998) Crystal growth and electrical properties of $Pb((Zn_{1/3}Nb_{2/3})_{0.91}Ti_{0.09})O_3$ single crystals produced by solution Bridgman method. J. Am. Ceram. Soc. 81: 2785-2788
41 Kobayashi T, Saitoh S, Harada K, Shimanuki S, Yamashita Y, (1998) Growth of large and homogeneous PZN-PT single crystals for medical ultrasonic array transducers. ISAF 98. Proc. 11th IEEE Int. Symp. Appl. Ferroel.: 235-238
42 Park SE, Shrout TR, Bridenbaugh P, Rottenberg J, Loiacono GM (1998) Electric field induced anisotropy in electrostrictive $Pb(Mg_{1/3}Nb_{2/3})O_3$ - $PbTiO_3$ crystals. Ferroelectrics 207: 519-526
43 Belegundu U, Du XH, Bhalla A, Uchino K (1999) Effect of electric field on domain formation in relaxor based $Pb(Zn_{1/3}Nb_{2/3})O_3$ - $PbTiO_3$ single crystals. Ferroelectrics, Lett. Sect. 26: 107-116

44 Belegundu U, Du XH, Cross LE, Uchino K (1999) In situ observation of domains in 0.9 $Pb(Zn_{1/3}Nb_{2/3})O_3$ - 0.1 $PbTiO_3$ single crystals. Ferroelectrics 221: 67-71
45 Dammak H, Lebon A, Calvarin G (1999) Kinetic study of PZN single crystal polarization under a static electric field. Ferroelectrics 235: 151-158
46 Durbin MK, Jacobs EW, Hicks JC, Park S-E (1999) In situ x-ray diffraction study of an electric field induced phase transition in the single crystal relaxor ferroelectric, 92% $Pb(Zn_{1/3}Nb_{2/3})O_3$ - 8% $PbTiO_3$. Appl. Phys. Lett. 74: 2848-2850
47 Hosono Y, Harada K, Shimanuki S, Saitoh S, Yamashita Y (1999) Crystal growth and mechanical properties of $Pb[(Zn_{1/3}Nb_{2/3})_{0.91}Ti_{0.09}]O_3$ single crystal produced by solution Bridgman method. Jpn. J. Appl. Phys., Part 1: 38(9B): 5512-5515
48 Kighelman, Z, Damjanovic D, Seifert A, Hiboux S, Sagalowicz L, Setter N (1999) Relaxor $Pb(Mg_{1/3}Nb_{2/3})O_3$ thin films and their electromechanical properties. Integr. Ferroelectrics 25: 125-133
49 Liu SF, Park SE, Shrout TR, Cross LE (1999) Electric field dependence of piezoelectric properties for rhombohedral 0.955 $Pb(Zn_{1/3}Nb_{2/3})O_3$ - 0.045 $PbTiO_3$ single crystals. J. Appl. Phys. 85: 2810-2814
50 Luo H, Xu G, Wang P, Yin Z (1999) Growth and characterization of relaxor ferroelectric PMNT single crystals. Ferroelectrics 231: 97-102
51 Paik DS, Park SE, Shrout TR, Hackenberger W (1999) Dielectric and piezoelectric properties of perovskite materials at cryogenic temperatures. J. Mater. Sci. 34: 469-473
52 Yanagisawa K, Rendon-Angeles JC, Kanai H, Yamashita Y (1999) Stability and single crystal growth of dielectric materials containing lead under hydrothermal conditions. J. Eur. Ceram. Soc. 19: 1033-1036
53 Ye ZG, Dong M, Zhang L (1999) Domain structure and phase transitions in relaxor-based piezo-/ferroelectric (1-x) $Pb(Zn_{1/3}Nb_{2/3})O_3$ – x $PbTiO_3$ single crystals. Ferroelectrics 229: 223-232
54 Yu H, Randall CA (1999) Dendritic domain configurations in $Pb(Zn_{1/3}Nb_{2/3})O_3$ - $PbTiO_3$ single crystals. J.Appl. Phys. 86: 5733-5738
55 Cross LE (1994) Relaxor ferroelectrics: An overview. Ferroelectrics 151: 305-320
56 Uchino K, Furuta A (1992) Destruction mechanism of multilayer ceramic actuators. Proc. 8th IEEE Int. Symp. Appl. Ferroel.: 195 – 198
57 Furuta A, Uchino K (1994) Destruction mechanism of multilayer ceramic actuators: case of antiferroelectrics. Ferroelectrics 160: 277-285
58 Chen W, Lynch CS (1998) Model for simulating polarization switching and AF-F phase changes in ferroelectric ceramics. J. Intell. Mater. Syst. Struct. 9: 427-431
59 Liu T, Lynch CS (2003) Ferroelectric properties of [110], [001] and [111] poled relaxor single crystals: measurements and modelling. Acta mater. 51: 407-416
60 McLaughlin EA, Liu T, Lynch CS (2005) Relaxor ferroelectric PMN-32%PT crystals under stress and electric field loading: II-33-mode measurements. Acta mater. 53: 4001-4008
61 Liu T, Oates WS, Wan S, Lynch CS (2005) Crack initiation at electrode edges in PZN-4.5%PT single crystals. J. Intell. Mater. Syst. Struct. 16: 373-379
62 Liu T, Lynch CS (2006) Orientation dependence of nonlinearity and hysteresis in PZN-4.5%PT single crystals I: Unipolar response. J. Intell. Mater. Syst. Struct. 17: 953-957
63 Liu T, Lynch CS (2006) Orientation dependence of nonlinearity and hysteresis in PZN-4.5%PT single crystals II: Bipolar electromechanical response. J. Intell. Mater. Syst. Struct. 17: 931-937
64 McLaughlin EA, Liu T, Lynch CS (2004) Relaxor ferroelectric PMN-32%PT crystals under stress and electric field loading: I-32 mode measurements. Acta mater. 52: 3849-3857
65 Liu T (2004) Electromechanical Behaviour of Relaxor Ferroelectric Crystals. PhD Thesis, The Georgia Institute of Technology, Atlanta, GA, USA

66 Lines ME, Glass AM (1977) Principles and Applications of Ferroelectrics and related Materials. Clarendon Press, Oxford
67 Fett T, Thun G (1998) Determination of room-temperature tensile creep of PZT. J. Mater. Sci. Lett. 17: 1929-1932
68 Lupascu DC, Fedosov S, Verdier C, von Seggern H, Rödel J (2004) Stretched Exponential Relaxation in Fatigued Lead-Zirconate-Titanate. J. Appl. Phys. 95: 1386-1390
69 Lohse O, Grossmann M, Boettger U, Bolten D, Waser R (2001) Relaxation mechanism of ferroelectric switching in $Pb(Zr,Ti)O_3$ thin films. J. Appl. Phys. 89: 2332-2336
70 Zhou D, Kamlah M (2005) Determination of room-temperature creep of soft lead zirconate titanate piezoceramics under static electric fields. J. Appl. Phys. 98: 104107
71 Lupascu DC (2004) Fatigue in ferroelectric ceramics and related issues. Springer, Heidelberg
72 Zhou D, Kamlah M, Munz D (2005) Effects of Bias Electric Fields on the Non-linear Ferroelastic Behaviour of Soft Lead Zirconate Titanate Piezoceramics. J. Am. Ceram. Soc. 88: 867–874
73 Morozov M (2005) Softening and hardening transitions in ferroelectric $Pb(Zr,Ti)O_3$ ceramics: PhD thesis, École Polytechnique Fédérale de Lausanne, Switzerland
74 Arlt G, Neumann H (1988) Internal Bias in Ferroelectric Ceramics: Origin and Time Dependence. Ferroelectrics 87: 109-120
75 Rosen CZ, Hiremath BV, Newnham R (1992) Key Papers in Physics, Piezoelectricity, Am. Inst. Phys., New York
76 Uchino K (1997) Piezoelectric Actuators and Ultrasonic Motors, Kluwer, Boston
77 Setter N, Ed. (2002) Piezoelectric Materials in Devices, Ceramics Laboratory, Lausanne
78 offmann MJ, Hammer M, Endriss A, Lupascu DC (2001) Correlation between microstructure, strain behaviour, and acoustic emission of soft PZT ceramics. Acta Mater. 49: 1301-1310
79 Jones JL (2007) The use of diffraction in the characterization of piezoelectric materials. J. Electroceram. 19: 67–79
80 Schönau K, Schmitt LA, Knapp M, Fuess H, Eichel R-A, Kungl H, Hoffmann M (2007) Nanodomain structure of $Pb[Zr_{1-x}Ti_x]O_3$ at its morphotropic phase boundary: Investigations from local to average structure. Phys. Rev. B75: 184117
81 Fett T, Munz D, Thun G (1999) Tensile and Bending Strength of Piezoelectric Ceramics. J. Mater. Sci. Lett. 18: 1899-1902
82 Munz D, Fett T (1999) Ceramics: Mechanical Properties, Failure Behaviour, Materials Selection. Springer, Heidelberg
83 Fett T, Munz D, Thun D (1998) Nonsymmetric deformation behaviour of PZT determined in bending tests. J. Am. Ceram. Soc. 81: 269-272
84 Balke N, Lupascu DC, Granzow T, Rödel J (2007) Fatigue of Lead Zirconate Titanate Ceramics II: Sesquipolar Loading. J. Am. Ceram. Soc. 90: 1088-1093
85 Chen W, Lynch CS (2001) Multiaxial constitutive behaviour of ferroelectric materials. J. Engin. Mater. Techn., Transa. ASME 123: 169-175
86 Shieh J, Huber JE, Fleck NA (2003) An evaluation of switching criteria for ferroelectrics under stress and electric field. Acta mater. 51: 6123-6137
87 Kounga-Njiwa AB, Lupascu DC, Rödel J (2004) Crack tip switching zone in ferroelectric ferroelastic materials. Acta Mater. 52: 4919-4927
88 Kounga-Njiwa AB, Aulbach E, Granzow T, Rödel J (2007) Influence of radial stress on the poling behaviour of lead zirconate titanate ceramics. Acta Mater. 55:675–680
89 Kounga AB, Granzow T, Aulbach E, Hinterstein M, Rödel J (2008) High-temperature poling of ferroelectrics. J. Appl. Phys. 104: 024116
90 Hall DA (2001) Review: Nonlinearity in piezoelectric ceramics. J. Mater. Sci. 36: 4575-4601

91 Bhattacharya K, Ravichandran G (2003) Ferroelectric perovskites for electromechanical actuation. Acta mater. 51: 5941–5960
92 Huber JE (2005) Micromechanical modelling of ferroelectrics. Curr. Opinion Solid State Mater. Sci. 9: 100-106
93 Landis CM (2004) Non-linear constitutive modelling of ferroelectrics. Curr. Opinion Solid State Mater. Sci. 8: 59-69
94 Robert G, Damjanovic D, Setter N, Turik AV (2001) Preisach modelling of piezoelectric nonlinearity in ferroelectric ceramics. J. Appl. Phys. 89: 5067-5074
95 Zhou X, Chattopadhyay A (2001) Hysteresis Behaviour and Modelling of Piezoceramic Actuators. Trans. ASME, J. Appl. Mech. 68: 270-277
96 Kamlah M, Tsakmakis C (1999) Phenomenological modelling of the non-linear electromechanical coupling in ferroelectrics. Int. J. Solids Struct. 36: 669-695
97 Huber JE, Fleck NA (2001) Multiaxial electrical switching of a ferroelectric: theory versus experiment. J. Mech. Phys. Solids 49: 785-811
98 McMeeking RM, Landis CM (2002) A phenomenological multiaxial constitutive law for switching in polycrystalline ferroelectric ceramics. Int. J. Engng. Sci. 40: 1553-1577
99 Landis CM (2002) Fully coupled, multiaxial, symmetric constitutive laws for polycrystalline ferroelectric ceramics. J. Mech. Phys. Solids 50: 127-152
100 Schröder J, Romanowski H (2005) A thermodynamically consistent mesoscopic model for transversely isotropic ferroelectric ceramics in a coordinate-invariant setting. Arch. Appl. Mech. 74: 863-877
101 Klinkel S (2006) A phenomenological constitutive model for ferroelastic and ferroelectric hysteresis effects in ferroelectric ceramics. Int. J. Solids Struct. 43: 7197-7222
102 Mehling V, Tsakmakis C, Gross D (2007) Phenomenological model for the macroscopical material behaviour of ferroelectric ceramics. J. Mech. Phys. Solids 55: 2106-2141
103 Hwang SC, Lynch CS, McMeeking RM (1995) Ferroelectric/ferroelastic interactions and a polarization switching model. Acta metall. mater. 43: 2073-2084
104 Delibas B, Arockiarajan A, Seemann W (2006) Rate dependent properties of perovskite type tetragonal piezoelectric materials using micromechanical model. Int. J. Solids Struct. 43: 697-712
105 Smith RC, Seelecke S, Dapino M, Ounaies Z (2006) A unified framework for modelling hysteresis in ferroic materials. J. Mech. Phys. Solids 54: 46-85
106 Menzel A, Arockiarajan A, Sivakumar SM (2008) Rate dependent properties of perovskite type tetragonal piezoelectric materials using micromechanical model. Smart Mater. Struct. 17: 15026
107 Zhang ZK, Fang DN, Soh AK (2006) A new criterion for domain-switching in ferroelectric materials. Mech. Mater. 38: 25-32
108 Shaikh MG, Phanish S, Sivakumar SM (2006) Domain switching criteria for ferroelectrics. Comp. Mater. Sci. 37: 178–186
109 Li FX, Fang DN, Soh AK (2004) An analytical axisymmetric model for the poling-history dependent behaviour of ferroelectric ceramics. Smart Mater. Struct. 13: 668-675
110 Li FX, Rajapakse RKND (2007) A constrained domain-switching model for polycrystalline ferroelectric ceramics. Part I: Model formulation and application to tetragonal materials. Acta mater. 55: 6472-6480
111 Ricinschi D, Okuyama M (2005) A nucleation-growth model for ferroelectric hysteresis loops with complete and partial switching. J. Eur. Ceram. Soc. 81: 2357–2361
112 Lo VC, Chung WY, Chow SC (2007) Simulation of electromechanical responses of ferroelectric ceramics driven by combined alternating electrical and mechanical loadings. J. Appl. Phys. 101: 114111
113 Wang J, Shi S-Q, Chen LQ, Li Y, Zhang T-Y (2004) Phase field simulations of ferroelectric/ferroelastic polarization switching. Acta mater. 52: 749-764

114 Soh AK, Song YC, Niz Y (2006) Phase Field Simulations of Hysteresis and Butterfly Loops in Ferroelectrics Subjected to Electro-Mechanical Coupled Loading. J. Am. Ceram. Soc. 89: 652–661
115 Arlt G (1990) Twinning in ferroelectric and ferroelastic ceramics: stress relief. J. Mater. Sci. 25: 2655-2666
116 Rödel J, Kreher WS (2000) Self-consistent modelling of non-linear effective properties of polycrystalline ferroelectric ceramics. Comp. Mater. Sci. 19: 123-132
117 Kreher WS, Rödel J (2002) Dissipative domain wall motion and macroscopic response of ferroelectric ceramics. Ferroelectrics 269: 51-56
118 Su Y, Weng GJ (2006) A polycrystal model for the anisotropic behaviour of a fully poled ferroelectric ceramic. J. Appl. Phys. 100: 114110-1-8
119 Rödel J, Kreher WS (2003) Modelling linear and nonlinear behaviour of polycrystalline ferroelectric ceramics. J. Eur. Ceram. Soc. 23: 2297-2306
120 Huber JE, Fleck NA, Landis CM, McMeeking RM (1999) A constitutive model for ferroelectrics. J. Mech. Phys. Solids 47: 1663-1697
121 Huber JE, Fleck NA (2004) Ferroelectric switching: a micromechanics model versus measured behaviour. Eur. J. Mech. A/Solids 23: 203-217
122 Zhou D, Kamlah M, Laskewitz B (2006). Multiaxial non-proportional polarization rotation tests of soft PZT piezoceramics under electric field loading. Proc. SPIE Vol. 6170, 617009, San Diego, CA, United States, Armstrong WD, Ed.
123 Pathak A, McMeeking RM (2008) Three-dimensional finite element simulations of ferroelectric polycrystals under electrical and mechanical loading. J. Mech. Phys. Solids 56: 663–683
124 Huo Y, Jiang Q (1997) Modelling of domain switching in polycrystalline ferroelectric ceramics. Smart Mater. Struct. 6: 441-447
125 Kamlah M, Jiang Q (1999) A constitutive model for ferroelectric PZT ceramics under uniaxial loading. Smart Mater. Struct. 8: 441-459
126 Lu W, Fang DN, Li CQ, Hwang KC (1999) Nonlinear electric-mechanical behaviour and micromechanical modelling of ferroelectric domain evolution. Acta mater. 47: 2913-2926
127 Li WF, Weng GJ (2002) A theory of ferroelectric hysteresis with a superimposed stress. J. Appl. Phys. 91: 3806–3815
128 Liu F, Li HJ, Wang TC (2008) Energy principle and nonlinear electric–mechanical behaviour of ferroelectric ceramics. Acta Mechanica 198: 147-170
129 Zhou D, Kamlah M (2006) Room-temperature creep of soft PZT under static electrical and compressive stress loading. Acta mater. 54: 1389-1396
130 Liu QD, Huber JE (2006) Creep in ferroelectrics due to unipolar electrical loading. J. Eur. Ceram. Soc. 26: 2799-2806
131 Kim S-J, Jiang Q (2002) A finite element model for rate-dependent behaviour of ferroelectric ceramics. Int. J. Solids Struct. 39: 1015-1030
132 Belov AY, Kreher WS (2007) Micromechanics of ferroelectrics: From domain walls to piezoceramic devices. Ferroelectrics 351: 79-87
133 Chong KB, Guiu F, Reece MJ (2008) Thermal activation of ferroelectric switching. J. Appl. Phys. 103: 14101
134 Belov AY, Kreher WS (2005) Viscoplastic behaviour of perovskite type ferroelectrics. Mater. Sci. Eng. B 118: 7-11
135 Belov AY, Kreher WS (2005) Viscoplastic models for ferroelectric ceramics. J. Europ. Ceram. Soc. 25: 2567-2571
136 Maugin GA, Pouget J, Drouot R, Collet B (1992) Nonlinear electromechanical couplings. Wiley, New York
137 Chen P, Peercy P (1979) One dimensional dynamic electromechanical constitutive relations of ferroelectric materials. Acta Mech. 31: 231–241

138 Chen PJ, Tucker TJ (1981) One dimensional polar mechanical and dielectric responses of the ferroelectric ceramic PZT 65/35 due to domain switching. Int. J. Eng. Sci. 19: 147–158
139 Bassiouny E, Ghaleb AF, Maugin GA (1988) Thermodynamical formulation for coupled electromechanical hysteresis effects–I. Basic equations. Int. J. Eng. Sci. 26: 1279–1295
140 Bassiouny E, Ghaleb AF, Maugin GA (1988) Thermodynamical formulation for coupled electromechanical hysteresis effects–II. Poling of ceramics. Int. J. Eng. Sci. 26: 1297–1306
141 Bassiouny E, Maugin GA (1989) Thermodynamical formulation for coupled electromechanical hysteresis effects–III. Parameter identification. Int. J. Eng. Sci. 27: 975–987
142 Landis CM, McMeeking RM (2001) A self-consistent constitutive model for switching in polycrystalline barium titanate. Ferroelectrics 255: 13–34
143 Hashin Z, Shtrikman S (1962) On some variational principles in anisotropic and nonhomogeneous elasticity. J. Mech. Phys. Sol. 10: 335-342
144 Walpole LJ (1966) On bounds for the overall elastic moduli of inhomogeneous system. J. Mech. Phys. Sol. 14: 151–162
145 Kröner E (1977) Bounds for effective elastic moduli of disordered materials. J. Mech. Phys. Sol. 25: 137–155
146 Willis JR (1966) Bounds and self-consistent estimates for the overall properties of anisotropic composites. J. Mech. Phys. Sol., 25:185–202
147 Francfort GA, Murat F (1986) Homogenization and optimal bounds in linear elasticity. Arch. Rat. Mech. Anal. 94: 307–334
148 Nemat-Nasser S, Hori M (1993) Micromechanics: Overall Properties of Heterogeneous Materials. North-Holland, London
149 Chen T (1994) Micromechanical estimates of the overall thermoelectroelastic moduli of multiphase fibrous composites. Int. J. Sol. Struct. 31: 3099–3111
150 Chen T (1993) Piezoelectric properties of multiphase fibrous composites: Some theoretical results. J. Mech. Phys. Sol. 41: 1781–1794
151 Dunn ML, Taya M (1993) Micromechanics predictions of the effective electroelastic moduli of piezoelectric composites. Int. J. Sol. Struct. 30: 161–175
152 Dunn ML,Taya M (1993) An analysis of piezoelectric composite materials containing ellipsoidal inhomogeneities. Proc. R. Soc. A 443(1918): 265–287
153 Benveniste Y (1993) Universal relations in piezoelectric composites with eigenstress and polarization fields, part I: Binary media: Local fields and effective behaviour. J. Appl. Mech. 60: 265–269
154 Benveniste Y (1993) Universal relations in piezoelectric composites with eigenstress and polarization fields, part II: Multiphase media-effective behaviour. J. Appl. Mech. 60: 270–275
155 Benveniste Y (1993) Exact results in the micromechanics of fibrous piezoelectric composites exhibiting pyroelectricity. Proc. R. Soc. A 441(1911): 59–81
156 Benveniste Y (1997) Piezoelectric inhomogeneity problems in anti-plane shear and inplane electric fields how to obtain the coupled fields from the uncoupled dielectric solution. Mech. Mater. 25: 59–65
157 Hori M, Nemat-Nasser S (1998) Universal bounds for effective piezoelectric moduli. Mech. Mat. 30: 1–19
158 Silva ECN, Fonseca JSO, Kikuchi N (1998) Optimal design of periodic piezocomposites. Comp. Meth. Appl. Mech. Eng. 159: 49–77
159 Silva ECN, Nishiwaki S, Fonseca JSO, Kikuchi N (1999) Optimization methods applied to material and flextensional actuator design using the homogenization method. Comp. Meth. Appl. Mech. Eng. 172: 241–271

160 Li Z, Wang C, Chen C (2003) Effective electromechanical properties of transversely isotropic piezoelectric ceramics with microvoids. Comp. Mat. Sci. 27: 381–392
161 Uetsuji Y, Nakamura Y, Ueda S, Nakamachi E (2004) Numerical investigation on ferroelectric properties of piezoelectric materials using a crystallographic homogenization method. Model. Sim. Mat. Sci. Eng. 12: S303–S317
162 Sanchez-Palencia E (1980) Non-Homogeneous Media and Vibration Theory (Lecture Notes in Physics vol. 172). Springer, Berlin
163 Hill R (1972) On constitutive macro-variables for heterogeneous solids at finite strain. Proc. R. Soc. A 326(1565): 131-147
164 Hill R (1985) On the micro-to-macro transition in constitutive analyses of elastoplastic response at finite strain. Proc. Cam. Phil. Soc. 98: 579-590
165 Suquet PM (1986) Homogenization Techniques for Composite Materials, Lecture Notes in Physics 272, chapter Elements of homogenization for inelastic solid mechanics, pp. 193–278. Springer-Verlag
166 Krawietz A (1986) Materialtheorie: Mathematische Beschreibung des phnomenologischen thermomechanischen Verhaltens. Springer-Verlag, Berlin
167 Miehe C, Schröder J, Schotte J (1999) Computational homogenization analysis in finite plasticity simulation of texture development in polycrystalline materials. Comp. Meth. Appl. Mech. Eng. 171: 387–418
168 Schröder (2000) Homogenisierungsmethoden der nichtlinearen Kontinuumsmechanik unter Beachtung von Instabilitäten. Bericht aus der Forschungsreihe des Instituts für Mechanik (Bauwesen), Lehrstuhl I, Universität Stuttgart
169 Schröder J, Keip M-A (2010) A framework for the two-scale homogenization of electromechanically coupled boundary value problems. In M Kuczma and K Wilmanski, editors, Computer Methods in Mechanics, 311-329, Springer
170 Schröder J, Keip M-A (2009) Multiscale Modeling of Electro-Mechanically Coupled Materials: Homogenization Procedure and Computation of Overall Moduli. In M Kuna and A Ricoeur, editors, IUTAM Symposium on Multiscale Modelling of Fatigue, Demage and Fracture in Smart Materials Systems, in press
171 Romanowski H, Schröder J (2005) Coordinate invariant modelling of the ferroelectric hysteresis within a thermodynamically consistent framework. A mesoscopic approach. In Y. Wang and K. Hutter, eds., Trends in Applications of Mathematics and Mechanics, pp. 419–428. Shaker Verlag, Aachen
172 Romanowski H (2006) Kontinuumsmechanische Modellierung ferroelektrischer Materialien im Rahmen der Invariantentheorie. PhD thesis, Institut für Mechanik, Fakultät Ingenieurwissenschaften, Abteilung Bauwissenschaften, Universität Duisburg-Essen
173 Schröder J, Gross D (2004) Invariant formulation of the electromechanical enthalpy function of transversely isotropic piezoelectric materials. Arch. Appl. Mech. 73: 533-552
174 Schröder J, Romanowski H (2005) A thermodynamically consistent mesoscopic model for transversely isotropic ceramics in a coordinate-invariant setting. Archive of Applied Mechanics 74, 863-877.
175 Schröder J, Romanowski H, Kurzhöfer I (2007) A computational meso-macro transition procedure for electro-mechanical coupled ceramics. In Schröder J, Lupascu D, Balzani D, eds., First Seminar on the Mechanics of Multifunctional Materials, Bad Honnef, Germany, May 7th - 10th 2007. Institut für Mechanik, Fakultät Ingenieurwissenschaften, Abteilung Bauwissenschaften, Universität Duisburg-Essen
176 Kurzhöfer I (2007) Mehrskalen-Modellierung polycrystalliner Ferroelektrika basierend auf diskreten Orientierungsverteilungsfunktionen. PhD thesis, Institut für Mechanik, Fakultät Ingenieurwissenschaften, Abteilung Bauwissenschaften, Universität Duisburg-Essen

177 Hill R (1963) Elastic properties of reinforced solids: Some theoretical principles. J. Mech. Phys. Sol. 11: 357-372

178 Boehler JP (1987) Introduction to the invariant formulation of anisotropic constitutive equations. In Applications of Tensor Functions in Solid Mechanics, Boehler JP, editor, pp. 3–65. Springer, Wien

179 Kamlah M, Böhle U (2001) Finite element analysis of piezoceramic components taking into account ferroelectric hysteresis behaviour. Int. J. Solids Struct. 38: 605-633

180 Pohanka RC, Freiman SW, Okazaki K, Tashiro S (1983) Fracture of piezoelectric materials. In Fracture Mechanics of Ceramics, ed. RC Bradt, AG Evans, DPH Hasselman, FF Lange, 5:353–64. New York: Plenum Press

181 Freiman SW, Pohanka RC (1989) Review of Mechanically Related Failures of Ceramic Capacitors and Capacitor Materials. J. Am. Ceram. Soc. 72: 2258-2263

182 Winzer SR, Shankar N, Ritter AP (1989) Designing cofired multilayer electrostrictive actuators for reliability. J. Am. Ceram. Soc. 72: 2246-2257

183 Cao HC, De Graef M, Evans AG (1993) Structure and properties at the ferroelectric/electrode interface between lead zirconate titanate and copper. J. Am. Ceram. Soc. 76(12): 3019-3023

184 Furuta A, Uchino K (1993) Dynamic observation of crack propagation in piezoelectric multilayer actuators. J. Am. Ceram. Soc. 76: 1615-1617

185 Aburatani, H., S. Harada, Uchino K, Furuta A, Fuda Y (1994). Destruction mechanisms in ceramic multilayer actuators. Jpn. J. Appl. Phys., Part 1, 33(5B): 3091-3094

186 Hao TH, Gong X, Suo Z (1996) Fracture mechanics for the design of ceramic multilayer actuators. J. Mech. Phys. Solids 44: 23-48

187 Fan J, Niu H, Lynch CS (1998) An experimental and theoretical study of electric field effects on cracked ceramics. J. Mater. Sci. Lett. 17: 189-192

188 Lynch CS (1998) Fracture of ferroelectric and relaxor electro-ceramics: influence of electric field. Acta mater. 46: 599-608

189 Ru CQ, Mao X, Epstein M (1998) Electric-field induced interfacial cracking in multilayer electrostrictive actuators. J. Mech. Phys. Solids 46: 1301-1318

190 Dos Santos e Lucato SL, Lupascu DC, Kamlah M, Rödel J, Lynch CS (2001) Constraint-induced crack initiation at electrode edges in piezoelectric ceramics. Acta mater. 49: 2751-2759

191 Dos Santos e Lucato SL, Bahr HA, Pham V-B, Lupascu DC, Balke H, Rödel J, Bahr U (2002) Electrically driven cracks in piezoelectric ceramics: Experiments and fracture mechanics analysis. J. Mech. Phys. Solids 50: 2333-2353

192 Maeng S-Y, Lee D-K, Choi J-W, Kim H-J, Kang C-Y, Sahn-Nahm, Yoon S-J (2005) Design and fabrication of multilayer actuator using floating electrode. Mater. Chem. Phys. 90: 405-410

193 Kusukawa K, Shiozaki Y (2007) Electric loading - Induced cracking behaviour at electrode edges in PZT ceramics. Key Engin. Mater. 352: 267-270

194 Westram I, Ricoeur A, Emrich A, Rödel J, Kuna M (2007) Fatigue crack growth law for ferroelectrics under cyclic electrical and combined electromechanical loading. J. Eur. Ceram. Soc. 27: 2485-2494

195 Megherhi MH, Dayton GO, Shrout TR, Mecholsky JJ (1990) Indentation fracture of lead magnesium niobate-based multilayer composite structures. J. Mater. Res. 5: 515-523

196 Davis JB, Cao HC, Bao G, Evans AG (1991). Fracture energy of interfaces. An elastic indentation technique. Acta metall. mater. 39: 1019-1024

197 Lynch CS, Yang W, Collier L, Suo Z, McMeeking RM (1995) Electric-field-induced cracking in ferroelectric ceramics. Ferroelectrics 166: 11-30

198 Shang JK, Tan X (2001) Indentation-induced domain switching in $Pb(Mg_{1/3} Nb_{2/3})O_3$ - $PbTiO_3$ crystal. Acta mater. 49: 2993-2999

199 Fang F, Yang W (2002) Indentation-induced cracking and 90 degrees domain switching pattern in barium titanate ferroelectric single crystals under different poling. Mater. Lett. 57: 198-202
200 Fang F, Li YH, Yang W (2005) Effect of poling directions on the electric-field-induced fatigue crack growth in barium titanate ferroelectric single crystals. Ferroelectrics 322: 11-21
201 Kolleck A, Schneider GA, Meschke FA (2000) R-curve behaviour of BaTiO3- and PZT ceramics under the influence of an electric field applied parallel to the crack front. Acta Mater. 48: 4099-4113
202 Dos Santos e Lucato SL (2003) Crack-growth-velocity-dependent R-curve behaviour in lead zirconate titanate. J. Am. Ceram. Soc. 86: 1037-1039
203 Felten F, Schneider GA, Knoblauch V (2005) In situ investigation of subcritical crack growth in lead zirconate titanate ceramics using atomic force microscopy. Acta mater. 53: 4051-4056
204 Kounga-Njiwa AB, Fett T, Lupascu DC, Rödel J (2006). Effect of geometry and electrical boundary conditions on R-curves for lead zirconate titanate ceramics. Engin. Fract. Mech. 73(3): 309-317
205 Oates WS, Lynch CS, Lupascu DC, Kounga-Njiva AB, Aulbach E, Rödel J (2004) Subcritical crack growth in lead zirconate titanate. J. Am. Ceram. Soc. 87: 1362-1364
206 Dos Santos e Lucato SL, Lupascu DC, Rödel J (2000) Effect of Poling Direction on R-Curve Behaviour in Lead Zirconate Titanate. J. Am. Ceram. Soc. 83: 424–426
207 Oates WS, Lynch CS, Kounga-Njiwa AB, Lupascu DC (2005) Anisotropic fracture behaviour in ferroelectric relaxor PZN-4.5%PT single crystals. J. Am. Ceram. Soc. 88: 1838-1844
208 Lupascu DC, Kreuzer M, Lucato SLS, Rödel J, Lynch CS (2001) A liquid crystal display of stress fields in ferroelectrics. Appl. Phys. Lett. 78: 2554-2556
209 Liu ZW, Fang DN, Xie HM, Lee JJ (2007) Study of effect of 90 degrees domain switching on ferroelectric ceramics fracture using the moire interferometry. Acta mater. 55: 3911-3922
210 ASTM E-399: Standard Test Method for Linear-Elastic Plane-Strain Fracture Toughness K_{Ic} of Metallic Materials
211 Waser R, Ed. (2005) Nanoelectronics and information technology, Wiley-VCH, Weinheim
212 Cao H, Evans AG (1994) Electric-field-induced fatigue crack growth in piezoelectrics. J. Am. Ceram. Soc. 77: 1783-1786
213 Weitzing H, Schneider GA, Steffens J, Hammer M, Hoffmann MJ (1999) Cyclic fatigue due to electric loading in ferroelectric ceramics. J. Eur. Ceram. Soc. 19: 1333-1337
214 Zhu T, Fang F, Yang W (1999) Fatigue crack growth in ferroelectric ceramics below the coercive field. J. Mat. Sci. Let. 18: 1025-1027
215 Zhu T, Yang W (1999) Fatigue crack growth in ferroelectrics driven by cyclic electric loading. Mech. Phys. Sol. 47: 81-9
216 Jiang LZ, Sun CT (1999) Crack growth behaviour in piezoceramics under cyclic loads. Ferroelectrics 233: 211-233
217 Shang JK, Tan X (2001) A maximum strain criterion for electric-field-induced fatigue crack propagation in ferroelectric ceramics. Mat. Sci. Eng. A301: 131-139
218 Liu B, Fang DN, Hwang KC (2002) Electric-field induced crack growth in ferroelectric ceramics. Mat. Lett. 54: 442-446
219 Shieh J, Huber JE, Fleck NA (2006) Fatigue crack growth in ferroelectrics under electrical loading. J. Eur. Ceram. Soc. 26: 95-109
220 Westram I, Laskewitz B, Lupascu DC, Kamlah M, Rödel J (2007) Electric-field induced crack initiation from a notch in a ferroelectric ceramic. J. Am. Ceram. Soc. 90: 2849-2854

221 Westram I, Oates WS, Lupascu DC, Rödel J, Lynch CS (2006) Mechanism of electric fatigue crack growth in lead zirconate titanate. Acta mater. 55: 301-312
222 Westram I, Kungl H, Hoffmann MJ, Rödel J (2008) Influence of crystal structure on crack propagation under cyclic electric loading in lead-zirconate-titanate. J. Eur. Ceram. Soc. 29: 425-430
223 Westram I (2006) Crack Propagation in $Pb(Zr,Ti)O_3$ under Cyclic Electric Loading. Ph. D. Thesis, Darmstadt University of Technology, Darmstadt, Germany
224 Salz RJ, Hoffman M, Westram I, Rödel J (2005) Cyclic Fatigue Crack Growth in PZT Under Mechanical Loading. J. Am. Ceram. Soc. 88: 1331-1333
225 Suo Z (1993) Models for breakdown-resistant dielectric and ferroelectric ceramics. J. Mech. Phys. Solids 41: 1155-1176
226 Yang W, Suo Z (1994) Cracking in ceramic actuators caused by electrostriction. J. Mech. Phys. Solids 42: 649-664
227 McMeeking RM, Hwang SC (1997) On the potential energy of a piezoelectric inclusion and the criterion for ferroelectric switching. Ferroelectrics 200: 151-173
228 Yang, W (2002) Mechatronic reliability: electric failures, mechanical electrical coupling, domain switching, mass flow instabilities. Springer, Berlin
229 Mao GZ, Fang DN (2004) Fatigue crack growth induced by domain switching under electromechanical load in ferroelectrics. Theoret. Appl. Fract. Mech. 41: 115-123
230 Zheng J, Takahashi S, Yoshikawa S, Uchino K, de Vries JWC (1996) Heat generation in Multilayer Piezoelectric Actuators. J. Am. Ceram. Soc. 79: 3193-3198

Chapter 11
The Elastic Properties of Ferroelectric Thin Films Measured Using Nanoindentation

C. Chima-Okereke, W. L. Roberts, A. J. Bushby, M. J. Reece

11.1 Introduction

Ferroelectric thin films are being used for a large number of applications such as sensors and actuators in MicroElectroMechanical Systems (MEMS) [1] and non-volatile memories [2]. For many of their applications as actuators, they are used in the form of cantilever or membrane devices. The operation of these devices is determined by their electromechanical properties, which includes their elastic properties [3]. These properties are difficult to measure in thin film form, and often designers resort to using the properties of bulk materials with the same composition. The properties of thin films can be quite different from bulk materials because of the high texture and residual stresses in the thin films. Many approaches have been explored to determine the electromechanical properties of thin films. This includes the following methods: tensile [4]; bending [4]; wafer bulge [5]; bending of cantilever beams [6, 7]; surface acoustic wave [8]; and nanoindentation [9, 10, 11]. The nanoindentation technique has the advantage of being non-destructive and has a spatial resolution of the order of a micrometer. Spherical indenters have been used to measure the elastic [12, 13], ferroelectric and piezoelectric [14, 15, 16, 17] properties of ferroelectric thin films. The effect of texture on the mechanical properties of thin films has also been studied using sharp indenters (Berkovich) [18, 19, 20, 21].

In this chapter, we review the use of nanoindentation to measure the elastic properties of ferroelectric thin films. This includes the analysis of nanoindentation data and the modelling of elastic properties. These are then applied to experimental results for $Pb(Zr,Ti)O_3$ (PZT) thin films.

Centre for Materials Research and School of Engineering and Materials Science,
Mile End Road, London E1 4NS.
Corresponding author – m.j.reece@qmul.ac.uk

11.2 Elastic Indentation Theory

To investigate the elastic properties of materials using indentation, a spherical indenter should be used to minimise plastic deformation. It is also necessary to use an indenter that is rigid in comparison to the material being investigated. Generally, a polished diamond indenter is used. In the ideal case of an isotropic elastic material, where no plastic deformation occurs, the elastic displacement, h_e, of a sphere of radius, R, into a flat surface under force, F, is given by the Hertzian relationship [4, 9, 10, 11, 22, 23, 24, 25].

$$h_e = \left(\frac{9}{16}\right)^{1/3}\left(\frac{F}{E^*}\right)^{2/3}\left(\frac{1}{R}\right)^{1/3} \tag{1}$$

E^* is the composite modulus of the contact resulting and is given by

$$\left(\frac{1}{E^*}\right) = \frac{(1-\nu_s^2)}{E_s} + \frac{(1-\nu_i^2)}{E_i} \tag{2}$$

where E is the Young's Modulus (if the material is isotropic), ν is Poisson's ratio and subscripts s and i denote the surface and the indenter respectively. Equation (2) defined in terms of the indentation modulus (E') is given by

$$\frac{1}{E^*} = \frac{1}{E'_s} + \frac{1}{E'_i} \tag{3}$$

The relationship between equations (2) and (3) ($E' = E/\left(1-\nu^2\right)$) is only valid for the cases where the material being indented is isotropic. For cases where the material is anisotropic, the relationship between indentation modulus and crystal elastic constants are more complex. These cases are discussed in section 5.

If the contact is completely elastic, and the indenter is spherical and rigid, the depth of penetration in contact with the sphere, h_c, is half of the total depth, h_{max},

$$h_c = \frac{h_{\max}}{2} \tag{4}$$

The radius of contact, a, is an important factor since it dictates the area over which the indenting force is applied, i.e., the indentation pressure. The ratio a/R expresses the geometry of the contact in non-dimensional form. The radius of the circle of contact for a rigid sphere is given by

$$a = \sqrt{Rh_e} \tag{5}$$

[10, 22, 26]

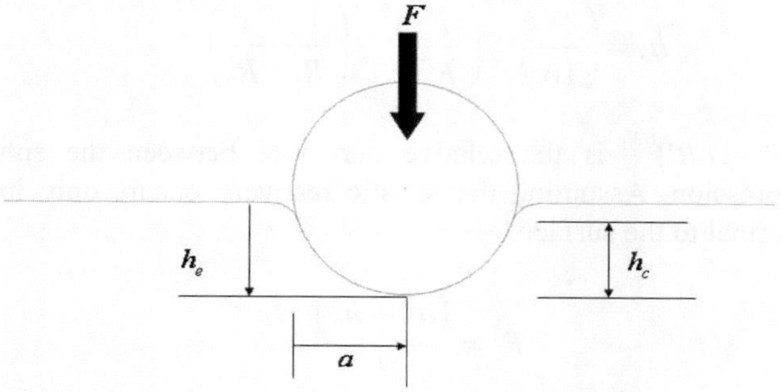

Fig. 11.1 Inden

Diamond indenters are never perfectly spherical due to the problem of polishing an anisotropic material into a spherical shape. So, in general R is a function of h_c. A calibration routine is used to find the indenter radius at a depth based on indenting reference materials of well-characterised elastic modulus, e.g., soda-lime glass. There are high contact stresses involved in nanoindentation, and these may cause the film to plastically deform. This needs to be incorporated into indentation theory, and we discuss how to do it in the next section.

11.3 Elastic-Plastic Indentation Theory

If plastic deformation occurs during indentation, the geometry of the indentation site changes. This needs to be reflected in the equations used to obtain the modulus. During indentation, there will be an elastic component to the penetration, h_e,. However, there will also be a residual depression, h_r, left after the indenter has been removed. The relationship between h_r , h_e, and the penetration at maximum force, h_{max}, is

$$h_e = h_{\max} - h_r \tag{6}$$

The depth of the indenter contact with the specimen becomes

$$h_c = (h_{\max} + h_r) / 2 \tag{7}$$

The radius of the circle of contact becomes

$$a^2 = 2Rh_c - h_c^2 \tag{8}$$

If the radius of curvature for the residual impression is R' [22]

$$h_e = \left(\frac{9}{16}\right)^{1/3}\left(\frac{F}{E^*}\right)^{2/3}\left(\frac{1}{R}-\frac{1}{R'}\right)^{1/3} \tag{9}$$

where $(1/R - 1/R')^{1/3}$ is the relative curvature between the sphere and the residual impression. Assuming that elastic recovery occurs only in the vertical direction, normal to the surface

$$R' = \frac{(a^2 + h_r^2)}{2h_r} \tag{10}$$

The modulus of the material can be calculated as

$$E^* = \left(\frac{3}{4}\right)\left(\frac{F}{h_e^{3/2}}\right)\left(\frac{1}{R}-\frac{1}{R'}\right)^{1/2} \tag{11}$$

The analytical solution may be used to obtain the indentation modulus of a material from the experimental force versus depth data collected during indentation. The two main techniques used to do this are discussed in the next section.

11.4 Evaluating Indentation Modulus from Spherical Indentation Force-Penetration Data

The two methods used to apply elastic-plastic spherical indentation theory to evaluate indentation modulus from experimental results are the Field and Swain and the Oliver and Pharr methods [10].

11.4.1 Field and Swain Method

The Field and Swain method [10] of analysis features a single partial unload from a maximum load. Its advantage is speed, convenience and the ability to measure elastic properties at various depths. The method assumes that unloading is completely elastic and no reverse plasticity takes place. The force and depth of a loading step are annotated as (F_1, h_1), and (F_2, h_2) denote the unloading step. After each loading step F_1, the indenter unloads to a fraction of this force, F_2. The force and depth are recorded at each point and the indenter continues to a higher F_1. The resulting graph contains two curves. One fully loaded curve F_1, h_1, and one partially unloaded curve F_2, h_2, see Fig. 11.2.

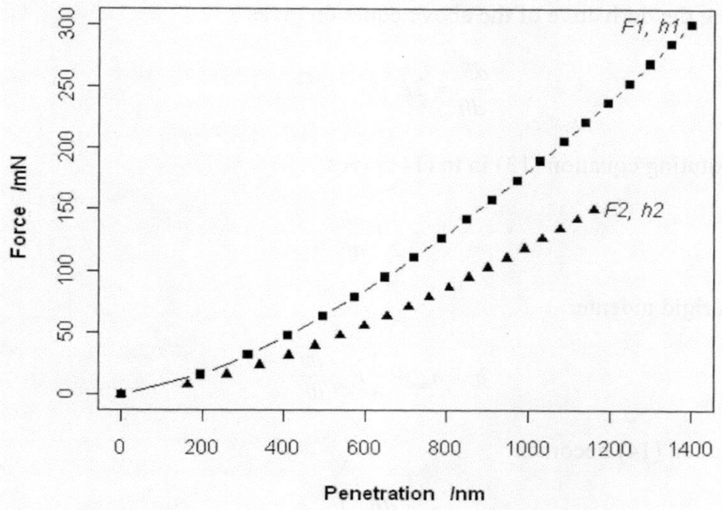

Fig. 11.2 Force penetration curve using the partial unloading technique.

When F_1 exceeds the yield point for the material, permanent deformation will occur and the two branches diverge. The residual depth is calculated as

$$h_r = \left[\frac{\left(h_2 \left(F_1 / F_2 \right)^{2/3} - h_1 \right)}{\left(F_1 / F_2 \right)^{2/3} - 1} \right] \tag{12}$$

The above equation assumes that the radius of the indenter R is the same at h_1 and h_2.

The residual penetration is used to calculate the contact depth, h_c, the radius of contact, a, and the radius of curvature of the surface, R', using equations (6) to (10). These are then in turn used to calculate the composite modulus of the indenter and surface response, E^*, using equation (11).

11.4.2 Oliver and Pharr

The Oliver and Pharr Method [10] uses the slope of the initial portion of the force-depth unloading curve to determine h_c. If R is taken to be the radius of the indenter, the Hertz equation given by equation (9) may be represented by

$$F = \frac{4}{3} E^* R^{1/2} h_e^{3/2} \tag{13}$$

Taking the derivative of the above equation gives

$$\frac{dF}{dh} = 2E^* R^{1/2} h_e^{1/2} \tag{14}$$

Substituting equation (13) in to (14) gives

$$h_e = \frac{3}{2} F \frac{dh}{dF} \tag{15}$$

For a rigid indenter

$$h_c = h_{max} - \frac{3}{4} F_{max} \frac{dh}{dF} \tag{16}$$

and equation (14) becomes

$$E^* = \frac{dF}{dh} \frac{1}{2a} \tag{17}$$

Fig. 11.3 shows a schematic of the elastic-plastic loading and the elastic unloading.

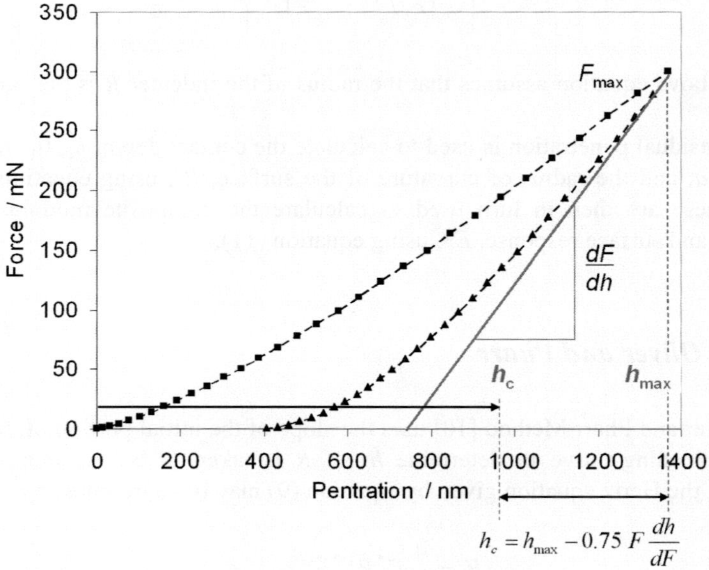

Fig. 11.3 Graph showing elastic-plastic indentation curves.

The Oliver and Pharr method is slower than the Field and Swain method as it requires a series of unloading points.

The methods mentioned in this section may be used to calculate the modulus of thin film-substrate complexes with depth. However, to measure the elastic modulus of a thin film requires a different approach. These are discussed in section 6.

11.5 Indentation of Anisotropic Materials

There are several analytical solutions for the indentation of different crystal symmetries. A solution for fully anisotropic materials has been published [27], but only transversely isotropic symmetries will be considered because it is the most relevant symmetry for ferroelectric thin films. A solution for the indentation modulus of transversely isotropic materials was published some time ago [28]. The stress-strain relationship of a planar isotropic half-space was used to calculate a force-depth relationship equation, which incorporated functions of the elastic constant of the material. The solution is shown below in equations (18) to (20).

$$\left.\begin{aligned} a &= \frac{s_{13}(s_{11}-s_{12})}{s_{11}s_{33}-s_{13}^{2}} \\ b &= \frac{s_{13}(s_{13}+s_{44})-s_{12}s_{33}}{s_{11}s_{33}-s_{13}^{2}} \\ c &= \frac{s_{13}(s_{11}-s_{12})+s_{11}s_{44}}{s_{11}s_{33}-s_{13}^{2}} \\ d &= \frac{s_{13}^{2}-s_{12}^{2}}{s_{11}s_{33}-s_{13}^{2}} \end{aligned}\right\} \tag{18}$$

$$2d\Omega_1^2 = a+c+[(a+c)^2-4d]^{1/2} \tag{19a}$$

$$2d\Omega_2^2 = a+c-[(a+c)^2-4d]^{1/2} \tag{19b}$$

$$\frac{(1-v^2)}{E} = -\frac{\Omega_1+\Omega_2}{2d^{1/2}(ac-d)}[(d-2bd+ac)s_{11}-(2d-bd-ac)s_{12}] \tag{20}$$

where s_{ij} are the compliance coefficients.

A solution for transverse isotropic piezoelectric materials has also been published [29], and is of the form

$$F = \frac{8}{3} K \frac{a^3}{R} \tag{21}$$

where K is a function of the elastic, dielectric, and piezoelectric properties of the materials. The solution presented here will be for the case when the material is unpoled. Consider a polynomial of sixth order given as a determinant a_{ij}, where

$$\det[a_{ij}] = 0 \tag{22}$$

and the coefficients of a_{ij} are given by

$$\left.\begin{aligned} &a_{11} = c_{44}k^2 - c_{11}, a_{12} = -a_{21} = (c_{13} + c_{44})k, \\ &a_{22} = c_{33}k^2 - c_{44}, a_{13} = -a_{31} = -(e_{31} + e_{15})k, \\ &a_{23} = -a_{32} = -e_{33}k^2 + e_{15}, \\ &a_{33} = \varepsilon_{33}k^2 - \varepsilon_{11} \end{aligned}\right\} \tag{23}$$

c_{ij} are the elastic constants, e_{ij} are the piezoelectric constants and ε_{ii} are the dielectric constants for transverse isotropy.

The equation resulting from the determinant in equation (22) may then be solved by numerical means to give six roots for k. Two of these roots are real, $k = \pm k_1$ and four are complex $k = \pm(\delta \pm \omega i)$, $i = \sqrt{-1}$. Where k_1 and δ are positive definite and ω is non-negative. These parameters are used to calculate M_1 and M_5 and hence the constant K in equation (21) using the following equations

$$\left.\begin{aligned} &\alpha_1 = a_{12}a_{23} - a_{13}a_{22}, \beta_1 = -a_{11}a_{23} - a_{12}a_{13} \\ &\gamma_1 = a_{11}a_{22} - a_{12}^2 \end{aligned}\right\} \tag{24}$$

$$\left.\begin{aligned} &\alpha_{21} + i\alpha_{22} = \alpha_1(\delta + i\omega), \beta_{21} + i\beta_{22} = \beta_1(\delta + i\omega) \\ &\gamma_{21} + i\gamma_{22} = \gamma_1(\delta + i\omega) \end{aligned}\right\} \tag{25}$$

$$\left.\begin{aligned} &m_1 = e_{15}\gamma_1 - c_{44}(k_1\alpha_1 + \beta_1) \\ &m_2 = e_{15}\gamma_{21} - c_{44}(\delta\alpha_{21} - \omega\alpha_{22} + \beta_{21}) \\ &m_3 = e_{15}\gamma_{22} - c_{44}(\delta\alpha_{22} - \omega\alpha_{21} + \beta_{22}) \end{aligned}\right\} \tag{26}$$

$$\left.\begin{aligned} &M_1 = \beta_1 - \beta_{22}\frac{m_1}{m_3} \\ &M_5 = \left(\frac{m_1}{k_1}\right) - \frac{m_3\delta - m_2\omega}{\delta^2 + \omega^2}\left(\frac{m_1}{m_2}\right) \end{aligned}\right\} \tag{27}$$

Hence equation (21) can be written as

$$F = \frac{8}{3}\left[\frac{M_5}{M_1}\right]\frac{a^3}{R} \tag{28}$$

11.6 Elastic Modulus of Isotropic Thin Films on Substrate

When an indenter penetrates the surface of a thin film on a substrate, the mechanical response is a combination of the film and the substrate. The measured modulus varies as a function of depth of penetration. As the depth of penetration increases, more of the mechanical contribution comes from the substrate. It is not always known what these relative contributions are. In order to measure film-only properties, a commonly used rule of thumb is to limit the indentation depth to less than 10 % of the film thickness. This is commonly known as the Bückle rule for measuring hardness in coatings [30, 31, 32, 33, 34, 35]. The extension of the relation, however, has never been proven for obtaining the indentation modulus of a thin film. Also, this rule is inaccurate since as soon as a measurement is obtained from a film, there is some response from the substrate [33, 36, 37]. Conventional methods consist of various extrapolation techniques on a modulus-depth graph, back to zero depth, to give the value for the film, [22, 37]. A more accurate measurement of a film modulus is obtained if a small indenter radius is used and the modulus difference between the film and substrate is large [33]. The composite film-substrate modulus indentation modulus E may generally be expressed as

$$E' = É_s + (É_f - É_s)\, ß\, (x) \tag{29}$$

or

$$E' = É_f + (É_s - É_f)\, Þ\, (x) \tag{30}$$

where $E' = E/(1-v^2)$ (for isotropic materials) and the subscripts s and f correspond to the substrate and film and $ß$ (or $Þ$) is a function of the relative penetration x, which is a ratio of the contact circle radius, a, to the coating thickness, t (i.e. $x = a/t$). Note that equations (29) and (30) are the same, just presented differently. Therefore, $Þ = 1 - ß$. Fig. 11.4 shows the effect of having a stiffer ($E_{s,1}$) and a more compliant ($E_{s,2}$) substrate than the film [37].

A problem for thin films is that it is difficult to obtain experimental data at low a/t values, and the projection back to zero a/t may not be linear. There are various methods of performing this extrapolation. Five of them have been reviewed [37], the linear, exponential, Gao, Doerner and Nix, and the reciprocal exponential function.

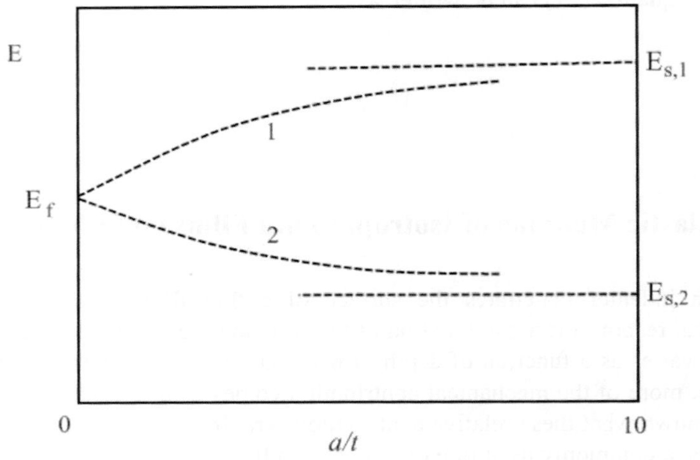

Fig. 11.4 Elastic modulus behaviour with increase of relative penetration into a coated specimen. E is the elastic modulus, and the subscripts f and s correspond to the film and substrate. Curve 1 describes a specimen with a more compliant film on a stiffer substrate *($E_f < E_s$)*; curve 2 shows the opposite case where $E_f > E_s$ [37].

11.6.1 Linear Function

The linear function is the simplest expression way of obtaining the indentation modulus of the film, and is given by

$$y = A + Bx \tag{31}$$

where $y = E'$, $x = a/t$, A and B are constants $Þ(x) = x$, $A = E'_f$, and $B = E'_s - E'_f$.

A and B are determined using linear regression. It was found that the linear function gives a satisfactory result for thick films, when the relative penetration (a/t) is small *(from $a/t \approx 1$ to zero)*. But, it is not valid for $a/t > 1$ [37].

11.6.2 Exponential Function

The exponential function for $\beta(x) = a/t$ is of the form

$$\beta_{\mathbf{exp}} = e^{-\Im x} \tag{32}$$

where $x = a/t$ and $\Im$ is a constant. Substituting equation (32) into equation (29), and taking natural logarithms gives the equation of a straight line. And $y = ln[abs$

(E' - E'ₛ)], A = ln [abs(E'ᶠ − E'ₛ)], B = -ℑ may be obtained using linear regression, which allows the film indentation modulus E'_f to be calculated. The exponential function was found to give unrealistic values at large depths of penetration.

11.6.3 Gao Function

This function is based on the derivation of the contact of a cylindrical indenter with a semi-infinite elastic body. This has a surface layer of a different material.

$$\beta_{Gao} = \frac{2}{\pi}\arctan\frac{1}{x} + \frac{1}{2\pi(1-\nu)}\times\left[(1-2\nu)\frac{1}{x}\ln\left(1+x^2\right)-\frac{x}{1+x^2}\right] \quad (33)$$

where $x = a/t$ and ν is the Poisson's ratio. Equation (29) may be written as

$$E' - E'_s = (E'_f - E'_s)\beta_{Gao} \quad (34)$$

which is in the form of a straight line

$$y = Bx \quad (35)$$

since $y = E' - E'_s$, and $x = \beta_{Gao}(x)$, the film modulus E'_f may be found by

$$E'_f = E'_s + B \quad (36)$$

11.6.4 Doerner and Nix Function

Doerner and Nix proposed the function below

$$\frac{1}{E} = \frac{1}{E_f} + \left(\frac{1}{E_s} - \frac{1}{E_f}\right)e^{-\Im(t/a)} \quad (37)$$

The Doerner and Nix function is an empirical relationship and the values $\Im$ and E_f are found by minimizing the function $Q = \sum\left(E_{meas} - E_{calc}\right)^2$, the sum of the squared difference between the measured modulus and the modulus obtained by using equation (37).

11.6.5 Reciprocal Exponential Function

The Reciprocal exponential function is given by

$$\frac{1}{E}=\frac{1}{E_s}+\left(\frac{1}{E_f}-\frac{1}{E_s}\right)e^{-\Im(a/t)} \quad (38)$$

Again, using logarithms, a straight line equation is obtained $x = a/t$, $y = ln[abs(1/E - 1/E_s)]$. Taking a linear fit of this curve give the values $A = ln[abs(1/E_f - 1/E_s)]$ and $B = -\Im$. The film modulus can then be obtained using $E_f = 1/[(1/E_s) + \kappa e^A]$ where $\kappa = +1$ if $E_f > E_s$, and $\kappa = -1$ if $E_f < E_s$.

It was concluded that the Gao function is the one that provides the best fittings of the experimental data, and that the linear regression is the only other function that should be used [37]. In addition, these methods assume that the film is mechanically isotropic and the Poisson's ratio of the film and substrate are similar, which may not always be so.

In the case of multilayered materials, the indentation modulus profile with depth is not readily normalised by using *a/t* as with a single film on a substrate. So, a different methodology needs to be employed, which is discussed in the next section.

11.7 Analytical Equations for Indentation of Multilayered Materials

Two authors [38, 39] have presented analytical solutions describing the force depth profile for the elastic spherical indentation of a single layer on a substrate, when the film and substrate are elastically isotropic. Kim's solution was for the case of a cylindrical indenter, and Hsueh for spherical indenters.

The indentation of the layered specimen may be described axisymmetrically by a cylindrical coordinate system with coordinates, r, θ, z, where r and z are the horizontal and vertical axes respectively. In this case, the surface of the specimen would be at $z = 0$ and h defines the deflection of this surface by the indenter. The derivation used by both authors is described below.

$$h=\int_0^\infty \frac{\partial h}{\partial z}dz \quad (39)$$

According to the Bousinesq result, the derivative of the axial displacement , h, with respect to z (i.e., the axial strain) is given by

$$\frac{\partial h}{\partial z} = \frac{F(1+v)}{2\pi E}[3r^2 z(r^2+z^2)^{-5/2} - (3-2v)z(r^2+z^2)^{-3/2}] \tag{40}$$

If the surface is coated with a film thickness t_f, the equation becomes

$$h = \int_0^{tf} \frac{\partial h}{\partial z} dz + \int_{tf}^{\infty} \frac{\partial h}{\partial z} dz \tag{41}$$

In order to obtain the equation for penetration of a sphere into a semi-infinite half space, it is necessary to integrate over the pressure distribution for a spherical indenter, q, and the circumference $2\pi r dr$ where $0 \geq r \leq a$ and

$$q = \frac{3F(a^2 - r^2)^{1/2}}{2\pi a^3} \tag{42}$$

for $0 \geq r \leq a$.

This method has been extended to include equations for a three-layer system [13]. In order to calculate the displacement at the surface of the film for a two layer system, equation (41) is extended to

$$h = \int_0^{t_1} \frac{\partial h}{\partial z} dz + \int_{t_1}^{t_1+t_2} \frac{\partial h}{\partial z} dz + \int_{t_1+t_2}^{\infty} \frac{\partial h}{\partial z} dz \tag{43}$$

where t_1 and t_2 are the thicknesses of the layers of the first and second layers respectively, and z is the vertical coordinate. For three layers,

$$h = \int_0^{t_1} \frac{\partial h}{\partial z} dz + \int_{t_1}^{t_1+t_2} \frac{\partial h}{\partial z} dz + \int_{t_1+t_2}^{t_1+t_2+t_3} \frac{\partial h}{\partial z} dz + \int_{t_1+t_2+t_3}^{\infty} \frac{\partial h}{\partial z} dz \tag{44}$$

So the general equation for a multilayer material with *n-1* layers may be written as

$$h = \sum_{i=1}^{n} \int_{\sum_{j=1}^{i} t_{j-1}}^{\sum_{j=1}^{i} t_j} \frac{\partial h}{\partial z} dz \quad \mathrm{i} = 1, 2, 3, \ldots, \mathrm{n}, \; t_0 = 0, \; t_n = \infty \tag{45}$$

The derivative of the axial displacement with respect to z (i.e., the axial strain) is given by equation (40). Substituting equation (40) into (44) gives

$$h = P\left(\int_0^{t_1} \frac{(1+\nu_1)\left(\frac{3r^2 z}{(r^2+z^2)^{5/2}} - \frac{(3-2\nu_1)z}{(r^2+z^2)^{3/2}} \right)}{2E_1\pi} dz + \int_{t_1}^{t_1+t_2} \frac{(1+\nu_2)\left(\frac{3r^2 z}{(r^2+z^2)^{5/2}} - \frac{(3-2\nu_2)z}{(r^2+z^2)^{3/2}} \right)}{2E_2\pi} dz + \int_{t_1+t_2}^{t_1+t_2+t_3} \frac{(1+\nu_3)\left(\frac{3r^2 z}{(r^2+z^2)^{5/2}} - \frac{(3-2\nu_3)z}{(r^2+z^2)^{3/2}} \right)}{2E_3\pi} dz + \int_{t_1+t_2+t_3}^{\infty} \frac{(1+\nu_4)\left(\frac{3r^2 z}{(r^2+z^2)^{5/2}} - \frac{(3-2\nu_4)z}{(r^2+z^2)^{3/2}} \right)}{2E_4\pi} dz \right) \tag{46}$$

where ν_1, ν_2, ν_3, ν_4, E_1, E_2, E_3, and E_4 are the Poisson's ratios and moduli of the first to the third layers and the substrate respectively. This gives an equation for the point loading surface displacement.

$$h = F\left(-\frac{1-\nu_1^2}{E_1\pi\sqrt{r^2}} - \frac{(1+\nu_1)(2r^2(-1+\nu_1)+t_1^2(-3+2\nu_1))}{2E_1\pi(r^2+t_1^2)^{3/2}} + \frac{(1+\nu_2)(2r^2(-1+\nu_2)+t_1^2(-3+2\nu_2))}{2E_2\pi(r^2+t_1^2)^{3/2}} - \frac{(1+\nu_2)(2r^2(-1+\nu_2)+t_2^2(-3+2\nu_2))}{2E_2\pi(r^2+t_2^2)^{3/2}} + \frac{(1+\nu_3)(2r^2(-1+\nu_3)+t_2^2(-3+2\nu_3))}{2E_3\pi(r^2+t_2^2)^{3/2}} - \frac{(1+\nu_3)(2r^2(-1+\nu_3)+t_3^2(-3+2\nu_3))}{2E_3\pi(r^2+t_3^2)^{3/2}} + \frac{(1+\nu_4)(2r^2(-1+\nu_4)+t_3^2(-3+2\nu_4))}{2E_4\pi(r^2+t_3^2)^{3/2}} \right) \tag{47}$$

The elastic penetration (h_e) of the indenter due to the pressure distribution q produced by a spherical indenter is given by

$$h_e = \int_0^a \frac{3r\sqrt{a^2 - r^2}}{a^3} h \, dr$$

$$\begin{aligned} h_e = \int_0^a \frac{3r\sqrt{a^2-r^2}}{a^3} F \Bigg\{ & -\frac{1-v_1^2}{E_1 \pi \sqrt{r^2}} - \frac{(1+v_1)(2r^2(-1+v_1)+t_1^2(-3+2v_1))}{2E_1\pi(r^2+t_1^2)^{3/2}} \\ & + \frac{(1+v_2)(2r^2(-1+v_2)+t_1^2(-3+2v_2))}{2E_2\pi(r^2+t_1^2)^{3/2}} - \frac{(1+v_2)(2r^2(-1+v_2)+t_2^2(-3+2v_2))}{2E_2\pi(r^2+t_2^2)^{3/2}} \\ & + \frac{(1+v_3)(2r^2(-1+v_3)+t_2^2(-3+2v_3))}{2E_3\pi(r^2+t_2^2)^{3/2}} - \frac{(1+v_3)(2r^2(-1+v_3)+t_3^2(-3+2v_3))}{2E_3\pi(r^2+t_3^2)^{3/2}} \\ & + \frac{(1+v_4)(2r^2(-1+v_4)+t_3^2(-3+2v_4))}{2E_4\pi(r^2+t_3^2)^{3/2}} \Bigg\} dr \end{aligned} \tag{48}$$

Integrating the above equation gives

$$\begin{aligned} h_e = \frac{3}{8} F \Bigg\{ & \frac{2(v_1^2-1)}{aE_1} - \left(\frac{\alpha}{a^3 E_1 \pi \sqrt{t_1^2}\sqrt{a^2+t_1^2}} \right) + \left(\frac{\beta}{a^3 E_2 \pi \sqrt{t_1^2}\sqrt{a^2+t_1^2}} \right) - \left(\frac{\delta}{a^3 E_2 \pi \sqrt{t_2^2}\sqrt{a^2+t_2^2}} \right) + \\ & \left(\frac{\gamma}{a^3 E_3 \pi \sqrt{t_2^2}\sqrt{a^2+t_2^2}} \right) - \left(\frac{\theta}{a^3 E_3 \pi \sqrt{t_3^2}\sqrt{a^2+t_3^2}} \right) + \left(\frac{\xi}{a^3 E_4 \pi \sqrt{t_3^2}\sqrt{a^2+t_3^2}} \right) \Bigg\} \end{aligned} \tag{49}$$

where

$$\alpha = (1+v_1)\left(a\left(a^2\pi\sqrt{t_1^2}\sqrt{1+\frac{t_1^2}{a^2}}(v_1-1) + \left(-4t_1^2\sqrt{a^2+t_1^2} + \pi t_1^3\sqrt{1+\frac{t_1^2}{a^2}} \right) v_1 \right) - 2t_1\sqrt{a^2+t_1^2}\left(a^2(v_1-1)+t_1^2 v_1\right) ArcCot\left[\frac{2at_1}{-a^2+t_1^2} \right] \right)$$

$$\beta = (1+v_2)\left(a\left(a^2\pi\sqrt{t_1^2}\sqrt{1+\frac{t_1^2}{a^2}}(v_2-1) + \left(-4t_1^2\sqrt{a^2+t_1^2} + \pi t_1^3\sqrt{1+\frac{t_1^2}{a^2}} \right) v_2 \right) - 2t_1\sqrt{a^2+t_1^2}\left(a^2(v_2-1)+t_1^2 v_2\right) ArcCot\left[\frac{2at_1}{-a^2+t_1^2} \right] \right)$$

$$\delta = (1+v_2)\left(a\left(a^2\pi\sqrt{t_2^2}\sqrt{1+\frac{t_2^2}{a^2}}(v_2-1) + \left(-4t_2^2\sqrt{a^2+t_2^2} + \pi t_2^3\sqrt{1+\frac{t_2^2}{a^2}} \right) v_2 \right) - 2t_2\sqrt{a^2+t_2^2}\left(a^2(v_2-1)+t_2^2 v_2\right) ArcCot\left[\frac{2at_2}{-a^2+t_2^2} \right] \right)$$

$$\gamma = (1+v_3)\left(a\left(a^2\pi\sqrt{t_2^2}\sqrt{1+\frac{t_2^2}{a^2}}(v_3-1) + \left(-4t_2^2\sqrt{a^2+t_2^2} + \pi t_2^3\sqrt{1+\frac{t_2^2}{a^2}} \right) v_3 \right) - 2t_2\sqrt{a^2+t_2^2}\left(a^2(v_3-1)+t_2^2 v_3\right) ArcCot\left[\frac{2at_2}{-a^2+t_2^2} \right] \right)$$

$$\theta = (1+v_3)\left(a\left(a^2\pi\sqrt{t_3^2}\sqrt{1+\frac{t_3^2}{a^2}}(v_3-1) + \left(-4t_3^2\sqrt{a^2+t_3^2} + \pi t_3^3\sqrt{1+\frac{t_3^2}{a^2}} \right) v_3 \right) - 2t_3\sqrt{a^2+t_3^2}\left(a^2(v_3-1)+t_3^2 v_3\right) ArcCot\left[\frac{2at_3}{-a^2+t_3^2} \right] \right)$$

$$\xi = (1+v_4)\left(a\left(a^2\pi\sqrt{t_3^2}\sqrt{1+\frac{t_3^2}{a^2}}(v_4-1) + \left(-4t_3^2\sqrt{a^2+t_3^2} + \pi t_3^3\sqrt{1+\frac{t_3^2}{a^2}} \right) v_4 \right) - 2t_3\sqrt{a^2+t_3^2}\left(a^2(v_4-1)+t_3^2 v_4\right) ArcCot\left[\frac{2at_3}{-a^2+t_3^2} \right] \right)$$

(50).

The above formula is the force *(F)* – penetration *(h_e)* relationship for a three-layer system indented with a sphere.

11.8 Indentation of Sub-Micron PZT 30/70 Thin Films

In this section, results are presented for the spherical indentation of thin films (<1 μm). At these thicknesses, the indentation modulus – *a/t* plots (indenter contact radius–PZT film thickness ratio) are far from linear because of the strong influence of the individual layers. It is therefore necessary to apply the analytical solution for isotropic, multilayer films presented in section 7. The effect of film thickness, indenter radii and poling state are presented.

11.8.1 Method

PZT films with 30/70 (30% $PbZrO_3$ and 70% $PbTiO_3$) composition were produced by Cranfield University using the sol-gel method [13]. A schematic diagram of the layered structure of the films, showing the various PZT film thicknesses, is shown in Fig. 11.5. The silicon wafer had [100] orientation. The PZT films had tetragonal crystal structure with [111] preferred orientation. The films were poled using the corona poling method [13]. This had the advantage of not using a metal top electrode, which would have affected the elastic response of the films.

The measurements were performed with a UMIS 2000 nanoindenter system using a diamond indenter. The indenters used were 5 μm, 10 μm, and 20 μm in radius. The effective elastic modulus and Poisson's ratio of the indenter were taken as 1150 GPa, and 0.07 respectively [12]. The Field and Swain method was used to determine the indentation modulus as a function of *a/t*. The contact radius, *a*, and indentation modulus, were determined at each of 40 load steps by partially unloading to a force fraction of 0.8 of the initial force. The specimens were each indented at 10 locations with one indent at each location. The depth data was corrected for machine stiffness, and the calculations took the variability of indenter radius with depth into account. More details may be found in a previous publication [13].

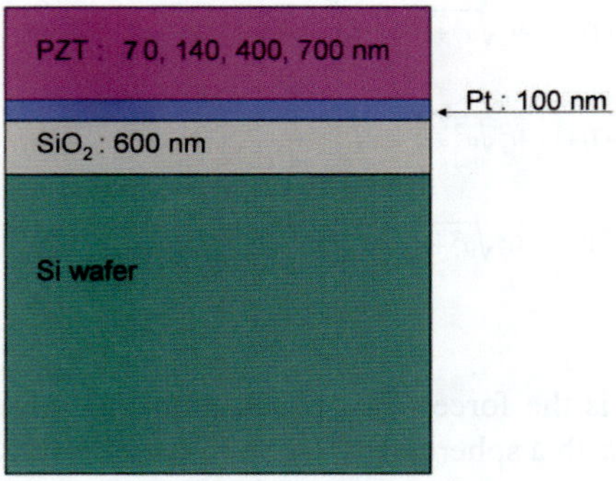

Fig. 11.5 Diagram of PZT thin film specimen showing the thickness of the various layers, and the four different PZT film thicknesses that were investigated.

Calculations using the analytical solution for indentation of multilayered materials introduced in section 11.7 will be compared to the experimental results. The values used in the analytical model are shown in Table 11.1.

Table 11.1 Elastic properties used in the axisymmetric model [40(i), 41(ii)]. The modulus used for Si and Pt in the simulation where the effective moduli calculated from indentation modulus derived from Swadener's indentation modulus solution [27] using available elastic coefficients [42, 43], and assuming the Possion's ratio in that particular orientation.

Material	Poisson's Ratio	Elastic Modulus /GPa
PZT [111]	0.3(ii)	120(ii)
Pt [111]	0.33	170
SiO_2	0.17(i)	65(i)
Si [100]	0.28	152.61

11.8.2 Results

11.8.2.1 Poled Versus Unpoled Films

Fig. 11.6 shows the indentation data for 700nm thick 30/70 PZT film in the poled and unpoled state with a 20 μm indenter. Within the scatter of the data, there is no apparent difference between the films. This is surprising since the electro-mechanical boundary conditions for poled bulk materials have a significant effect on the elastic response of piezoelectric materials [44]. In this case, highly electrically insulating diamond indenters were used suggesting open circuit conditions. So, the poled films would have been expected to be much stiffer than the unpoled films. A similar result is reported in Section 9, where data is presented for thicker unpoled and poled films. One explanation for these observations is that the electrical boundary conditions were effectively closed-circuit because of local charge transport through or across the films. There is a good fit between the analytical model and the experimental data

11.8.2.2 Indentation Modulus Profiles for Films of Various Thicknesses

Assuming that no plastic deformation occurred during the partial unload step, it was expected that the indentation modulus – *a/t* profile of the films would follow that described by the analytical equation. The modulus would start off from the film modulus at *a/t* = 0 and dip as a result of the relatively compliant and thick SiO_2 layer, before tending towards the modulus of the substrate. It was not anticipated that the effect of the Pt layer would be observed experimentally since it is relatively thin. This is generally what is observed in the results (Fig. 11.7 – Fig. 11.10). However, for the 700 nm PZT film, there is a significant discrepancy between what is expected and what is observed experimentally (Fig. 11.10). In this case the observed indentation modulus at higher *a/t* values is some 10 GPa higher than is predicted. The reason for this is not known.

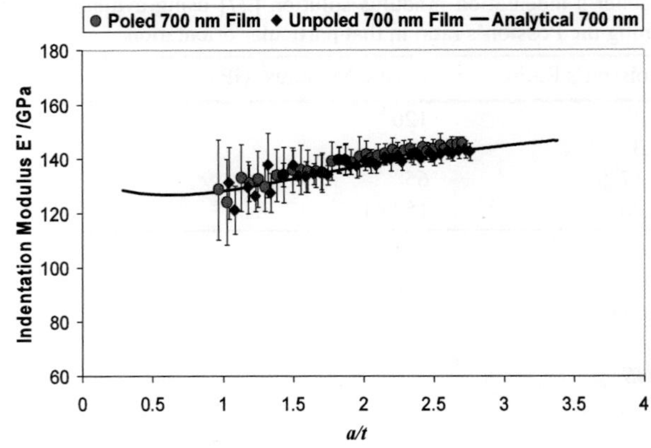

Fig. 11.6 Experimental and analytical results for Indentation Modulus vs. *a/t* for 700 nm poled and unpoled PZT films on Pt/SiO_2/Si substrate indented with a 20 μm indenter. The error bars are the standard deviation of the indentation modulus.

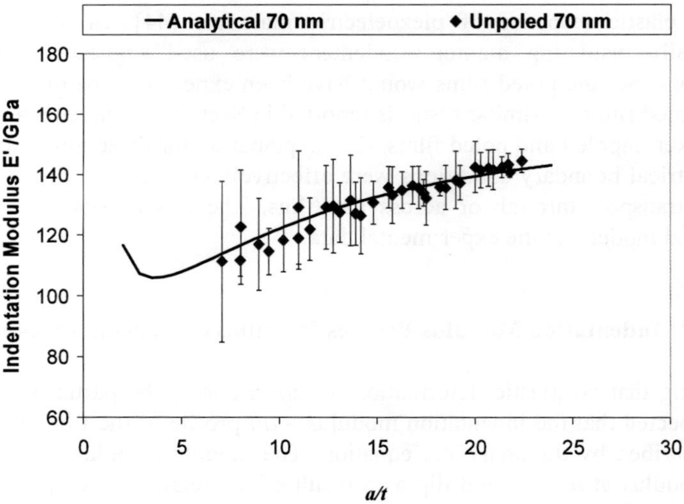

Fig. 11.7 Experimental and analytical results for Indentation Modulus versus *a/t* for 70 nm unpoled PZT film on Pt/SiO_2/Si substrate indented with a 10 μm indenter.

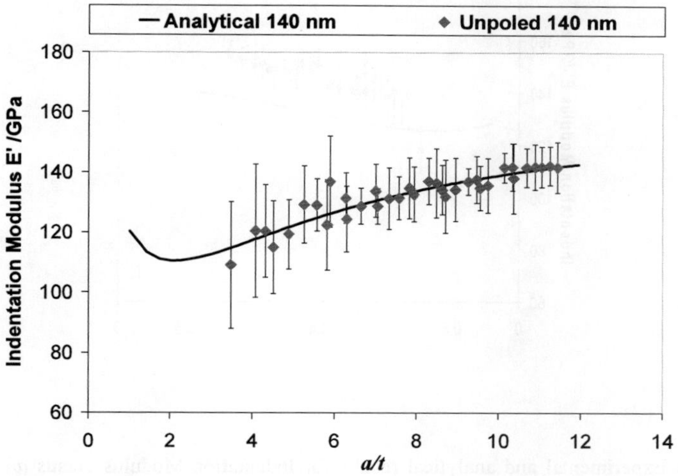

Fig. 11.8 Experimental and analytical results for Indentation Modulus versus *a/t* for 140 nm unpoled PZT film on $Pt/SiO_2/Si$ substrate indented with a 10 μm indenter.

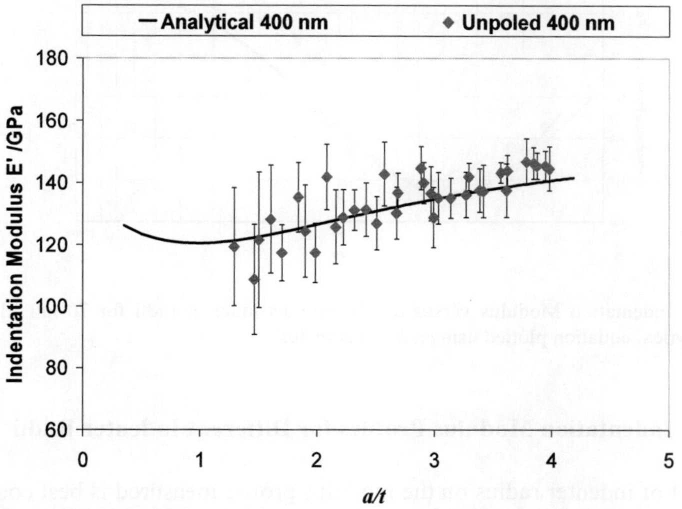

Fig. 11.9 Experimental and analytical results for Indentation Modulus versus *a/t* for 400 nm unpoled PZT film on $Pt/SiO_2/Si$ substrate indented with a 10 μm indenter.

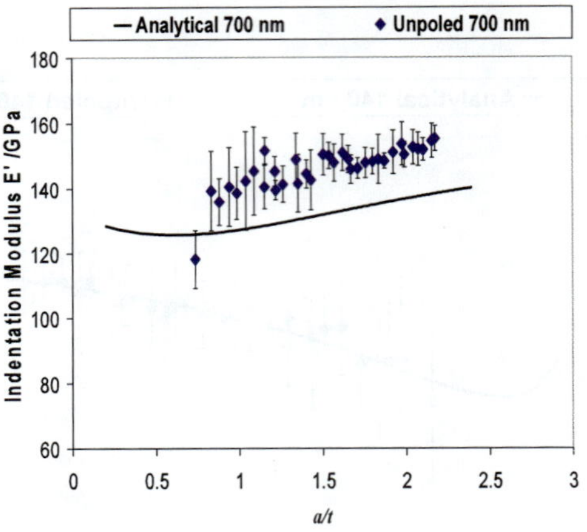

Fig. 11.10 Experimental and analytical results for Indentation Modulus versus *a/t* for 700 nm unpoled PZT film on $Pt/SiO_2/Si$ substrate indented with a 10 μm indenter.

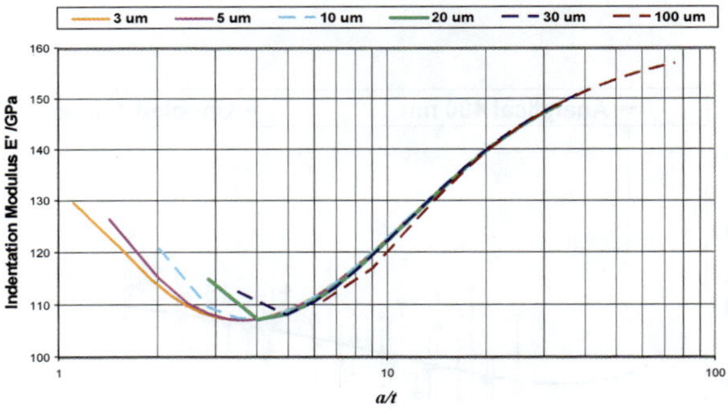

Fig. 11.11 Indentation Modulus versus *a/t* for various indenter radii for 70 nm film obtained using analytical equation plotted using a *log(a/t)* scale.

11.8.2.3 Indentation Modulus Profiles for Different Indenter Radii

The effect of indenter radius on the modulus profile measured is best compared by plotting the output from the analytical equation for different indenter radii for a film of the same thickness (Fig. 11.11). The thinnest film (70 nm) was chosen so that the effect of indenter radius is clearly marked. At very high values of *a/t*, all of the indenters behave similarly because the mechanical properties of the substrate dominate. However, at lower penetration depths, the effect of the dissimilar stress fields in the film layers, with their differing properties, take over and the behaviour of each indenter is unique. At any given penetration, the smaller radius indenters will measure properties closer to the surface of the film-substrate heterostructure.

In this section, the experimental data for multilayer thin films (<1 μm) was compared with the theoretically predicted behaviour calculated, by using an analytical equation [13]. The results for different film thicknesses and indenter radii show a good fit. This gives confidence that the analytical solution can simulate the elastic response of thin films. However, the modelling was performed using effective isotropic elastic properties for textured films. In the next section, the anisotropic nature of the films will be considered.

11.9 Indentation of Thick Films (> 1 μM)

11.9.1 Single Crystal Elastic Coefficients of PZT

In this section, results for thicker PZT films (>1 μm) are considered, where the indentation modulus versus depth of penetration tends towards a linear behaviour. Experimental and theoretical values for the elastic coefficients of single crystal PZT are used to model the behaviour of textured films. This contrasts with the approach that was used in the previous section where extrapolated values for the effective indentation modulus of PZT were used for the modelling. Since there is no analytical solution available for an anisotropic multilayer system, the isotropic solution described in Section.7 is used with theoretically calculated indentation moduli for bulk textured PZT. These are estimated using the Conway [28] and Swadener [27] solutions for anisotropic spherical indentation.

There exists no experimental single crystal elastic data for PZT because of the inability to grow large enough crystals. The experimental data that does exist is for poled polycrystalline ceramics. Berlincourt [45] determined by resonance methods the elastic coefficients for different PZTs with different Zr/Ti compositions. The values obtained represent the statistical average of the single crystal coefficients of the individual crystallites. The effective modulus is therefore dependent on the degree of poling (texture) and electrical boundary conditions. The data has the additional complication that it is not simply the intrinsic elastic response, but has an extrinsic contribution from the ferroelastic deformation produced by the movement of the ferroelectric / ferroelastic domain walls [46]. It is therefore not possible using the experimental data for ceramics to obtain useful elastic coefficients for the modelling of the thin films, which are highly textured. We can, however, use the *ab-initio* data for 50-50 PZT reported by Heifets and Cohen [47].

In this section, a method is presented for estimating the elastic indentation response of 50-50 PZT films with different crystallographic texture. This will be compared with experimental data for thick 52-48 PZT films. The stiffness and compliance matrices for non-polarised, tetragonal, single crystal, 50/50 PZT calculated by Heifets and Cohen [47], where stiffness, c, is in units of GPa and compliance, s, is in units of 1000GPa^{-1} are as follows

$$c = \begin{pmatrix} 250 & 84.5 & 83.6 & 0 & 0 & 0 \\ 84.5 & 250 & 83.6 & 0 & 0 & 0 \\ 83.6 & 83.6 & 92 & 0 & 0 & 0 \\ 0 & 0 & 0 & 24.6 & 0 & 0 \\ 0 & 0 & 0 & 0 & 24.6 & 0 \\ 0 & 0 & 0 & 0 & 0 & 84.1 \end{pmatrix}$$

$$s = \begin{pmatrix} 5.76 & -0.28 & -4.98 & 0 & 0 & 0 \\ -0.28 & 5.76 & -4.98 & 0 & 0 & 0 \\ -4.98 & -4.98 & 19.93 & 0 & 0 & 0 \\ 0 & 0 & 0 & 40.65 & 0 & 0 \\ 0 & 0 & 0 & 0 & 40.65 & 0 \\ 0 & 0 & 0 & 0 & 0 & 11.89 \end{pmatrix}$$

11.9.2 Estimation of Elastic Properties of Textured PZT

In order to simulate textured films, the assumption is made that the film is purely oriented in one orientation. The matrix for a particular orientation can be obtained by the appropriate rotations of the matrices above as described by Auld [48]. The matrices for the tetragonal single crystal [111] orientation are given below

$$c_{[111]} = \begin{pmatrix} 127.84 & 83.45 & 100.88 & 0 & 31.45 & 0 \\ 83.45 & 251.35 & 83.3 & 0 & -0.21 & 0 \\ 100.88 & 83.3 & 180.98 & 0 & 43.67 & 0 \\ 0 & 0 & 0 & 63.64 & 0 & 27.41 \\ 31.45 & -0.21 & 43.67 & 0 & 41.88 & 0 \\ 0 & 0 & 0 & 27.41 & 0 & 43.98 \end{pmatrix}$$

$$s_{[111]}=\begin{pmatrix} 16.31 & -3.39 & -6.10 & 0 & -5.91 & 0 \\ -3.40 & 5.71 & -1.82 & 0 & 4.48 & 0 \\ -6.10 & -1.82 & 11.57 & 0 & -7.50 & 0 \\ 0 & 0 & 0 & 21.60 & 0 & -13.47 \\ -5.91 & 4.480 & -7.50 & 0 & 36.16 & 0 \\ 0 & 0 & 0 & -13.47 & 0 & 31.13 \end{pmatrix}$$

If a film is completely oriented in one direction but random in the transverse directions, it is transverse isotropic. A tetragonal symmetry possesses six or seven elastic constants, whereas a transversely isotropic symmetry has five. The films investigated in this study are highly textured. This was achieved by the use of a seeding layer and appropriate heat treatments [49]. The transversely isotropic state of a film can be created by rotating a set matrix about a chosen axis, and then averaging the strain. This can be applied to any crystal system. The exact consideration of such a system would require details of individual grain sizes that interact with one another on a microscopic scale. Since we do not have this information, a generalised treatment can be made with the assumption of homogeneity. The relation between stress and strain is given by

$$\varepsilon_i = s_{ij}\sigma_j \tag{52}$$

The average strain in the material is calculated after rotating the vertical (polar) axis by a full 360° in steps of 1° the strains at each orientation are summed and then divided by the number of orientations.

Performing a rotation about a crystallographic axis, the matrix effectively becomes transverse isotropic, i.e., there are five elastic coefficients. The s_{66} and c_{66} coefficients are related to the s_{12} and s_{11} (also the c_{12} and c_{11}) coefficients respectively. It must be noted that there is a small error of <1% associated with the s_{66} and the c_{66} through the relationships of

$$s_{66} = 2(s_{11} - s_{12}) \text{ and } c_{66} = \tfrac{1}{2}(c_{11} - c_{12}) \text{ respectively.}$$

[001] orientation transverse isotropic compliance matrix for tetragonal PZT 50/50 calculated with data from Heifets and Cohen [47] in 1000 GPa^{-1}

$$s_{[001\text{-trans}]}=\begin{pmatrix} 5.736 & -0.26 & -4.98 & 0 & 0 & 0 \\ -0.26 & 5.736 & -4.98 & 0 & 0 & 0 \\ -4.98 & -4.98 & 19.915 & 0 & 0 & 0 \\ 0 & 0 & 0 & 40.65 & 0 & 0 \\ 0 & 0 & 0 & 0 & 40.65 & 0 \\ 0 & 0 & 0 & 0 & 0 & 11.98 \end{pmatrix}$$

[100] orientation transverse isotropic compliance matrix for tetragonal PZT 50/50 calculated with data from Heifets and Cohen [47] in 1000 GPa^{-1}

$$S_{[100\text{-trans}]}=\begin{pmatrix} 13.465 & -5.605 & -2.63 & 0 & 0 & 0 \\ -5.605 & 13.465 & -2.63 & 0 & 0 & 0 \\ -2.63 & -2.63 & 5.76 & 0 & 0 & 0 \\ 0 & 0 & 0 & 26.271 & 0 & 0 \\ 0 & 0 & 0 & 0 & 26.271 & 0 \\ 0 & 0 & 0 & 0 & 0 & 38.14 \end{pmatrix}$$

[111] orientation transverse isotropic compliance matrix for tetragonal PZT 50/50 calculated with data from Heifets and Cohen [47] in 1000 GPa^{-1}

$$S_{[111\text{-trans}]}=\begin{pmatrix} 11.16 & -3.578 & -4.021 & 0 & 0 & 0 \\ -3.578 & 11.16 & -4.021 & 0 & 0 & 0 \\ -4.021 & -4.021 & 11.88 & 0 & 0 & 0 \\ 0 & 0 & 0 & 29.092 & 0 & 0 \\ & 0 & 0 & 0 & 29.092 & 0 \\ 0 & 0 & 0 & 0 & 0 & 29.476 \end{pmatrix}$$

11.9.3 Indentation Modulus of Textured Bulk PZT

Following the strain averaging, the transverse isotropic compliance and stiffness matrices were used in conjunction with Conway's [28] and Swadener and Pharr's [27] solutions respectively. The results of the calculated indentation moduli are shown in Table 11.2. The two methods are in good agreement with each other. The values from Table 11.2 are used in the multilayer isotropic model developed by Chima-Okereke *et al.* [13]. The indentation moduli from Table 11.2 are first converted into the effective isotropic elastic modulus using the relation $M=E/(1-\nu^2)$ for isotropic materials where M is the indentation modulus. A Poisson's ratio ν of 0.33 was used. This was estimated using the Hill averaging method [50].

$$\nu=\frac{1}{2}\left[1-\frac{3G}{3K+G}\right],\ \frac{1}{E}=\frac{1}{3G}+\frac{1}{9K} \qquad (53)$$

Where K is the bulk modulus and G is the shear modulus calculated by the Hill averaging method [50] and using the Heifets and Cohen [47] Single crystal data for 50-50 PZT.

Table 11.2 Indentation modulus values obtained using different anisotropic solutions in units of GPa.

Orientation	Conway/GPa	Swadener and Pharr/GPa
[001]	72.3	72.2
[100]	149.5	149.5
[111]	99.3	99.3

11.9.4 Indentation Modulus Profiles for Textured PZT Films

Indentation experiments were carried out on sol-gel prepared 52-48 PZT films spun at 3000 rpm, pyrolysed at 300°C and of [100] and [001] preferred orientation (Table 11.3). The film thicknesses were 1.5 and 2.9 μm. The experimental conditions were the same as those described in section 8.1 Three rows of 25 indentations were made in a straight line with at least 50 μm spacing between them. The standard deviation between the 75 indents was used to estimate the error bars. In all of the figures, the analytically modelled response of [001, 100] and [111] films is included for comparison.

Table 11.3 XRD characterisation of texture.

	(001) %	(100) %	(111) %
1.5 μm unpoled	35	65	-
1.5 μm poled	46.4	43.6	-
2.9 μm unpoled	30.8	68.69	1
2.9 μm poled	56.1	33.3	10.6

Fig. 11.12 shows the experimental data for the 1.5 and 2.9 μm thick, unpoled 52-48 PZT films. The experimental data lies between the predicted lines for the [100] and [001] orientated films as would be expected based on their mixed [100] and [001] texture. The data for the thicker film lies above that of the thinner film as would be expected.

Fig. 11.13. shows the experimental data for a 1.5 μm unpoled film obtained with different indenter radii. Within the scatter, the data overlaps as predicted by the analytical plots for $a/t > 0.1$.

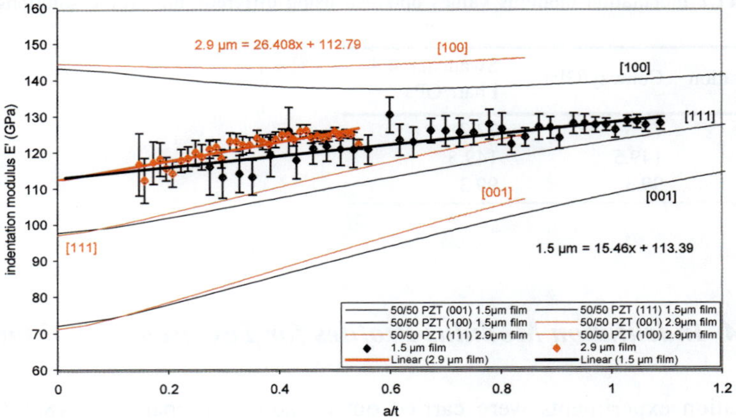

Fig. 11.12 Indentation data for unpoled 52-48 PZT films as a function of film thickness of 1.5 µm and 2.9 µm obtained with a 5 µm radius indenter.

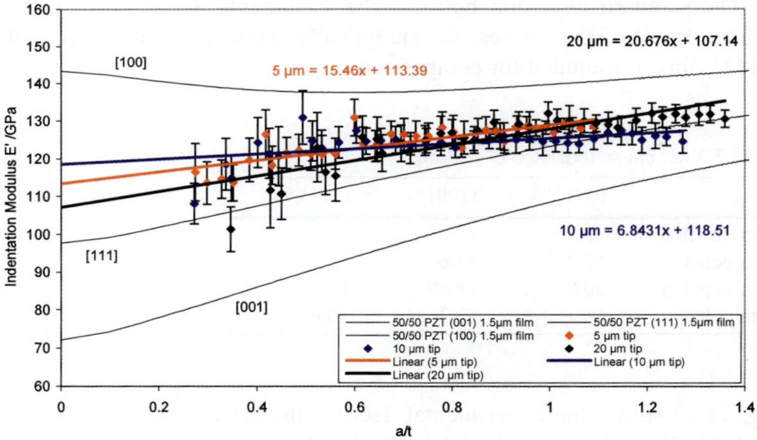

Fig. 11.13 Indentation data for unpoled 52-48 PZT film of 1.5 µm thickness as a function of indenter radius of 5 µm, 10 µm and 20 µm.

Fig. 11.14 compares the indentation behaviours of unpoled and poled 1.5 µm thick films. The data for the films show a similar trend. This is similar to the results for unpoled and poled thin 30-70 PZT films (Fig. 11.6). The differences in texture between the unpoled and poled films is small (Table 11.3) and would not produce a significant difference in the indentation modulus of the films, considering the scatter of the experimental data. As commented on earlier, the fact that the unpoled and poled behaviours are similar suggests that the electrical boundary conditions were effectively closed-circuit.

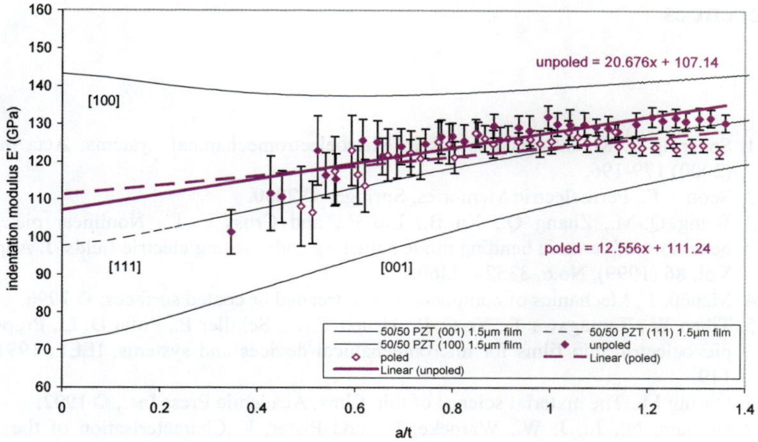

Fig. 11.14 Indentation data for 52-48 composition films as a function of polarity (poled versus unpoled).

11.10 Conclusions

The nanoindentation method can provide elastic data for submicrometer ferroelectric thin films with a spatial resolution of the order of micrometers. An analytical method has been developed that allows the response of a multilayer system to be predicted. The experimental data shows the trends predicted for different film thicknesses, orientations and indenter radii. To fully characterise the elastic properties of a textured ferroelectric film requires the determination of five elastic coefficients. This is beyond the capability of the experimental analytical techniques. However, by the use of *ab-initio* elastic data and modelling, it is possible to gain quantitative information about the elastic properties of ferroelectric thin films.

Acknowledgement

We would like acknowledge the assistance of Dr Luc Vandeperre, Imperial College, London, with the averaging method for the transverse isotropic calculations. We would like to thank our collaborators at Cranfield University, Dr Qi Zhang, Roger Whatmore and Silvana Corkovic for the preparation and characterisation of the films. Part of the work was supported by an EPSRC grant GR/S45034/01.

References

1. Spearing S. M., Material Issues in Microelectromechanical Systems; Acta mater. 48 (2000) 179-196.
2. Scott J. F., Ferroelectric Memories, Springer © 2000.
3. Wang Q.-M., Zhang Q., Xu B., Liu R., and Cross E. L., Nonlinear piezoelectric behaviour of ceramic bending mode actuators under strong electric fields, J. Appl. Phys. Vol. 86 (1999), No.6, 3352 – 3360.
4. Menĉík J., Mechanics of components with treaded or coated surfaces, © 1996.
5. Tjhen W., Tamagawa T., Ye C.-P., Hsueh C.-C., Schiller P., Polla D. L., Properties of piezoelectric thin films for micromechanical devices and systems, IEEE, 1991, 114 – 119.
6. Ohring M., The material science of thin films, Academic Press Inc., © 1992.
7. Hossain, N., Ju, J. W., Warneke, B., and Pister, K.,Characterisation of the Young's Modulus of CMOS Thin Films, Mechanical Properties of Structural Films, ASTM STP 1413, C. L. Muhlstein and S. B. Brown, Eds., American Society for Testing and Materials, West Conshohocken, PA, Online, Available: www.astm.org/STP/1413/1413_15, 1 July 2001.
8. Schneider D., Siemroth P., Schülke T., Berthold J, Schultrich B., Schneider H. H., Ohr R., Petereit B., and Hillgers H.; Quality control of ultra-thin and super-hard coatings by laser-acoustics, Surface and Coating Technology 153 (2002) 252-260.
9. Fischer-Cripps A. C., Simulation of sub-micron indentation tests with spherical and Berkovich indenters; J. Mater. Res., Vol. 16, No. 7,(2001) 2149- 2157.
10. Fischer-Cripps A. C., Study of analysis methods of depth-sensing indentation test data for spherical indenters; J. Mater. Res., Vol. 16, No. 6,(2001) 1579 – 1584.
11. Fischer-Cripps A. C., Use of combined elastic modulus in the analysis of depth-sensing indentation data; J. Mater. Res., Vol. 16, No. 11, (2001) 3050 – 3052.
12. Algueró M., Bushby A. J., and Reece M. J., Direct measurement of mechanical properties of (Pb,La)TiO3 ferroelectric thin films using nanoindentation techniques; J. Mater. Res., Vol. 16, No.4, (2001), 993-1002.
13. C. Chima-Okereke, A.J. Bushby, M.J. Reece, R.W. Whatmore and Q. Zhang, Experimental, analytical, and finite element analyses of nanoindentation of multilayer PZT/Pt/SiO2 thin film systems on silicon wafers, J. Mater. Res., Vol. 21, No. 2, (2006), 409 – 419.
14. Algueró M., Bushby A. J., and Reece M. J., Poyato R., Ricote J.,Calzada M. L., Pardo L., Stress-induced depolarisation of (Pb,La)TiO3 ferroelectric thin films by nanoindentation; Appl. Phys. Lett., Vol. 79, No. 23, (2001), 3830 – 3832.
15. Algueró M., Bushby A. J., and Reece M. J., Anelastic deformation of Pb(Zr,Ti)O3 thin films by non-180o ferroelectric domain wall movements during nanoindentation; Appl. Phys. Lett., Vol. 81, No. 3, (2002), 421 – 423.
16. V. Koval, M.J. Reece, A.J. Bushby Enhanced Ferroelectric Loop Asymmetry of Lead Zirconate Thin Films, J Applied Physics, 101, 0241131-8, (2007).
17. V. Koval, M.J. Reece and A.J. Bushby, Ferroelectric / Ferroelastic Behaviour and Piezoelectric Response of PZT Thin Films Under IndentationJ Applied Physics, 97, 074301-1 -7 (2005).
18. Delobelle, P. Guillon, O., Fribourg-Blanc, E., Soyer, C., Cattan, E., and Rèmiens, D., Appl. Phys. Lett., 85, 22, (2004), 5185-5187.
19. Delobelle, P., Fribourg-Blanc, E., and Rèmiens, D., Thin Solid Films, 515, (2006), 1385-1393.

20. Delobelle, P., Wang, G. S., Fribourg-Blanc, E., and Rèmiens, D., Surface & Coatings Technology, 201, (2006), 3155-3162.
21. Delobelle, P., Wang, G. S., Fribourg-Blanc, E., and Rèmiens, D., Journal of the European Ceramic Society, 27, (2007), 223-230.
22. Bushby A. J., Nanoindentation using spherical indenters; Non-destructive testing and evaluation, Vol. 17, (2001), 213-234.
23. Herbert E. G., Pharr G. M., Oliver W. C., Lucas B. N., Hay J. L., On the measurement of stress-strain curves by spherical indentation; Thin Solid Films, 398 – 399 (2001) 331 – 335.
24. Johnson K. L., Contact Mechanics, Cambridge University Press © 1985.
25. Martin M., Taylor M., Fundamental relations used in nanoindentation: Critical examination based on experimental measurements; J. Mater. Res., Vol. 17, No. 9 (2002) 2227 – 2234.
26. Fischer-Cripps A. C., Methods of correction for analysis of depth-sensing indentation test data for spherical indenter; J. Mater. Res., Vol. 16, No. 8, (2001) 2244 – 2250.
27. Swadener J. G., and Pharr G. M., Indentation of elastically anisotropic half-spaces by cones and parabolae of revolution; Phil. Mag. A, Vol. 81, No.2, (2001) 447-466.
28. Conway H. D., Farnham K. A., Ku T. C., The indentation of a Transversely isotropic half space by a rigid sphere; Journal of Applied Mechanics, Vol. 34 (1967), No. 2, 491 – 492.
29. Ramamurty U., Sridhar S., Giannakoplous A. E., and Suresh S., An experimental study of spherical indentation on piezoelectric materials; Acta Mater., Vol. 47, No. 8, 1999, 2417-2430.
30. Bull S. J., Korsunsky A. M., Mechanical properties of thin carbon overcoats, Tribology International Vol. 31, No. 9, 547-551, 1998.
31. Chen X., Vlassak J. J., Numerical study on the measurement of thin film mechanical properties by means of nanoindentation; J. Mater. Res. Vol. 16, No. 10 (2001) 2974 – 2982.
32. Chudoba T., Schwarzer N., Richter F., Determination of elastic properties of thin films by indentation measurement with a spherical indenter; Surface and Coatings Technology 127 (2000) 9-17.
33. Chudoba T., Schwarzer N., Richter F., Steps towards mechanical modelling of layered systems; Surface and Coatings Technology, Vol. 154 (2002) 140-151.
34. Saha R., Nix W. D., Effects of substrate on the determination of thin film mechanical properties by nanoindentation; Acta Mater., Vol. 50 (2002), 23 – 38.
35. Tsui T. Y., Vlassak J., Nix W. D., Indentation plastic displacement field; Part I. The case of soft films on hard substrates; J. Mater. Res., Vol. 14, No. 6, (1999) 2196 – 2203.
36. Kouitat-Njiwa R., Jürgen von Stebut, Boundary element numerical analysis of elastic indentation of a sphere into a bi-layer material, International Journal of Mechanical Sciences, Vol. 45, (2003) 317 – 324.
37. Menčík J., Munz D., Quandt E., Weppelmann E.R., and Swain M.V., Determination of elastic modulus of thin layers using nanoindentation; J. Mater. Res., Vol. 12, No. 9, Sep 1997, 2475-2484.
38. Kim M. T., Influence of substrates on the elastic reaction of films for the microindentation tests; Thin Solid Films, Vol. 283, (1996), 12 – 16.
39. Hsueh C-H. ,and Miranda P., Master curves for Hertzian indentation on coating/substrate systems, J. Mater. Res., Vol.19, No.1, (2004), 94-100.
40. Kim J. – K., Kim N. – K., Park B. – O., Effects of ultrasound on microstructure and electrical properties of Pb (Zr0.5 Ti0.5) O3 thin films prepared by sol–gel method; Materials Letters 39 (1999) 280 – 286.

41. Algueró M., Bushby A. J., Hvizdos P., Reece M. J., Whatmore R. W., Zhang Q., Mechanical and electromechanical properties of PZT sol-gel thin films measured by nanoindentation, Integrated Ferroelectrics, Vol. 41, (2001), Part ¼, 53 – 62.
42. Brantely W.A., Calculated elastic constants for stress problems associated with semiconductor devices, J.Appl.Phys. Vol.44, no.1 pp 534-535, 1973.
43. Lide D. R. (Editor-in-Chief), Handbook of chemistry and physics, *1st edition, 2000-2001, CRC Press.
44. Jaffe B., Cook W. R., Jaffe H., Piezoelectric Ceramics.
45. Berlincourt D. A., Cmolik C., Jaffe H., Proceedings of IRE, 48, 200 – 209, 1960.
46. Damjanovic D., Ferroelectric, dielectric and piezoelectric properties of ferroelectric thin films and ceramics; Rep. Prog. Phys. 61 (1998) 1267-1324.
47. Heifets E. and Cohen R. E., Ab initio study of elastic properties of Pb(Zr,Ti)O3, Fundamental Physics of Ferroelectrics, AIP Conference Proceedings(NY) 626, (2002), p150-159.
48. Auld B. A., Acoustic fields and waves in solids, Vol. 1, 1973.
49. Marshall, J.M. & Corkovic, S. & Zhang, Q. & Whatmore, R.W. & Chima-Okereke, C. & Roberts, W.L. & Bushby, A.J. & Reece, M.J. (2006) "The electromechanical properties of highly 100.oriented Pb(Zr0.52Ti0.48)O3, PZT.thin films" , Integrated Ferroelectrics, vol. 80, page 77-85.
50. Hill, R., *Proc. Phys. Soc.*, A55, p349-354, 1952

Chapter 12
Pyroelectricity in Polycrystalline Ferroelectrics

R. Jiménez, B. Jiménez.

12.1 Introduction

12.1.1 History

Pyroelectricity (from Greek *Pyro* (fire) and *electricity*) is the electrical potential created in certain materials when they are heated. As a result of a change in temperature, positive and negative charges accumulate or move to opposite ends of the material and hence, an electrical potential is established.

The first reference to pyroelectric effect is by Theophrastus in 314 BC, who noted that tourmaline becomes charged because it attracted bits of straw and ash when heated. Tourmaline's properties were reintroduced in Europe in 1707 by Johann George Schmidt, who also noted the attractive properties of the mineral when heated. Pyroelectricity was first described by Louis Lemery in 1717. In 1747, Linnaeus first related the phenomenon to electricity, although this was not proven until 1756 by Franz Ulrich Thodor Aepinus. In 1824, Sir David Brewster gave the effect the name it has today. William Thomson in 1878 and Voight in 1897 helped develop a theory for the processes behind pyroelectricity. Pierre Curie and his brother, Jacques Curie, studied pyroelectricity in the 1880s, leading to their discovery of some of the mechanisms behind piezoelectricity.

Crystallographic studies have allowed the identification of those structures that show pyroelectricity. They are summarised as follows [1]: "Of the 32 crystal classes, 21 are non-centrosymmetric (not having a centre of symmetry). Of these, 20 exhibit direct piezoelectricity, the remaining one being the cubic class 432. Ten of these are polar (i.e., spontaneously polarise), having a dipole in their unit cell, and exhibit pyroelectricity. If this dipole can be reversed by the application of an

Instituto de Ciencia de Materiales de Madrid. Consejo Superior de Investigaciones Científicas, 28049, Madrid, Spain.
riqjim@icmm.csic.es

electric field, the material is said to be ferroelectric". All polar crystals are pyroelectric, so the 10 polar crystal classes are sometimes referred to as the pyroelectric classes.

Piezoelectric Crystal Classes (point groups) are : 1, 2, m, 222, mm2, 4, -4, 422, 4mm, -42m, 3, 32, 3m, 6, -6, 622, 6mm, -62m, 23, -43m.

Pyroelectric point groups are: 1, 2, m, mm2, 3, 3m, 4, 4mm, 6, 6mm.

To understand the piezoelectric properties of ceramics was necessary for the development of texture point groups. These textures consist of aggregates of crystallites randomly arrayed in a plane, but possessing elements of order in a direction normal to that plane. Of the seven groups that have been described by Shubnikov [2], two (∞ and ∞ mm) satisfy the above symmetry requirements and may exhibit pyroelectricity: they possess an infinite-fold rotational axis and an infinite-fold rotational axis at the intersection of infinity of mirror planes, respectively.

Although pyroelectric effect was first discovered in minerals such as tourmaline and much later in ferroelectrics and other ionic crystals, there are materials whose pyroelectric properties do not depend on a specific crystalline structure but on engineering of the material. We refer specifically to polymers and semiconductors. Even some biological tissues as bone, tendon and some woods have pyroelectric properties.

To have a wider vision of materials with pyroelectric properties, we believe it interesting to introduce the concept of electret [3]. Electret (*elektr-* from "electricity" and *-et* from "magnet") is a dielectric material that has a quasi-permanent electric charge or dipole polarisation. An electret generates internal and external electric fields, and is the electrostatic equivalent of a permanent magnet. Oliver Heaviside coined this term in 1885. Materials with electret properties were, however, already studied since the early years of the 18th century. One particular example is the electrophorus, a device consisting of a slab with electret properties and a separate metal plate. The electrophorus was originally invented by Johan Carl Wilcke in Sweden and again by Alessandro Volta in Italy.

Cellular space charge electrets, with internal bipolar charges at the voids, provide a new class of electret materials that mimic ferroelectrics. Hence, they are known as ferroelectret. Ferroelectrets display strong piezoelectricity, comparable to ceramic piezoelectric materials. Some dielectric materials are capable of acting both ways. An overview on the history and bright future of Electret Science may be found in the work of Gerhard–Multhaupt [4, 5].

Pyroelectricity is definitely not a new concept, but its research and application continues to this day. There are more than twenty guides to pyroelectricity and pyroelectric materials, which are quite useful in following scientific development over the years in the knowledge of pyroelectric materials and the mechanisms behind this phenomenon. These guides cover the references of the most significant works on pyroelectricity from 1970 to today [6]. We want to mention the interesting works of Lang [7, 8] concerning the history of the pyroelectricity and its natural existence in minerals, plants and animals giving an idea on its widespread presence in nature.

12.1.2 Pyroelectric Materials

The evolution of the pyroelectricity in the science of materials began when ferroelectric materials burst upon the field of technological applications. This must be attributed to its reversible spontaneous electrical polarisation that exhibits a great sensitivity to small changes of temperature in a margin about 200°C below the structural phase transition temperature.

Materials displaying pyroelectric properties may appear under different physical forms:

- Single or poly-crystalline: ferroelectric single crystals and ceramics
- Polymers and electrets
- Crystalline thin films.

12.1.2.1 Ferroelectric Single Crystals and Ceramics

Table 12.1 Some well known pyroelectric materials are listed [9].

Material		$\gamma(10^{-8}C\cdot cm^{-2}K^{-1})_{300K}$	$\Re_V\ (10^{-10}V\cdot W^{-1})$
$BaTiO_3$	(S.C.)	7.00	1.10
GASH	(S.C.)	0.10	2.00
$LiNbO_3$	(S.C.)	0.40	0.50
$LiTaO_3$	(S.C.)	1.70	1.30
$NaNO_2$	(S.C.)	0.40	2.50
PL_xZT	(ceramic)	3.50—17.00	2.00
$Sr_x\ Ba_{1-x}Nb_2O_6$	(S.C.)	4.20—28.00	0.30
TGS	(S.C.)	3.50	8.50
PVF_2	(thick film)	0.24	

(S.C.= Single Crystal. γ= pyroelectric coefficient. $\Re_v$ = responsivity. These parameters will be explained in section 4). PVF_2 is a polymer. GASH: Guanidine Aluminium Sulphate. TGS: Triglycine Sulphate.

A more comprehensive list of pyroelectric materials may be found in the work of Muralt [10].

Among ferroelectric single crystals, we emphasise the organic compound triglycine sulphate TGS [11] due to its high pyroelectric coefficient at room temperature and the ease of growing that makes it a good candidate for use in technological applications. Many of the first vidicon applications (see section 4 for an explanation) use TGS single crystals [12]. The inorganic Lithium tanatalate $LiTaO_3$ is a crystal exhibiting both piezoelectric and pyroelectric properties, which has been used in infrared sensor devices [13].

Beside the ferroelectric single crystals, polycrystalline ferroelectrics have received most of the attention when the use of these materials in technological applications becomes widespread [14, 15]. Material engineering allows for designing solid solutions with improved pyroelectric properties. It is possible today to design stocks in ferroelectric compounds that constitute the base of their technological applications. Concerning semiconductor materials, thin film technology has allowed for designing artificial pyroelectric materials based on these compounds [16]. We must mention the recent appearance of pyroelectric and semiconductor granular systems operating in the terahertz range of frequencies [17, 18]. This is due to the coupling between electromagnetic waves and movable electrons and/or hole islands that should exist in these materials.

12.1.2.2 Polymers and Electrets

Most of the work on polymers has been interpreted by assuming that they are electrets that can be included in one of the two following groups.

1. Real-charge electrets which contain excess charge of one or both polarities, either on the dielectric's surfaces (a surface charge), or within the dielectric's volume (a space charge).
2. Oriented-dipole electrets contain oriented (aligned) dipoles. Ferroelectric materials are a variant of these.

Materials with extremely high resistivity, such as Teflon [19] may retain excess charge for many hundreds of years. Most of the commercially produced electrets are based on fluoropolymers (e.g., Amorphous Teflon) machined to thin films, polyvinyl fluorides, derivatives of phenylpyrazine, and cobalt phthalicuanine.

Non-crystalline polymeric materials have acquired great importance in technological applications because they can develop pyro and piezoelectric properties after specific treatment [20, 21, 22, 23]. However, among the most usual polymers, we find the well-known semi-crystalline PVDF and the copolymer P(VDF-TrFE). In these materials, during the poling process, charges are injected from the electrodes into the film. These charges are caught in Coulomb traps at the surfaces of the poled crystallites causing a mutual stabilisation of charges and dipoles. This is the reason for the remnant polarisation after switching off the electric field. The binding energy of the charges is equal to the activation energy of the polarisation, which means that setting free of charges from the traps is equivalent to the destruction of polarisation.

12.1.2.3 Crystalline Thin Films

Significant new insights into the pyroelectric behaviour of ferroelectrics have been extracted from the development in the methods of preparation of materials in thin film form: laser ablation, ion beam deposition, metal-organic deposition, spinning deposition, etc. [24, 25, 26] The deposition on semiconductor substrates has allowed the integration of pyroelectric materials in micro and nanotechnologies, creating high expectations for promising technological applications of these materials. All of the known inorganic ferroelectric compounds today can be prepared in the form of polycrystalline thin film.

In the work of Muralt [10], the reader can find a large list of pyroelectric thin film materials grouped in the following classes:

- Thin films on silicon substrates.
- Thin films grown epitaxially on crystalline substrates.
- Thin films suitable for induced pyroelectricity.

In the past decade, ferroelectric thin films have attracted increasing attention for application in non-volatile random access memories due to their bi-stable polarisation [27, 28]. These devices are required to have long-term stability and reliability under various operating conditions.

Concerning semiconductor materials, progress has been made in creating artificial pyroelectric materials, usually in the form of a thin film, as Gallium nitride (GaN), Gallium nitrate ($GaNO_3$) [29].

The thermoelectric response of semi-amorphous thin films [30, 31] semiconducting (superconducting) Y–Ba–Cu–O thin films [16] was investigated by illumination with 150 ps optical pulses at a wavelength of 1064 nm and with a continuous wave at 493 nm ~argon laser. The measured unbiased voltage response was consistent with pyroelectricity.

12.2 Pyroelectric Effect

12.2.1 Background on Pyroelectricity

The requirement for a material to be pyroelectric is the existence of spontaneous electrical polarisation, either of intrinsic character or induced by an electric field that varies with changing temperature.

Under normal circumstances, even polar materials do not display a net dipole moment. As a consequence, there are no electric dipole equivalents of bar magnets because the intrinsic dipole moment is neutralised by "free" electric charge that builds up on the surface by internal conduction or from the ambient atmosphere. Polar materials and crystals only reveal their nature when perturbed in such a way that the balance between the dipoles and the compensating surface charges is momentarily lost.

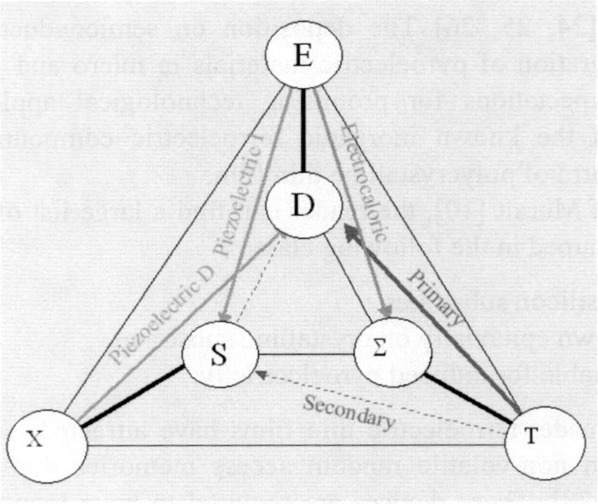

Fig. 12.1 The thermodinamically reversible interactions that may occur among the thermal, electrical and mechanical properties of a crystal (Heckmann Diagram). E = electric field, X = mechanical stress, T = temperature, D = electric displacement, S = Strain, Σ = entropy.

Pyroelectricity represents one of the linear relationships between mechanical, electrical and thermal variables as laid out in the Heckmann Diagram, Fig. 12.1 [32].

The change of the vector of spontaneous polarisation P with temperature T defines the pyroelectric coefficient:

$$\gamma = \frac{\Delta P}{\Delta T} = \frac{dP}{dT} \tag{1}$$

From the relations that can be obtained between the extensive variables in the inner triangle of the Heckmann Diagram, and the intensive thermodynamic variables in the outer triangle (Fig. 12.1), we take the extensive variable, the electric displacement, D, as perfect differential, according to the methods of classic thermodynamics. Then we can write [33]

$$dS_i = (s_{ij})^{E,T} dX_j + (d_{im})^T dE_m + (\alpha_i)^E dT \tag{2}$$

$$dD_n = (d_{nj})^T dX_j + (\varepsilon_{nm})^{X,T} dE_m + (\gamma_n)^X dT \tag{3}$$

Where, S = Strain, X = Stress, E = Electric field, T = Temperature, D = P + εE = Electric displacement, s_{ij} = Elastic compliance, d_{im} = Piezoelectric coefficient, α_i = Thermal expansion coefficient, γ_n = Pyroelectric coefficient

The equations (2), (3) become

$$dS_i = (s_{ij})^{E,T} dX_j + (\alpha_i)^E dT \tag{4}$$

$$dP_n = (d_{nj})^T dX_j + (\gamma_n)^X dT \tag{5}$$

In a stress free conditions (X = 0) the pyroelectric effect is $dP_n = dP = (\gamma_n)^X dT$.

In a clamped crystal, dS = 0 it is convenient to use the strain as independent variable resulting in the equations

$$dX_i = (c_{ij})^{E,T} dS_j - (e_{im})^T dE_m + (\lambda_i)^E dT \tag{6}$$

$$dD_n = (e_{nj})^T dS_j + (\varepsilon_{nm})^{S,T} dE_m + (\gamma_n)^S dT \tag{7}$$

Where: e_{ni} is the piezoelectric constant, λ_I the thermal stress coefficient and γ^S the pyroelectric coefficient at constant strain

At constant E, dE = 0, dD = dP and we have the relation between γ^X and γ^S

$$dP = (e_{nj})^T dS_j + (\gamma_n)^S dT \tag{8}$$

By taking into account expression (4), we have

$$dP = (e_{nj})^T \left[(s_{ij})^{E,T} dX_i + (\alpha_i)^E dT \right] + (\gamma_n)^S dT \tag{9}$$

At constant stress, dX = 0. Therefore, we can write

$$(e_{nj})^T + (\alpha_i)^E dT + (\gamma_n)^S dT = (\gamma_n)^X dT \tag{10}$$

and the pyroelectric coefficients at constant stress, γ^X and constant strain, γ^S, are related in the form :

$$\gamma^X = \gamma^S + e_{ni}\alpha_i \tag{11}$$

Experimentally, it is difficult to clamp the crystal to measure the primary γ^S coefficient. Hence, it is calculated by the equation (11) if piezoelectric, elastic, thermal expansion and γ^X are known. For planar symmetry (∞mm, 6mm) we can write

$$\gamma^X = \gamma^S + 2e_{31}\alpha_1 + e_{33}\alpha_3 \tag{12}$$

The pyroelectric coefficient measured at constant stress is often called the total pyroelectric coefficient and is the sum of the pyroelectric coefficient at constant strain (primary pyroelectric effect) and the piezoelectric contribution from thermal expansion (secondary pyroelectric effect). As an example of the importance of the primary pyroelectric coefficient on the total one, the measured pyroelectric coefficient and the calculated primary coefficient for a $BaTiO_3$ ceramic are included in Fig. 12.2 it may be verified that the primary pyroelectric coefficient is the main contribution to the total pyroelectric effect in this material [34].

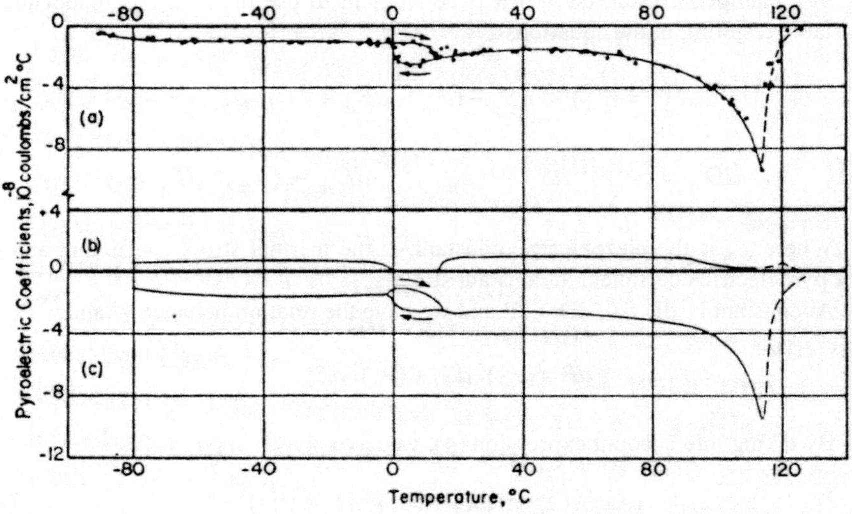

Fig. 12.2 Pyroelectric coefficient vs. temperature in $BaTiO_3$; (a) total pyroelectric coefficient γ^X, (b) secondary pyroelectric coefficient (γ''), (c) primary pyroelectric coefficient (γ^S) Reproduced with permission from reference 34.

The above derivation of the pyroelectric coefficient is for the case of zero electric field. In the case of non-zero electric field (internal or external), the contribution to the total pyroelectric coefficient coming from the evolution of the sample dielectric constant with temperature must be included. This extra contribution is due to the induced polarisation. The total pyroelectric coefficient measured at constant stress becomes:

$$\gamma^X = \left(\frac{\partial P_S}{\partial T}\right)_{E=0} + \frac{\partial \varepsilon}{\partial T} E \tag{13}$$

Where E is the electric field and ε is the real part of the material permittivity.

This correction to the pyroelectric coefficient means that by applying an external electric field to the sample, the pyroelectric coefficient can be modified.

The nature of the polarisation that appears in (8) and (9) is also an open question. In the case of zero electric field, some authors identify such a polarisation as the remnant polarisation (P_r) instead of the spontaneous polarisation P_s.

12.2.2 Pyroelectricity Fundamentals in Thin Films

The physical conditions of a pyroelectric material in the form of thin film are different from those of bulk material. For this reason, the electrical properties as defined for the bulk material will be influenced significantly by the boundary conditions such as limited thickness, microstructure, nature of the substrate, and semiconductor-metal-dielectric interfaces.

There are several ways to address the theoretical modelling of the pyroelectric effect in thin films, depending on the deposition conditions and thermal treatment of the deposited film. The lattice mismatch between the film and the substrate is one the most important factors in the case of epitaxial films. Therefore, it could be said that every thin film preparation condition should have its own theoretical treatment. In general, films can be considered as partially clamped by the substrate in the plane and partially free along its thickness. Therefore, by the use of specific boundary conditions, we can derive the pyroelectric coefficients of films [33].

Assuming that the film is free to expand along its thickness, but only to contract or expand coupled with the substrate in the plane of the film, we shall use the above derivation of the pyroelectric coefficient for bulk material with specific boundary conditions [33].

We first try the case when the temperature of the substrate is constant. The strains in the film are $S_1 = S_2 = 0$ and the thermal stresses $X_1 = X_2 \neq 0$ and $X_3 = 0$. Under these conditions, we shall have

$$0 = dS_1 = dS_2 = (s_{11} + s_{12})dX_1 + \alpha_1 dT \tag{14}$$

$$dS_3 = 2s_{31}dX_1 + \alpha_3 dT \tag{15}$$

$$dP_3 = 2d_{31}dX_1 + \gamma_3^X dT \tag{16}$$

with $s_{31} = s_{32}$ and $d_{31} = d_{32}$. From these equations we obtain

$$dX_1 = -\alpha_1 \frac{dT}{(s_{11} + s_{12})} \tag{17}$$

$$dP_3 = -\left(\alpha_1 \frac{2d_{31}}{(s_{11} + s_{12})}\right)dT + \gamma_3^X dT \tag{18}$$

The planar clamped pyroelectric coefficient will be

$$\gamma^{pc} = \frac{dP_3}{dT} = \gamma_3^X - \frac{2d_{31}\alpha_1}{(s_{11} + s_{12})} \tag{19}$$

Secondly, we shall consider the case when the temperature of the substrate changes and the stress X_1 induced by the substrate can be neglected (stress-free condition).

In this case, the deformation in the substrate is directly translated to the film: $dS_1 = dS_2 = \alpha_{1s}dT$, being α_{1s} the thermal expansion coefficient of the substrate. Then we have

$$dS_1 = dS_2 = \alpha_{1s}dT = (s_{11} + s_{12})dX_1 + \alpha_1 dT \tag{20}$$

and the pyroelectric coefficient will be

$$\gamma^{pc} = \frac{dP_3}{dT} = \gamma_3^X - \frac{2d_{31}(\alpha_1 - \alpha_{1s})}{(s_{11} + s_{12})} \tag{21}$$

P_3 is the polarisation normal to the film surface, $\gamma^X{}_3$ is the primary pyroelectric coefficient, α_1 and α_{1S} are the thermal expansion coefficient of the pyroelectric film and the substrate, respectively. d_{31} is the piezoelectric coefficient in the direction perpendicular to the polar axis and s_{11} and s_{12} are the elastic compliance constants of the ferroelectric thin film.

If the film and the substrate are at the same temperature, the pyroelectric effect can be considered as that obtained in a "dc" or quasi-static experiment where the substrate undergoes a uniform temperature change.

The physical meaning and the relative contribution to the total pyroelectric coefficient of the primary and secondary pyroelectric effects measured in polycrystalline ferroelectric thin film can be found in the work of C.P. Ye [35].

In the considerations given above, changes in phase diagrams due to the effect of the strain are not considered and the thin film is supposed to have the usual ferro-paraelectric phase transition at a given temperature.

12.2.2.1 Heterostructure-Induced Polarisation Effects in Ferroelectric Thin Films

The heterostructure formed after the deposition of the film on a substrate means that the film has a free surface. Another interface with the substrate introduces specific electrical and mechanical boundary conditions, and will be different from the interface attached to the substrate and the interface facing the preparation chamber environment. This asymmetry can produce effects on the polarisation state of the material that affect the pyroelectric properties. In some cases, a self-polarisation can be established and it may originate a pyroelectric response without previous poling in both polycrystalline and epitaxial thin films. This self-polarisation is due to the appearance of a layer in the film, usually close to one of the interfaces, strongly poled in one direction.

Some authors have proposed that in polycrystalline thin films, the driving force for self-polarisation is the built–in electric field produced by the Schottky barrier in the contact ferroelectric and bottom electrode, which is capable of polarising the sample when cooling from the paraelectric phase [36]. Other authors claim that this effect is triggered by space charge formation due to oxygen vacancy accumulation at grain boundaries [37]. The defective surface state of the thin film produces an increase in the internal field [38]. The residual stress of the films is also used to explain the apparition of self-polarisation [39, 40]. The relaxation of the stress in the film can produce the appearance of built–in electric field due to the flexoelectric effect that can give rise to the self-polarisation [41, 42]. Also, the deposition of a top electrode can produce an effect related to the self-polarisation [43, 44].

Hence, the pyroelectric coefficient in ferroelectric thin films can be high at room temperature due to the self-polarisation, with good figures of merit for applications. But, the physical origin of its high value is difficult to understand yet. This polarisation enhancement seems to be related to the characteristics of the heterostructure formed, which can be used to tailor the pyroelectric properties of films.

We shall try some cases where special physical conditions of thin film materials (substrate, electrodes, structure, etc.) can produce particular properties that sometimes provide advantages for their technological applications. In the case of epitaxial thin films, it has been possible to perform a theoretical treatment with a specific set of known boundary conditions that can be introduced in the Landau-Ginzburg-Devonshire (LGD) thermodynamic formalism.

The effect of the strain imposed by the heterostructure in the ferroelectric thin film can produce changes in the phase transitions. Thus, the phase diagram of the integrated ferroelectric film becomes different to that of the free "bulk" material. Current physical models based on the phenomenological theory or first principles calculations have been developed to understand the effects of the misfit strain that originates from both the lattice mismatch and the different thermal expansion coefficients between film and substrate. Misfit dislocations, which lead to the strain and depolarisation field, decrease in the case of superconducting electrodes onto epitaxial ferroelectric thin films. The LGD formalism provides a phase transition sequence in the thin film that is defined as follows

<u>Phase **p**</u>: Paraelectric phase at high temperature without polarisation that transforms from cubic to tetragonal symmetry due to the clamping effect.

<u>Phase **c**:</u> Tetragonal ferroelectric phase for high compression strains.

<u>Phase **aa**</u>: Orthorhombic ferroelectric phase for high tensile strains.

<u>Phase **r**</u>: This phase is monoclinic at low temperature and low strains with polarisation in-plane (x-y) and out- plane (z)

A thermodynamic formalism has been developed by A. Sharma *et al.* [45], to calculate the pyroelectric coefficients of epitaxial thin films on single crystalline substrates as a function of the film thickness, by taking into account the considerations given above. Expressions for spontaneous polarisation and pyroelectric coefficient of **c** and **r** phases are given.

A treatment of the polarisation states in epitaxial ferroelectric thin films has also been performed by M.D Glinchuk *et al.* [46]. The work constitutes a complete treatment of the problem applied to an ideal ferroelectric thin film in the single domain case. The results of this model are really interesting because they predict effects like self-polarisation and, better still, the existence of an electret state below a critical thickness. The transition from the paraelectric phase on lowering the temperature can lead to any of four different ferroelectric phases depending on the misfit strain imposed on the film.

The authors [46] studied the phase diagrams of thin films with self-polarised phase by taking bulk and surface contributions of the Landau-Ginsburg-Devonshire free energy and calculating the three components of polarisation. These include the lattice mismatch, misfit dislocations and surface piezoelectric effects. In the strained films, the surface piezoelectric effect induces an internal electric field E_m normal to the surface, producing an increase of diffusivity and shifts the phase transition. This E_m can be the origin of the self-polarisation.

Below a critical thickness, this E_m induces an electret-like polar state with non-switchable polarisation. The existence of the electret–like state implies that in very thin films, the centrosymmetric phase cannot happen and so pyroelectric and piezoelectric effects can be found despite the phase not being switchable. The critical thickness of the existence of the electrect-like state depends on the sign and magnitude of the misfit strain.

The authors take into account free energy expression with bulk and surface terms $\Delta G = \Delta G_V + \Delta G_S$ in the form of powers of polarisation components (including odd power terms) with coefficients depending on the film thickness, misfit strain, temperature, etc. [46] The surface free energy contribution (ΔG_S) takes into account the surface tension and strain. It includes terms as:

$$(\Delta G_S)z \approx \left(\delta_z/L\right)\left[P_z\left(L/2\right)+P_{m1}\right]^2/\lambda_{z1}+\left[P_z\left(-L/2\right)+P_{m2}\right]^2/\lambda_{z2} \tag{22}$$

Where $P_{mi} = u^i{}_{m,i}e_{31}$ is the polarisation due to the misfit strains $u^i{}_{m,i}$ (i= 1 free upper surface and i=2 film-substrate boundary) via piezoelectric coefficient e_{31}. $\lambda_{Z1,2} = (\delta_z e^2{}_{31})/\mu_{1,2}$ being μ the surface tension coefficient. L is the thin film thickness.

The odd power terms of P_z included in the free energy expression come from the induced internal built-in electric field

$$E_{ind} = 2\pi(P_{m1} + P_{m2})\xi(h) \tag{23}$$

h =L/2L$_z$ is a dimensionless parameter, L the thickness of the film and $L_z \approx (\delta_z/4\pi)^{1/2}$ the characteristic length. $\xi(h)$ is a function that takes into account the polarisation state along the thickness of the film (46).

This internal electric field, due to misfit strain and piezoelectricity in the vicinity of the surface, should lead to the self-polarisation phenomenon that provokes asymmetry in the total polarisation P_z

$$P_Z \neq - P_Z$$

For $P_Z \neq 0$ the study determines a critical temperature $T^Z_{cr}(h,U_m)$, a critical thickness $h_{cr}(T>T_C^Z,U_m)$ (where the paraelectric phase loses its stability independent of the phase transition order, being $U_m = u_m^1 = u_m^{(2)}$) and a phase transition temperature :

$$T_C^Z(Um) = T_C + \frac{2Q_{12}U_m}{\alpha_x(s_{11}+s_{12})} \tag{24}$$

Q_{12} is the electrostriction coefficient.

The model allows us to calculate the evolution of the permittivity along the z axis (ε_{zz}) as a function of temperature for different values of the parameter h calculated as a constant misfit strain of Um= 0.005, (Fig. 12.3). The calculated pyroelectric coefficient is included in Fig. 12.4 for the same conditions. In Fig. 12.5a, the calculated phase diagram as a function of h for the same material is presented in the case of an external field imposed to compensate the internal one; in Fig. 12.5b the same calculation but without external field is presented. The value of strain U_m depends on the film thickness because of misfit dislocations appearance at some critical thickness $h_d = 1/|U_m|$

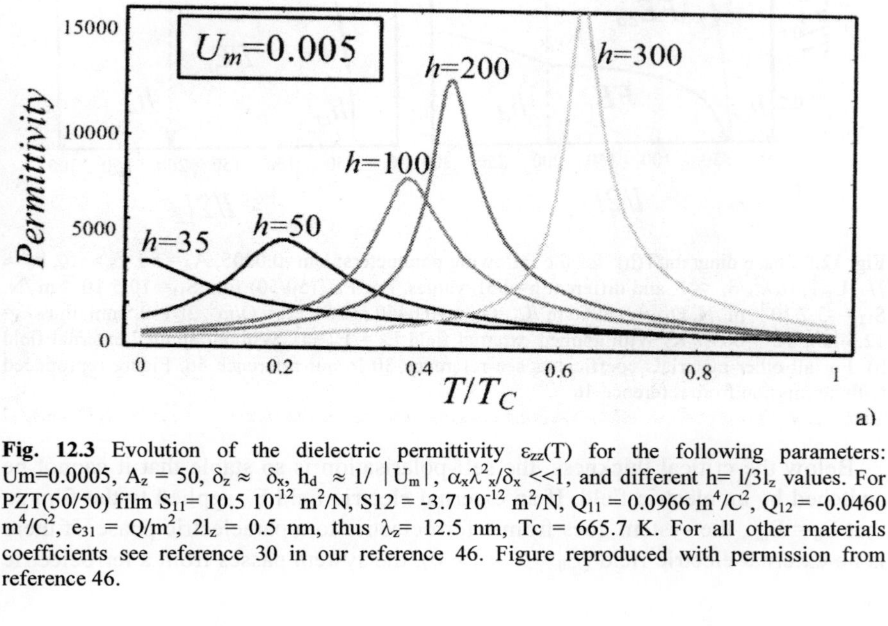

Fig. 12.3 Evolution of the dielectric permittivity $\varepsilon_{zz}(T)$ for the following parameters: Um=0.0005, $A_z = 50$, $\delta_z \approx \delta_x$, $h_d \approx 1/|U_m|$, $\alpha_x\lambda^2_x/\delta_x \ll 1$, and different h= $l/3l_z$ values. For PZT(50/50) film S_{11}= 10.5 10^{-12} m^2/N, S12 = -3.7 10^{-12} m^2/N, Q_{11} = 0.0966 m^4/C^2, Q_{12} = -0.0460 m^4/C^2, e_{31} = Q/m^2, $2l_z$ = 0.5 nm, thus λ_z= 12.5 nm, Tc = 665.7 K. For all other materials coefficients see reference 30 in our reference 46. Figure reproduced with permission from reference 46.

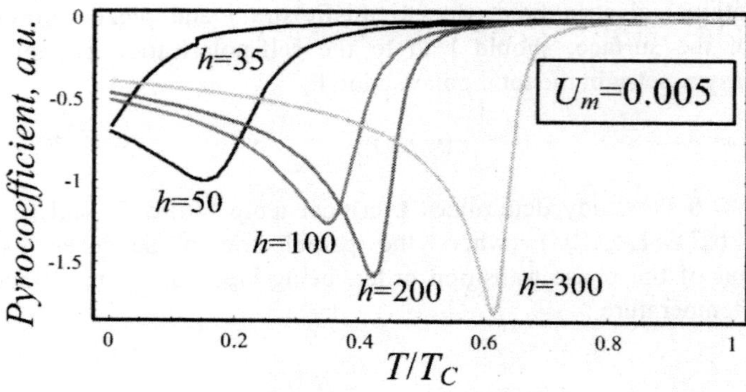

Fig. 12.4 Evolution of the Pyroelectric coefficient γ(T) for the following parameters: Um=0.0005, $A_z = 50$, $\delta_z \approx \delta_x$, $h_d \approx 1/|U_m|$, $\alpha_x\lambda_x^2/\delta_x \ll 1$, and different h= $l/3l_z$ values. For PZT(50/50) film S_{11}= 10.5 10^{-12} m^2/N, S12 = -3.7 10^{-12} m^2/N, Q_{11} = 0.0966 m^4/C^2, Q_{12} = -0.0460 m^4/C^2, e_{31} = Q/m^2, $2l_z$ = 0.5 nm, thus λ_z= 12.5 nm, Tc = 665.7 K. For all other materials coefficients see reference 30 in our reference 46. Figure reproduced with permission from reference 46.

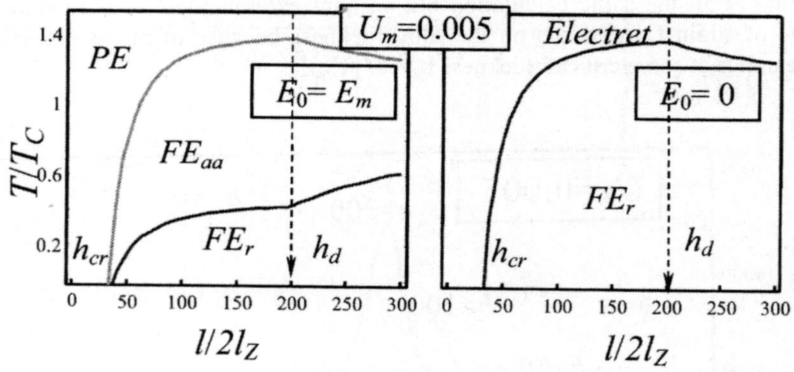

Fig. 12.5 Phase diagram T(h) for the following parameters: Um=0.0005, $A_z = 50$, $\delta_z \approx \delta_x$, $h_d \approx 1/|U_m|$, $\alpha_x\lambda_x^2/\delta_x \ll 1$, and different h= $l/3l_z$ values. For PZT(50/50) film S_{11}= 10.5 10^{-12} m^2/N, S_{12} = -3.7 10^{-12} m^2/N, Q_{11} = 0.0966 m^4/C^2, Q_{12} = -0.0460 m^4/C^2, e_{31} = Q/m^2, $2l_z$ = 0.5 nm, thus λ_z= 12.5 nm, Tc = 665.7 K. With applied external field $E_0 = E_m$ a). Without applied external field b). For all other materials coefficients see reference 30 in our reference 46. Figure reproduced with permission from reference 46.

Below the critical thickness, the self-polarisation is so stable that it cannot be switched by an electric field. If an external electric field E_0 applied to the film so that $E_0 = E_m$, the system goes from a ferroelectric to a paraelectric phase. If there is no external electric field ($E_0 = 0$, $E_m \neq 0$), the system passes from a ferroelectric phase to an electret-like polar state. The main condition for self-polarisation appearance is $E_m(h) > E_{0c}(h)$= thermodynamic coercive field.

In the case of a PZT (50/50) thin film, the calculated phase diagram at $E_0 = 0$ due to misfit strain gives the following: For compressive strain ($U_m < 0$), the electret-like phase transforms into FE_C phase. For tensile strain ($U_m > 0$), the electret-like phase can transform into FE_{aa} and then into $FE_{r\ phase}$ depending on the film thickness.

The existence of polarisation states in the as-prepared films leading to self-polarisation is interesting from the point of view of applied pyroelectricity. The preparation of strongly self-polarised pyroelectric capacitors saves poling time. Also, the stability of this polarisation is high, which is good for the device stability. Furthermore, the existence of intermediate (not fully poled) self-polarisation states should be known in advance in order to pole the material capacitor along this preferential polarisation direction and obtain larger and more stable pyroelectric coefficients.

Experimental data obtained in BTO epitaxial thin films [47] can be analysed qualitatively in terms of the theoretical models. The calculations [48, 49] show that the paraelectric ferroelectric phase transition is of the second order and its temperature is always higher than the Curie-Weiss temperature (Tc ≈ 400 K) of a free BTO crystal. Both features are in good agreement with the experimental observations and indicate, as suggested by the in-plane misfit strain–phase diagram calculated through LGD theory, that BTO film is effectively strained but weakly as Tc of the film remains close to the bulk one. To estimate the in-plane misfit strain in the film, the authors first determine the out-of-plane strain $S_\perp$ by using the following equation:

$$S_\perp(T) = S_\perp(T_G) + (\alpha_{film} - \alpha_{BTO})(T - T_G) \tag{25}$$

where $\alpha_{film} = 17.0\text{x}10^{-6}\ K^{-1}$ and $\alpha_{BTO} = 9.6\text{x}10^{-6}\ K^{-1}$ are the thermal expansion coefficients of the film and of the unstrained bulk BTO, respectively. T_G is the growing film temperature.

If the lattice misfit strain is relaxed at the processing temperature, e.g., by the creation of dislocations, $S_\perp(T_G)$ is small and the second term, related to the strains produced by the difference of the expansion coefficients of film and substrate, dominates. After the out-of-plane strain expression is determined, in the limit of the linear elasticity, the in-plane strain $S_{||}$ may be determined by

$$S_{||}(T) \approx [(\nu - 1)/2\nu]S_\perp(T) \tag{26}$$

where ν is the BTO Poisson coefficient (ν=0.35).

The in-plane strain calculated using expression (26), just above the phase transition temperature ($T_{C(film)} \cong 420K$), is equal to $S_{||}(420K) \approx 0.34\%$. For this tensile strain the LGD theory predicts a transition temperature of 550K, so that the ferroelectric phase should be of **aa** type and the polarisation will be confined in the plane of the film. The out-of-plane polarisation should be equal to zero, contrary to the experimental results. This indicates that the LGD theory does not model correctly the thermal behaviour of thin films, which instead behave as if they were in a monoclinic phase.

In the case of polycrystalline thin films, the main contribution to the stress comes from the difference in thermal expansion coefficients of the film and the substrate. This also produces a mismatch strain that can produce effects similar to the lattice misfit strain, which in this case is less important. See reference [45] for a comment on and a free energy expression for a textured polycrystalline thin film.

12.3 Measurement Methods

Due to the experimental difficulty of obtaining the constant strain boundary condition, the measurements of pyroelectric properties are performed under constant stress at zero external electric field. These boundary conditions are assumed in almost all the measuring methods.

A possible classification of the experimental methods to obtain the pyroelectric coefficient or the polarisation is based on the different types of a thermal wave that can be used to change the sample temperature. In the "Source book of Pyroelectricity" S. Lang [50] made a classification of the measurement methods in two categories: "qualitative" and "quantitative", depending on the approximation to a quantitative value of the pyroelectric coefficient. In this chapter, we are concerned with the "quantitative" methods because the "qualitative" ones are more devoted to demonstrating the existence of pyroelectricity, usually to assess the non-existence of a centre of symmetry in the sample. But that is not much in use these days. The term "quantitative" should not be taken strictly, as the accuracy of the different methods varies largely.

The "quantitative" methods are divided into two categories: "static" and "dynamic" methods.

The "static" methods consist of producing a "heat burst" that changes the sample temperature slightly. The charge developed by the pyroelectric effect is compensated by a capacitor in series with the sample, during the temperature relaxation [51, 52]. This technique yields accurate pyroelectric coefficients but is difficult to be implemented in a wide range of temperatures. Nevertheless, the large accuracy of this measurement may be used to confirm the results obtained with other methods in the same temperature range.

The "dynamic" methods are based on producing a continuous temperature change in the sample. Among the different "dynamic" methods, a subdivision may be made by taking into account the way of changing the temperature: "The temperature is increased at a constant slope along a large temperature interval" or "The temperature of the sample fluctuates around a central value and changing the temperature derivative sign with time". According with this, the subdivision may be described as: a) "constant sign temperature slope" methods and b) "oscillating" methods.

12.3.1 "Constant Sign Temperature Slope" Methods

R.L. Byer published one of the first descriptions of the "constant sign slope" method and C. B. Roundy [13] also called it the "direct" method. In this work, a scanning temperature rate of 1°C min^{-1} was used. The diagram of the measurement set-up and its equivalent circuit are included in Fig. 12.6 a and b, respectively. As may be deduced from the plots, the measurement method is simple and effective. If a constant temperature slope is maintained, the current is directly proportional to the pyroelectric coefficient.

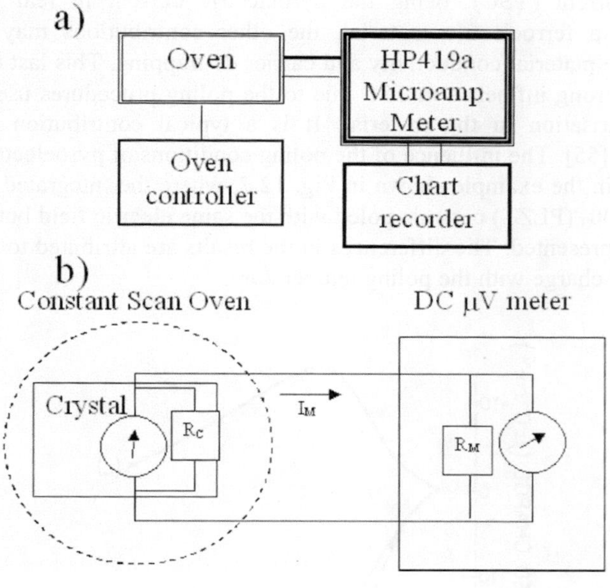

Fig. 12.6 (a) Measurement set-up for measuring the pyroelectric coefficient by the "direct" method. (b) Equivalent circuit. Adapted from reference 13. R_c id the crystal leakage resistance, I_M is the current flowing and R_M is the meter inpur resistance.

Based on this measurement method, S.B. Lang and F. Steckel [53], developed a method that allowed the measurement of the pyroelectric current, the DC dielectric constant, and the volume resistivity of the pyroelectric material, by changing the magnitude of the shunt resistor used to convert the pyroelectric current to voltage. The equivalent measuring circuit is similar to that of the "direct" method, see Fig. 12.6b. By changing the shunt resistance (in series with the sample) and the charge time, it was possible to obtain different curves, and therefore different properties, of the sample under test. As a difference with the "direct" method proposed by Byer and Roundy that maintains a constant slope [13], in the set-up of Lang and Steckel [53], the heating power is fixed in the beginning of the experiment not to reach an initial temperature slope larger than 1 or 2°C·min^{-1}. On increasing T, the slope decreases. So, knowledge of the temperature derivative during the experiment time is necessary to obtain the pyroelectric coefficient.

The "direct" method has also been used to study the phase transitions in ferroelectric materials. As an example of a system devoted to this purpose, the one described in [54] is convenient because it allowed poling the sample with an electrical field and then in the same set-up, the measurement of the polarisation change with the temperature.

It has to be considered that the electric current collected is the thermo stimulated current (TSC), being the pyroelectric current in real pyroelectric materials. In a ferroelectric material, the other contributions may be due to changes in the material conductivity and carrier de-trapping. This last contribution may have a strong influence in TSC due to the poling procedures used to induce remnant polarisation in the material. It is a typical contribution in polymer ferroelectrics [55]. The influence of the poling conditions of pyroelectric materials may be seen in the example shown in Fig. 12.7, where the integrated charge of a (Pb,La)(Zr,Ti)0_3 (PLZT) ceramic poled with the same electric field but at different conditions is presented. The differences in the results are attributed to the increase in the trapped charge with the poling temperature.

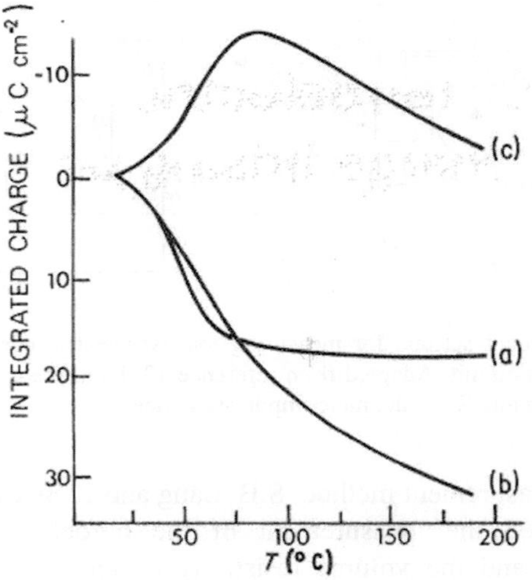

Fig. 12.7 P-T diagrams of PLZT. (a) sample poled at 20°C. (b) sample poled at 200°C. (c) sample poled at 200°C, cooled to room temperature and then electrically switched. Figure reproduced with permission from Alemany C, Jiménez B, Mendiola J, Maurer E. (1984) "Aging of (Pb, La)(Zr, Ti)O_3 ferroelectric ceramics and the space-charge arising on hot poling" . J. Mat. Science. 19(8), 2555.

The "direct" methods cannot distinguish among the different contributions to the TSC. Thus, to minimise the effects of the sample conductivity on the pyroelectric current, it is necessary to perform measurements on heating and on cooling the sample. But, the de-trapping contribution cannot be corrected with this procedure due to its irreversible character.

One of the consequences is that the evolution of the polarisation (at zero applied electric field), as a function of the temperature, is hard to be followed by these pyroelectric measurement methods. This is because the non-pyroelectric contributions are significant if the material resistivity is not very high. In Fig. 12.8, the P versus T curves for a ceramic that was poled at different values of the poling electric field are plotted. The shape of the curves resembles that of a diffuse phase transition, but a strong increase of the polarisation is observed on increasing the poling field. The polarisation increase reflects the effect of the charge trapped in the sample during the poling that is released close to the phase transition temperature where extended defects like domain boundaries disappear.

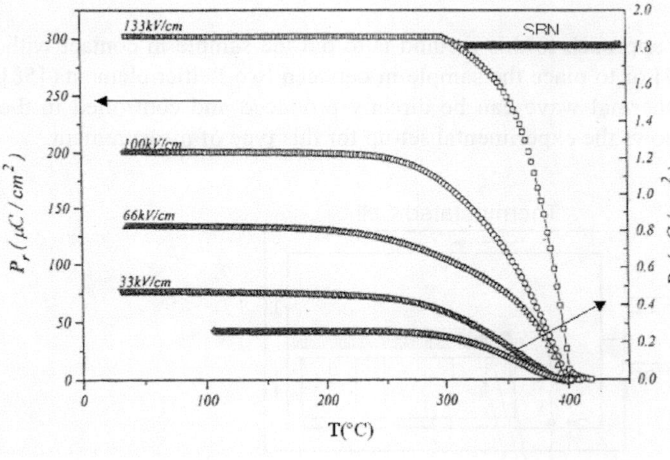

Fig. 12.8 P_r –T diagram as a function of the poling field applied to a $Bi_2SrNb_2O_9$ (SBN) ceramics sample. The right axis corresponds to the measurement without previous poling of the sample. Adapted from Durán-Martín P. PhD thesis (1997). "Propiedades ferroeléctricas de materiales cerámicos con estructura tipo Aurivillius de composiciones basadas en $Bi_2SrNb_2O_9$". Dpto de Fiscia de Materiales UAM.

12.3.2 Oscillating Methods

It can be demonstrated that most of the non-pyroelectric contributions to TSC (I_{np}), in a small temperature interval, may be considered to have a linear dependence with T (I_{np} α *R*T, being R a proportionality constant) as long as the pyroelectric current is proportional to the local temperature derivative with time (dT/dt). Hence, the TSC (I_p) may be described in a small ΔT as [56]:

$$i_P = i_0 + RT + \gamma A \frac{dT}{dt} \tag{27}$$

Where the constant term (i_0) represents a constant, non-pyroelectric, temperature- independent current and A is the electrode area.

The dynamic oscillating methods can reduce the influence of other non-pyroelectric contributions to the TSC as well as separate the pyroelectric contribution from those. For example, in the case of induced sinusoidal temperature changes in the sample, the true pyroelectric current is 90° out of phase with the temperature change. The contributions due to sample conductivity or charge de-trapping are in-phase with the temperature change [56].

12.3.2.1 Sinusoidal Thermal Waves Method

A practical approach to this method is to put the sample in contact with a Peltier element [57] or to place the sample in between two Peltier elements [58]. Then, a sinusoidal thermal wave can be directly produced and controlled in the sample. Fig. 12.9 shows the experimental set-up for this type of measurement.

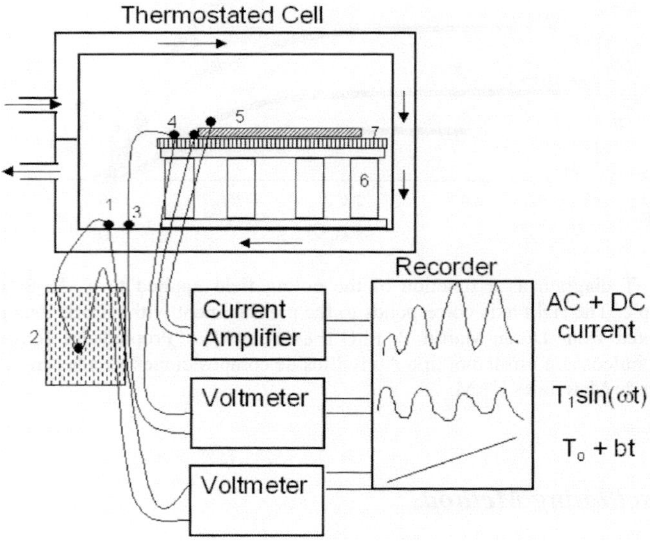

Fig. 12.9 Set-up for measuring pyroelectric coefficients with sinusoidal temperature waves ($T_1\sin(\omega t)$). (1),(2),(3) and (4) are copper-constantan thermocouples that control the temperature changes in the chamber; (5) is the sample, (6) is the Peltier cooler element and (7) is the brass thermal conductor supporting the sample. Figure adapted from reference 57.

Heating with two Peltier elements on both sides of the sample produces a better heat distribution than with a single element. This minimises the effect of temperature gradients, and avoids contributions from tertiary pyroelectricity [58]. An important advantage of this experimental set-up is that it allows the study of the temperature distribution inside the sample. This makes possible the experimental determination of the maximum frequency of the temperature excitation that produces a homogeneous temperature distribution inside the sample. For sinusoidal homogeneous heating, the pyroelectric current is proportional to the amplitude (T_0) and frequency (ω_0) of the thermal wave:

$$I_0 = \omega_0 \gamma A T_0 \tag{28}$$

where γ is the pyroelectric coefficient and A is the area of the electrode. On increasing the frequency, the validity of equation (28) vanishes. The maximum frequency to obtain a homogenous temperature in the sample is [58]:

$$\omega << \frac{1}{2\alpha}, \quad \alpha = \frac{K}{L^2} \tag{29}$$

where K is the thermal diffusivity and L the sample thickness.

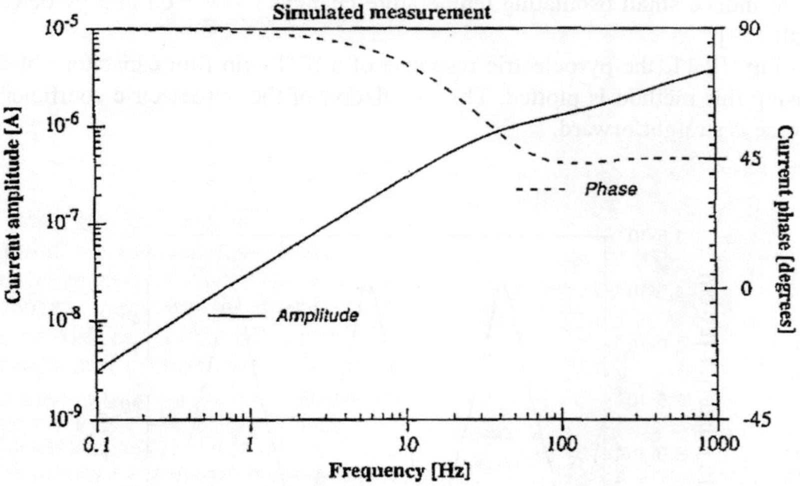

Fig. 12.10 Simulation of the evolution of the pyroelectric current amplitude and phase with the frequency of the sinusoidal thermal wave used for the measurements. . In this simulation the following values were used: L = 100 μm, $K = 2.5\text{x}10^{-7}$ m^2 s^{-1}, $A = 4.9\text{x}10^{-4}$ m^2, $\gamma = 1\text{x}10^{-5}$ C m^{-2} K^{-1}, T_0 = 1K. Reproduced with permission from reference 58.

To show the evolution of the pyroelectric current with the frequency of the thermal wave used for the measurement, a simulation has been made for a $LiTaO_3$ sample of 0.1 mm thickness and K = 2.5 10^{-7} m^2 s^{-1} (Fig. 12.10). The rest of the parameters used in the calculation are included in the figure caption. It may be seen that above the so called cut-off frequency (8 Hz), the amplitude of the pyroelectric current no longer shows a linear correlation with the frequency of the thermal wave, and the phase goes from 90º to 45º. As stated earlier, equation (28) is no longer valid for high frequencies, which implies a non-homogeneous heating of the sample for those frequencies. But, if a low enough heating frequency is used, it is possible to calculate the pyroelectric coefficient at a fixed temperature of reference. Or to introduce a temperature ramp superimposed to the sinusoidal wave that gives rise to the following temperature change in the sample:

$$T = T_0 + bt + T_a \sin(\omega t) \tag{30}$$

where T_0 is the initial temperature, b is the slope of the temperature ramp and ω is the frequency of the sinusoidal thermal wave with amplitude T_a of the Peltier element. Using this ramp the, pyroelectric coefficient can be calculated continuously on a selected temperature range.

12.3.2.2 Triangular Thermal Waves Method

Temperature triangular thermal waves have been used experimentally for a long time to induce small oscillating temperature changes (ΔT≈ 1°C) in a pyroelectric sample [59].

In Fig. 12.11, the pyroelectric response of a SBT thin film capacitor obtained by using this method is plotted. The calculation of the pyroelectric coefficient in this case is straightforward.

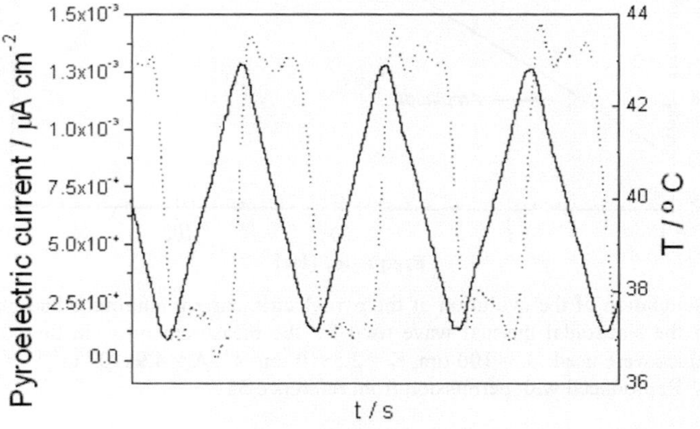

Fig. 12.11 Pyroelectric current response (dashed line) to a triangular thermal wave (continuous line) of a $SrBi_{2.2}Ta_2O_9$ thin film sample. ΔT = 5 °C, f= $3x10^{-3}$ Hz.

Among the different experimental set-ups used, the one designed by Daglish M. [60] is among those that provide the most accurate results. They can be considered as a real quantitative estimation (a precision of 1% is expected) of the total pyroelectric coefficient. This set-up could be used as a possible standard measuring method. It combines the ability of a Peltier element to produce accurate temperature waves with the temperature homogeneity of He as the heating gas. The use of large surface to thickness ratio samples improves the thermal homogeneity achieved in the experiment. The frequencies of the oscillating temperature signal can be varied between 10 Hz and 100 mHz. The equipment could be operated in two modes: fixed reference temperature mode or ramped reference temperature mode. In the second mode, the sample experienced a constant linear rate of temperature change, superimposed to the temperature oscillation. The pyroelectric coefficient can be calculated continuously over the selected temperature range.

12.3.2.3 Heating by Light Methods

One of the first dynamic oscillating methods was described by A.G. Chynoweth [61]. The idea was to provide large rates of temperature changes by using a focussed light beam from a tungsten projection lamp that is chopped to produce light and dark pulses on the sample, which is maintained at a reference temperature T. As a consequence, the pyroelectric current is much higher than other contributions to the TSC.

Due to the strong difficulty of knowing the true temperature change and the temperature gradient within the sample, the results given for the pyroelectric coefficient should be taken as "non-quantitative". This term is used to distinguish from the other "qualitative" methods explained at the beginning of this section. In this case, the determination of the sample temperature requires the knowledge of several parameters to obtain the correct value. For example, in stationary cases, sinusoidally modulated radiation power absorbed $W_w = W_0 e^{i\omega t}$, the stationary solution for the temperature modulation amplitude (T_ω) is [10]:

$$|T_\omega| = \frac{\eta W_\omega}{G\sqrt{1+\omega^2 \tau_{th}^2}} \quad (31)$$

where τ_{th} is H/G the thermal time constant, H is the heat capacity, G is the thermal conductivity and η is the emissivity of the top electrode.

The pyroelectric current as a function of frequency ω can be derived as [10]:

$$J_\omega = \gamma A \omega |T_\omega| \quad (32)$$

Thermal conductivity is not known for many materials, so the error in the temperature determination can produce large errors on the pyroelectric coefficient determination.

The use of light sources like LASER for heating the sample allows the use of higher frequencies than in previously described methods. But, it induces non-homogeneous heating in the samples, see equation (29). Taking advantage of the controlled non-homogenous heating produced, new measurement methods of the pyroelectric coefficient are developed. These are also able to map the polarisation changes within the sample. These methods are known as Laser Intensity Modulation Methods (LIMM), and will be treated in the next section.

As mentioned earlier, the most serious difficulty when using oscillating methods is to ensure a homogeneous heating of the sample. The presence of temperature gradients within the sample can produce tertiary pyroelectricity that may affect the evaluation of the true pyroelectric coefficient. However, these methods have the advantage of distinguishing the pyroelectric current from the total TSC. This allows the application of a DC bias during the measurements and obtains pyroelectric coefficients in these conditions.

12.3.2.4 "Noisy" Temperature Derivative Based Method

Recently, a new method has been proposed which is a simplification of the ramped oscillating methods used in light heating or the generation of AC thermal waves in the samples [62]. The idea consists of applying equation (27) to the TSC delivered by the material under a temperature variation with a nominal slope. The heating element does not produce an increase of the temperature of the material at a constant rate for small time intervals. Instead, dT/dt show fluctuations that can be rather large even in producing changes in the sign of the derivative. This is the origin of the noise in the TSC signal in most dynamic measurements of the pyroelectric coefficient, due to the term

$$\gamma A \frac{dT}{dt}$$

in equation (27). However, it should be considered that the response of the material to large fluctuations in the temperature change rate must be mostly pyroelectric in nature. This is because the non-pyroelectric components to the response do not depend on dT/dt . Moreover, it can be considered that in small time intervals, the sample is subject to a free-form thermal wave. So, we can use the material response to these dT/dt fluctuations present in dynamic heating measurements to obtain information on the pyroelectric properties of the material.

With the proposed method, it is possible to extract an estimation of γ as a function of temperature by non-linear least square fitting of equation (27) in a small temperature interval to keep the validity of the equation. This means the term RT in equation (27) is valid and γ is assumed as constant. Then, after integration of the pyroelectric coefficient over the whole range of the temperature interval, the variation of the ferroelectric polarisation with temperature may be obtained by using the calculation method proposed by Lang [62].

It is worth commenting that this method may be considered as a pseudostatic experiment. After the Fourier transform of the local derivative (after subtraction of the average component), the usual period of oscillation that results is close to 0.05 Hz. This assures a homogeneous heating of the sample. But, in order to keep equation (27) valid, the physical properties of the sample must not change too fast with temperature. The developed method becomes useful for the study of materials with diffuse phase transitions, where the physical properties vary slowly, as is usually the case in ceramics and thin films.

The experimental set-up for this method is shown in Fig. 12.12. This set-up is similar to that used for "dynamic" measurements at constant temperature rate ("direct" method, see Fig. 12.6). Fig. 12.13 presents a typical heating profile obtained with the experimental set-up. The average temperature slope is 1.7 °C min^{-1}; the regression factor R= 0.9997 is quite good. But, as may be observed in the inset of Fig. 12.13, the instantaneous derivative is not constant and strongly fluctuates around the nominal one producing what we can call a "noisy derivative".

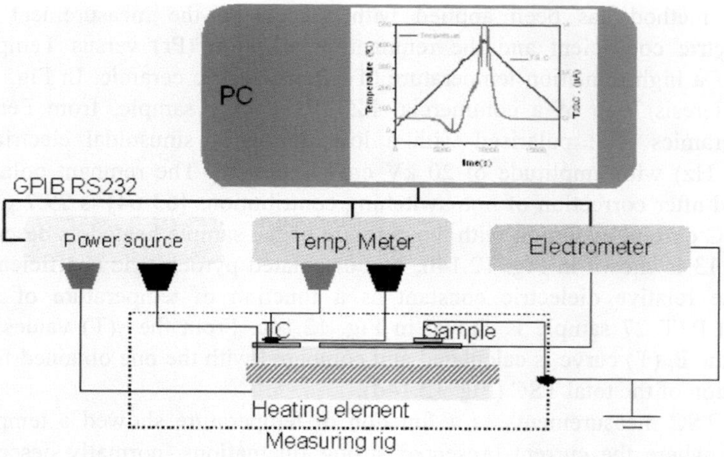

Fig. 12.12 Experimental set-up for the measurement of the pyroelectric coefficient based on the "noisy " method temperature derivative. The system is similar to that proposed in reference [13] for the "direct" method.

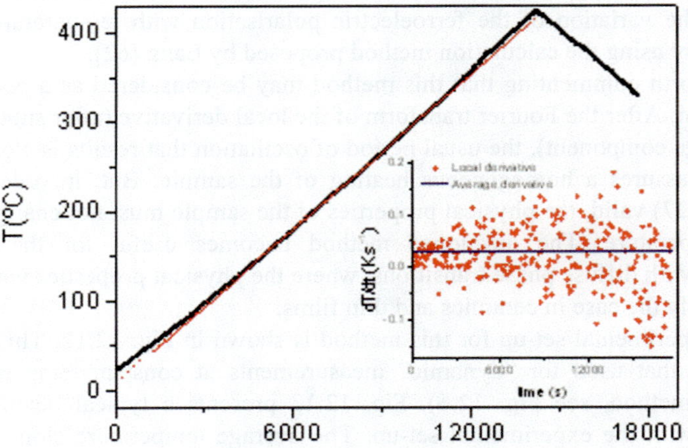

Fig. 12.13 Temperature variation used to measure the TSC in polar samples. In the figure the result of the fitting of the T vs. t data is shown . The inset is the local or instantaneous derivative of the temperature. It can be seen that the oscillations in the temperature derivative increases with increasing the temperature.

This method has been applied with success to the measurement of the pyroelectric coefficient and the remnant polarisation (Pr) versus Temperature curve of a high transition temperature (Tc) ferroelectric ceramic. In Fig. 12.14a, the hysteresis loop of a commercial PZT27 ceramic sample, from Ferroperm Piezoceramics A/S, polarised with a low frequency sinusoidal electric wave (f=0.01 Hz) with amplitude of 20 kV cm^{-1} is plotted. The remnant polarisation obtained after correction of non-switching contributions [63 64] is 29.7 μC cm^{-2}. The TSC current evolution with temperature of the sample heated as described in Fig. 12.13 is shown in Fig. 12.14b. The calculated pyroelectric coefficient along with the relative dielectric constant as a function of temperature of another unpoled PZT 27 sample is shown in Fig. 12.14c. From the γ(T) values of this graph, the P_r (T) curve is calculated and compared with the one obtained from the integration of the total TSC (Fig. 12.14d).

The TSC measurements as a function of temperature showed a temperature interval where the current presented strong fluctuations, normally described as "noise". Above the ferroelectric-to-paraelectric transition temperature (Tc), the fluctuations almost disappear. No significant fluctuations are is observed in the cooling run. In the inset of Fig. 12.14b, it can be seen in detail the TSC current evolution in a small temperature interval showing oscillating changes with temperature variation. In the temperature range between room temperature and Tc, the pyroelectric component of the TSC is proportional to the instantaneous derivative of the temperature. This fluctuates around its average value as shown in the inset of Fig. 12.13. Above Tc, the pyroelectric contribution to the TSC disappears. But, the temperature derivative continues to be "noisy". The disappearance of the fluctuations / "noise" in the TSC is directly related to the loss of the pyroelectric effect at the ferro-paraelectric phase transition temperature.

To derive γ(T), a non-linear least square fitting of the TSC (t) data to equation (27) is applied. The time interval used corresponds to a change in temperature close to 1K. The resulting γ(T) *curve* showed the typical shape of an almost diffuse phase transition, Fig. 12.14c. The temperature of the maximum (337ºC) is slightly lower than the Tc obtained from the dielectric constant curve (344ºC). Fig. 12.14c. The total polarisation of the sample calculated from the pyroelectric coefficient is 32 μC cm^{-2} . This is in good agreement with the value measured in the hysteresis loop, Fig. 12.14d. On the other hand, polarisation obtained from the total TSC current increases on increasing the temperature mainly due to the contribution of sample conductivity and the release of the trapped charge carriers due to the disappearance of ferroelectric domain boundaries, Fig. 12.14d.

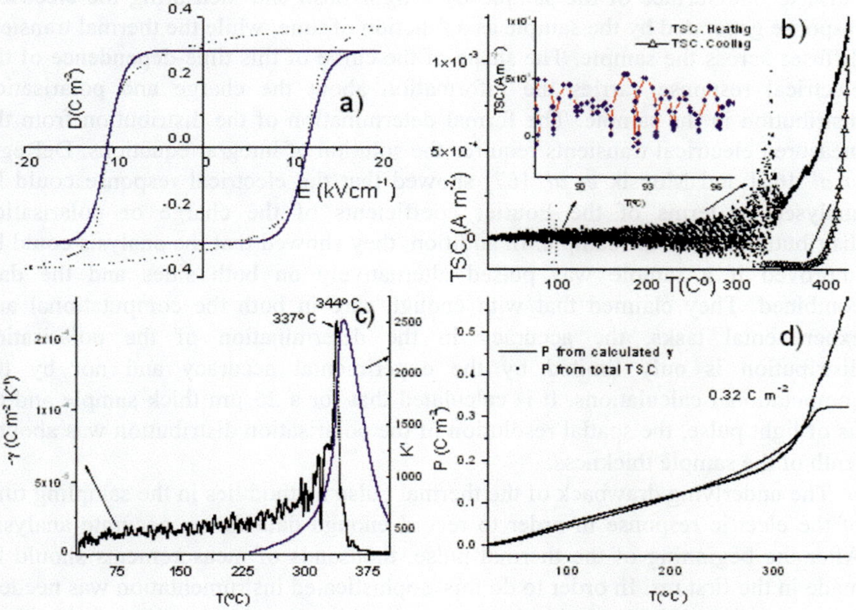

Fig. 12.14 a Experimental ferroelectric hysteresis loop of the PZT27ceramic from FERROPERM measured at 0.01 Hz (dotted line) and corresponding corrected switching loop after elimination of all non-switching contributions (solid line) .**b** TSC from the sample heated with a thermal wave similar to that showed in Figure 13. The TSC measured on cooling is displaced by an amount of – 1.25 10^{-4} A·m^{-2} for clarity. The inset is a zoom on a small temperature interval showing the strong current fluctuations.**c** Temperature variation of the calculated pyroelectric coefficient γ (•) and the relative dielectric constant measured at 10 kHz (blue line) showing the good agreement between both magnitudes. The temperatures at maximum of both magnitudes are shown in the plot. **d** Polarization versus temperature curves, calculated from the γ(T) curve (solid line) and calculated from the integration of the measured TSC (•).

12.3.3 Evaluation of the Polarisation Distribution Through Pyroelectric Effect Based Methods

For the measurement of charge and polarisation distributions, various experimental techniques are available, which are based either on the piezoelectric or on the pyroelectric effect. The use of fast heating or high frequency thermal waves produces non-homogeneous heating of the sample that can be used to determine the polarisation distribution in the sample. Thickness profile, for example. The polarisation profiling methods based on the pyroelectric effect are implemented in the time or the frequency domains.

The response of the sample to a thermal pulse is the base of the methods in the time domain. One of the first papers that used the thermal pulse method to determine the charge or polarisation distributions across thin dielectric samples was published by Collins [65]. The experiment consists in applying a thermal pulse to one surface of the sample by a light flash and measuring the electrical response generated by the sample as a function of time, while the thermal transient diffuses across the sample. The shape of the curve of this time-dependence of the electrical response carries the information about the charge and polarisation distribution in the sample. The formal determination of the distribution from the measured electrical transients requires the solution of integral equations. DeReggi *et al.* [66] and Mopsik *et al.* [67] showed that the electrical response could be analysed in terms of the Fourier coefficients of the charge or polarisation distribution across the sample. In addition, they showed that the analysis could be improved if a sample was pulsed alternatively on both sides and the data combined. They claimed that with enough care in both the computational and experimental tasks, the accuracy in the determination of the polarisation distribution is only limited by the experimental accuracy and not by the computational calculations. It is calculated that for a 25 μm thick sample and 60 μs of light pulse, the spatial resolution of the polarisation distribution was about a tenth of the sample thickness.

The underlying drawback of the thermal pulse method lies in the sampling time of the electric response in order to record enough data for an accurate analysis. After the beginning of the thermal pulse, thousands of measurements should be made in the first ms. In order to do this, sophisticated instrumentation was needed, increasing the difficulty in implementing the time domain technique at that time, the end of the 1970s.

The experimental difficulties pushed the scientific community to develop easier methods in the frequency domain. One of the first publications on polarisation profiling using periodic heating was that of Lang and Das Gupta [68]. In this case, a thin polymer film coated with evaporated opaque electrodes on both surfaces is freely suspended in an evacuated chamber containing a window through which radiant energy can be admitted. Each surface of the sample, in turn, is exposed to a periodically modulated radiant energy source such as a Laser. The absorbed energy produces temperature waves that are attenuated and retarded in phase as they propagate through the thickness of the sample. Because of the attenuation of the waves, a non-uniform thermal force acts on the spatially distributed dipoles or space charges to produce pyroelectric current.

Because of use of a laser source, the technique was named Laser intensity modulation method (LIMM). The laser presents many advantages like high modulation frequencies and good focussing, although for low frequency ranges as those needed to study thicker samples, modulated incandescent lamps can be used [68]. Many papers have been published using the basis of this technique on different types of materials (see [69, 70] and references therein).

These days, the implementation of the LIMM method is possible in most characterisation laboratories. A schematic of a LIMM set-up is included in Fig. 12.15.

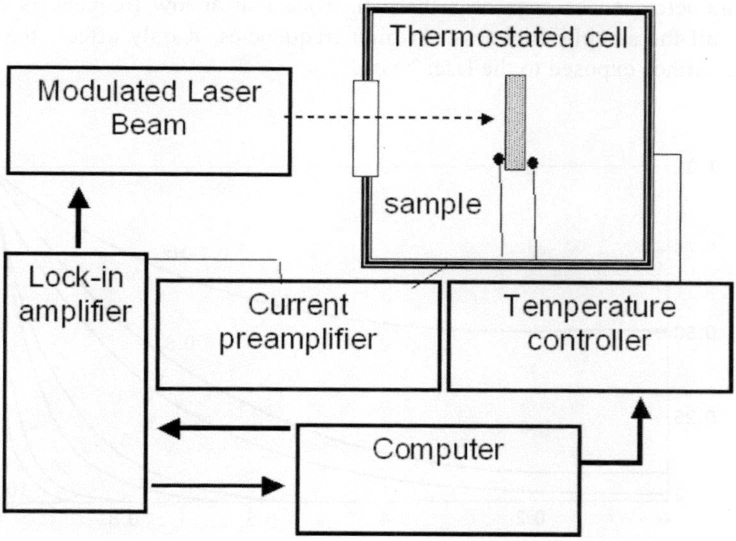

Fig. 12.15 Block diagram of the experimental set-up for Light intensity modulation method (LIMM). Adapted from reference [70].

12.3.3.1 A Short Introduction to the Basics of LIMM

The first assumption to be made is one-dimensional heating propagation in the sample. This is true as long as the area irradiated on the sample surface is much larger than its thickness. Therefore, the temperature evolution inside the sample will be only expressed as a function of the distance to the surface, x. T_ω (x, t) for a given frequency of the irradiation source, ω, can then be expressed [71] as the real part of:

$$T_{\omega}(x,t) = T_0(x)e^{i\omega t} \tag{33}$$

the spatial and temporal distribution of the thermal flux in the sample, on a first approximation is:

$$T_{\omega}(x,t) = \frac{q_0 \eta (1-i)\cosh[D(1+i)x]}{2gD\, senh[D(1+i)L]} e^{i\omega t} \tag{34}$$

where q_0 is the maximum irradiation intensity, η is the electrode absorption, g is the thermal conductivity of the sample, L is the distance between electrodes and D is thermal diffusion length.

Performing the time derivative of the thermal flux in the sample, we found the "thermal force" acting in the sample (in Fig. 12.16 it is plotted in normalised values as a function of position in the sample at different modulation frequencies). This parameter can be seen as a thermal probe that at low frequencies extends through all the sample thickness. At high frequencies, it only affects the region near the surface exposed to the laser beam.

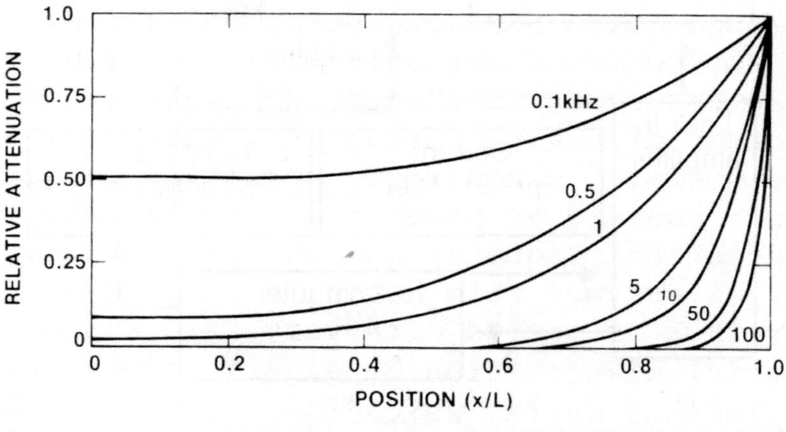

Fig. 12.16 Normalized curves of dΔT/dt "thermal force" versus x/L for different frequencies of the irradiation source. Reproduced with permission from reference. [68].

The polarisation process in the ferroelectric samples produces, besides the ferroelectric polarisation, an induced polarisation coming from the bound charge.

The thermal profile interacts separately with the bound or space charge polarisation distribution and with that associated to the ferroelectric polarisation.

In the case of a short circuit configuration, the TSC generated can be expressed after Collins [67, 68, 72] as:

$$I(\omega) = \frac{\omega}{D senh(D(1+i)L}(B_P(1+i)) \times$$

$$\int_o^L P(x)\cosh\left[D(1+i)x\right]dx + \frac{B_P}{D}\int_o^L p(x)senh\left[D(1+i)x\right]dx \qquad (35)$$

where B_p and B_P correspond to:

$$B_P = \frac{A.q_0\eta(\alpha_P + \alpha_x - \alpha_\varepsilon)}{2gL} \qquad B_\rho = \frac{A.q_0\eta(\alpha_x - \alpha_\varepsilon)}{2gL}$$

In this equations α_p, α_x and α_ε correspond to the coefficients of variation with the temperature of the polarisation, sample dimension (thermal expansion) and dielectric permittivity, respectively.

If the polarisation is completely compensated by space charge, the equation can be simplified to that of Ploss *et al.* [73]:

$$I(\omega) = \gamma_P \frac{A}{x_2 - x_1}\int_{z_1}^{2} d_{PS}(x)\frac{\partial T(x,t)}{\partial t}dx, \quad d_{PS} = \frac{P(x)}{P_S} \qquad (36)$$

Where P_s is the saturation polarisation and γ_P is the pyroelectric coefficient at constant stress. Due to the numerous parameters involved in this equation, which are difficult to determine, just the relative value of $d_{ps}(x)$ is computed.

The equation of the thermo-stimulated current at each frequency is a Fredholm integral equation of the first kind. This is a poorly posed problem because a large number of polarisation distributions $d_{ps}(x)$ will satisfy the equation within the experimental error.

The first evaluation procedures for LIMM were approximations to the pyroelectric functions by a sum of trial functions [74]. A Fourier procedure was later implemented in combination with a deconvolution procedure [75]. Other approaches included the constrained regularisation method developed by Provencher [76, 77] and the Tikhonov regularisation, adapted to the pyroelectric spectra [78]. But, it can be said that the most frequent evaluation method used is based on the scanning function [73].

As a guide to finding a solution to equation (35) close to the physically correct one, Provencher [76] proposed the "principle of parsimony": "*of all solutions choose the simplest one, the solutions that reveals the least amount of detail or information that was not already known or expected*". Using this principle and a polynomial regularisation method, Lang [70] developed an evaluation procedure to find a stable and accurate solution for the LIMM equation (36). The problem of finding the correct solution of the integral equation is also found in the time domain methods (thermal pulse method).

12.3.3.2 New Methods Based on LIMM

Many different experimental techniques have been developed by implementing modifications of the original LIMM. For example, the Surface laser intensity modulation method (SLIMM), developed by Lang [79]. In this case, for the analysis of the SLIMM results, it is assumed that the sample behaves thermally as a semi-infinite solid. This method is most appropriate for the determination of the polarisation distribution near the surfaces of the ferroelectric samples.

Another modification of the LIMM method is based on the focussing of the laser beam. Related to the focussing of the laser beam, in 1976, previously to the LIMM development, A. Hadni *et al.* [80] reported the development of a laser-scanning microscope that was able to map the pyroelectric response of the material in real time. It is not a LIMM technique in a strict sense because it uses a fixed chopper frequency for the laser beam, between 0.5 -200 kHz, and not modulation in the amplitude. The microscope can analyse the sample surface fast, collecting the pyroelectric voltage response in each point at a single frequency. Using this technique, the authors were able to follow the evolution of the ferroelectric domains with temperature in a TGS crystal. In Fig. 12.17, the map of the pyroelectric response of a 170μm x 130μm TGS plate area at two chopping frequencies is presented. The pictures clearly show the domain boundary as well as the effect of the chopping frequency on the pyroelectric image resolution. Later, both the surface scanning and the laser beam focussing developed in this work were incorporated into the LIMM methods.

Recently, the FLIMM technique has permitted the mapping of the polarisation in three dimensions [81]. This technique was developed initially for the observation of the space charge polarisation profile on dielectrics. FLIMM requires a good three–dimensional modelling of the spatial evolution of the thermal gradient, because the focused beam spot on the surface can be smaller than the sample thickness. This prevents the one-dimensional approximation to the evolution of the thermal gradient used in LIMM. The laser beam can be focused from 1μm to several millimetres. Using this focused laser, a three-dimensional mapping of the polarisation with a maximum surface resolution of 10μm x 10μm and an in-depth resolution of 1μm can be obtained. The technique presents a drawback: the diffusion of the thermal flux as a function of frequency. This produces a lowering in the in-plane spatial resolution when lower frequencies are used.

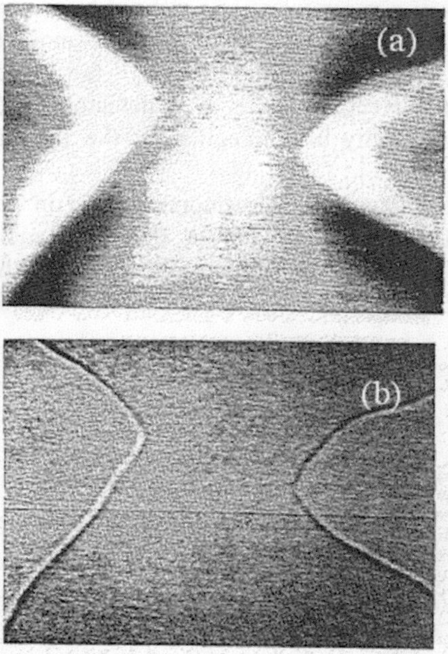

Fig. 12.17 Map of the pyroelectric response of a 170 μm x 130 μm TGS plate area at a chopping frequency of (a) ν = 0.5 kHz and (b) ν = 2000 KHz. The images correspond to the 90° out-ofphase detection mode. Reproduced with permission from [80].

12.4 Applications of the Pyroelectric Effect

The parameters that define the quality of pyroelectric material for technological applications is based on the following parameters: Responsivity ($\Re$), Noise Equivalent Power (NEP), Detectivity (D*) and Figure of Merit (FM).

The response of a pyroelectric material to an applied thermal radiation is known as responsivity. It is defined as the specific response in voltage or current per watt of power radiation [10].

Responsivity in current:

$$\Re_I\left(\frac{I}{W}\right)=\frac{\eta\gamma\omega A}{G_{th}\left(1+\omega^2\tau_{th}^2\right)^{1/2}} \tag{37}$$

Responsivity in voltage:

$$\Re_V\left(\frac{V}{W}\right)=\frac{\eta\gamma\omega AR_p}{G_{th}G_{el}\left(1+\omega^2\tau_{th}^2\right)^{1/2}\left(1+\omega^2\tau_{el}^2\right)^{1/2}} \tag{38}$$

In these expressions: γ = pyroelectric coefficient. η = fraction of the incident radiation that is converted into heat (emissivity). ω = modulation frequency of the incident radiation. $\tau_{th} = H/G_{th}$ = thermal time. G_{th}= thermal conductance. H = heat capacity. τ_{el} = RpCp = electrical time. A = detector surface. In figure 19, the evolution of the responsivity as a function of the thermal wave frequency is plotted.

The performance of infrared (IR) detectors depends on the signal-to-noise ratio, S/N, being the minimal detectable power, the noise equivalent power (NEP), defined as $\nu_D/\Re_V$. ν_D is the noise voltage generated by the conductance of the element, i.e., $\omega C\tan\delta$ [81]. Therefore, ν_D is equivalent to the Johnson noise in a resistor:

$$\nu_D = \left(\frac{4kT}{\omega C \tan\delta}\right)^{\frac{1}{2}} \tag{39}$$

being C, tanδ, ω, k, T, the element capacity, the loss tangent, the frequency, the Boltzman constant and the temperature respectively: This noise gives an equivalent noise current of

$$J_D = (4kT\omega C \tan\delta)^{\frac{1}{2}} \tag{40}$$

Therefore, the NEP can also be expressed for unit bandwidth as a function of the responsivity in current:

$$NEP = \frac{J_D}{\Re_I} = \frac{G_{th}\left(1+\omega^2\tau_{th}^2\right)^{\frac{1}{2}}\left(4kT\omega C\tan\delta\right)^{\frac{1}{2}}}{\eta\gamma\omega A} \tag{41}$$

The detectivity limited by Johnson noise current is defined for any bandwidth as:

$$D^* = \frac{A^{\frac{1}{2}}B^{\frac{1}{2}}}{NEP} = \frac{\gamma\eta\omega A^{\frac{3}{2}}B^{\frac{1}{2}}}{G_{th}\left(1+\omega^2\tau_{th}^2\right)^{\frac{1}{2}}\left(4kT\omega C\tan\delta\right)^{\frac{1}{2}}} \tag{42}$$

A, is the detector surface. B, is the bandwidth.

An important parameter for the evaluation of detectors is the figure of merit (FM) [10] of the pyroelectric material that relates dielectric and pyroelectric parameters for an optimal Signal to Noise ratio, S/N.

We include the most outstanding figure of merit in current:

$$FM_I = \frac{\gamma}{c_p} \tag{43}$$

Figure of merit in voltage:

$$FM_{V(bulk)} = \frac{\gamma}{k'\varepsilon_0 c_p} \quad (44)$$

$$FM_{V(thin-film)} = \frac{\gamma}{k'\varepsilon_0} \quad (45)$$

Figure of merit of detectivity:

$$FM_{V(bulk)} = \frac{\gamma}{c_p (k'\varepsilon_0 \tan\delta)^{1/2}} \quad (46)$$

Where c_p =Volume specific heat, k' = relative dielectric constant.

Technological applications of pyroelectric materials are mainly related to thermal radiation detection. Thermal detectors measure infrared (IR) radiation through the temperature changes induced by the absorption of radiation in the 10 μm wavelength range. They are used for contact-less temperature measurements, security detectors, human presence sensors and thermal imaging. If the detector is connected to a high impedance amplifier, it can be represented as an equivalent circuit formed by a capacitor, a current generator, and a shunt resistance as shown in Fig. 12.18.

The IR pyroelectric detectors used in technological devices [82] are classified in:

1. Single element detectors.
2. Compensated detectors:

Compensation is obtained by connecting two oppositely polarised detector elements either in series or in parallel, but with only one of the elements exposed to the infrared radiation to be detected. This arrangement is preferable in environments where the ambient temperature is fluctuant and the noise signal level is large. With this configuration, the noise coming from the piezoelectric response (microphone noise) is strongly suppressed.

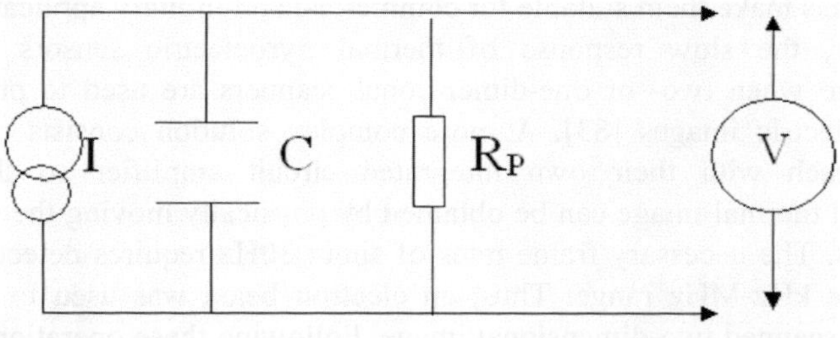

Fig. 12.18 Equivalent circuit of a simple pyroelectric detector. Adapted from reference 82.

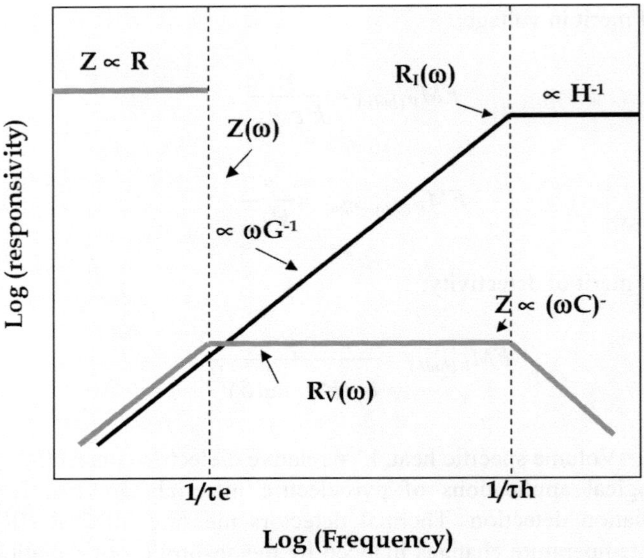

Fig. 12.19 Log–log graph of the current and voltage responsivity vs. the frequency of the thermal excitation wave, together with the impedance for the typical thinfilm situation where the thermal time constant is much shorter than the electrical one. Thermal wavelength effects are neglected. Adapted from [10].

3. Multi-element arrays:

It is advantageous for thermal imaging or laser beam profile monitoring applications. The small size elements of around 100 μm^2 are photolithographically etched on the pyroelectric bulk material.

Pyroelectric thermal detectors present important advantages over other detectors such as photovoltaic and photoconductive devices. They can be operated over a wide spectral range: from the far infrared above 8 μm, at least for wavelengths through visible radiation, down to X-rays. Their sensitivity at room temperature is comparable to a cooled photoconductive detector and do not require a bias supply. At room temperature, the fundamental limits of thermal IR detectors are not different from those of cooled quantum detectors. These characteristics make them suitable for commercial and military applications.

However, the slow response of thermal pyroelectric sensors becomes a disadvantage when two- or one-dimensional scanners are used to produce real-time pyroelectric images [83]. A more complex solution consists of arrays of detectors each with their own integrated circuit amplifier so that a two-dimensional thermal image can be obtained by physically moving the whole array of detectors. The necessary frame rates of about 30Hz requires detector response times in the kHz-MHz range. Thus, an electron beam was used to produce an electrically scanned two-dimensional image. Following these operation principles, a device named vidicon has been developed [84, 85]. In the case of the use of a

TGS crystal as the pyroelectric element, the pyroelectric vidicon can detect thermal differences of 30°C. In Fig. 12.20, the schematic sketch of a vidicon is presented.

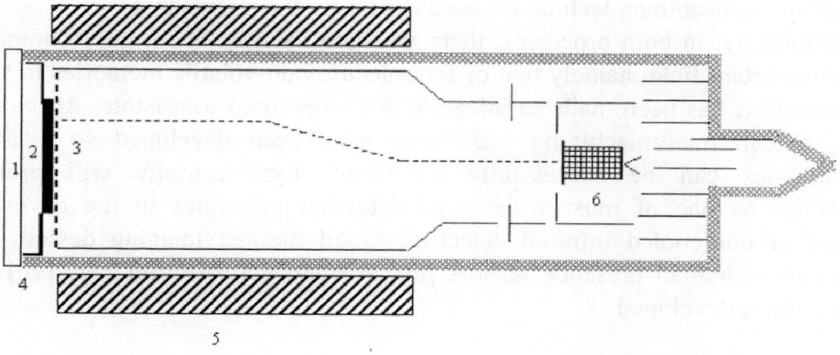

Fig. 12.20 Schematic sketch of a pyroelectric vidicon. (1) Germanium plate. (2) Target mount and TGS crystal. (3) Mesh. (4) Indium seal and target connection. (5) Focus and deflector coils. (6) Cathode.

Most usual applications of pyroelectric detectors are headed towards:

- Fire and intruder alarms
- Pollution monitoring and gas analysis (CO_2) sensors.
- Radiometers.
- Thermal imaging.
- Laser detectors: (CO_2, HCN) or any CW laser beam, provided that it is modulated at a suitable frequency.

The performance of infrared (IR) detectors applied to imaging is usually described by the noise-equivalent (target) temperature difference (NETD), i.e., the smallest detectable temperature difference of a target whose image covers the whole detector placed in the focal plane of an imaging optic. The NETD is defined for a number of optics f = 1 (number of phase levels of a Talbot array illuminator) and for video frequencies in the range 25-60Hz. The relation between NETD and NEP [83, 10], can be written as

$$NETD = M.\,NEP = MB^{1/2}(D^*)^{-1} \quad (47)$$

The M constant contains: pixel area, f-number of the optics and dP/dT term. $B^{1/2}$ is the frequency bandwidth.

The thinner the pyroelectric target crystal, the larger the video read-out signal, which means a minimisation of the heat capacity and heat conduction of an element detector. The one- or two-dimensional detector arrays can be obtained by micromachining ceramic or single crystal wafers to reduce the NETD [86]. Below 50 μm, decreasing the wafer thickness becomes increasingly difficult. However, silicon micromachining techniques enable smaller and thinner pixels as it happens with micro-bolometres of VO_x and amorphous silicon [87].

The key question for pyroelectric detectors is whether they can compete in the race for downscaling and integration in silicon technology. The answer is yes, as they are compatible with the CMOS circuitry due to capacitor condition. Besides, pyroelectric materials are available in film form, which is compatible with the silicon micromachining techniques used to prepare the microsensors.

Fortunately, in both processes, there was good synergy with an economically more important field, namely that of ferroelectric non-volatile memories in which a great effort has been made to integrate thin films onto transistors. At the same time, silicon micromachining techniques have been developed so that both technologies can be successfully combined. Pyroelectricity will probably crystallise as one of most widespread detection principles in the developing market of non-cooled infrared detectors. Good thermal imaging devices [88], detection of human presence sensors [89, 90] and gas spectrometers [90] have already been developed.

12.5 Emerging Applications

12.5.1 Special (Emerging) Applications

In the past few years, new possibilities of technological application of the pyroelectric effect are emerging. These applications are based on the utilisation of the electrical field, $E = 4\pi P/\varepsilon\varepsilon_0$, due to the electrical spontaneous polarisation of pyroelectric materials existing in uniaxial single crystal elements, or in grains of pyroelectric-semiconductor granular systems.

Following a change of temperature in a pyroelectric material [91], a depolarising electric field appears inside the pyroelectric, as well as outside and at the edges. The latter is called edge depolarising electric field (EDEF). The EDEF extends outwards to a distance of the order of magnitude of the pyroelectric element width. This strong EDEF (10^4 -10^6 V/cm), when penetrating into the surrounding medium, creates a variety of physical effects:

- Inducing electrical current in a semiconductor and affecting its resistance.

The pyroelectric itself is not a part of the electrical, current-carrying, circuit. The sole role of the pyroelectric in this structure is to create the EDEF. This principle constitutes the basis of pyroelectric (AlGaN)/semiconductor (GaN) granular systems in terahertz technology [91, 92].

- Accelerating charged and neutral particles in vacuum or in a gas. The EDEF allows the acceleration of electrons, ions or neutral polarisable particles and controlling their motion in vacuum and in gases [91, 93]. A scheme of the proposed device can be found in Fig. 12.21. The pyroelectric crystals have been used to ionise gas and accelerate ions to energies of up to 200keV at room temperature [94, 95] (a scheme of the neutron generator system can be found in Fig. 12.22).

- Generating electromagnetic waves, modifying optical characteristics by electro-optical and photoelastic effects, generating piezoelectric deformation and more.
- The EDEF of a needle-shaped pyroelectric probe can be used to control nanostructures, and to investigate the characteristics of nano-sized systems. Since ferroelectricity does not disappear at a crystal size of about 40 Å [91], it is reasonable to presume that pyroelectricity will do it as well at such a dimension.

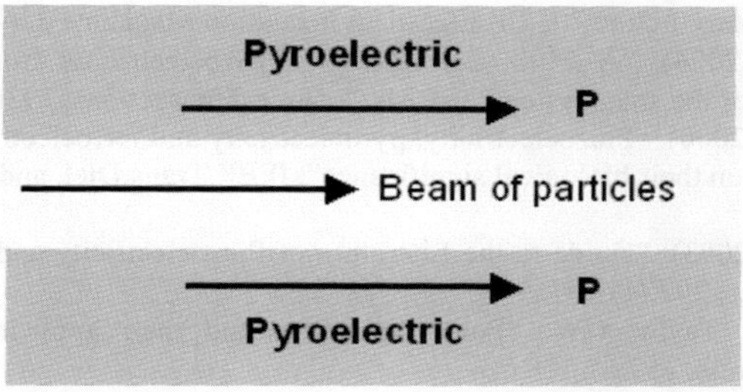

Fig. 12.21 Schematic outline of a EDEF-based particle accelerator. Adapted from reference 94.

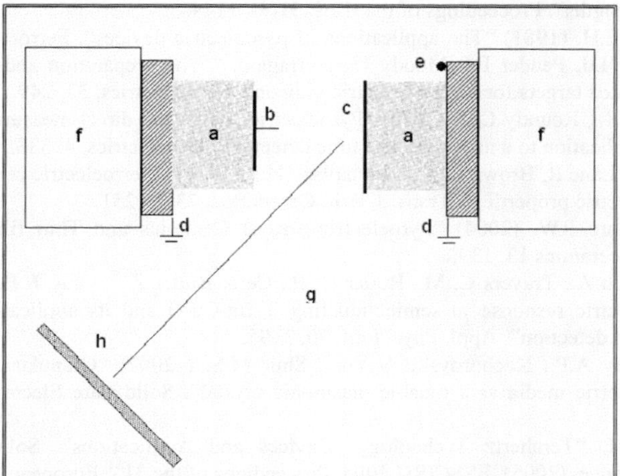

Fig. 12.22 Setup for paired-crystal neutron generation experiments. (a) 20mm diameter x 10 mm thick $LiTaO_3$ crystal. (b) 3 mm long cat whisker tip with 70nm apex mounted on a 13 mm diameter 1mm thick Cu disk. (c) Deuterated polystyrene target . (d) thermoelectric heater cooler. (e) Thermocouple. (f) Cu heat sink. (g) Vacuum chamber filled with 1.2 mTorr D_2 gas. (h) 3''x3'' EJ-301 detector at 45° angle to crystals.

References

1. Nye JF (1957) "The Physical Properties of Crystals". Clarendon Press. Oxford.
2. Shubnikov AV, Zheludev IS, Konstantinova VP. (1958) "Etude des textures piezoelectriques" Dunod, Paris .
3. Jefimenko OD, Walker DK, (1973) "Volume charge distribution in carnauba wax electrets", J. Appl. Phys. 44, 3459.
4. Multhaupt RG, Faria RM, Oliveira ON, Giacometti JA. (2006)"Useful effects of space charge and dipole polarisation: Recent developments in polymer electrets and organic semiconductors (Editorial)" IEEE Transactions on Dielectrics and Electrical Insulation 13, 939
5. Multhaupt RG. (2006) "Electret Science Has a Long History and a Bright Interdisciplinary Future " IEEE Electrical Insulation Magazine 22(4), 50.
6. Lang S.B. (2004). "A 2400 year history of pyroelectricity: from ancient Greece to exploration of the solar system" British Ceramic Transactions, 193(7), 65.
7. Lang. S.B. (2000). "Piezoelectricity, pyroelectricity and ferroelectricity in biomaterials - Speculation on their biological significance". IEEE Trans.Diel. and Electrical Insulation, 7,(4), 466.
8. Lang S.B. (2003). "Guide to the Literature of Piezoelectricity and Pyroelectricity. 20." Ferroelectrics, 297,(1) 107.
9. Burfoot JC, Taylor GW. "Polar dielectrics and their applications" University of California Press. 1979.
10. Muralt P. (2001)."Micromachined infrared detectors based on pyroelectric thin films" Rep. Prog. Phys 64, 1339.
11. Steier WH, Yamashita EE. (1963) "A Pyroelectric Effect Detector For Submillimetre Wavelengths". Proceedings of the IEEE.51(8), 1144.
12. Putley E.H. (1981) "The applications of pyroelectric devices". Ferroelectrics, 33, 207. Warner DJ, Pedder DJ, Moody IS, Burrage J. " The preparation and performance of reticulated targets for the pyroelectric vidicon". Ferroelectrics, 33, 249.
13. Byer R.L., Roundy C.B. (1972) "Pyroelectric coefficient direct measurement technique and application to a nsec response time detector" Ferroelectrics, 4, 333.
14. Luff D Lane R, Brown K.R., Marshallsay H.J. (1974). "Ferroelectric ceramics with high pyroelectric properties". Trans. J. Brit. Ceram.Soc. 73(7), 251.
15. Whatmore RW. (2004) "Pyroelectric arrays: Ceramics and Thin films". Journal of Electroceramics 13, 139.
16. Jahanzeb A., Travers C. M., Butler D. P., Çelik-Butler Z., . Gray J. E. (1997) "Strong pyroelectric response in semiconducting Y-Ba-Cu-O and its application to uncooled infrared detection". Appl. Phys. Lett. 70, 3495.
17. Dmitriev A.P., Kachorovskii V.Yu. , Shur M.S. (2007) "Granular semiconductor / pyroelectric media as a tunable plasmonic crystal". Solid-state Electronic vol. 51 (5), 812.
18. Shur M. "Terahertz Technology: Devices and Applications". Solid-State Circuits Conference, (2005). ESSCIRC 2005. Proceedings of the 31ST European 13.
19. 29.- Schwodiauer R, Gogonea SB, Bauer S, Heitz J, Arenholz E, Bauerle D. (1998)." Charge stability of pulsed-laser deposited polytetrafluoroethylene film electrets". Appl. Phys. Lett. 73(20);2941.
20. Pfister G, Abkowitz M, Crystal R.G. (1973) " Pyroelectricity in polyvinylidene Fluoride". J. Appl. Phys. 44 (5), 2064.

21. Nix E.L, Nanayakkara J, Davies G. R, Ward I. M. (1988) "Primary and secondary pyroelectricity in highly oriented polyvinylidene Fluoride". J. Polymer Science B: Polymer physics, 26, 127.
22. Bauer S, Lang S B. (1996). "Pyroelectric Polymer Electrets" .IEEE Transations on Dielectrics and Electrical Insulation, 3(5), 647.
23. Lang SB, Muensit S.(2006). "Review of some lesser known applications of piezoelectric and pyroelectric polymers" Appl. Phys. A 85, 125..
24. Auciello O, Kingon AI, Krupanidi SB (1996) "Sputter synthesis of ferroelectric thin films and heterostructures". MRS Bull., 21(6);25. Auciello O, Ramesh R. (1996) "Laser- Ablation deposition and characterization of ferroelectric capacitors for non-volatile memories". MRS Bull., 21(6); 31 .
25. DeKeijser M, Dormans GJM. (1996) "Chemical vapour deposition of electroceramic thin films" MRS Bull., 21(6); 37.
26. Tuttle BA, Schwartz RW (1996) "Solution deposition of ferroelectric thin films" MRS Bull., 21(6);49.
27. Scott JF, DeAraujo CAP. (1989). "Ferroelectric memories" Science, 246 (4936), 1400.
28. DeAraujo CAP, Cuchiaro JD, McmillanLD, Scott MC, Scott JF. (1995) "Fatigue free ferroelectric capacitors with platinum electrodes". Nature 374 (6523):627.
29. Shur M S. (2005).. "GaN – based devices". Electron Devices 2005, Spanish Conference on. 15.
30. Frenkel AI, Ehre D, Lyahovitskaya V, Kanner L, Wachtel E, Lubomirsky I. (2007)"Origin of polarity in amorphous $SrTiO_3$" Phys. Rev. Lett. 99(21) 215502.
31. Lyahovitskaya V, Zon I, Feldman Y, Cohen SR, Tagantsev AK, Lubomirsky I. (2003) "Pyroelectricity in highly stressed quasi-amorphous thin films". Adv. Mat. 15(21),1826.
32. Lang SB. (2005) "Pyroelectricity: from ancient curiosity to modern imaging tool". Physics today. August, 31
33. Zook J.D. Liu S.T. J. (1978) "Pyroelectric effect in thin film". J. Appl. Phys. 49(8), 4604.
34. Perls TA, Diesel J, Dobrov WI. (1958) "Primary Pyroelectricity in Barium Titanate Ceramics" J. Appl. Phys 19, 1279.
35. Ye CP, Tamagawa T, Polla DL. (1991) "Experimental Studies on Primary and Secondary Pyroelectric Effects in $Pb(Zr_xTi_{1-x})O_3$, $PbTiO_3$, and ZnO Thin-Films" J. Appl. Phys, 70(10), 5536.
36. Kholkin A, Brooks K, Taylor D, Hiboux S, Setter N. (1998)."Self-polarisation effect in $Pb(Zr,Ti)O_3$ thin films" Integrated Ferroelectrics, 22,(1-4) 1045.
37. Gerlach G, Shuchaneck G, Khöler R. Sandner T, Pasmini P, Krawietz R, Pompe W, Frey J, Jost O, Schönecker A. (1999) "Properties of sputter and sol-gel deposited PZT thin films for sensor and actuator applications: Preparation, stress and space charge distribution, self poling" Ferroelectrics 230(1-4), 411.
38. Bratkovsky AM, Levanyuk AP. (2005) "Smearing of phase transition due to a surface effect or a bulk inhomogeneity in ferroelectric nanostructures "Phys. Rev. Lett. 94, 107601.
39. Yamaka E, Watanabe H, Kimura H, Kanaya H, Ohkuma H. (1988) "Structural, Ferroelectric, and Pyroelectric Properties of Highly C-axis Oriented $Pb_{1-X}Ca_XTiO_3$ Thin-Film Grown by radio-frequency Magnetron Sputtering "J. Vac. Sci Technology A, 6(5), 2921.
40. Sviridov E, Sem I, Alyoshin V, Biryukov S, Dudkevich V. (1995) "Ferroelectric film self-polarisation " Mater res. Symp. Proc. 361, 141.
41. Catalan G, Noheda B, McAneney J, Sinnamon LJ, Gregg JM. (2005) "Strain gradients in epitaxial ferroelectrics" Phys. Rev. B,72 (2), 0201021-4.
42. Ma WH, Cross LE. (2006) "Flexoelectricity of barium titanate" Appl. Phys. Lett. 88, 232902.

43. Jiménez R, Alemany C, Mendiola J. (2002) "Top electrode induced self-polarisation in CSD processed SBT thin films" Ferroelectrics. 268 ,551.
44. Wang B, Kwok K, Chan H, Choy C. (2004)" Internal field and self-polarisation in sol-gel-derived lead zirconate titanate films" Applied Physics A. 79 ; 643.
45. Sharma A, Ban ZG, Alpay SP, Mantese JV. (2004) "Pyroelectric response of ferroelectric thin films" J. Appl. Phys. 95(7), 3618.
46. Glinchuk MD, Morozovska AN, Eliseev EA. (2006) "Ferroelectric thin films phase diagrams with self-polarised phase and electret state "J. Appl. Phys, 99, 114102.
47. Dkhil B, Defay E, Guillan J. (2007) "Strains in BaTiO3 thin film deposited onto Pt-coated Si substrate". Applied. Phys. Lett. 90, 022908.
48. Pertsev NA, Zembilgotov ZG, Tagantsev AK. (1999) "Equilibrium states and phase transitions in epitaxial ferroelectric thin films". Ferroelectrics, 223(1-4), 79.
49. Junquera J, Ghosez P.(2003) "Critical thickness for ferroelectricity in perovskite ultrathin films".Nature.422 (6931), 506.
50. Lang S.B. "Source book of pyroelectricity". Gordon and Breach, New York, 1974.
51. Ackermann W. (1915) "Observations on pyroelectricity, in its dependency on temperature.". Annalen der Physik, 46(2),197.
52. Gavrilova ND. (1965) "Study of the temperature dependence of the pyroelectric coefficients of crystals by the static method" Soviet Physics- Crystallography. 10(3), 278
53. Lang SB, Steckel F. (1965) "Method for measurement of pyroelectric coefficient DC dielectric constant and volume resistivity of a polar material" Rev. Scientific Inst. 36(7), 929.
54. Bernard M, Briot R., Grange G. (1980)" Experimental and theoretical study of polarisation of niobium doped PZT ceramics" J. Solid State Chem. 31(3) 369.
55. Weise W, Keith T, von Malm N, von Seggern H (2005) "Trap concentration dependence of thermally stimulated currents in small molecule organic materials" Phys.Rev. B. 72(4), 045202.
56. Sharp JE, Garn LE. (1982) "Use of low-frequency sinusoidal temperature waves to separate pyroelectric currents from non-pyroelectric currents. Part I. Theory" J. Appl. Phys. 53(12), 8974
57. Sharp JE, Garn LE. (1982) "Use of low-frequency sinusoidal temperature waves to separate pyroelectric currents from non-pyroelectric currents. Part II. Experiment" J. Appl. Phys. 53(12), 8980.
58. Dias C, Simon M, Quad R, Das-Gupta. DK. (1993) "Measurement of the pyroelectric coefficient in composites using a temperature–modulated excitation" J. Phys D: Appl. Phys. 26 ,106
59. Jimenez R, Ramos P; Calzada ML, Mendiola J. (1998) "Pyroelectricity in lead titanate thin films" Bol. Soc. Esp. Ceram. Vid. 37 (2-3) 117.
60. Daglish M. (1998) "A dynamic method for determining the pyroelectric response of thin films " Integrated. Ferroelectrics. 22(1-4) , 993.
61. Chynoweth AG. (1956) "Dynamic method for measuring the pyroelectric effect with special reference to Barium titanate" J. Appl. Phys. 27(1), 78.
62. Jiménez R, Hungría T, Castro A. Jiménez-Riobóo R. (2008) "Phase transitions in $Na_{1-x}Li_xNbO_3$ solid solution ceramics studied by a new pyroelectric current based method". J. Phys. D: Appl. Phys. 41 065408.
63. Lang S.B., Rice LH, Saw SA. (1969) " Pyroelectric effect in Barium Titanate ceramic". J. Appl. Phys. 40(11), 4335.
64. Jiménez R, Alemany C, Calzada ML, González A, Ricote J, Mendiola. (2002) "Processing effects on the microstructure and ferroelectric properties of strontium bismuth tantalate thin films " J. Appl phys A, 75(5), 607.
65. Collins RE. (1975) "Distribution of charge in electrets" Appl Phys. Letters. 26(12), 675.

66. DeReggi AS, Guttman CM, Mopsik FI, Davis GT, Broadhurst MG. (1978) "Determination of charge or polarisation distribution across polymer electrets by thermal pulse method and Fourier analysis" Phys Rev. Lett. 40(6), 413.
67. Mopsik FI, DeReggi AS. (1982) "Numerical evaluation of dielectric polarisation distribution from thermal pulse data" J. Appl. Phys. 53(6) 4333.
68. Lang SB, Das Gupta DK. (1981) "A technique for determining the polarisation distribution in thin polymer electrets using periodic heating". Ferroelectrics. 39(1-4) 1249.
69. DeFrutos J, Jiménez B. (1990) "Study of the spatial distribution of the polarisation in ferroelectric ceramics by means of low frequency sinusoidal thermal waves." Ferroelectrics 109 101.
70. Lang SB. (2004) "Laser intensity modulation method (LIMM): Review of the fundamentals and a new method for data analysis" IEEE transactions on dielectrics and electrical insulation. 11(1), 3.
71. Frutos J, Ph. D Thesis. (1992) "Estudio de distribución de polarización en materiales ferroeléctricos por medio de ondas térmicas sinusoidales de baja frecuencia". UAM, Madrid, Spain.
72. Collins RE (1976) "Analysis of spatial distribution of charges and dipoles in electrets by a transient heating technique" J. Appl. Phys. 47(11), 4804.
73. Ploss B, Emmerich R, Bauer S.(1992) " Thermal wave probing of pyroelectric distributions in the surface region of ferroelectric materials. A new method for the analysis" J. Appl. Phys. 72(11), 5363.
74. Lang SB, DasGupta DK. (1986) "Laser Intensity modulation Method: A technique for determination of spatial distributions of polarisation and space charge in polymer electrets" J. Appl. Phys 59(6), 2151.
75. Bauer S, Ploss B. (1991) "Polarisation distribution of thermally poled PVDF films measured with a heat wave method (LIMM)" Ferroelectrics 118(1-4), 363.
76. Provencher SW. (1982) "A constrained regularization method for inverting data represented by linear algebraic or integral equations". Comput. Phys. Communications. 27(3), 213.
77. Lang SB. (1991) "Laser Intensity modulation Method (LIMM): Experimental techniques. Theory and solution of the integral equation" Ferroelectrics. 118 (1-4), 343.
78. Bloss P, Steffen M, Schäfer H. (1994) "LIMM Investigations of thermally poled PVDF and FEP samples" 8th International symposium on electrets (ISE 8) Proceedings, (Paris).
79. Lang SB. (1998) "An analysis of the integral equation of the surface laser intensity modulation method using the constrained regularization method" IEEE. Trans. Dielectr. Electr. Insul. 5(1), 70.
80. Hadni A, Bassia JM, Gerbaux X, Thomas R. (1976) "Laser Scanning method for pyroelectric display in real time". Applied Optics. 15(9), 2150.
81. Marty-Dessus D, Berquez L, Petre A, Franchesi JL. (2002) "Space charge cartography by FLIMM: a three-dimensional approach" J. Phys D: Appl. Phys. 35, 3249.
82. Porter S.G. (1981) "A brief guide to pyroelectric detectors". Ferroelectrics 33, 193.
83. Kruse PW. (1995) "A comparison of the limits to the performance of thermal and photon detector imaging arrays". Infrared Phys. Technol. 36, 209.
84. Tompsett MF. (1997) "Pyroelectric Vidicon. Uncooled Infrared Imaging Arrays and Systems". Ed P W Kruse and D D Skatrud (New York: Academic Press), 219.
85. Watton R. (1978) "Thermal properties of reticulated layers: Analytical solutions and design parameters" Infrared Physics 18 (2), 73.
86. Hanson CM, Beratan HR, Belcher JF, Udayakumar KR, Soch K. (1998) "Advances in monolithic ferroelectric uncooled IRFPA technology". Infrared detectors and focal plane Arrays V. (3379), 60.

87. Barcerak, R S. Uncooled infrared sensors: rapid growth an future perspective". Infrared Detectors and Focal Plane Arrays . Proceedings of photo-optical instrumentation engineers (SPIE), 3379, 60.
88. Togami Y, Okuyama M, Hamakawa Y, Kimata M, Denda M. (1991) "Pyroelectric infrared image sensors operable at room temperature". Optoelectronic Devices Technolo. 6, 205.
89. Fujii S, Kamata T, Hayashi S, Tomita Y, Takayama R, Hirao T, Nakayama T, Deguchi T. "Pyroelectric Linear Infrared Sensors made of La-Modified PbTiO3 Thin Films and their Applications". (1995) ". Orlando, FL.:SPIE Optical Engineering Press.
90. Willing B, Kohli M, Muralt P, Oehler O. (1998) "Thin film pyroelectric array as a detector for a infrared gas spectrometre". Infrared Phys. Technol. 39, 443.
91. Dmitriev AP, Kachorovskii VY, Shur MS. (2007) "Granular semiconductor./pyroelectric media as a tunable plasmonic crystal". Solid-State Electronics, 51, 812.
92. Shur M. (2005) "Terahertz technology: devices and applications". Proceedings of ESSCIRC, Grenoble, France , 13.
93. Naranjo B, Gimzewskl JK, Putterman S. (2005) "Observation of nuclear fusion driven by a pyroelectric crystal". Nature, 434, 28l.
94. Geuther J, Danon Y, Saglime F. (2006) "Nuclear Reaction Induced by a Pyroelectric Accelerator". Phys. Rev. Letters, 96, 054803.
95. Sandomirsky V, Schlesinger Y, Levin R. (2006) "The Edge Electric Field of a Pyroelectric and its Applications". J. Appl. Phys. 100(11), 11372.

Chapter 13
Properties of Ferro-Piezoelectric Ceramic Materials in the Linear Range: Determination from Impedance Measurements at Resonance

L. Pardo[1], K. Brebøl[2]

Abstract

A critical review of the methods of characterization in the linear range of piezoceramics of industrial interest from impedance measurements at resonance is presented, from the 1961 standards to the most recent fitting and iterative methods. This review takes into account the need to determine the material losses and introduces the Alemany *et al.* method. This is shown as the only method systematically applied to date in the determination of the full matrix of elasto-piezo-dielectric coefficients, including all losses, of a poled ferroelectric ceramic, from three samples and four resonances, in an accurate and consistent way.

Matrix material characterization, including losses, of a soft, Navy-II type, commercial PZT ceramic (PZ27), carried out by the Alemany *et al.* iterative method, allows us to perform the accurate modelling by FEA of shear samples, study of mode coupling and optimization of shear geometries for the purpose of material characterization.

---

[1] Instituto de Ciencia de Materiales de Madrid (ICMM-CSIC). Cantoblanco 28049 - Madrid (Spain) lpardo@icmm.csic.es
[2] Limiel ApS. DK - 4772 Langebæk. (Denmark) kbreboel@post3.tele.dk

13.1 The Resonance Method in the Determination of the Properties of Ferro-Piezoelectric Ceramics in the Linear Range

The characterization in the linear range of ferro-piezoelectric bulk ceramic materials, from impedance measurements at their electromechanical resonances, has been used since the early times of the development of these materials in the 1960s [1, 2], and has become a widespread practice. In the resonance method, a low voltage a.c. signal is used to excite an elastic wave in the piezoceramic via the inverse piezoelectric effect. A resonance condition, which occurs at a critical frequency depending on the sample dimensions and properties, is monitored by measuring the impedance or admittance of the sample as a function of the frequency. The obtained experimental curve, the resonance spectrum, is analyzed to determine the material properties. The simple experimental set-up, just involving an impedance analyzer and an appropriated sample holder, needed to apply this method has contributed to its widespread use in research and industrial laboratories and to the standardization of the technique. The standard methods of measurements have been issued a long time ago and revised several times. The following is not a self-explanatory or exhaustive review of the method, for which the reader is addressed to the wider classical literature [1, 2, 3, 4, 5] on it. But, it aims to provide him or her with the key ideas and references to understand the power and difficulties of this method, as well as the present state of the art on its application.

13.1.1 Properties of Ferro-Piezoelectric Ceramics

The modelling and design of new piezoelectric devices by, among other numerical methods, the finite element analysis, relies on the accuracy of the dielectric, piezoelectric and elastic coefficients of the active material used. This is frequently an anisotropic ferroelectric polycrystal, commonly referred to as ferroelectric ceramic, piezoelectric ceramic or piezoceramic. The polarization in the randomly oriented grains of an as-sintered ferroelectric ceramic may be oriented in the closest crystallographic direction to an external electric field, in the so-called "poling" process (Fig. 13.1a). This gives as a result, an induced anisotropy of the properties and a piezoelectric effect [1].

An accurate description of piezoceramics involves the evaluation of the dielectric, piezoelectric and mechanical losses. This accounts for the out of phase material response to the input signal, which is not always accomplished despite of their important role in the material performance. Losses in piezoceramics have inconvenient consequences for positioning actuator applications, since they lead to hysteresis in the field-induced strain, and for resonance applications, such ultrasonic motors, since they are the cause of heat generation. On the other hand, they may be an advantage for force sensors and acoustic transducers, since they widen the frequency band for receiving signals.

The description of the material parameters by complex values (P* =P′-iP′′) is a convenient way to separately account for the dielectric, piezoelectric and mechanical losses (tanδ= P′′/ P′). The origin of the losses in ferroelectric ceramics has been analyzed in numerous works [6, 7, 8, 9, 10]. While the dielectric losses are related to the ionic and the ferroelectric nature of these compounds, the mechanical losses in piezoceramics arise from crystal lattice defects, microstructure (grain boundaries, porosity) and ferroelastic domain wall motions, and the piezoelectric losses from the coupling of all such effects.

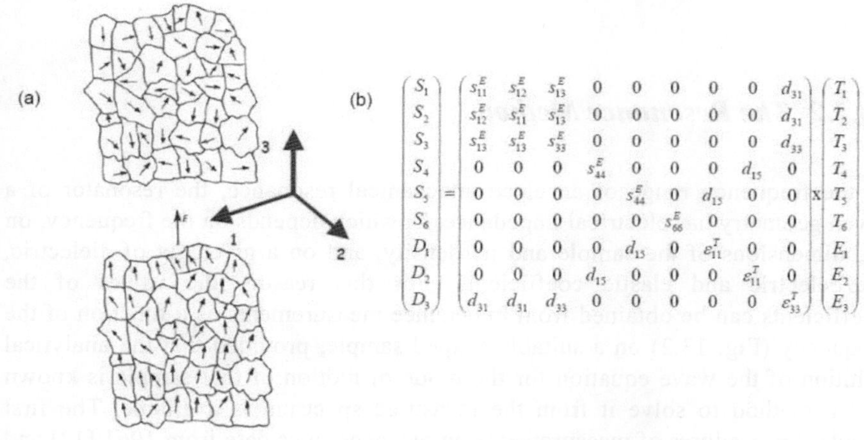

Fig. 13.1 a Schematic view of grains of a ferroelectric ceramic before (up) and after (down) the poling process. The arrow in each grain represents the net polarization of the ferroelectric domains oriented at different angles. The 3 direction is arbitrarily chosen as the one of the applied electric field. **b** Matrix of elasto-piezo-dielectric coefficients of a poled ferroelectric ceramic (S=mechanical strain, s=elastic compliance, D=electric displacement, d=piezoelectric coefficient, ε=dielectric permittivity, T=mechanical stress, E=electric field).

Poled ferroelectric ceramics in their normal operating range show substantially linear relations between the stress (T_{ij}) and the strain (S_{ij}), which are tensor magnitudes, on the one hand and between the electric field (E_i) and the dielectric displacement (D_i), which are vector magnitudes, on the other. Besides, the piezoelectric coefficients provide relations between mechanical and electrical magnitudes [1]. The constitutive equations may take various forms. This relationship may be written by making use of different sets of coefficients in reduced matrix form. One of these frequently used sets is:

$$S = s^E T + d_t E \quad (1)$$

$$D = d T + \varepsilon^T E$$

where *s* is the matrix of elastic compliance, *d* is the matrix of piezoelectric charge coefficients and d_t its transposed (changing rows by columns) matrix, and ε^T the dielectric permittivity matrix. Superindex E means constant (zero) electric field and superindex T means constant (zero) mechanical stress. Thus, there are various sets of parameters, related among them, that characterize a piezoceramic. Piezoceramics, which have fibre (6mm) symmetry [11] induced by poling, are characterized by only five independent elastic constants (note that $s_{66}^E = 2\ (s_{11}^E - s_{12}^E)$), three independent piezoelectric coefficients and two independent dielectric coefficients (Fig. 13.1b).

13.1.2 The Resonance Method

In the frequency range of an electromechanical resonance, the resonator of a given geometry has electrical impedance, Z, which depends on the frequency, on the dimensions of the sample and its density, and on a given set of dielectric, piezoelectric and elastic coefficients. For this reason, the values of the coefficients can be obtained from impedance measurements as a function of the frequency (Fig. 13.2) on a suitable shaped sample, provided that the analytical solution of the wave equation for the mode of motion of that sample is known and a method to solve it from the measured spectrum is available. The first standard procedures of measurements on piezoceramics date from 1961 [12] and were issued by the North American "Institute of Radio Engineers (IRE)". The aim of this standard was to adapt previous definitions, relations and measurement methods, developed for piezoelectric crystals in general, to the characteristics of those new materials.

The electromechanical coupling factor, k, is defined as the square root of the energy transformed by the resonator divided by the total energy input, and expresses the ability of the resonator to transform electric energy into mechanical energy, and vice-versa. The resonator can be described by an LCR (inductance, capacitance, resistance) resonant circuit. As is well known, the complex impedance of the resonator is Z= R + iX, where X is the reactance. The mechanical quality factor, Q_m, is defined as the expression of the internal mechanical damping of the resonator and by the ratio between the circuit reactance and resistance. The higher the mechanical quality factor, the lower the mechanical losses of the resonator.

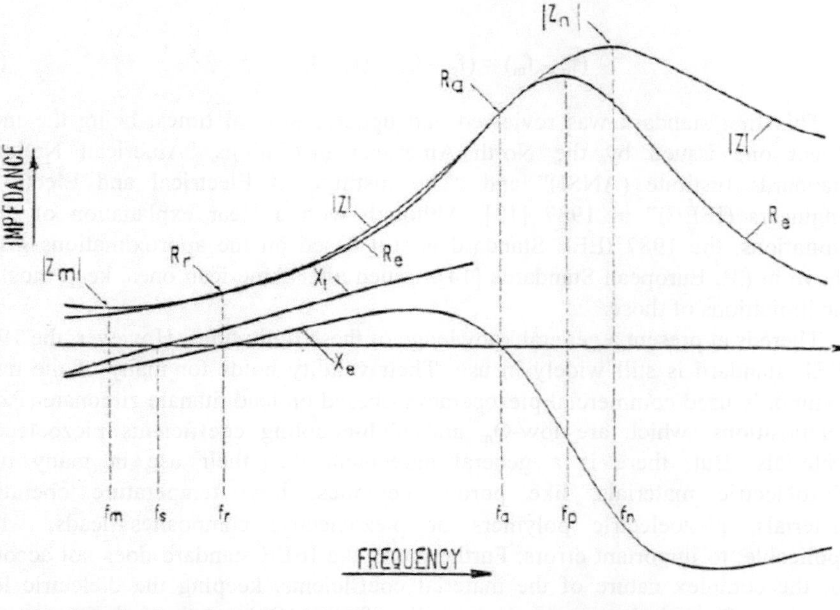

Fig. 13.2 Complex impedance (modulus, |Z|; real, resistance(R), and imaginary, reactance (X), components) dependence on frequency in the vicinity of an electromechanical resonance of a piezoelectric ceramic.

The complex admittance of the resonator is defined as the reciprocal of the impedance and as Y=G+iB, where G is the conductance and B the susceptance. Standards define f_n and f_m as the frequencies of maximum and minimum impedance (Fig. 13.2), respectively, in the neighbourhood of the resonance, and also define the procedures (the so-called transmission circuit method) to measure the resonator impedance and determine these frequencies. Coupling coefficients and quality factors of the resonator are calculated from the difference between these two frequencies. Standards also define other pairs of frequencies: fs and fp, the series and parallel frequencies, as the frequencies for maximum G and maximum R, respectively, and fr and fa, the resonance and antiresonance frequencies, as the frequencies of B=0 and X=0, respectively. The values of these frequencies are such that (Fig. 13.2):

$$(f_n - f_m) > (f_p - f_s) > (f_a - f_r) \quad (2)$$

The main piezoceramic characteristics considered in the development of the 1961 standard were the high symmetry (6mm), leading to simpler relations than for most piezoelectric crystals, high electromechanical coupling factors, low mechanical quality factors and noticeable dielectric loss factor ($\tan\delta = \varepsilon''/\varepsilon'$). If such conditions are fulfilled, the following approximations are valid:

$$(f_n - f_m) \approx (f_p - f_s) \approx (f_a - f_r) \tag{3}$$

This first standard was reviewed and updated several times, being the most recent one issued by the North American institutions "American National Standards Institute (ANSI)" and "The Institute of Electrical and Electronic Engineers (IEEE)" in 1987 [13]. Although with a clear explanation of their limitations, the 1987 IEEE Standard is still based on the approximations given above in (3). European Standards [14], issued after American ones, keep most of the limitations of those.

There is at present a general knowledge of these limitations. However, the 1987 IEEE standard is still widely in use. Their validity holds for many of the most commonly used commercial piezoceramics based on lead titanate zirconate (PZT) compositions, which are low-Q_m and high-coupling coefficients piezoelectric materials. But, there is a general agreement that their use in many new piezoelectric materials, like porous ceramics, high temperature operating materials, piezoelectric polymers or piezoelectric composites leads, when applicable, to important errors. Furthermore, the IEEE standard does not account for the complex nature of the material coefficients, keeping the dielectric loss factor ($\tan\delta$) and the mechanical quality factor (Q_m) as the only parameters accounting for the losses in materials.

The dielectric permittivity is "traditionally" obtained from the sample capacitance measured at 1kHz. This is not a good practice. The reason is the change in permittivity, elastic compliance and piezoelectric response with frequency, a dispersion, which is more pronounced in ferroelectric materials. Variations between such permittivity value at 1kHz and that included in the analytical solution of the wave equation corresponding to a resonance taking place at a given frequency (from 100kHz to 10MHz) are, thus, assumed. Recommendation for the use of such low frequency, also included in the 1987 IEEE Standard for ε^T_{ij}, is made on the basis of avoiding interference with the capacitance measurement of the lowest resonances of plates, usually taking place at $f \sim 100$kHz.

However, the 1987 IEEE Standard contributions to the characterization of piezoelectric ceramics are remarkable. The standard states the resonator geometries, the relationships among their dimensions for the validity of the equations used in the IEEE Standard, and all the relationships among the coefficients needed to get all the independent coefficients for piezoceramics.

13.1.3 Iterative Methods in the Complex Characterization of Piezoceramics

One of the first attempts to treat losses in piezoceramics by considering complex coefficients is due to Holland and dates from 1967 [15]. At that time, Holland and EerNisse [16] proposed a gain-bandwidth method for length extensional mode of thickness poled bars that was valid for moderate losses and low electromechanical coupling coefficient materials. Since then, many papers on the topic have been published, some of which will be referred to later.

An iterative method is a way of solving a mathematical problem, here a non-linear equation, by a series of approximations, which obtains a more accurate solution by using the preceding approximation as a starting point. Alternatively, fitting methods are used in the comparison with the experimental impedance spectrum to find empirically the appropriated values for the analyzed portion of the spectrum.

Smits [17] published the first iterative procedure for the accurate determination of complex materials coefficients. The major drawback of the method is that it requires a skilled operator, because the judicious choice of the three frequencies for measurement of Z is needed, in order to avoid the determination of constants with large errors. From such Z values, a system of non-linear equations is established and solved in the corresponding coefficients by an iterative procedure.

After the development and extension of the use in science of personal computers, these methods became realizable in a number of laboratories. The first automation and the extension to complex parameters of the procedure by Meitzler *et al.* [18], previously adopted in the 1987 IEEE standard [13], for a piezoelectric resonator in the radial mode of a thin disk, was published by Sherrit *et al.* [19]. This is the most mathematically complex resonance mode used for materials characterization. A linear interpolation is used in the 1987 IEEE standard, for the data of the Poisson ratio, σ^P, as a function of the ratio of the first overtone to the fundamental resonance frequencies, $f^{(2)}{}_s/f^{(1)}{}_s$, and in the determination of σ^P. In [19], these authors also introduced a refinement of the standard method for σ^P by a polynomial fit, easier to implement in terms of automation and more accurate. Besides, Sherrit *et al.* also proposed a non-iterative evaluation method, and applied it to thickness extensional plate resonators [20]. This method implicitly predetermines that the phase of the electromechanical coupling factor is only dependent on mechanical loss. It is only applicable for materials with low dielectric and piezoelectric losses. Sherrit *et al.* accomplished as well the first complete characterization in a complex form by the resonance method of a commercial ceramic [21].

They used four geometries and five resonance modes, shown in Table 13.1. For that purpose they used a non-iterative method for the radial mode of a thickness-poled thin disk resonator and the Smits method for the rest of the modes.

Alemany *et al.* [22, 23] developed an automatic iterative method, based on the Smits model, that overcomes its main drawback by the automation of the choice of the frequencies for the measurement of Z (or Y depending on the resonance mode). The Alemany *et al.* method has been also applied to the determination of the parameters from overtone resonances in the radial and thickness modes of thin disks

[24]. It thus accounts for a determination of the dispersion of the complex material parameters, though in a discrete instead of a continuous way. This method is the only iterative method that was applied to all resonance modes needed for the full matrix characterization of piezoceramics (Table 13.1). The systematic use of the Alemany *et al.* method on the three sample shapes and four resonance modes of Table 13.1, have recently led to the full complex characterization of a soft, Navy II-type, lead circonate titanate, PZT, commercial piezoceramic (PZ27 Ferroperm Piezoceramics A/S) [25] and a high sensitivity Mn-doped Morphotropic Phase Boundary $0.655Pb(Mg_{1/3}Nb_{2/3})O_3$-$0.345PbTiO_3$ ceramic [26].

In 1997, Kwok *et al.* [27] published a comparative study, applied to thickness extensional plate resonators. The methods compared were: a non-linear regression method proposed by the authors (the Gauss-Newton fitting method), the 1987 IEEE standard method and a commercial software (the Piezoelectric Resonance Analysis Program (PRAP), based on Smits [17] and Sherrit *et al.* [20] works). Together with other moderate loss materials (copolymers, lead metaniobate ceramics and composites), they characterized polyvinylidene fluoride (PVDF), a high-loss piezoelectric polymer. They found the already mentioned limitations of such methods, i.e., that 1987 IEEE standard, Smits and Sherrit *et al.* methods, although requiring skilled operators, are valid methods for materials with moderate losses but fail in the characterization of high-loss materials. As for any other fitting method, the proposed non-linear regression method has the drawback of being sensitive to the choice of the data segment used. Each calculated material parameter represents an average within the frequency range where data points are used for the calculation. The narrower the fitting range, the closer are the calculated average values to the actual values of the parameters.

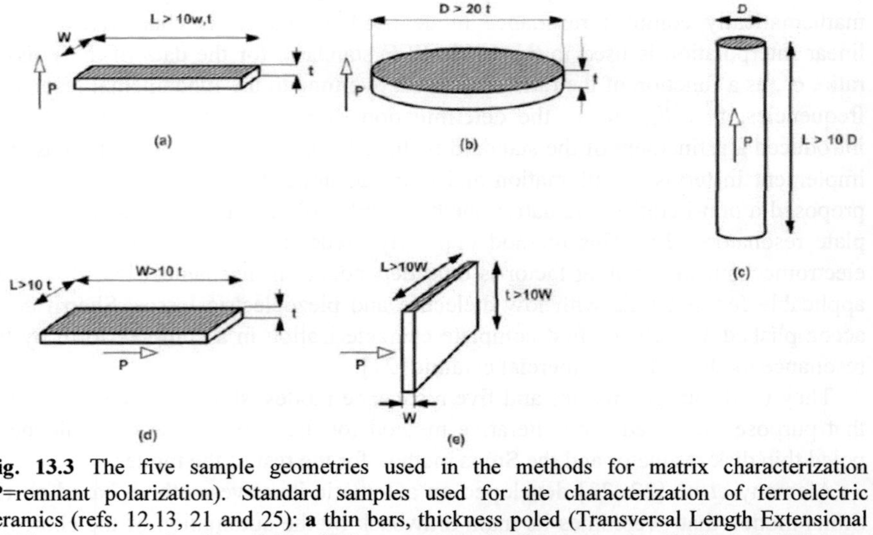

Fig. 13.3 The five sample geometries used in the methods for matrix characterization (P=remnant polarization). Standard samples used for the characterization of ferroelectric ceramics (refs. 12,13, 21 and 25): **a** thin bars, thickness poled (Transversal Length Extensional (TLE) resonator), **b** thin disks, thickness poled (Radial resonator (RD) and Thickness Extensional (TE) resonator), **c** long bars, length poled (Length Extensional resonator (LE)) and **d** shear plates, length poled (Thickness Shear resonator (TS)). Sample **e** is a shear plate, thickness poled, (Non-standard Thickness Shear resonator) used in this work.

Table 13.1 Sample shapes, with the required dimensional ratios, corresponding resonance modes and directly obtained coefficients for each mode, which are needed for the determination of the full characteristic matrices of a poled ferroelectric ceramic according to the standards [refs. 12,13 and 21] and by Alemany *et al.* method with one of the two types of shear resonators [refs. 25 and 55]. The coefficients that are not displayed in this Table are calculated from those showed here using the relationships that exist among them [refs. 13, 21 and 25].

Sample geometry	**Resonance mode** (Standards)	**Coefficients** (Standards)	**Resonance mode** (Alemany *et al.* method)	**Directly obtained coefficients** (Alemany *et al.* method)
thin bars, thickness poled (Length > 10. width, thickness)	transverse length extensional mode of thin plates (TLE)	• d_{31}, ε^{T}_{33}, s^{E}_{11} • k_{31}		
thin disks, thickness poled. (Diameter>20.thickness)	radial mode (RD)	• d_{31}, ε^{T}_{33}, s^{E}_{11}, s^{E}_{12} • k_p	radial mode	• d_{31}, ε^{T}_{33}, s^{E}_{11}, s^{E}_{12} • k_p, k_{31}
	thickness extensional mode (TE)	• k_t, c^{D}_{33}	thickness extensional mode of thin disks	• h_{33}, ε^{S}_{33}, c^{D}_{33} • k_t
long bars, length poled. (Length > 10.Diameter)	length extensional mode (LE)	• d_{33}, ε^{T}_{33}, s^{E}_{33} • k_{33}	length extensional mode	• g_{33}, ε^{T}_{33} , s^{D}_{33} • k_{33}
shear plates, length poled. (Length, width> 10.thickness)	thickness shear mode of thin plates (TS)	• d_{15} , ε^{T}_{11}, s^{E}_{55} • k_{15}	thickness shear mode of thin plates [25]	• h_{15}, ε^{S}_{11}, c^{D}_{55} • k_{15}
shear plates, thickness poled. (Length, width> 10.thickn/ess)			second thickness shear mode of thin plates [55]	• e_{15}, ε^{S}_{11}, s^{E}_{55} • k_{15}

13.1.4 *Iterative Automatic Method Developed by C. Alemany et al. at CSIC*

In this method, the coefficients are calculated as the solution – obtained by an automatic iterative numerical method of the set of non-linear equations that results when experimental impedance, (or admittance) data are introduced into the analytical solution of the wave equation for the mode of motion of the sample. Such impedance or admittance data are taken at four given frequencies in the neighbourhood of the resonance. The determination of such frequencies is also automatic. This analytical solution exists for all the four resonance modes (Table 13.1) needed for the determination of the complete set of parameters and given in Fig. 13.3. The impedance for the thickness resonance of a disk poled and excited along its thickness is given by the equation:

$$Z = R + iX = -i\frac{t}{2\pi f S \varepsilon_{33}^{s}} + i\frac{h_{33}^{2}}{2\pi^{2} S f^{2} c_{33}^{D}\sqrt{\frac{\rho}{c_{33}^{D}}}}\tan\left(\pi f t\sqrt{\frac{\rho}{c_{33}^{D}}}\right) \tag{4}$$

where S = π(D/2)^2 (Fig. 13.3b) is the electroded surface area and ρ is the ceramic density. The admittance for the radial resonance of a disk poled and excited along its thickness is given by:

$$Y = G + iB = i\frac{2\pi^{2} f D^{2}}{4t}\left(\varepsilon_{33}^{T} + 2d_{31}^{2}\frac{c_{11}^{P}}{\frac{1}{2 - \Im_{1}\left[\pi f D\sqrt{\frac{\rho}{c_{11}^{P}}}\right]} - \frac{1}{1+\sigma^{P}}}\right) \tag{5}$$

where t is the thickness of the disk (Fig. 13.3b), σ^P is the Poisson ratio, $\sigma^P = - s_{12}^E / s_{11}^E$, c_{11}^P is a relationship among constants, $c_{11}^P = s^E_{11}\,[(s^E_{11})^2 - (s^E_{12})^2]^{-1}$, and $\Im$ is the Onoe function defined as:

$$\Im_{1}[z] = \frac{zJ_{o}(z)}{J_{1}(z)} \tag{6}$$

being Jo and J_1 the Bessel functions of the first kind and zero and first orders, respectively.

The impedance for the shear resonance of a plate poled across its length and excited along its thickness is

$$Z = R + iX = -i\frac{t}{2\pi f S \varepsilon_{11}^{s}} + i\frac{h_{15}^{2}}{2\pi^{2} S f^{2} c_{55}^{D}\sqrt{\frac{\rho}{c_{55}^{D}}}}\tan\left(\pi f t\sqrt{\frac{\rho}{c_{55}^{D}}}\right) \tag{7}$$

where S = Lw (Fig. 13.3d) is the electroded surface area.

The impedance for the length resonance of a bar poled and excited along its length is given by:

$$Z = R + iX = -i\frac{L}{2\pi f S}\left(\frac{1}{\varepsilon_{33}^{T}} + \frac{g_{33}^{2}}{s_{33}^{D}}\right) + i\frac{g_{33}^{2}}{2\pi^{2} S f^{2} s_{33}^{D}\sqrt{\rho s_{33}^{D}}}\tan\left(\pi f L\sqrt{\rho s_{33}^{D}}\right) \quad (8)$$

where $S = \pi(D/2)^2$ (Fig. 13.3c) is the electroded surface area.

Only the dimensions and density of the sample, together with the value of the complex admittance or impedance at such four frequencies around the resonance, are required to get the dielectric, elastic and piezoelectric complex coefficients that rules in each mode of resonance. In the radial mode of thin disks, it is additionally needed to know the value of $f^{(2)}{}_s$ of the first overtone. In practice, an experimental data file of absolute values of admittance, $|Y_i|$, and its phase angle, θ_i, at each frequency, f_i, is obtained in a frequency interval around the resonance. From these values, the corresponding values of conductance, $G_i=|Y|_i\cos\theta_i$, and resistance $R_i = |Y|_i^{-1}\cos\theta_i$ are obtained.

Two of the four frequencies involved in the calculation, f_s and f_p, are determined by location of the maximum values of R_i and G_i in the measured interval, and the values of complex impedance at such frequencies introduced in the system of non-linear equations to be solved, in the so-called central iteration of the method. The determination of the other two frequencies involved in the calculation of the central iteration, f_1 and f_2, constitutes a second iterative process, called peripheral iteration of the method, which finishes when the convergence of f_2 fulfils the criteria: $|f_2(\text{final})-f_2(\text{initial})|<0.05\%$. The details concerning the two iterations of the Alemany *et al.* method are given in [22] and [23].

Contrary to what it is done in the fitting methods, both the G and the R profiles are reconstructed here just as a quality criteria of the results obtained by the method, since the solution of the equation system is unique and cannot be further modified. Such reconstruction is done by insertion of the obtained complex coefficients in the analytical solution of the wave equation of the given resonance mode and calculation of R and G as a function of frequency. When the experimental resonance spectrum is free of spurious resonances, thus corresponding to a single resonator, the regression factor between the experimental and reconstructed values is usually above 0.99, which underlines the quality of the method.

In the first publications on the method [22, 23], the complex characterization of a number of commercial piezoceramics (Ferroperm Piezoceramics A/S, Denmark), with high (PZ27, lead circonate titanate) and low (PZ34, modified lead titanate, and PZ45, bismuth niobate) coupling coefficients and with high (PZ27) and low (PZ35, lead metaniobate) mechanical quality factors, was accomplished. Although still fully valid as described in references [22] and [23], some of the calculation details given in the first works have benefited from the long experience of application of the method at ICMM-CSIC to a wide number of ceramic materials [28, 29, 30]. It has been made easier to use in the research and industrial characterization and quality control laboratories. The first versions of the software were controlling the connected impedance analyzer, primarily to adjust the four frequencies for the measurements at each iteration.

Later versions can also work on already made datasets, in order to accept data from any kind of equipment, and have been extensively used in the study of high temperature piezoelectric ceramics [30, 31, 32] where calculations takes place after all the measurements were done.

13.2 Complementary use of Finite Element Analysis and Laser Interferometry to the Characterization of Piezoceramics from Impedance Measurements at Resonance

The interest that the topic of the characterization of piezoelectric materials from the analysis of the electromechanical resonance still drags nowadays is shown by the new publications on fitting or iterative methods. To date, these are commonly applied to length extensional mode of thickness poled bars [33, 34, 35] or planar mode of disks [36]. However, due to the possibility of their being widely used for a variety of geometries [37], it can be expected that these methods will gradually extend to other useful modes of resonance for characterization of piezoceramics. At this point, the authors have considered that a critical review of the advantages and drawbacks of the method, at the light of the valuable information that complementary techniques provide, was necessary.

13.2.1 Finite Element Analysis for the Matrix Characterization of Piezoceramics

Widely used for device modelling, the Finite Element Analysis (FEA) has been scarcely used in the science of materials. The characterization of piezoceramics is a field in which this tool is beginning to receive some attention [38, 39, 40]. However, the strong potential of this tool in the study, for example, of the modes of motion of the material resonators [41, 42] remains yet scarcely exploited for characterization purposes. One of the important reasons for this is the scarce number of complete sets of data, including all losses, for the piezoceramics of industrial interest. The modelling of piezoelectrics has been restricted to considering those as materials without losses [43] and isotropic media [44]. As the determination of matrix data on piezoceramics seems to be a field of intense activity, it is also foreseen that the potential of the FEA modelling technique will extend to the field of material science.

13.2.1.1 Matrix Characterization Based on Standard Shear Samples

The availability of the full set of coefficients, including all losses, has allowed the modelling by 3-D FEA of PZ27 (Ferroperm Piezoceramics A/S, Denmark) piezoceramic items of the three geometries and four resonance modes (Table 13.1) used for the matrix characterization [45, 46]. Here, we choose to work with the following coefficients: s_{11}^E, s_{12}^E, s_{13}^E, s_{33}^E, s_{44}^E and s_{66}^E (note that $s_{66}^E = 2\ (s_{11}^E - s_{12}^E)$), where superindex E means constant electric field, d_{13}, d_{33} and d_{15}; ε^S_{11} and ε^S_{33}, where superindex S means constant strain, as they are those requested by the FEA software used. The three-dimensional FEA modelling was done using ATILA software [47]. Originally developed for modelling sonar transducers, this program has the ability to include piezoelectric materials defined by a full data set of complex variables. The three-dimensional harmonic analysis used here yields the impedance values in a given interval of frequencies, from which the resonance and antiresonance frequencies and the electromechanical coupling factors can be obtained.

The three sample shapes modelled here were those of the samples used to determine the set of dielectric, elastic and piezoelectric complex coefficients [25], namely: (1) a thin ceramic disk of diameter D=37,80 mm and thickness t=0,76 mm, thickness poled, from which planar mode and thickness mode where measured, (2) a long bar, length poled, of D=2mm and length L= 20mm, from which the length extensional mode was measured and (3) an in plane poled standard shear plate of thickness t=0.59mm, width for poling w=5.9mm and length L=5.9mm, from which a shear mode was measured. The measured R and G peaks for such resonance modes of this samples are shown in Fig. 13.4 (a-d).

The mesh used in each resonator simulation was selected so as to have a minimum of five nodes per wavelength, except the shear element, which were divided into a 30x30x3 mesh consisting of 2700 hexagonal 20 node elements. This fine mesh was used to include the secondary resonances found in the experimental spectrum, which has a wavelength much smaller than the one of the main resonance. The number of frequencies analyzed was chosen to achieve a compromise between the computing time and the required resolution of the calculated spectrum, depending on its complexity. The disk and the long bar were simulated as rotationally symmetric items, which results in calculation times of the order of a few minutes for the whole frequency sweep in a PC with a Pentium IV, 3GHz, processor. The shear elements were simulated as full three-dimensional items, resulting in 15 minutes calculation time for each discrete frequency point. The modelled R and G curves are also shown for the four modes of resonance in Fig. 13.4 (a-d).

This work had the double purpose of assessing the validity of the material constants, by comparing modelled and experimental impedance data, and studying the modes of resonance of the three standard samples used in the characterization.

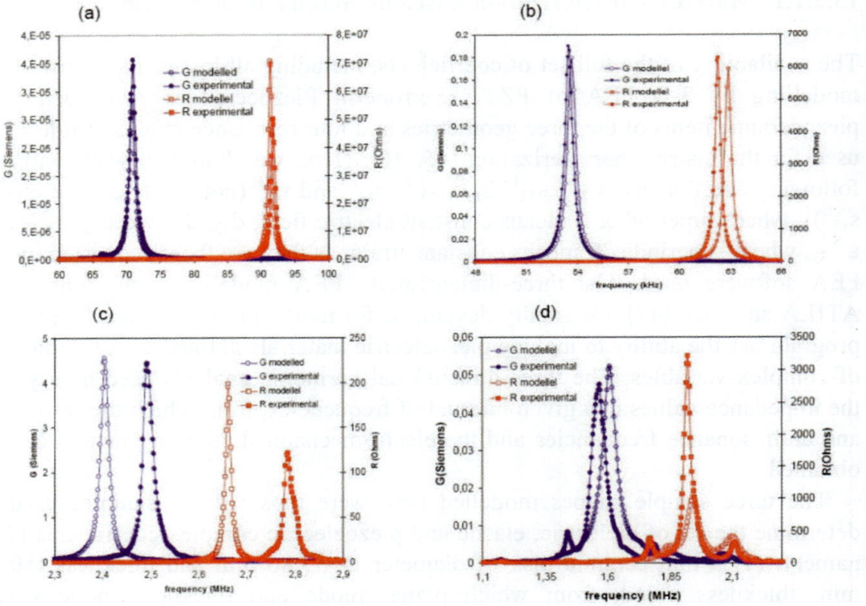

Fig. 13.4 R and G peaks (full symbols are experimental values and open symbols are FEA modelling results using the matrix of material coefficients in [25]) corresponding to the four modes of resonance needed for the full characterization of PZ27 ceramic: **a** length extensional mode of a rod, length poled, **b** radial mode of a disk, thickness poled, **c** thickness mode of a disk, thickness poled, and **d** thickness shear mode of a plate, length poled.

It can be observed that the good agreement between the measured and FEA modelled R and G peaks for both the length extensional resonance of the bar (Fig. 13.4(a)) and the radial resonance of the disk (Fig. 13.4(b)), both of which are purely extensional modes. Fig. 13.4(c) shows that for the thickness resonance mode of the thin ceramic disk, the measured resonance and antiresonance frequencies differ from the frequencies obtained by FEA and those measured (deviation <5%). The height of the R and G FEA modelled peaks is, once again, in reasonably good agreement with experimental values.

Secondary satellite resonance peaks (Fig. 13.4(d)) are commonly observed in the resonance spectrum of in-plane poled shear samples, independently of the composition and dimensions of the shear sample, even while keeping the standard dimensional ratios L, w≥10t. For this reason, the method of using electrical resonance and the antiresonance frequencies to determine the shear coupling coefficient is not recommended by the IRE [12] and IEEE standards [13]. Instead, a dielectric measurement method was chosen using the relationship $\varepsilon^S_{11}= \varepsilon^T_{11}(1-k^2_{15})$. It is easy to obtain the free permittivity value, ε^T_{11}, by measuring the capacitance at a low frequency (<1 kHz), which is well below the fundamental resonance. However, the clamped permittivity, ε^S_{11}, at very high frequencies may not be easily accessible because of the influence of those resonances and/or their higher harmonics.

Such satellite peaks are sometimes explained as a consequence of material inhomogeneity, either compositional, microstructural or due to inhomogeneous poling caused by edge effects. When comparing measured and modelled peaks for the in-plane poled standard shear plate (Fig. 13.4d), it is noticeable that the modelled spectrum also shows such secondary peaks. The results of the FEA analysis, corresponding to an elastically, dielectrically and piezoelectrically homogeneous item of PZ27, show that the current explanation of these peaks is not correct and that they can also be modelled at frequencies in good agreement with the measured ones. Secondary peaks seem to be most likely due to the occurrence of non purely-shear modes inherently to such a standard sample geometry and dimensional ratios.

FEA modelling of the mode of motion reveals that, for this particular shear sample, for one of the satellite modes at f=1790 kHz, the mode of motion resembles a composition of a thickness shear and an asymmetric Lamb wave modes of motion with perpendicular propagation directions. For the other satellite at f=1440 kHz, the mode of motion is even more complex [45]. FEA results also show that, at the main resonance, f =1572 kHz, there is an inhomogeneous mode of motion (Fig. 13.5 (a-c)) and the shear displacement (Fig. 13.5a) in the centre of the sample is much higher than the one at the edges of the standard shear item.

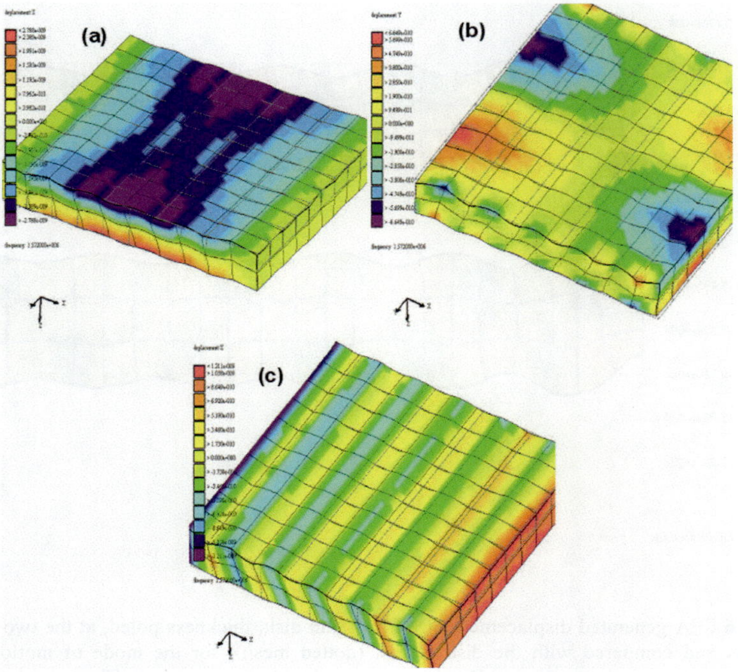

Fig. 13.5 FEA modelled displacement at the main resonance of the PZ27, length poled, shear plate (1572 kHz for 0.59×5.9×5.9 mm sample). The grey-level code corresponds to the values of the: **a** in-plane shear displacement X, **b** in-plane shear displacement Y and **c** out-of-plane displacement Z. Material parameters used correspond to the matrix of [25].

The piezoelectric shear coefficient, d_{15}, reported for PZ27, d_{15}=(396-26i) pC.N^{-1}, from the iterative calculation at the resonance of the standard sample [25] is lower than the actual value quoted by the manufacturer (d_{15}=500 pC.N^{-1}), which is obtained from direct measurements on accelerometers working with shear elements [48]. A similar result was obtained [21] for Motorola 3203HD piezoceramic, later corrected by the authors [49]. The resonance method with such standard shear geometry gave k_{15} = 61%, whereas the manufacturer states k_{15} =72% as catalogue value [50]. Other authors [51] stated that there is an aspect ratio dependence of the coefficients obtained with the standard shear sample,

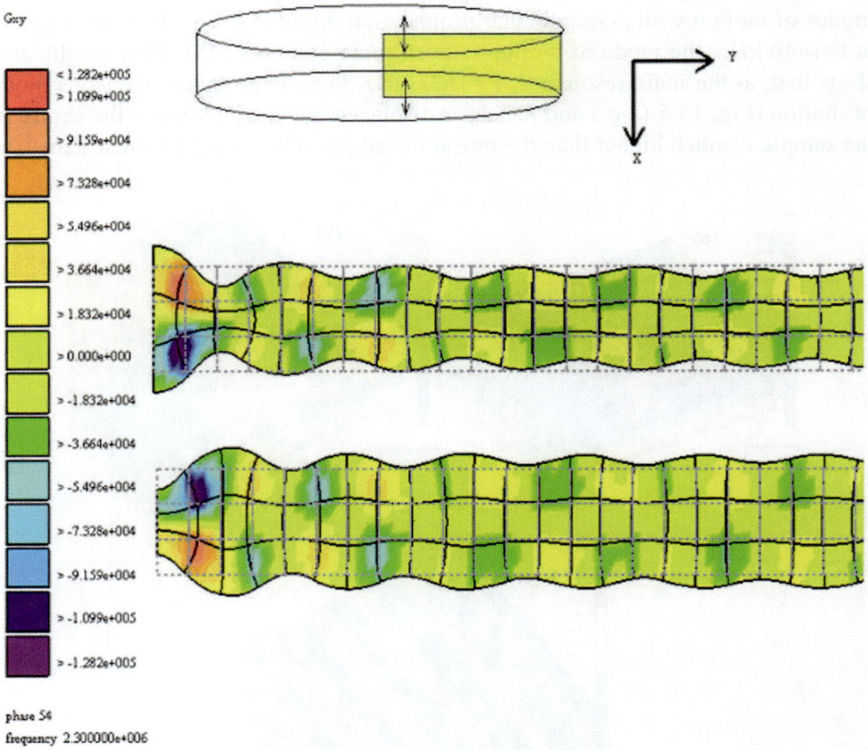

Fig. 13.6 FEA generated displacement of a PZ27 thin disk, thickness poled, at the two extreme positions and compared with the disk at rest (dotted mesh), for the mode of motion at the thickness extensional resonance. Displacement is highly exaggerated. The grey-level code indicates the intensity of the shear stress in the XY plane. The inset shows the section of the disk that the FEA modelling displays.

These results also reveal dynamic clamping as the source of the underestimation of the material parameters when using standard shear items that explain the insufficient agreement between measured and FEA modelled R and G peaks in Fig. 13.4d.

Fig. 13.6 shows that the mode of motion of the disk's displacement frequency does not follow a simple piston like scheme [52]. On the contrary, shear contributions to the displacement at the thickness mode are non-negligible. Thus, the low values of the shear coefficients obtained from the standard sample result in differences of the frequencies determined by FEA with respect to the experimental ones. These results lead to the conclusion that a procedure revision is needed to determine complex shear coefficients. This is to be used for piezoceramic elements modelling, from resonance measurements to shear samples. Here, FEA has been a useful tool in studying the reliability of the complex matrix characterization by comparison of modelled and experimental complex electrical impedance.

13.2.1.2 Matrix Characterization Based on Thickness-Poled Shear Samples

The non-standard, thickness-poled shear sample (Fig. 13.3e) considered in this work for the refinement of the matrix characterization is that of the second thickness shear resonance mode, already described by Berlincourt [2], and which was later analyzed by other authors [51, 53, 54]. The following formula [51] is used as the analytical solution of the wave equation for the admittance at the second Thickness Shear resonance mode of motion, when the dimensional ratio given by t>> w (Fig. 13.3) is fulfilled:

$$Y = G + iB = i\frac{2\pi fwL\varepsilon_{11}^{s}}{t} + i\frac{2we_{15}^{2}}{t}\sqrt{\frac{s_{55}^{E}}{\rho}}\tan\left(\pi fL\sqrt{\rho s_{55}^{E}}\right) \qquad (9)$$

As usual w,L and t (Fig. 13.3e) are the dimensions of the sample, and ρ the material density. In the method used for the shear sample, the coefficients e_{15}, $\varepsilon^{S}{}_{11}$ and $s^{E}{}_{55}$ (note that $s^{E}{}_{55} = s^{E}{}_{44} = 1/c^{E}{}_{55}$) are calculated (Table 13.1) by solving – using an automatic iterative numerical method as previously described – the set of non-linear equations that results when experimental complex admittance data are introduced into this analytical expression. Such admittance data are needed at four frequencies in the neighbourhood of the resonance and antiresonance peaks, the determination of such frequencies also being automatic. The shear results obtained from this mode of resonance in samples of the commercial piezoceramic under study, PZ27, were used to get a refined set of characteristic dielectric, elastic and piezoelectric complex coefficients of this material [55]. The value of the real part of the piezoelectric shear coefficient d_{15} obtained (524 – 21i) is higher than the one obtained from measurements on standard shear samples. This is in agreement with the value derived from measurement of actual shear samples in devices [48].

The three-dimensional FEA modelling of the R and G peaks of the new set of samples is shown in Fig. 13.7 (a-d), now including the thickness-poled non-standard shear plate here suggested, with actual dimensions of thickness for the poling t=0.9mm, w=8.97mm and L=8.1mm. The FEA modelling shown in figures 7 (a-d) was accomplished using the refined matrix of material coefficients [55] and is shown again in comparison with the experimental spectra of the four resonance modes.

For the length extensional and radial modes (Fig. 13.7a and b), the agreement between FEA modelled and measured spectra is similar to the one previously obtained (Fig. 13.4a and b), since these modes are not affected by shear coefficients. However, the results obtained from the refined matrix are better for the thickness mode of the thin disk and the shear mode of the thickness-poled shear plate (Figures 7(c) and (d)) than those obtained with the original one (Fig. 13.4b and (d)) for these modes involving shear strains. On the other hand, unwanted satellite resonances were also observed in the spectrum of the thickness-poled shear sample (Fig. 13.7d). These satellites are again reproduced in the FEA modelled spectrum (Fig. 13.7d) and are, thus, intrinsic modes associated with the geometry and dimensional ratio.

The strong coupling of resonance modes and the dependence on the sample geometry of the resonance spectrum of the thickness-poled shear plates is well known[51, 53, 54]. For other geometries, like the thickness-poled thin disk (Fig. 13.3c), for which two resonances are expected, one linked to the thickness (t) and the other to the diameter (D), the simplest way to avoid coupling of resonance modes is to make these two dimensions significantly different. This takes the corresponding resonance frequencies well apart from each other. This is the standard procedure [12, 13, 14].

The first authors that faced the problem of establishing the condition for an uncoupled resonance in the thickness poled shear plate [51], stated that a ratio between lateral dimensions (L, t) and thickness for poling (w) higher than 5:1 is enough to obtain a well-defined shear resonance spectrum and reliable material parameters. Other authors [53] could not find satisfactory experimental results with such aspect ratio. Even for samples with aspect ratio of width between measuring electrodes (t) to thickness for poling (w) of t/w= 27:1, a coupling of the main shear resonance mode with other mode was observed (Fig. 13.8).

In [53], the authors recommended a minimum ratio of 20:1 to determine material parameters from impedance measurements. More recently [54], it was shown, once again, that even for a ratio 20:1 extraneous resonances, such as longitudinal resonances, appears close to the main thickness shear resonance mode, but with reduced influence on this so that reproducible measurements can take place. It is also stated [54] that ratios between 5:1 and 10:1 can lead to errors of 25% in the determination of the material properties from impedance measurements.

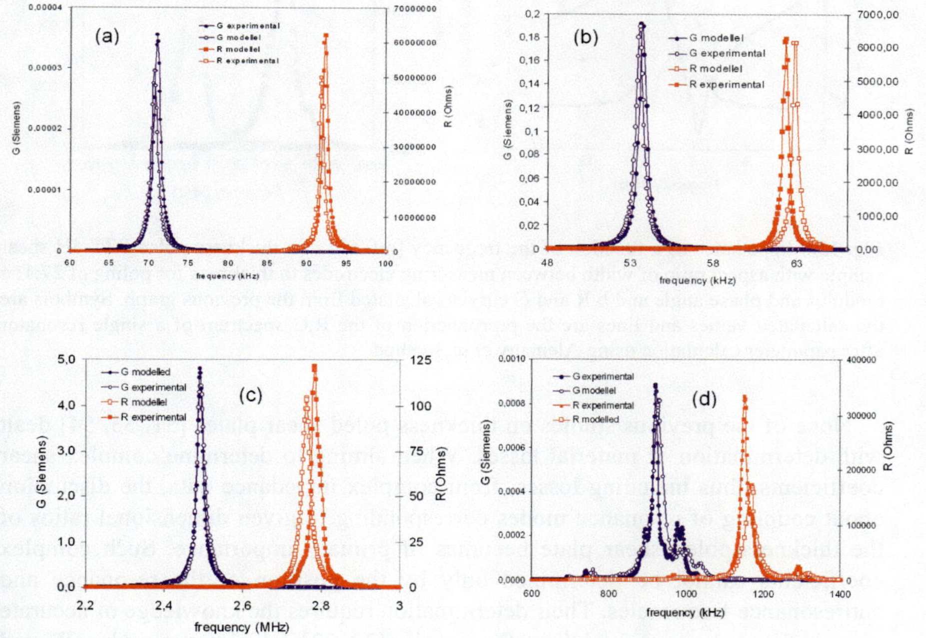

Fig. 13.7 R and G peaks (full symbols are experimental values and open symbols are FEA modelling results using the refined matrix of material coefficients in [55]) corresponding to the four modes of resonance needed for the full characterization of PZ27 ceramic: **a** length extensional mode of a rod, length poled, **b** radial mode of a disk, thickness poled, **c** thickness mode of a disk, thickness poled, and **d** thickness shear mode of a plate, thickness poled.

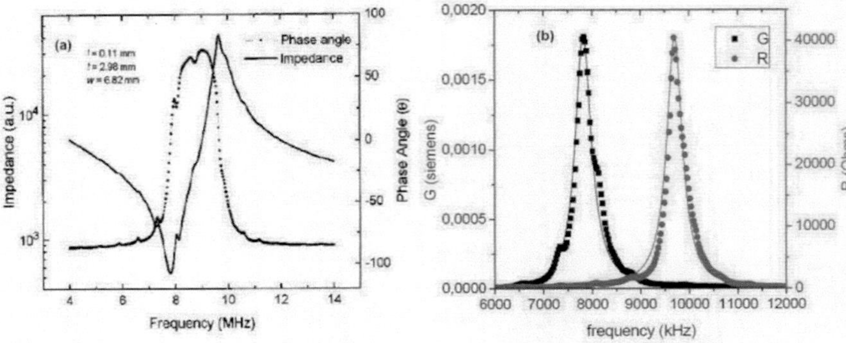

Fig. 13.8 Impedance as a function of the frequency [ref. 53] of a thickness-poled PZT-5H shear sample with aspect ratio of width between measuring electrodes to thickness for poling of 27:1: **a** modulus and phase angle and **b** R and G curves calculated from the previous graph. Symbols are the calculated values and lines are the reproduction of the R,G spectrum of a single resonator after parameter calculation using Alemany *et al.* method.

None of the previous studies on thickness-poled shear plates [51, 53, 54] dealt with determination of material losses. When aiming to determine complex shear coefficients, thus including losses, from complex impedance data, the discussion about coupling of resonance modes corresponding to given dimensional ratios of the thickness-poled shear plate becomes of primary importance. Such complex coefficients cannot be determined only by the position of the resonance and antiresonance frequencies. Their determination requires the knowledge of accurate values of impedance around these frequencies [22, 23], which must not be affected by mode coupling. This state of affairs seemed to be in contradiction with the high quality of the results of the reproduction by FEA of the spectra of PZ27 samples (Fig. 13.7 (a-d)) from thickness-poled shear samples, which had an aspect ratio of t/w=10:1 [55].

In addition to the mentioned contradiction and the lack of agreement among the authors who worked on the topic, the discussion about the resonance modes of the thickness-poled shear plates lacks experimental data that can provide a clear description of the unwanted spurious resonance modes. In order to elucidate this, we studied non-standard thickness-poled shear piezoceramic plates of PZ27 of 15x15mm and 2.0, 1.5 and 1.0mm thickness, with aspect ratio t/w=7.5:1, 10:1 and 15:1, respectively (Fig. 13.3e). The measured R and G spectra for these samples, showed in Fig. 13.9(a-c), reveals mode coupling in all them. Such coupling is less perturbing to the main resonance as the aspect ratio increased, in agreement with previous literature. Resemblance of the two spectra (Fig. 13.7d and Fig. 13.9b, found for the two samples with the same aspect ratio (10:1), shows again that the satellite peaks are associated with the natural resonances of a given geometry.

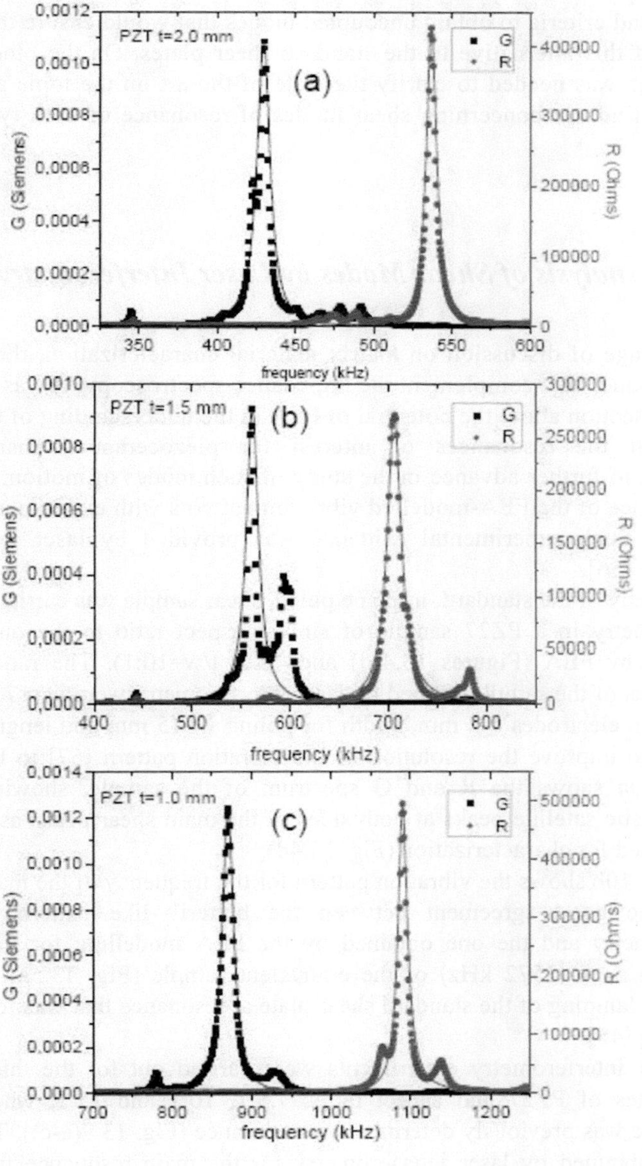

Fig. 13.9 Experimental spectra of resonance of non-standard, thickness-poled, shear piezoceramic plates of PZ27 of 15x15mm and **a** 2.0mm, **b** 1.5mm and **c** 1.0 mm thickness, with aspect ratio t/w=7.5:1, 10:1 and 15:1, respectively. Symbols are the experimental values and lines are the reconstruction after the shear parameters calculation by the Alemany *et al.* method.

A more detailed experimental knowledge of the resonance modes of the thickness-poled shear plate was needed at this stage of the problem, on the one hand to find criteria to obtain uncoupled modes that would ensure the validity of the use of this alternative to the standard shear plates. On the other hand, this knowledge was needed to clarify the state of the art on the topic and to verify the FEA findings concerning shear modes of resonance of both types of shear plates.

13.2.2 Analysis of Shear Modes by Laser Interferometry

At this stage of discussion on matrix material characterization, there is a need for techniques that complement the impedance spectroscopy. It was stated in the previous section about the potential of FEA in the understanding of the modes of motion at the resonances of interest for piezoceramics characterization. However, to further advance in the study of such modes of motion, proof of the concordance of the FEA-modelled vibration patterns with experimental data was required. Such experimental evidence was provided by laser interferometry technique [56].

The study of the standard, in-plane poled, shear sample was carried out by laser interferometry in a PZ27 sample of similar aspect ratio to the one previously modelled by FEA (Figures 13.4(d) and 13.5; t/w=10:1). The ratio and actual dimensions of the sample studied by laser interferometry were w/t=7.5:1, distance in between electrodes t=2 mm, width for poling w=15 mm and length L=15 mm, in order to improve the resolution of the vibration pattern [57] to be measured. Fig. 13.10a shows the R and G spectrum of the sample, showing the same characteristic satellite peaks at both sides of the main shear peaks as those of the sample used for characterization (Fig. 13.4d).

Fig. 13.10b shows the vibration pattern for the frequency of the main peak (444 kHz). The good agreement between the butterfly-like pattern obtained by interferometry and the one obtained by the FEA modelling for the equivalent resonance peak (1572 kHz) of the equivalent sample (Fig. 13.5a), confirm the dynamic clamping of the standard shear plate at resonance that was found by FEA modelling [45].

Similar interferometry experiments were carried out for the thickness-poled shear plates of PZ27 and aspect ratios 7.5:1, 10:1 and 15:1, whose complex impedance was previously determined at resonance (Fig. 13.9(a-c)). The vibration patterns obtained by laser interferometry for the main resonance frequency are shown in Fig. 13.11(a-c). These samples were modelled by FEA, on the basis of the refined matrix of coefficients [55]. The results obtained for the out-of-plane displacement at the same frequencies as the measurements is also shown in Fig. 13.11(d-f).

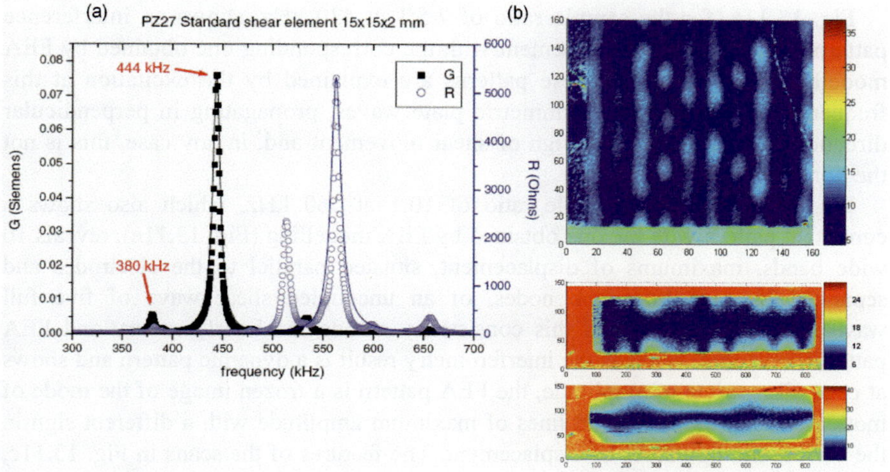

Fig. 13.10 a Spectrum for the standard, in-plane poled, shear sample of PZ27 and 15x15x2mm. Symbols are the experimental values and lines are the reconstruction after the shear parameters calculation by Alemany *et al.* method. **b** Laser interferometry pattern of the plate surface at 444 kHz for: (top) the major surface of the shear plate, (middle) the lateral horizontal surface of the plate and (bottom) the lateral vertical surface. The x and y-axes indicates the number of motor steps for the scan. Step for the scan at the top is 0.1 mm, whereas for lateral surfaces is 0.02 mm. The grey-level code indicates the value of the displacement perpendicular to the scanned surface in a.u. /1 V driving voltage.

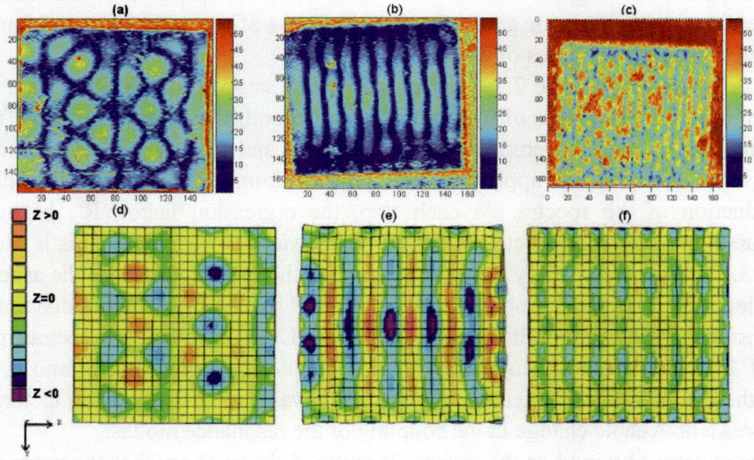

Fig. 13.11 Upper row: normal to the surface displacements scans measured by laser interferometry on one of the major surfaces of the, non-standard, thickness poled PZ27 shear plates: **a** measured at 430 kHz for the 7.5:1 ratio sample, **b** measured at 560 kHz for the 10:1 ratio sample and **c** measured at 869 kHz for the 15:1 sample. The x and y-axes indicates the number of motor steps for the scan. Step for the scan is 0.1mm. The grey-level code indicates the displacement perpendicular to the scanned surface in a.u. /1V driving voltage. Lower row: FEA modelled patterns of the out-of-plane displacement (Z direction) for the same samples at **d** 437 kHz, **e** 570 kHz and **f** 870 kHz, frequencies of the maximum of the FEA modelled G peaks.

Fig. 13.11a, for the sample ratio of 7.5:1 at 430 kHz, shows an interference pattern, which is in good agreement with the corresponding one obtained by FEA modelling (Fig. 13.11d). These patterns are explained by the excitation at this frequency of two standing symmetric plate waves, propagating in perpendicular directions. There is barely a sign of shear movement and, in any case, this is not the dominant mode.

Fig. 13.11b, for the sample ratio of 10:1 at 560 kHz, which also shows a consistent pattern with the one obtained by FEA modelling (Fig. 13.11e), reveals 10 wide bands, maximums of displacement, situated parallel to the electrodes and separated by nine lines, the nodes, of an uncoupled shear wave of five full wavelengths. To understand this consistency among the interferometry and FEA patterns, note that, whereas the interferometry result is a dynamic pattern and shows at once all maxima of amplitude, the FEA pattern is a frozen image of the mode of motion and shows nodes and lines of maximum amplitude with a different sign in the corresponding values of displacement. The features of the scans in Fig. 13.11c, and Fig. 13.11f, for the sample ratio of 15:1 at 869 kHz, show 16 bands separated by 15 nodes, corresponding to eight full wavelengths. The ratio between the numbers of wavelengths (5:8) fits reasonably well with the inverse of the ratio between the thicknesses of the last two samples (1.5:1.0), indicating that for both samples this is a thickness-driven shear mode of motion. For both thickness-poled samples, the vibration pattern is homogeneous, in contrast to the ones found for the dynamically-clamped standard, length-poled, shear samples (Fig. 13.5a and Fig. 13.10b). The fact that, even for the highest aspect ratio, 15:1, mode coupling is found means that inherently to this resonator geometry, the thickness of the sample is simultaneously driving two types of modes of resonance.

In order to test this, the 7.5:1 aspect ratio PZ27 sample was reduced from 2.00 to 1.00 mm thickness in steps of 0.02mm, varying smoothly the ratio from 7.5:1 to 15:1. The impedance spectra were recorded at each step and the Alemany *et al.* automatic iterative method [55] was applied to determine the complex coefficients that allow the reproduction of the spectra. At each step, the regression factor, R^2, between the measured and the reconstructed R and G peaks was also recorded. This is plotted in Fig. 13.12 and shows clearly the cyclic nature of the phenomena with the aspect ratio variation [58]. This most probably was not noticed by previous authors due to the fine dimensional scale at which it takes place. Fig. 13.12 also shows two spectra (numbers 1 and 2) that correspond to relatively close values of dimensional ratio and that, nevertheless, presents a noticeable change in the value of the regression factor, which indicates a noticeable change in the coupling of the resonance modes.

The spectra obtained as the sample is thinned down show that the more intense peaks, corresponding to the electrically-driven thickness-shear mode, move towards higher frequencies. When changing frequency, this mode excites mechanically the different overtones of plate waves (n, n+1, n+2, etc., in Fig. 13.12), which have fixed frequencies, as the lateral dimensions of the plate remain unchanged. Due to this mechanism, shear and plate waves couple necessarily in a periodic manner, no matter how high the aspect ratio (Fig. 13.8 and Fig. 13.9c). Such a phenomenon explains the different results found by different authors.

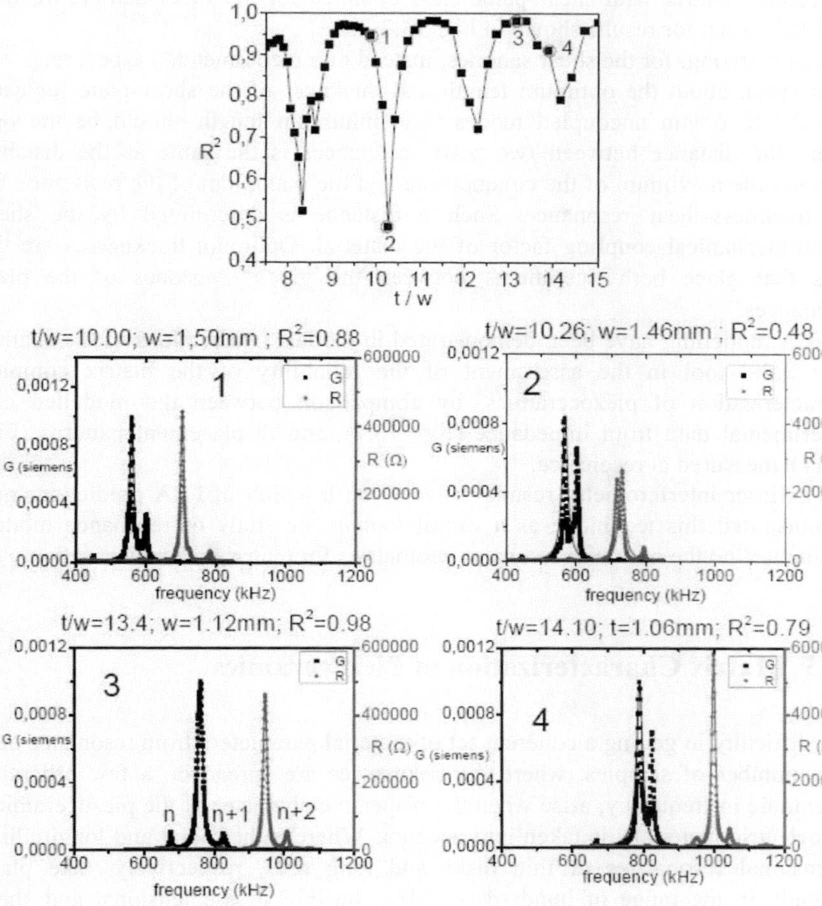

Fig. 13.12 The upper graph shows the regression factor of the reconstructed resonance spectrum after Alemany *et al.* calculation to the measured one (R^2), as a function of the aspect ratio of the PZ27 thickness-poled shear sample. Below are shown the spectra corresponding to the four points marked in the R^2(t/w) graph.

Therefore, despite the inherent coupling between modes regardless of the dimensional ratio of the shear plate, we found a number of dimensional ratios below 15:1 for which an accurate determination of the shear properties of the material can be carried out. The FEA simulations of Fig. 13.11 were made using a refined matrix of coefficients with the shear parameters obtained for the PZ27 sample of ratio 13.4:1, for which the maximum of R^2 was obtained in the studied interval (Fig. 13.12). The results of such FEA modelling have the same features, and differ only in minor details like position of the frequency for the G and R maxima within <1%. This is with respect to those simulations carried out using the refined matrix with shear parameters obtained for the PZ27 sample of ratio 10:1 [55], used for results shown in Fig. 13.7.

Summarizing, for the shear samples, instead of a recommended aspect ratio, we must speak about the optimum length and thickness of the shear plate for each material to obtain uncoupled modes. The minimum length should be the one where the distance between two plate resonances is the same as the distance between the maximum of the conductance and the maximum of the resistance for the thickness-shear resonance. Such a distance is determined by the shear electromechanical coupling factor of the material. Optimum thicknesses are the ones that place both maximums between the given overtones of the plate resonances.

FEA modelling have been demonstrated in the last two sections of the chapter as a valid tool in the assessment of the reliability of the matrix complex characterization of piezoceramics, by comparison between the modelled and experimental data from impedance (Fig. 13.7), and displacement patterns (Fig. 13.11), measured at resonance.

The laser interferometry results provide confirmation of FEA predictions and demonstrated this technique as a useful tool in the study of resonance modes, aiming to find the optimum resonator geometries for material characterization.

13.3 Matrix Characterization of Piezoceramics

The difficulty in getting a coherent set of material parameters from resonance data on a number of samples, where the resonances are spread on a few orders of magnitude in frequency, arise when the dispersive character of the piezoceramics, ferroelectric materials, is taken into account. Whereas the radial and longitudinal extensional resonances of thin disks and long rods, respectively, take place typically in the range of hundreds of kHz, the thickness-extensional and shear resonances of thin disks and shear plates can be found in the range of few MHz. Besides, there is a dependence of the piezoceramic properties on the polarization level achieved that affects the consistency of the results obtained from length-poled resonators, the long bar and the standard in-plane poled shear plate, and the thickness poled resonators, like the thin disk or plates, due to edge effects. Despite all these, our results show that with due care in the selection of samples for certain materials, such an accurate and coherent set of parameters can be obtained from complex impedance data.

13.3.1 State of the Art of the Matrix Characterization of Bulk Piezoceramics

The resonance method has received revisions and criticisms, for the above mentioned reasons, and alternative experimental methods to obtain the full ensemble of coefficients of bulk piezoceramics have been explored.

Disregarding the lossy character of the piezoelectric ceramics as the standard procedure, combined methods of resonance and ultrasonic spectroscopy have been developed [59, 60] and recently applied to the matrix characterization of high sensitivity 0.7 $Pb(Mg_{1/3}Nb_{2/3})O_3$ -0.3$PbTiO_3$ ceramics [61].

A recent practical revision of alternative methods [62] for matrix characterization has selected two methods that aim to provide all set of data with the great advantage of using only one sample. One of them, purely acoustic, is the measurement of transmission coefficient of an ultrasonic plane wave, generated in water and which propagates through a piezoelectric plate, as a function of frequency and incidence angle. This is a technique that presents a higher experimental complexity than the resonance method, since it requires the use of rather large samples (79.8×79.8×2.025 mm^3 in [62]) fixed on a motorized rotation stage and immersed into a large water tank. The full tensor characterization of PZ27 piezoelectric plate [63] was reported and compared with characterizations using other methods. This study reveals the problems of the technique in the determination of the piezoelectric shear coefficient and all the material losses.

The second technique analyzed [62] was the resonance ultrasound spectroscopy [64](RUS) applied to the study of 0.65 $Pb(Mg_{1/3}Nb_{2/3})O_3$0.35$PbTiO_3$ ceramic using only one sample (10x10x10 mm^2). This has also been recently applied to a 0.88 $Pb(Zn_{1/3}Nb_{2/3})O_3$-0.12 $PbTiO_3$ single crystal [65]. This technique uses a piezoelectric cube electroded on two faces and placed on the impedance test clip fixture of an impedance analyzer. This analyzer is used as a broadband excitation source to generate free resonances in the cube and to visualize its electrical input impedance. A laser interferometer is used to detect the displacement field in the normal direction to the surface for the excited resonances. The fitting procedure to the analytical model that describes the materials properties used to determine them, is performed on resonant frequencies of the cube step by step, by study of the different modes sequentially. This rather complex, second technique has not yet been reported to determine the material losses.

A series of works were also recently published to determine the matrix of properties, again not taking into account the losses, from resonance method, by using IEEE standards [13], combined with methods from literature in Chinese [66] and Japanese standards [67]. $BiScO_3$-$PbTiO_3$ [68] high sensitivity high temperature piezoceramics and lead-free ceramics based on $(K_{0.5}Na_{0.5})NbO_3$ [69, 70, 71] have been characterized recently in this way.

To date there is only one [72] recent work, apart from those using the Alemany method [25, 26], that focuses on the full complex characterization, and the temperature dependence of the obtained coefficients, of both a soft and a hard PZT compositions, making use of the PRAP commercial program. None of the works mentioned in this section applies any kind of reliability or consistency criteria to the set of material parameters obtained.

13.3.2 Matrix Characterization of Piezoceramics from Resonance Using Alemany et al. Method and Thickness-Poled Shear Samples

In addition to dispersion and differences in the polarization achieved in the different piezoceramic resonators, the length poling process may be a problem for some ceramics. This creates an additional difficulty when characterizing piezoceramics using the resonance method. Using insufficiently-poled samples for various reasons, the resonance spectrum of in-plane poled standard shear samples may show double peaks that cannot be treated as a single resonance, using formula (7). This was observed for PZT ceramics with closed porosity and high sensitivity PMT-PT ceramics, preventing their full characterization from resonance data. The use of the non-standard, thickness-poled, shear plate provided, by better poling and better control of the mode coupling, an accurate characterization of the shear parameters, demonstrated for both materials [73, 74]. Additionally, this non-standard shear sample allows characterization from identically, thickness-poled, thin disks and shear plates. Though still imperfect due to the need of the use of the long rod, the method presented here allows the consistency of the data obtained and proved the accuracy of such data by allowing accurate FEA modelling of piezoelectric samples in a wide range of frequencies.

Summary

A number of issues on the topic of the ferro-piezoceramics matrix characterization, in the linear range from complex impedance at resonance, were analyzed in this chapter and illustrated for a soft, Navy-II type, commercial PZT (PZ27). This was possible because of the complementary use of Finite Element Analysis (FEA) and Laser Interferometry.

A critical review of the methods for characterization available to date was presented. The Alemany *et al.* iterative method applied to complex material characterization in the linear range from complex impedance measurements was assessed as the most systematic among those methods available, giving both accurate and consistent piezoceramic characterization within the limits of the dispersive character of the material, and from the lowest number of resonators and the most consistently poled set of samples.

FEA was used as a reliability criteria of accuracy and consistency of the parameters obtained, from three samples and four modes of resonance, using the Alemany *at al.* automatic iterative method. The agreement between FEA modelling, based on the mentioned set of parameters, and both the complex electrical impedance measurements and the vibration patterns obtained by laser interferometry, was presented for PZ27.

The advantages of the proposed thickness-poled shear plate for piezoceramics with poling problems and, in general, by the use two identically poled samples, thin disks and shear plates, thickness poled, was pointed out.

Besides, the modes of motion of standard, in-plane poled, and thickness-poled shear plates were studied in the reduction of intrinsic mode coupling, required for accurate characterization including losses. The dynamic clamping of the standard, length-poled, shear plate main resonance was found to cause the underestimation of the piezoelectric shear coefficients, which recommends the use of an alternative shear sample. The periodic character of the mode coupling in thickness-poled shear plates was also demonstrated, enlightening the choice of the best dimensional ratios for complex characterization.

Acknowledgements

The authors wish to thank Ms. W.W. Wolny and Dr. E. Ringaard from Ferroperm Piezoceramics A/S (Kvistgaard, Denmark) for the ceramic samples used. Mr. A. García (contract holder at ICMM-CSIC) is greatly acknowledged for the technical support and the large number of electrical measurements he carried out for this work. The authors are also indebted to the late Dr. C. Alemany for the generous implementation of the automatic iterative method to the second thickness shear geometry and thoughtful discussions. His chess player-like perseverance in the experimental work along his scientific career was a permanent inspiration. The financial support of the EU 6FP Network of Excellence MIND (NoE 515757-2) is gratefully acknowledged.

References

1. "Piezoelectric Ceramics". B. Jaffe, W.R. Cook and H. Jaffe. Academic Press. London, 1971.
2. "Piezoelectric and Piezomagnetic Materials and Their Function in Transducers". D.A. Berlincourt, D.R. Curran and H. Jaffe in "Physical Acoustics" vol.1 Part A. Ed W.P. Mason, Academic Press, 1964.

3. "Electroceramics. Materials, Properties and Applications". A.J. Moulson and J.M. Herbert. Chapman and Hall. London, 1990.
4. "Piezoelectricity". W.G. Cady. McGraw-Hill Book Co. Inc. London, 1946.
5. "Ultrasonic Transducer Materials". O.E. Mattiat. Plenum Press. New York – London, 1971.
6. K.H. Haerdtl. "Electrical and mechanical losses in ferroelectric ceramics". Ceram Int. 8, 121 (1982).
7. K. Uchino and S. Hirose."Loss mechanisms in piezoelectrics: How to measure different losses separately" IEEE Trans. Ultrasonics, Ferroelectrics and Frequency Control 48, 307(2001).
8. G. Robert, D. Damjanovic and N. Setter. "Piezoelectric hysteresis analysis and loss separation". J. Appl. Phys. 90(9), 4668(2001).
9. A.V. Mezheritsky. "Elastic, Dielectric and Piezoelectric Losses in Piezoceramics: How It Works All Together" IEEE Trans. Ultrasonics, Ferroelectrics and Frequency Control, 51(6), 695-707 (2004).
10. S. Sherrit. "Complex Material Coefficients and Energy Ratios for Lossy Piezoelectric Materials". IEEE Trans. Ultrasonics, Ferroelectrics and Frequency Control, 55(11), 2479-2483 (2008)
11. "Physical Properties of Crystals. Their representation by tensors and matrices". J.F. Nye. Oxford at the Clarendon Press. London, 1957.
12. "IRE standard on piezoelectric crystals: Measurements of piezoelectric ceramics, 1961", Proc. IRE, 49(7), 1161-1169 (1961).
13. "IEEE Standard on piezoelectricity". ANSI/IEEE Std. 176-1987.
14. Piezoelectric properties of ceramic materials and components. Part 2: methods of measurement – Low power European Standard CENELEC, EN 50324-2 (2002).
15. R. Holland. "Representation of dielectric, elastic and piezoelectric losses by complex coefficients". IEEE Trans.Sonics Ultrason. SU-14(1),18 (1967).
16. R. Holland and E. EerNisse. "Accurate measurement of coefficients in a ferroelectric ceramic". IEEE Trans.Sonics Ultrason. SU-16(4),173 (1969).
17. J.G. Smits. "Iterative method for accurate determination of the real and imaginary parts of the materials coefficients of piezoelectric ceramics". IEEE Trans. Sonics Ultrason. SU-23(6),393-402(1976).
18. A.H. Meitzler, H.M. O´Bryan, Jr., and H.F. Tiersten. "Definitions and measurements of radial mode coupling factors in piezoelectric ceramic materials with large variations in Poisson´s ratio" IEEE Trans.Sonics Ultrason. SU-20(3), 233 (1973).
19. S. Sherrit, N. Gauthier, H. D. Wiederick and B.K. Mukherjee. "Accurate evaluation of the real and imaginary material constants for a piezoelectric resonator in the radial mode". Ferroelectrics 119, 17 (1991).
20. S. Sherrit, H. D. Wiederick and B.K. Mukherjee. "Non-iterative Evaluation of the Real and Imaginary Material Constants of Piezoelectric Resonators". Ferroelectrics 134, 111 (1992).
21. S. Sherrit, H.D. Wiederick and B.K. Mukherjee "A complete characterization of the piezoelectric, dielectric and elastic properties of Motorola PZT3203HD including losses and dispersion". Medical Imaging 1997: Ultrasonic Transducer Engineering, SPIE Proceedings, 3037, 158 (1997).
22. C. Alemany, L. Pardo, B. Jiménez, F. Carmona, J. Mendiola and A.M. González. "Automatic iterative evaluation of complex material constants in piezoelectric ceramics". J. Phys. D: Appl. Phys. 27, 148 (1994).

23. C. Alemany, A.M. González, L. Pardo, B. Jiménez, F. Carmona and J. Mendiola. "Automatic determination of complex constants of piezoelectric lossy materials in the radial mode". J. Phys. D: Appl.Phys. 28(5), 945 (1995).
24. A.M. Gonzalez and C. Alemany. "Determination of the frequency dependence of characteristic constants in lossy piezoelectric materials". J. Phys. D: Appl.Phys. 29, 2476 (1996).
25. M. Algueró, C. Alemany, L. Pardo and A. M. Gonzalez. "Method for Obtaining the Full Set of Linear Electric, Mechanical and Electromechanical coefficients and All Related Losses of a Piezoelectric Ceramic". J. Am. Ceram. Soc. 87(2), 209 (2004).
26. M. Algueró, C. Alemany, L. Pardo and M.P. Thi. "Piezoelectric Resonances, Linear Coefficients and Losses of Morphotropic Phase Boundary $Pb(Mg_{1/3}Nb_{2/3})O_3$-$PbTiO_3$ Ceramics". J.Am.Ceram.Soc. 88(10), 2780-2787 (2005).
27. K.W. Kwok, H.L.W. Chan and C.L.Choy. "Evaluation of the Material Parameters of Piezoelectric Materials by various methods". IEEE Trans. Ultrasonics, Ferroelectrics and Frequency Control 44(4), 733 (1997).
28. J. Ricote, C. Alemany and L. Pardo. "Microstructural effects on dielectric and piezoelectric behaviour of calcium modified lead titanate ceramics". J. Mater.Res.10 (12), 3194 (1995).
29. L. Pardo, A. Castro, P. Millán, C. Alemany, R. Jiménez and B. Jiménez. "$(Bi_3TiNbO_9)_x(SrBi_2Nb_2O_9)_{1-x}$ Aurivillius type structure piezoelectric ceramics obtained from mechanochemically activated oxides". Acta Materialia 48(9), 2421-2428 (2000).
30. A. Moure, M. Algueró, L. Pardo, E. Ringgaard and A. Pedersen. "Microstructure and temperature dependence of properties of morphotropic phase boundary $Bi(Mg_{1/2}Ti_{1/2})O_3$-$PbTiO_3$ piezoceramics processed by conventional routes". J. Eur. Ceram. Soc. 27, 237-245 (2007).
31. L. Pardo, P.Duran-Martín, J.P. Mercurio, L. Nibou and B. Jiménez. "Temperature behaviour of structural, dielectric and piezoelectric properties of sol-gel processed ceramics of the system $LiNbO_3$-$NaNbO_3$". J.Phys.Chem.Solids 58(9), 1335 (1997).
32. A. Moure, C. Alemany and L. Pardo. "Temperature dependence of piezoelectric, electromechanical and elastic parameters of Aurivillius-type structure piezoceramics with optimized texture and microstructure". IEEE Trans. Ultrasonic Ferroelectrics and Frequency Control, 52(4), 570-577 (2005).
33. T. Tsurumi, Y.B. Kil, K. Nagatoh, H. Kakemoto and S. Wada. "Intrinsic Elastic, Dielectric, and Piezoelectric Losses in Lead Zirconate Titanate Ceramics Determined by an Immittance-Fitting Method" J.Am.Ceram.Soc. 85(8), 1993 (2002).
34. X. Hong Du, Q. M. Wang and K. Uchino. "Accurate Determination of Complex Materials Coefficients of Piezoelectric Resonators" IEEE Trans. Ultrasonics, Ferroelectrics and Frequency Control 50(3), 312 (2003).
35. M. Maeda, H. Hashimoto, I. Suzuki "Measurements of complex materials constants of piezoelectrics: extensional vibrational mode of a rectangular bar". J.Phys.D: Appl. Phys. 36 (2), 176-180 (2003).
36. M. Maeda, H. Hiroyuki, M. Washihira and S. Ikuo "Measurements of complex materials constants of piezoelectric ceramics: Radial vibrational mode of a ceramic disk". J.Eur.Ceram.Soc., 28(1), 133-138 (2008).
37. A. M. Gonzalez, J. de Frutos and M.C. Duro. "Procedure for the characterization of piezoelectric samples in Non-standard Resonant Modes". J.Eur.Ceram.Soc. 19 , 1285-1288 (1999).

38. D.D. Ebenezer "Determination of Complex Coefficients of Radially Polarized Piezoelectric Ceramic Cylindrical Shells Using Thin Shell Theory". IEEE Trans. Ultrasonics, Ferroelectrics and Frequency Control 51 (10), 1209 (2004).
39. R. Ramesh and D.D. Ebenezer "Analysis of axially polarized piezoelectric ceramic rings". Ferroelectrics 323, 17-23 (2005).
40. M.A.B. Andrade, E.C.N. Silva, F. Buiochi and J.C. Adamowski. "Characterization of piezoelectric materials by using an optimization algorithm". 2007 International Congress of Ultrasonics. Session R25.
41. C.H, Huang, Y.Y. Chen. "Transverse Vibration Analysis for Piezoceramic Rectangular Plates Using Ritz's Method with Equivalent Constants". IEEE Trans. Ultrasonics, Ferroelectrics and Frequency Control 53 (2), 1209 (2006).
42. H.Y. Lin and C.C. Ma"The Influence of Electrode Designs on the Resonant Vibrations for Square Piezoelectric Plates". IEEE Trans. Ultrasonics, Ferroelectrics and Frequency Control 53 (5), 825 (2006).
43. A. Barzegar, D. Damjanovic and N. Setter. "The effect of boundary conditions and sample aspect ratio on apparent d_{33} piezoelectric coefficients determined by direct quasistatic method "IEEE Trans. Ultrasonics, Ferroelectrics and Frequency Control 51(3), 262-270 (2004).
44. R. Steinhausen, T. Hauke, H. Beige, S. Seifert, U. Lange, D. Sporn, S. Gebhardt and A. Schönecker. "Characterization and modelling of ferroelectric thin films and 1-3 composites". Bol. Soc. Esp. Ceram. Vidrio 41(1), 158-165 (2002).
45. L. Pardo, M. Algueró and K. Brebøl. " Resonance modes in Standard piezoceramic shear geometry: a discussion based on Finite Element Analysis". *J. Physique IV France* 128 207 (2005).
46. L. Pardo, M. Algueró and K. Brebøl. "Resonance modes in Standard characterization of piezoceramics: a discussion based on Finite Element Analysis". Ferroelectrics 336, 181 (2006).
47. ATILA User´s Manual, Institut Superieur d´Electronique du Nord, Acoustics Laboratory (1997).
48. http://www.ferroperm-piezo.com
49. D.J. Powell, G.L. Wojcik, C.S. Desilets, T.R. Gururaja, K. Guggenberge, S. Sherrit, B.K. Mukherjee. "Incremental "model-build-test'' validation exercise for a 1-d biomedical ultrasonic imaging array" in Proc. IEEE Ultrasonics Symposium, 1997, pp.1669-1674.
50. http://www.ctscorp.com/components/Datasheets/PZT.pdf
51. N. Aurelle, D. Roche, C. Richard and P. Gonnard. "Sample aspect ratio influence on the shear coefficients measurements of a piezoelectric bar". *Proc.* 9th IEEE Iternacional Symposium on Applications of Ferroelectrics (ISAF). Eds. R.K. Pandey, M. Liu and A. Safari. The Pennsylvania State University, University Park, PA, USA. August 7-August 10, 1994. pp. 162-165.
52. S. Ueha, S. Sakuma, and E. Mori, "Measurements of vibration velocity distributions and mode analysis in thin disks of $Pb(Zr{\cdot}Ti)O_3$". J. Acoustic Soc. Am. 75(5), 1842–1847 (1983).
53. W. Cao, S. Zhu and B. Jiang. "Analysis of the shear modes in a piezoelectric vibrator". J.Appl.Phys. 83 , 4415 (1998).
54. T. Comyn T, A.W. Tavernor. "Aspect ratio dependence of d_{15} measurements in Motorola 33203 material". J.Eur.Ceram.Soc. 21 , 12 (2001).
55. L. Pardo, M. Algueró and K. Brebøl. "A non-Standard shear sample for the matrix characterization of piezoceramics and its validation study by finite element analysis". J.Phys.D: Appl.Phys. 40 2162 (2007).
56. C.B. Scruby and L.E. Drain. Laser Ultrasonics. Techniques and Applications. Ed. Adam Hilger, Bristol. ISBN 0-7503-0050-7 (1990).

57. L. Pardo, F. Montero De Espinosa and K. Brebøl. "Study by laser interferometry of the resonance modes of the shear plate used in the Standards characterization of piezoceramics". J. Electroceramics 19(4), 437-442 (2007)
58. L. Pardo, F. Montero De Espinosa, A. García and K. Brebøl. "Choosing the best geometries for the linear characterization of lossy piezoceramics: study of the thickness poled shear plate". Appl. Phys. Lett. 92, 172907 (2008).
59. H. Wang, W. Jiang and W. Cao." Characterization of lead zirconate titanate piezoceramic using high frequency ultrasonic spectroscopy". J.Appl.Phys. 85 (12) (1999)
60. H. Wang and W. Cao. "Determination of full set material constants from phase velocities". J. App. Phys. 92(8), 4578- 4583(2002).
61. H. Wang, B. Jiang, T. R. Shrout and W. Cao "Electromechanical Properties of Fine-Grain, 0.7 $Pb(Mg_{1/3}Nb_{2/3})O_3$-0.3$PbTiO_3$ Ceramics". IEEE Trans. Ultrasonics, Ferroelectrics and Frequency Control, 51(7), 907 (2004).
62. G. Feuillard , E. Le Clezio, F. Levassort, L. P. Tran Huu Hue, T. Delaunay and M. Lethiecq. "New tools for electromechanical characterisation of piezoceramics" J. Electroceramics 19(4), 419 (2007)
63. E. Le Clezio and A. Shuvalov, "Transmission of acoustic waves through piezoelectric plates: Modelling and experiment," in Proc. IEEE Ultrason. Symp., 2004, pp. 553–556.
64. I. Ohno, "Rectangular parallelepiped resonance method for piezoelectric crystals and elastic constants of alphaquartz," Phys. Chem. Minerals. 17, 371 (1990).
65. T. Delaunay, E.L.Clezio, M. Guennou, H. Dammak, M.P.Thi and G. Feuillard "Full Tensorial characterization of PZN-12%PT single crystal by resonant ultrasound spectroscopy". IEEE Trans. Ultrasonics, Ferroelectrics and Frequency Control, 55 (2), 476 - 488 (2008).
66. Y. Li, Z. K. Qin, and Z. G. Zhou, "Measurements of Piezoelectric and Ferroelectric Materials". Beijing, China: Science Press, 1984. (in Chinese)
67. Electrical Measurement method for piezoelectric ceramic elements, EMAS-6100 (Standard of Electronic Materials Manufacture Association Japan), April 1993.
68. S. Zhang, E. F. Alberta, R. E. Eitel, C. A. Randall and T. R. Shrout, "Elastic, Piezoelectric, and Dielectric Characterization of Modified $BiScO_3$-$PbTiO_3$ Ceramics". IEEE Trans. Ultrasonics, Ferroelectrics and Frequency Control, 52(11), 2131(2005).
69. S. Zhang, R. Xia, T. R. Shrout, G. Zang and J. Wang. "Piezoelectric properties in perovskite 0.948$K_{0.5}Na_{0.5}NbO_3$–0.052$LiSbO_3$ lead-free ceramics". J.Appl.Phys. 100, 104108 (2006)
70. S.Zhang, R.Xia, T.R. Shrout, G. Zang, J. Wang. "Characterization of lead free $(K_{0.5}Na_{0.5})NbO_3$–$LiSbO_3$ piezoceramic". Solid State Communications 141 675–679 (2007).
71. S. Zhang, T. R. Shrout, H. Nagata, Y. Hiruma and T. Takenaka. "Piezoelectric Properties in $(K_{0.5}Bi_{0.5})TiO_3$-$(Na_{0.5}Bi_{0.5})TiO_3$-$BaTiO_3$ Lead-Free Ceramics". IEEE Trans. Ultrasonics, Ferroelectrics and Frequency Control, 54(5), 910 (2007).
72. R. G. Sabat, B.K. Mukherjee, W. Ren and G. Yang. "Temperature dependence of the complete material coefficients matrix of soft and hard doped piezoelectric lead zirconate titanate ceramics". J. Appl. Phys. 101, 064111 (2007)
73. L. Pardo, A. García, K. Brebøl, D. Piazza and C. Galassi "Matrix Characterization, Including Losses, of Porous PZT based ceramics with Morphotropic Phase Boundary Composition: Key Issues" J. Electroceramics 19(4), 413-418 (2007)
74. L. Pardo, M. Algueró and K. Brebøl. "Iterative method in the characterization of piezoceramics of industrial interest". Advances in Science and Technology 45 , 2448-2458 (2006).

Chapter 14
Domain Engineered Piezoelectric Resonators

Jiří Erhart

14.1 Introduction

Today, many dielectric materials are used in applications for their piezoelectric properties. Among all piezoelectric materials, ferroelectrics built an important subgroup. Ferroelectricity in such materials is a typical phenomenon resulting from the spontaneous existence of permanent electric dipole moments in their structure that can be reversed by application of an electric field. From the structural point of view, ferroelectricity is a structural phase transition from paraelectric (also called parent or higher symmetry) to ferroelectric (lower symmetry) phase. Crystallographic groups are in group-subgroup relationship for para- and ferroelectric phases. Spontaneous dipole moment presence requires some symmetry constraints, i.e., the existence of singular polar axis in the crystallographic symmetry group (e.g., $4mm$ group with one four-fold polar axis). Symmetry constraints are the same as for pyroelectricity, but dipole moment must spontaneously exist in ferroelectric material. Both symmetry groups for para- and ferroelectric phase together specify unique structure–*ferroelectric species* (e.g., $m\bar{3}m \rightarrow 4mm$ for tetragonal $BaTiO_3$ crystal). For a complete list of all ferroelectric and ferroelastic species see [1, 2, 3].

Ferroelectric phase allows for the existence of several energetically equivalent variants with the different spontaneous dipole moment orientation. Such equivalent states are called *domain states*. Similarly, spatially connected volume with the same dipole moment orientation belongs to one *domain*. Crystal volume may be typically divided into many lamellar domains with two or more domain states alternated (see Fig. 14.1). Each domain state is characterized by its tensor properties like spontaneous strain, piezoelectric, elastic, dielectric properties etc. For more details on different options for various properties in allowed domain states, see e.g. [3]. In ferroelectrics, spontaneous polarization is also accompanied by the spontaneous

Department of Physics and International Centre for Piezoelectric Research, Technical University of Liberec, Studentská 2, CZ-461 17 Liberec 1, Czech Republic strain. Ferroelastic domain states differ in their spontaneous strain tensor, which is symmetrical second order tensor, contrary to the vector character of spontaneous polarization. Different spontaneous polarization orientations (i.e., different ferroelectric domain states) may correspond to the same spontaneous strain tensor (i.e., to the same ferroelastic domain state). Based on various tensor properties in neighbouring domains, we can distinguish different domain states in the material. This is a principle of polarizing microscopy visualization for ferroelectric and ferroelastic domains – either based on the optical activity, or on the refractive index indicatrix. Other methods of domain visualization are based on spontaneous strain (e.g., laser interferometry [4]) or piezoelectric coefficient (e.g., piezoresponse Atomic Force Microscopy – AFM [5]).

Two neighbouring ferroelectric domains are connected at common interface– i.e., at *domain wall* (DW). For our symmetry analysis purpose, we can assume DW only as an infinitely thin plate. In reality, it is a volume region with spontaneous polarization gradient from one dipole moment orientation in the first domain to the second one. The type of DW is commonly described by an angle between dipole moment orientations in the neighbouring domains. 180° or 90^odomain walls in $BaTiO_3$ crystal are the typical examples of such notation. The angle between spontaneous polarizations in the neighbouring domains could be arbitrary; its value is given by the symmetry for each ferroelectric species. DW plane orientation could be either arbitrary (W_∞-walls, e.g., 180° walls of arbitrary orientation and irregular shape in tetragonal $BaTiO_3$), or fixed with respect to the crystallographic lattice of crystal (W_f-walls, e.g., $(110)_C$ wall in tetragonal $BaTiO_3$), or dependent on the spontaneous strain values (so called *strange walls*, S-walls, e.g., $(11K)_C$, K=0.38 in orthorhombic $BaTiO_3$), or do not exist at all in an equilibrium domain structure. For detailed discussion on different DW types and their possible combinations, see e.g., [1, 2]. If the normal component of spontaneous dipole moment is equal for both domains joined at DW, such DW is called *neutral* and it does not carry additional electric charge. In the case of spontaneous polarization normal discontinuity, Maxwell's equation requires charged interface between domains and such DW must be therefore *charged*. For schematic explanation of described terms, see Fig. 14.2.

A typical domain structure is a result of crystal growth conditions (thermal stresses, chemical composition and temperature gradients etc.) or subsequent mechanical, electrical and thermal treatment of ferroelectric crystal. To some extent, desired domain structure can be engineered to the specific domain volume and type distribution by external fields–mainly by the most easily controlled electrical field. Spontaneous dipole moments are switchable among their existing domain states by a properly oriented electric field. Specific external field could prefer either single domain, or multidomain configuration depending on its crystallographic orientation with respect to the allowed spontaneous dipole moment possibilities. For theoretical crystallographic analysis on external field effects on domain structure stabilization, see e.g., [6]. Although the most typical domain structure configuration is twinned lamellar structure, some examples were reported previously also for multidomain configuration–see e.g., [7, 8] in $KNbO_3$

as-grown crystal. Schematic picture of multidomain configuration reported in [7] is in Fig. 14.3. Note the S-walls with orientation close to $(1;\pm0.3;1)_C$ non-crystallographic orientation between two domain states.

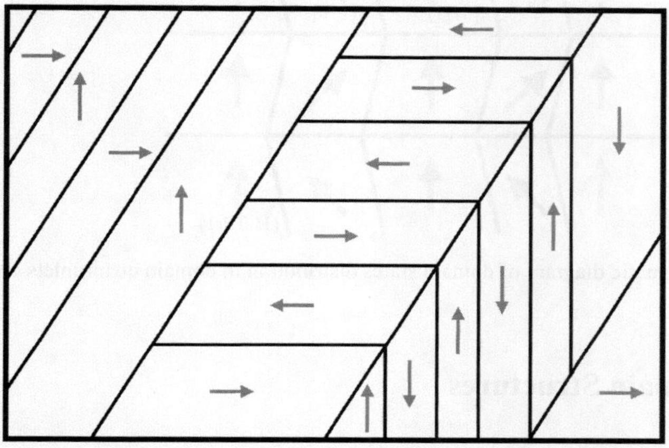

Fig. 14.1 Schematic picture of typical domain lamellar structure in crystal. Arrows indicate orientations of in-plane projections of spontaneous polarization vectors.

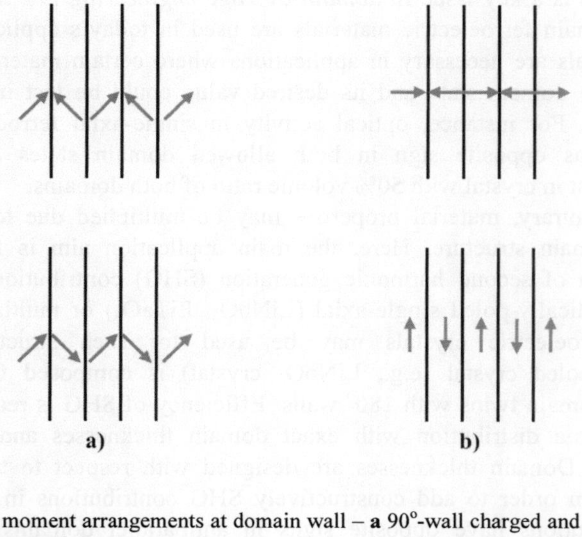

Fig. 14.2 Dipole moment arrangements at domain wall – **a** 90°-wall charged and neutral, **b** 180°-wall charged and neutral.

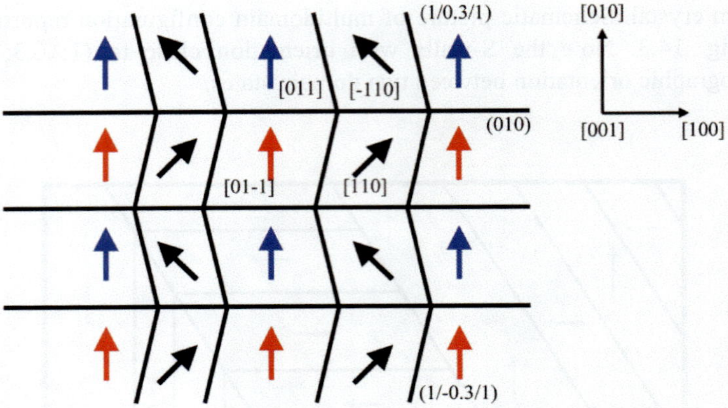

Fig. 14.3 Schematic diagram of domain states distribution in domain quadruplets observed in [7].

14.2 Domain Structures

Domain structure configuration is characterized not only by the number and type of specific domain states, but also by their space distribution. These are two main features, which may be designed in ferroelectric materials. The design of domain volume distribution is specific for *domain-geometry engineering* and design of domain states is a key issue in *domain-average engineering* [9]. Single domain and multidomain ferroelectric materials are used in today's applications. Single domain crystals are necessary in applications where certain material property is unique in the domain state and its desired value could be lost in multidomain configuration. For instance, optical activity in single-axial ferroelectric crystal $Pb_5Ge_3O_{11}$ has opposite sign in both allowed domain states and could be effectively lost in crystal with 50% volume ratio of both domains.

On the contrary, material properties may be multiplied due to the properly designed domain structure. Here, the main application aim is for instance a multiplication of second harmonic generation (SHG) contribution in nonlinear optics. Periodically-poled single-axial ($LiNbO_3$, $LiTaO_3$) or multiaxial ($BaTiO_3$, $KNbO_3$) ferroelectric crystals may be used for such structures. Typical periodically-poled crystal (e.g., $LiNbO_3$ crystal) is composed from regularly alternating domain twins with 180° walls. Efficiency of SHG is reached by exact domain volume distribution with exact domain thicknesses and parallel DW arrangement. Domain thicknesses are designed with respect to the light wave propagation in order to add constructively SHG contributions in each domain. SHG contributions have opposite signs in antiparallel domains in 180° DW structure in a single-axial ferroelectric crystal like $LiNbO_3$. The typical thickness of domain lamella is in the range 1-10μm. Sophisticated poling strategies (pulse poling with back-switching field [10], differential vector poling [11] etc.) were developed for periodic poling in $LiNbO_3$, $LiTaO_3$ or $KNbO_3$, KTA and other materials. However, the main external poling field is pulsed electric field. Periodically-poled optical structures are applicable for multiplication of laser light frequency with the aim of resulting blue laser beam.

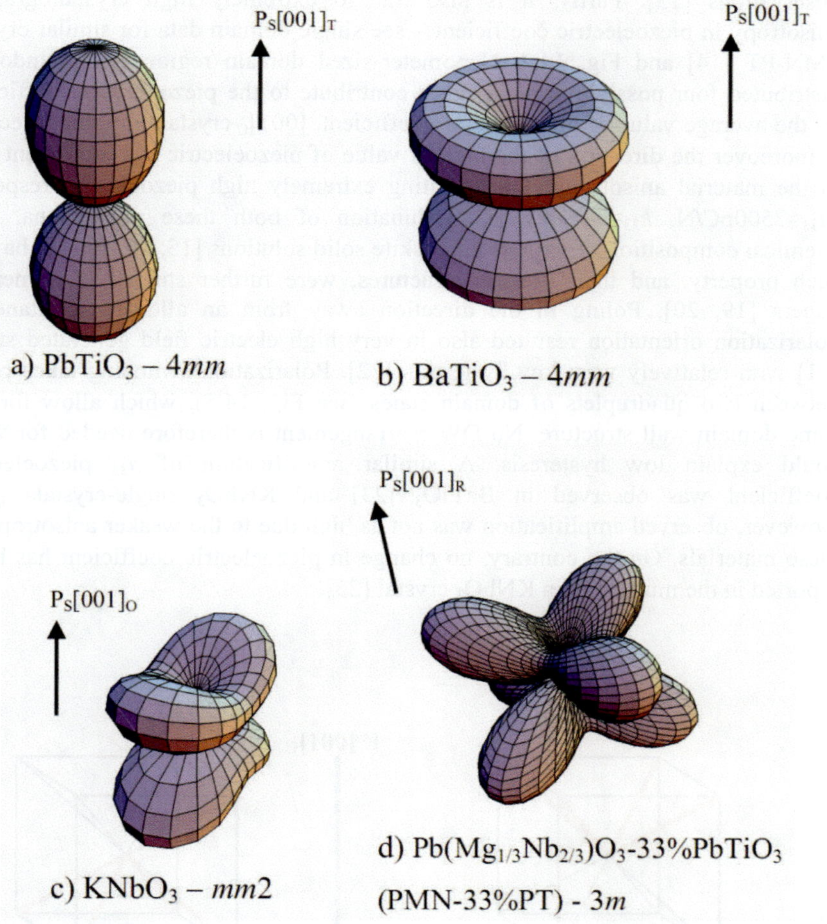

Fig. 14.4 Crystallographic anisotropy (d_{33}-surface) for piezoelectric coefficient in **a** $PbTiO_3$, **b** $BaTiO_3$, **c** $KNbO_3$ and **d** PMN-33%PT crystals.

Another strategy of using multidomain ferroelectric materials appeared relative recently in the early 1980s [12]. Multiaxial ferroelectric solid solution crystal $Pb(Zn_{1/3}Nb_{2/3})O_3$-$PbTiO_3$ (PZN-PT) was accidentally poled in the direction different from any spontaneous polarization direction allowed by the symmetry of its ferroelectric species. For instance, perovskite rhombohedral PZN-PT crystal was poled along a $[001]_C$ – direction. Such poling prefers equally quadruplet domain states – $[111]_C, [\bar{1}11]_C, [1\bar{1}1]_C, [\bar{1}\bar{1}1]_C$ – and it resulted in extremely high piezoelectric coefficient, almost one order of magnitude higher than any applied piezoelectric material before that discovery. Two decades of intensive research on this phenomenon showed that it is due partly to the relaxor nature of PZN-PT crystal leading to very fine domain structure in nanometer range – see AFM observations [13]. Partly, it is also due to extremely high crystallographic anisotropy in piezoelectric coefficient – see single domain data for similar crystal PMN-PT [14] and Fig. 14.4. Nanometer-sized domain regions with randomly distributed four possible domain states contribute to the piezoelectric coefficient by the average value of piezoelectric coefficient. $[001]_C$ crystallographic direction is moreover the direction of the highest value of piezoelectric d_{33} coefficient due to the material anisotropy. The resulting extremely high piezoelectric response (d_{33}≈2500pC/N, k_{33}≈95%) is a combination of both these phenomena. The chemical composition of various perovskite solid solutions [15, 16, 17, 18] having such property, and their domain structures, were further studied in numerous papers [19, 20]. Poling in the direction away from an allowed spontaneous polarization orientation resulted also in very high electric field generated strain [21] with relatively very low hysteresis [22]. Polarization switching takes place between two quadruplets of domain states (see Fig. 14.5), which allow for the same domain wall structure. No DW rearrangement is therefore needed for what could explain low hysteresis. A similar amplification of d_{33} piezoelectric coefficient was observed in $BaTiO_3$ [23] and $KNbO_3$ single-crystals [24]. However, observed amplification was not as high due to the weaker anisotropy in these materials. On the contrary, no change in piezoelectric coefficient has been reported in the multidomain $KNbO_3$ crystal [25].

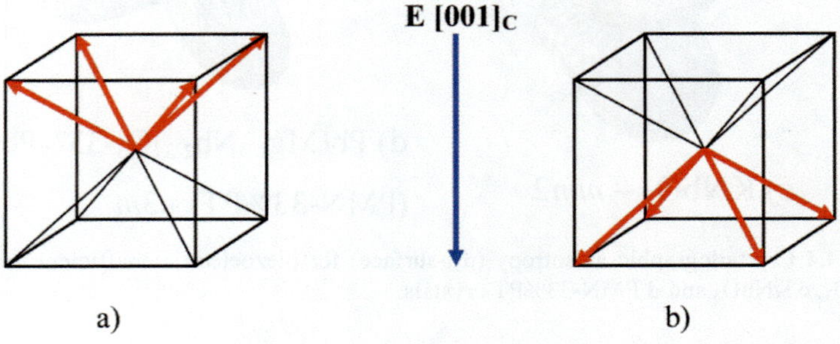

Fig. 14.5 Two domain states quadruplets in domain-engineered rhombohedral PZN-PT or PMN-PT single crystals (ferroelectric species $m\bar{3}m \rightarrow 4mm$). Domain states **a** before and **b** after electric field application in $[001]_C$-direction.

Until now, minor attention has been paid to the other applications of multidomain configurations in ferroelectrics. Piezoelectric resonators are one of further possible applications. After the first pioneering remark on the possibility of vibration mode design [26], several authors addressed domain engineering for piezoelectric resonators [27, 28, 29, 30]. The converse piezoelectric effect is typically used for generating vibrations of the elastic body (i.e., in piezoelectric resonator) in an alternating electric field of desired frequency. Strains in the resonator depend on the value of piezoelectric coefficient and could therefore be engineered by the domain structure. Vibration modes not allowed in the homogeneously-poled resonator could be generated in the properly designed non-homogeneously poled one.

The idea of this domain engineering in piezoelectric resonators for contour extensional mode [26] was experimentally studied in partially twinned $LiNbO_3$ crystal plates with head-to-head (or tail-to-tail) configuration [27]. The creation of a bending vibration mode and the shift of fundamental resonance frequency for thickness-extensional (TE) vibration mode in thin plate were experimentally demonstrated in an impedance spectrum. For 50% volume ratio of both antiparallel domains (180° DW structure), the fundamental resonance shifted from about 7MHz to the value approximately doubled. The resonator vibrated at thickness mode overtone, which is not allowed in homogeneously-poled crystal (see Fig. 14.6). Moreover, a completely new bending mode (low frequency mode) appeared as a result of domain engineering in twinned crystal.

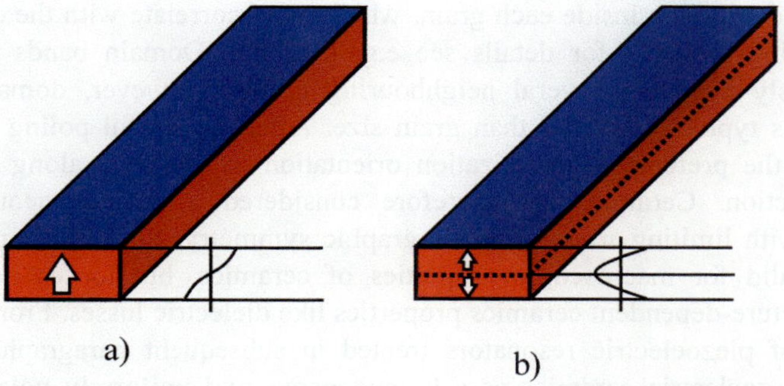

Fig. 14.6 $LiNbO_3$ thin plate – **a** homogeneously poled, **b** domain engineered with 50% volume ratio of both domains. Vibration modes for TE vibrations are schematically shown in transversal displacements.

The same idea was demonstrated in periodically poled $LiNbO_3$ crystal with very fine domain structure (domain lamella thicknesses were 2-10μm) in [28, 29, 30]. Keeping the same macroscopic dimensions for piezoelectric resonator, the fundamental resonance frequency has been shifted from 1-10MHz range into 500MHz-1GHz range depending on the domain thickness. This technique is potentially applicable for very high frequency resonators. However, it is limited to the materials with geometrically well-defined domain walls.

Although an electric field is almost exclusively used for domain manipulation, some authors also reported mechanical stress for multidomain crystal de-twinning [31, 32]. Uniaxial mechanical stress of 10-20MPa oriented along $[100]_C$-direction in $BaTiO_3$ (ferroelectric species $m\bar{3}m \rightarrow 4mm$) crystals was used for the preference of c-domain state (polarization oriented in $[001]_C$-direction) with respect to a-domain state (polarization oriented in $[100]_C$-direction). After such treatment, the crystal was free of 90° domain walls. 180° domain walls were removed by electric field poling along crystal c-axis. Similarly, thermal treatment was reported for de-twinning of $Bi_4Ti_3O_{12}$ crystal (ferroelectric species $4/mmm \rightarrow m_{xy}$) [33].

Thermal treatment was used for domain engineering in quartz crystals (SiO_2, ferroelectroelastic and ferrobielastic second order ferroics). Local temperature gradients created by laser beam irradiation produced thermal stresses that resulted in quartz twinning (non-ferroelectric; left-hand and right-hand quartz variants) [34, 35]. However, mechanical or thermal stresses could be controlled less precisely than an electric field.

Domains in single crystals are of the typical size from nanometre range up to the macroscopic dimensions, for instance in single domain commercially produced $LiNbO_3$ crystals. A somewhat different situation takes place in polycrystalline materials like piezoelectric (ferroelectric) ceramics, for instance PZT ceramics. Polycrystal is composed from tiny grains of the typical dimension from 1μm to 10μm, or even smaller in nano-grained ceramics. Domain structure is a mesoscopic structure inside each grain, which may correlate with the domains in neighbouring grains – for details see e.g. [36, 37]. Domain bands sometimes continuously penetrate several neighbouring grains. However, domain size in ceramics is typically smaller than grain size. After successful poling process in ceramics, the preferential polarization orientation in grains is along the poling field direction. Ceramics are therefore considered as homogeneously poled material with limiting ∞mm crystallographic symmetry. Such assumption is, of course, valid for macroscopic properties of ceramics, but looses its sense in microstructure-dependent ceramics properties like dielectric losses. From the point of view of piezoelectric resonators treated in subsequent paragraphs, we will assume piezoelectric ceramics as a homogeneous and uniformly poled material with macroscopic tensor properties belonging to ∞mm symmetry group. Under the domain states in a domain-engineered ceramics resonator, we will understand macroscopic "effective" domain states, i.e., each domain state represented by an "average" polarization orientation over all grains inside the domain's volume.

14.3 Domain Engineering for Piezoelectric Resonators

A sophisticated technique for DW engineering was developed by Wada *et al.* [38] in the poling process of normal ferroelectrics $BaTiO_3$ crystal in $[111]_C$-direction. The authors succeeded in engineering the DW density in crystal (i.e., also in domain thickness engineering) by simultaneous application of special temperature heating-cooling cycle in the vicinity of Curie temperature and an electric field. The sample is heated slightly above Curie temperature and cooled with applied electric field in several steps to room temperature. Such combined thermal and electrical treatment resulted in the lamellar domain structure with different DW density, i.e., with domain thicknesses in the range from 6.5μm, 13.3 μm and over 40μm. Effective piezoelectric d_{31} coefficient (in crystallographic $[1\bar{1}0]_C$-direction) was measured by the resonant method. It was reported that d_{31} coefficient is dependent on the domain thickness and increases with increasing DW density (i.e., with decreasing domain thickness). d_{31} coefficient in domain engineered $[111]_C$-oriented $BaTiO_3$ bar with regular domain thickness of 6.5μm was reported almost three times higher (i.e., 180pC/N) than the theoretical value of d_{31} coefficient in $[1\bar{1}0]_C$-direction in single domain crystal (i.e., 62pC/N). This phenomenon has not yet been fully explained. It opens an interesting possibility for increasing an effective piezoelectric coefficient in fine-grained $BaTiO_3$ ceramics. Domains are mesoscopic structures inside individual ceramics grains. By decreasing the average size of grains (e.g., in nanograin ceramics), we can control the domain size at the same time. If we could control domain wall density at the same time, a highly piezoelectric lead-free $BaTiO_3$ ceramics might result. However, ceramics grain size is also limited by certain minimum size, otherwise ferroelectricity is lost.

Effective electromechanical properties and their temperature dependence for regularly twinned ferroelectric crystal were theoretically studied for $PbTiO_3$ crystal [39] and $BaTiO_3$ crystal [40]. Single-domain material data and their temperature dependences have been either calculated from the thermodynamic potential for $PbTiO_3$ [41], or taken from the experiment for $BaTiO_3$ [42] in their tetragonal phases up to the Curie temperature. Material properties for regularly twinned crystals (volume ratio 50% for both domains) were calculated using simple property averaging scheme [43]. Different combinations of twin domain states were compared with respect to their temperature dependence. Tensor components for different electromechanical properties show an opposite sign in both domain states building domain twins. Their temperature coefficients have opposite signs and they might therefore compete in twinned crystals, resulting in a decrease of temperature coefficient for some tensor components of an effective property in a twinned crystal. Based on the results [39, 40], we can theoretically design a proper resonator's shape and its vibration mode with reduced resonant frequency temperature coefficient. It means design of proper domain wall orientation and density with respect to crystal orientation and available domain engineering technology. However, practical realization of such a piezoelectric resonator is a key issue and is still highly questionable despite the great effort invested in domain engineering studies in the past decades.

14.4 Twin-Domain Piezoelectric Ceramics Resonators

In this section, we will address a systematic study on the domain engineered piezoelectric ceramics resonators. Electric impedance/admittance, resonance frequency and its temperature coefficient is derived for each resonator and various vibration modes. Temperature coefficients for homogeneously poled (i.e., single-domain) ceramics resonators were already published [44].

14.4.1 Length-Extensional Modes of Thin Bars

For the length-extensional (LE) mode of thin bar, we have two different options for ceramics resonator:

- bar poled along its thickness (also called d_{31}-mode)
- bar poled along its length (also called d_{33}-mode).

Standards on piezoelectric materials [45, 46] list the following formula for admittance/impedance for such resonators in LE vibration mode

- longitudinally-poled bar

$$Z = \frac{1}{j\omega\left(\varepsilon_{33}^{T}\dfrac{bw}{l}\right)\left(1-k_{33}^{2}\right)}\left\{1-k_{33}^{2}\frac{\tan(\eta)}{\eta}\right\} \tag{1}$$

$$\eta = \frac{1}{2}kl, \quad k = 2\pi f\sqrt{\rho s_{33}^{D}}, \quad k_{33}^{2} = \frac{d_{33}^{2}}{s_{33}^{E}\varepsilon_{33}^{T}}$$

- thickness-poled bar

$$Y = j\omega\left(\varepsilon_{33}^{T}\frac{wl}{b}\right)\left\{1-k_{31}^{2}+k_{31}^{2}\frac{\tan(\eta)}{\eta}\right\} \tag{2}$$

$$\eta = \frac{1}{2}kl, \quad k = 2\pi f\sqrt{\rho s_{11}^{E}}, \quad k_{31}^{2} = \frac{d_{31}^{2}}{s_{11}^{E}\varepsilon_{33}^{T}}$$

where *l*, *w*, and *b* are length, width and thickness of the bar resonator. Both admittance/impedance functions were derived for thin bar, i.e., its length *l* is much bigger than the other two dimensions *w*, *b*. Boundary conditions included mechanically free ends of bar, i.e., mechanical stress equal zero at the bar ends. Resonance condition is simple for thickness-poled bar and is not dependent on electromechanical coupling coefficient k_{31}

$$\eta_r = \frac{1}{2}k_r l = \frac{\pi}{2}, \frac{3\pi}{2}, \frac{5\pi}{2}, \ldots \tag{3}$$

But, it is solution of transcendental equation employing electromechanical coupling coefficient k_{33} in the case of longitudinally-poled bar

$$1 - k_{33}^2 \frac{\tan(\eta_r)}{\eta_r} = 0 \tag{4}$$

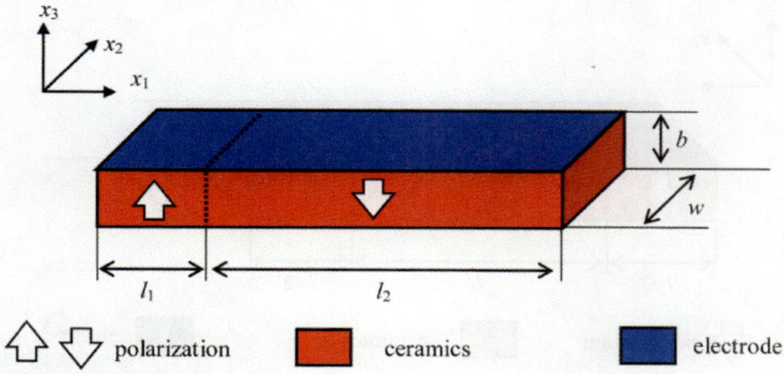

Fig. 14.7 Length-extensional domain engineered bar resonator – twin domain structure, thickness-poling.

In non-homogeneously poled ceramics bars, we have basically two choices of different domain states – either poling along bar length, or along bar thickness. A bar resonator's volume could therefore be divided into two or more domains with alternating domain (antiparallel) states. Admittance for twinned thickness-poled bar resonator working in LE vibration mode (Fig. 14.7) could be derived as

$$Y = j\omega\left(\varepsilon_{33}^T \frac{wl}{b}\right)\left\{1 - k_{31}^2 + k_{31}^2 \frac{\tan(\eta)}{\eta} - k_{31}^2 \frac{4(1-\cos kl_1)(1-\cos kl_2)}{2\eta\sin(2\eta)}\right\} \tag{5}$$

with the variables meaning the same as above. Boundary conditions include not only mechanically free bar ends, but also displacement and normal stress continuity at the domain interface. Resonance condition is again very simple, giving rise to harmonics for the resonance frequency overtones

$$\eta_r = \frac{1}{2}k_r l = \frac{\pi}{2}, \frac{3\pi}{2}, \frac{5\pi}{2}, \ldots \text{ or } 2\eta_r = k_r l = \pi, 2\pi, 3\pi, 4\pi, \ldots \tag{6}$$

It is interesting to note that the resonance frequency does not depend on the specific volume division into two domains, i.e., on the values of lengths l_1 and l_2. However, new resonance frequencies appear as a result of resonator's domain structure as seen in the second part of Eq. (6). A similar phenomenon has been observed in $LiNbO_3$ domain engineered crystal [28, 29]. Effective symmetry of this twinned resonator is lowered to *m*, contrary to *mm*2 symmetry for homogeneously-poled bar. In special cases of equal domain volumes (i.e., $l_1 = l_2 = l/2$), effective resonator's symmetry is 2/*m*, which is not a subgroup of the *mm*2 symmetry group. However, effective symmetry is always a subgroup of materials symmetry ∞*mm* in all cases, as required by Neuman's principle.

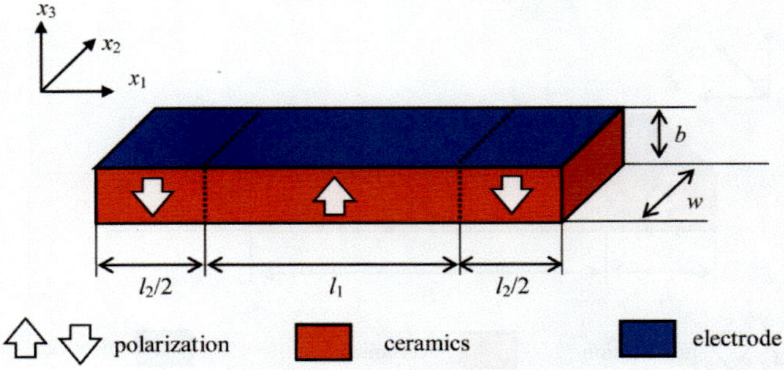

Fig. 14.8 Symmetrical length-extensional domain engineered bar resonator – three domain structure, thickness-poling.

Keeping the same *mm*2 effective symmetry for domain engineered resonator, we can design three-domain structure independently from the actual dimensions l_1 and l_2 (Fig. 14.8). Under similar assumptions and boundary conditions as for twinned thickness-poled bar, we can express the resonator's admittance as

$$Y = j\omega\left(\varepsilon_{33}^T \frac{wl}{b}\right)\left\{1 - k_{31}^2 + k_{31}^2 \frac{\tan(\eta)}{\eta} - k_{31}^2 \frac{8\sin\frac{1}{2}kl_1\left(1-\cos\frac{1}{2}kl_2\right)}{2\eta\cos\eta}\right\} \tag{7}$$

Note that this formula is not symmetrical with respect to interchange of domain lengths l_1 and l_2. Resonance condition is simple

$$\eta_r = \frac{1}{2}k_r l = \frac{\pi}{2}, \frac{3\pi}{2}, \frac{5\pi}{2}, \ldots \tag{8}$$

No additional resonance frequencies dependent on the actual domain lengths l_1 and l_2 appear in the impedance spectrum with respect to the homogeneously-poled bar resonator.

The thin bar poled along thickness could also be easily engineered by subsequent poling to a larger number of domains with alternating antiparallel polarizations. The fundamental resonance frequency for a multidomain resonator with equal size domains is in harmonics with the fundamental resonance frequency for the homogeneously-poled resonator of the same dimensions, as seen from the admittance function [47].

$$Y_N = j\omega\left(\varepsilon_{33}^T \frac{lw}{b}\right)\left\{1 - k_{31}^2 + k_{31}^2 \frac{\tan\left(\frac{1}{N}\eta\right)}{\frac{1}{N}\eta}\right\}, \quad N = 1,2,\ldots,5\,. \tag{9}$$

$$\eta = \frac{1}{2}kl, \quad k = 2\pi f\sqrt{\rho s_{11}^E}$$

However, even-numbered harmonics are also allowed in the resonator with even number of domains of equal length. Note that even-numbered harmonics are not allowed in the homogeneously-poled resonator with the same dimensions. A multidomain structure with very thin domains of equal dimensions may substantially increase the resonance frequency for the LE mode. The practical processing of piezoelectric resonators is limited to certain dimension (e.g., by 30μm thickness for quartz plates) and the introduction of a tiny intrinsic domain structure might solve the problem. We can shift the fundamental resonance to very high frequency by domain engineering, retaining the resonator's macroscopic dimensions. This is potentially applicable for a practical resonator's design. Non-equal domain dimensions are a source of potentially complicated impedance spectrum as seen from Eqs. (5) or (7). The exact volume distribution of domains with planar domain walls may be a challenging task for the domain engineering technology in some technically applied piezoelectric materials.

The linear temperature coefficient (at certain temperature T_0) for the resonance frequency is defined [48] by its first temperature derivative as

$$TK(f_r) = \frac{1}{f_r}\left(\frac{\partial f_r}{\partial T}\right)_{T=T_0} \tag{10}$$

We can calculate this coefficient by the first derivative of resonance condition for domain engineered bar resonator:

- twinned or three-domain thickness-poled resonator (from Eq. (6) or (8))

$$TK(f_r) = TK(\eta_r) + \frac{1}{2}\alpha_{33} - \frac{1}{2}TK(s_{11}^E),\ TK(\eta_r) = 0 \tag{11}$$

Homogeneously-poled as well as domain engineered thickness-poled resonators studied here have their resonance frequency temperature coefficient independent from the domain lengths l_1 and l_2.

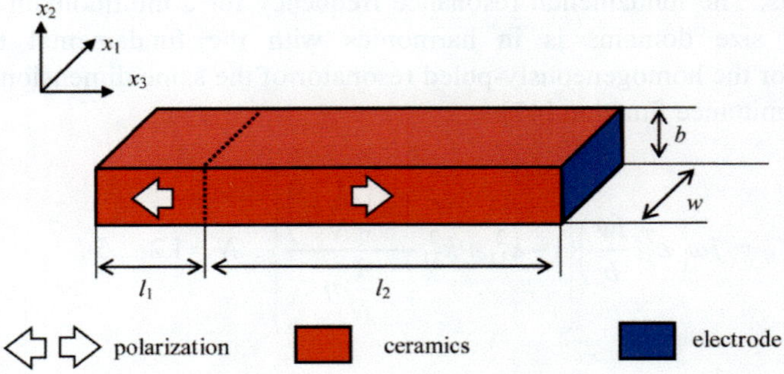

Fig. 14.9 Length-extensional domain engineered bar resonator – twin domain structure, longitudinal poling.

Similarly, we also derived an impedance function for twinned longitudinally-poled thin bar working in LE vibration mode (see Fig. 14.9)

$$Z = \frac{1}{j\omega\left(\varepsilon_{33}^T \frac{bw}{l}\right)\left(1-k_{33}^2\right)}\left\{1 - k_{33}^2\frac{\tan(\eta)}{\eta} + k_{33}^2\frac{4(1-\cos kl_1)(1-\cos kl_2)}{2\eta\sin(2\eta)}\right\} \tag{12}$$

$$\eta = \frac{1}{2}kl, \quad k = 2\pi f\sqrt{\rho s_{33}^D}$$

Boundary conditions include mechanically free resonator's ends and continuity of electrical displacement, and mechanical stress and displacement at the interface between domains. In this case, resonance condition is a transcendental equation

$$1 - k_{33}^2\frac{\tan(\eta_r)}{\eta_r} + k_{33}^2\frac{4(1-\cos k_r l_1)(1-\cos k_r l_2)}{2\eta_r\sin(2\eta_r)} = 0 \tag{13}$$

similarly to a homogeneously-poled bar. Resonance condition (13) differs from Eq. (4) by the last term and is (symmetrically) dependent on actual domain dimensions l_1 and l_2. In a special case of equal domain volumes (i.e., $l_1 = l_2 = 0.5l$), we can get resonance condition in the form

$$1 - k_{33}^2 \frac{\tan(\frac{1}{2}\eta_r)}{\frac{1}{2}\eta_r} = 0 \tag{14}$$

Resonance frequency is twice as high as in the case of a single domain resonator and no other resonance frequencies are present due to the domain different thicknesses. In the case of non-equal domain dimensions l_1 and l_2, new resonance frequencies may appear in the impedance spectrum. Contrary to the thickness-poled bar, such resonance frequencies depend also on the actual domain dimensions l_1 and l_2. Further impedance spectrum analysis could be done only in numerical calculations for specific material due to the transcendental equation (13). Temperature coefficient for resonance frequency could be derived for twinned resonator as

$$TK(f_r) = TK(\eta_r) + \alpha_{11} - \frac{1}{2}\alpha_{33} - \frac{1}{2}TK(s_{33}^D) \tag{15}$$

$$TK(\eta_r) = \frac{2}{k_{33}^2} TK(k_{33}) \left\{ \frac{4}{2\eta_r \sin 2\eta_r} \left[k_r l_1 \sin k_r l_1 (1 - \cos k_r l_2) + k_r l_2 \sin k_r l_2 (1 - \cos k_r l_1) \right] + \right.$$
$$\left. + \frac{1}{k_{33}^2} - 2 + \frac{2\eta_r \cos 2\eta_r}{k_{33}^2 \sin 2\eta_r} \right\}^{-1}$$

and it could be tuned to a certain extent by the domain dimensions l_1 and l_2. Numerical values cannot be calculated in general. They depend on specific material properties, i.e., on electromechanical coupling coefficient k_{33} and thermal expansion coefficients α_{11} and α_{33}.

14.4.2 Thickness-Extensional Mode of Thin Plate

Thin ceramics plates, homogeneously-poled along their thickness, vibrate in a thickness-extensional (TE) mode. Planar displacements are clamped due to the resonator shape and therefore this vibration mode is also one-dimensional (displacement only in thickness direction). Impedance function [46] for this resonator is

$$Z = \frac{1}{j\omega\left(\varepsilon_{33}^S \frac{lw}{b}\right)} \left\{ 1 - k_t^2 \frac{\tan(\eta)}{\eta} \right\} \tag{16}$$

$$\eta = \frac{1}{2}kb, \quad k = 2\pi f\sqrt{\frac{\rho}{c_{33}^D}}, \quad k_t^2 = \frac{e_{33}^2}{c_{33}^D \varepsilon_{33}^S}$$

This formula was derived under the assumption of a thin plate. Boundary conditions include mechanical stress equal to zero at the plate main surfaces. Resonance frequency is a solution of the transcendental equation

$$1 - k_t^2 \frac{\tan(\eta_r)}{\eta_r} = 0 \tag{17}$$

and it depends on the thickness electromechanical coupling factor k_t.

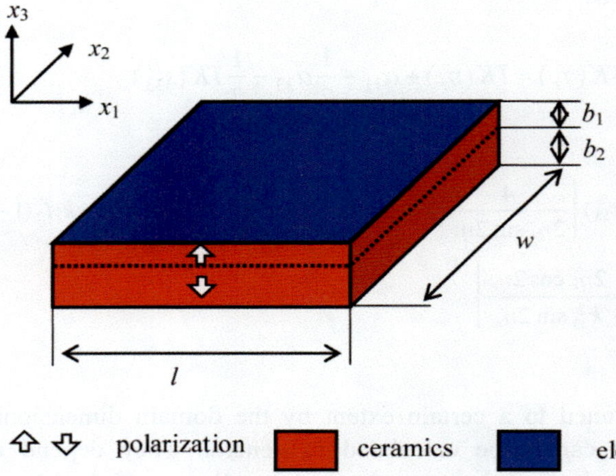

Fig. 14.10 Thickness-extensional domain engineered plate resonator – twin domain structure, thickness-poling.

For thickness-poled ceramics resonator, we can design a twin domain engineered structure (Fig. 14.10). However, a practical realization of this structure is questionable. Such a domain structure was observed in yttrium-doped $LiNbO_3$ single crystals as a result of doping and crystal growth conditions [30]. Impedance function for twinned ceramics resonator (Fig. 14.10) could be expressed as

$$Z = \frac{1}{j\omega\left(\varepsilon_{33}^S \frac{lw}{b}\right)}\left\{1 - k_t^2 \frac{\tan(\eta)}{\eta} + k_t^2 \frac{4(1-\cos kb_1)(1-\cos kb_2)}{2\eta \sin(2\eta)}\right\} \tag{18}$$

Boundary conditions include zero mechanical stress on the main resonator faces and continuity of normal mechanical stress, displacement and electric displacement at the domain interface. Resonance frequency is again a solution of transcendental equation

$$1-k_t^2\frac{\tan(\eta_r)}{\eta_r}+k_t^2\frac{4\left(1-\cos k_r b_1\right)\left(1-\cos k_r b_2\right)}{2\eta_r\sin(2\eta_r)}=0 \quad (19)$$

and it is dependent also on the actual domain thicknesses b_1 and b_2. This domain engineered resonator has the same effective symmetry ∞mm as a homogeneously-poled one, only in case of circular shape. In the case of a rectangular shape, its symmetry is $mm2$ and is $4mm$ for square plate resonator (and $4/mmm$ for square plate resonator and equal domain thicknesses). In the special case of 50% volume ratio of both domains, i.e., $b_1 = b_2 = 0.5b$, the impedance function reduces to

$$Z=\frac{1}{j\omega\left(\varepsilon_{33}^S\frac{lw}{b}\right)}\left\{1-k_t^2\frac{\tan(\frac{1}{2}\eta)}{\frac{1}{2}\eta}\right\} \quad (20)$$

Fundamental resonance frequency in this special case is second harmonics of the fundamental resonance frequency for TE mode for a homogeneously-poled resonator. Note that such a resonance frequency is not allowed for a homogeneously-poled resonator. Domain engineering could be therefore used for multiplication of fundamental resonance frequency. By introducing a tiny domain structure, we can substantially increase the fundamental resonance frequency for the TE vibration mode. In the case of non-equal domain thicknesses b_1 and b_2, a new resonance frequency might appear in the impedance spectrum due to the domain engineering.

Temperature coefficient for the resonance frequency could be derived as

$$TK(f_r)=TK(\eta_r)+\alpha_{11}-\frac{1}{2}\alpha_{33}+\frac{1}{2}TK(c_{33}^D) \quad (21)$$

$$TK(\eta_r)=\frac{2}{k_t^2}TK(k_t)\left\{\frac{4}{2\eta_r\sin 2\eta_r}\left[k_r b_1\sin k_r b_1(1-\cos k_r b_2)+k_r b_2\sin k_r b_2(1-\cos k_r b_1)\right]+\right.$$
$$\left.+\frac{1}{k_t^2}-2+\frac{2\eta_r\cos 2\eta_r}{k_t^2\sin 2\eta_r}\right\}^{-1}$$

and it can be tuned by the domain thicknesses b_1 and b_2 in a certain range. The perspective of this vibration mode in a domain engineered resonator is applicable to resonance frequency multiplication.

14.4.3 Thickness-Shear Mode of Thin Plate

Thickness-shear (TS) vibration mode is another possibility of one-dimensional solution for thin plate vibrations. We have generally two different options for TS mode resonator (or d_{15}-mode):

- longitudinally-poled plate (or thin bar) with electrodes deposited on the main resonator's faces
- thickness-poled plate (or thin bar) with electrodes deposited on side faces (i.e., on width x thickness face)

The first option is a resonator recommended by piezoelectric standards [45, 46] for material property measurement. The second resonator type is not recommended by standard, but it could also serve as an alternative for TS mode measurement of material data in case the thickness-poling is more desirable than the longitudinal one.

In the case of a homogeneously-poled ceramics resonator, we can derive impedance function for longitudinally-poled TS resonator as

$$Z = \frac{1}{j\omega\left(\varepsilon_{11}^{T}\frac{lw}{b}\right)\left(1-k_{15}^{2}\right)}\left\{1-k_{15}^{2}\frac{\tan(\eta)}{\eta}\right\} \tag{22}$$

$$\eta = \frac{1}{2}kb, \quad k = 2\pi f\sqrt{\rho s_{55}^{D}}, \quad k_{15}^{2} = \frac{d_{15}^{2}}{s_{55}^{E}\varepsilon_{11}^{T}}$$

This formula has been derived under the assumption of thin plate. Boundary conditions include zero mechanical stress at the main (and electroded) faces of resonator. Resonance frequency is a solution of transcendental equation

$$1-k_{15}^{2}\frac{\tan(\eta_{r})}{\eta_{r}} = 0 \tag{23}$$

and it depends on the value of shear electromechanical coupling factor k_{15}.

Impedance function for TS domain engineered resonator (Fig. 14.11, longitudinal poling) could be expressed as

$$Z = \frac{1}{j\omega\left(\varepsilon_{11}^{T}\frac{lw}{b}\right)\left(1-k_{15}^{2}\right)}\left\{1-k_{15}^{2}\frac{\tan(\eta)}{\eta}+k_{15}^{2}\frac{4\left(1-\cos kb_{1}\right)\left(1-\cos kb_{2}\right)}{2\eta\sin(2\eta)}\right\} \tag{24}$$

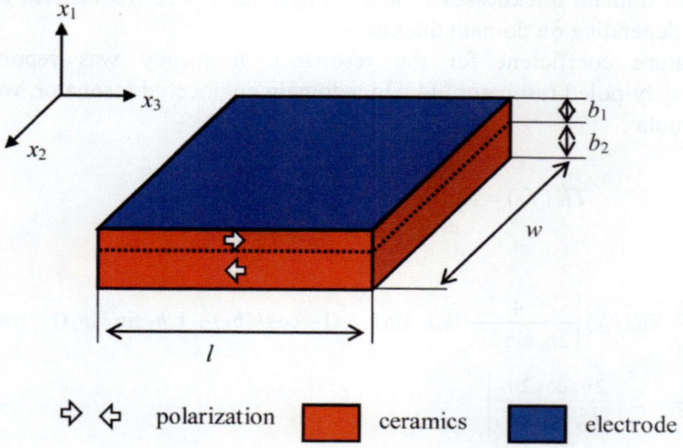

Fig. 14.11 Thickness-shear domain engineered plate resonator – twin domain structure, longitudinal poling.

Boundary conditions include zero mechanical stress at the main resonator faces and continuity conditions for mechanical stress, displacement and electrical displacement at the domain interface. Resonance frequency is a solution of transcendental equation

$$1-k_{15}^2\frac{\tan(\eta_r)}{\eta_r}+k_{15}^2\frac{4(1-\cos k_r b_1)(1-\cos k_r b_2)}{2\eta_r\sin(2\eta_r)}=0 \tag{25}$$

and it depends not only on the electromechanical coupling factor k_{15}, but also on the domain thicknesses b_1 and b_2. Effective symmetry of such domain engineered resonator is *m* in case of rectangular shape, but it could belong even to 2/*m* symmetry for special case of equal domain thicknesses b_1 and b_2. When the volume ratio of both domains is equal, i.e. $b_1 = b_2 = 0.5b$, the impedance function is reduced to

$$Z=\frac{1}{j\omega\left(\varepsilon_{11}^T\frac{lw}{b}\right)(1-k_{15}^2)}\left\{1-k_{15}^2\frac{\tan(\frac{1}{2}\eta)}{\frac{1}{2}\eta}\right\} \tag{26}$$

Fundamental resonance frequency for this special case is twice as high as the fundamental resonance frequency for a homogeneously-poled resonator from the same material and with the same dimensions. Note that such a resonance mode is not allowed in the homogeneously-poled resonator. Domain engineering could multiply the fundamental resonance frequency in such a resonator's type. In case of non-equal domain thicknesses b_1 and b_2, new resonance frequencies could be introduced depending on domain thicknesses.

Temperature coefficient for the resonance frequency was reported for homogeneously-poled resonator [44]. In a domain engineered resonator, we obtain similar formula

$$TK(f_r) = TK(\eta_r) + \frac{1}{2}\alpha_{33} - \frac{1}{2}TK(s_{55}^D) \tag{27}$$

$$TK(\eta_r) = \frac{2}{k_{15}^2}TK(k_{15})\left\{\frac{4}{2\eta_r \sin 2\eta_r}\left[k_r b_1 \sin k_r b_1 (1-\cos k_r b_2) + k_r b_2 \sin k_r b_2 (1-\cos k_r b_1)\right] + \right.$$
$$\left. + \frac{1}{k_{15}^2} - 2 + \frac{2\eta_r \cos 2\eta_r}{k_{15}^2 \sin 2\eta_r}\right\}^{-1}$$

which is dependent on actual domain thicknesses b_1 and b_2. Temperature coefficient could be also tuned in a certain range by a proper choice of domain thicknesses.

A non-standard TS resonator is based on a thickness-poled ceramics plate with electrodes on side faces. Admittance function for such resonator is

$$Y = j\omega\left(\varepsilon_{11}^T \frac{wb}{l}\right)\left\{1 - k_{15}^2 + k_{15}^2 \frac{\tan(\eta)}{\eta}\right\} \tag{28}$$

$$\eta = \frac{1}{2}kb, \quad k = 2\pi f\sqrt{\rho s_{55}^E}, \quad k_{15}^2 = \frac{d_{15}^2}{s_{55}^E \varepsilon_{11}^T}$$

similarly to LE vibrating thin bar. Boundary conditions include zero mechanical stress on the main resonator faces. Resonance condition is the same as Eq. (3), but for different frequency constant η

$$\eta_r = \frac{1}{2}k_r b = \frac{\pi}{2}, \frac{3\pi}{2}, \frac{5\pi}{2}, \ldots \tag{29}$$

and resonance frequency overtones are harmonics (contrary to antiresonance frequencies). Temperature coefficient for resonant frequency for homogeneously-poled non-standard TS resonator is

$$TK(f_r) = TK(\eta_r) + \frac{1}{2}\alpha_{11} - \frac{1}{2}\alpha_{33} - \frac{1}{2}TK(s_{55}^E), \quad TK(\eta_r) = 0 \tag{30}$$

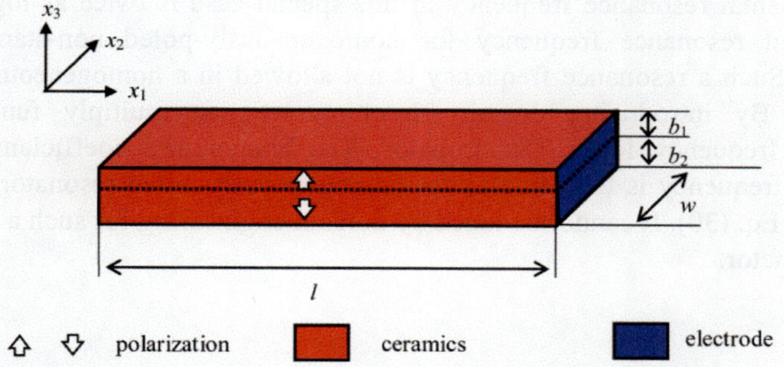

Fig. 14.12 Thickness-shear domain engineered plate (bar) resonator – twin domain structure, thickness-poling.

Similar to a TE domain engineered resonator, a TS resonator could also be designed with twin domain structure in a thickness-poled thin plate (Fig. 14.12). However, the practical realization of such domain configuration is questionable and has not been reported so far. Admittance function could be derived in the form

$$Y = j\omega\left(\varepsilon_{11}^{T}\frac{wb}{l}\right)\left\{1-k_{15}^{2}+k_{15}^{2}\frac{\tan(\eta)}{\eta}-k_{15}^{2}\frac{4(1-\cos kb_1)(1-\cos kb_2)}{2\eta\sin(2\eta)}\right\} \quad (31)$$

Boundary conditions include zero mechanical stress on the main resonator faces and continuity of mechanical stress, displacement and the same electric field inside both domains. This domain engineered resonator exhibits *mm*2 effective symmetry in case of rectangular shape. Symmetry may be the higher *mmm* in case of equal domain thicknesses b_1 and b_2. Resonance frequency is a solution of simple condition

$$\eta_r = \frac{1}{2}k_r b = \frac{\pi}{2}, \frac{3\pi}{2}, \frac{5\pi}{2}, \ldots \text{ or } 2\eta_r = k_r b = \pi, 2\pi, 3\pi, 4\pi, \ldots \quad (32)$$

which is not dependent on actual domain thicknesses b_1 and b_2. Domain engineering introduces new resonance frequencies in the admittance spectrum (e.g. at $\eta_r = \pi$) independently from domain thicknesses, but does not change the fundamental resonance frequency. In the special case of 50% volume ratio of both domains, i.e., $b_1 = b_2 = 0.5b$, admittance function is reduced to

$$Y = j\omega\left(\varepsilon_{11}^{T}\frac{wb}{l}\right)\left\{1-k_{15}^{2}+k_{15}^{2}\frac{\tan(\frac{1}{2}\eta)}{\frac{1}{2}\eta}\right\} \quad (33)$$

Fundamental resonance frequency in this special case is twice as high as the fundamental resonance frequency for homogeneously-poled non-standard TS resonator. Such a resonance frequency is not allowed in a homogeneously-poled TS plate. By introducing domain structure, we can multiply fundamental resonance frequency for a TS resonator. The temperature coefficient for the resonance frequency is the same as for homogeneously-poled resonator, i.e., the same as in Eq. (30). It cannot be tuned by domain engineering for such a vibration mode resonator.

14.4.4 Contour-Extensional Mode of Thin Disc

A thin ceramics disc is another example of one-dimensional solution of equation of motion for a piezoelectric ceramics resonator. If homogeneously-poled along its thickness, it is isotropic in plane perpendicular to the poling direction and it could vibrate in a radially-extensional mode (RE) due to its circular shape. The exact solution of RE vibrations for a homogeneously-poled disc has been published [49] with resulting admittance function

$$Y = j\omega\left(\varepsilon_{33}^{P}\frac{\pi r_2^2}{2b}\right)\left[\frac{k_p^2}{1-k_p^2}\frac{(1+\sigma^P)J_1(\eta_2)}{\left[(1-\sigma^P)J_1(\eta_2)-\eta_2 J_0(\eta_2)\right]}-1\right] \tag{34}$$

$$\eta_2 = 2\pi f r_2\sqrt{\frac{\rho}{c_{11}^P}},\ k_p^2 = \frac{2d_{31}^2}{\varepsilon_{33}^T\left(s_{11}^E+s_{12}^E\right)},\ c_{11}^P = c_{11}^E - \frac{\left(c_{13}^E\right)^2}{c_{33}^E},\ \sigma^P = -\frac{s_{12}^E}{s_{11}^E},\ \varepsilon_{33}^P = \varepsilon_{33}^S + \frac{e_{33}^2}{c_{33}^E}$$

and J_0, J_1 are zero- and first-order Bessel's functions of the first kind and r_2 is a disc radius. Boundary conditions include zero radial displacement in the disc centre and zero mechanical stress at the outer disc diameter. Resonance frequency is a solution of transcendental equation

$$(1-\sigma^P)J_1(\eta_{2r}) - \eta_{2r}J_0(\eta_{2r}) = 0 \tag{35}$$

Resonance (as well as antiresonance) frequency overtones are not harmonics. The temperature coefficient for the resonant frequency in homogeneously-poled disc has been derived [44] as

$$TK(f_r) = TK(\eta_{2r}) + \frac{1}{2}\alpha_{33} + \frac{1}{2}TK(c_{11}^P),\ TK(\eta_{2r}) = \frac{\sigma^P TK(\sigma^P)}{\eta_{2r}^2 + \left(\sigma^P\right)^2 - 1}. \tag{36}$$

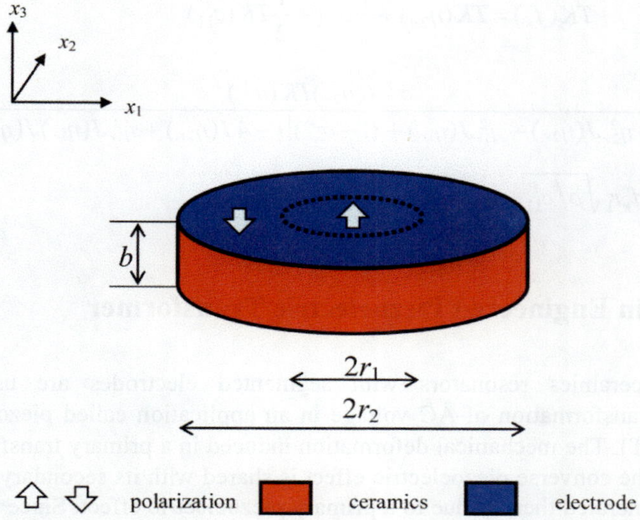

Fig. 14.13 Contour-extensional domain engineered disc resonator – twin domain structure, thickness-poling.

A twin domain structure could be introduced into this resonator by a circular and axially symmetrical domain segment (Fig. 14.13) poled in an antiparallel direction. For details on the exact solution for twin-domain disc resonator, see [47]. We will report here on the main results only. Independent from the size of internal domain, the effective symmetry of such a resonator is the same as for a homogeneously-poled one, i.e., ∞mm. Resonance frequency is a solution of transcendental equation [47]

$$Y_0(\eta_{1r})\left[Y(\eta_{1r})-J(\eta_{1r})\right]J_0\left(\eta_{2r}\right)\left[1-(1-\sigma^P)J\left(\eta_{2r}\right)\right]=0 \tag{37}$$

where $J(x)\equiv\dfrac{J_1(x)}{xJ_0(x)},\quad Y(x)\equiv\dfrac{Y_1(x)}{xY_0(x)}$.

Resonance frequency could be defined by the radius of disc as well as by inner domain radius, but not by both radii as seen from Eq. (37). Domain engineering therefore may introduce new resonance frequencies into the admittance spectrum. According to the published analysis [47], the fundamental resonance of thin disc could not be changed by the domain engineering on thin disc in an RE vibration mode. Also, the temperature coefficient for the fundamental resonance frequency is independent from the inner domain radius. Higher resonance overtones could however be tuned by the domain structure inside an RE resonator. Their temperature coefficient is [47]

$$TK(f_r) = TK(\eta_{2r}) + \frac{1}{2}\alpha_{33} + \frac{1}{2}TK(c_{11}^P) \tag{38}$$

$$TK(\eta_{2r}) = \frac{\sigma^P J(\eta_{2r}) TK(\sigma^P)}{2 + \eta_{2r}^2 J(\eta_{2r}) - \eta_{1r}^2 J(\eta_{1r}) + (1-\sigma^P)\left[1 - 4J(\eta_{2r}) + \eta_{1r}^2 J(\eta_{1r}) J(\eta_{2r})\right]}$$

where $\eta_{1r} = 2\pi f_r r_1 \sqrt{\rho / c_{11}^P}$, $\eta_{2r} = 2\pi f_r r_2 \sqrt{\rho / c_{11}^P}$.

14.5 Domain Engineered Piezoelectric Transformer

Piezoelectric ceramics resonators with segmented electrodes are used for piezoelectric transformation of AC voltage in an application called piezoelectric transformer (PT). The mechanical deformation induced in a primary transformer's circuit due to the converse piezoelectric effect is shared with its secondary circuit. The voltage generated there is due to a primary piezoelectric effect. Since the first idea of the Rosen-type transformer, there are many different types of PTs realized on various resonator shapes and by using different vibration modes. Electric voltage is transformed without the generation of a magnetic field harmful to magnetic records, etc. The main transformer parameters are a step-up (or step-down) ratio, and the efficiency of electric energy transformation. We will not present an exhaustive review on PTs, but we will show how domain engineering could help improve PT parameters.

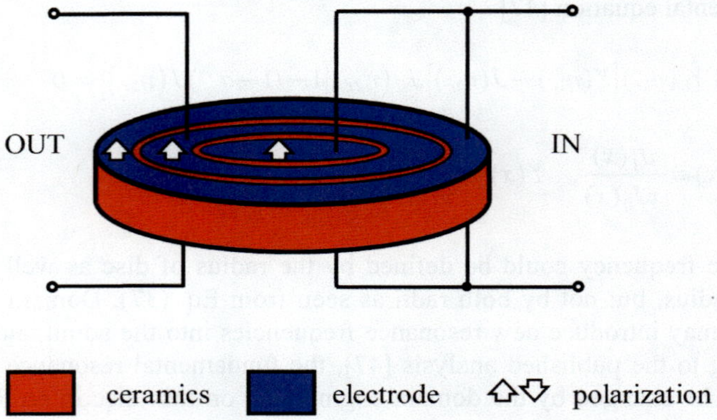

Fig. 14.14 Homogeneously poled „double ring-dot" piezoelectric ceramics transformer. IN = primary circuit, OUT = secondary circuit.

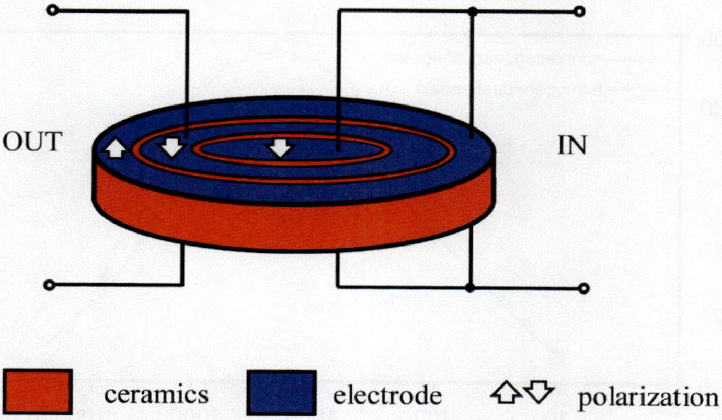

Fig. 14.15 Domain- engineered „double ring-dot“ piezoelectric ceramics transformer. IN = primary circuit, OUT = secondary circuit.

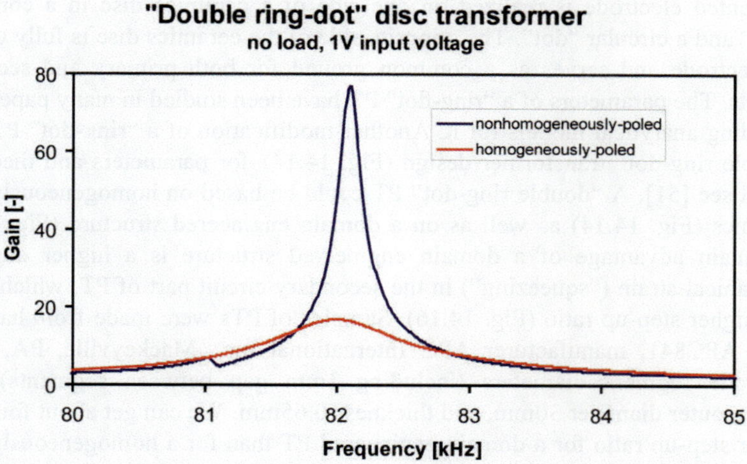

Fig. 14.16 Step-up ratio (gain) for homogeneously-poled and domain-engineered piezoelectric ceramics transformer.

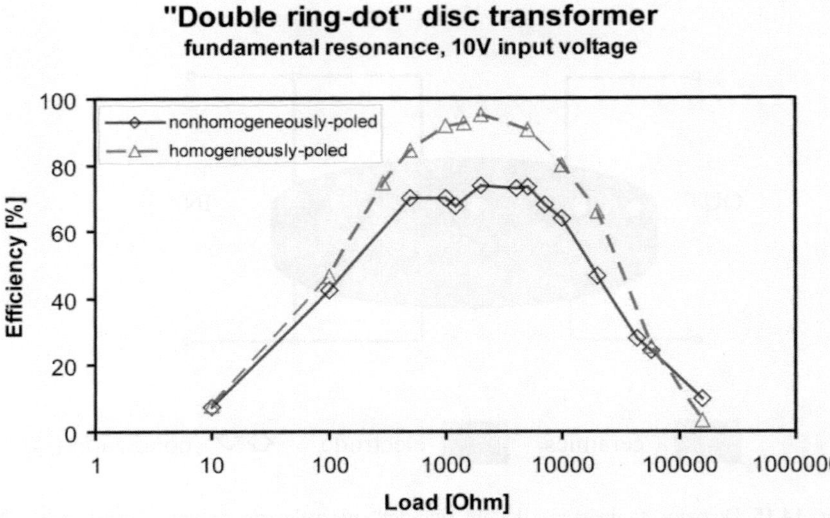

Fig. 14.17. Efficiency for homogeneously-poled and domain-engineered piezoelectric ceramics transformer.

One of the typical PT designs is the so-called "ring-dot" transformer. A segmented electrode is realized on one side of a ceramics disc in a concentric "ring" and a circular "dot". The opposite side of the ceramics disc is fully covered by electrode and serves as a common ground for both primary and secondary circuits. The parameters of a "ring-dot" PT have been studied in many papers, [50] including analytical models for it. Another modification of a "ring-dot" PT is the "double ring-dot" transformer design (Fig. 14.14)–for parameters and theoretical model see [51]. A "double ring-dot" PT could be based on homogeneously-poled ceramics (Fig. 14.14) as well as on a domain engineered structure (Fig. 14.15). The main advantage of a domain engineered structure is a higher degree of mechanical strain ("squeezing") in the secondary circuit part of PT, which results in a higher step-up ratio (Fig. 14.16). Samples of PTs were made from hard PZT (type APC841, manufacturer APC International Inc., Mackeyville, PA, USA), electrode segment diameters (including 1mm gap between segments) 8mm, 16mm, outer diameter 30mm, and thickness 0.65mm. We can get about four times higher step-up ratio for a domain engineered PT than for a homogeneously-poled one. However, PT efficiency of energy transfer is smaller due to the non-homogeneous internal domain structure (Fig. 14.17). The idea of using ferroelectric domains for the improvement of ceramics resonator properties could help for transformers of specific type and electrode design.

14.6 Conclusions

In our brief contribution, we discussed issues of domain engineering applied for piezoelectric resonators. Twin domain structures are compared with the homogeneously-poled ceramics resonators in resonance frequency (fundamental frequency or overtones) and temperature dependence of resonance frequency. We identified the possibilities of domain engineering application with respect to a tuning resonance spectrum and resonance frequency temperature dependence of some twinned resonators with typical configuration (i.e., shape, electrode pattern and poling direction). Finally, we also showed experimental data for an example of piezoelectric ceramics resonator–i.e., domain engineered piezoelectric transformer of a "double ring-dot" design. Domain engineering could improve its step-up ratio significantly, but it could also decrease its efficiency at the same time.

Although the existence of ferroelectric domains is known for many decades, domain engineering studies for resonators are not frequent. The demand for new applications in electronics and progress in ferroelectric domain engineering technology will certainly stimulate further research in the field–especially in the application of non-180° DW structures, not reported up to now. Domain engineering in a nanoscale could help to solve the problem of resonators with high fundamental frequency; they are limited by its physical size in the current piezoelectric resonator technology.

Acknowledgements

The author would like to thank Prof Fousek and Prof Janovec, who stimulated ferroelectric domain studies in our Department of Physics at TU Liberec, and Petr Půlpán for performing measurements on piezoelectric transformers. Financial support from the Grant Agency of the Czech Republic under project No. 202/06/0411 is acknowledged.

References

1. J.Fousek, V.Janovec: The orientation of domain walls in twinned ferroelectric crystals, J.Appl.Phys. 40, 1 (1969) 135-142.
2. J.Sapriel: Domain-wall orientations in ferroelastics, Phys.Rev. B 12, 11 (1975) 5128-5140.

3. J.Erhart: Domain wall orientations in ferroelastics and ferroelectrics, Phase Transitions 77, 12 (2004) 989-1074.
4. K.G.Desmukh, S.G.Ingle: Interferometric studies of domain structures in potassium niobate single crystals, J.Phys. D: Appl.Phys. 4 (1971) 124-132.
5. L.M.Eng, M.Abplanalp, P.Günter: Ferroelectric domain switching in tri-glycine sulphate and barium-titanate bulk single crystals by scanning force microscopy, Appl.Phys. A 66 (1998) S679-S683.
6. J.Fuksa, V.Janovec: Macroscopic symmetries and domain configurations of engineered domain structures, J.Phys.: Condens. Matter 14 (2002) 3795-3812.
7. E.Wiesendanger: Domain structures in orthorhombic $KNbO_3$ and characterisation of single domain crystals, Czech.J.Phys. B 23 (1973) 91-99.
8. Li Lian, T.C.Chong, H.Kumagai, M.Hirano, Lu Taijing, S.C.Ng: Temperature evolution of domains in potassium niobate single crystals, J. Appl. Phys. 80, 1 (1996) 376-381.
9. J.Fousek, D.B.Litvin, L.E.Cross: Domain geometry engineering and domain average engineering of ferroics, J.Phys.: Condens. Matter 13 (2001) L33-L38.
10. V.Ya.Shur, E.L.Rumyantsev, E.V.Nikolaeva, E.I.Shishkin, R.G.Batchko, G.D.Miller, M.M.Fejer, R.L.Byer: Regular ferroelectric domain array in lithium niobate crystals for nonlinear optic applications, Ferroelectrics 236 (2000) 129-144.
11. J.Hirohashi, K.Yamada, H.Kamiyo, S.Shichijyo: Artificial Fabrication of 60° Domain Structures in $KNbO_3$ Single Crystals, J.Korean Phys.Soc. 42 (2003) S1248-S1251.
12. J. Kuwata, K. Uchino, S. Nomura: Dielectric and piezoelectric properties of $0.91Pb(Zn_{1/3}Nb_{2/3})O_3$-$0.09PbTiO_3$ single crystals, Jpn.J.Appl.Phys. 21, 9 (1982) 1298-1302.
13. M.Abplanalp, D.Barošová, P.Bridenbaugh, J.Erhart, J.Fousek, P.Günter, J.Nosek, M.Šulc: Domain structures in PZN-8%PT and PMN-29%PT single crystals studied by scanning force microscopy, J.Appl.Phys. 91, 6 (2002) 3797-3805.
14. R.Zhang, B.Jiang, W.Cao: Single-domain properties of $0.67Pb(Mg_{1/3}Nb_{2/3})O_3$–$0.33PbTiO_3$ single crystals under electric field bias, Appl.Phys.Lett. 82, 5 (2003) 787-789.
15. Y.Yamashita, Y.Hosono, K.Harada, N.Ichinose: Effect of molecular mass B-site ions on electromechanical coupling factors of lead-based perovskite piezoelectric materials, Jpn.J.Appl.Phys. 39, Part 1, 9B (2000) 5593-5596.
16. R.E.Eitel, C.A.Randall, T.R.Shrout, P.W.Rehrig, W.Hackenberger, Seung-Eek Park: New High Temperature Morphotropic Phase Boundary Piezoelectrics Based on $Bi(Me)O_3$–$PbTiO_3$ Ceramics, Jpn. J. Appl. Phys. 40, Part 1, 10 (2001) 5999–6002.
17. Y.Yamashita, Y.Hosono, K.Harada, N.Yasuda: Present and Future of Piezoelectric Single Crystals and the Importance of B-Site Cations for High Piezoelectric Response, IEEE Trans. UFFC 49, 2 (2002) 184-192.
18. S.Zhang, C.A.Randall, T.R.Shrout: Recent Developments in High Curie Temperature Perovskite Single Crystals, IEEE Trans. UFFC 52, 4 (2005) 564-569.
19. S.Wada, S.-E.Park, L.E.Cross, T.R.Shrout: Engineered domain configuration in rhombohedral PZN-PT single crystals and their ferroelectric related properties, Ferroelectrics 221, 1-4 (1999) 147-155.
20. Z.-G.Ye, M.Dong: Morphotropic domain structures and phase transitions in relaxor-based piezo-/ferroelectric $(1-x)Pb(Mg_{1/3}Nb_{2/3})O_3$-$xPbTiO_3$ single crystals, J.Appl.Phys. 87, 5 (2000) 2312-2319.
21. S.-E. Park,T. R. Shrout: Ultrahigh strain and piezoelectric behaviour in relaxor-based ferroelectric single crystals, J.Appl.Phys. 82, 4 (1997) 1804-1811.
22. K.Takemura, M.Ozgul, V.Bornand, S.Trolier-McKinstry, C.A.Randall: Fatigue anisotropy in single crystal $Pb(Zn_{1/3}Nb_{2/3})O_3$–$PbTiO_3$, J.Appl.Phys. 88, 12 (2000) 7272-7277.

23. S.Wada, S.Suzuki, T.Noma, T.Suzuki, M.Osada, M.Kakihana, S.E.Park, L.E.Cross, T.R.Shrout: Enhanced piezoelectric property of barium titanate single crystals with engineered domain configurations, Jpn.J.Appl.Phys. 38, Part 1, 9B (1999) 5505-5511.
24. S. Wada, A. Seike, T.Tsurumi: Poling Treatment and Piezoelectric Properties of Potassium Niobate Ferroelectric Single Crystals, Jpn.J.Appl. Phys. 40, Part 1, 9B (2001) 5690–5697.
25. K.Nakamura, T.Tokiwa, Y.Kawamura: Domain structures in $KNbO_3$ crystals and their piezoelectric properties, J.Appl.Phys. 91, 11 (2002) 9272-9276.
26. R.E.Newnham, L.E.Cross: Secondary ferroics and domain-divided piezoelectrics, Ferroelectrics 10 (1976) 269-276.
27. V.D.Kugel, G.Rosenman, D. Shur: Piezoelectric properties of bidomain $LiNbO_3$ crystals, J. Appl. Phys. 78, 9 (1995) 5592-5596.
28. Y.-Y.Zhu, N.-B.Ming: Ultrasonic excitation and propagation in an acoustic superlattice, J.Appl.Phys. 72, 3 (1992) 904-914.
29. Y.-Y.Zhu, S.-N.Zhu, Y.-Q.Qin, N.-B.Ming: Further studies on ultrasonic excitation in an acoustic superlattice, J.Appl.Phys. 79, 5 (1996) 2221-2224.
30. N.B.Ming, J.F.Hong, D.Feng: The growth striations and ferroelectric domain structures in Czochralski-grown $LiNbO_3$ single crystals, J.Mater.Sci. 17 (1982) 1663-1670.
31. P.G.Schunemann, T.M.Pollak, Y.Yang, Y.-Y.Teng, C.Wong: Effects of feed material and annealing atmosphere on the properties of photorefractive barium titanate crystals, J. Opt. Soc. Am. B 5, 8 (1988) 1702-1710.
32. S.Ajimura, K.Tomomatsu, O.Nakao, A.Kurosaka, H.Tominaga, O.Fukuda: Photorefractive effect of $BaTiO_3$ single crystals grown in inert atmospheres, J. Opt. Soc. Am. B 9, 9 (1992) 1609-1613.
33. M.M.Hopkins, A.Miller: Preparation of poled, twin-free crystals of ferroelectric bismuth titanate, $Bi_4Ti_3O_{12}$, Ferroelectrics 1 (1970) 37-42.
34. S.Noge, T.Uno: Formation of Artificial Twinning Quartz Plate with *x*-axis Inversion Area by Laser Beam Irradiation, Jpn.J.Appl.Phys. 38, Part 1, 7A (1999) 4250–4253.
35. S.Noge, T.Uno: Twinning of a quartz plate at low temperature using a laser beam, Jpn.J.Appl.Phys. 39, Part 1, 5B (2000) 3056-3059.
36. G.Arlt, P.Sasko: Domain configuration and equilibrium size of domains in $BaTiO_3$ ceramics, J.Appl.Phys. 51, 9 (1980) 4956-4960.
37. Sang-Beom Kim, Doh-Yeon Kim: Stabilization and Memory of the Domain Structures in Barium Titanate Ceramics: Microstructural Observation, J. Am. Ceram. Soc. 83, 6 (2000) 1495–1498.
38. S.Wada, K.Yako, H.Kakemoto, J.Erhart, T.Tsurumi: Enhanced piezoelectric property of $BaTiO_3$ single crystals with the different domain sizes, Key Engineering Materials 269 (2004) 19-22.
39. J.Erhart: Theoretical calculation of the temperature dependence for the material coefficients of the domain-engineered ferroelectric crystals, Ferroelectrics 292 (2003) 71-81.
40. J.Erhart, S.Wada: Theoretical calculation of the resonant frequency temperature dependence for domain-engineered piezoelectric resonators, Materials Science and Engineering B 120 (2005) 175-180.
41. M.J.Haun, E.Furman, S.J.Jang, H.A. McKinstry, L.E.Cross: Thermodynamic theory of $PbTiO_3$, J.Appl.Phys. 62, 8 (1987) 3331-3338.
42. A. Schaefer, H. Schmitt, A. Dörr: Elastic and piezoelectric coefficients of TSSG barium titanate single crystals, Ferroelectrics 69 (1986) 253–266.
43. J.Erhart, W.Cao: Effective symmetry and physical properties of twinned perovskite ferroelectric single crystals, J. Mater. Res. 16, 2 (2001) 570-577.

44. J. Erhart, L. Rusin, L.Seifert: Resonant frequency temperature coefficients for the piezoelectric resonators working in various vibration modes, Journal of Electroceramics 19 (2007) 403-406.
45. IRE Standards on Piezoelectric Crystals: Measurements of Piezoelectric Ceramics, IEEE Std. 179-1961 (R1971).
46. IEEE Standard on Piezoelectricity, ANSI/IEEE Std. 176-1987
47. J.Erhart, M.Franclíková and L.Rusin: Piezoelectric resonators with engineered domain structures, Ferroelectrics 376, 1 (2008) 99-115.
48. J.Zelenka: *Piezoelectric resonators* (Elsevier, Amsterdam 1986).
49. A.H.Meitzler, H.M.O'Bryan, Jr., H.F.Tiersten: Definition and measurement of radial mode coupling factors in piezoelectric ceramic materials with large variations in Poisson's ratio, IEEE Trans. Sonics Ultrason. SU-20, 3 (1973) 233-239.
50. P.Půlpán, J.Erhart: Transformation ratio of "ring-dot" planar piezoelectric transformer, Sensors and Actuators A 140 (2007) 215-224.
51. P.Půlpán, J.Erhart, O.Štípek: Modelling of piezoelectric transformers, Ferroelectrics 351 (2007) 204-215.

Chapter 15
Non-Linear Behaviour of Piezoelectric Ceramics

Alfons Albareda, Rafel Pérez

15.1 Introduction

By their ferroelectric nature, ferro-piezoelectric ceramics have a non-linear behaviour. This nature is reinforced by their notably complex grain and domain structure, which leads to a less linear behaviour than is expected for a single crystal.

The dielectric, piezoelectric and elastic behaviour of a piezoceramic can first be described by its constitutive equations, which express a linear relation between the applied fields (either electric or stress field) and the deformation produced by these fields (either electric displacement or elastic strain). However, its behaviour is only linear when the applied fields are low enough. So, it is necessary to find the best way to relate the quantities E, D, T and S when the fields are too large. It is also necessary to search for the most suitable measurement methods in order to characterize the material under these new conditions [1].

The study of non-linear behaviour has a double objective: on the one hand, it is necessary to describe correctly the behaviour under the application of high signal in order to design high power devices adequately. On the other hand, it enables us to understand better the processes that take place inside the material by taking advantage of the fact that these processes are in some way manifested in the non-linear behaviour [2].

It is important for the user of the material to know its behaviour fully. It is known that when large fields are applied, the coefficients do not maintain their values since the permittivity values, the elastic compliance and the piezoelectric coefficient become larger, as do their corresponding losses. The application of a sinusoidal excitation produces no sinusoidal response; high order harmonics appear, and the application of two signals of different frequency may produce a

---

Applied Physics Department. Universitat Politècnica de Catalunya. 08034 Barcelona. Spain
+34934016086
+34934016090
alfons@fa.upc.edu rafel@fa.upc.edu

response whose frequency is equal to the sum or the difference of the former (intermodulation). When the excitation signal exceeds a certain level, the hysteresis phenomenon appears, and thus there is more than one response to a single excitation. This depends on the fields previously applied to the sample. Simultaneous application of a DC electric field or a DC stress can also alter the response to an AC excitation (tunability) [3, 4, 5].

However, high excitation fields not only produce non-linear effects, they also cause material heating which alters their characteristics, so that both effects are usually produced simultaneously. The excitation by short bursts enables the amplitude to be notably increased by minimizing heating, but without preventing the non-linear effects. This fact is often used in non-linear characterization, because it enables the strictly non-linear effects to be fully distinguished from the thermal ones [6].

Non-linear behaviour must be described by many more coefficients than the linear one, so it is expected that much more information can be obtained from them. Thus, non-linear analysis is a powerful tool for obtaining (or for completing) knowledge of some aspects of the complex internal dynamics of the ceramics. Taking into account that the non-linear effects are related to the domain wall motion, it provides the possibility of analyzing such behaviour [7, 8, 9, 10].

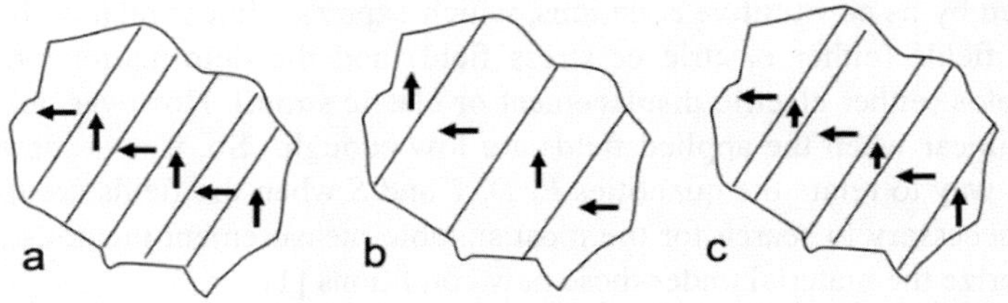

Fig. 15.1 The domain distribution **b** has the same ratio as **a** (in volume) but different wall extension while **c** has a different ratio than **a**, but the wall extension is identical.

The change in size and polarization of the crystallographic cell by the application of an electric field or a mechanical stress constitutes the so-called intrinsic effect. However, the major contribution often comes from the domain wall displacement, or extrinsic effect, which is more dependent on the amplitude of the field than the intrinsic effect.

While the intrinsic effect is manifested according to the volume of the domains oriented in each direction of the space, and therefore depends on the polarization of the whole material, the extrinsic effect depends on the extent of the domain walls and on their orientation [7, 11]. As an example, in the three configurations described in Fig. 15.1 it is expected that a) and b) must have the same intrinsic response, while a) and c) must have the same extrinsic response.

In a ceramic with a given composition, the addition of small quantities of strange ions, either of the same or of different valence than the ions they substitute, can produce a change in its linear and non-linear behaviour. On one hand, it can cause a phase change, which will fully modify its structure, and on the other it can modify the domain structure, dramatically changing the electrical conductivity. This modifies the speed at which the system returns to equilibrium. Finally, it can favour or hinder either the nucleation of new domains or the motion of their walls [9, 12].

The domain structure is rather complex and is in general far from being in a state of stable equilibrium. Rather, it is often in a metastable equilibrium, or out of equilibrium, but with a very slow evolution. This means that it is subject to a process of ageing, so that its properties change with time due to its slow evolution towards the equilibrium state, and also to fatigue processes, which occur when the application of intense fields changes the structure in a significant way [10, 13].

The action of a field or stress equal to or higher than the coercive field produces the ferroelectric switching, which consists in diverse phenomena of a non-linear nature, such as the annihilation and coalescence of existing domains or the nucleation of new ones. The application of a sub-coercive field produces phenomena that are moderately non-linear, such as the displacement of domain walls, which is usually accompanied by the creation of mechanical stress and space charge, or the interaction between the walls and impurities, vacancies, defects and lattice dislocations [14, 15].

15.1.1 Methods for Non-Linear Characterization

In order to characterize the materials by focusing on their non-linear aspects, different methods can be considered depending on the level of the excitation signal. Moreover, due to their piezoelectric nature, there are diverse aspects that can be analyzed: dielectric, elastic, and direct and converse piezoelectric behaviour.

For the lowest excitation level, their behaviour is practically linear. So, the coefficients barely undergo change, and the most notable effect is the generation of high order harmonics. Due to the problems posed by the non-linearity of the experimental system itself, intermodulation techniques are frequently preferred [3, 16], in which two different frequencies are applied and a signal is obtained whose frequency is related to them.

For higher amplitudes, it is possible to analyze how the coefficient changes with the amplitude, and if the noise is low enough, to analyze its evolution with time [10]. If the excitation is electrical, the dependence of the dielectric constant and the converse piezoelectric coefficient can be obtained in terms of the amplitude [17, 18, 19], while a mechanical excitation enables the direct piezoelectric coefficient to be obtained [20, 21]. In both cases, it is advisable to work well below the first resonant frequency in order to guarantee that no other fields except the external one are present, and that the field is uniform and known.

In these cases, the Rayleigh model can often be applied [22], and it fits better or worse depending on the type of material, for which reason it is often taken as a reference. According to this model, the increment of the losses is proportional to the increment of the coefficients ε or d, and these are proportional to the amplitude of the applied field.

These measurements can be complemented by the simultaneous application of a bias field. Although the superposition of a DC electric field to the AC excitation is an option that allows the information about the material to be extended, the AC mechanical excitation must be accompanied by a compression stress greater than the amplitude, because tensile forces are not usually allowed. It is also possible to apply a DC bias stress superposed to an AC electric field [21].

When the amplitude of the applied electric field is high enough, the ferroelectric ceramic switches its polarization and then appear some processes that at low fields are rare or inexistent. By measuring the electrical charge, it is possible to obtain the hysteresis loop, and we can also obtain the piezoelectric loop (butterfly) if the changes in the sample size can be measured. In this case, the dissipated energy is proportional to the number of loops completed, so a low frequency must be applied in order to avoid sample heating (less than 1 Hz).

Due to the high quality factor of the ceramics, the resonance phenomena can be used to apply a great mechanical stress without applying mechanical excitation to the sample. In this case the frequency cannot be freely chosen, since it is determined by the size of the sample. The higher the stress amplitude, the greater the compliance and the losses, so that resonance frequency falls and the bandwidth become broader, facts that can be measured by observing the change in the electrical impedance. This measurement method, which is relatively easy to apply, enables the non-linear elastic behaviour to be analyzed, although interpretation involves two difficulties: the indirect evaluation of the stress and the fact that it is not uniform.

In this chapter, we analyze the non-linear behaviour of the piezoelectric ceramics, whether under electrical excitation (dielectric and converse piezoelectric effects), under mechanical excitation (direct piezoelectric effect), or by taking advantage of the resonance phenomenon in order to apply high stress (elastic effect). The last part of the chapter is devoted to the phenomenological models that allow us to interpret some of the quoted behaviours, and finally, some theoretical considerations are made about the appearance on the non-linearity, the domain structure and the role of the dopants.

15.2 Dielectric and Converse Piezoelectric Behaviour

The non-linear dielectric behaviour can be obtained by observing the electrical charge induced at the electrodes when an AC electric field of given amplitude is applied. The range of amplitudes and the frequency must be decided *a priori*. The frequency must be somewhat lower than the lowest resonant frequency in order to avoid the stress caused by inertia. As a consequence of the application of the field, the sample will suffer a mechanical deformation due to the converse piezoelectric effect. So, its edges will move with a certain velocity, which can be measured. It is therefore advisable to carry out both measurements simultaneously.

These analyses can be performed at different levels: at an elementary level the mean permittivity can be obtained as a ratio between the amplitudes of the main frequency of the electrical displacement D and the electrical field E [19, 23]. More detailed information can be obtained by also looking at the amplitude and phase of the high order harmonics generated. [24]. Finally, it is possible to register the evolution of the fields D and E along the time, in order to obtain $D(E)$ along a complete loop [10, 23]. In any case, such measurements must be performed for different amplitudes E_0.

At the first level, it is convenient to represent either the real ε' as the imaginary ε'' parts of the permittivity as a function of the amplitude of the applied field. If both functions are similar, the plot $\varepsilon''(\varepsilon')$ must be linear. When the Rayleigh model can be applied, both relations $\varepsilon'(E_0)$ and $\varepsilon''(E_0)$ must be linear, and $\varepsilon''(\varepsilon')$ is plotted as a straight line of slope 0,43 [1, 19]. A method for deeply analyzing its behaviour consists in measuring the second and third harmonic of the function $D(t)$ with their related phases. By assuming that the applied field has the form $E_0 \cdot \sin(\omega t)$, the measurement of the components $D_2 \cdot \sin(2\omega t)$, $D'_2 \cdot \cos(2\omega t)$, $D_3 \cdot \sin(3\omega t)$, $D'_3 \cdot \cos(3\omega t)$ indicates the dependence between the field D and the powers E^2 or E^3 of the instantaneous field E applied [25].

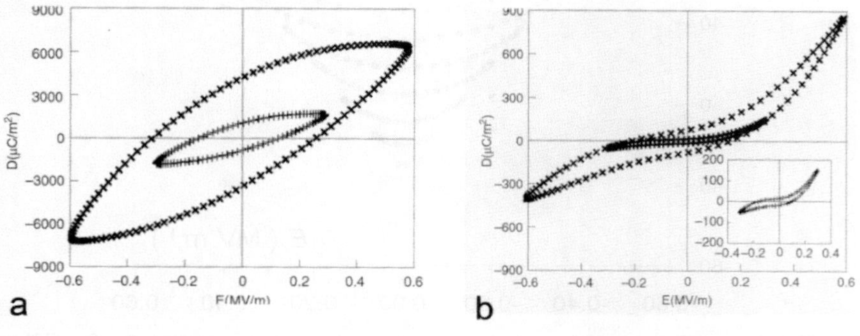

Fig. 15.2 Minor hysteresis loops after removing the linear part of D, for different field amplitudes. **a** Soft PZT. **b** Hard PZT.

A hysteresis loop is obtained when $D(E)$ is plotted if the amplitude is greater than the coercive field; however, with smaller fields, minor loops can be obtained, whose shape tends to an ellipse when the amplitude falls to zero. These loops can be modified by subtracting the linear part of D (the field D expected if its behaviour were the same as at null field). The loops obtained reflect the non-linear behaviour and appear to be rather different in soft than in hard materials, showing that the Rayleigh model is only valid in the former (Fig. 15.2).

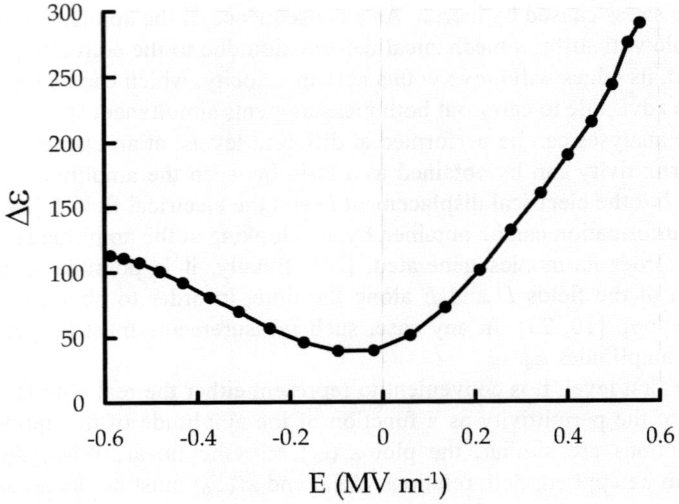

Fig. 15.3 Non-linear increment ε_{NL} of the dielectric constant for a hard ceramic PZT4.

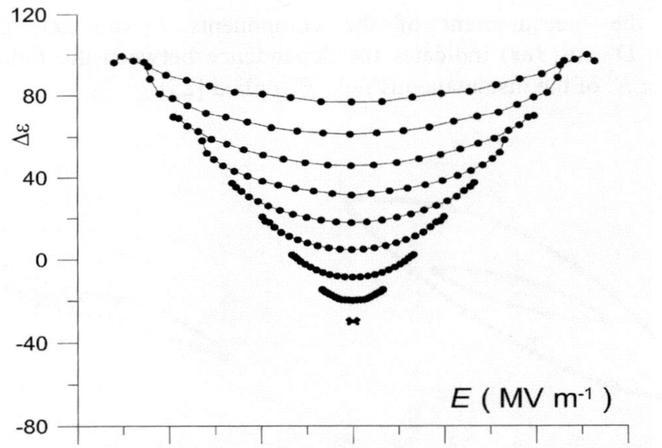

Fig. 15.4 Symmetrical part of non-linear increment of dielectric constant ε_{NL} for different amplitudes (soft PZT). It depends both on the amplitude and on the instantaneous field.

Due to the hysteresis, there are two values of D for each value of E. Thus, the mean value of both can be computed and the instantaneous permittivity defined as the derivative of this mean value with respect to the instantaneous field E. In view of the result, which gives the dependence of the non-linear permittivity in terms of the instantaneous field, we see that it depends not only on the instantaneous field, but also on its amplitude (Fig. 15.3, Fig. 15.4).

15.2.1 Experimental Method

In order to analyze the non-linear dielectric behaviour at the elementary level, the Schering bridge can be used, which gives directly both the real and imaginary parts of the dielectric constants. Signals of different amplitude should be applied over a short period in order to avoid overheating [26].

Notwithstanding, the Sawyer-Tower bridge can be used to give the electrical charge on the electrodes, so the value of D is computed and the shape of the loop obtained. This can also be done by measuring the current I, so that the charge is then obtained by numerical integration.

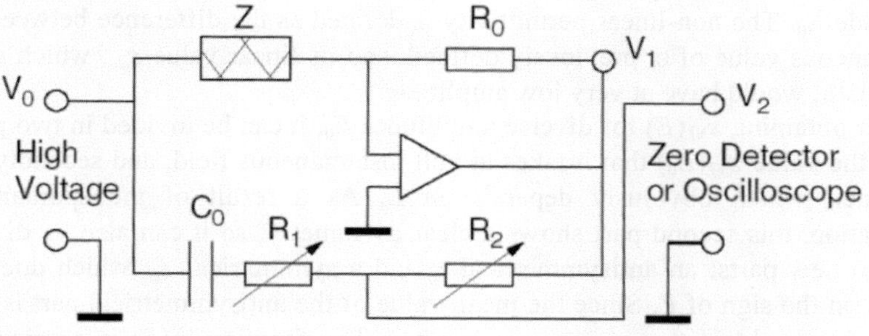

Fig. 15.5 Capacitance bridge for the measurement of non-linear permittivity. Once it has been balanced at low level signal, the output V_2 allows us to reconstruct the non-linear current.

These methods do not have sufficient resolution if the non-linear effect is low, so it is necessary to compensate for the linear part of the current by exclusively showing the non-linear behaviour. This can be achieved by the use of a modified compensation capacitance bridge [23]. The bridge, shown in Fig. 15.5, adjusted to zero while a low amplitude signal is applied, and since the amplitude is low the sample behaviour is linear. A high voltage is then applied without re-adjusting R_1 and R_2 values. So, thanks to the large inequality between the arms, the signal error accurately describes the value of the non-linear current (the increment of the current due to the non-linearity) as a time function.

It is necessary to work in a pulsed mode rather than a continuous mode in order to avoid sample heating. A set of some consecutive oscillations is applied, from which we must not consider the first three because the stationary state has yet to be attained, and the others are averaged. Although the burst mode should be used at high voltage, it is not recommended at low voltage because in continuous mode a lock-in amplifier can be used to distinguish easily between the signal and the noise.

It is recommended to use disc-shaped samples with a diameter of the order of 1 cm. By doing so, the resonant frequency is of the order of 200 kHz. So, measurements can be performed up to 10 kHz without trouble.

By attaching the disc at its centre, the edge moves with some radial speed, which can be measured by Laser interferometry. This procedure enables us to obtain the mechanical radial displacement of the edge and the mean value of the strain S perpendicular to the axis, so the converse piezoelectric coefficient d_{31} can be measured.

The mean value of D_{NL} of the sample is obtained by integration of such a signal, and it can be plotted versus field E. For each value of E there are two different values of D_{NL}, so we can compute the mean value $\check{D}$ and ΔD, which is equal to half the difference of such a pair of values. We define ε_{NL} as the derivative of $\check{D}$ with respect to E, and we analyze its dependence on E and on the amplitude E_0. The non-linear permittivity is defined as the difference between the instantaneous value of ε, previously defined, and its linear value ε_{lin}, which is the value that it would have at very low amplitude.

After obtaining $\varepsilon_{NL}(E)$ for diverse amplitudes E_0, it can be divided in two parts: firstly, the value $\varepsilon_{NL}(E_0)$ that it takes at null instantaneous field, and secondly, the remainder, which obviously depends on E. As a result of the spontaneous polarization, this second part shows a clear asymmetry, so it can also be divided into two new parts: an antisymmetrical ε_γ and a symmetrical ε_β, which does not depend on the sign of E. Since the mean value of the antisymmetrical part is null, $\langle\varepsilon_\gamma\rangle = 0$, we obtain that $\langle\varepsilon_{NL}\rangle = \varepsilon_\alpha + \langle\varepsilon_\beta\rangle$. The function $\varepsilon_\beta(E)$ is obtained by averaging the values of ε_{NL} that correspond to $-E$ and $+E$:

$$\varepsilon_\beta(E) = \tfrac{1}{2}\left(\varepsilon_{NL}(E) + \varepsilon_{NL}(-E)\right) - \varepsilon_\alpha \qquad (1)$$

So, the mean overall value of ε is the sum of three terms: ε_L, ε_α and $\langle\varepsilon_\beta\rangle$.

15.2.2 Results Obtained

The dependence of ε_{NL} and ε'' on the amplitude E_0 reveals a simple behaviour in soft PZT ceramics [27]: a linear dependence is observed in both cases, so the plot of ε'' versus ε_{NL} is also a straight line whose slope is approximately equal to $0.42 = 4/3\pi$. This is exactly the behaviour predicted by the Rayleigh model. This fact can also be observed in the hysteresis loop, which in this case consists of two parabolic sections. There are discrepancies only at high amplitudes, and these can be attributed to the appearance of other mechanisms that probably produce the ferroelectric switching when the field is higher than its coercive value. Although it is assumed that non-linear behaviour begins at a threshold field, experimental data shows that it has a very low value.

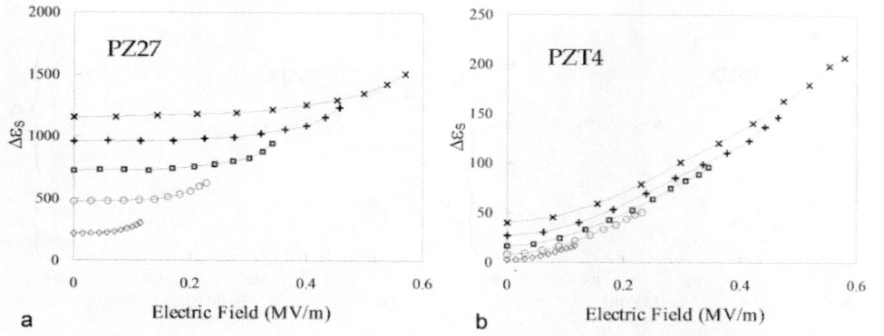

Fig. 15.6 Value of $\Delta\varepsilon$ (symmetrical part of ε_{NL}) versus the electrical field. **a** Soft PZT. **b** Hard PZT.

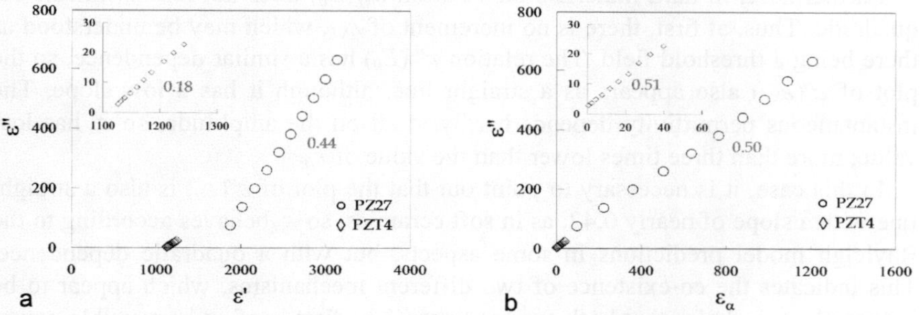

Fig. 15.7 Plot of imaginary part of dielectric constant **a** versus real part, **b** versus ε_{α}. Note that in **b** the slopes are similar for both materials.

Although either the real ε′ or the imaginary part ε″ of the permittivity rise with the increase of the amplitude [18, 19, 23], it is observed that this effect is much greater in soft than in hard ceramics, and while the dependence is linear in the former case it becomes rather quadratic in the latter. However, in both cases the dependence of $\varepsilon''(E_0)$ and $\varepsilon'(E_0)$ are of the same kind (Fig. 15.6). This fact is clearly shown when $\varepsilon''(\varepsilon')$ is plotted, and a straight line is formed in both cases (Fig. 15.7).

A further notable difference is observed between these materials: while in soft ones ε_{α} is greater than ε_{β}, in hard ones just the opposite occurs (Fig. 15.8).

With regard to the instantaneous permittivity of the soft materials, further details may be noted: For given amplitude, the value of ε_{NL} depends slightly on the instantaneous field, and this is only noticeable when the field values are extreme. Then, the value of $<\varepsilon_{\beta}>$ is small, although it is not completely null, as predicted by the Rayleigh model.

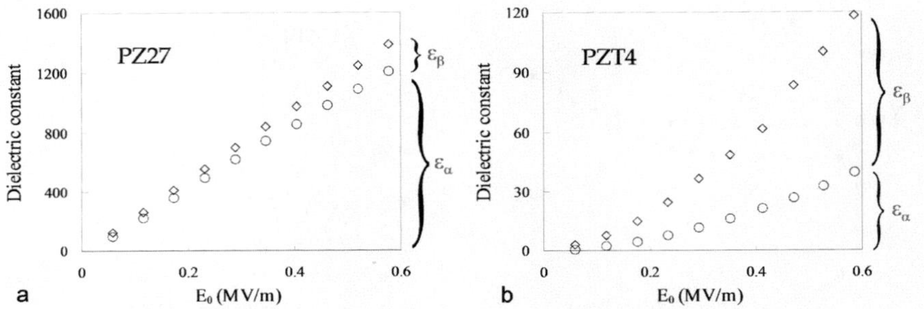

Fig. 15.8 Plot of ε_α and ($\varepsilon_\alpha + \varepsilon_\beta$) versus the amplitude of the field. **a** Soft PZT (the relation is linear, and $\varepsilon_\alpha > \varepsilon_\beta$). **b** Hard PZT (the relation is quadratic, and $\varepsilon_\alpha < \varepsilon_\beta$).

Furthermore, in hard materials the relation $\varepsilon_{NL}(E_0)$ does not remain linear, but quadratic. Thus, at first, there is no increment of ε_{NL}, which may be understood as there being a threshold field. The relation $\varepsilon''(E_0)$ has a similar dependence, so the plot of $\varepsilon''(\varepsilon_{NL})$ also appears as a straight line, although it has a low slope. The instantaneous permittivity depends hardly at all on the amplitude, so ε_α has low value; more than three times lower than the value of ε_β.

In this case, it is necessary to point out that the plot of $\varepsilon''(\varepsilon_\alpha)$ is also a straight line with a slope of nearly 0.42, as in soft ceramics, so ε_α behaves according to the Rayleigh model predictions in some aspects, but with a quadratic dependence. This indicates the co-existence of two different mechanisms, which appear to be independent, and also that both are not linear. The first is of an irreversible nature and depends on E_0, and is responsible for ε_α and ε''; while the second one, which is reversible, depends on E and is expressed through ε_β.

The measurement of $d(E_0)$ at different amplitudes shows that its dependence on E is similar to that of $\varepsilon_{NL}(E_0)$, in any case. So, it is reasonable to plot d versus ε_{NL}, and a linear relation is expected in both soft and hard samples. It transpires that not only a straight line is obtained for all amplitudes, but that this line crosses near the origin, and that it has the same slope for all the materials derived from PZT, whether they are hard or soft, providing they are reasonably well poled [28].

Although dopants will strongly condition both linear and non-linear behaviour, it is clearly shown that the slope of $d(\varepsilon')$ (Fig. 15.9) does not depend on the dopants. If those effects are produced by the wall motion, and thus by the switch of a certain volume of the material, it is reasonable to expect that this ratio will be equal to the ratio between the spontaneous mechanical strain and polarization.

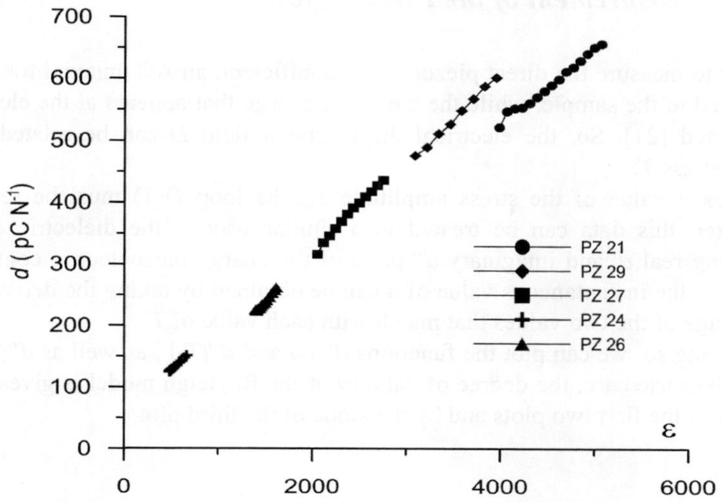

Fig. 15.9 Experimental plot of converse piezoelectric coefficient versus dielectric constant, for different well poled PZT samples. Note that the slope is similar in both soft and hard ceramics.

15.2.3 Anisotropy

In a poled ceramic, the non-linear dielectric effect depends on the direction of the applied field, which can be either parallel with or perpendicular to the poling direction. As described in [29], in rhombohedral ceramics it is observed that the effect is more pronounced in the normal direction than in the parallel one, mainly in acceptor doped ceramics. On the other hand, it is interesting to compare such results with those obtained over non-poled ceramics.

15.3 Direct Piezoelectric Behaviour

In dielectric measurements, the action of an electrical field over the wall motion is analyzed. However, direct piezoelectric measurements enable us to ascertain the effect of external stress [30]. It is necessary to take into account that 180° walls cannot be moved by applying stress, and that non-180° walls will only be moved by some shear stress, as will be remarked later.

Moreover, the stress T is described by a second rank tensor, while the electric field is described by a first rank tensor. It is for this reason that the effects caused by the lack of orientation of a grain with respect to the sample will be somewhat different in both cases.

15.3.1 Measurement of the Direct Effect

In order to measure the direct piezoelectric coefficient, an AC uniaxial force must be applied to the sample, while the electrical charge that appears at the electrodes is collected [21]. So, the electrical displacement field D can be related to the applied stress T.

For each value of the stress amplitude T_0, the loop $D(T)$ must be recorded. Thereafter, this data can be treated in a similar way to the dielectric data by computing real d' and imaginary d'' parts of the charge piezoelectric coefficient. Moreover, the instantaneous value of d can be obtained by taking the derivative of the average of the two values that match with each value of T.

By doing so, we can plot the functions $d'(T_0)$ and $d''(T_0)$, as well as $d''(d')$. As in the dielectric case, the degree of validity of the Rayleigh model is given by the linearity of the first two plots and by the slope of the third plot.

15.3.2 Experimental Method

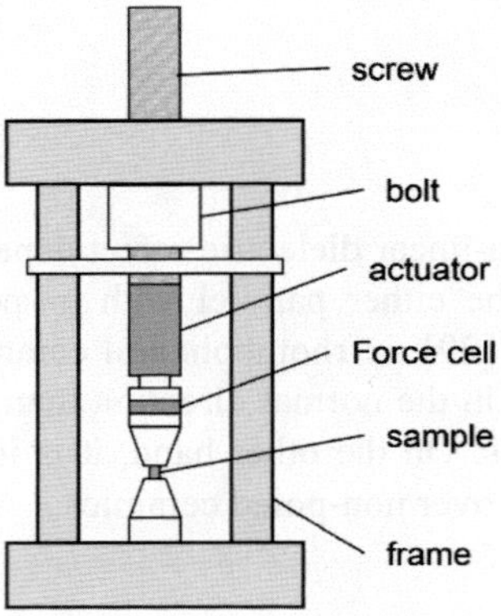

Fig. 15.10 Experimental set-up for measuring non-linear direct piezoelectric coefficient.

The best way to perform such a measurement consists in taking a solid frame to which some elements are attached: the sample, a charge cell for measuring the force, and a piezoelectric transducer for applying it. As the maximum displacement of the transducer is always small (some tenths of microns), the frame must be rigid enough to obtain a sufficiently large force. A screw is added in order to fasten all the system and to give an initial pre-stress (Fig. 15.10).

Both the sample and the charge cell need their own charge integrators in order to measure the respective variations of charge and force.

Since the sample cannot be subjected to tensile stress, a DC bias stress with a value greater than the amplitude of the AC stress must be applied. As a consequence, the equilibrium of the sample is altered, so some internal stress and space charges appear, which take some time to relax and disappear. Furthermore, the DC field can change the AC response quite significantly, because they have a similar order of magnitude.

When different DC bias T_B are applied, the value of the coefficient d at low amplitudes decreases as the compression rises, and tends to stabilize for a stress higher than 30 MPa. It is therefore necessary to control both variables. In order to see the influence of the AC amplitude on d, different amplitudes must be applied with the same bias, and measures must only be taken some time after the continuous stress has been applied.

15.3.3 Results

Unlike the dielectric behaviour, a linear dependence of d on the stress amplitude T_0 is found in both soft and hard PZT materials [31]. So, in this respect, the Rayleigh model is achieved in a universal mode.

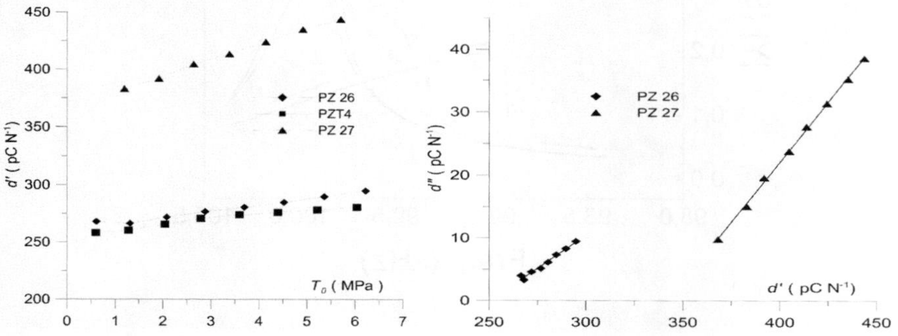

Fig. 15.11 a Direct piezoelectric coefficient versus field amplitude. **b** Imaginary versus real part of the piezoelectric coefficient. Note that the slope is 0.4 or 0.2 depending on whether the material is soft or hard.

In the analysis of the relation between d'' and d', a linear relation is found, but with a different slope. While the soft ceramics show a slope of 0.37 near the theoretical value 0.42, hard ceramics have a rather small slope, of the order of 0.2 (Fig. 15.11).

The dependence of d on the instantaneous stress is far less than in the dielectric case, which is consistent with the fact that the behaviour is closer to the Rayleigh model.

15.4 Resonance Measurements

An alteration of linear resonator behaviour occurs and non-linear phenomena appear when piezoelectric devices are used in resonance with high excitation signal, as happens in the power devices [32, 33, 34, 35, 36, 37, 38]. In many cases, this behaviour alters the normal operation of devices. Thus, it is important to have a non-linear characterization that can foresee what will occur.

This non-linearity is due to the great strains and stresses that the resonator undergoes in the neighbourhood of the resonance, and for this reason we will deal in general with non-linear elastic behaviour. This does not rule out the existence of other piezoelectric and dielectric non-linear behaviours, which can be superposed to the non-linear elastic one.

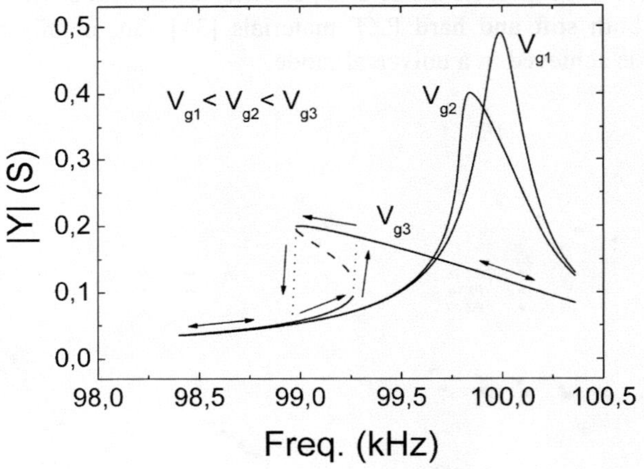

Fig. 15.12 Admittance modulus representation near resonance for different input levels. Frequency hysteresis at high level amplitude.

Among the most important non-linear elastic alterations that may occur in the resonance, the following phenomena that appear when the excitation signal is increased may be emphasized (Fig. 15.12):

- Increase of mechanical losses.
- Amplitude-frequency shift effect.
- Appearance of frequency hysteresis when a frequency sweep at constant voltage is applied.
- Harmonic generation at frequencies different to the excitation signal.

If these non-linear properties of the materials and the devices are known accurately, the operating limits of the resonators can be determined. It is also important to determine when the frequency hysteresis takes place, because sweepings in decreasing frequencies must be carried out so that the resonance can always be reached [38].

To carry out the material non-linear characterization, it is necessary to take piezoelectric resonators with the suitable forms to be able to characterize the non-linearity of the relevant elastic tensorial coefficients: s_{33}, s_{11}, etc. Generally, it is assumed that the constitutive equations are valid, but it is accepted that the coefficients cease to be constant and become a function of the excitation level amplitude.

An important factor in non-linear behaviour characterization of piezoelectric resonators is the prevention of increases in the sample temperature when the measurements are made at high level signal. Both phenomena can be superposed: non-linearity and thermal effects. In order to prevent this, it is necessary to design an appropriate measurement method.

The best way to avoid the overheating of the resonators is by the use of burst signals that are more limited in time. The advantages of this characterization system with burst excitation will be analyzed.

Firstly, the different resonance non-linear measurement and characterization methods are presented, and later measurements based on burst signals are analyzed. The method described is useful for the characterization of those non-linear coefficients corresponding to the vibration mode used, but the non-linear characterization developed for a resonator mode can be generalized to any other vibration mode.

The use of the motional impedance plane is also described: it provides a better understanding of each measurement method, as well as the sweep at constant frequency and the hysteresis phenomenon.

15.4.1 Resonance at High-Level: Measurement Methods

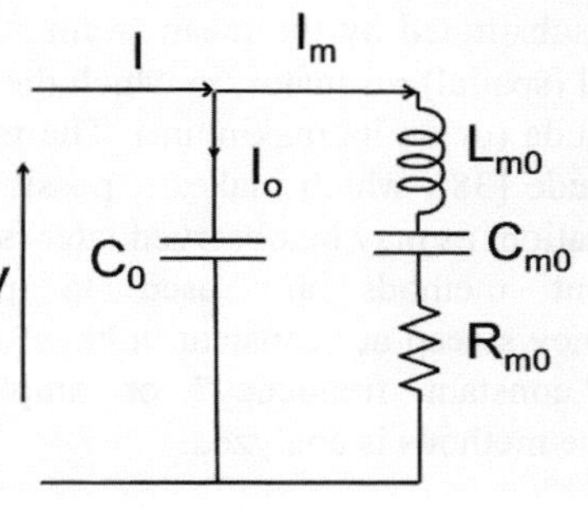

Fig. 15.13 Equivalent electric circuit of a resonator.

The most usual resonator non-linear characterization consists in the measurement of the increase in mechanical losses, $\Delta tan\delta_m$, and the resonance frequency decrease$\Delta\omega_s$, versus the amplitude of the excitation signal [38, 39]. This characterization is based on a lumped model, as in the circuit shown in Fig. 15.13: a motional branch L_{m0}, C_{m0}, R_{m0} (subindex 0 is used for low signal, or linear, magnitudes) in parallel with the electric branch C_0. When a high excitation level is used, the new equivalent circuit changes: the new discrete elements of the motional branch become L_m, C_m, R_m.

A way of obtaining this characterization is by the resistance measurement R (or inverse of the maximum admittance modulus $|Y_{max}|$) at resonance, which is proportional to the losses, and the reactance variation measurement, ΔX, due to the signal level increase (Fig. 15.13). This reactance increase is proportional to the resonance angular frequency decrease $\Delta\omega_s$:

$$\frac{\Delta R_m}{\overline{Z}} = \Delta \tan \delta_m = \Delta(\frac{1}{Q_m}) = \Psi(I_m), \tag{2}$$

$$\frac{\Delta X_m}{\overline{Z}} = -2\frac{\Delta\omega_s}{\omega_{s0}} = \Psi'(I_m), \tag{3}$$

$$\text{where } \overline{Z} = L_{m0}\omega_{s0}, \tag{4}$$

where L_{m0} is the motional inductance, from the low signal equivalent circuit. ω_s and ω_{s0} are the resonant frequencies at high and low signal level. Q_m is the mechanical quality factor. Ψ, Ψ' are functions of the motional current I_m. We assume that these non-linear resistance and reactance correspond, in the equivalent circuit, to the new elements L_m, C_m, R_m, with non-linear values that depend on the motional current I_m.

In these equations, the mechanical losses and the resonant frequency variations are related with the excitation signal level through the motional current I_m. The motional current is proportional to the spatial mean strain $<S>$ that exists in the resonator, and is the cause of the non-linear behaviour of these frequencies. When the motional current is substituted by the mean strain $<S>$, the lumped model is also substituted by a real (spatial) resonator, in which the non-linearity depends on the average of a magnitude (or on its maximum). The mean strain $<S>$ could be the independent magnitude [38], which makes it possible to obtain a good non-linear elastic characterization, as may be observed later (see Sect. 15.4.3.1).

Several measurement methods are used to perform the non-linear characterization: frequency sweep at "constant voltage", at "constant current", or amplitude sweep at "constant frequency" or amplitude sweep in "strict resonance". Each of these methods is analyzed.

15.4.1.1 *Frequency Sweep at Constant Voltage*

The standard system for carrying out the non-linear characterization is the use of an impedance analyzer making frequency sweeps at different voltage amplitude levels, obtaining for each level the resistance value at resonance and the resonant frequency (Fig. 15.12). This is the measurement system at *"constant voltage"*. Both the mechanical losses values $\tan\delta_m$, and the resonant frequency f_s, are obtained at each voltage amplitude level of the applied signal [26, 35, 37, 38, 39].

The non-linear effects are due to the high strains and stresses that appear at resonance, so it is necessary to carry out the electrical current measurement in order to obtain the motional current I_m and the mean strain $<S>$, for a radial disc resonator $<S>=<S_r+S_\theta>$:

$$<D_m>=<D_3>-\varepsilon_{33}^P E_3 = e_{31}^P <S>=\frac{I_m}{j\omega a}. \tag{5}$$

For other vibration modes, thickness, length extensional and shear, the corresponding tensorial coefficients must be considered. This measurement system is used to study the elastic non-linearities, and the characterization obtained shows that the non-linear contributions, both to the mechanical losses and to resonant frequency, depend exclusively on the motional current or mean strain $<S>$.

One of the problems of this method is that in these measurements one complete frequency sweep is carried out to obtain only one point on the representation of the non-linear characterization curve (Eqs. 2, 3). Therefore, it is necessary to make as many frequency sweeps as points on this curve, with the corresponding temperature increase. It is also necessary to make the frequency sweep only in decreasing frequencies, in order to ensure that the resonance is always reached, even if the hysteresis phenomenon appears.

15.4.1.2 *Frequency Sweep at Constant Current*

Hirose [40, 41] proposed carrying out the measurements at *"constant motional current"* [39]. This author uses a feedback electronic system that generates a constant current signal, after compensating the electrical capacitance C_0 (Fig. 15.13), to ensure that the motional current flowing through the resonator remains constant. The main advantage of this system is that the frequency hysteresis disappears, enabling frequency sweep at increasing or decreasing frequencies to be made. Furthermore, these measurements show impedance resonance $Z(\omega)$, $Y(\omega)$, curves that are symmetric in relation to the resonant frequency, even at high level.

A drawback of this system is the use of a rather sophisticated, non-standard generator. However, the main problem of this measurement method is that the power applied to the resonator is always very high. For all the frequencies of a sweep, the power applied is the same as that used only at resonance f_s in the previous method.

15.4.1.3 Amplitude Sweep at Constant Frequency

A further measurement method developed in [37, 38] consists in making an amplitude sweep at *"constant frequency"*. The measures of resistance and reactance increase, for the non-linear characterization depend only on the motional current and not on the frequency. In this system, an amplitude sweep is carried out for a fixed frequency near the resonance, and the measurements of the resistance R, reactance X and current I are made in order to obtain the resonator non-linear characterization (Eqs. 2, 3).

As pointed out in Sect. 15.4.3.1, the curves $R(I_m)$ and $\Delta X(I_m)$ are frequency independent. Thus, it is possible to obtain the non-linear laws by this method. Therefore, in the case of frequency independent laws, only one amplitude sweep is necessary to complete the non-linear characterization.

The measures can be made simultaneously with the vibration velocity measurements v, on a resonator characteristic point, for example on an edge point, where this velocity v is proportional to the mean strain $<S>$. The velocity measurement can easily be made by using a laser vibrometer. The vibration velocity measurement informs us directly of the mean strain $<S>$, making the non-linear measurement of piezoelectric coefficients d, e possible. For a disc of radius ρ, in radial resonance:

$$< S >=< S_r + S_\theta >= \frac{2u(\rho)}{\rho} = \frac{2v(\rho)}{\rho\omega}. \tag{6}$$

15.4.1.4 Amplitude Sweep at Constant Motional Reactance. Strict Resonance

A final improvement in the measurement system consists in making the amplitude sweep at constant motional reactance, in the resonance X_m=0, instead of at constant frequency. To do this, it is necessary to change the frequency and the amplitude in order to ensure the X_m=constant condition.

The main advantage of this system is that no hypothesis about the non-linear behaviour being independent from frequency is necessary. It also improves the characterization possibilities and provides a better comparison with other classical non-linear measurement systems.

In this case, it is necessary to know the electric capacitance C_0^P in order to calculate the electric branch admittance Y_0 (it is assumed that this electric admittance is independent of the amplitude level in resonance). For each measure, the impedance $\boldsymbol{Z}$ is calculated as well as the motional impedance $\boldsymbol{Z_m}$. This measurement system is detailed in [42]. For each frequency, it is necessary to obtain measurements points by increasing the excitation level, with the motional reactance negative and positive (Fig. 15.14). An interpolation between these measurements points enables us to obtain for each frequency the amplitude level corresponding to the resonance, $X_m=0$.

In this process, it is necessary to prevent the hysteresis phenomenon, because the interpolation would otherwise be incorrect.

As a conclusion about the different measurement systems, it is important to observe that the constant current method has the following disadvantage: the overheating of the resonators due to the high power used. The constant voltage method is not useful for characterizing non-linearity, since it requires multiple frequency sweeps and also because the independent variable, the current, changes throughout each sweep. Finally, the amplitude sweep at constant frequency, or in the strict resonance, is a better method for obtaining non-linear characterization.

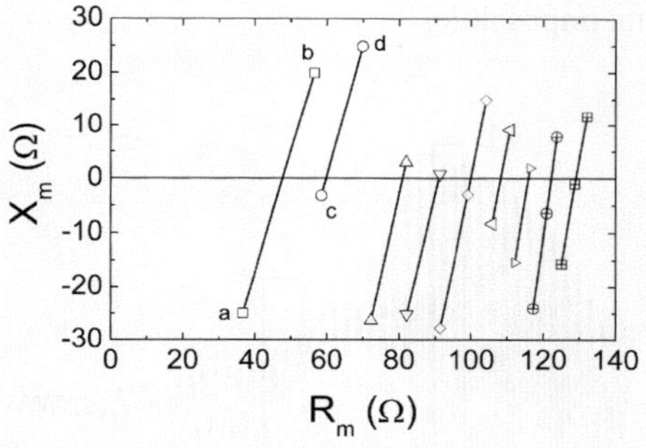

Fig. 15.14 Measurements points in the motional impedance plane for obtaining the strict resonance values at X_m=0 by interpolation.

15.4.2 Burst Measurements

As previously stated, an important factor in non-linear characterization at high power is taking the measurements while preventing resonator overheating [35]. The best solution is the use of burst signals, since this prevents temperature effects and is perfectly compatible with the use of an oscilloscope for making these measurements, as well as for measuring the vibration velocity at a point of resonator by a laser vibrometer.

To this end, the first option is to use a burst with a high signal level and to take the measurements in the free response at the transient state produced in the tail of the burst, when the excitation signal is finished (Fig. 15.15a) [6, 43, 44, 45, 46]. From the decay curve, the time constant $\tau(I)$ is calculated from the enveloping amplitude signal for different instants of the burst tail. The oscillation period T and frequency f at the same instants are also calculated. From all these measurements, the losses $tan\delta_m(I)$, and shift frequency $\Delta f/f_0\ (I)$ (Eq. 3) versus the current level can be obtained by using a lumped model:

$$\tan\delta_m = \frac{1}{Q_m} = \frac{1}{2\pi f\tau} \ . \qquad (7)$$

This measurement system can be completed by velocity measures with a laser vibrometer *v(I)*.

Measurements of this type have been used successfully [45, 47] to characterize materials with high mechanical quality factor by working at the first resonance mode. However, when it is used to characterize 1-3 composites [6, 43, 44], serious measurement difficulties are encountered. Indeed, in this case it is easy to find several resonant modes, with close frequency values. When the resonator is in the free oscillation interval, a beat phenomenon may easily occur between the different frequencies, because several vibration modes are excited. These beat effects, which are reflected in the enveloping signal, make the correct time constant measurements impossible.

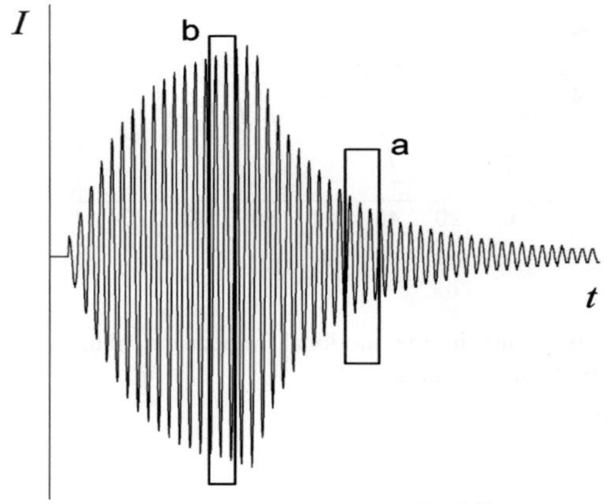

Fig. 15.15 Burst signal current. **a** Free vibrations in the tail, **b** steady-state vibrations before the end of the burst excitation.

15.4.2.1 *Steady-State Vibrations*

The solution for preventing thermal effects on the resonators is the use of burst signal excitation, but by carrying out the measurements in the time interval of the steady-state vibration zone (Fig. 15.15b). To this end, after the beginning of the burst excitation, a prudent delay is left to ensure that the transient state is finished and that the steady-state is achieved. At this moment, the signals voltage *V(t)*, current *I(t)* and velocity *v(t)*, are captured for a cycle number *N*, just before the end of the burst excitation signal. A high *N* value ensures accuracy by an average taken at the excitation frequency.

This method prevents noise and obtains the relative phase between signals: fundamental harmonic component measurements similar to those carried out in a lock-in amplifier are made by synchronous detection. With this acquisition data system, it is also possible to prevent the beating phenomenon. The experimental system is described in [42].

These results constitute the real and imaginary part of the complex voltage ***V***, the current ***I*** and the velocity ***v***. This method allows us to construct the complex impedance ***Z***=***V/I***=*R*+*jX*. In the first step, the non-linear behaviour (Eqs. 2, 3) can be obtained from the dependences *R(I)*, *X(I)* and *v(I)*.

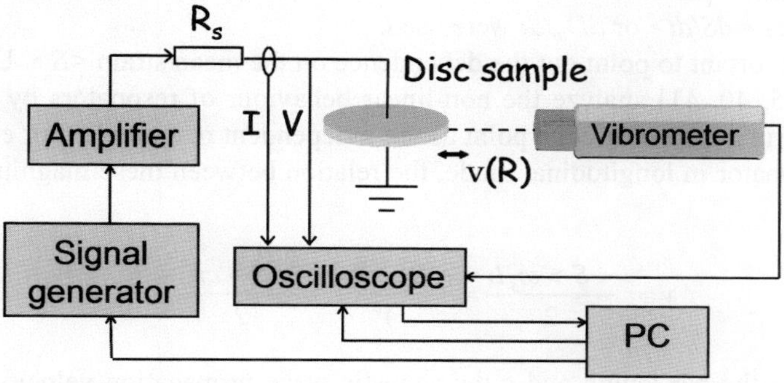

Fig. 15.16 Experimental block diagram.

Fig. 15.16 shows the experimental system. A digital oscilloscope acquire the time dependent signals of voltage *V(t)*, current *I(t)* (by a current probe) and velocity *v(t)*, obtained by a vibrometer laser.

We may conclude that the use of burst signals is preferable to prevent overheating, especially in the steady-state zone, where special data treatment is easily used to prevent noise.

15.4.3 Non-Linear Elastic Characterization

To make the non-linear characterization, it is necessary to transform the direct measures into other more appropriate magnitudes. Firstly, since non-linearity is associated with the mechanical or motional parameters of the resonator, it is necessary to modify the magnitudes *R*, *X*, for the corresponding motional magnitudes, R_m, X_m, which the resonator would have if the dielectric contribution C_0^P were discounted (Fig. 15.13). For the disc resonators, this capacitance C_0^P corresponds to the dielectric contribution in the frequency interval between the radial resonance and thickness resonance. For other vibration modes, this capacitance will be taken similarly.

In the same way, the total current *I* allow the calculus of the motional current $I_{m,}$ to be used as the independent variable. In the second step, it is necessary to find a magnitude that is independent of the resonator dimension. A first possibility is the use of the electric displacement $D_m = I_m / A\,\omega$ corresponding to the motional current, which is directly proportional to the mean strain $<S>$. The important question then arises as to whether the preferable magnitude is the mean strain $<S>$ or its time derivative $<dS/dt>$; that is to say, whether magnitude D_m or dD_m/dt is more appropriate. Non-linear characterization of resonators with the same material and different dimensions has been carried out [38]. It is observed that the best magnitude to explain the non-linearity is $<S>$ or D_m, because the dependence of the non-linear parameters on the resonator frequency is much lower than if the magnitudes $<dS/dt>$ or dD_m/dt were used.

It is important to point out the dependence on the mean strain $<S>$. Uchino and Hirose [35, 40, 41] analyze the non-linear behaviour of resonators by taking the velocity v in a resonator edge point as the independent magnitude. For example, in a bar resonator in longitudinal mode, the relation between these magnitudes v and $<S>$, is:

$$v = \frac{<S>\omega_s L}{2} = \frac{<S>\omega_s \lambda}{4} = \frac{<S> c\pi}{2} , \tag{8}$$

where L is the bar length and c the acoustic wave propagation velocity, which is near constant, with low variations for different materials. For other resonant modes, the frequency is always proportional to the inverse of the dimension associated to the resonator mode.

Thus the velocity vibration v and the main strain $<S>$ are proportional magnitudes, and are equivalent to taking either of them as the independent magnitude. While the vibration velocity is important in applications, and more especially in resonant devices, the mean strain $<S>$, or the mean stress $<T>$, are more closely related to the material non-linearity and enable ceramic properties to be characterized.

Assuming that the independent magnitude responsible for the non-linear behaviour is the mean stress $<T>$ rather than the mean strain $<S>$, it is necessary to use the magnitude I', proportional to the mean stress $<T>$:

$$<D'> = <D_3> - \varepsilon_{33}^T E_3 = d_{31} <T> = \frac{I'}{j\omega A} , \tag{9}$$

which for a radial resonator corresponds to the use of a dielectric capacitance $C_0 = C_0^T$ [48]:

$$I' = V/Z' = V(1/Z - Y_0) , \tag{10}$$

$$\text{where} \quad Y_0 = j\omega C_0^T . \tag{11}$$

15.4.3.1 *Motional Impedance Plane*

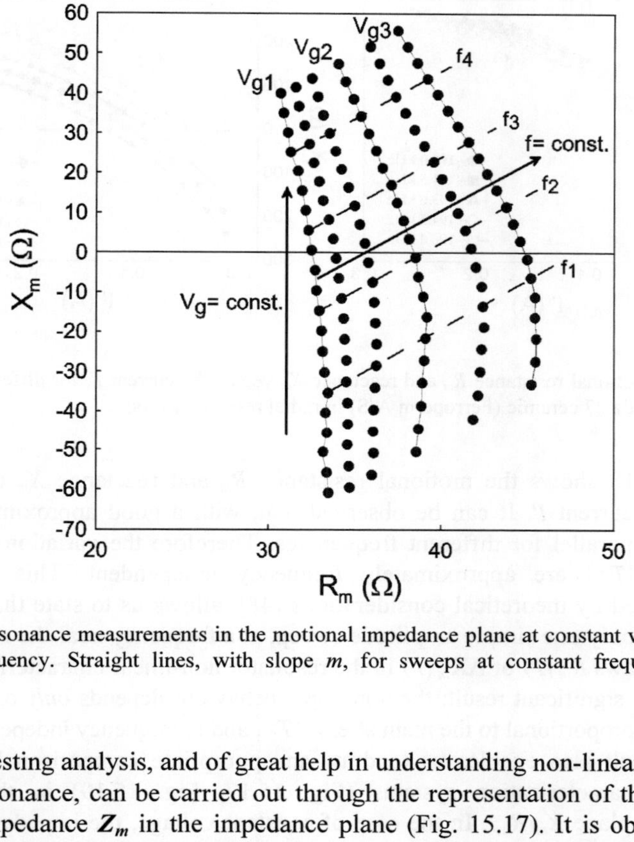

Fig. 15.17 Resonance measurements in the motional impedance plane at constant voltage and at constant frequency. Straight lines, with slope *m*, for sweeps at constant frequency. PXE5 ceramic.

An interesting analysis, and of great help in understanding non-linear behaviour near the resonance, can be carried out through the representation of the motional complex impedance $\boldsymbol{Z_m}$ in the impedance plane (Fig. 15.17). It is observed that when the amplitude level at constant frequency increases, the motional resistance R_m increases when the losses increase, and the motional reactance X_m also increases. In the impedance plane, the points corresponding to an amplitude sweep at constant frequency f_1 are aligned on a straight line with a slope m. This observation is in agreement with the consideration that the cause of elastic non-linear phenomena is the high strains or stress amplitudes. For other close frequencies, this representation is similar on another straight line nearly parallel to the line above. When the frequency of this second sweep is higher than the first sweep, $f_2>f_1$, the second straight line corresponds to higher reactances X_m. This representation shows that the non-linearity due to the increase in amplitude implies a reactance increase. Thus, in order to reach the resonance, that is to say $X_m=0$, it is necessary to use a lower frequency with a high level signal. One may therefore conclude that by increasing the excitation amplitude, there is equivalence between the decreases or shift frequency of resonance f_s and the motional reactance increase ΔX_m [37], a relation previously pointed out in Eq. (3).

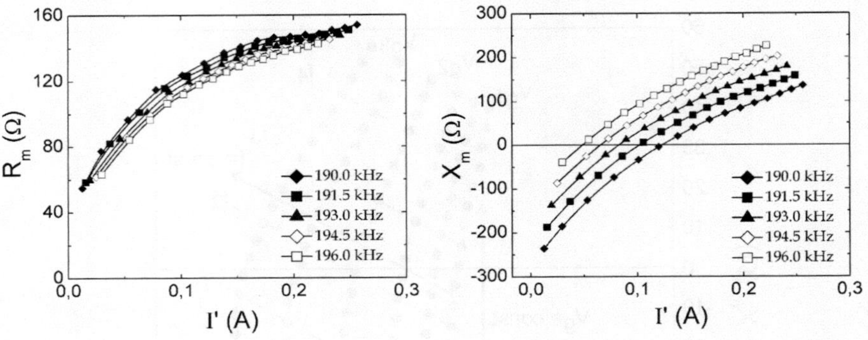

Fig. 15.18 Motional resistance R_m and reactance X_m versus the current I', for different frequency sweeps. Soft Pz 27 ceramic (Ferroperm A/S) in radial resonance mode.

Fig. 15.18 shows the motional resistance R_m and reactance X_m dependences versus the current I'. It can be observed that, with a good approximation, these curves are parallel for different frequencies. Therefore the variations $\Delta R_m(<T>)$ and $\Delta X_m(<T>)$ are approximately frequency independent. This result, also demonstrated by theoretical considerations [48], allows us to state that it suffices to perform only one simple amplitude sweep, at a frequency near the resonance, to obtain the laws $R_m (I')$ or $\Delta X_m (I')$ in the resonator non-linear characterization.

This is a significant result: the non-linear behaviour depends *only* on the current I', which is proportional to the main stress $<T>$, and is frequency independent.

This result also enables us to draw an appropriate comparison between the different characterization systems (Fig. 15.17, Fig. 15.19) in the motional impedance plane $X_m(R_m)$. In the case of constant voltage, the impedance curve is bent with higher resistance near the resonance (Fig. 15.17), because at these frequencies the total resonator impedance is minimum and the current I' is maximum, so the resistance is also maximum. This representation is unable to show that the independent magnitude is the current I'.

For a characterization at constant current [40, 41], the non-linearity produces ΔR_m and ΔX_m that are constant throughout the entire sweep, so the curve in this $\boldsymbol{Z_m}$ plane is approximately a straight line, parallel to the corresponding one at low signal (Fig. 15.19). The straight line at low signal is displaced ΔR_m horizontally and ΔX_m vertically, quantities that are constant across the entire sweep, since the current I is constant (also I'). The fact that in this case there is no bending of the curve in this impedance plane ensures the disappearance of the hysteresis phenomenon [37, 38].

Finally, when the amplitude sweep is carried out at constant frequency, the measured points are aligned on a straight line with a slope m, as mentioned above. The straight lines are parallel for different frequencies and are longer when they are closer to the resonance $X=0$ at high signal.

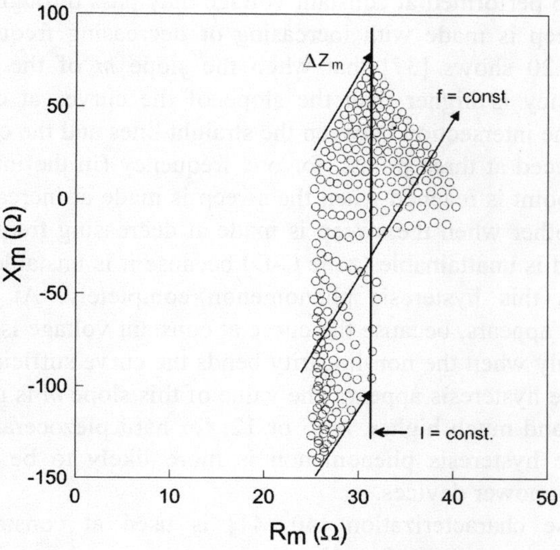

Fig. 15.19 Measurements in the motional impedance plane. Straight lines at constant frequency. Vertical straight line for sweep frequencies at constant current: at high level the impedance increase ΔZ_m is frequency independent.

15.4.3.2 *Hysteresis Phenomenon*

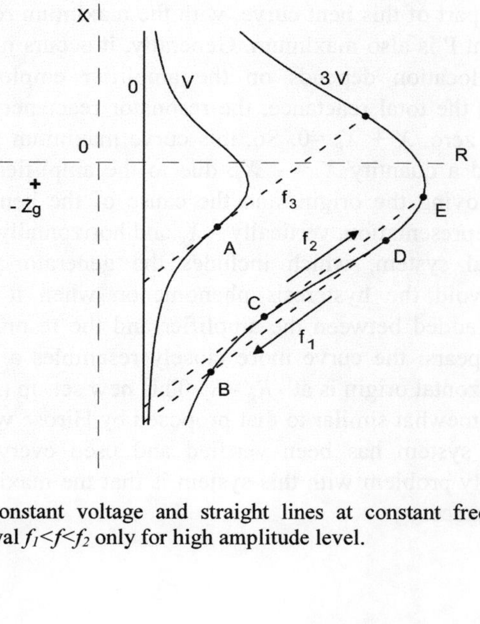

Fig. 15.20 Curves at constant voltage and straight lines at constant frequency. Hysteresis phenomenon in the interval $f_1<f<f_2$ only for high amplitude level.

The use of the motional impedance plane $\boldsymbol{Z_m}$ also allows us to characterize the hysteresis phenomenon. When the resonator is excited at high signal level, a frequency sweep performed at constant voltage may pass through a different path than if this sweep is made with increasing or decreasing frequencies (see Fig. 15.12). Fig. 15.20 shows [37] that when the slope m of the straight lines at constant frequency is higher than the slope of the curves at constant voltage, dX_m/dR_m, then the intersection between the straight lines and the curves at constant voltage is produced at three points, for one frequency (in the interval $f_1<f<f_2$, in Fig. 15.20). A point is reached when the sweep is made at increasing frequencies (zone B-C); another when the sweep is made at decreasing frequencies (zone E-D), and the third is unattainable (zone C-D) because it is unstable. Thus the slope m characterizes this hysteresis phenomenon completely. At low signal, the hysteresis never appears, because the curve at constant voltage is approximately a straight line. Only when the non-linearity bends the curve sufficiently, at constant voltage, does the hysteresis appear. The value of this slope m is near m=2 for soft PZT ceramics, and much higher, m=6 or 12, for hard piezoceramics. This result implies that the hysteresis phenomenon is more likely to be reached in hard materials used in power devices.

If the Hirose characterization [40, 41] is used at constant current, this phenomenon does not occur, since the sweep curves remain vertical straight lines (Fig. 15.19) and the intersection with the straight lines at constant frequency always gives only one point.

It has been observed that the hysteresis zone always occurs in the lower zone ($X_m<0$, frequencies lower than the resonance) of the bent curve at constant voltage. The resonance is always reached with decreasing frequencies. When amplitude sweeps are used at constant frequency, a zone of the impedance plane $\boldsymbol{Z_m}$, without measured points (see Fig. 15.19) has also been observed [37].

The extreme part of this bent curve, with the maximum resistance R_m, is found where the current I' is also maximum. Generally, it occurs just at resonance, X=0. However, this location depends on the amplifier employed. The current is maximum when the total reactance, the resonator reactance X plus the amplifier reactance X_g, is zero, $X + X_g = 0$. So, this curve maximum at constant voltage is vertically moved a quantity $X = -X_g$, due to the amplifier reactance [37]. It is equivalent to moving the origin, and the cause of the bending on the constant voltage sweep representation, vertically $-X_g$ and horizontally $-R_g$. This analysis of the experimental system, which includes the generator-amplifier impedance, allows us to avoid the hysteresis phenomenon when it appears. If a series resistance R_s is added between the amplifier and the resonator (Fig. 15.16), the hysteresis disappears: the curve more closely resembles a vertical straight line, because the horizontal origin is at $-R_g - R_s$. This new set-up (Fig. 15.16) makes the configuration somewhat similar to that proposed by Hirose with a constant current generator. This system has been verified and used every time the hysteresis appears. The only problem with this system is that the maximum current reached is lower at resonance.

The use of the motional impedance plane is important for understanding the different measurement methods, especially the constant frequency and strict resonance methods, and is able to foresee the frequency hysteresis phenomenon.

15.4.4 Elastic Non-Linear Behaviour

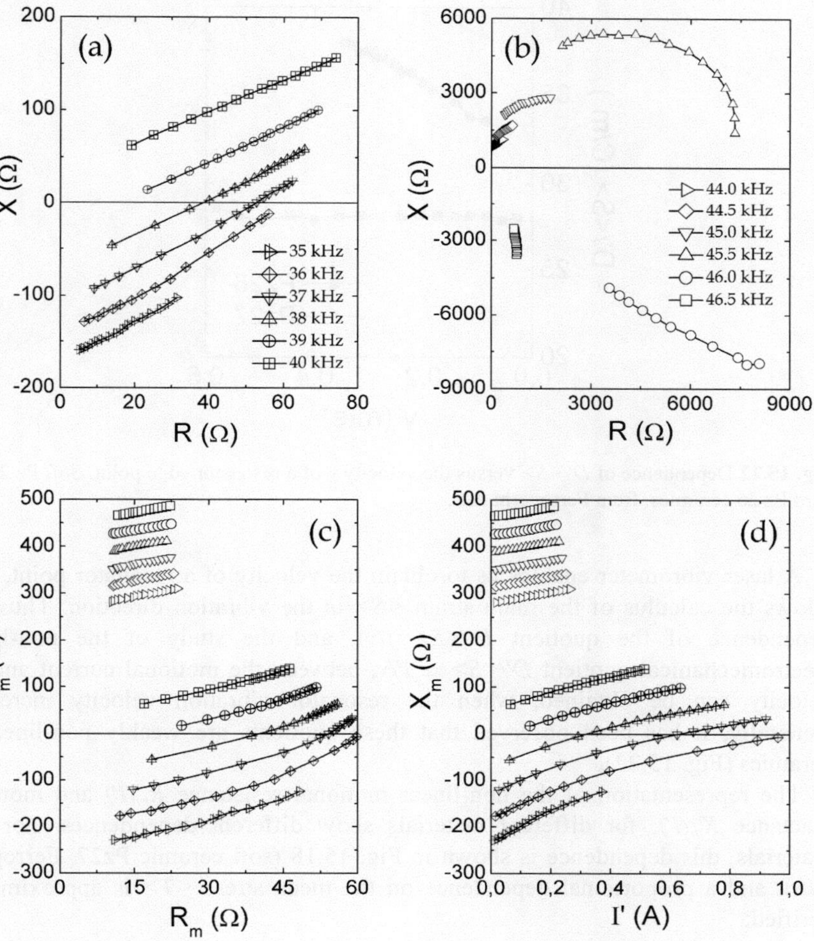

Fig. 15.21 Resonance **a** and anti-resonance **b** measurements at constant frequencies, in the impedance plane. **c** The same measurements in the motional impedance plane. **d** Motional reactance versus current *I'*. Soft PZT-5A ceramic (Morgan Matroc Ltd).

The high performance of the above results is shown in Fig. 15.21 [48]. The resonator total impedance near the resonance (Fig. 15.21a) and near the anti-resonance (Fig. 15.21b) are shown. It is observed that when the excitation level increases, the anti-resonance is reached from lower frequencies, due to the reactance non-linear increase. However, all these curves become parallel straight lines when they are represented in the motional impedance plane $\boldsymbol{Z_m}$ (Fig. 15.21c), or when they are analyzed in terms of the current I', proportional to the mean stress $<T>$ (Fig. 15.21d).

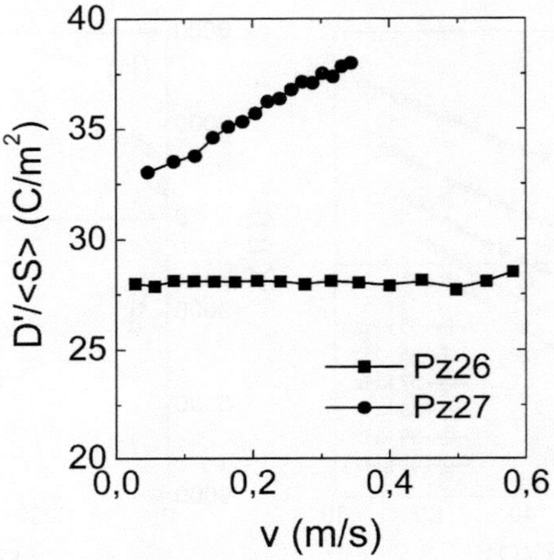

Fig. 15.22 Dependence of $D'/<S>$ versus the velocity v of a resonator edge point. Soft Pz 27 and hard Pz 26 ceramics, from Ferroperm A/S.

A laser vibrometer enables us to obtain the velocity of a resonator point, so it allows the calculus of the main strain $<S>$ in the vibration direction. Thus, the dependence of the quotient $D'/<S>=f(v)$, and the study of the non-linear electromechanical quotient $D'/<S> \propto I'/v$, between the motional current and the velocity can be obtained, when the resonator vibration velocity increases. Generally, it has been observed that these quotients are weakly non-linear in ceramics (Fig. 15.22).

The representation of the non-linear motional resistance $R_m(I')$ and motional reactance $X_m(I')$, for different materials show different dependences. For soft materials, this dependence is shown in Fig. 15.18 (soft ceramic Pz27, Ferroperm A/S), and a proportional dependence on the mean stress $<T>$ is approximately verified:

$$\Delta \tan \delta_m = r_1 \cdot <T> \, , \tag{12}$$

$$\Delta f_s / f_{s0} = x_1 \cdot <T> . \tag{13}$$

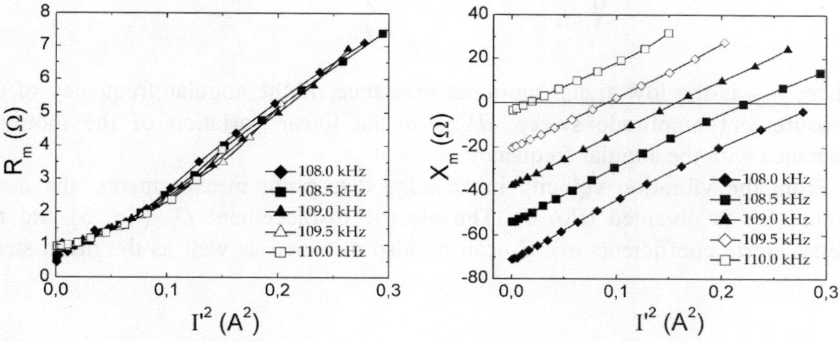

Fig. 15.23 Motional resistance R_m and reactance X_m versus the square of current I'. Hard Pz 26.

For hard ceramic materials, Fig. 15.23, the dependence versus the stress $<T>$ ceases to be linear and becomes approximately quadratic:

$$\Delta \tan \delta_m = r_2 \cdot <T>^2 \ , \tag{14}$$

$$\Delta f_s / f_{s0} = x_2 \cdot <T>^2 \, . \tag{15}$$

These dependences, either linear or quadratic, have also been observed in the same ceramic materials when a non-linear dielectric characterization is performed at low frequency (typically at 1 kHz). For soft materials, the non-linear dielectric permittivity shows a linear dependence with the electric field amplitude $\Delta\varepsilon = \alpha \cdot E_0$, in agreement with the Rayleigh model (see Sect. 15.2.2) [28, 49, 50].

Finally, these magnitudes R_m, X_m, I', v, must subsequently be modified in order to obtain other magnitudes closer to the constitutive equations and to the ceramic elastic coefficients. The mechanical losses tangent $tan(\delta_m)$ is used instead of R_m, the shift resonance frequency $\Delta f_s/f_{s0}$ instead of the motional reactance, and the mean strain $<T>$ instead of I'.

The main strain $<S>$ can be obtained by the current I_m or by the laser measurements of v. The main stress $<T>$ is obtained from I' (Eq. 9, in the case of a disc resonator in the radial mode).

After obtaining the direct measurements R, X, I, v, as well as the modified magnitudes R_m, X_m, I', D', D_m, $\Delta f/f$, $<S>$, it is necessary to make some transformations to analyze the elastic and piezoelectric coefficient variations. From the linear elastic coefficient c_{110}^P and the decrease of resonant frequency (increase of motional reactance), it is possible to obtain the high signal stiffness c_{11}^P, assuming that the Poisson ratio σ remains constant:

$$\sqrt{\frac{c_{110}^{P}}{c_{11}^{P}}} = 1 + \frac{(X_m - X_{m0})/\omega}{dX_{m0}/d\omega}, \qquad (16)$$

where X_{m0} is the low signal motional reactance, ω the angular frequency of the measurement amplitude sweep, $dX_{m0}/d\omega$ the linear variation of the motional reactance with the angular frequency.

From the vibration velocity in an edge disc point measurements, the mean strain <S> is obtained (Eq. 6). The electric displacement D_m (Eq. 5) and the piezoelectric coefficients e_{31}, d_{31} can be also obtained as well as the mean stress $<T>$:

$$e_{31} = \frac{D_m}{<S>}, \qquad (17)$$

$$d_{31} = \frac{e_{31}}{c_{11}^{P}(1+\sigma)}, \qquad (18)$$

$$<T> = c_{11}^{E}(1+\sigma) <S> - 2e_{31}^{P} E_3 . \qquad (19)$$

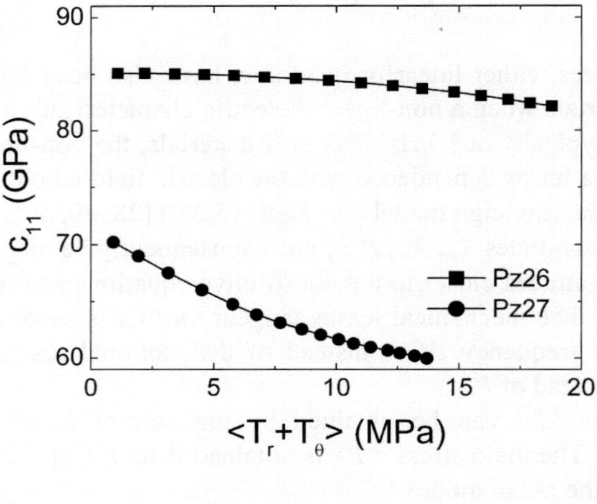

Fig. 15.24 Stiffness coefficient c_{11} versus the mean stress $<T>$. Soft Pz 27 and hard Pz 26 ceramics.

With all these magnitudes, it is possible to analyze these coefficient non-linear variations with the mean stress, assuming that the Poisson coefficient σ remains constant. Fig. 15.24 shows the dependence between the elastic c_{11} coefficient and the mean stress for two different materials. The relation between the piezoelectric coefficient d_{31} and the elastic s_{11} allows an interpretation of these results, with the result that the extrinsic coefficient $\Delta d_{31}/\Delta s_{11}$ is higher than the low signal linear coefficient d_{31}/s_{11} [48] (see Sect. 15.5.1).

The proposed independent variable is the mean stress $<T>$ or the mean strain $<S>$. The advantage of the use of the mean stress is that the non-linear relations are verified both in resonance and in the anti-resonance. But, the drawback is the necessary hypothesis to obtain this mean stress. When the stress is obtained (Eq. 19) in disc resonators, it is necessary to assume that the Poisson coefficient σ is constant. In other resonant modes, for example in bar resonators, this hypothesis is not necessary, and the calculus of the mean stress $<T>$ is easier. Otherwise, the use of the mean strain $<S>$ as the independent variable through the Eq. (6) allows its direct measurement (it is also possible to obtain the mean strain from Eq. (5) with the hypothesis that e_{31}^{P} is constant, if the velocity vibration has not been measured). In this case, the non-linear behaviour depends weakly on the frequency between resonance and anti-resonance.

15.5 Phenomenological Models

Given that the complexity of the phenomena makes their interpretation difficult, some phenomenological models have been proposed that do not take into consideration the deep nature of the problem, but enable us to understand some facts and establish relations between the phenomena observed.

First of all, rather than consider the complex internal structure, it seems reasonable to limit the degrees of freedom of the system to two: the polarization P and the strain S [51]. At high amplitude, the expression of the energy as a function of such variables does not have to be quadratic, but it can be described by functions smooth enough to allow a series development. In equilibrium, however, both forces (E field and stress T) must be null, so this development has not first-order term. As a consequence, if this model holds, the dependence of any coefficient on amplitude will be quadratic, although other terms ought to be included if the amplitude were very high. Moreover, the coefficients would only depend on the instantaneous value of the field, and not on the peak value.

A first glance at the experimental results [10, 18, 20, 22, 23, 24, 30, 52, 53] reveals that there are many cases in which this model cannot be applied, because the coefficients depend to some extent on the amplitude of the applied field, and this dependence is often linear. Only some materials, such as hard PZT, depend on the square of applied field, although they always show some small dependence on the amplitude.

A more realistic model takes into account that, given the complexity of the system, the set of successive states traversed by the system through a complete loop does not have to coincide either totally or partially with the states obtained when an excitation of another amplitude is applied. So, the instantaneous behaviour, throughout a complete loop, depends on the amplitude. Consequently, we must assume that dielectric, piezoelectric and elastic coefficients depend explicitly on the amplitude of the applied field (or stress). If the system is modelled by a random potential, we obtain behaviour analogous to that described by Rayleigh for ferromagnetic materials [20]:

$$D=\left(d_0+\alpha_d T_0\right)T\pm{}^{\alpha_d}\!/_2\left(T_0^2-T^2\right) \tag{20}$$

In this expression, it can be observed that the piezoelectric coefficient depends exclusively on the amplitude of the stress, and that such dependence is linear. When the amplitude tends to zero, a non-null coefficient d_0 persists.

The same reasons can be applied to the dielectric behaviour, so a similar relation is expected to hold [18, 54]:

$$D=\varepsilon_0\left[\left(\varepsilon+\alpha_\varepsilon E_0\right)E\pm{}^{\alpha_\varepsilon}\!/_2\left(E_0^2-E^2\right)\right] \tag{21}$$

The second term of this equation depends on the upward or downward direction towards which the field is evolving, so that $D(E)$ encloses a certain area, denoting the existence of energy losses. It must also be noted that the coefficient of this second term is the same as that appearing in the first term, and that the energy lost during a loop is proportional to the square of the field amplitude. Thus the plots of ε' and ε'' as a function of the amplitude would indicate agreement with the Rayleigh relation: the plot $\varepsilon'(E_0)$ must be linear, and $\varepsilon''(\varepsilon')$ must not only be linear but its slope must also be equal to 0.42. Something similar will occur with the piezoelectric coefficient [24].

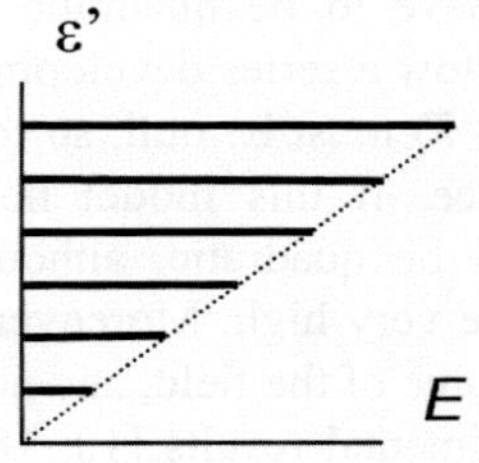

Fig. 15.25 Rayleigh model: theoretical dependence of ε on instantaneous and amplitude field.

The experimental measure of the dielectric constant as a function of the instantaneous field allows us to do a deeper analysis. By applying such a point of view, a material that totally agrees with the Rayleigh model would behave as shown in Fig. 15.25. Actually, soft materials behave like Fig. 15.1a and hard ones like Fig. 15.1b. So we conclude that the Rayleigh model never fits the experiments exactly, although it fits better in soft ceramics than in hard ones. Let us assume the existence of two parallel mechanisms without any interaction between them. Let us also assume that the resultant D is the sum of both contributions. We will then approach the real behaviour in all cases.

The split of ε_{NL} into ε_{α} and ε_{β} can provide us with further information. The plot of ε_{α} versus the amplitude shows us two types of dependence: linear, according to the Rayleigh model, which is a characteristic of soft ceramics, and quasi-quadratic, characteristic of hard ones. However, the plot of ε'' versus ε_{α} makes a linear dependence evident, with a slope of 0.42 in almost all cases. By analyzing samples with different degrees of ageing or poling, it can be shown that ε_{α} is sensitive to such conditions, while ε_{β} is nearly indifferent to them. We conclude that there is a mechanism related to ε_{α}, which clearly has an irreversible and extrinsic nature, because it is related to the evolving domain structure. Furthermore, energy losses are closely related to the increment of dielectric constant, which is common to all materials, although the types of dependence with the field were diverse [19].

The plot of the mean value of ε_{β} versus the amplitude shows the same aspect than ε_{α}, but the relation between both parts is very different: while in soft materials the term ε_{β} is residual, in hard ones it becomes the main term. This term usually depends on the instantaneous field through a power law, whose exponent usually has a high value in soft materials but is nearly two in hard ones, where a low dependence on the amplitude is often found. This is just the behaviour expected from the former model. The fact that losses are always proportional to ε_{α} leads us to believe that there are no significant losses related to ε_{β}, which is in accordance with that model. Thus, the mechanism related to ε_{β} probably has a reversible nature.

This scheme is not followed by the direct piezoelectric behaviour. Under the action of a field, the non-linear part of the coefficient depends linearly on the stress and no dependence on the instantaneous stress appears (or it is very low). This may be because the field or stress tensors are of different rank, or because in the latter case the use of a bias DC field cannot be avoided. Although the slope of the plot $d''(d')$ is close to the theoretical value 0.42 in soft materials, it is reduced by a half in hard ones. This is the only fact that is not in accordance with the Rayleigh model [21].

At a phenomenological level, a model has been proposed [55, 56], initially developed by Preisach for ferromagnetism, which on the one hand enables us to explain the behaviour expressed by the Rayleigh law, and on the other seeks to explain the behaviour in situations where the Rayleigh law does not work, so it can be taken as a general tool.

The model consists of the assumption that the sample behaves as a set of hysteretic systems, all connected in parallel, so that each element is defined by a coercive field E_{C} and a bias field E_{B}, and thus the element switches when the external electric field is greater than $(E_{B} + E_{C})$ in the upward direction, or less than $(E_{B} - E_{C})$ in the downward direction. The overall system is determined by a two-dimensional distribution function $P(E_{C}, E_{B})$ that tells how many elements there are for each pair of values. So, the sequence of the successive states that the system goes through depends on the extreme values of the applied field. The ensemble of elements that switch through a complete loop correspond to the elements belonging to a square triangle in the space (E_C, E_B), whose area is proportional to the square of the amplitude of the applied field (Fig. 15.26).

If the distribution function is uniform, the Rayleigh relation between the electrical displacement vector D and the electric field E is obtained. So, the behaviour of soft ceramics is interpreted in a reasonable way.

The same model can be used to understand the direct piezoelectric effect by assuming that there are some elements that can be switched by the action of a mechanical stress of value $T_B \pm T_C$. In this case, the distribution is defined in the space (T_C, T_B), which does not have to be similar to the distribution used for the dielectric properties, which is defined on the space (E_C, E_B). Then, the behaviours due to the action of a field or a stress do not have to be similar, although the Preisach model was correct in both cases.

If the distribution is not uniform, the Rayleigh equation no longer holds. So, it is reasonable to interpret hard ceramic behaviour through some type of distribution [54, 56]. The dependence of the dielectric constant with the amplitude is determined by the first derivative of P versus E_C, as well as by the second derivative of P versus E_B. Thereafter, the relation between real and imaginary components of ε is determined by the dependence on E_C, so that the relation must be a decreasing function in order to model the low losses of hard materials. It is not easy to find a distribution that fits the entire behaviour of such materials. Thus, this behaviour probably cannot be interpreted by accepting the Preisach model as an exclusive mechanism. This supports the opinion that other mechanisms may exist.

Nevertheless, the fact that a uniform distribution leads to the Rayleigh equation explains why this model works well in many cases.

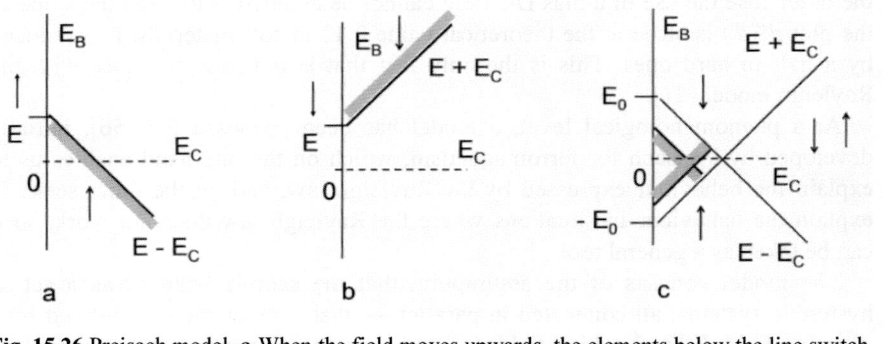

Fig. 15.26 Preisach model. **a** When the field moves upwards, the elements below the line switch. **b** If the field moves downwards, those that switch are above the line. **c** With an alternating field, the elements that are inside the triangle in the plane (E_C, E_B) switch.

15.5.1 Theoretical Considerations

Although such models enable us to describe ceramic behaviour more or less successfully, which is useful for the device designer, it is still necessary to understand the origin of these phenomena, how they are related to the structure, and finally how they depend on the composition and on the thermal or mechanical actions undergone by the sample up to this moment.

The non-linear effects observed in ceramics are mainly due to the alteration of their properties, which are described by defined coefficients. Although non-linear effects must appear when a solid changes its size, the alteration of the properties produced in this way is proportional to the variation of its size. However, non-linear effects observed in ceramics are usually two orders of magnitude larger, so that the size effect is negligible in this case, although it can be important in single crystals.

Any dielectric, piezoelectric or elastic phenomenon that implies the alteration of the electrical polarization, and of the mechanical strain of the material, can be produced by two different mechanisms: first, the intrinsic one, which consists in the deformation of each crystalline cell, so it is the only one found in a perfect crystal without twins; secondly, the extrinsic one, which is related to some other cause, either to the domain wall motion, and which causes some cells to change the domain to which they belong, or to the motion and re-orientation of defects.

Furthermore, it is necessary to distinguish between linear processes, in which the response is strictly proportional to the cause, and non-linear ones, which are defined as those in which the coefficients depend on the excitation amplitude.

Thirdly, it is necessary to consider the case of reversible processes, where the system returns to exactly the same state it had before excitation was applied (by passing through the same intermediate states); and irreversible processes, if it returns by a different way, so an energy loss and an entropy gain are unavoidably produced.

This triple classification may lead us to assume that there are eight different types of processes, but in practice only four of them are relevant: the intrinsic one (which is nearly linear and reversible), the linear extrinsic one (which is also reversible), the non-linear extrinsic reversible (which is given by the fraction β), and finally the non-linear extrinsic irreversible one (fraction α).

There is common agreement that the non-linear effects are due exclusively to the extrinsic effects, as may be deduced by the fact that they depend completely on the dopants and the defects they contain, although this fact should be subject to verification [57]. The deformation of a crystal cell does not have to be linear with the applied field; however, its degree of non-linearity is usually much less than that observed in ceramics. So, it is not worth taking into consideration.

The fact that a good linearity is maintained between D and the strain S, even when their dependence on the field is in no way linear, is clearly illustrative of the extrinsic nature of the process. In effect, wall displacement produces in the sample changes of charge and shear that are proportional to the volume switched and to the values of spontaneous polarization and strain, respectively. This guarantees proportionality between the increments of D and S, even when the relation between the wall displacement and the field is no longer linear.

There are two aspects that distinguish intrinsic and extrinsic contributions. The domain wall mobility depends on the temperature, and falls to zero when it approaches the absolute temperature. By also assuming that the intrinsic behaviour depends barely on the temperature, the difference between the coefficients measured at room temperature and at absolute zero must correspond to the extrinsic linear part [11, 58]. The other aspect only concerns the piezoelectric coefficients: as the cells of both domains are identical and differ only in their orientation, their volume must be the same, so there is no volume change in purely extrinsic processes. Then, the hydrostatic piezoelectric coefficient $d_h = d_{33}+2d_{31}$ is due exclusively to the intrinsic effect, because it must be null in an extrinsic process.

In ferroelectric perovskites, domain walls can be of 180° or non-180° (90° in the tetragonal case, and 71° or 109° in the rhombohedral case). The movement of 180° walls implies a change in the direction of polarization but no change in the strain, thereby contributing to the electrical displacement D by altering only the dielectric effect, but not the piezoelectric or the elastic effects. On the other hand, the movement of non-180° walls produces a change in both fields D and S and contributes to all three effects. A simultaneous change of dielectric and piezoelectric coefficients is a sign that non-linearity is produced by the non-180° wall movement.

This can easily be observed by plotting the reverse piezoelectric coefficient as a function of the permittivity for different amplitudes of the excitation field. No matter what type of dependence the coefficients might have on the amplitude, it can be seen that in this plot the experimental points are aligned along a straight line. Its slope g_{ext} corresponds to the relation $\Delta S/\Delta D$, where ΔD is the difference of the spontaneous polarization of the possible domains and ΔS is the difference of the spontaneous strains, which shows the extrinsic character of the process. This depends on the cell rather than on the defects. So, a good concordance is observed between the slopes of different materials, whether they are soft or hard [28].

In reality, however, things are not so simple, because ceramics are composed of random oriented grains containing domains whose polarization distribution depends on how the material has been poled. Since the tensors d and ε have different rank, the effect of the disparity of orientations cannot be the same for both. Then, the sign of the slope g_{ext} of the plot $d(\varepsilon)$ depends on the poling direction, and the slope must be null for non-poled ceramics. For the same reason, even though a sample has been well-poled, the slope will be less than $\Delta S/\Delta D$, according to a factor that depends on the distribution function of the domain walls [7].

On the plot $d(\varepsilon)$, an extra point corresponding to a very low temperature and low amplitude measurement can be added. In a first approximation, it corresponds to the intrinsic behaviour. In this case, it depends on the distribution of domains, taking into account only their volumes, and regardless of the amount, extension or orientation of the walls. As can be seen in Fig. 15.27, it is possible to discern clearly the intrinsic, linear extrinsic and non-linear extrinsic contributions, while the reversible ones cannot be distinguished from the irreversible.

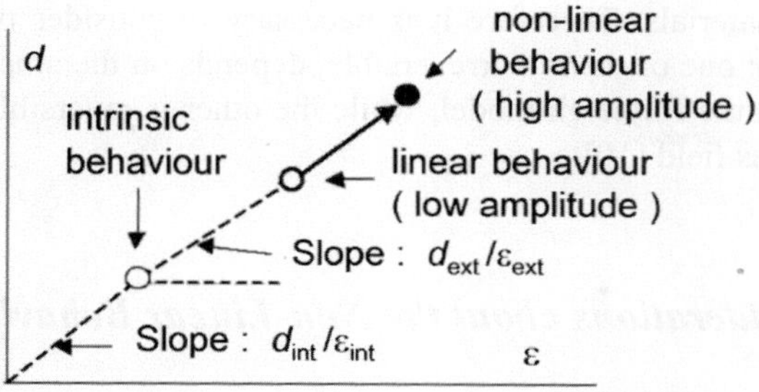

Fig. 15.27 Plot of converse d versus ε, determined at low frequency (there is no stress). The first point can only be obtained at low temperature and at low amplitude, and the second one at room temperature but at low amplitude. For the third one, we apply high amplitude.

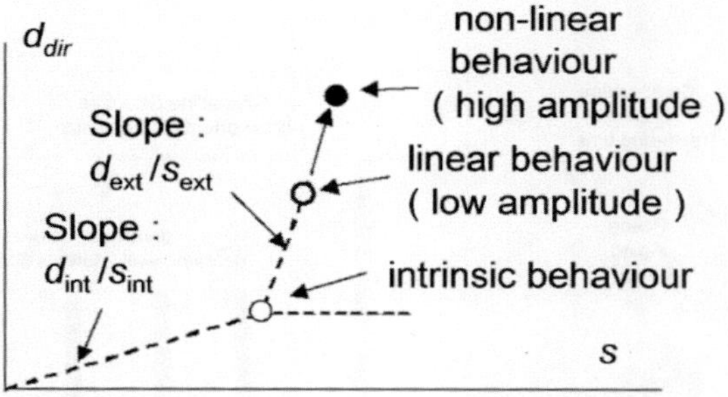

Fig. 15.28 Plot of d versus the compliance s, obtained for a disc in radial resonance mode, so the field E is nearly null.

Likewise, the direct piezoelectric coefficient can be plotted versus the compliance. In this case, a slope e_{ext} lower than $\Delta D/\Delta S$ is obtained (Fig. 15.28). The product of the slope of this plot e_{ext} and the slope of the previous one g_{ext} would be equal to 1 if the domains were well oriented, so the theoretical coupling constant of the extrinsic part must be 1. However, in practice, this is not the case. The result $k_{ext}^2 = g_{ext}e_{ext} < 1$ reflects the dispersion of the orientation of the grains, and thus the quality of the poling process.

In order to distinguish the reversible from the irreversible part, both being extrinsic and non-linear, it is necessary to analyse the signal as a function of the instantaneous applied field. Indeed, although in the plot $\varepsilon''(\varepsilon')$ the increment of ε'' is exclusively due to the irreversible part, the increment of ε' depends on both.

When both terms are measured in a partially poled sample by altering its domain structure, and ageing is observed, it can be seen that the distinction between reversible (ε_β) and irreversible (ε_α) is not arbitrary at all: the reversible part remains unaltered, while the reversible part (ε_β) shows considerable changes [10].

Furthermore, the plot $\varepsilon''(\varepsilon_\alpha)$ has the slope expected by the Rayleigh equation even in hard materials. Therefore it is necessary to consider two different non-linear processes: one of them is irreversible, depends on the amplitude of the field and agrees with the Rayleigh model, while the other is reversible and depends on the instantaneous field [10].

15.5.2 Considerations about the Non-Linear Behaviour

In order to understand the linear and non-linear behaviour of the piezoelectric ceramics, several factors must be taken into account. In broad terms, the manufacture and later treatment determine the crystalline and domain structures of the material, which in turn condition the dielectric and piezoelectric behaviour when an electric field or a mechanical stress is applied.

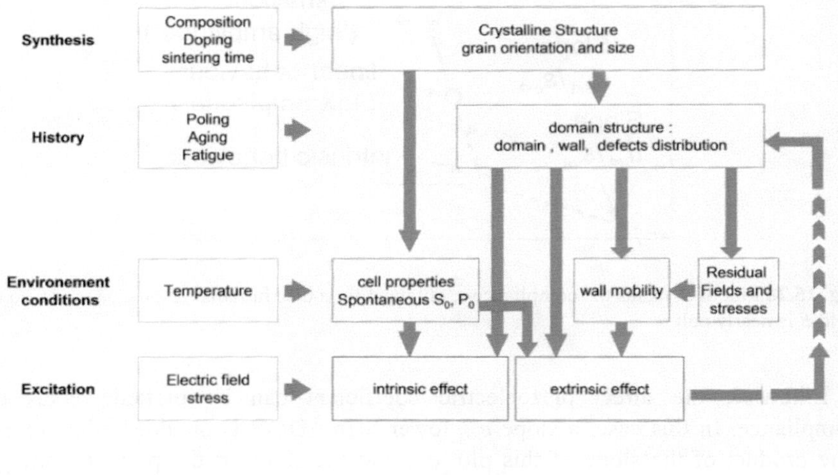

Fig. 15.29 Four level effects on a piezoelectric ceramic.

In Fig. 15.29, one may see some of the effects and relations that can be found in a ceramic. Composition is defined in the manufacturing process (for example, the Titanium-Zirconium ratio in a PZT ceramic), as well as the type and concentration of doping ions or dopants added. Other factors such as temperature and sintering time are also controlled. Altogether, this determines the crystalline structure, the Curie temperature and the grain size, as well as grain orientation if the ceramic has been textured.

The subsequent poling process, as well as fatigue and ageing, establish the current state of the sample, which is defined by the distribution of domains and domain walls (for each possible structure), as well as by defect orientation.

Mechanical stresses and electric fields appear when there is change of domain configuration, so a fully stable equilibrium is never attained. At a given temperature, crystal cells have a given spontaneous strain and polarization, so the material has definite intrinsic dielectric, piezoelectric and elastic properties. Furthermore, both domain walls and defects have specific mobilities.

The intrinsic effect depends on the amplitude of the applied field, on the particular coefficients of a cell and on the domain distribution. The extrinsic effect is proportional to the values of spontaneous S_0 and P_0, and depends on the amplitude of the applied field (linearly or not) according to the mobility and the domain wall distribution. This is altered by the defect distribution and by residual mechanical stresses.

Wall motion in turn modifies the domain distribution by changing the intrinsic effect [59] and by producing a non-linear response.

From the foregoing, it can be deduced that there are two key aspects for understanding non-linear behaviour: domain structure and domain wall mobility.

15.5.3 On the Domain Structure

The crystallographic directions are well defined inside a grain. Thus, the different domains can be identified by the directions in which the vectors of spontaneous polarization ***P*** are pointing (six in tetragonal case, eight in the rhombohedral case). There is only half the number of spontaneous strain possibilities, because the strain is the same for two domains whose polarizations differ by 180°.

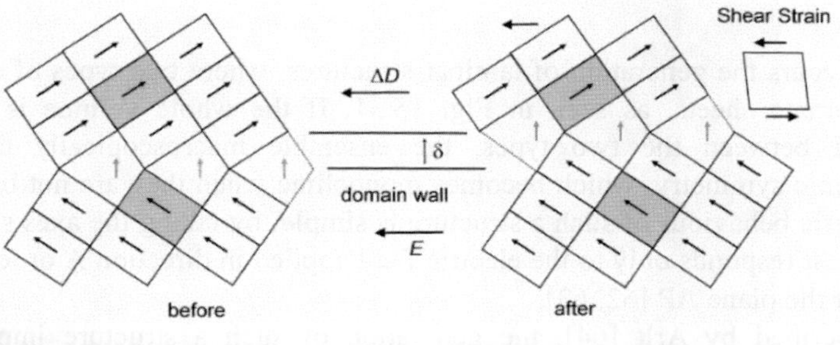

Fig. 15.30 Displacement of a domain wall: when a line of cells switches, so that they become a part of the lower domain, the domain wall moves upwards, while the upper domain moves to the left (by causing a shear strain) and the change of the mean electrical displacement *D* points towards the left.

The wall connecting two domains cannot adopt just any orientation, since it must fulfil two conditions: the continuity of the component of *P* normal to the wall, and the continuity of the lattice periodicity along the wall surface [60, 61].

This can be achieved in two different ways. In the first, the vectors P of both domains are pointing to form a 180° angle, so the wall must be parallel to such a direction (180° wall). Or, they form a different angle defined by the crystalline system, in which case, the wall must be perpendicular to the bisectrix of that direction (90° in tetragonal system, 71° or 109° in rhombohedral one) (Fig. 15.30). In the second case, the wall must be flat, because its orientation is perfectly defined by the poling directions. This is not so in the first case, where the walls may have a cylindrical shape, and so may adopt an orientation containing the polarization direction, which is common to both domains.

While the 180° wall movement only produces a change of the mean value of D in the same direction of P_s, the displacement of a non-180° wall generates an increment of D in the direction of the difference of the two vectors P_s involved, as well as a shear strain over the sagital plane.

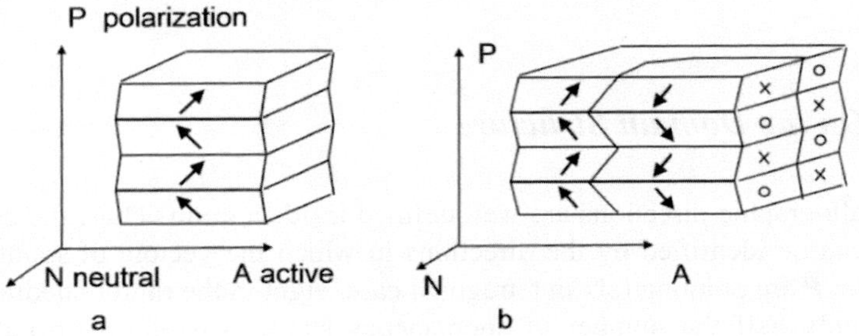

Fig. 15.31 a Laminar structure formed by alternating domains. Notation used for the three coordinate axes. **b** A 180° lacuna inside a 90° twin structure. If the contour mechanical and electrical contour conditions are fulfilled, it forms a tube in the direction of vector P.

This favours the generation of laminar structures, where two types of domains lie in alternate sheets, as seen in Fig. 15.31. If the whole volume is equally distributed between the two types, the ensemble macroscopically takes an orthorhombic symmetry, which becomes monoclinic when they are not balanced. The extrinsic behaviour of such a structure is simple: by taking the axes shown in Fig. 15.31, it responds only to the electric field applied in direction A or to a shear stress over the plane AP [62, 63].

As described by Arlt [64], the generation of such a structure implies the creation of mechanical stresses near the grain boundary that depend on the domain thickness [65]. A state of equilibrium is thus established in which thickness is proportional to the square root of the width, which frequently coincides with the grain size. As a result, the grain size determines the domain density, and indirectly, the extrinsic response. However, if the grain size is too small it cannot be subdivided into domains, while in big grains more complex structures are produced, which are formed by four domain types. There is therefore an optimum size for practical ceramic grains, which usually measures several micrometers.

Within these laminar structures formed by domains separated by non-180° walls, lacunae can exist in which polarization has been inverted. Then, the borders of such lacunae are formed by 180° walls, so that they acquire the shape of cylinders with irregular bases. These structures can be clearly identified by direct observation, and it usually is observed that there are fewer 180° walls than non-180° ones.

Each grain has its own orientation with respect to the applied external field, so the effective field E (in direction A) is different for each grain, and something similar occurs with the effective stress T. It must be taken into account that E and T are tensors of different rank, so the relation between effective and external field is not the same as the relation between the corresponding stresses. Furthermore, because of the wall motion, an increment of the effective D and S occurs, and they appear outside the grain in a different form depending on the rank of their tensors [66].

Moreover, it is necessary to consider the case where the crystalline phase is not pure. It is known that the dielectric and piezoelectric properties are optimized near the morphotropic phase boundary MPB. It has been often supposed that the two phases co-exist, so in addition to the domain walls there are also walls between phases. This increases the probability of finding well-oriented walls by improving the extrinsic phenomena. It has recently been discovered that in the phase diagram there is a monoclinic phase that occupies a narrow band between the other two phases [67, 68]. According to this, it is assumed that in a monoclinic phase the displacement of the ions inside the cell is easier than inside the other phases. Therefore, the increase of the piezoelectric effect in the MPB would be of an intrinsic rather than an extrinsic nature. Nevertheless, it must be take into account that, although symmetry imposes no restriction on the direction of polarisation, this does not necessarily imply that this direction can be easily changed. Since both intrinsic and extrinsic phenomena depend on different distributions (either domains or domain walls), this is an additional fact to be considered in order to discern which of the models is more appropriate.

15.5.4 On the Role of the Dopants

By assuming that the most important extrinsic contribution under the effect of a sub-coercive field comes from the motion of non-180° domain walls, the problem of the cause of the non-linear behaviour arises. If the applied fields are not strong enough to change significantly either the area or the orientation of the domain walls, we must search for the cause in their anomalous mobility.

Both the linear and non-linear behaviour depend greatly on the type of dopants the material contains, and it is important to know how this influence is exerted. It is customary to consider that dopants have a direct influence on mobility by hindering wall motion, and the effect is not so great when the excitation is high (pinning effect) [12, 25]. While some dopants (acceptors) hinder the motion, others can improve them (donors).

However, it is necessary to distinguish between reversible and irreversible extrinsic contribution. By its very nature, a pinning mechanism is irreversible and non-linear. This may account for one aspect of the behaviour, but not others. At low field, the walls can undergo a displacement proportional to the excitation field (elastic type behaviour), while at high field this relation may be of a non-proportional type (non-linear reversible extrinsic effect). In some cases, the observed effect seems to be the sum of two effects, since some walls move in a reversible form, while the others do so irreversibly.

The direct effect of dopants on mobility may not be the only mechanism to significantly determine its behaviour. In addition to those cases in which dopants can modify the crystalline structure, the possibility that they may have an influence on the domain structure must also be taken into account, and that this may in turn modify mobility. The motion of a wall produces new stresses in the domain border, which may generate a restoring force that could explain the elastic behaviour at low level signal.

We must also remember that impurities strongly influence electrical conductivity. So, a low conductivity makes electrical relaxation slower when the system needs to re-adapt due to change. The presence of an electrical force will have an influence on the quasi-equilibrium state, which could be different from that attained by a more conductive material.

The interaction between defects and walls depends on their nature and on their relative orientation. For instance, acceptor dopants produce defects (dopant-oxygen vacancy) oriented in the direction (1 0 0) that interacts with domains in tetragonal materials, which are oriented in the same directions. However, donor dopants have no preferential orientation, although it has been observed that they interact differently in tetragonal than in rhombohedral materials.

Even if there were a well-established relation between the field and the motion of a simple laminar structure, two problems would still remain.

Firstly, in order to model the behaviour, it is convenient to have (or at least, to assume) a distribution function that shows us the wall domain area as a function of the angles defining their orientation (domain wall distribution). With an isotropic distribution, as found in a non-poled ceramic, a null piezoelectric response should be found.

Secondly, we do not assume significant interaction between grains, which is somewhat unreasonable. They are not arranged electrically in parallel, but rather some of them are in series, so a phenomenon analogous to the Maxwell-Wagner effect can be produced [69]. The dispersion in the orientation of the grains implies that they do not all expand at the same rate, thereby generating a mechanical stress field both in the direction parallel with the applied field and perpendicular to it. This has an influence on the neighbour grains, setting up an interaction between intrinsic and extrinsic effects.

References

1. Bassiri-Gharb N, Fujii I, Hong E, Trolier-McKinstry S, Taylor D V, Damjanovic D (2007) Domain wall contributions to the properties of piezoelectric thin films. J. Electroceram 19, pp 49-67
2. Cross L E (1995) Ferroelectric materials for electromechanical transducer applications. Jpn. J. Appl. Phys. 34, pp 2525-2532
3. Albareda A, Pérez R, Villar J L, Minguella E, Gorri J A (1997) Intermodulation measurement of nonlinearities in piezoceramic resonators. Rev. Sci. Instrum. 68 no 8, pp 3143 - 3149
4. Ikeda T (1990) Fundamentals of Piezoelectricity. Oxford Univ. Pr., Oxford
5. Maiti T, Guo R, Bhalla A S (2007) Enhanced electric field tunable dielectric properties of $BaZr_xTi_{1-x}O_3$ relaxor ferroelectrics. Applied Phys. Letters 90, 182901
6. Albareda A, Casals JA, Pérez R, Montero de Espinosa F (2002) Nonlinear measurements of high-power 1-3 piezo-air-transducers with burst excitation. Ferroelectrics 273, pp 47-52
7. Pérez R, Albareda A, García JE, Tiana J, Ringgaard E, Wolny WW (2004) Extrinsic contribution to the non-linearity in a PZT disc. J. Phys. D: Appl. Phys. 37, pp 2648-2654
8. Albareda A, Pérez R, García J E, Ochoa D A, Gomis V, Eiras J A (2007) Influence of donor and acceptor substitutions on the extrinsic behaviour of PZT piezoceramics. J. Eur. Ceram. Soc. 27, pp 4025-4028
9. García J E, Pérez R, Albareda A, Eiras J A (2007) Non-linear dielectric and piezoelectric response in undoped and Nb5+ or Fe3+ doped PZT ceramic system. J. Eur. Ceram. Soc. 27, pp 4029-4032
10. García J E, Pérez R, Albareda A (2005) Contribution of reversible processes to the non-linear dielectric response in hard lead zirconate titanate ceramics. J. Phys: Condens. Matter 17, pp 7143-7150
11. Zhang Q M, Wang H, Kim N, Cross L E (1994) Direct evaluation of domain-wall and intrinsic contributions to the dielectric and piezoelectric response and their temperature dependence on lead zirconate-titanate ceramics. J. Appl. Phys. 75, pp 454-459
12. Lente M H, Picinin A, Rino J P, Eiras J A (2004) 90° domain wall relaxation and frequency dependence of the coercive field in the ferroelectric switching process. J. Appl. Phys. 95, pp 2646-2653
13. Lupascu D C (2004) Fatigue of ferroelectric ceramics and related issues. Hull R, Osgood R M, Parisi J, Warlimont H (eds), Springer Series in Materials Science no 61, Berlin Heidelberg, Germany
14. Uchino K (1999) Recent trend of piezoelectric actuator developments. Proc. 1999 Int. Symp. Micromechatronics & Human Science, IEEE-MHS, pp 3-9
15. Robert G, Damjanovic D, Setter N (2001) Piezoelectric hysteresis analysis and loss separation. J. Appl. Phys. 90, pp 4668-4675
16. Albareda A, Kayombo J H, Gorri J A (2001) Nonlinear direct and indirect third harmonic generation in piezoelectric resonators by the intermodulation method. Rev. Sci. Instrum. 72 no 6, pp 2742 - 2749
17. Hall D A, Stevenson P J (1996) Field induced destabilization of hard PZT ceramics. Ferroelectrics 223, pp 309-318

18. Stevenson P J, Hall D A (1999) The effect of grain size on the high field dielectric properties of hard PZT ceramics. Ferroelectrics 228, pp 139–58
19. Pérez R, García J E, Albareda A (2002) Relation between nonlinear dielectric behaviour and alterations of domain structure in a piezoelectric ceramic. Ferroelectrics 273, pp 205 - 210
20. Demartin M, Damjanovic D (1996) Dependence of the direct piezoelectric effect in coarse and fine grain barium titanate ceramics on dynamic and static pressure. Appl. Phys. Lett. 68 no 21, pp 3046-3048
21. Pérez R, Albareda A, Pérez E, García J E, Tiana J, Gorri J A (2006) No linealidad del efecto piezoeléctrico directo d_{33} en cerámicas PZT. Bol. Soc. Esp. Ceram. V. 45 no 3. pp 197-201
22. Taylor D V, Damjanovic D (1997) Evidence of domain wall contribution to the dielectric permittivity in PZT thin films at sub-switching fields. J. Appl. Phys. 82, pp 1973-1975
23. García J E, Pérez R, Albareda A (2001) High electric field measurement of dielectric constant and losses of ferroelectric ceramics. J. Phys. D: Appl. Phys. 34, pp 3279 - 3284
24. Eitel R E, Shrout T R, Randall C A (2006) Nonlinear contributions to the dielectric permittivity and converse piezoelectric coefficient in piezoelectric ceramics. J. Appl. Phys. 99, 124110
25. Bassiri-Gharb N, Trolier-McKinstry S, Damjanovic D (2006) Piezoelectric nonlinearity in ferroelectric thin films. J. Appl. Phys. 100, 044107
26. European Standard (2002) Piezoelectric properties of ceramic materials and components—Part 3: Methods of measurement—High power. CENELEC, 50324-3
27. Hall D A (2001) Nonlinearity in piezoelectric ceramics. J. Mat. Science 36, pp 4575-4601
28. Andersen B, Ringgaard E, Bove T, Albareda A, Pérez R (2000) Performance of Piezoelectric Ceramic Multilayer Components Based on Hard and Soft PZT. Proc. Actuators'00, pp 419-422
29. García J E, Pérez R, Albareda A, Eiras J A (2007) Extrinsic response anisotropy in ferroelectric perovskite polycrystals. Solid State Commun. 144, pp 23-26
30. Damjanovic D (1997) Stress and frequency dependence of the direct piezoelectric effect in ferroelectric ceramics. J. Appl. Phys. 82 no 4, pp 1788-1797
31. Davis M, Damjanovic D, Setter N (2004) Pyroelectric properties of $(1-x)Pb(Mg_{1/3}Nb_{2/3})O_3$-$xPbTiO_3$ and $(1-x)Pb(Zn_{1/3}Nb_{2/3})O_3$-$xPbTiO_3$ single crystals measured using a dynamic method. J. Appl. Phys. 96 no 5, pp 2811-2815
32. Planat M, Hauden D (1985) Nonlinear properties of bulk and surface acoustic waves in piezoelecric crystals. In: Taylor GW, Gagnepain JJ, Meeker TR, Nakamura T, Shuvalov LA (eds) Piezoelectricity. Gordon and Breach, New York, pp 277-296
33. Beige H, Shmidt G (1985) Electromechanical resonances for investigating linear and nonlinear properties of dielectrics. In: Taylor GW, Gagnepain JJ, Meeker TR, Nakamura T, Shuvalov LA (eds) Piezoelectricity. Gordon and Breach, New York, pp 93-103
34. Holland R, Eernisse EP (1969) Accurate measurement of coefficients in a ferroelectric ceramic. IEEE Trans. on Sonics and Ultrason. SU-16 no. 4, pp 173-181
35. Priya S, Viehland D, Vazquez Carazo A, Ryu J, Uchino K (2001) High-power resonant measurements of piezoelectric materials: importance of elastic nonlinearities. J. Appl. Phys, vol. 90 no. 3, pp 1469-1479
36. Guyomar D, Aurelle N, Richard C, Gonnard P, Eyraud L (1994) Non Linearities in Langevin Transducers. Proc IEEE Int.Ultrasonics Symposium, Cannes, vol 2, pp 925-928
37. Pérez R, Albareda A (1996) Analysis of non-linear effects in a piezoelectric resonator. J. Acoust. Soc. Amer. 100 no. 6, pp 3561-3570

38. Albareda A, Gonnard P, Perrin V, Briot R, Guyomar D (2000) Characterization of the Mechanical Nonlinear Behaviour of Piezoelectric Ceramics. IEEE Trans. Ultrason., Ferroelect., Freq. Contr. 47 no. 4, pp 844-853
39. Blackburn JF, Cain M (2006) Nonlinear piezoelectric resonance: A theoretically rigorous approach to constant I-V measurements. J. Appl. Phys. 100, 114101
40. Uchino K (1997) Piezoelectric actuators and ultrasonic motors. Tuller HL (eds), Kluwer Acad. Pub., Norwell MA
41. Hirose S, Takahashi S, Uchino K, Aoyagi M, Tomikawa Y (1995) Measuring methods for high-power characteristics of piezoelectric materials. Proc. Mater. for Smart Systems, Mater. Res. Soc. 360, pp 15-20
42. Albareda A, Pérez R, Casals JA, García JE, Ochoa DA (2007) Optimization of elastic nonlinear behaviour measurements of ceramic piezoelectric resonators with burst excitation. IEEE Trans. Ultrason., Ferroelect., Freq. Contr. 54 no 10, pp 2175-2188
43. Albareda A, Casals JA, Pérez R, Montero de Espinosa F (2001) Nonlinear measurements of piezocomposite transducers with burst excitation. Proc. 12th-IEEE ISAF'00, pp 979-982
44. Casals JA, Albareda A, Pérez R, García JE, Minguella E, Montero de Espinosa F (2003) Non-linear Characterization with Burst Excitation of 1-3 Piezocomposite Transducers. Ultrasonics 41, no 4, pp 307-311
45. Takahashi S, Sasaki Y, Umeda M, Nakamura K, Ueha S (2001) Nonlinear Behaviour in Piezoelectric Ceramic Transducers. Proc. 12th IEEE Int. Symp. Appl. Ferroelect., pp 11-16
46. Umeda M, Nakamura K, Ueha S (1998) The measurement of high power characteristics for a piezoelectric transducer based on the electrical transient response. Jpn. J. Appl. Phys. 37, pp 5322-5325
47. Blackburn JF, Cain M (2007) Non-linear piezoelectric resonance analysis using burst mode: a rigorous solution. J. Phys. D: Appl. Phys. 40, pp 227-233
48. Albareda A, Pérez R, García JE, Ochoa DA (2007) Non-linear elastic phenomena near the radial antiresonance frequency in piezoceramic discs. J. of Electroceramics 19, pp 427-431
49. Hall D A (1999) Rayleigh behaviour and the threshold field in ferroelectric ceramics. Ferroelectrics 223, pp 319-328
50. Damjanovic D, Demartin M (1997) Contribution of the irreversible displacement of domain walls to the piezoelectric effect in barium titanate and lead zirconate titanate ceramics. J. Phys.: Condens. Matter. 9, pp 4943-4953
51. Pérez R, García J E, Albareda A (2001) Nonlinear Dielectric Behaviour of Piezoelectric Ceramics. Proc. IEEE-ISAF'00, pp 443-446
52. Damjanovic D, Demartin M (1996) The Rayleigh law in piezoelectric ceramics. J. Phys. D : Appl. Phys. 29, pp 2057-2060
53. García J E, Pérez R, Albareda A (2002) Manifestación de la estructura de dominios en el comportamiento dieléctrico no lineal de una cerámica piezoeléctrica. Bol. Soc. Esp. Ceram. V. 41 no 1, pp 75-79
54. Robert G, Damjanovic D, Setter N (2000) Preisach modelling of ferroelectric pinched loops , Appl. Phys. Lett. Vol. 77, pp 4413-4415
55. Robert G, Damjanovic D, Setter N (2001) Preisach distribution function approach to piezoelectric nonlinearity and hysteresis. J. Appl. Phys. 90, pp 2459-2464
56. Turik S A, Reznitchenko L A, Rybjanets A N, Dudkina S I, Turik A V, Yesis A A (2005) Preisach model and simulation of the converse piezoelectric coefficient in ferroelectric ceramics. J. Appl. Phys. 97, 064102
57. Cross L E (2000) Domain and phase change contributions to response in high strain piezoelectric actuators. AIP Conf. Proc. 535, pp 1-15

58. García J E, Gomis V, Pérez R, Albareda A, Eiras J A (2007) Unexpected dielectric response in lead zirconate titanate ceramics: the role of ferroelectric domain wall pinning effects. Appl. Phys. Letter 91, 042902
59. Trolier-McKinstry S, Bassiri-Gharb N, Damjanovic D (2006) Piezoelectric nonlinearity due to motion of 180° domain walls in ferroelectric materials at subcoercive fields: A dynamic poling mode. Appl. Phys. Lett. 88, 202901
60. Fousek J, Janovec V (1969) Orientation of domain walls in twinned ferroelectric crystals. J. Appl. Phys. 40, pp 135-142
61. Mueller V, Zhang Q M (1998) Nonlinearity and scaling behaviour in donor-doped lead zirconate titanate piezoceramic. Appl. Phys. Lett. 72, pp 2692-2694
62. Pérez R, García J E, Albareda A, Ochoa, D A (2007) Extrinsic effects in twinned ferroelectric polycrystals. J. Appl. Phys. 102, 044117
63. Chaplya P M, Carman G P (2001) Dielectric and piezoelectric response of lead zirconate–lead titanate at high electric and mechanical loads in terms of non-180° domain wall motion. J. Appl. Phys. 90, pp 5278-5286
64. Arlt G, Hennings D, With G (1985) Dielectric properties of fine-grained barium titanate ceramics. J. Appl. Phys. 58, pp 1619-1625
65. Arlt G, Sasko P (1980) Domain configuration and equilibrium size of domains in $BaTiO_3$ ceramics J. Appl. Phys. 51, pp 4956-4960
66. Pérez R, Albareda A, García J E, Casals J A (2004) Relación entre los comportamientos no lineales dieléctrico y mecánico en cerámicas piezoeléctricas de PZT. Bol. Soc. Esp. Ceram. V. 43 no 3, pp 658-662
67. Noheda B, Cox D E, Shirane G, Gonzalo J A, Cross L E, Park S-E (1999) A monoclinic ferroelectric phase in the $Pb(Zr_{1-x}Ti_x)O_3$ solid solution. Appl. Phys. Lett. 74, pp 2059-2061
68. Noheda B, Gonzalo J A, Cross L E, Guo R, Park S-E, Cox D E, Shirane G (2000) Tetragonal-to-monoclinic phase transition in a ferroelectric perovskite: The structure of $PbZr_{0.52}Ti_{0.48}O_3$. Phys. Rev. B 61, pp 8687-8695
69. Damjanovic D, Demartin M, Duran Martin P, Voisard C, Setter N (2001) Maxwell–Wagner piezoelectric relaxation in ferroelectric heterostructures. J. Appl. Phys. 90, pp 5708-5712

Chapter 16
Piezoelectric Transducers for Structural Health Monitoring: Modelling and Imaging

Yago Gómez-Ullate Ricón, Francisco Montero de Espinosa Freijo

16.1 Introduction

The use of ultrasonic Lamb waves [1, 2, 3] is emerging as one of the most effective techniques for damage detection in aeronautical structures [4]. The advantage of using these waves is evident as they can propagate over large distances; thus, avoiding moving the transducer over the whole structure as is case with conventional point-by-point measurement techniques. These waves can be excited and detected by a variety of methods, i.e., interdigital transducers (IDTs), fine point contact transducers, air-coupled ultrasonic transducers, laser-generation methods, and wedge transducers. However, among these methods, embedded piezoelectric array transducers are an effective method for the Non Destructive Inspection of panel structures made of metallic or composite materials. Piezoelectric materials (PZTs) are particularly attractive for damage detection as long as they can act simultaneously as transmitters and receivers. Moreover, these ceramics are available as small plates of different thickness which can be cut to sensors of the desired geometry. Piezoelectric materials produce strain as a result of an electric signal excitation and also display the converse effect where an applied strain produces an electrical signal [5]. These sensors can be bonded or embedded on the structure to be analyzed. Its reduced thickness, low weight and low cost make them useful when designing an integrated damage monitoring system. The degradation of the piezoceramic materials under the influence of mechanical cyclic loading is an important area of work that has also been widely investigated [6]. Mechanical stress, electrical stress and temperature are the key factors in determining the long-term durability of the piezoelectric material. Many studies have been conducted to determine degradation of piezoelectric elements based on the above parameters.

Instituto de Acústica, Serrano 144, 28006 Madrid

The design of ultrasonic transducers for application in an ultrasonic, nondestructive evaluation (NDE) system to facilitate in-service condition monitoring of plate-type structures is of considerable interest to materials engineers [7]. The integration of a permanently installed transducer into a structure, either embedded into or bonded onto the surface, can be used to provide periodic information pertaining to the health of the structure. Consequently, the ultrasonic transducer will need to be lightweight, unobtrusive with respect to the thickness of the plate under investigation, and conformable to facilitate operation in nonplanar structures. The development of a Structural Health Monitoring (SHM) system using Lamb waves requires the use of these piezoceramics [8, 9, 10]. With the use of few of these transducers in array configuration, a large area of the structure can be monitored, and then a 2D image can be implemented with the received interrogating Lamb wave signals [11].

Principal sources of damage in aluminium or composite material plate-like structures are crack propagation and delamination under fatigue loading [12, 13, 14, 15]. Currently, there exist several techniques for inspecting these structures, such as aircraft panels, but these are time consuming and expensive, and require the aircraft involved to be taken out of service. A convenient inspection technique for the interrogation of large structures involves ultrasonic Lamb wave propagation. The monitoring of aeronautical structures with Lamb waves requires the use of thin piezoelectric transducers integrated into the structure [16, 17, 18]. This has the advantages of long-range propagation, sensitivity to internal flaws and whole-thickness coverage.

This chapter presents the study, design and development of a damage detection system for plate-like structures. Lamb waves are generated and detected by several piezoceramic sensors and actuators bonded to the structure (a multi-transducer system). The piezoelectric elements used in the experiments were made from lead zirconate titanate (PZT), and were bonded onto the aluminium specimen using a Loctite® room temperature cure structural adhesive.

By propagating over large distances with low attenuation, these waves provide information about the integrity of the structure monitored. The developed system allows a 2D image of the inspected structure to be obtained, and thus enable defects to be detected.

The developed prototype is a low-cost system, designed to be easily and permanently integrated into critical structural locations so as to facilitate life-cycle management decisions.

16.2 Lamb Wave Dispersion Curves

Piezoelectric transducers are suitable for the generation and reception of Lamb waves in plate-like structures made of materials such as aluminium or carbon fibre composites. The problem arrives with their dispersive nature. That is, for a given frequency, multiple modes can exist, making defect identification difficult. The choice of the best frequency (and consequently the Lamb mode) is always a big issue in a Structural Health Monitoring system. The dispersion curves give us an idea of the various existing modes and their velocities for each frequency of excitation. Therefore, it becomes necessary to choose the optimal frequency excitation and the appropriate actuator location to obtain the modes more suited to defect detection.

16.2.1 Experimental Dispersion Curves

The experimental dispersion curves were obtained for the case of metallic (aluminium) and composite (Carbon Fibre Reinforced, CFR) structures. PZT actuators (Ferroperm® PZ 27, 4 MHz) were used in this work because of their high force output at relatively low voltages, and their good response qualities at both low and high frequencies.

16.2.1.1 Flat Aluminium Plate

A 1.1 mm thick, 1200x1200 mm aluminium plate was instrumented with one piezoceramic 0.5 mm thick, 7x7 mm square. A small drop of instant adhesive was placed on the centre of the plate and then the piezoceramic was glued on by pressing firmly for a few seconds. The piezoceramic, when excited, resonated in its thickness mode to generate an omnidirectional Lamb wave. A 5052 Panametrics® Pulser-Receiver was used to excite the actuator with a broadband signal. An optical vibrometer (Polytec® OFV 5000) controlled by a 3D computerized system, measured the propagating Lamb modes, capturing the displacement signals from each millimetre over a distance of 50 mm. A 3D stage with 0.1 mm resolution moved the vibrometer head along the three Cartesian directions. A LabVIEW® program, controlling a Tektronix® TDS 220 Oscilloscope, was developed to digitize the measured signals and record them in a computer for post-processing. The collected signals were 2500 points length and 64 signals averaged. The experimental dispersion curves were obtained by applying to the data collected a 2D-FFT algorithm implemented in a MATLAB® program. A schematic diagram of the experimental set-up is shown in Fig. 16.1.

Fig. 16.2 shows a plot of the wave number versus the frequency for the symmetric and antisymmetric Lamb wave modes. The dispersion curves obtained confirm the effective generation of Lamb waves. It can be observed that the first five Lamb propagating modes are present. Note that, at a given frequency value, several Lamb modes may be present. The branches corresponding to each Lamb mode are clearly identifiable. At low frequencies, below 1 MHz, only two Lamb modes are present: the A0 mode and the S0 mode. Frequency excitation ranges above 1,5 MHz would produce other modes making lamb mode selection difficult. Consequently, a frequency range up to 500 kHz has been chosen for this study.

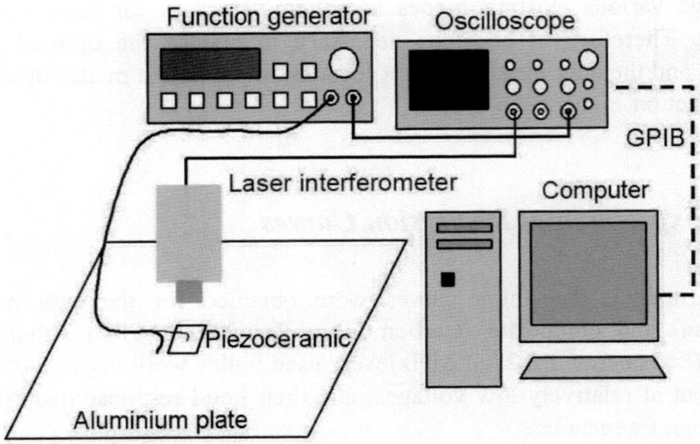

Fig. 16.1 Schematic diagram of the experimental set-up.

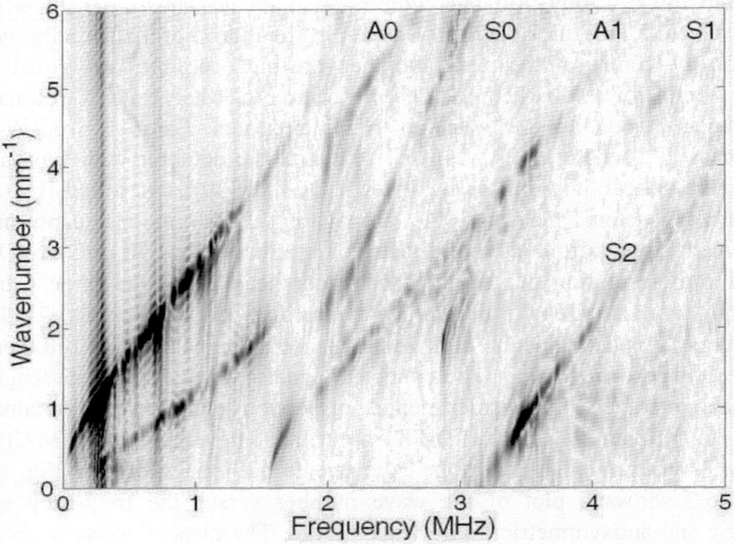

Fig. 16.2 Experimental dispersion curves for an aluminium plate of thickness 1.1 mm.

A detailed part of the experimental dispersion curves is plotted in Fig. 16.3. The X-axis has been set to be 0 to 500 kHz. The first two fundamental modes are present, the antisymmetric mode A0 (superior branch) and the symmetrical mode S0 (inferior branch).

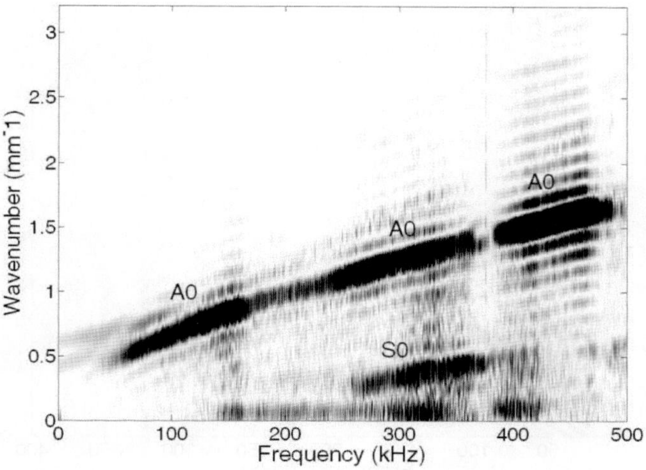

Fig. 16.3 Detailed part of the symmetric and antisymmetric modes showing the frequency region of interest of the experimental dispersion curves.

16.2.1.2 Carbon Fibre Plate

The experimental dispersion curves of a carbon fibre (CFR) plate were obtained following the same procedure as in the case of the aluminium plate (see Fig. 16.4). The tested specimen was a 804 mm x 624 mm x 2.4 mm simple flat plate built of eight layers of 0/90 fabric (Hexcel Composites S.L.) and four layers of unidirectional tape AS4/8552 material (Hexcel Composites S.L.) with the following ply arrangement $(0/90, 0/90, 0, 0, 0/90, 0/90)_{sym}$, where the principal stiffness direction is aligned with the longitudinal axis of the transducer. It can be observed that, for the case of a CFR plate, only the antisymmetric mode is present.

16.3 Design, Manufacture and Installation of a Flexible Linear Array

In order to facilitate the installation of the piezoelectric actuators in the structure to be analyzed, a linear flexible array was designed and built. The array substrate used was a specially designed flexible kapton-cooper circuit as shown in Fig. 16.5. The individual elements (eight in all) composing the array are square piezoceramics (American Piezoceramic Inc. APC85) 0.5 mm thick, 7x7 mm. These transducers can be effectively used as both sensors and actuators. The piezoceramics were gently bonded to the kapton surface with glue by the face, with the two electrodes leaving free the ground face which will be in direct contact with the inspected plates.

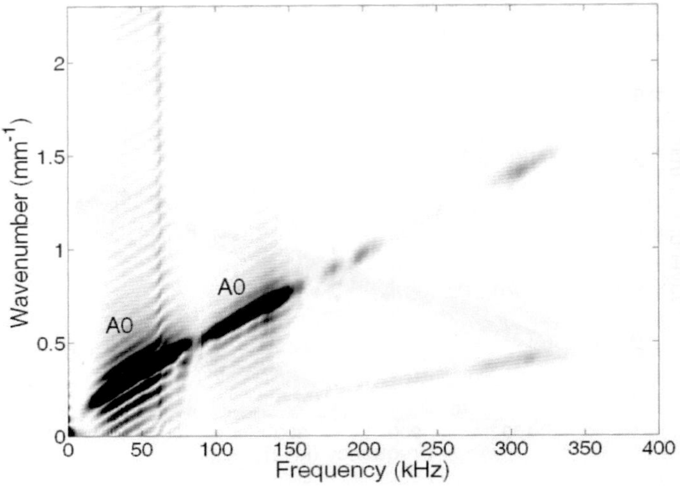

Fig. 16.4 Experimental dispersion curves for a carbon fibre reinforced plate with uniform thickness.

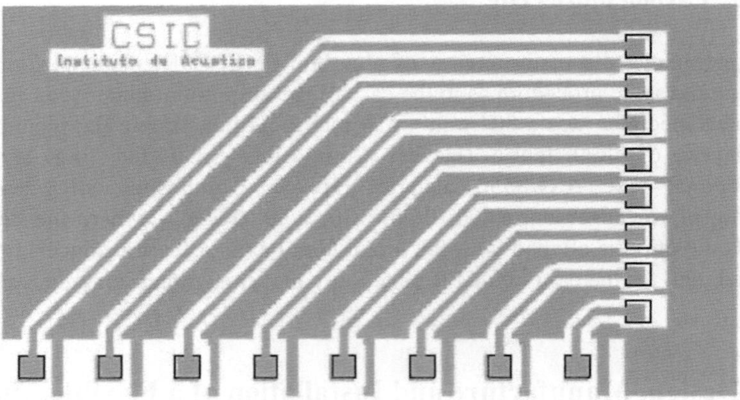

Fig. 16.5 Layout of the array flexible support.

The piezoceramic elements are arranged in a linear shape and separated a distance of $d \cong \lambda/2$, where λ is the wavelength associated with the guided wave propagating along the plate-like structure. As shown before, by exciting electrically the structure formed by a square piezoceramic 7x7x0.5 mm bonded to an aluminium plate, 1.1 mm thick, several Lamb wave modes are generated. A frequency of 300 kHz is obtained as the first optimal excitation frequency for the fundamental symmetric mode S0 with its corresponding wave velocity c = 5440 m/s and wavelength λ = 18mm. Similarly, for the case of the fundamental antisymmetric mode A0, an optimal frequency of 100 kHz, with a propagation velocity c = 2000 m/s, has been obtained. Hence, the spacing for the individual elements of the array was selected d = 9mm.

16.3.1 Study of the Diffraction Pattern of Piezoceramic Elements Attached to Aluminium Plates

In order to interpret the images obtained with an ultrasonic NDE system, as the one presented in this work, it is necessary to study the diffraction pattern of the transducers used. The diffraction of ultrasonic in fluid media and solids is well established. This is not the case for the diffraction of thin ultrasonic actuators attached to solids emitting plate waves. The theoretical study is complex and it is not yet solved. An alternative way to study the diffraction pattern of these transducers is the use of a commercial diffraction simulation program (FIELD). A modified acoustic field model was used to predict as first approximation the diffracted elastic field. The model, based in the Rayleigh integral for longitudinal waves in fluids, was modified to take into account that the waves were confined along a plate and so, there is no geometrical dispersion. The square piezoceramic bonded to the border of an aluminium plate was approximated to a rectangular emitter of equal aperture size and infinite lateral dimensions (see Fig. 16.6). In this way, the effect of diffraction in the X-axis due to the aperture is eliminated, generating a non-diffracted wave in that dimension and therefore "confined" as is the case of the plate wave.

With this approximation and using the piezoceramic dimensions and the propagating wavelength of the Lamb modes, the diffraction lobe in the XY plane was simulated. Fig. 16.7 shows the results obtained for a single piezoceramic element emitting the A0 mode (100 kHz).

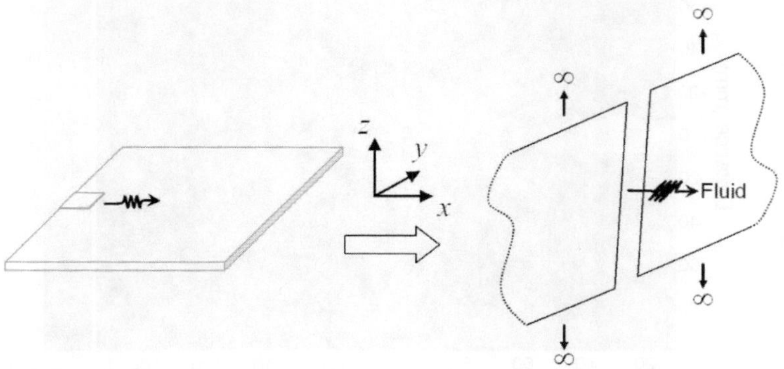

Fig. 16.6 Schematic diagram of the model approximation used to simulate the diffraction pattern.

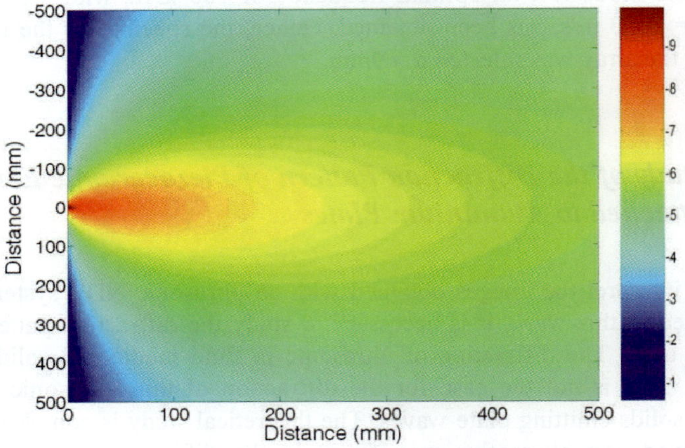

Fig. 16.7 Simulation results of the diffraction pattern for a single piezoceramic element.

In order to validate the simulation, the diffraction of a single array element was measured using a laser vibrometer (POLYTEC® OFV 5000). The laser senses the out-of-plane vibration. A computerized stage with displacement accuracy lower than 0.05 mm was used, scanning the entire plate. The diffraction for the antisymmetric A0 mode (100 kHz) of a single piezoceramic is shown in Fig. 16.8. It can be appreciated that, except for the appearance of the lobes due to the lateral sides of the piezoceramic, the simulations agree quite well with experimental results.

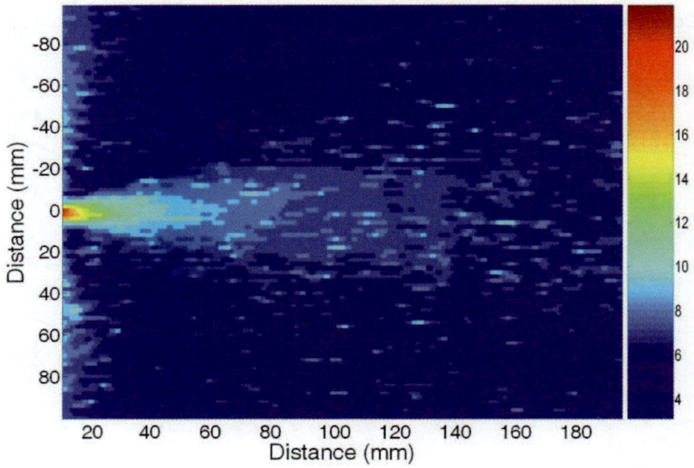

Fig. 16.8 Experimental results of the diffraction pattern for a single piezoceramic element.

After simulating a single transducer, the same program was used to simulate the diffraction pattern for a linear array composed of eight piezoceramic elements (see Fig. 16.9). The simulations were compared with the interferometric measurements (see Fig. 16.10). Good agreement is obtained between the simulations and the experimental measurements. These results confirm the validity of the approximation made in the simulations.

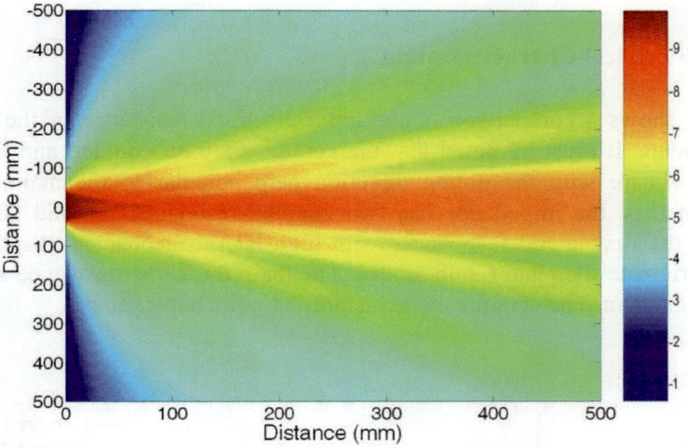

Fig. 16.9 Simulation results of the diffraction pattern for an eight elements linear array.

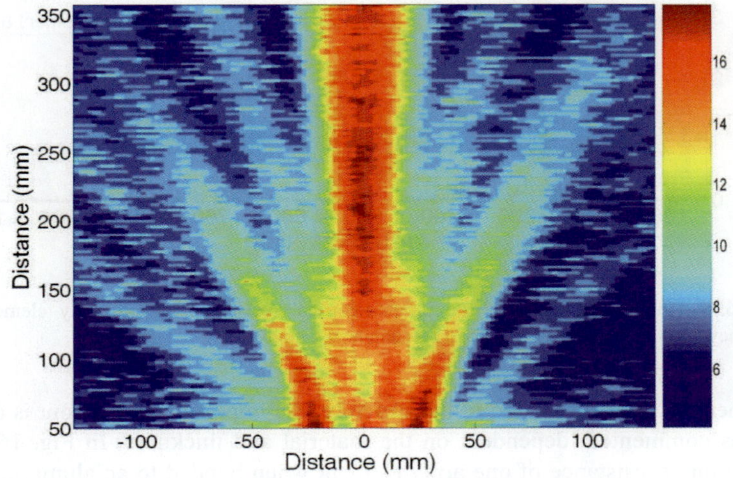

Fig. 16.10 Experimental results of the diffraction pattern for an eight elements linear array.

16.3.2 Characterization of the Array

The characterization of the flexible array included electrical impedance and crosstalk individual measurements. The array pulse cannot be measured and characterized because it must be bonded to the structure to be inspected. The working resonance and the final performance are extremely dependent on the structure (material, thickness).

16.3.2.1 Electrical Characterization

Fig. 16.11 shows a comparison of the input electrical resistance of the flexible array elements. It must be noted that the ceramics are air-coupled and the main surface is square with the poling direction perpendicular to the main surface. Because of that, the impedance has two coupled resonances around 280 KHz. These are the first lateral resonances with high lateral mechanical displacement that will originate the plate Lamb modes. The main thickness resonance around 4 MHz is not shown. The ceramic elements present good homogeneity.

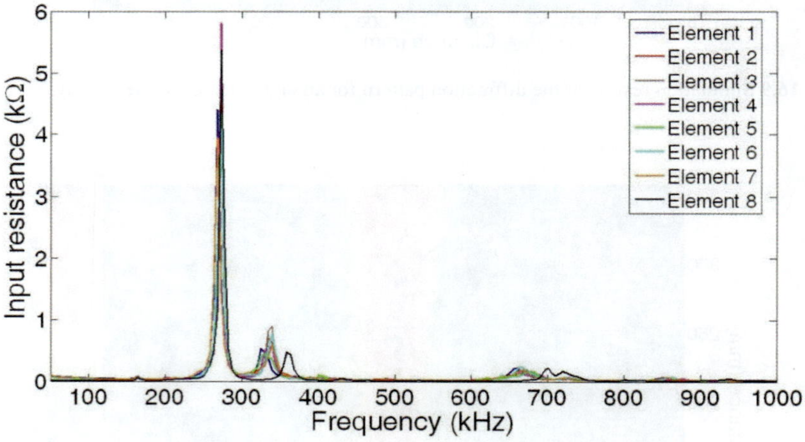

Fig. 16.11 Homogeneity test of the input electrical resistance of the array elements. The frequency range around the first lateral resonance is displayed.

When the array is bonded to the structure, the impedance behaviour is different and, as commented, dependent on the material and thickness. In Fig. 16.12, the typical input resistance of one array element when bonded to an aluminium plate, 1.1 mm thick, is shown. A shift in frequency can be observed, with the main resonance around 300 kHz.

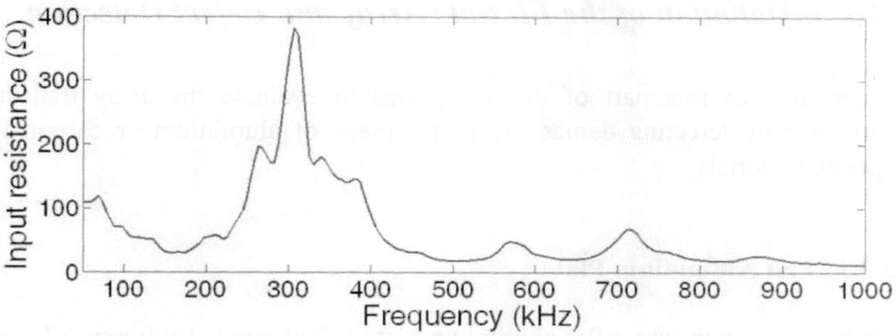

Fig. 16.12 Input electrical impedance of one array element bonded to an aluminium plate 1.1 mm thick.

16.3.2.2 Mechanical Characterization

When the array is air-coupled, no cross coupling exists between the array elements. Nevertheless, when glued, the mechanical cross coupling is high. This is the reason of the big death zone observed in the images performed with the array. The mechanical cross-coupling measured at the elements from the second to the eighth, when the first ceramic is excited with a broad band electrical signal, is depicted in
Fig. **16.13**. The test was made with a SONATEST® MASTERSCAN 330 in through-transmission mode (dual mode).

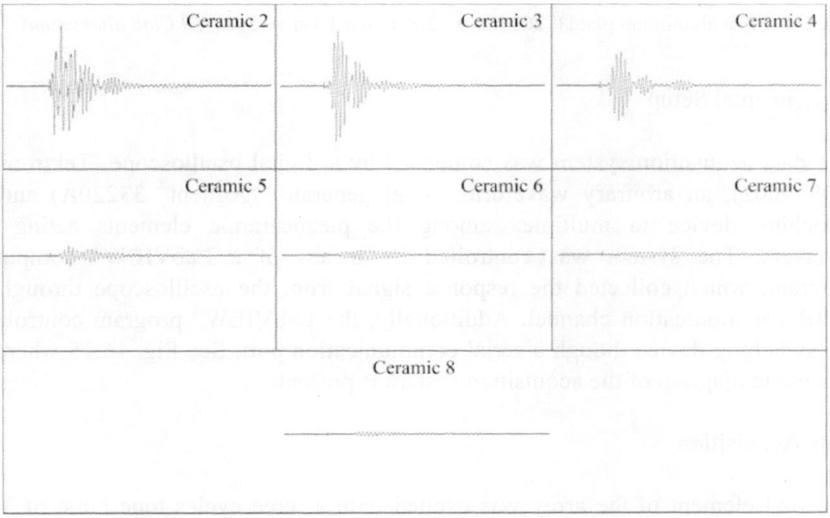

Fig. 16.13 Recorded mechanical cross-coupling of the glued array. From up-left to down- right, the received signals from the second to the eighth ceramic. First ceramic is acting as emitter.

16.3.3 Installation of the Flexible Array and Defect Detection

The objective of this part of the study was to evaluate the array transducer performance in detecting damage in plates made of aluminium or carbon fibre composite materials.

16.3.3.1 Flat Aluminium Plate

The tested specimen was a flat aluminium plate (1200 mm x 1200 mm x 1.1 mm, *2024-T3 Clad aluminium*), see Fig. 16.14. The flexible array was surface bonded at the middle of one end of the plate.

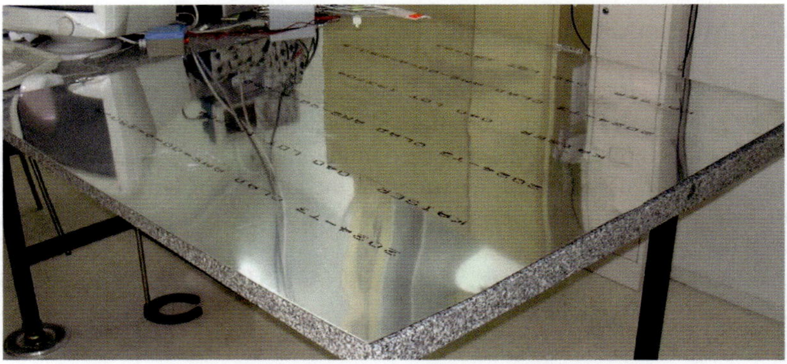

Fig. 16.14 Flat aluminium plate (1200 mm x 1200 mm x 1.1 mm, *2024-T3 Clad aluminium*).

Experimental Setup

The data acquisition system was composed by a digital oscilloscope (Tektronix® TDS 2002), an arbitrary waveform signal generator (Agilent® 33220A) and a switching device to multiplex among the piezoceramic elements acting as receivers. The system was controlled by means of a LabVIEW® computer program, which collected the response signal from the oscilloscope through a GPIB communication channel. Additionally, the LabVIEW® program controlled the switching device though a serial communication port. See Fig. 16.15 where a schematic diagram of the acquisition system is presented.

Data Acquisition

The first element of the array was excited with a three cycles tone burst of 360 kHz receiving sequentially with all the sensors including itself. This process was repeated for the rest of the elements composing the array. A multiplexer unit controlled with a LabVIEW® program was used to switch between the active emitting and receiving channels. The developed program recorded the data from the digital oscilloscope and stored it in a file. The resulting data was then processed using the Synthetic Aperture Focusing Technique (SAFT). The SAFT analysis is an imaging process that increases the signal-to-noise ratio by numerically focusing the acoustic fields [19].

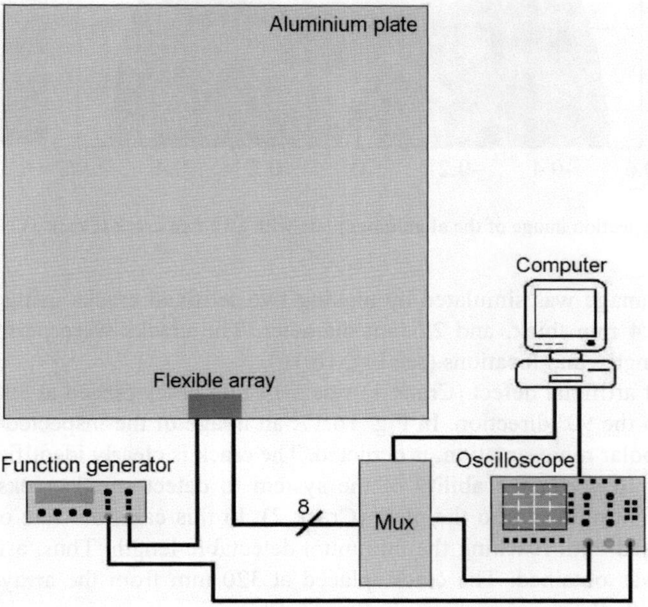

Fig. 16.15 Schematic diagram of the data acquisition system.

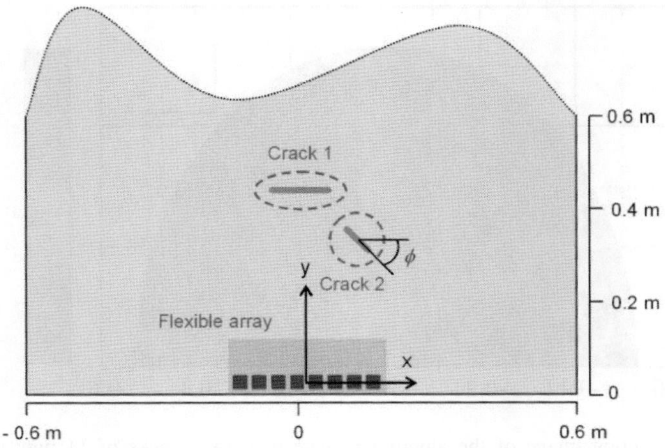

Fig. 16.16 Schematic diagram of the portion of the aluminium plate with defects.

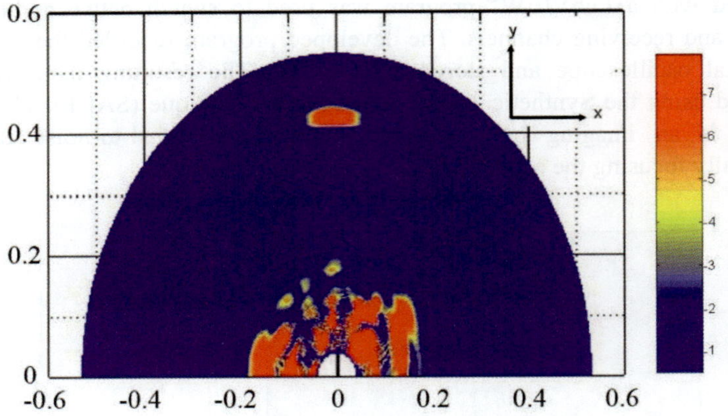

Fig. 16.17 Inspection image of the aluminium plate with a 14 mm crack (Crack 1).

Panel damage was simulated by making two artificial cracks using a diamond cut disk, 0.4 mm thick, and 22 mm diameter. The cracks were performed with different lengths and locations (see Fig. 16.16).

The first artificial defect (Crack 1) was a 14 mm crack placed at 300 mm from the array in the 90° direction. In Fig. 16.17, an image of the inspected aluminium plate, in a polar representation, is depicted. The crack is clearly identifiable.

In order to verify the ability of the system to detect small cracks, a second defect was introduced into the plate (Crack 2). In this case, the size of the crack was increased until reaching the minimum detectable length. Thus, a final length of 8 mm was obtained. The crack, placed at 320 mm from the array in the 75° direction with 45° slope, is shown in Fig. 16.18.

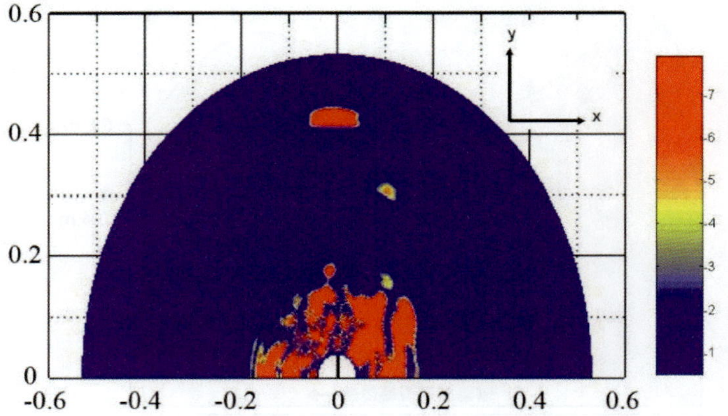

Fig. 16.18 Inspection image of the aluminium plate with two cracks of 14 mm and 8 mm, respectively (Crack 1, Crack 2).

16.3.3.2 Carbon Fibre Plate

The tested specimen was the CFR plate used for the calculation of the dispersion curves described in the previous section (see Fig. 16.19a). Panel damage was simulated by making a delamination using a 1.54 kg, 20 mm in diameter, drop weight mobile impactor. The impacts were performed with different energy levels (5, 10, 15 and 20 J). After each impact, the damage extent was monitored using conventional pulse-eco techniques. A delamination with a final diameter of 22.5 mm was used in the experiments. Most of the dead zone, due to the mechanical cross-coupling between the elements of the array, has been removed for a better representation. The presence of the delamination and the border of the plate can be observed at 400 and 600 mm respectively (see Fig. 16.19b).

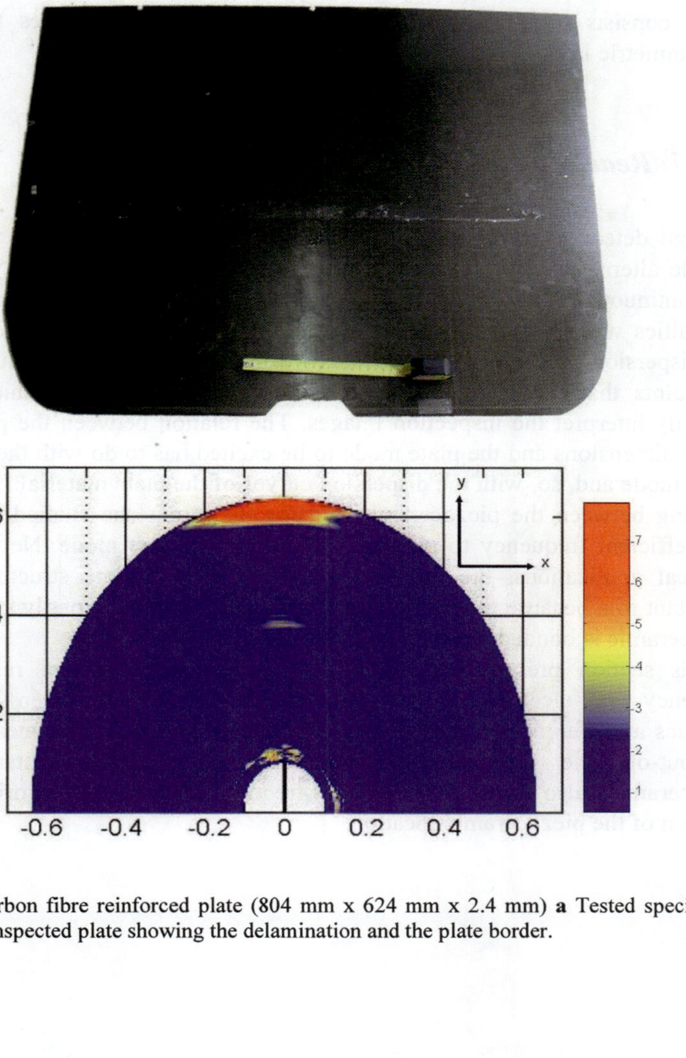

Fig. 16.19 Carbon fibre reinforced plate (804 mm x 624 mm x 2.4 mm) **a** Tested specimen **b** Image of the inspected plate showing the delamination and the plate border.

16.4 Study of Crosstalk Reduction in Linear Piezoelectric Arrays for Imaging in Structural Health Monitoring Applications

The performances of transducers can be strongly affected when the piezoceramic is bonded close to a plate border. This is because of the finite dimensions of the plate structures. In that case, a large cross-coupling between the different piezoelectric array elements appears. This effect increases the image "dead zone". This cross-coupling has been minimized by separating the array from the border. However, when the separation distance is in the order of the wavelength or so, unwanted frequency band-gaps may appear in the frequency transducer band because of the reactive effect of the reflected wave, making difficult the choice of the optimal excitation frequency of the system. A final improvement is proposed which consists of the use of 2-2 piezocomposite square plates that kill the antisymmetric lateral cross-coupling [20].

16.4.1 Reactive Effect of the Plate Border

Damage detection techniques based on the propagation of Lamb waves are a reliable alternative to conventional methods and offer the possibility of a quick and continuous method of inspection of plate structures. However, there are some difficulties when designing such a system. The presence of multiple modes and the dispersion phenomenon associated with Lamb waves introduce physical constraints that must be taken into account to design the techniques and to correctly interpret the inspection images. The relation between the piezoceramic lateral dimensions and the plate mode to be excited has to do with the wavelength of the mode and, so, with the dispersion curves of the plate material. The dynamic coupling between the piezoceramic and the plate must be studied to know the most efficient frequency to propagate the desired Lamb mode. Nevertheless, in practical applications, the limited dimensions of the plate structures play an important role because the transducer performances can be strongly affected if the piezoceramic is bonded close to a plate border [21].

This section presents the finite element and experimental results of the frequency and electric response of the plate-ceramic set when bonding the ceramics at different positions from the border of the plate. The measurement of the out-of-plane mechanical displacement at a certain distance of the piezoceramics also shows that different frequency spectrums are originated as a function of the piezoceramic location.

16.4.1.1 Finite Element Study

In order to study the dynamic coupling of the plate-ceramic set, a finite element analysis has been performed. The simulation results were validated with experimental measurements. First, the mechanical out-of-plane displacement for different positions of the piezoceramic (emitter) from the aluminium plate border was simulated. Next, the electrical response of the plate-ceramic set was obtained. The frequency spectrum of the mentioned results gives information about the behaviour of the transducer at different frequencies and helps in determining the optimal excitation frequency. The study of the reactance introduced by the plate border on the transducer answer shows that when it is on the order of the wavelength, this distance is critical.

Modelling the Mechanical Displacement

A 2D model of the plate-ceramic set was developed using the commercial simulation program PZFlex® (Weidlinger Associates Inc, Los Altos, CA. USA) [22], see Fig. 16.20. PZFlex® has multiple element and material types available, including fully coupled piezoelectric materials, and isotropic and anisotropic elastic solids, in both 2D and 3D [23]. The model was developed for each position of the piezoceramic. In order to minimize the processing time, the length of the plate in the wave propagation direction has been reduced with respect to the dimensions of the plates used in the previous sections. Absorbing boundaries were also set on the side of the plate more distant from the ceramic, thus assuming a plate large enough that the reflections from the border do not affect the piezoceramic. The mesh was chosen small enough so that the element size was significantly smaller than the wavelength under study. A ceramic, with the same dimensions and material properties as in the experiments, was directly surface-mounted on the plate to generate the different Lamb wave modes. The simulation outputs were the out-of-plane displacement and the electrical impedance of the plate-ceramic set.

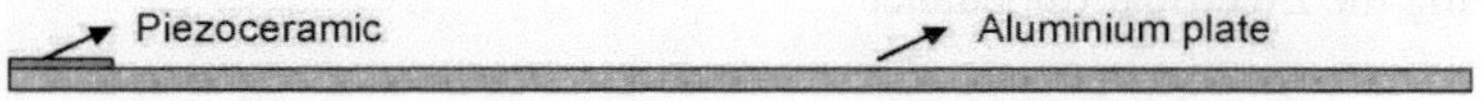

Fig. 16.20 Schematic diagram of the 2D model developed.

For each position of the piezoceramic, the out-of-plane displacement was simulated and its corresponding power spectrum was calculated. These ceramic positions (0 mm, 3.5 mm and 7 mm from the border of the plate) are related to the wavelength of the Lamb wave response at the spectrum central frequency for this piezoceramic geometry, $\lambda \approx 14$ mm. The wavelength of the A0 and S0 modes is similar because of their different velocity. A broadband signal was used to excite the piezoceramic.

Applying the Fast Fourier Transform to the simulated data, the power spectrum S_{xx} is calculated, which is the squared modulus of the Fourier transform defined as

$$X(f) = \int_{-\infty}^{\infty} x(t) e^{2\pi jft} dt \quad (1)$$

and which gives an indication about the frequency contents of the analyzed signals.

In the next figures, the simulated out-of-plane displacement signals (Fig. 16.21a, Fig. 16.22a and Fig. 16.23a) and its corresponding power spectrum (Fig. 16.21b, Fig. 16.22b and Fig. 16.23b) are plotted. Taking as reference the power spectrum of the piezoceramic attached to the border, it is found that several frequency band-gaps appear when the piezoceramic is being separated from the 0 mm position. It must be borne in mind that the laser interferometer measures the displacement perpendicular to the plate. This means that the modes having a preferential on-plane displacement (S0 modes) are under-considered when compared to those with a preferential out-of-plane displacement (A0 modes). Nevertheless, the effect of the plate border reactance is observed in both the low frequency region (antisymmetric mode, 100 kHz) and in the high frequency region (symmetric mode, 300 kHz).

The reactive effect of the plate border can be observed by comparing the amplitude spectrum shown in Fig. 16.21b with the corresponding Fig. 16.22b (3.5 mm separation). In this, the presence of frequency band-gaps in the region of 100 kHz and 200 kHz is evident. These band-gaps are still present in Fig. 16.23b (7 mm separation). In this case, the sharp decrease in amplitude around 300 kHz is remarkable and so is the presence of two new band-gaps around 50 kHz and above 400 kHz.

A wrong placement of the piezoceramics on the plate can affect negatively the mode selection, being ineffective when exciting the desired modes. As a consequence, for an appropriate system design, these band-gaps must be taken in consideration to optimize the Lamb mode generation.

Modelling the Electrical Impedance

The other strategy adopted to study the effect of the plate border on the excitation frequencies was the analysis of the electrical impedance of the plate-ceramic set. A 3D model, for each position of the piezoceramic, has been performed.

Fig. 16.24 shows the simulation results of the electrical impedance for each ceramic position. It can be observed that the higher value for the resistance is obtained when the ceramic is placed on the border of the plate (0 mm). When exciting the ceramic, the reflections produced on the border contribute to the Lamb wave mode propagating to the middle of the plate. This is due to their coincidence in phase. In the case of a separation distance of 7 mm from the border, $\lambda / 2$, we observe that the resistance is almost the same as in the case of

0 mm. The small decrease in amplitude is due to diffraction losses of the piezoceramic rear wave contribution. Finally, for a separation distance of 3.5 mm, the resistance has the minimum value. In this case, the reflecting wave arriving from the border is opposite in phase to the wave propagating to the middle of the plate, resulting in a large decrease in amplitude. This effect is well known and extensively used to enlarge the frequency bandwidth of resonant transducer: the $\lambda / 4$ matching. The changes in frequency for the maximum resonance must then be taken into account when tuning the best frequency to obtain the maximum amplitude to excite the system.

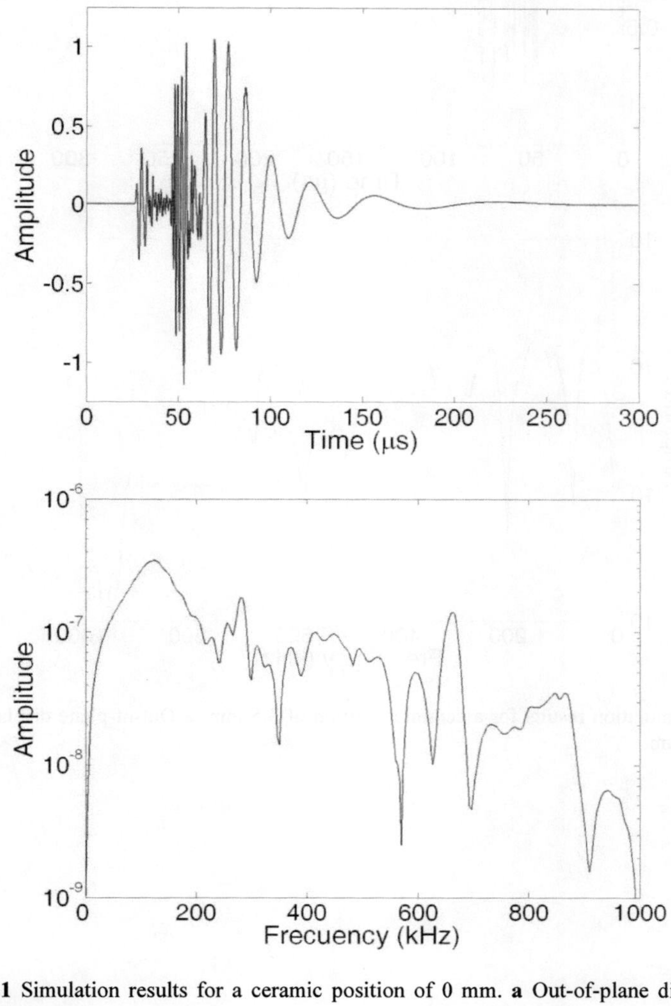

Fig. 16.21 Simulation results for a ceramic position of 0 mm. **a** Out-of-plane displacement **b** Power Spectrum.

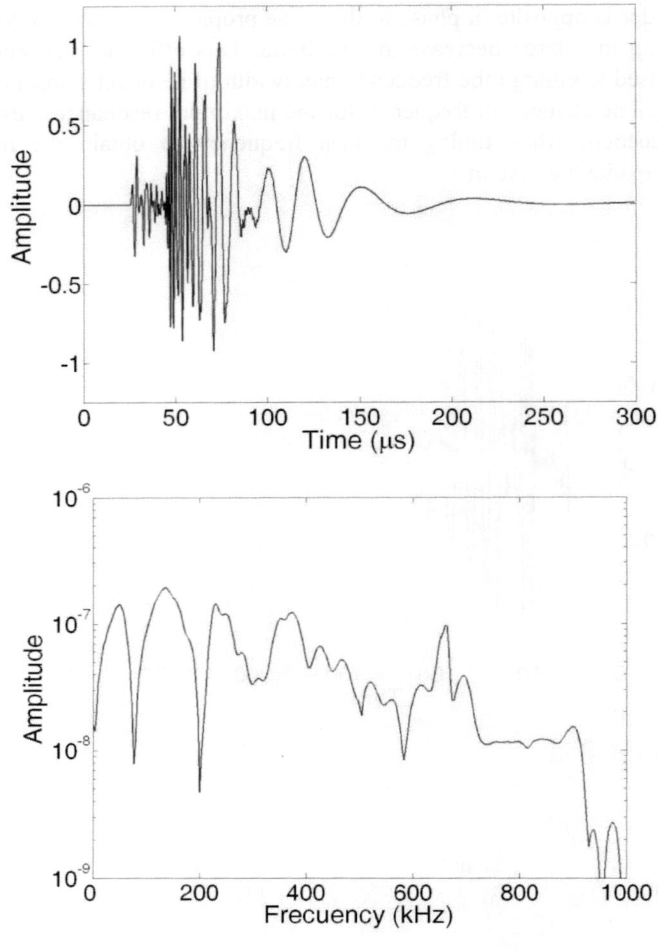

Fig. 16.22 Simulation results for a ceramic position of 3.5 mm. **a** Out-of-plane displacement **b** Power spectrum.

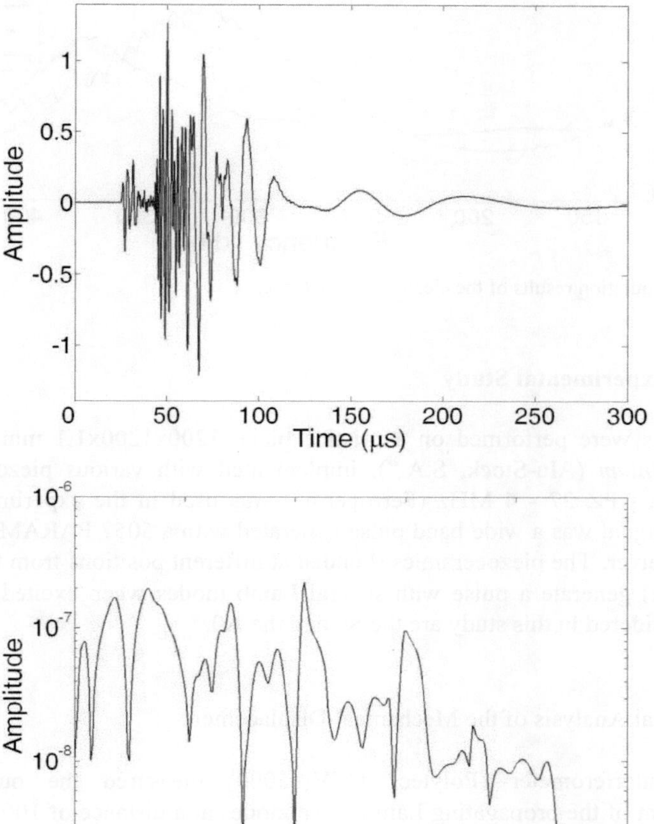

Fig. 16.23 Simulation results for a ceramic position of 7 mm. **a** Out-of-plane displacement **b** Out-of-plane displacement.

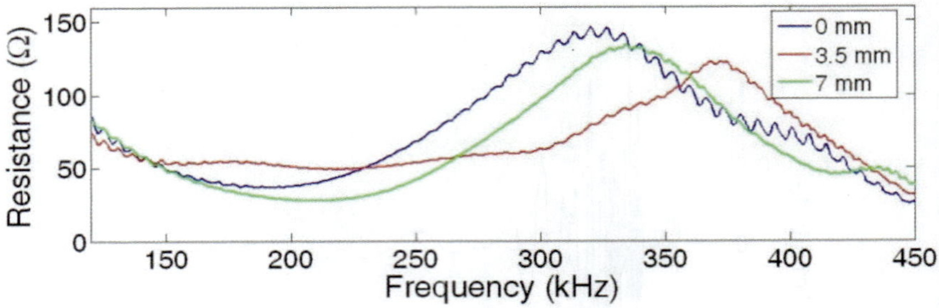

Fig. 16.24 Simulation results of the electrical impedance.

16.4.1.2 Experimental Study

Experiments were performed on a pristine plate, 1200x1200x1.1 mm, *2024-T3 Clad aluminium* (Alu-Stock, S.A.®), implemented with various piezoceramics; 7x7x0.5mm - PZ 27 - 4 MHz (Ferroperm®) was used in the experiments. The excitation signal was a wide band pulse generated with a 5052 PARAMETRICS® Pulser-Receiver. The piezoceramics (bonded at different positions from the border of the plate) generate a pulse with several Lamb modes when excited. The two modes considered in this study are the S0 and the A0.

Experimental Analysis of the Mechanical Displacement

A laser interferometer (Polytec® OFV 5000) measured the out-of-plane displacement of the propagating Lamb wave modes at a distance of 100 mm from the border of the plate. The time signal was digitized in a Tektronix® oscilloscope and recorder in a computer. Next, the power spectrum of the collected signals (2500 points length and 64 signals averaged) was calculated using the FFT. This allows us to obtain the frequency response of the plate-ceramic set, and thus to determine the optimal excitation frequency of the system.

The experimental results of the mechanical displacements are shown in the next figures Fig. 16.25a, Fig. 16.26a and Fig. 16.27a, together with the corresponding power spectrums Fig. 16.25b, Fig. 16.26b and Fig. 16.27b.

The comparison between experimental measurements (Fig. 16.25, Fig. 16.26 and Fig. 16.27) and simulation results (Fig. 16.21, Fig. 16.22 and Fig. 16.23) show a good agreement. The frequency band-gap around 100 kHz can be observed for a separation distance of 3.5 mm (Fig. 16.26b), while around 200 kHz the amplitude decrease is less than in the simulation results. In the case of a separation distance of 7 mm, the same attenuation peaks predicted in the simulations can be observed in the experimental results shown in Fig. 16.27b.

Experimental Analysis of the Electrical Impedance

The electrical impedance for each position of the plate-ceramic set was measured experimentally with an HP® 4194 Impedance/Gain-Phase Analyzer. Because of the piezoelectric nature of the transducer, the resonant frequencies and the frequency behaviour of the system can be deduced from the electrical impedance measurements.

Again, good agreement is found between the simulations (Fig. 16.24) and the experimental results (Fig. 16.28).

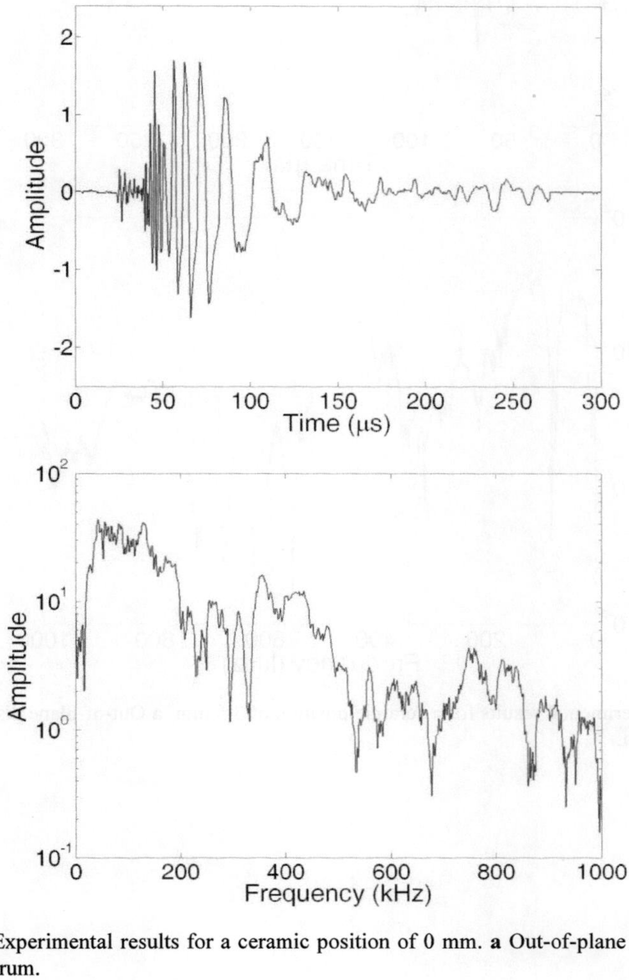

Fig. 16.25 Experimental results for a ceramic position of 0 mm. **a** Out-of-plane displacement **b** Power spectrum.

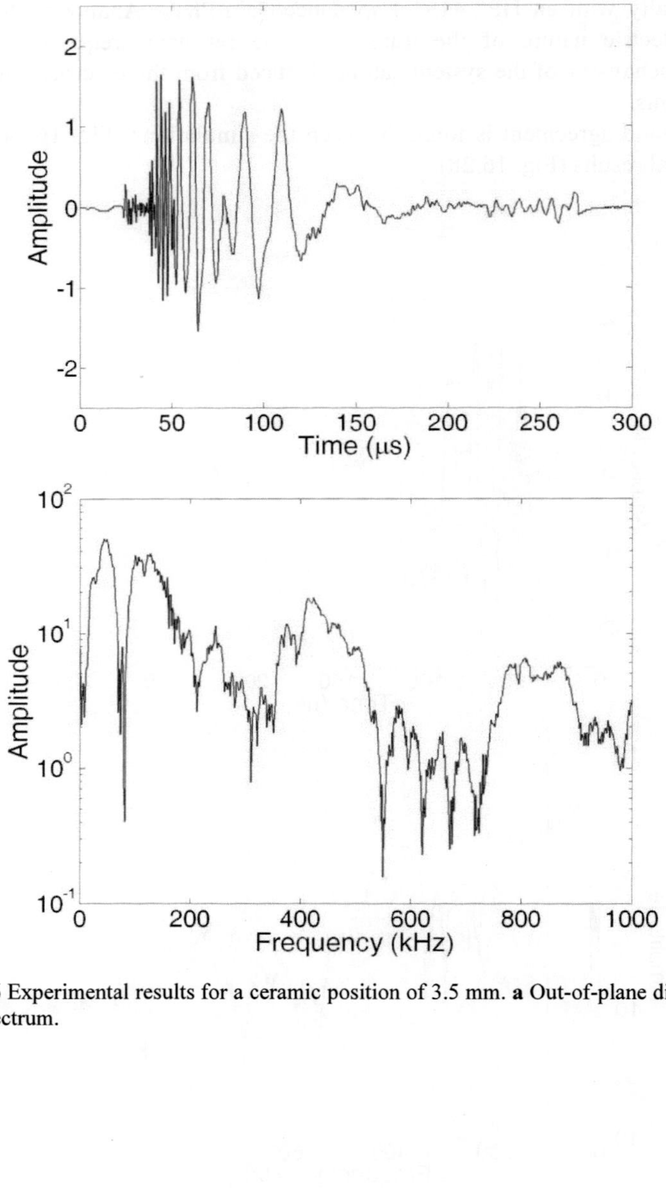

Fig. 16.26 Experimental results for a ceramic position of 3.5 mm. **a** Out-of-plane displacement **b** Power spectrum.

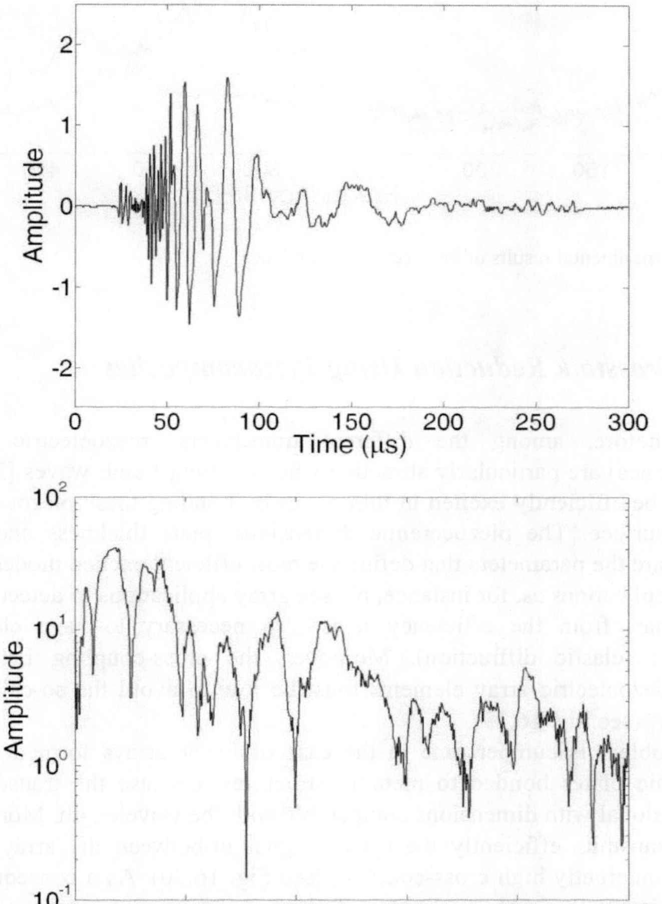

Fig. 16.27 Experimental results for a ceramic position of 7 mm. **a** Out-of-plane displacement **b** Power spectrum.

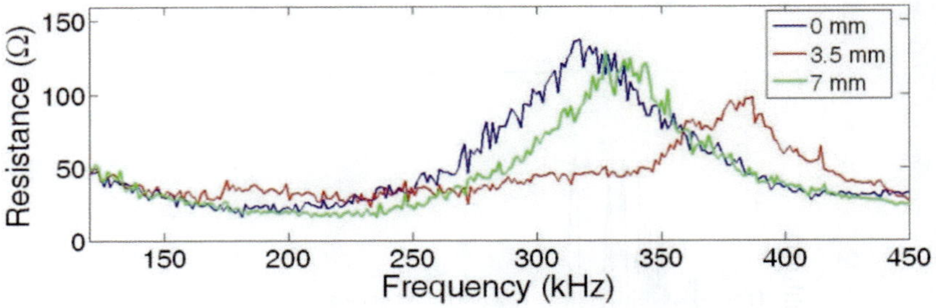

Fig. 16.28 Experimental results of the electrical impedance.

16.4.2 Crosstalk Reduction Using Piezocomposites

As said before, among the different transducers, piezoelectric materials (piezoceramics) are particularly attractive when exciting Lamb waves [24]. These waves can be efficiently excited in thin plates by bonding these piezoceramics to the plate surface. The piezoceramic dimensions, plate thickness and material properties are the parameters that define the most efficient excited modes. In some practical applications as, for instance, phased array applications to detect structural defects, apart from the efficiency it is also necessary to have clean wave propagation (elastic diffraction). Moreover, the cross-coupling between the different piezoelectric array elements must be low to avoid the so-called image "dead zone" (see Fig. 16.29).

This problem is cumbersome in the case of linear arrays formed by square piezoceramic plates bonded to metallic structures, because the transducers are two-dimensional with dimensions comparable with the wavelength. Moreover, the structure transmits efficiently the elastic signal in-between the array elements producing inherently high cross-coupling (see Fig. 16.30). As a consequence, the diffraction is not as simple as the one of piezoelectric array transducers in fluids. All these factors must be taken into account when designing a phased array system for thin metallic plates.

With these transducers, when using a conventional piezoceramic to excite a Lamb wave, several modes are generated. At the frequency excitation ranges used in this work, at least two Lamb modes are present, the fundamental symmetric S0 and antisymmetric A0 modes. The analysis of the received signals when having a multimodal Lamb wave response is complex, making the identification of defects difficult. The use of 2-2 piezocomposites is proposed in this work as an effective method to carry out the Lamb wave mode selection, decreasing the cross-coupling of inter-elements [25].

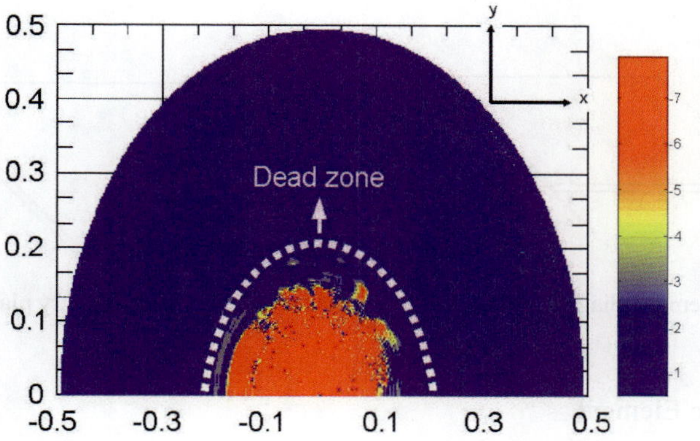

Fig. 16.29 Polar representation of the inspected aluminium plate with dead zone.

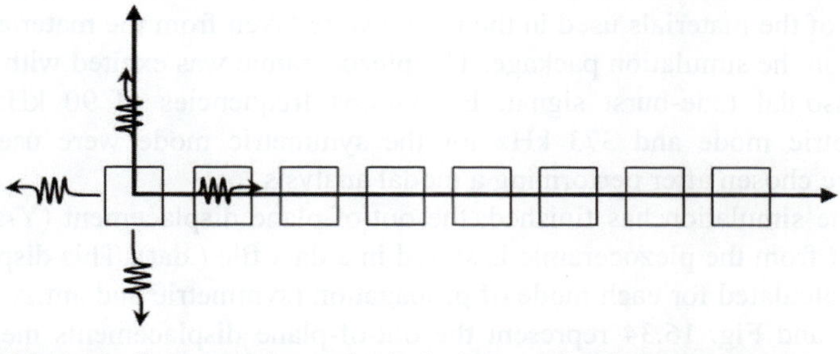

Fig. 16.30 Schematic diagram of the mechanical cross-coupling between elements.

16.4.2.1 Finite Element Study

Simulations were conducted in two steps. First, Lamb waves were generated on an aluminium plate by bonding a conventional piezoceramic on to its surface. The presence of the two fundamental modes, A0 and S0, was then verified. Next, two different piezocomposites were bonded to the same aluminium plate in order to achieve the Lamb wave mode selection.

Several 3D models using the commercial simulation program PZFlex® were developed. The metallic plate dimensions were set large enough to avoid unwanted reflections from the edges of the plate. Two symmetry planes were also set to the models to minimize the simulation time. That is, only a quarter of the plate-ceramic set was modelled and then symmetry conditions were applied to obtain the full model (Fig. 16.31). The aluminium plate was modelled as an isotropic solid, while the piezoelectric plates considered the full anisotropic material properties. Single point integration was used, and mesh density was at least twenty elements per wavelength of interest. Regular element spacing was chosen throughout the model.

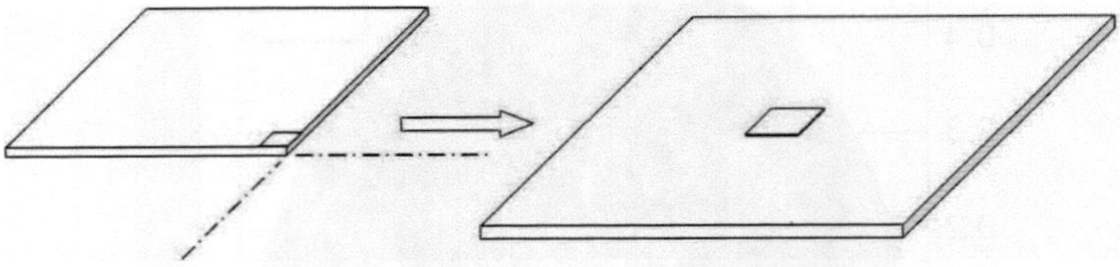

Fig. 16.31 Schematic diagram of the plate-ceramic model with the symmetry planes.

Piezoceramic Element

The first model consisted of a single piezoceramic element attached to the surface of an aluminium plate. The piezoceramic dimensions were 7x7x0.5 mm and the thickness of the aluminium plate was 1.1 mm. The mechanical and electrical properties of the materials used in the model were taken from the materials library contained on the simulation package. The piezoceramic was excited with an eight-cycle sinusoidal tone-burst signal. Excitations frequencies of 90 kHz for the antisymmetric mode and 373 kHz for the symmetric mode were used. These values were chosen after performing a modal analysis.

Once the simulation has finished, the out-of-plane displacement (Y-axis) at 1 mm distant from the piezoceramic is stored in a data file (.dat). This displacement has been calculated for each mode of propagation (symmetric and antisymmetric). Fig. 16.33 and Fig. 16.34 represent the out-of-plane displacements measured at both lateral sides of the emitter (X-axis and Z-axis). The coordinate system used in the piezoceramic simulation is shown in Fig. 16.32. As it can be seen in Fig. 16.33, the signals obtained for the antisymmetric mode (A0-90 kHz) have the same displacement amplitude. In the case of the symmetric mode (S0-373 kHz), the amplitude of displacement follows the same behaviour (see Fig. 16.34). It can be concluded that the piezoceramic exhibits the same amplitude displacements in both propagation axis. This confirms the crosstalk generation due to the propagation of theses modes along the linear array axis.

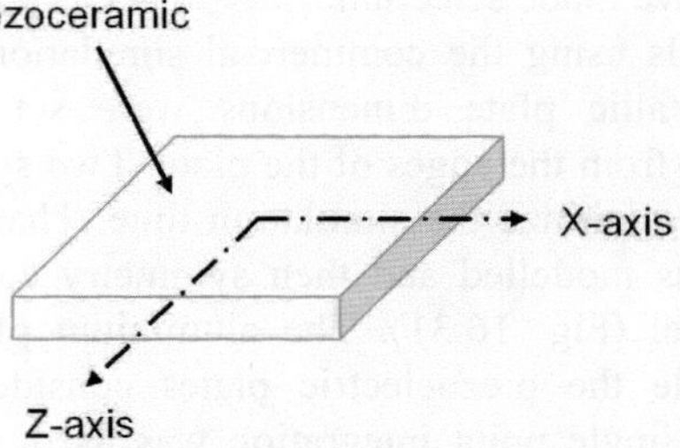

Fig. 16.32 Schematic diagram of the coordinate system for the simulated piezoceramic.

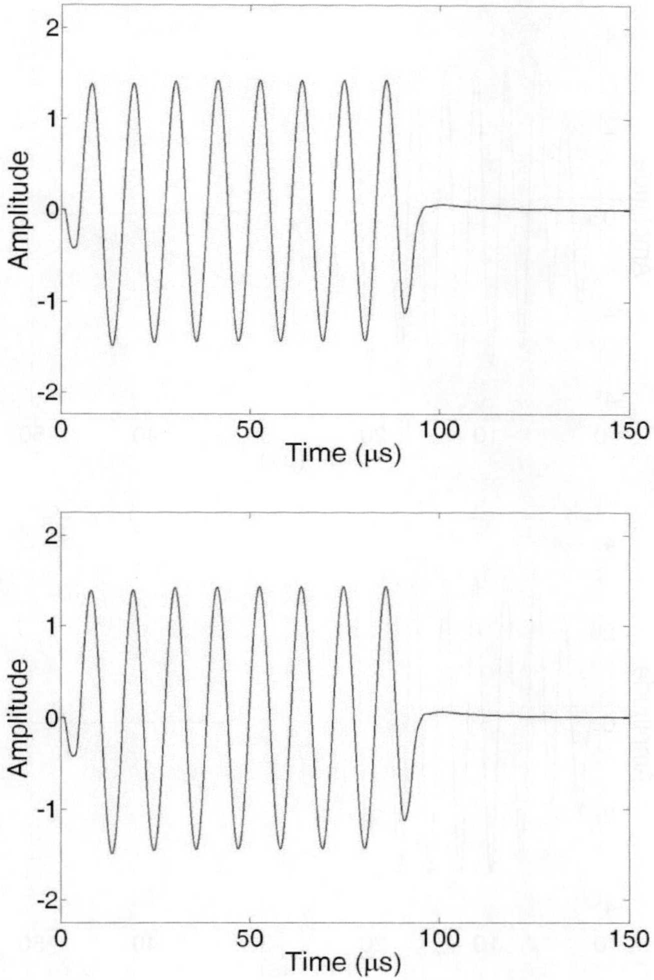

Fig. 16.33 Simulation results of the out-of-plane displacement for the case of a piezoceramic. Results for the antisymmetric mode (90 kHz) calculated at 1 mm from the emitter. **a** Z-axis **b** X-axis.

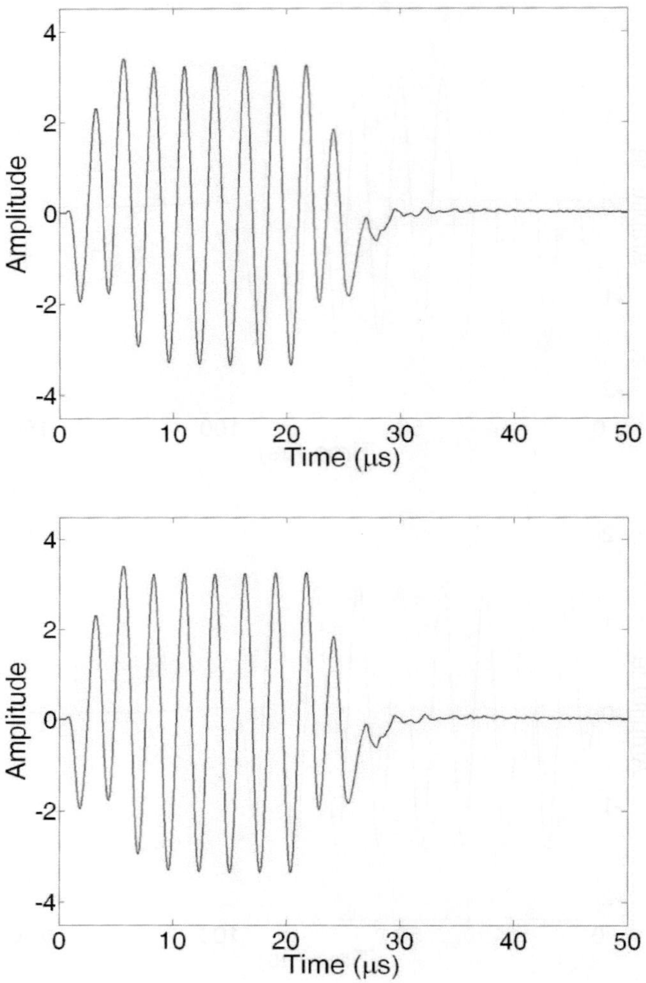

Fig. 16.34 Simulation results of the out-of-plane displacement for the case of a piezoceramic. Results for the symmetric mode (373 kHz) calculated at 1 mm from the emitter. **a** Z-axis **b** X-axis.

Piezocomposite

Piezoelectric composites (also called piezocomposites) are piezoelectrically active materials composed by a ceramic component and a passive, usually polymer, component [26, 27]. In order to verify the mode selection a 2-2 piezocomposite was constructed with the same dimensions of the piezoceramic element (7x7x0.5 mm). The piezocomposite was modelled intercalating longitudinal piezoceramic elements (0.2x7x0.5 mm) with a soft polymer (0.05x7x0.5 mm) until reaching the ceramic dimensions, see Fig. 16.35. The piezocomposite electrode thickness was assumed to be negligible. The mechanical properties of the soft polymer used are listed in Table 1.

Table 1 Mechanical properties of the soft polymer.

Vantico® HY956/CY208	
Young's modulus (m s^{-2} Kg)	1.8486×10^9
Poisson's coefficient	0.4188
Density (Kg m^{-3})	1165

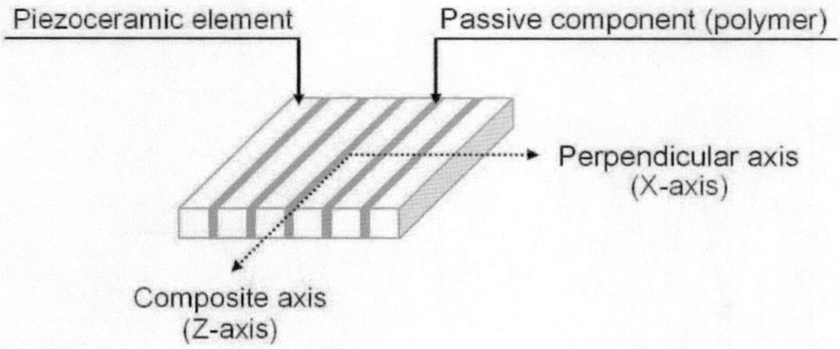

Fig. 16.35 Schematic diagram of the piezocomposite with the coordinate system.

As is the case of the piezoceramic, the out-of-plane displacement was simulated along the composite axis (Z-axis) and perpendicular axis (X-axis). This displacement was calculated at 1 mm from the piezocomposite. Frequencies of 90 kHz for the antisymmetric mode and 373 kHz for the symmetric were used to excite the piezocomposite. Fig. 16.36 and Fig. 16.37 show the simulated out-of-plane displacements when using the soft polymer composite. The propagation of the antisymmetric mode in the perpendicular axis (see Fig. 16.36b) is dramatically damped, while along the composite direction the amplitude remains unaltered (see Fig. 16.36a). This decrease in amplitude is due to the mechanical displacements originated by the cuts introduced in the piezoceramic material and the mismatch between the piezoceramic and polymer. In the case of the symmetric mode, the amplitude of displacement in both axes is of the same magnitude (see Fig. 16.37).

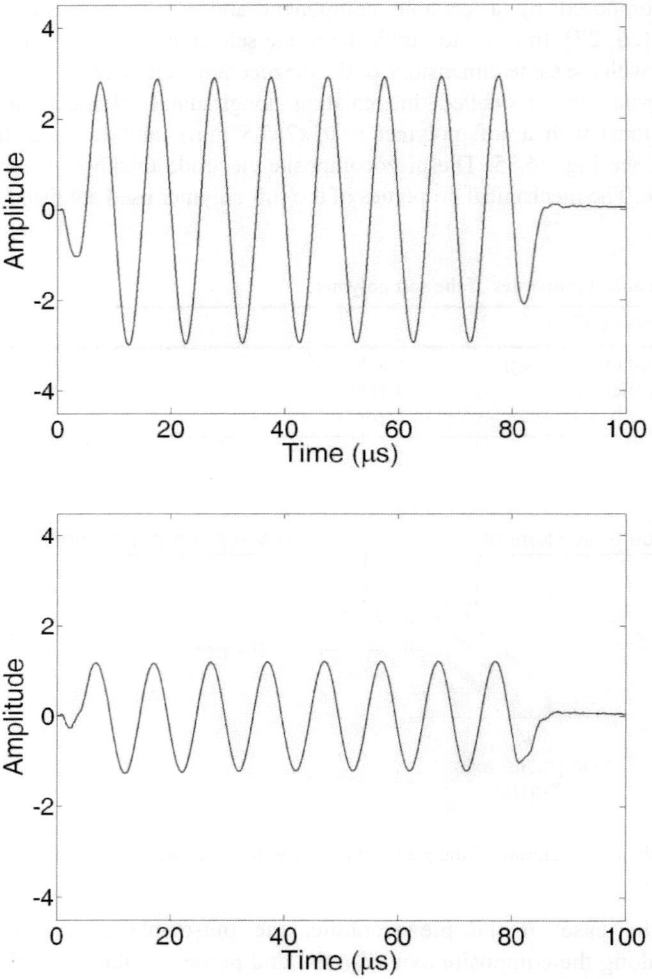

Fig. 16.36 Simulation results of the out-of-plane displacement for the case of a piezocomposite. Results for the antisymmetric mode (90 kHz) calculated at 1 mm from the emitter. **a** Composite axis (Z-axis) **b** Perpendicular axis (X-axis).

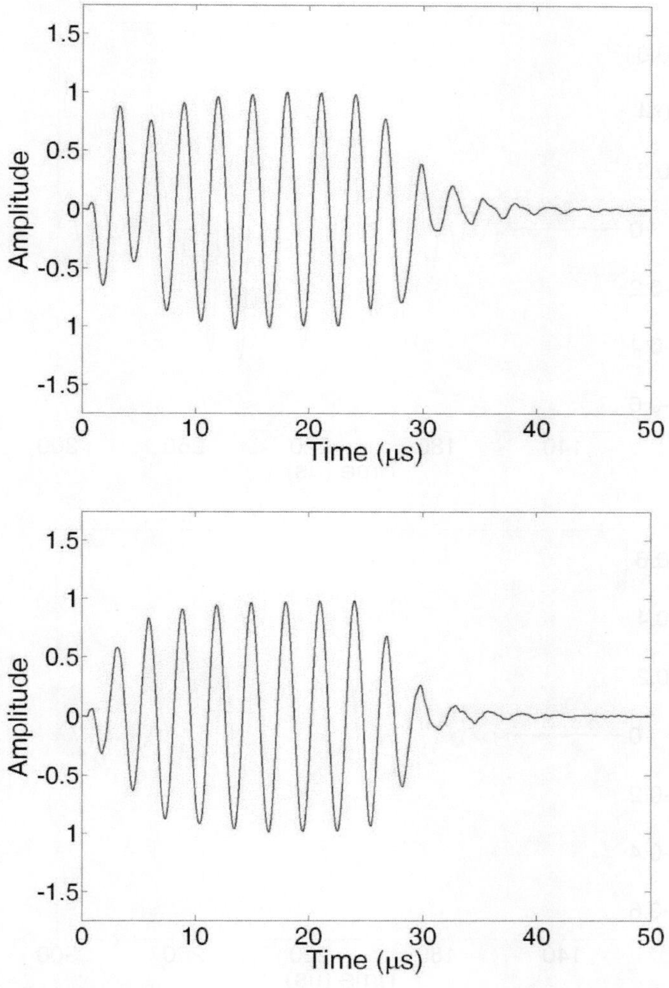

Fig. 16.37 Simulation results of the out-of-plane displacement for the case of a piezocomposite. Results for the symmetric mode (373 kHz) calculated at 1 mm from the emitter. **a** Composite axis (Z-axis) **b** Perpendicular axis (X-axis).

To test the mode cancellation at distances of several wavelengths, the amplitude of displacement was calculated at 300 mm from the composite edge following the X-axis and Z-axis. Fig. 16.38 and Fig. 16.39 show the results. The cancellations shown in Fig. 16.36b for the antisymmetric mode (1 mm from the piezocomposite) are also observed at the far field (see Fig. 16.38). No amplitude decrease is observed for the symmetric mode shown in Fig. 16.39.

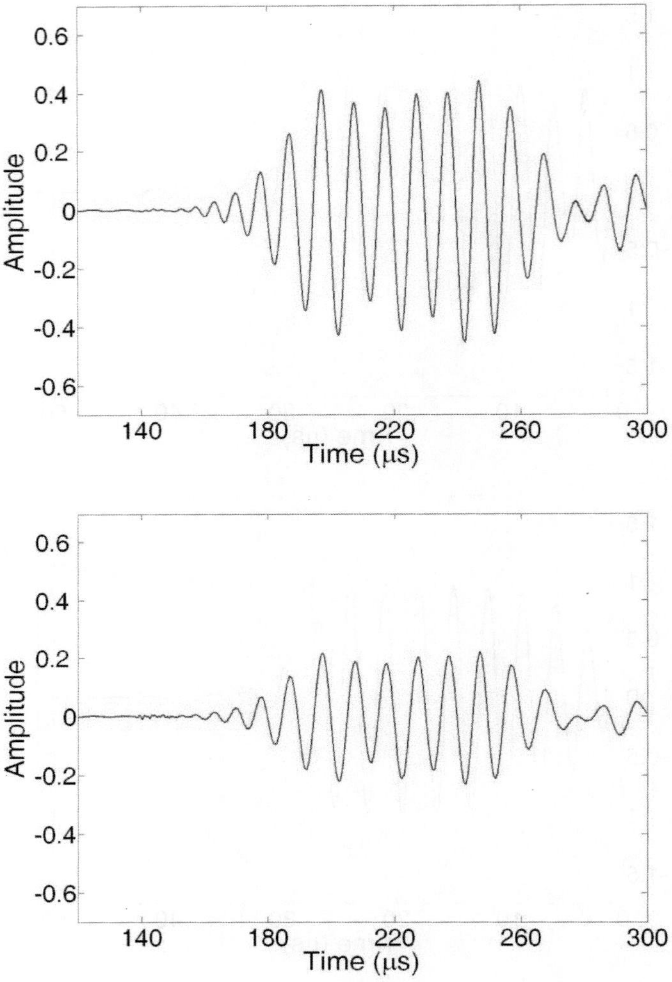

Fig. 16.38 Simulation results of the out-of-plane displacement for the case of a piezocomposite. Results for the antisymmetric mode (90 kHz) calculated at 300 mm from the emitter. **a** Composite axis (Z-axis) **b** Perpendicular axis (X-axis).

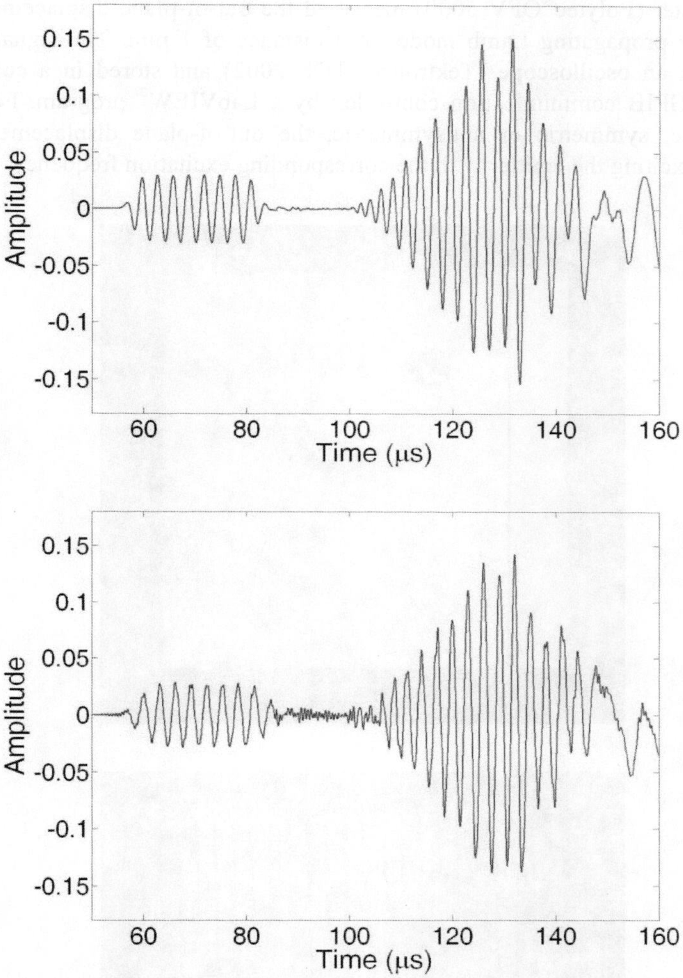

Fig. 16.39 Simulation results of the out-of-plane displacement for the case of a piezocomposite. Results for the symmetric mode (373 kHz) calculated at 300 mm from the emitter. **a** Composite axis (Z-axis) **b** Perpendicular axis (X-axis).

16.4.2.2 Experimental Study

The simulations were validated with experimental data. For this purpose, experimental measurements were performed in a pristine aluminium plate of dimensions 1200x1200x1.1 mm. The emitter (piezoceramic-piezocomposite, Fig. 16.40) was glued to the middle of the plate using an instant adhesive. The emitter was driven with an 8-cycle, 20 V_{pp}, sinusoidal tone-burst signal generated with an

Agilent 33220A Function/Arbitrary waveform generator. An optical interferometer (Polytec®OFV 5000) measured the out-of-plane displacement (Y-axis) of the propagating Lamb modes at a distance of 1 mm. The signals were digitized in an oscilloscope (Tektronix® TDS 2002) and stored in a computed through a GPIB communication controlled by a LabVIEW® program. For each Lamb mode, symmetric or antisymmetric, the out-of-plane displacement was measured exciting the emitter with the corresponding excitation frequency.

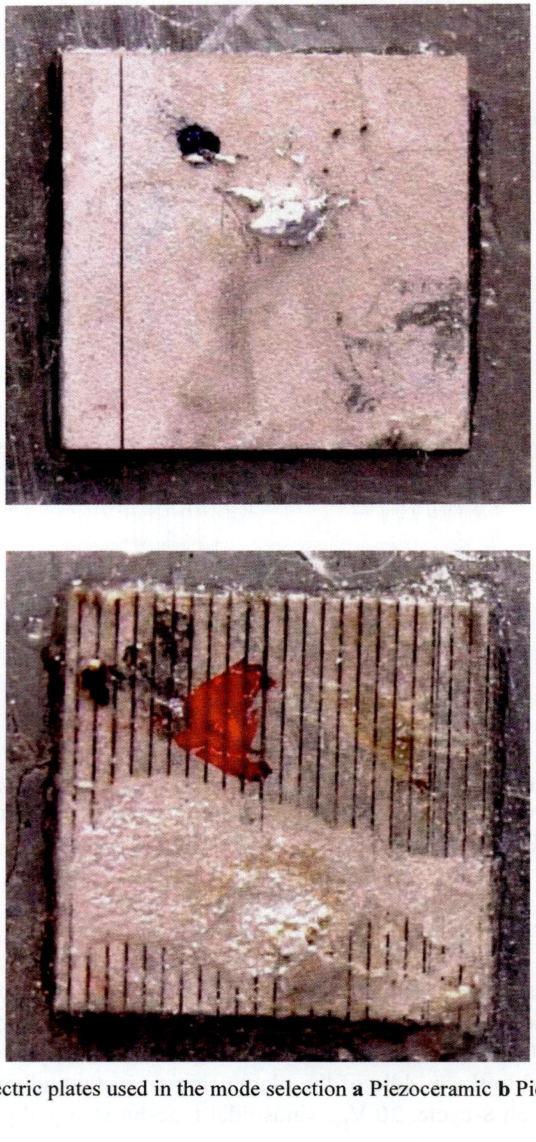

Fig. 16.40 Piezoelectric plates used in the mode selection **a** Piezoceramic **b** Piezocomposite.

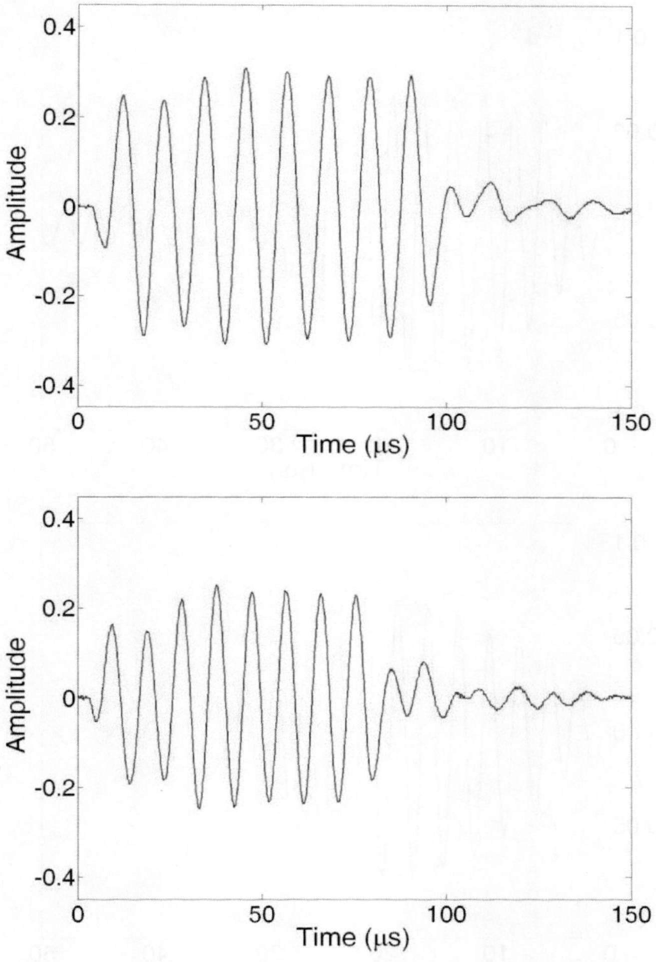

Fig. 16.41 Experimental results of the out-of-plane displacement for the case of a piezoceramic. Results for the antisymmetric mode (90 kHz) measured at 1 mm from the emitter. **a** Z-axis **b** X-axis.

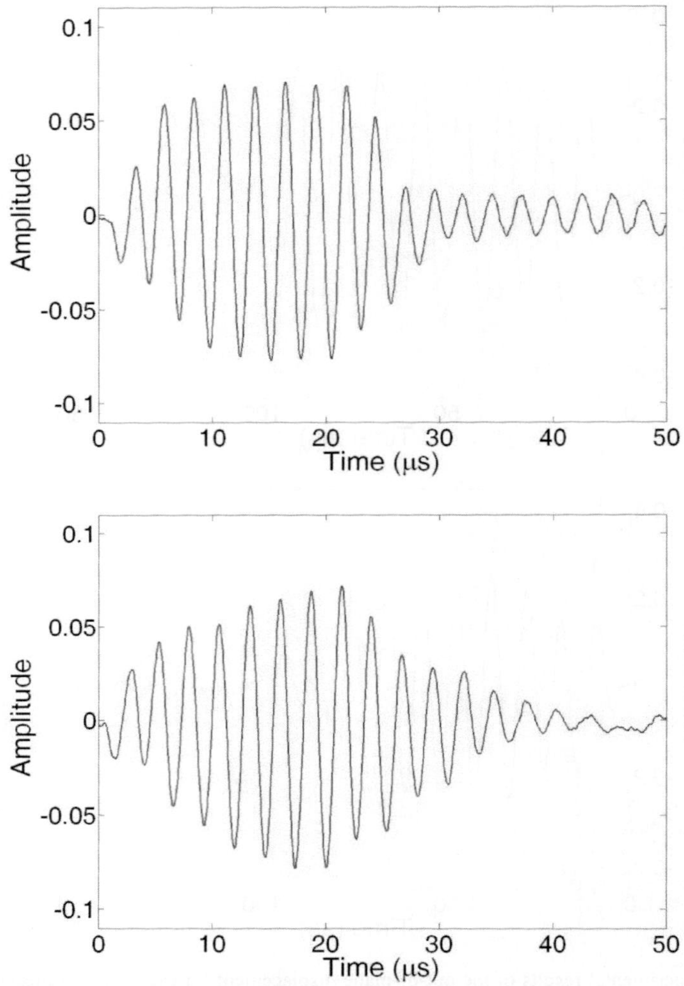

Fig. 16.42 Experimental results of the out-of-plane displacement for the case of a piezoceramic. Results for the symmetric mode (373 kHz) measured at 1 mm from the emitter. **a** Z-axis **b** X-axis.

Piezoceramic

A piezoceramic (Ferroperm® PZ 27, 7x7x0.5 mm) was used for the first measurement (see Fig. 16.40a). The displacement signals measured with the interferometer are shown in Fig. 16.41 and Fig. 16.42. As can be seen, the experimental measurements fit the simulation results presented in Fig. 16.33 and Fig. 16.34. The differences in amplitude of displacement between the X-axis and Z-axis may be due to small variations in the intensity of the signal reflected by the plate.

Piezocomposite

The piezocomposite was fabricated following the well known dice-and-filling technique using a K&S® dicing system [28, 29]. Metallization was made by sputtering using a Baltzer® SCD 050 station. Fig. 16.43 shows a flexible array fabricated with the individual piezocomposite elements of Fig. 16.40b. A soft polymer, Eccogel 1365, was used as the passive component to construct the 2-2 piezocomposite. As before, the piezocomposite was bonded to the aluminium plate and the out-of-plane displacement was measured with the interferometer. The measurement point distance was 1 mm from the edge of the composite plate. The displacement was measured at both axes of the composite. Once again, the displacement for the antisymmetric mode decreases in amplitude along the perpendicular axis (see Fig. 16.44b). The displacement amplitudes for the symmetric mode are not affected (see Fig. 16.45).

Finally, measurements with soft polymer composite were done to test the mode cancellation at far distance (see Fig. 16.46 and Fig. 16.47). The experimental results show good correlation with the simulations.

Fig. 16.43 Flexible array of piezocomposites 2-2 used for the mode selection.

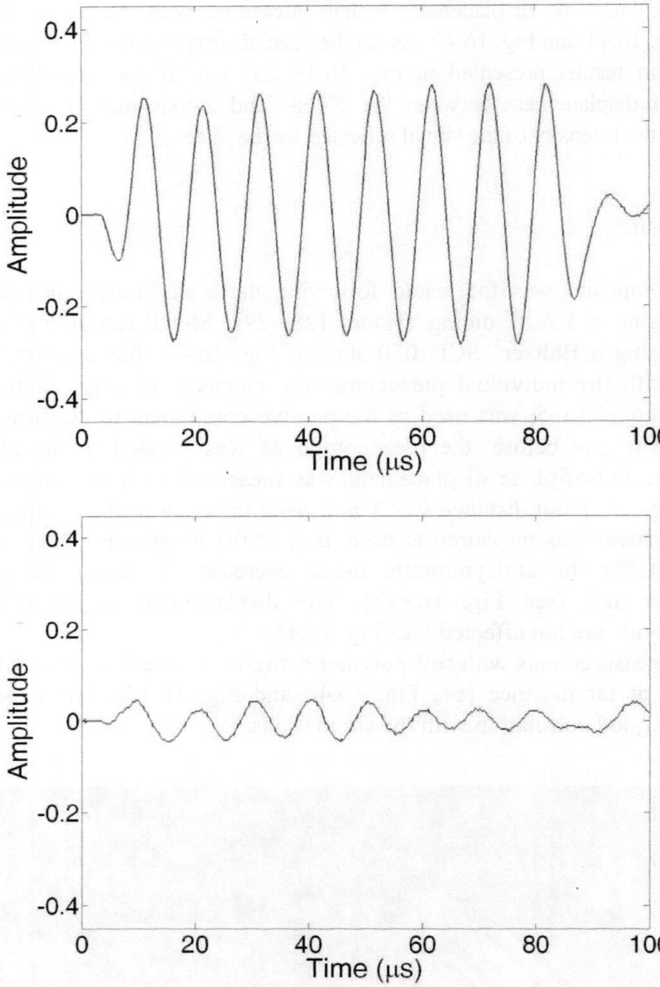

Fig. 16.44 Experimental results of the out-of-plane displacement for the case of a piezocomposite. Results for the antisymmetric mode (90 kHz) measured at 1 mm from the emitter. **a** Composite axis (Z-axis) **b** Perpendicular axis (X-axis).

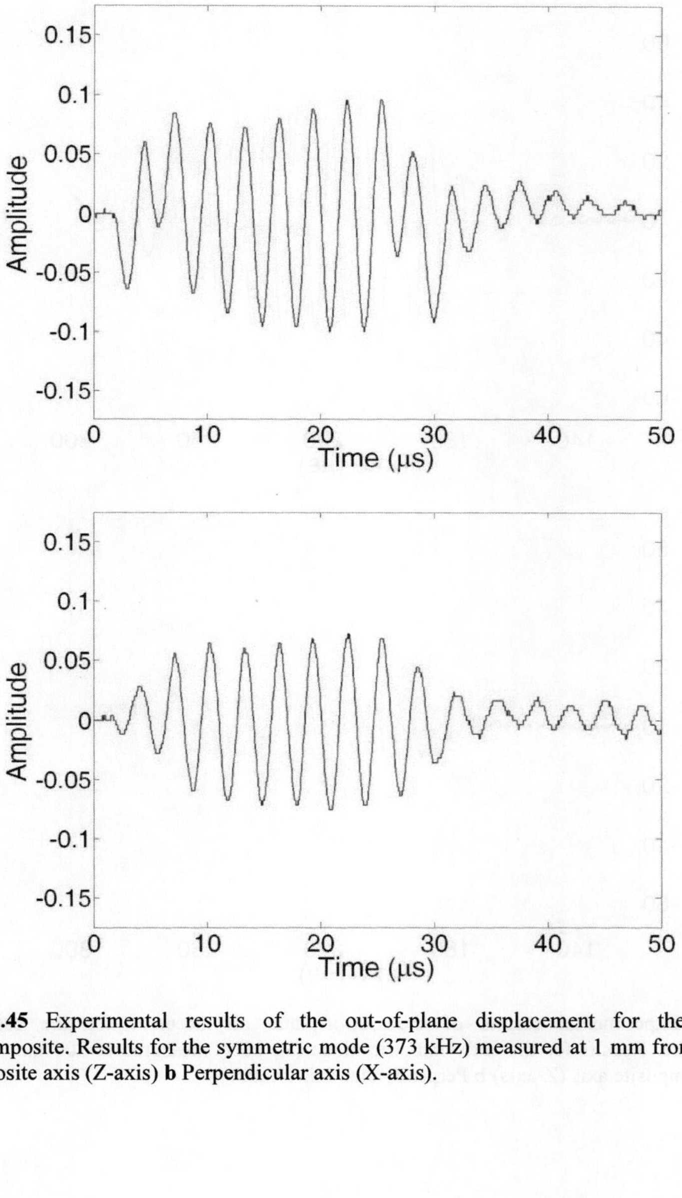

Fig. 16.45 Experimental results of the out-of-plane displacement for the case of a piezocomposite. Results for the symmetric mode (373 kHz) measured at 1 mm from the emitter. **a** Composite axis (Z-axis) **b** Perpendicular axis (X-axis).

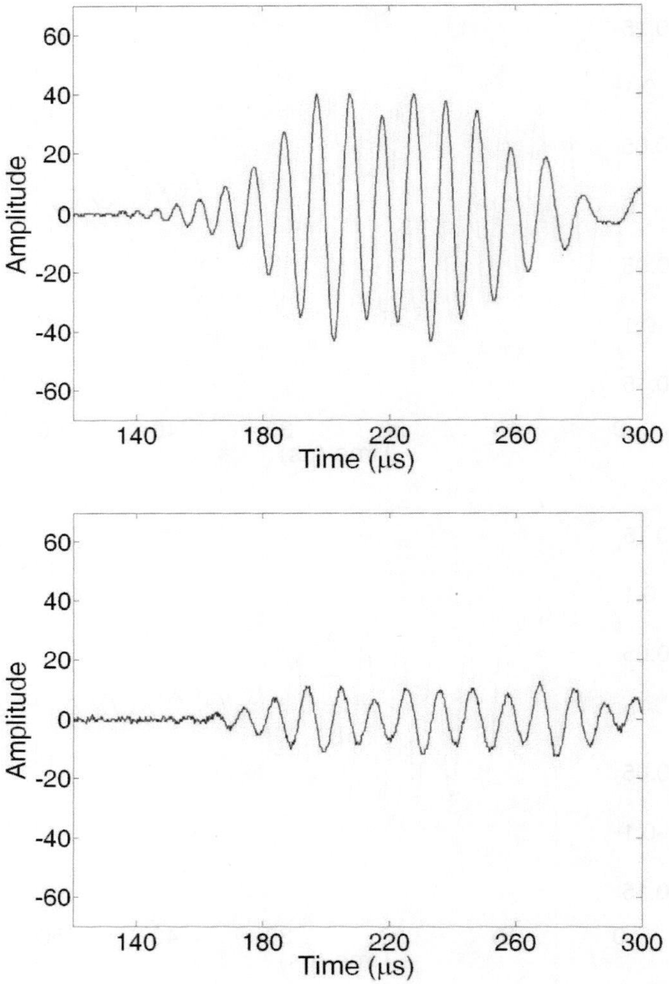

Fig. 16.46 Experimental results of the out-of-plane displacement for the case of a piezocomposite. Results for the antisymmetric mode (90 kHz) measured at 300 mm from the emitter. **a** Composite axis (Z-axis) **b** Perpendicular axis (X-axis).

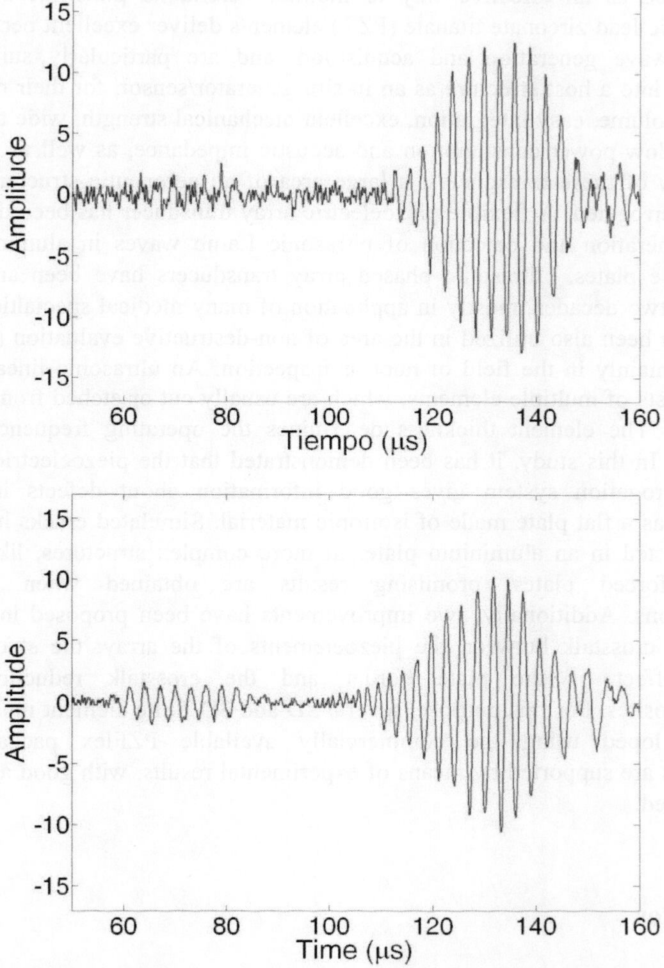

Fig. 16.47 Experimental results of the out-of-plane displacement for the case of a piezocomposite. Results for the symmetric mode (373 kHz) measured at 300 mm from the emitter. **a** Composite axis (Z-axis) **b** Perpendicular axis (X-axis).

16.5 Conclusions

The use of ferroelectric materials, such as piezoelectric transducers, has been demonstrated as an effective way to monitor aeronautic plate-like structures. Piezoelectric lead zirconate titanate (PZT) elements deliver excellent performance in Lamb wave generation and acquisition, and are particularly suitable for integration into a host structure as an in-situ generator/sensor, for their negligible mass and volume, easy integration, excellent mechanical strength, wide frequency responses, low power consumption and acoustic impedance, as well as low cost. With a few of these transducers, a large area of an aeronautic structure can be rapidly interrogated. A flexible piezoelectric array transducer has been developed for the generation and detection of ultrasonic Lamb waves in aluminium and carbon fibre plates. Ultrasonic phased array transducers have been around for more than two decades, mostly in application of many medical specialties. These arrays have been also utilized in the area of non-destructive evaluation (NDE) of materials, mainly in the field of nuclear inspection. An ultrasonic linear phased array consists of multiple elements, which are usually cut or etched from a single PZT plate. The element thickness determines the operating frequency of the transducer. In this study, it has been demonstrated that the piezoelectric phased-array interrogation system gives good information about defects in simple specimens as a flat plate made of isotropic material. Simulated cracks have been easily detected in an aluminium plate. In more complex structures, like carbon fibre reinforced plates, promising results are obtained when detecting delaminations. Additionally, two improvements have been proposed in order to reduce the crosstalk between the piezoelements of the array: the study of the reactive effect of the plate border and the crosstalk reduction using piezocomposites. For this purpose, several 2D and 3D finite element models have been developed using the commercially available PZFlex package. The simulations are supported by means of experimental results, with good agreement demonstrated.

References

1. Viktorov, I. A., Rayleigh and Lamb Waves - Physical Theory and Applications, Plenum Press, NY, 1967.
2. Lamb, H., "On Waves in an Elastic Plate" Proceedings of the Royal Society of London. Series A, Containing Papers of a Mathematical and Physical Character, vol.93, n.651, 293-312, 1917.
3. Rose, J.L., Ultrasonic Waves in Solid Media, Cambridge University Press, 1999.
4. Staszewski, W.J., Boller, C., Tomlinson, G.R., Health Monitoring of Aerospace Structures. John Wiley & Sons, Chichester, UK., pp. 167-169, 2003.

5. Brissaud, M. "Characterization of piezoceramics" IEEE Transactions on Ultrasonics, Ferroelectrics and Frequency Control, Volume: 38, Issue: 6. On page(s): 603-617 Nov 1991.
6. Hoon, T.H., Cheam, D.W., et al., "Mechanical Degradation of the Piezoelectric Elements" The 5th International Workshop Structural Health Monitoring, Stanford University, 2005.
7. Azar, L., Shi, Y., & Wooh, S.-C., "Beam focusing behaviour of linear phased arrays" NDT and E International, 33(3): p. 189-198, 2000.
8. Barnoncel, D., Osmont, D. & Dupont, M., "Health Monitoring of Sandwich Plates with Real Impact Damages Using PZT Devices" Structural Health Monitoring 2003, Fu-Kuo Chang, ed., DEStech Pub, 871-878, 2003.
9. Cawley, P., Simonetti, F., "Structural health monitoring using guided waves - Potential and challenges" In Fu-Kuo Chang, editor. Structural Health Monitoring, DEStech Pub. Inc., Lancaster, pages 503-510, 2005.
10. Díaz Valdés, S.H. & Soutis, C., "Health Monitoring of Composites using Lamb Waves generated by Piezoelectric Devices" Plastics, Rubber and Composites, vol. 29, 475-481, 2000.
11. Giurgiutiu, V., "Lamb Wave Generation with Piezoelectric Wafer Active Sensors for Structural Health Monitoring" Proceedings of the SPIE 10th Annual International Symposium on Smart Structures and Materials and 8th Annual International Symposium on NDE for Health Monitoring and Diagnostics, March 2-6, San Diego, CA, paper 5056-7, 2003.
12. Grondel, S., Delebarre, C., Assaad, J., Dupuis, J.-P., Reithler, L., "Fatigue crack monitoring of riveted aluminium strap joints by Lamb wave analysis and acoustic emission measurement techniques" NDT and E International 35 (3), pp. 137-146, 2002.
13. Ihn, J.-B. & Chang, F.-K., "Multicrack Growth Monitoring at Riveted Lap Joints Using Piezoelectric Patches" 7th Annual International Symposium on Non-destructive Evaluation for Health Monitoring and Diagnostics, March 17-21, San Diego, CA, Proceedings of the SPIE, vol. 4702, 29-40, 2002.
14. Diamanti, K., Hodgkinson, J.M. & Soutis, C., "Damage Detection of Composite Laminates Using PZT Generated Lamb Waves" Proceedings of the 1st European Workshop on Structural Health Monitoring, July 10-12, Paris, France, 398-405, 2002.
15. Su, Z. & Ye, L., "Fundamental Lamb Mode-based Delamination Detection for CF/CP Composite Laminates Using Distributed Piezoelectrics" Structural Health Monitoring - An International Journal, vol. 3, No. 1, 43-68, 2004.
16. Gachagan, A., G. Hayward, et al., "A flexible piezoelectric transducer design for efficient generation and reception of ultrasonic lamb waves" IEEE Transactions on Ultrasonics, Ferroelectrics, and Frequency Control, 52(7): 1175-1182, 2005.
17. Moetakef, M.A., Joshi, S.P. & Lawrence, K.L., "Elastic Wave Generation by Piezoceramic Patches" AIAA Journal, vol. 34, 2110- 2117, 1996.
18. Nieuwenhuis, J. H., N. Jr., et al., "Generation and detection of guided waves using PZT wafer transducers" IEEE Transactions on Ultrasonics, Ferroelectrics, and Frequency Control, 52(11): 2103-2111,2005.
19. Nagai, K., "A new synthetic-aperture focusing method for ultrasonic B-scan imaging by the Fourier transform" IEEE Transactions on Sonics and Ultrasonics, SU-32 (4) 531–536, 1985.
20. Ricon, Y.G.U. & Freijo, F.Md.E. "Study of Crosstalk Reduction in Linear Array for Imaging in SHM Applications" Ultrasonics Symposium, 2006. IEEE , vol., no., pp.1790-1793, 2006.
21. Gómez-Ullate, Y. & Montero de Espinosa Freijo, F., "Lamb waves generation in thin plates using bonded piezoceramics: reactive effect of the plate border" Structural Health

Monitoring 2006: Proceedings of the Third European Workshop, Conference Centre, Granada, Spain, 2006.

22. PZFLEX User's manual, Version 1-J.10.0, December 2005.
23. Gómez-Ullate Ricón, Y. & de Espinosa Freijo, F.M., "Piezoelectric modelling using a time domain finite element program" Journal of the European Ceramic Society, 27 (13-15), pp. 4153-4157, 2007.
24. Gómez-Ullate, Y., Saletes, I. & Montero de Espinosa, F., "Lamb waves generation in plates using glued piezoceramics" Boletín de la Sociedad Española de Cerámica y Vidrio, 45 (3), pp. 188-191, 2006.
25. Gómez-Ullate Ricón, Y., Chinchurreta Segovia, F.J. & Montero de Espinosa Freijo, F., "Lamb waves generation on thin plates using piezocomposites" Journal of the European Ceramic Society, 27 (13-15), pp. 4147-4151, 2007.
26. Reynolds, P., Hyslop, J., Hayward, G., "Analysis of spurious resonances in single and multi-element piezocomposite ultrasonic transducers" Proceedings of the IEEE Ultrasonics Symposium. 2 pp. 1652-1653, 2003.
27. Reynolds, P., Pereyra, V., "Application of optimisation techniques to finite element analysis of piezocomposite devices" Proceedings - IEEE Ultrasonics Symposium. 1 pp. 634-637, 2004.
28. Savakus, H. P., K. A. Klicker, Newnham, R. E., "PZT-Epoxy Piezoelectric Transducers: A Simplified Fabrication Procedure" Materials Research Bulletin. vol. 18, p.p. 677-680, 1981.
29. Smith, W.A., "The Role of Piezocomposites in Ultrasonic Transducers" Ultrasonics Symposium Proceedings. 2, pp. 755-766, 1989.

Index